T0324183

Seamounts: Ecology, Fisheries & Conservation

Fish and Aquatic Resources Series

Series Editor: Tony J. Pitcher

Director, Fisheries Centre, University of British Columbia, Canada

The *Blackwell Publishing Fish and Aquatic Resources Series* is an initiative aimed at providing key books in this fast-moving field, published to a high international standard.

The Series includes books that review major themes and issues in the science of fishes and the interdisciplinary study of their exploitation in human fisheries. Volumes in the Series combine a broad geographical scope with in-depth focus on concepts, research frontiers and analytical frameworks. These books will be of interest to research workers in the biology, zoology, ichthyology, ecology and physiology of fish and the economics, anthropology, sociology and all aspects of fisheries. They will also appeal to non-specialists such as those with a commercial or industrial stake in fisheries.

It is the aim of the editorial team that books in the *Blackwell Publishing Fish and Aquatic Resources Series* should adhere to the highest academic standards through being fully peer reviewed and edited by specialists in the field. The Series books are produced by Blackwell Publishing in a prestigious and distinctive format. The Series Editor, Professor Tony J. Pitcher is an experienced international author, and founding editor of the leading journal in the field of fish and fisheries.

The Series Editor and Publisher at Blackwell Publishing, Nigel Balmforth, will be pleased to discuss suggestions, advise on scope, and provide evaluations of proposals for books intended for the Series. Please see contact details listed below.

Titles currently included in the Series (Full details at www.blackwellfish.com)

1. *Effects of Fishing on Marine Ecosystems and Communities* (S. Hall) 1999
2. *Salmonid Fishes* (Edited by Y. Altukhov *et al.*) 2000
3. *Percid Fishes* (J. Craig) 2000
4. *Fisheries Oceanography* (Edited by P. Harrison and T. Parsons) 2000
5. *Sustainable Fishery Systems* (A. Charles) 2000
6. *Krill* (Edited by I. Everson) 2000
7. *Tropical Estuarine Fishes* (S. Blaber) 2000
8. *Recreational Fisheries* (Edited by T.J. Pitcher and C.E. Hollingworth) 2002
9. *Flatfishes* (Edited by R. Gibson) 2005
10. *Fisheries Acoustics* (J. Simmonds and D.N. MacLennan) 2005
11. *Fish Cognition and Behavior* (Edited by C. Brown, K. Laland and J. Krause) 2006
12. *Seamounts* (Edited by T.J. Pitcher, T. Morato, P.J.B. Hart, M.R. Clark, N. Haggan and R.S. Santos) 2007
13. *Sharks of the Open Ocean* (M. Camhi, E. Pikitch and E. Babcock) 2008

For further information concerning books in the series, please contact:

Nigel Balmforth, Professional Division, Blackwell Publishing, 9600 Garsington Road, Oxford OX4 2DQ, UK
Tel: +44 (0) 1865 476501; Fax: +44 (0) 1865 471501
Email: nigel.balmforth@oxon.blackwellpublishing.com

Seamounts: Ecology, Fisheries & Conservation

Edited by

Tony J. Pitcher
Fisheries Centre, University of British Columbia, Vancouver, Canada

Telmo Morato
Department of Oceanography and Fisheries, University of the Azores, Horta, Portugal

Paul J.B. Hart
Department of Biology, University of Leicester, United Kingdom

Malcolm R. Clark
National Institute of Water and Atmospheric Research, Wellington, New Zealand

Nigel Haggan
Fisheries Centre, University of British Columbia, Vancouver, Canada

Ricardo S. Santos
Department of Oceanography and Fisheries, University of the Azores, Horta, Portugal

Blackwell Publishing editorial offices:
Blackwell Publishing Ltd, 9600 Garsington Road, Oxford OX4 2DQ, UK
Tel: +44 (0)1865 776868
Blackwell Publishing Professional, 2121 State Avenue, Ames, Iowa 50014-8300, USA
Tel: +1 515 292 0140
Blackwell Publishing Asia Pty Ltd, 550 Swanston Street, Carlton, Victoria 3053, Australia
Tel: +61 (0)3 8359 1011

Cover Credits: Les Gallagher/FishPics & *Imag*Dop

First published 2007 by Blackwell Publishing Ltd

ISBN: 978-1-4051-3343-2

Library of Congress Cataloging-in-Publication Data

Seamounts : ecology, fisheries & conservation / edited by Tony J. Pitcher . . . [et al.].
p. cm. – (Fish and aquatic resources series ; 12)
Includes bibliographical references and index.
ISBN-13: 978-1-4051-3343-2 (hardback : alk. paper)
1. Seamounts. I. Pitcher, T. J.

GC87.6.S4S414 2007
577.7–dc22

2007022245

A catalogue record for this title is available from the British Library

Set in 10/13 Times by Charon Tec Ltd (A Macmillan Company), Chennai, India
www.charontec.com
Printed and bound in Singapore by C.O.S. Printers Pte Ltd

The publisher's policy is to use permanent paper from mills that operate a sustainable forestry policy, and which has been manufactured from pulp processed using acid-free and elementary chlorine-free practices. Furthermore, the publisher ensures that the text paper and cover board used have met acceptable environmental accreditation standards.

For further information on Blackwell Publishing, visit our website:
www.blackwellpublishing.com

Contents

List of Contributors

Javier Arístegui
Facultad de Ciencias del Mar
Universidad de Las Palmas de Gran Canaria
Spain
E-mail: jaristegui@dbio.ulpgc.es

Amy Baco
Woods Hole Oceanographic Institution
Biology Department
214 Redfield, Woods Hole
Massachusetts, 02543
USA
E-mail: abaco@whoi.edu

Igor Bashmachnikov
Department of Oceanography and Fisheries
University of the Azores
PT-9901-862, Horta
Portugal
E-mail: igorb@notes.horta.uac.pt

Georgy Z. Beck-Bulat
Pacific Research Institute of Fisheries and Oceanography (TINRO)

Karen A. Bjorndal
Vladivostok, Russian Federation
Archie Carr Center for Sea Turtle Research and Department of Zoology
University of Florida
PO Box 118525
Gainesville, FL 32611-8525
USA
E-mail: kab@zoology.ufl.edu

Alan B. Bolten
Archie Carr Center for Sea Turtle Research and Department of Zoology
University of Florida
PO Box 118525
Gainesville, FL 32611-8525
USA
E-mail: abb@zoo.ufl.edu

Paul E. Brewin
San Diego Supercomputer Center
9500 Gilman Drive, La Jolla
CA 92093-0505
USA
E-mail: pebrewin@sdsc.edu

Cathy Bulman
Commonwealth Scientific and Industrial Research Organisation (CSIRO)
Division of Marine and Atmospheric Research
GPO Box 1538
Hobart, Tasmania, 7001
Australia
E-mail: cathy.bulman@csiro.au

William W. Cheung
Fisheries Centre
Aquatic Ecosystems Research Laboratory (AERL)
2202 Main Mall
The University of British Columbia
Vancouver, BC, V6T 1Z4
Canada
E-mail: w.cheung@fisheries.ubc.ca

Sabine Christiansen
Global Environmental Conservation Organization (WWF International)
North-East Atlantic Program
Am Güthpol 11, D-28757 Bremen
Germany
E-mail: christiansen@wwfneap.org

Malcolm Clarke
Rua do Porto 18, S. João
PT-9930 Lajes dos Pico, Açores
Portugal
E-mail: malcolmclarke@sapo.pt

Malcolm R. Clark
National Institute of Water and Atmospheric Research
Private Bag 14-901
Wellington
New Zealand
E-mail: m.clark@niwa.co.nz

John F. Dower
Department of Biology and School of Earth and Ocean Sciences
PO Box 3020
Stn CSC, University of Victoria
Victoria, BC, V8W 3N5
Canada
E-mail: dower@uvic.ca

Beth Fulton
Commonwealth Scientific and Industrial Research Organisation (CSIRO)
Division of Marine and Atmospheric Research
GPO Box 1538
Hobart, Tasmania, 7001
Australia
E-mail: Beth.Fulton@csiro.au

Amatzia Genin
The Inter-university Institute for Marine Sciences of Eilat and The Hebrew University of Jerusalem
PO Box 469
Eilat 88103
Israel
E-mail: amatzia@vms.huji.ac.il

Kristina M. Gjerde
The World Conservation Union (IUCN)
Global Marine Program
Konstancin-Chylice, 05-510
Poland
E-mail: kgjerde@it.com.pl

John D.M. Gordon
Scottish Association for Marine Science (SAMS)
Dunstaffnage Marine Laboratory
Oban, PA37 1QA
Scotland
UK
E-mail: dmg@sams.ac.uk

Huw Griffiths
British Antarctic Survey
High Cross
Madingley Road, Cambridge
Cambridgeshire, CB3 0ET
UK
E-mail: hjg@bac.ac.uk

R. Dean Grubbs
Hawai'i Institute of Marine Biology
PO Box 1346
Kane'ohe, HI 96744
USA
E-mail: rgrubbs@hawaii.edu

Susan Gubbay
Bamford Cottages
Upton Bishop
Ross-on-Wye, HR9 7TT
UK
E-mail: sgubbay@mayhill.wyenet.co.uk

Nigel Haggan
1777, East 7th Avenue
Vancouver, BC, V5N, 1S1
Canada
E-mail: n.haggan@fisheries.ubc.ca

Jason M. Hall-Spencer
School of Biological Sciences
University of Plymouth
Drake's Circus, Plymouth
Devon, PL4 8AA
UK
E-mail: jason.hall-spencer@plymouth.ac.uk

Paul J.B. Hart
Department of Biology
University of Leicester
Leicester LE1 7RH
UK
E-mail: pbh@le.ac.uk

Thomas Hart
British Antarctic Survey
High Cross, Madingley Road, Cambridge
Cambridgeshire, CB3 0ET
UK
E-mail: thha@bas.ac.uk

Kim N. Holland
Hawai'i Institute of Marine Biology
PO Box 1346
Kane'ohe, HI 96744
USA
E-mail: kholland@hawaii.edu

Alexander F. Kakora
Polar Research Institute of Marine
Fisheries and Oceanography (PINRO)
Murmansk, Russian Federation

Kristin Kaschner
Sea Around Us Project, Fisheries Centre
Aquatic Ecosystems Research Laboratory
(AERL)
2202 Main Mall
The University of British Columbia
Vancouver, BC, V6T 1Z4
Canada
Present: Forschungs- und
Technologiezentrum Westküste
Hafentörn, 25761 Büsum
Germany
E-mail: k.kaschner@fisheries.ubc.ca

Adrian Kitchingman
23/38 Chapman St
North Melbourne
Victoria, 3051
Australia
E-mail: a.kitchingman@pgrad.unimelb.
edu.au

Anthony J. Koslow
Commonwealth Scientific and Industrial
Research Organisation (CSIRO)
Private Bag 5
Wembley 6913, Perth
Australia
E-mail: Tony.Koslow@csiro.au

Nikolai N. Kukharev
Southern Scientific Research Fisheries of
the Marine Fisheries and Oceanography
(YugNIRO)
Kerch
Ukraine
E-mail: fish@kerch.com.ua

Sherman Lai
Fisheries Centre
Aquatic Ecosystems Research Laboratory
(AERL)
2202 Main Mall
The University of British Columbia
Vancouver, BC, V6T 1Z4
Canada
E-mail: s.lai@fisheries.ubc.ca

Feodor Litvinov
Atlantic Scientific Research Institute
of Marine Fisheries and Oceanography
(AtlantNIRO)
5, Dmitry Donskoy Str
Kaliningrad, 236000
Russia
E-mail: flit@atlant.baltnet.ru

Helder Marques da Silva
Department of Oceanography and Fisheries
University of the Azores
PT-9901-862, Horta
Portugal
E-mail: helder@notes.horta.uac.pt

Ana Martins
Department of Oceanography and Fisheries
University of the Azores
PT-9901-862, Horta
Portugal
E-mail: anamartins@notes.horta.uac.pt

Helen R. Martins
Department of Oceanography and Fisheries
University of the Azores
PT-9901-862, Horta
Portugal
E-mail: hrmartins@oma.pt

Gui Menezes
Department of Oceanography and Fisheries
University of the Azores
PT-9901-862, Horta
Portugal
E-mail: gui@notes.horta.uac.pt

Telmo Morato
Department of Oceanography and Fisheries
University of the Azores
PT-9901-862, Horta
Portugal
E-mail: telmo@notes.horta.uac.pt

Daniel Pauly
Fisheries Centre
Aquatic Ecosystems Research Laboratory (AERL)
2202 Main Mall
The University of British Columbia
Vancouver, BC, V6T 1Z4
Canada
E-mail: d.pauly@fisheries.ubc.ca

Mário Rui Pinho
Department of Oceanography and Fisheries
University of the Azores
PT-9901-862, Horta
Portugal
E-mail: maiuka@notes.horta.uac.pt

Tony J. Pitcher
Fisheries Centre
Aquatic Ecosystems Research Laboratory (AERL)
2202 Main Mall
The University of British Columbia
Vancouver, BC, V6T 1Z4
Canada
E-mail: pitcher.t@gmail.com

Filipe M. Porteiro
Department of Oceanography and Fisheries
University of the Azores
PT-9901-862, Horta
Portugal
E-mail: filipe@notes.horta.uac.pt

P. Keith Probert
Department of Marine Science
University of Otago
PO Box 56, Dunedin
New Zealand
E-mail: keith.probert@stonebow.otago.ac.nz

Bertrand Richer de Forges
Institut de Recherche pour le Développement (IRD)
Centre de Nouméa (UMR7138 UPMC)
101 Promenade Roger Laroque
Anse Vata
BP A5 – 98848 NOUMEA Cedex
New Caledonia
Bertrand.Richer-de-Forges@
E-mail: noumea.ird.nc

Brian Riewald†
Archie Carr Center for Sea Turtle Research
and Department of Zoology

University of Florida
PO Box 118525
Gainesville, FL 32611-8525
USA
([†]deceased)

Alex D. Rogers
British Antarctic Survey
High Cross, Madingley Road, Cambridge
Cambridgeshire, CB3 0ET
UK
Present: Institute of Zoology
Zoological Society of London
Regents Park, London, NW1 4RY
UK
E-mail: Alex.rogers@ioz.ac.uk

Sarah Samadi
Institut de Recherche pour le Développement (IRD)
213 rue La Fayette
75 480 PARIS Cedex 10
France
E-mail: Sarah@mnhn.fr

Marco A. Santos
Department of Oceanography and Fisheries
University of the Azores
PT-9901-862, Horta
Portugal
E-mail: msantos@notes.horta.uac.pt

Ricardo S. Santos
Department of Oceanography and Fisheries
University of the Azores
PT-9901-862 Horta
Portugal
E-mail: ricardo@notes.horta.uac.pt

Thomas Schlacher
University of the Sunshine Coast
Maroochydore DC Queensland, 4558
Australia
E-mail: tschlach@usc.edu.au

Karen I. Stocks
San Diego Supercomputer Center
9500 Gilman Drive
La Jolla, CA 92093-0505
USA
E-mail: kstocks@sdsc.edu

Tracey Sutton
Harbor Branch Oceanographic Institution (HBOI)
5600 US 1 North Fort Pierce
Florida, 34946
USA
E-mail: TSutton@HBOI.edu

David R. Thompson
National Institute of Water and Atmospheric Research Ltd
Private Bag 14-901
Kilbirnie, Wellington
New Zealand
E-mail: d.thompson@niwa.co.nz

Vladimir I. Vinnichenko
Polar Research Institute of Marine Fisheries and Oceanography (PINRO)
Murmansk, Russian Federation
E-mail: vinn@pinro.ru

Reg Watson
Fisheries Centre
Aquatic Ecosystems Research Laboratory (AERL)
2202 Main Mall
The University of British Columbia
Vancouver, BC, V6T 1Z4
Canada
E-mail: r.watson@fisheries.ubc.ca

Paul Wessel
Department of Geology and Geophysics

School of Ocean and Earth Science and Technology
University of Hawaii
1680 East-West Road
Honolulu, HI 96822
USA
E-mail: pwessel@hawaii.edu

Martin White
Department of Earth and Ocean Sciences
National University of Ireland, Galway
Ireland
E-mail: Martin.White@nuigalway.ie

Series Editor's Foreword

> The toilsome fishermen . . . in tiny barks . . . shudder at the terrors awful to behold of the grim sea, even the seamonsters* which encounter them when they traverse the secret places of the deep.
>
> Oppian, Halieutica, c.173 AD (translated A.W. Mair 1928)

Seamounts have been known, but their nature not fully revealed to humans, for a long time; Oppian's 'secret places of the deep' were also familiar to Haida fishermen from the remote western islands in the northwest Pacific. Darwin (1839) mentioned them as places that attract seabirds, and in the tropics, become the foundations of coral atolls. Today, seamounts – the term was first used in the 1930s – and the organisms that live on them are still surprisingly mysterious: the Editors mention that they are, 'some of the least understood habitats on the planet'. This book attempts to draw back the veil with a clear synopsis of what is known about seamounts world wide. In fact, what is not known about seamounts would make a longer book: recently adopted by the conservation community as islands of biodiversity, in fact there turns out to be only a small amount of hard scientific evidence supporting this view. Even the numbers and locations of seamounts remain largely unknown, and this is rather surprising as they are evidently quite hulking things, averaging 5 km high and 15 km across. Now you might think that a nation that will merrily send a cruise missile thousands of kilometres in the direction of an erstwhile terrorists' mobile phone would have very good information about our planet's topology, so you will be surprised to learn that a US nuclear submarine recently ran full tilt into a seamount in the Pacific.

This book follows a logical progression from geological and physical processes, ecology, biology and biogeography, to exploitation, management and conservation concerns. In 21 Chapters written by 57 of the world's leading seamount experts, the book reviews all aspects of their geology, ecology, biology, exploitation, conservation and management. In Part I of this book, several detection and estimation techniques for tallying seamounts are reviewed, along with a history of seamount research. Seamounts are ubiquitous undersea mountains rising from the ocean seafloor that do not reach the surface. There are likely many hundreds of thousands of seamounts, they are usually formed from volcanoes in the deep sea and are defined by oceanographers as independent features that rise to at least 0.5 km above the seafloor, although smaller features may have the same origin.

---

*Ketos, large sea creatures, meaning whales, dolphins, seals, sharks and tunas.

Some seamounts support and attract rich living communities and may be important nodes of biodiversity. Part II of this book covers biophysical coupling, the basis of a long-standing controversy over higher seamount productivity. Seamounts are certainly important 'way stations' for many migratory fish, cetaceans, some seabirds and cephalopods, but many of the fundamental ecological processes that maintain seamount communities are poorly understood. For example, the elegant geophysics described herein shows how spinning Taylor columns may form over seamounts, but they are an evanescent phenomenon and the precise influences of local current patterns on nutrient upwelling, entrainment of water masses and the possible enhancement of primary and secondary production are often uncertain. Complex seamount food webs generally seem to depend on advected, trapped food supplies. These factors determine just how much a seamount may attract feeding visitors like tuna and whales. Part III covers the biology and ecology of seamount organisms.

Most seamount fish, corals and sponges are very long lived, making their recovery slow and hence the impact of serial overfishing on seamount ecosystems raises serious concerns; a history of intensive boom and bust fisheries has depleted fish populations and likely damaged or destroyed associated benthic communities. Part IV presents synoptic views of seamount ecology and their fisheries, mainly focused on how modelling can improve our insight into seamount ecology. The final part of this book, Part V, presents material on the exploitation, management and conservation of seamount ecosystems. Some data from former soviet distant water fishing fleets that targeted seamounts is brought together in the book for the first time. The book also reviews a few small-scale fisheries that have laudably sustainable fisheries on seamounts. The book closes with reviews of ongoing but as yet only partially successful efforts to conserve seamount biota, and presents a synoptic evaluation framework that might be employed to tally progress in research and conservation.

The editors recognize this book as a 'triumph of collaboration' among researchers from the Department of Oceanography and Fisheries in the Azores, Portugal, the University of Leicester in the UK, the National Institute of Water and Atmospheric Research in New Zealand and the University of British Columbia's Fisheries Centre in Canada. The book was planned by many of the authors while attending a stimulating workshop held in the beautiful Azores Islands in May 2005: the list of far-sighted and generous sponsors of this venture are fully acknowledged in the Preface, but principal among them I would like to mention the Census of Marine Life project on Seamounts and the Regional Government of the Azores.

As Series Editor I am confident that this book represents a unique and fresh synthesis of our knowledge of seamounts and their biota, that it will become an essential reference work on the topic, and trust also that it may provide a stimulus for further groundbreaking research on these intriguing 'secret places of the deep'.

Professor Tony J. Pitcher
Editor, Blackwell Publishing Fish and Aquatic Resources Series
Fisheries Centre, University of British Columbia, Vancouver, Canada
July 2007

References

Mair, A.W. (1928) *Oppian, Colluthus, Tryphiodorus*, 635 pp. Loeb Classical Library, Heinemann, UK.

Darwin, C. (1839) *Journal of Researches into the Geology and Natural History of the Various Countries Visited by H.M.S. Beagle*. Henry Colburn, London, UK (Chapter 1, St. Jago – Cape de Verd Islands).

Series rationale

Fish researchers (aka fish freaks) like to explain, to the bemused bystander, how fish have evolved an astonishing array of adaptations; so much so that it can be difficult for them to comprehend why anyone would study anything else. Yet, at the same time, fish are among the last wild creatures on our planet that are hunted by humans for sport or food. As a consequence, today we recognize that the reconciliation of exploitation with the conservation of biodiversity provides a major challenge to our current scientific knowledge and expertise. Even evaluating the tradeoffs that are needed is a difficult task. Moreover, solving this pivotal issue calls for a multidisciplinary consilience of fish physiology, biology and ecology with social sciences such as economics and anthropology in order to probe the frontiers of applied science. In addition to food, recreation (and inspiration for us fish freaks), it has, moreover, recently been realized that fish are essential components of aquatic ecosystems that provide vital services to human communities. Sadly, virtually all sectors of the stunning biodiversity of fishes are at risk from human activities. In freshwater, for example, the largest mass extinction event since the end of the dinosaurs has occurred as the introduced Nile perch in Lake Victoria eliminated over 100 species of endemic haplochromine fish. But, at the same time, precious food and income from the Nile perch fishery was created in a miserably poor region. In the oceans, we have barely begun to understand the profound changes that have accompanied a vast expansion of human fishing over the past 100 years. The Blackwell Series on Fish and Aquatic Resources is an initiative aimed at providing key, peer-reviewed texts in this fast-moving field.

Preface

The world ocean has somewhere between 10 000 and 100 000 seamounts more than 1 km tall and as many as 1 000 000 features over 100 m tall. These are some of the least understood habitats on the planet. Large seamounts, particularly those close to or within the photic zone, support and attract rich biotic communities and are important for the status of marine food webs and biodiversity. Intensive boom-and-bust fisheries have depleted fish populations and damaged or destroyed associated benthic communities. Most seamount fish and the corals and sponges of the benthos are extremely long lived, making their recovery uncertain, or at best very slow. Seamounts are important 'way stations' for many migratory fish, cetaceans seabirds and cephalopods. Other oceanic or 'seamount-associated' species may simply come to 'raid the larder'. The abundance of birds and fish was the first indication that humans had of the existence of seamounts, which, in places like the Azores, still support sustainable artisanal fisheries. The impact of overfishing on seamount ecosystems and the abundance of seamount visitors raises serious concerns.

Many of the fundamental ecological processes that maintain seamount communities are still poorly understood: the influences of local current patterns on upwelling of nutrients, the entrainment of water masses and the enhancement of primary and secondary production; recruitment to the seamount ecosystem; complex food web structure depending on advective or trapped food supplies; transient feeding by visitors like tuna and whales and the integration of different environmental compartments stratified by depth. By bringing together international experts on seamount ecosystems and their fisheries to create a fresh synthesis, this book aims to address these issues and lead the way to an improved insight into seamount ecology and identify measures necessary to conserve their biodiversity and integrity.

Themes running throughout the book are: recruitment and vulnerability of seamount organisms, impacts on seamount biodiversity and the sustainability and economic basis of seamount fisheries.

Seamounts are underwater volcanoes, most of them extinct, but there has been some controversy over a more precise definition. We therefore provide the following definition and characterization of seamounts.

Definition of a seamount

Traditionally, geologists have defined seamounts as undersea mountains whose summits rise more than 1000 m above the surrounding seafloor and that, to first order, exhibit a conical shape with a circular, elliptical or more elongate base. Most seamounts are volcanic in origin. Initially, a seamount was defined as a 'large isolated elevation characteristically of conical form' (Murray, 1941), where 'large' would only later be quantified to mean ⩾1000 m (Menard, 1964). When the Davidson Seamount was named in 1938, the US Board on Geographic Names stated, 'The Generic term "seamount" is here used for the first time, and is applied to submarine elevations of mountain form whose character and depth are such that the existing terms bank, shoal, pinnacle, etc., are not appropriate'. As understanding of the geologic processes that form seamounts and their distribution has improved, the strict 1000 m relief limitation was relaxed and the geological literature now routinely applies the term 'seamount' to much smaller structures (down to a few tens of metres). Studies of seamount populations reveal that their size–frequency distributions are continuous with no obvious break. Thus, seamounts do not have a clear lower-size limit, making any size-based criteria for defining them arbitrary. Consequently, the term 'seamount' has been applied more generally to topographic 'hill' elevations regardless of size and relief (e.g., Epp and Smoot, 1989; Rogers, 1994).

As this book is more biological than geological, we adopt a more functional approach and define as a seamount any topographically distinct seafloor feature that is at ⩾100 m but which does not break the sea surface. We exclude large banks and shoals (e.g., Georges Bank, Porcupine Bank) as well as topographic features on continental shelves as dealt with elsewhere in the literature and different from true seamounts in size (in the case of large banks and shoals) and proximity to other shallow topography (in the case of topographic features on the continental shelf).

We classify individual seamounts on the basis of summit depth using functional criteria important in regulating biological productivity rather than arbitrary depth limits. We define shallow seamounts as those that penetrate the euphotic zone; intermediate seamounts as those that are shallower than the daytime depth of the deep scattering layer (but which do not reach the euphotic zone) and deep seamounts as those with summits below the deep scattering layer. Oceanic islands, many of which have the same origins as seamounts, share many common features and ecological effects on their submerged slopes.

Finally, we classify seamounts as large or small, depending on whether the heights exceed 1500 m (regardless of depth). This height separation is useful in isolating large seamounts, whose global distribution is well resolved by satellite altimetry, from small seamounts whose distribution must be inferred from local, acoustic mapping and therefore remains poorly sampled.

Origin of this book

At the International Deep Sea Conference held in Queenstown, New Zealand, in December 2003 (FAO, 2005) it was evident that seamount fisheries have moved far ahead of our knowledge of their ecology, oceanography and levels of sustainable exploitation. Such serious depletions were reported worldwide that Keynote speakers and others called for concerted international action to protect seamounts from further depredations, especially by distant water fleets. In the Azores, a good degree of protection already exists, seamount fisheries are exclusively local and small scale, bottom trawling has never been used and officially banned in 2002. The Azores also host a large amount of seamount research and so provide a logical base from which to launch a world-wide effort to develop better insight of impacts on seamount systems, and seek ways to improve the management of their fisheries. The content and structure of this book grew from a workshop of over 40 seamount experts who gathered for a week at the Old Whaling Station in Horta on the lovely island of Faial, Azores in May 2005 (Fig. 1). All of the Chapters have been peer reviewed. The workshop and the book have been made possible by a number of generous sponsors, who are listed with their logos in the acknowledgements section, but we wish to offer our especial thanks to the CenSeam project of the Census of Marine Life and the Regional Government of the Azores.

Parts of this book

This book follows a logical progression from geological and physical processes, ecology, biology and biogeography, to exploitation, management and conservation concerns. Part I introduces and characterizes seamounts. In Chapter 1 Paul Wessel covers the geological origins of seamounts and their geomorphological characteristics. In Chapter 2, Adrian Kitchingman, Sherman Lai, Telmo Morato and Daniel Pauly review counts of the numbers and locations of seamounts worldwide, and present a new estimate. In Chapter 3, Paul Brewin, Karen Stocks and Gui Menezes set out a brief history of scientific research on seamounts.

Part II covers biophysical coupling on seamounts, the crux of controversy over their much-discussed higher productivity. Physical processes and seamount productivity are reviewed in Chapter 4 by Martin White, Igor Bashmachnikov, Javier Arístegui and Ana Martins, showing how basic geo-physics builds the templates of water masses that drive biological processes on seamounts. Chapter 5 by Amatzia Genin and John Dower, covers plankton dynamics laying out the arguments and evidence for and against the enhancement of primary production, the trapping of food for filter feeding organisms or vertically migrating plankton of higher trophic levels, including the mesopelagic organisms that are the subject of Chapter 6 by Filipe Porteiro and Tracey Sutton.

Part III covers the biology and ecology of seamount organisms. Chapter 7 on seamount benthos by Sarah Samadi, Thomas Schlacher and Bertrand Richer de Forges, includes an extensive discussion of endemism, an oft-mentioned feature of seamount faunas. Deepwater corals on seamounts, one of the fauna most threatened by trawling, are covered in Chapter 8 by Alex Rogers, Amy Baco, Huw Griffiths, Tom Hart and Jason Hall-Spencer. In Chapter 9, Telmo Morato and Malcolm Clark discuss the characterization and life histories of seamount fishes; while Chapter 10 describes the activities of large pelagic fish, especially tuna, on seamounts by Kim Holland and Dean Grubbs, with a second section on pelagic sharks by Feodor Litvinov. The almost unrecognized role of cephalopods at seamounts is analysed by Malcolm Clarke in Chapter 11. This part concludes with a three-section Chapter 12 on evidence about air-breathing visitors to seamounts: marine mammals by Kristin Kaschner; turtles by Marco Santos, Alan Bolten, Helen Martins, Brian Riewald and Karen Bjorndal; and seabirds by David Thompson.

Part IV presents three synoptic views of seamount ecology. In Chapter 13, Karen Stocks and Paul Hart review the biogeography and biodiversity of seamounts in the light of biogeographic theory and samples of world-wide plankton. The many components of food webs at seamounts are systematized by Tony Pitcher and Cathy Bulman in Chapter 14. In Chapter 15, Beth Fulton, Telmo Morato and Tony Pitcher cover the range of approaches with which seamount ecosystems and their fisheries may be modelled.

Finally, Part V presents six views of exploitation, management and conservation of seamount ecosystems. Chapter 16 by Helder Marques da Silva and Mário Rui Pinho documents the range of artisanal and small-scale fisheries on seamounts. In Chapter 17, Malcolm Clark, Vladimir Vinnichenko, John Gordon, Georgy Beck-Bulat, Nikolai Kukharev and Alexander Kakora document large-scale distant-water fisheries on seamounts, including much material previously available only in Russian. The world catch from seamount fisheries is estimated using a novel technique in Chapter 18 by Reg Watson, Adrian Kitchingman and William Cheung. In Chapter 19, Malcolm Clark and Tony Koslow examine evidence for the range of impacts that fisheries may have on seamount ecosystems. Chapter 20 by Keith Probert, Sabine Christiansen, Kristina Gjerde, Susan Gubbay and Ricardo Santos reviews the full panoply of concerns of the management and conservation of seamount ecosystems. This book closes with a final synoptic chapter by the editors presenting a synthesis of the major issues for seamount ecology, fisheries and conservation.

Supplemental material backing several chapters of this book may be found on the web site: www.seamountsbook.info. At the same website, material from the Horta workshop, newsletters and other material from the 3-year project that produced this book may also be found.

References

Epp, D. and N.C. Smoot (1989) Distribution of seamounts in the North Atlantic. *Nature*, 337, 254–7.

FAO (2005) *Report on DEEP SEA 2003, An International Conference on Governance and Management of Deep-Sea Fisheries*, Queenstown, New Zealand, 1–5 December 2003. FAO, Rome, Italy. *FAO Fisheries Report 772*, 84 pp.

Menard, H.W. (1964) *Marine Geology of the Pacific*. McGraw-Hill, New York.

Murray, H.W. (1941) Submarine mountains in the Gulf of Alaska. *Bulletin of the Geological Society of America*, 52, 333–62.

Rogers, A.D. (1994) The biology of seamounts. *Advances in Marine Biology*, 30, 305–50.

Fig. 1 Participants at the May 2005 seamounts workshop held at the Old Whaling Factory, Horta, Faial, Azores, Portugal: 1 Keith Probert; 2 Amatzia Genin; 3 Adrian Kitchingman; 4 Ashley Rowden; 5 Ricardo S. Santos; 6 Martin White; 7 Mireille Consalvey; 8 Helder M. Silva; 9 Cathy Bulman; 10 Mário R. Pinho; 11 Karen Stocks; 12 Tony J. Pitcher; 13 Gui Menezes; 14 Chuck Hollingworth†; 15 Kristin Kaschner; 16 Paul Hart; 17 Malcolm R. Clark; 18 Tony Koslow; 19 Marco Santos; 20 Sabine Christiansen; 21 John Dower; 22 Jason Hall-Spencer; 23 Susan Gubbay; 24 André Couto; 25 John Gordon; 26 Helen R. Martins; 27 Malcolm Clarke; 28 Cláudia Silveira; 29 Filipe Porteiro; 30 Alex Rogers; 31 Paul Wessel; 32 Paul E. Brewin; 33 Ana Martins; 34 Sarah Samadi; 35 Dean Grubbs; 36 Telmo Morato; 37 Thomas Schlacher.

(†deceased)

Publisher's Acknowledgement

Blackwell Publishing would like to record thanks to the following bodies, whose generous funding has allowed this book to be published at a price much reduced from the usual model for this level of publication:

- The Regional Government of the Azores, Portugal (Regional Directorate for Fisheries, Regional Secretary for the Environment and the Sea, and Regional Directorate for Science and Technology)
- The Census of Marine Life Seamounts Project
- The MARMAC project under the InterReg IIIB/FEDER programme

Acknowledgements

This book is a triumph of collaboration between several research institutions around the globe. The editorial board of researchers from the Fisheries Centre in Canada, the Department of Oceanography and Fisheries in the Portuguese islands of the Azores, the University of Leicester in the UK and the National Institute of Water and Atmospheric Research in New Zealand, joined forces to create a fresh synthesis on seamount ecology, their fisheries and conservation. Getting most of the 56 authors from 16 different countries together in a workshop held in the Faial island, Horta, Azores in May 2005, was no easy task. We would like to thank the three eminent scientists who peer-reviewed the chapters. We have many institutions to thank for the funding, inspiration and support that made it possible. We are very grateful to the Regional Government of the Azores through their departments for fisheries (Regional Directorate for Fisheries), for environment and the sea (Regional Secretary for the Environment and the Sea) and for research (Regional Directorate for Science and Technology), to the CoML – CenSeam project (Census of Marine Life – Seamounts), and to the MARMAC project under the InterReg IIIB/FEDER programme, for their generous funding support to the Azores workshop and book. We also thank the Luso-American Foundation (FLAD), the Portuguese National Science Foundation (FCT – Fundação para a Ciência e Tecnologia) and the OASIS project (an European Union funded research initiative) for additional funding. The editors and authors are also grateful to the organizers of the Faial workshop, mainly the Department of Oceanography and Fisheries (DOP) of the University of the Azores, the IMAR (Institute of Marine Research) and the OMA (Observatory of the Sea of the Azores). The logos of major and other sponsors of this book and workshop are presented with our thanks on the next page. On behalf of all editors and authors, we wish to thank Nigel Balmforth and Kate Nuttall at Blackwell Publishers for accepting our challenge of publishing a book on seamounts.

Major sponsors

Secretaria Regional da Educação e Ciência

Direcção Regional da Ciência e Tecnologia

Sponsors

Part I

Introduction and Characterization of Seamounts

Chapter 1
Seamount characteristics

Paul Wessel

Abstract

Seamounts have a volcanic origin from magmatic intrusions through the oceanic crust and are associated with spreading centres, mid-plate hotspots comprising upwelling mantle plumes, and ocean subduction zones. Factors that control the formation of seamounts include water depth, chemistry of the source, age of the seafloor, and proximity to areas of mantle upwelling. Our knowledge of the geomorphology of seamounts is based on underwater observations, acoustic and tomographic mapping; factors responsible for geomorphological features include magma supply, plate stresses, and crustal fabric. The spatial and temporal distribution of seamounts is discussed using studies and techniques at a variety of scales, and an analysis of sizes, frequencies, geographical patterns, and correlations with other features, such as island arcs or lines, is presented. Age determinations (geochronology) for seamounts are based on a limited set of dated rock samples or inferred indirectly from the seamounts' size and location in the context of the age of the underlying seafloor. I review what happens to seamounts after they have been formed, including plate tectonic considerations, subsidence due to thermal cooling of the lithosphere, and changes in the environment due to latitudinal motion. Seamounts may be used as geological markers, tracers of intraplate volcanism, and delineators of absolute motion over the mantle, and there is a growing awareness of their influence on tidal dissipation and global circulation. Volcanic islands are a subset of seamounts that currently exceed sea level, and may sustain unique flora and fauna, providing natural laboratories for ecology, evolution, and cultural diversification. Eroded seamounts once above sea level are special cases that encourage coral formation in temperate zones.

Introduction

Seamounts are some of the most ubiquitous landforms on Earth and are present in uneven densities in all ocean basins. Being volcanic in nature, seamounts are mostly found on oceanic crust and to a lesser extent on extended continental crust. They are generated near mid-ocean spreading ridges, in plate interiors over upwelling plumes (hotspots), and in island arc convergent settings. While some technical definitions of a seamount state that the feature must have an elevation exceeding 1000 m above the seabed (Menard, 1964), there are no geological reasons to separate smaller seamounts from their taller counterparts using an arbitrary cut-off height (see Preface for definitions of seamounts). Seamounts form by effusive eruptions of lava on the deep seafloor and exhibit strong axial symmetry in the early stages of formation. If magma is available and the seafloor is mechanically strong, the seamount may grow larger, eventually losing its axi-symmetrical shape due to more complex distribution of stresses that often lead to the development of rift zones and fissure eruptions. Volcanic oceanic islands form small subsets of large seamounts that have breached sea level. While they exhibit some features that set them apart from seamounts, such as exposure to erosion when above sea level, this is a temporary situation that depends on relative sea level, which changes as the islands age, subside, and eventually drown, or as eustatic sea level changes through time.

While the smallest seamounts can only be imaged by high-resolution, multi-beam echo-sounders mounted on surface ships, larger seamounts may be observed indirectly from the way they perturb the Earth's gravity field, causing the equipotential surface (known as the geoid) to exhibit local 'bumps' that can be attributed to the seamounts below (see Box 1.1 and Fig. 1.1). Over oceanic areas, the geoid roughly

Box 1.1 Counting seamounts using satellite altimetry

Global estimates of seamount abundances (see Chapter 2; Wessel, 2001; Kitchingman and Lai, 2004) are based on measurements originating with satellite altimetry, specifically the Geosat/ERS-1 gridded products (for details see Sandwell and Smith, 1997; Smith and Sandwell, 1997). The radar onboard the satellite directly measures the height above the ocean surface. Simultaneously, several ground stations track the satellite's orbit. With this information, one can determine the height of the ocean above the Earth's reference ellipsoid; this anomaly is known as the geoid and reflects variations in the Earth's gravitational field (Fig. 1.1). Basic potential field theory is used to convert the geoid anomaly into a free-air gravity anomaly. Over the oceans, the main cause of such gravity anomalies are lateral density variations related to undulations of the seafloor. Because basalt is much denser than seawater, the gravity field over a seamount is locally enhanced. However, large seamounts deform the seafloor giving rise to gravity anomalies. Conveniently, these anomalies tend to be of longer wavelengths than those due to the seamount itself, making it possible to study the relationship between seafloor topography and observed gravity in a limited band of wavelengths. Smith and Sandwell (1997) used this technique to predict relative seafloor relief given the observed gravity anomalies and then added in regional depths based on available ship track bathymetry measurements. The resulting gridded bathymetry product is the best global solution as it uses actual bathymetry measurements to calibrate the predictions. However, because satellite altimetry has limited resolution, it is unable to portray the shortest wavelengths (<15 km) faithfully. Further resolution is lost because of the physical law of 'upward continuation', i.e., the gravitational signal originating at the water/rock interface decays with distance away from the seafloor. Upward continuation affects the shortest wavelengths the most, thus at the sea surface much of the short-wavelength information have decayed below the noise level. Finally, the methodology has some minor limitations, as there will always be some spectral overlap between the two sources of gravity anomalies mentioned above. Overall, this means that features such as the smallest (and deepest) seamounts and abyssal hill fabric cannot be well resolved in the presently available satellite altimetry. While an improved altimeter may obtain better data and thus improve the instrument's resolution, the degradation by upward continuation will remain a fact of nature, preventing a global mapping of all small seamounts from space. It is likely that tens of thousands of seamounts in the 1–1.5 km range could be reliably detected if improved satellite altimetry were to become available in the future (Sandwell *et al.*, 2002).

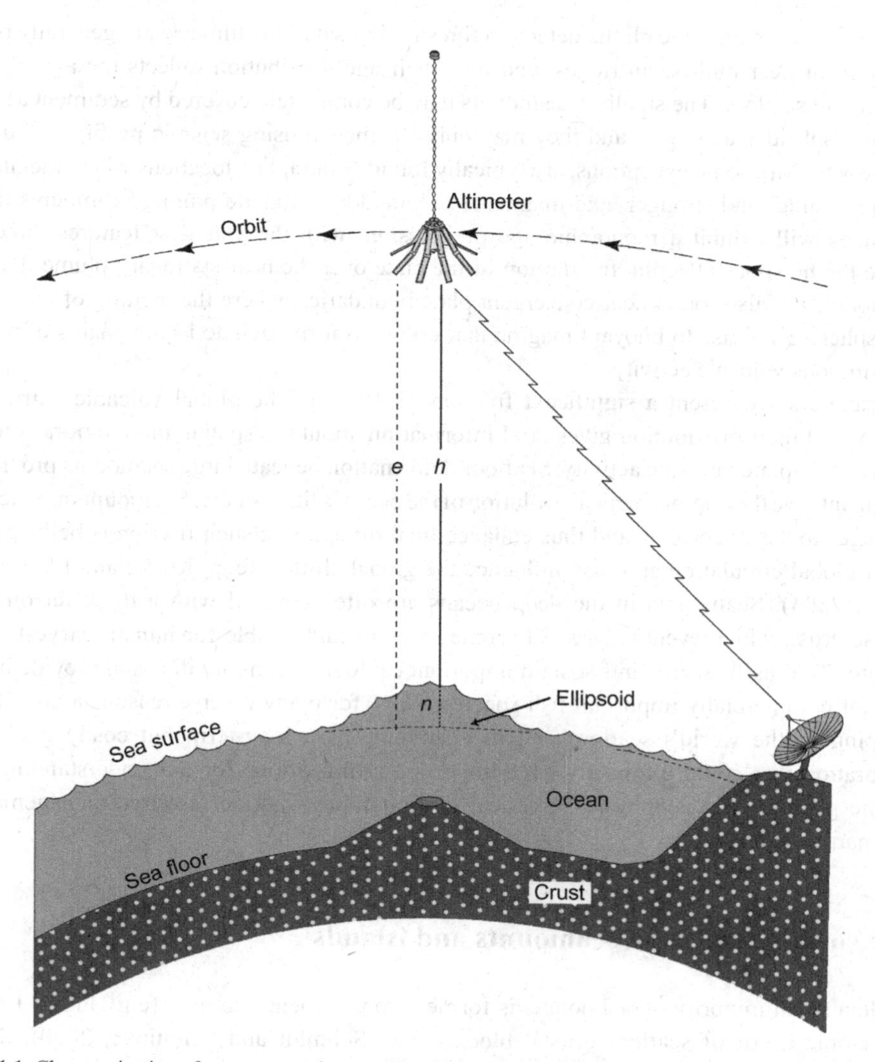

Fig. 1.1 Characterization of seamounts using satellite altimetry. An orbiting altimeter sends radar pulses towards the ocean and detects the returns. Given orbit geometry, pulse travel times, and ellipsoid parameters, geoid height (n) may be estimated. Since n is a representation of the gravity anomaly over the seamount, it is thus possible to infer the location and approximate dimension of the feature indirectly from the gravity anomaly (reproduced from Sandwell and Smith, 1997 by permission of American Geophysical Union).

corresponds to the sea surface; hence, orbiting satellites carrying radar altimeters capable of measuring sea-surface height can map these anomalies. Physical laws can then be used to convert the geoid signal into a likely seamount shape and size. As acoustic mapping improves and spatial coverage increases, additional smaller seamounts are being explored, making it apparent that the distribution of seamount sizes follows a power-law relationship, which itself implies that the number of seamounts with height $\geqslant 100$ m may exceed 1 000 000. The distribution of seamounts among the world's ocean basins is not uniform, and the preponderance of features is located in the Pacific Ocean. Few seamounts are found on the smallest tectonic plates, such as the Cocos and Philippine Sea plates.

Small seamounts beneath the detection threshold of satellite altimetry are generally produced at or near mid-ocean ridges, and thus their age distribution reflects the age of the underlying seafloor. The smallest seamounts may be completely covered by sediment as the seafloor subsides and ages, and they may only be imaged using seismic profiling. Larger seamounts, with some exceptions, are typically found at intraplate locations where the lithosphere is older and stronger, and magma is provided by a mantle plume. Seamounts thus produced will exhibit a monotonic age progression, with the youngest features directly above the hotspot, reflecting the motion of the plate over the near-stationary plume. Large seamounts are also found near convergent plate boundaries, where the melting of recycled lithosphere gives rise to buoyant magma that erupts to form arcuate island chains of near-synchronous volcanic activity.

Seamounts represent a significant fraction (5–10%) of the global volcanic extrusive budget and their distribution gives vital information about the spatial and temporal variations in intraplate volcanic activity. Seafloor deformation beneath large seamounts provides insight into the thermomechanical evolution of the oceanic lithosphere. Seamounts also act as obstacles to water currents and thus enhance tidal dissipation; such friction is believed to affect global circulation and may influence the global climate (e.g., Kunze and Llewellyn Smith, 2004). Seamounts in the deep oceans are often covered with a thick ferromanganese crust, which eventually will become economically viable for human harvest (see Chapter 7). Finally, seamounts sustain important ecological communities and provide habitats for commercially important fish species. Thus, for many diverse reasons a complete mapping of the world's seafloor and its seamounts is a long-term but costly goal for exploration that could ultimately have important ramifications for our understanding of marine geology and geophysics, physical oceanography, marine resource management, and marine ecology.

The volcanic origin of seamounts and islands

While a small minority of seamounts is formed from serpentinite mud (e.g., Fryer, 1992) or tectonic uplift of seafloor crustal blocks (e.g., Schmidt and Schminke, 2000), their presence is limited to the forearc regions of subduction zones. The vast majority of seamounts, however, are constructional aggregates composed of basalt, reflecting their volcanic origin. While sharing a similar material source, seamounts are typically formed in one of three distinct tectonic settings, each imparting unique tectonic characteristics to their offspring. The main factors that separate these origins and the relative importance of these classes of seamounts are discussed below.

Intraplate seamounts

It seems clear that the majority of the larger seamounts found in the ocean basins were formed in an intraplate setting, far from the presence of active plate boundaries. Because of their frequent alignment into linear, sub-parallel chains that appear to correlate with the direction of current and past plate motions (e.g., Duncan and Clague, 1985; Wessel and Kroenke, 1997; Koppers *et al.*, 2001; O'Neill *et al.*, 2005), the most widely accepted origin of such seamounts remains the 'hotspot hypothesis'. Proposed by Wilson (1963) and further elaborated by Morgan (1971), the hotspot hypothesis states that these seamounts form above more or less stationary mantle plumes or hotspots in the Earth's mantle. As the plates move over these sources, the seamounts thus formed are carried away from the source of magma and hence cease to be active. The net result is the formation of a line of extinct volcanoes that exhibit a monotonic age progression reflecting the history of past plate motion.

Since the early formulation of the hotspot hypothesis, hotspots have been proposed to explain numerous sites of unusual volcanic activity (e.g., Burke and Wilson, 1976; Sleep, 1990; Clouard and Bonneville, 2001). Nevertheless, despite tantalizing results from Iceland (Wolfe *et al.*, 1997), conclusive imaging of the underlying mantle plumes using seismic tomography remains elusive (Nataf, 2000). For instance, the archetypal strong plume believed to have formed the Hawaii-Emperor seamount chain and currently assumed to underlie the Hawaii hotspot at the southeast end of the Big Island of Hawaii is not well resolved, whereas other, lesser hotspots (e.g., Easter and Ascension Islands, Azores) appear as stronger manifestations in the tomographic images (Montelli *et al.*, 2004). Although the simple age progressions predicted by the hotspot hypothesis are confirmed for several seamount chains (such as Hawaii-Emperor and Louisville chains), others exhibit a more complex age pattern that casts some doubt on the hotspot theory being the only explanation for such volcanism (e.g., Anderson *et al.*, 1992; McNutt *et al.*, 1997; Dickinson, 1998).

With some exceptions, seamounts formed by hotspot volcanism tend to grow into the largest (Fig. 1.2). In particular, hotspot seamounts formed on old, and hence thicker and stronger, oceanic lithosphere can in some cases reach almost 10 km from base on the seafloor to the tallest island peak, making Mauna Kea, one of the five volcanoes that form the Big Island of Hawaii, the tallest mountain on Earth. Given the smallest features (~100 m) considered seamounts in this volume, the size of observed seamounts spans almost three orders of magnitude. Because large seamounts often penetrate the euphotic zone, they have traditionally been the main focus of ecological studies, despite being a small subset of all seamounts globally. After all, seamounts formed on oceanic crust must reach at least 2.5 km just to match the typical mid-ocean ridge depth, and since most larger seamounts were formed in even deeper water, only truly large seamounts will have a shallow-water presence.

A few large intraplate seamounts are isolated and clearly not associated with any hotspot seamount chain, e.g., Shimada seamount in the Pacific (Gardner and Blakely, 1984)

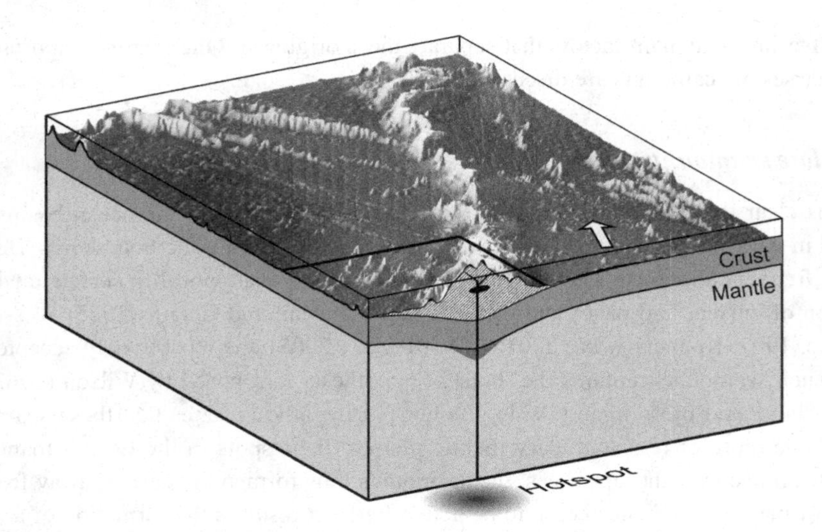

Fig. 1.2 Intraplate seamount and island formation over the Hawaii hotspot. Given a thick and strong lithosphere (~90 million years old), intraplate seamounts can grow very tall and even breach sea level to form oceanic islands. The large volcanic piles deform the lithosphere, which responds by flexure. The plume beneath the plate feeds the active volcanoes by a network of feeder dikes; some magma may pond beneath the crust as well (Watts and ten Brink, 1989). As the plate motion carries the volcanoes away from the hotspot (arrow indicates current direction of motion), they cease to be active and form a linear seamount/island chain.

and Vesteris seamount in the north Atlantic (Haase and Devey, 1994). Because of their large size they must have formed on relatively thick and strong crust, which makes it likely that the available magma intruded into preexisting zones of weakness in the lithosphere, probably made more vulnerable by extensional stresses related to plate motion changes (e.g., Sager and Keating, 1984). It is possible that their isolation makes them unique in that no nearby seamounts have served as stepping stones for the proliferation of marine species.

Mid-ocean ridge seamounts

Most seamounts are small and are believed to have formed near a divergent plate boundary (Batiza and Vanko, 1983; Lonsdale, 1983; Fornari *et al.*, 1988; Jaroslow *et al.*, 2000). Here, excess amounts of magma find their way through the thin and fractured crust to form small, sub-circular seamounts of short stature, often just a few tens to hundreds of metres in elevation (Smith and Cann, 1990). Occasionally, somewhat larger seamounts can be formed (Fig. 1.3). It is likely that most, if not all, smaller seamounts were formed in this tectonic environment as the thickness of the lithosphere rapidly increases away from the mid-ocean ridge, making the penetration of small amounts of magma from an increasingly deeper source more difficult. Indeed, seamount production rates decrease with increasing crustal age and lithospheric thickness, being highest close to the ridge axis (e.g., Batiza, 1982). At fast-spreading ridges, such as the East Pacific Rise, small seamounts form on the flanks of the ridge where the crust is just 0.2–0.3 million years old, and their abundance correlates with spreading rate (White *et al.*, 1998). In contrast, at slow-spreading ridges, such as the Mid-Atlantic Ridge, small seamounts appear to be produced almost exclusively within the median valley (e.g., Smith and Cann, 1990). Studies have shown that many of these new seamounts undergo extensive tectonic deformation by normal faulting, reducing their original heights considerably (Jaroslow *et al.*, 2000). It is also clear that because of increased sediment coverage on older seafloor (e.g., Ludwig and Houtz, 1979), the smallest and most numerous seamounts with height less than 100 m are likely to be buried after a few tens of millions of years.

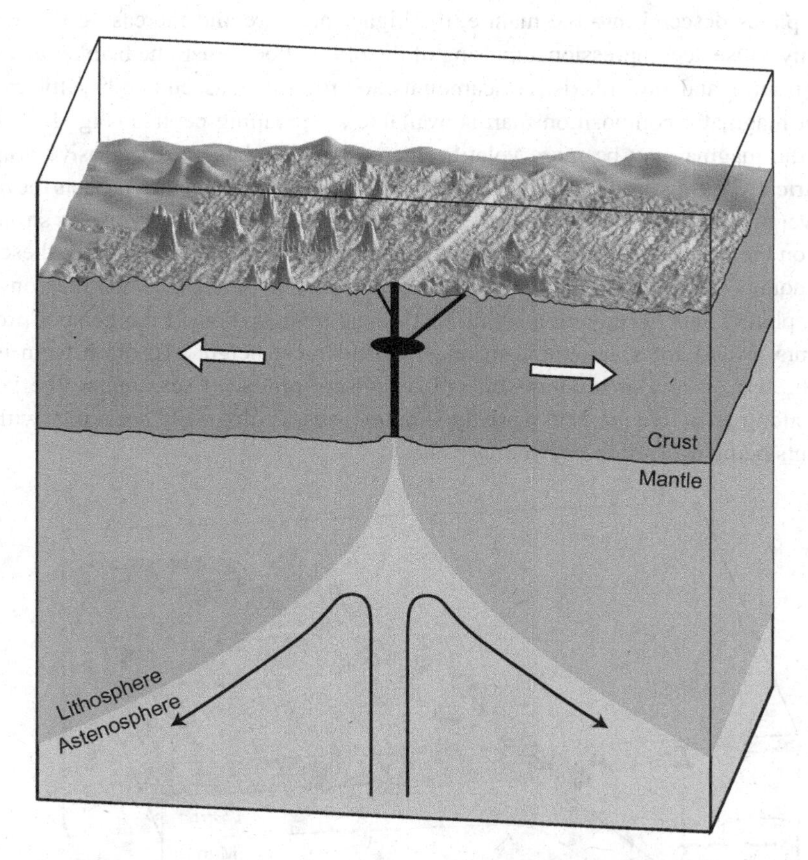

Fig. 1.3 Seamount formation near the East Pacific Rise. Here, the Pacific and Nazca plates are separating at rates exceeding 10 cm/year. The thin and weak plates cannot sustain very large volcanoes and typically only smaller cones are found. In this setting, excessive magma at depth is diverted into feeder dikes that reach the surface on the ridge flank, forming small volcanoes.

Whereas this class of seamounts is important to geologists interested in the study of underwater eruption mechanisms and the cumulative magmatic budget of mid-ocean ridge systems, their small sizes virtually guarantee that their summits will never reach shallower water and become as important ecologically as their taller, hotspot-produced counterparts. However, the ecological role of deep, small seamounts is still poorly understood, although they can have large impacts on the doming of watermasses and consequent upwelling of deep nutrients (see Chapter 4). In few instances where mid-ocean ridge seamounts do grow tall, a hotspot plume such as those beneath Iceland or the Azores is believed to provide the extra magma.

Island arc seamounts

Distinctly different from the two other classes of seamounts, island arc seamounts form at subduction zones where one oceanic plate is being forced beneath the other. As subducting plates descend into the mantle, the higher pressure and increasing temperatures eventually cause decompressional melting of the old seafloor crust, the blanket of wet sediments (if any), and any preexisting seamounts, to produce an ascending basaltic melt of a different magmatic composition than is available at spreading centres (Fig. 1.4). In particular, the magma may be more volatile, increasing the chance of explosive eruptions. Geometrically, the distribution of these island arc seamounts and islands reflects the trend of the convergent plate boundaries, and the overall plate tectonic geometry places strong constraints on the evolution of such seamounts (e.g., Fryer, 1996). Geographically, these island arc seamounts are found in the relatively narrow collision zones between the converging tectonic plates, thus occupying a small area of the total seafloor. Like hotspot-produced seamounts, island arc seamounts can reach considerable height and often form islands, hence the term island arcs. However, unlike hotspot-produced seamounts, the volcanic activity along an active arc is essentially simultaneous, geologically speaking, with older seamounts being overprinted by younger ones.

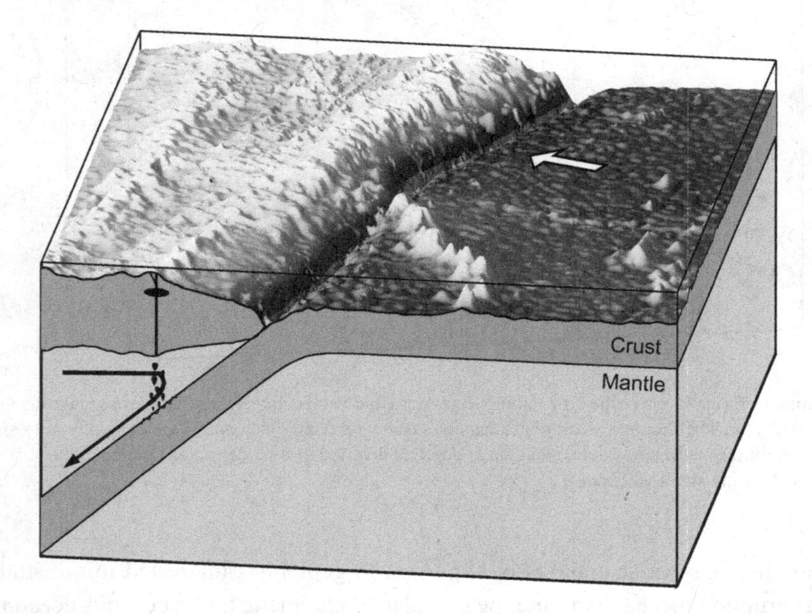

Fig. 1.4 Island arc formation in the Kermadec trench. The subducting Pacific plate and its sediments will melt at depth and form rising diapirs, eventually erupting to create an arc of volcanoes that parallels the subduction zone. Note that the oldest part of the Louisville seamount chain (another intraplate chain) is currently being subducted, possibly affecting the trench geometry.

Morphology and the evolution of seamounts

Seamounts are born deep below the sea surface, where a considerable weight of water ('overburden') is pressing down upon the embryonic volcanoes (Fig. 1.5a). Following a pathway of preexisting cracks or weaknesses, buoyant magma finds its way to the surface. Because seamounts are constructed predominantly on oceanic crust that typically formed at mid-ocean ridge depths of ~2.5 km and subsequently began subsiding as the crust became colder and denser, eventually reaching depths of 5–6 km (e.g., Hillier and Watts, 2004), an emerging seamount will be exposed to a water pressure of 25–50 MPa. Consequently, volcanic gases within the magma body cannot expand and extrusive flows are generally effusive. The cooling effect of seawater also affects the shape of the volcano, allowing construction of steeper flanks that are generally possible once the volcano builds up above sea level (e.g., Macdonald *et al.*, 1986). At first, the seamount is typically fed from a central vent, resulting in an almost circular feature. Many seamounts, in particular those formed near the mid-ocean ridge from small amounts of excess magma, do not develop beyond this stage. However, hotspot seamounts will generally undergo a more complex evolution as developing stresses within the crust and the flanks of the seamount become larger and start to dictate the further development of the volcano (Borgia *et al.*, 2000; Mitchell, 2001).

During the deep-water stage, the high water pressure will impose non-explosive eruptions, which take the form of lava flows or pillow basalts. Circular symmetry is largely maintained, and many develop summit craters. If adequate magma supply is available and the seamount is allowed to grow taller, gravitational stresses in the flanks, possibly enhanced by flexural stresses transmitted from the increasingly deformed subsurface, will favour the development of rift zones. Such rift zones (well developed in the Hawaiian Islands) will break the circular symmetry and lead to the construction of long ridges from fissure eruptions (e.g., Dieterich, 1988). As the summit of the seamount approaches sea level (Fig. 1.5b), water pressure can no longer keep gases locked up in the magma and explosive eruptions become more common (e.g., Macdonald *et al.*, 1986). As sea level is breached, the extrusive products tend to be finer-grained, more vesicular, and structurally less resistant to modifications by the environment. This inherent weakness allows steady erosion to shape the islands, augmented by large landslides as has been observed on the flanks of many large volcanic islands (e.g., Holcomb and Searle, 1991). The combined effect of rift zones, erosion, and landslides is to modify the basic circular shape of seamounts into stellate forms (e.g., Mitchell, 2001).

Once the island is well established, the life of the volcano enters the shield-building stage, during which large flows of *aa* and *pahoehoe* lava are extruded, gradually building up the island (Fig. 1.5c). When active construction finally wanes, the island no longer regenerates to keep up with the destructive forces of erosion (Fig. 1.5d). Combined with long-term thermal subsidence of the seafloor and isostatic adjustments (e.g., Moore, 1987), this brings the summit area back to sea level, where wave erosion flattens the summit to produce a flat-topped guyot (Clague, 1996). As pointed out by Darwin (1842), coral growth will tend to keep up with the subsidence rate, capping many volcanic islands with a thick coral reef layer before subsidence eventually drowns the seamount altogether. Complex interplays between eustatic sea level changes, vertical isostatic adjustments,

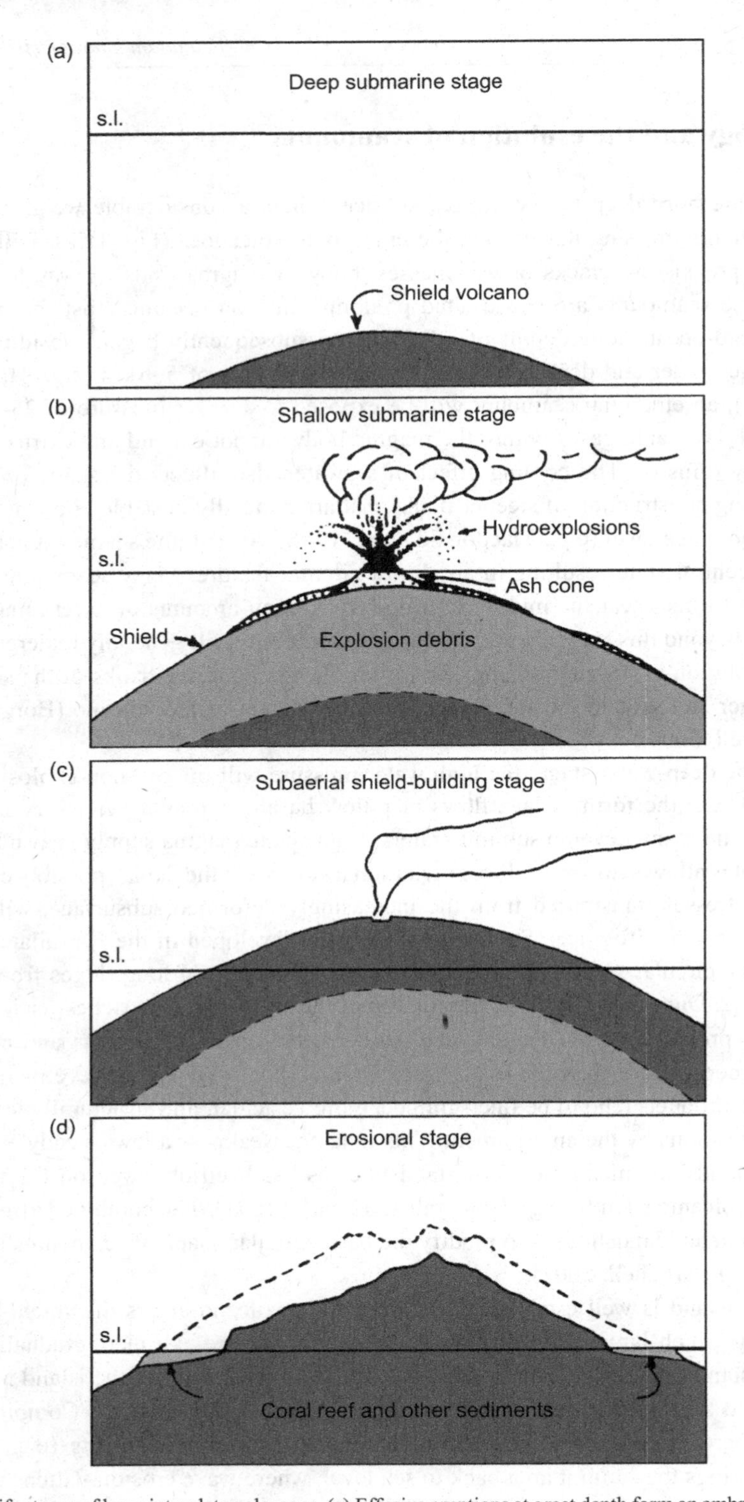

Fig. 1.5 Life stages of large intraplate volcanoes. (a) Effusive eruptions at great depth form an embryonic volcano. (b) As the volcano approaches sea level, the lower water pressure allows gases to expand leading to explosive eruptions, forming more vesicular lavas and volcanoclastic deposits. (c) Once above sea level, effusive eruptions return to produce large volcanic shields that constitute the bulk of the island. (d) After the volcano becomes extinct, erosion and subsidence eventually bring the volcano down to sea level. Fringing reefs may form and reef growth may match subsidence for a while before the seamount drowns, leaving a flat-topped guyot (redrawn from Macdonald *et al.*, 1986 with permission of the University of Hawaii Press).

and climate changes at the seamount caused by plate motion, result in a wide variety of seamounts, some with fringing reefs, others with lagoons with calcareous sediments, while others never developed a coral cap and may have drowned long ago (e.g., Grigg and Epp, 1989; Moore *et al.*, 1996; Flood, 2001).

Research has shown that while many flat-topped seamounts are guyots that obtained their shape as a consequence of wave erosion, others never reached sea level (Fig. 1.6) and must have developed their truncated shape by other means (e.g., Simkin, 1972). While it is unlikely that a single mechanism can explain the shape of all seamounts, it is believed that the presence of a shallow magma chamber within the volcano may modulate the growth of the seamount (e.g., Barone and Ryan, 1990; Clague *et al.*, 2000). This insight is being corroborated by geophysical investigations that have inferred the presence of fossil magma chambers beneath several islands and atolls in French Polynesia on the basis of their anomalous contributions to the residual gravity field (Clouard *et al.*, 2000).

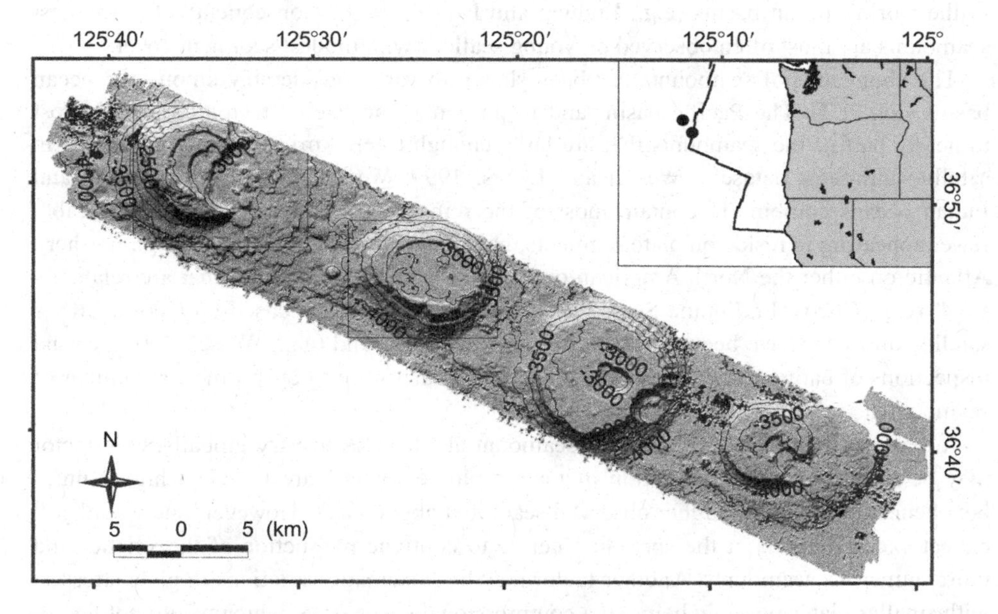

Fig. 1.6 Taney seamounts located on the flank of the Juan de Fuca ridge. While near-ridge seamounts are generally small, occasionally somewhat larger seamounts can form, such as these seamounts that reach ~1–2 km in height. Despite their truncated appearances, these seamounts were never at sea level but obtained their shape by filling in collapsed calderas (modified from Clague *et al.*, 2000 by permission of American Geophysical Union). Inset: Map of Oregon and Washington State coastline showing location of ridge.

The distribution of seamounts

Seamounts are distributed both in space (geographically) and time (temporally), and studies of these variations have provided key insights into several factors that control their formation.

Spatial distribution

Studies of the distribution of seamounts in the world's oceans have determined that their number varies considerably between ocean basins, they tend to form both linear and random constellations, and their sizes and distributions provide invaluable information about their origins. From single-beam echo-sounder profiles, *via* multi-beam surveys, to deflections of the vertical and bathymetry grids inferred from satellite altimetry studies have found that the distribution of seamounts over a wide range of sizes is well approximated by an exponential or a power-law model (Jordan *et al.*, 1983; Smith and Jordan, 1987, 1988; Craig and Sandwell, 1988; Wessel and Lyons, 1997; Wessel, 2001; Kitchingman and Lai, 2004). Such models reflect the observation that most seamounts are small, and by extrapolating the power-law trends obtained for larger seamounts observed in global altimetry-derived grids there may be perhaps as many as 100 000–200 000 seamounts reaching heights of 1 km or more (see Box 1.1 and Chapter 2). Further extrapolation down to the smallest seamount sizes observed (a few tens of metres) suggests seamount populations reaching into millions, but at the same time the majority of such small seamounts will likely become buried by sediments, given the typical thickness (100–200 m) of sediment in the world's ocean basins (e.g., Ludwig and Houtz, 1979). Consequently, the smallest seamounts are most often observed on young seafloor with modest sediment cover.

The abundance of seamounts has been shown to vary considerably among the ocean basins (Fig. 1.7). The Pacific basin (and in particular the Pacific tectonic plate) is host to nearly half of the seamounts that are large enough (>~2 km) to be well expressed in satellite altimetry datasets (Wessel and Lyons, 1997; Wessel, 2001). The Atlantic and Indian oceans combine to contain most of the remaining seamounts, with considerably fewer appearing to reside on plate segments that are located at high latitude (e.g., northern Atlantic on either the North American or Eurasian plates) or on plates that are relatively small (e.g., Cocos, Philippine Sea). While global investigations based on Geosat/ERS-1 satellite altimetry have been limited to the ±72 latitude band (e.g., Wessel, 2001), casual inspections of bathymetry data for the polar ocean basins reveal only a modest number of seamounts.

It is not entirely clear what causes seamount abundances to vary globally. One factor may be the underlying distribution of mantle plumes, which are found in higher numbers beneath plates with higher observed seamount abundances. However, one would still expect excess magma at the spreading centre to continue production of the smaller and more numerous seamounts. Another factor may be systematic variations in plate stresses, with smaller plates possibly being in a compressional stress state, which would not favour the intrusion of magma through the seafloor. Indeed, the incremental construction of oceanic plates, which thicken by the addition of new material freezing onto their undersides, leads to a stress profile that implies a compressional stress regime in the crust (Parmentier and Haxby, 1986; Denlinger and Savage, 1989). Smaller plates are also less likely to have a directional regional stress dominate, and often, as in the case of the Cocos plate, exhibit relatively young, and hence light, seafloor being subducted, thus being associated with only a marginal negative buoyancy force. In contrast, the large Pacific plate, in particular the equatorial region, appears to be under tension from the slab pull forces at the distant subduction zones, as evidenced by widespread extensional volcanism that is neither associated with hotspots nor mid-ocean ridges (Winterer and Sandwell, 1987; Sandwell *et al.*, 1995; Searle *et al.*, 1995). Finally, plates that move the fastest over the underlying mantle appear to have the highest seamount abundances provided they share at least one spreading plate boundary.

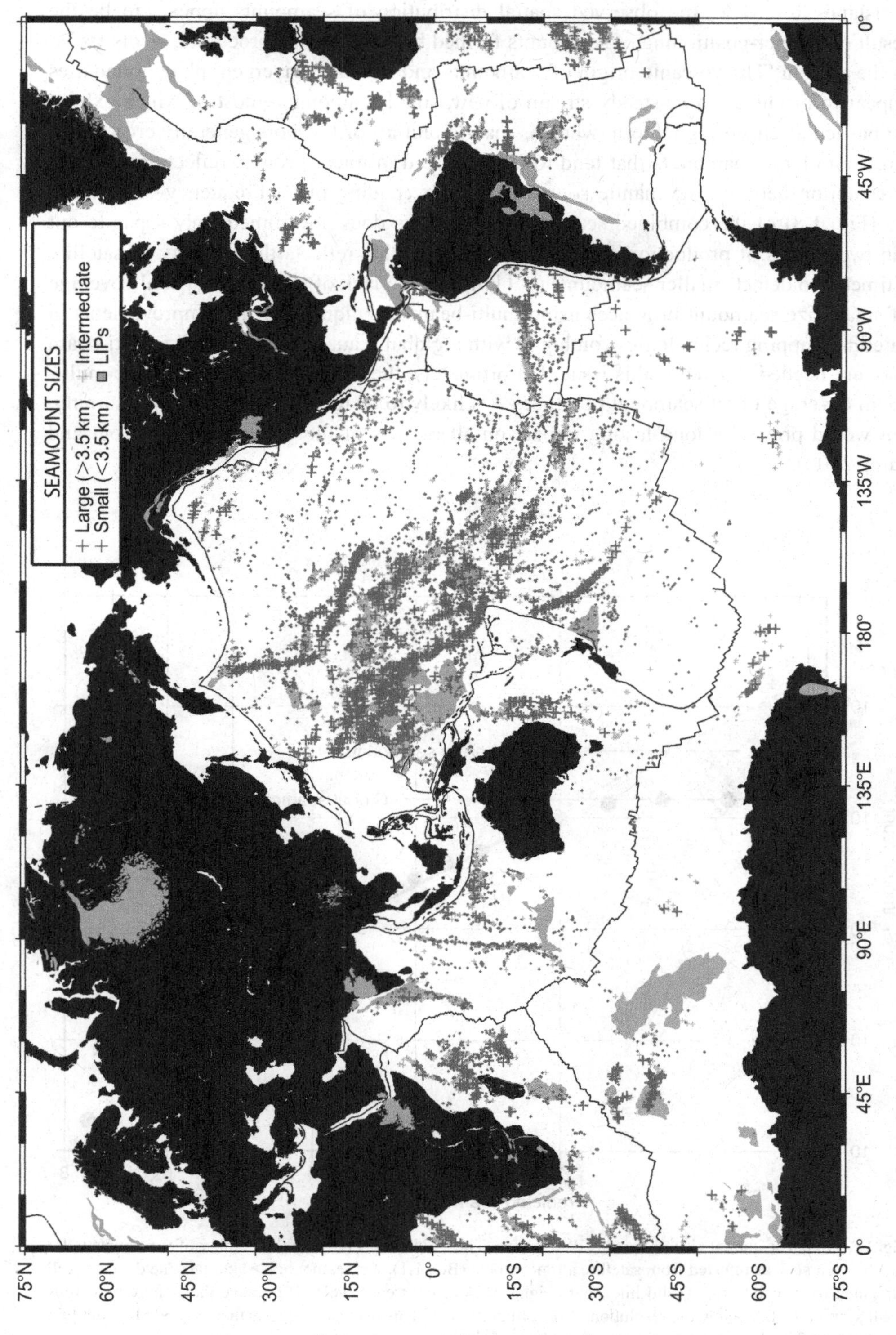

Fig. 1.7 Global distribution of seamounts as identified by Wessel (2001) in a vertical gravity gradient grid (in units of mGal/m or Eötvös) derived from the Geosat/ERS-1 satellite altimetry (Sandwell and Smith, 1997). Red crosses are large seamounts (>120 Eötvös, generally >3.5 km tall), blue are small seamounts (<60 Eötvös, generally <2.5 km tall), whereas green crosses are of intermediate size. The majority of seamounts can be found in the Pacific basin, with the remainder divided between the Atlantic and the Indian oceans. Large igneous provinces are outlined in orange (e.g., Coffin and Eldholm, 1994); these are often associated with seamount provinces.

Island arcs aside, the observed spatial distribution of seamounts appears to be the result of a super-positioning of seamounts formed by two separate processes, as discussed in the section 'The volcanic origin of seamounts and islands'. Divergent plate boundaries appear to produce a near-steady stream of new, small seamounts, most of which exhibit no particular clustering pattern, whereas mantle plumes or hotspots generally create both small and large seamounts that tend to be organized in linear groups, reflecting the relative motion between the mantle source and the overriding plate. Frequency–size analysis (Fig. 1.8) of the combined seamount populations does not immediately separate out the two modes of production, but it is possible that this reflects the inability of satellite altimetry to detect smaller seamounts (<1 km) and the lack of significant spatial coverage of small-size seamount provinces using multi-beam techniques. Further improvements in satellite mapping technologies combined with a global reanalysis of available multi-beam data are needed to resolve this issue. Unfortunately, the ideal situation – complete multi-beam coverage of all seamount provinces – is likely to be too costly and time consuming, but would provide a long-lasting impact on all aspects of the ocean sciences (Vogt and Jung, 2000).

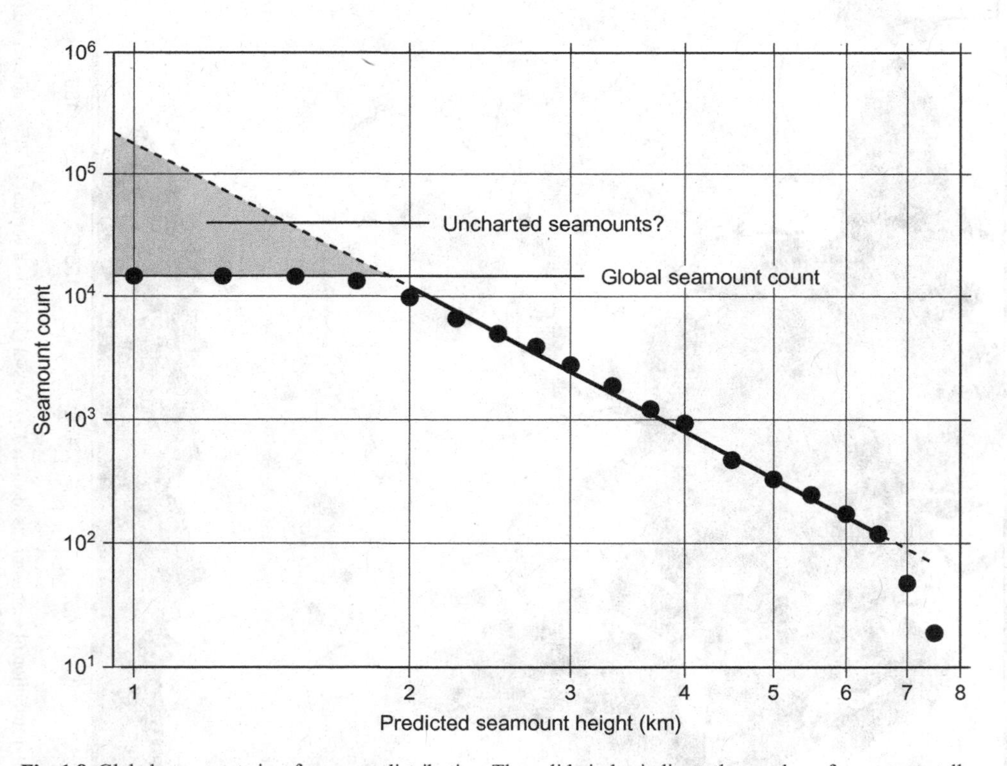

Fig. 1.8 Global seamount size–frequency distribution. The solid circles indicate the number of seamounts taller than a given size, as inferred from satellite altimeter data (Box 1.1). For seamounts >2 km tall, the data are well explained by a scaling rule (solid line). For heights <2 km, the trend levels off because these more numerous small seamounts fall below the resolution of the altimetry data. This analysis suggests that only ~15 000 out of a potential of ~200 000 seamounts greater than 1 km in height have been mapped.

Temporal distribution

Despite the lack of absolute ages except for a tiny subset of islands and seamounts, enough has been inferred from available age data, the constraints furnished from seafloor age and the size of seamounts to conclude that the production of seamounts is not strictly steady state. The evidence points to periods characterized by enhanced production rates. This appears to be particularly clear during the Cretaceous (146–65 million years ago), during which time Pacific seamount production rate was almost twice as high as today, resulting in numerous large intraplate seamounts now present in the West Pacific. It is probably no coincidence that this period also saw the formation of several large oceanic plateaus, such as the Ontong Java, Manahiki, Shatsky, and the Mid-Pacific Mountains (Larson, 1991; Coffin and Eldholm, 1994). While the majority of these large seamounts are not dated and hence cannot be directly correlated with plateau formation, the available dates (Koppers *et al.*, 1998) as well as predicted ages based on the size–frequency distribution appear to place the extra production of seamounts to about the same time as the plateau formation (Cazenave and Royer, 2001; Wessel, 2001).

The ages of seamounts

Seamounts are among the youngest geological features on Earth, reflecting the youthfulness of the oceans in general and the regenerative processes of plate tectonics in particular. However, only a few seamounts are currently volcanically active and they tend to be restricted to (i) the very youngest volcanoes of hotspot island chains (such as Hawaii, Samoa, Reunion, and others); (ii) various places along active island arcs; and (iii) newly formed smaller seamounts associated with mid-ocean ridges (e.g., Scheirer *et al.*, 1996). Many volcanic islands and a few seamounts have been dated using radiometric techniques, yielding absolute ages with a fairly narrow uncertainty band (Clouard and Bonneville, 2001). However, only a tiny fraction of all seamounts has been dated with such techniques and, given the time and cost of acquiring suitable samples and performing the careful laboratory procedures necessary for reliable age determinations (e.g., Koppers *et al.*, 2003), other techniques are necessary to study the population at large. Several different lines of enquiry allow an indirect assessment of the ages of seamounts. Beyond the obvious geological interest in dating seamounts, the ages of seamounts may be a significant parameter in the study of the evolution of endemic ecologies at seamounts.

The age of the seafloor

Unlike seamounts, the underlying seafloor has been well dated as a result of the near-global availability of marine magnetic anomalies, resulting from the frequent reversal of the Earth's magnetic field (e.g., Cande *et al.*, 1989). Detected by magnetometers towed behind surface ships and by low-flying aircraft, linear magnetic anomalies, called isochrones, have been mapped throughout the world's oceans, and assigned absolute ages by careful calibration with radiometrically dated lava flows both on land and at sea. Apart from seafloor formed during the Cretaceous Superchron (a time of few magnetic reversals: e.g., Larson and Olson, 1991), we now know the age of the seafloor almost everywhere (Müller *et al.*, 1997). Because seamounts are necessarily formed after the creation of the underlying seafloor, such ages also serve as upper age boundaries for seamounts. While the oldest seafloor is close to 200 million years, the mean seafloor age is only ~80 million years; hence, most of the seafloor features, and thus most seamounts, are young, geologically speaking.

Implied age progressions

Seamount chains believed to have formed over a mantle plume are expected to exhibit a simple, monotonic age progression away from the active hotspot or site of the most recent volcanic activity. Hence, plate motion models calibrated from available age data can be used to predict the ages of undated seamounts in linear seamount chains. Such estimates can be fairly reliable, in particular in younger seamount chains, and furthermore provide useful tests for the hotspot hypothesis. However, as we go back in time beyond ~70 million years such models become increasingly less reliable due to the scarcity of dates and copolar linear chains (Koppers *et al.*, 2001; Kroenke *et al.*, 2004) as well as the observed complexity of age trends in the older ocean basins (Koppers and Staudigel, 2005).

Isostatic implications

Seamounts represent long-term loads on the oceanic lithosphere, which deforms mechanically by elastic flexure (e.g., Watts, 2001). The exact nature of the flexural deformation depends on the age of the lithosphere when the load was emplaced, with younger, and thus warmer and thinner, lithosphere deforming at a shorter wavelength than older, and thus colder and thicker, lithosphere. These different states of isostatic compensation can be inferred by studying the gravity field over mapped seamounts and, by equating short-wavelength flexural deformation with a young plate at the time of seamount formation, an approximate age of the seafloor at the time the seamount was emplaced can be deduced (and by knowing the seafloor age we thus obtain the present seamount age). Although the resolution of such ages is in most cases only adequate to classify a seamount as having formed either in an 'on-ridge' or an 'off-ridge' setting, such information can help to determine if the seamount formed in an intraplate setting and thus most likely was formed over a hotspot (e.g., Ribe and Watts, 1982; Watts *et al.*, 2006).

Size–age relationship

Statistical studies reveal that small seamounts form predominantly on young and thin oceanic crust, whereas the largest seamounts form on old and thick lithosphere (Epp, 1984; Wessel, 2001). No single physical explanation, such as constant hydrostatic head of magma columns (e.g., Eaton and Murata, 1960; Vogt, 1974) or limiting stresses in the flanks of seamounts (e.g., ten Brink, 1991; McGovern and Solomon, 1993; Wessel, 2001), has yet been conclusively shown to cause this observed trend. Nevertheless, most of the proposed mechanisms predict a similar (and strictly monotonic) relationship between the seamount size and the age of the seafloor at the time of seamount formation.

Consequently, one may attempt to use the observed size of a seamount as a proxy for its age at the time of formation, and with seafloor ages being generally available, a predicted age can be obtained for most undated seamounts. Whereas such predictions are subject to large errors, they may nevertheless serve as reconnaissance ages, which may be useful in testing theories of seamount formation, intraplate volcanic magnetic budgets, and plate tectonic evolution (Wessel, 1997, 2001), and thereby provide upper age estimates for the time available for endemic species to have evolved.

The impact of seamounts

Seamounts are important in several contexts, including geological, oceanographic, and biological/ecological, and possibly in the mining for mineral resources. To date, most explorations of seamounts have had a geological focus, but other aspects (such as biological) are becoming more common. The main impacts that seamounts have had on these different fields are outlined below.

Geological impact

Seamounts are windows to the mantle that allow scientists to study the nature of erupting magma in detail. While overwhelmingly basaltic, minor changes in the chemical and isotopic composition of lavas can be mapped. Such patterns can be used to make inferences about the source, such as the depth of the magma and its ultimate composition (e.g., Janney *et al.*, 2000). The extrusive volume represented by seamounts is a significant fraction of the entire crustal production, perhaps as much as 10% (Batiza, 1982). Both spatial and temporal variations in this intraplate volcanic budget shed light on the plate tectonic evolution of the ocean basins and the mechanisms by which the Earth gets rid of its excess heat. The alignment of seamounts into chains provides a means to decipher the motion of the tectonic plates over long geological intervals, enabling a better understanding of the climatic changes experienced at islands than simply follows from latitudinal migration of plates carrying seamount provinces on their backs (e.g., Kroenke, 1996). Many seamounts have active hydrothermal convection systems that may have a significant effect on element cycles involving seawater (Batiza, 2001). They also participate in the dissipation of residual heat from the formation of both seamount and seafloor (e.g., Harris *et al.*, 2004). Finally, seamounts and islands act as measuring sticks for relative sea level variations, which can have both eustatic and tectonic components (Douglas, 1991; Tushingham and Peltier, 1992).

Oceanographic impact

The bathymetry of the ocean floor influences the ocean circulation in several ways (see Chapter 4). Most importantly, first-order bathymetric features such as ridges and plateaus steer the currents and in places provide barriers that prevent deep waters from mixing with warmer, shallower waters. However, there is a growing awareness that smaller-scale bathymetry, such as seamounts, may play a formerly overlooked role in the turbulent mixing of the oceans (e.g., Kunze and Llewellyn Smith, 2004). In fact, measurements suggest that mixing around a shallow seamount is many orders of magnitude more vigorous than in areas far from seamounts (Lueck and Mudge, 1997). The precise understanding of how the Earth's climate will evolve depends on how quickly heat and carbon dioxide can penetrate into the deep oceans, and assumed rates of vertical mixing can considerably affect model predictions. Further mapping of seamounts and the inclusion of an accurate representation of the seafloor into models of global ocean circulation will be necessary to improve predictions of climate (e.g., Jayne *et al.*, 2004).

Ecological impact

Representing obstacles to flow, seamounts induce local currents, which can enhance upwelling around the seamount (see Chapters 4 and 5). As this may bring up nutrients from the deeper ocean, primary productivity is enhanced, supporting a wide variety of life (Rogers, 1994). The importance of seamounts as habitats for biological communities and, in particular, fish is being increasingly recognized. Islands, a subset of seamounts that currently exceed sea level, also sustain unique flora and fauna, and provide natural laboratories for the study of marine ecology, evolution, and cultural diversification (e.g., Price and Clague, 2002).

Mineral resources impact

Older seamounts may accumulate a ferromanganese oxide crust that is enriched in the elements cobalt, copper, manganese, and sulphur, typically occurring at depths exceeding 3 km (e.g., Friedrich and Schmitz-Wiechowski, 1980; Grigg *et al.*, 1987). It has been estimated that the total cumulative amounts of such marine mineral resources might exceed the amounts currently available on land. So far, the cost of harvesting deep ocean nodules and crusts has been prohibitive. However, rising prices associated with depletion of similar, terrestrial resources will likely make deep ocean resources more attractive, especially since the bulk of these are outside the exclusive economic zones of the adjacent countries (e.g., Baker *et al.*, 2001).

References

Anderson, D.L., Tanimoto, T. and Zhang, Y.-S. (1992) Plate tectonics and hotspots: the third dimension. *Science*, 256, 1645–51.

Baker, C.M., Bett, B.J., Billett, D.S.M. and Rogers, A.D. (2001) An environmental perspective. In: *The Status of Natural Resources on the High-Seas* (ed. Wwf/Iucn), pp. 1–68. World Wide Fund for Nature and International Union for Conservation of Nature and Natural Resources, Gland, Switzerland.

Barone, A.M. and Ryan, W.B.F. (1990) Single plume model for asynchronous formation of the Lamont seamounts and adjacent East Pacific Rise terrains. *Journal of Geophysical Research*, 95, 10801–27.

Batiza, R. (1982) Abundances, distribution, and sizes of volcanoes in the Pacific Ocean and implications for the origin of non-hotspot volcanoes. *Earth and Planetary Science Letters*, 60, 195–206.

Batiza, R. (2001) Seamounts and off-ridge volcanism. In: *Encyclopedia of Ocean Sciences* (eds. Steele, J., Thorpe, S. and Turekian, K.), pp. 2696–708. Academic Press, San Diego, CA.

Batiza, R. and Vanko, D. (1983) Volcanic development of small oceanic central volcanoes on the flanks of the East Pacific Rise inferred from narrow-beam echo-sounder surveys. *Marine Geology*, 54, 53–90.

Borgia, A., Delaney, P.T. and Denlinger, R.P. (2000) Spreading volcanoes. *Annual Review of Earth and Planetary Sciences*, 28, 539–70.

Burke, K.C. and Wilson, J.T. (1976) Hot spots on the Earth's surface. *Journal of Geophysical Research*, 93, 7690–708.

Cande, S.C., Labrecque, J.L., Larson, R.L. *et al.* (1989) *Magnetic Lineations of the World's Ocean Basins*. American Association of Petroleum Geologists, Tulsa, OK.

Cazenave, A. and Royer, J.Y. (2001) Applications to marine geophysics. In: *Satellite Altimetry and Earth Sciences* (eds. Fu, L.-L. and Cazenave, A.), pp. 407–39. Academic Press, San Diego, CA.

Clague, D.A. (1996) The growth and subsidence of the Hawaiian-Emperor volcanic chain. In: *The Origin and Evolution of Pacific Island Biotas* (eds. Keast, A. and Miller, S.E.), pp. 35–50. SPB Academic Publishing, Amsterdam.

Clague, D.A., Reynolds, J.R. and Davis, A.S. (2000) Near-ridge seamounts chains in the northeastern Pacific Ocean. *Journal of Geophysical Research*, 105, 16541–61.

Clouard, V. and Bonneville, A. (2001) How many Pacific hotspots are fed by deep-mantle plumes? *Geology*, 29, 695–8.

Clouard, V., Bonneville, A. and Barsczus, H.G. (2000) Size and depth of ancient magma reservoirs under atolls and islands of French Polynesia using gravity data. *Journal of Geophysical Research*, 105, 8173–91.

Coffin, M.F. and Eldholm, O. (1994) Large igneous provinces: crustal structure, dimensions, and external consequences. *Reviews in Geophysics*, 32, 1–36.

Craig, C.H. and Sandwell, D.T. (1988) Global distribution of seamounts from Seasat profiles. *Journal of Geophysical Research*, 93, 10408–20.

Darwin, C. (1842) *On the Distribution of Coral Reefs with Reference to the Theory of Their Formation*. Smith, Elder and Co, London.

Denlinger, R.P. and Savage, W.Z. (1989) Thermal stresses due to cooling of a viscoelastic oceanic lithosphere. *Journal of Geophysical Research*, 94, 744–52.

Dickinson, W.R. (1998) Geomorphology and geodynamics of the Cook-Austral island-seamount chain in the south Pacific Ocean: implications for hotspots and plumes. *International Geology Review*, 40, 1039–75.

Dieterich, J.H. (1988) Growth and persistence of Hawaiian volcanic rift zones. *Journal of Geophysical Research*, 93, 4258–70.

Douglas, B.C. (1991) Global sea level rise. *Journal of Geophysical Research*, 96, 6981–92.

Duncan, R.A. and Clague, D.A. (1985) Pacific plate motion recorded by linear volcanic chains. In: *The Ocean Basins and Margins* (eds. Nairn, A.E.M., Stehli, F.G. and Uyeda, S.), pp. 89–121. Plenum Press, New York.

Eaton, J.P. and Murata, K.J. (1960) How volcanoes grow. *Science*, 132, 925–38.

Epp, D. (1984) Implications of volcano and swell heights for thinning of the lithosphere by hotspots. *Journal of Geophysical Research*, 89, 9991–6.

Flood, P.G. (2001) The 'Darwin Point' of Pacific Ocean atolls and guyots: a reappraisal. *Palaeogeography, Palaeoclimatology, Palaeoecology*, 175, 147.

Fornari, D.J., Perfit, M.R., Allan, J.F. *et al.* (1988) Geochemical and structural studies of the Lamont seamounts: seamounts as indicators of mantle processes. *Earth and Planetary Science Letters*, 89, 63–83.

Friedrich, G. and Schmitz-Wiechowski, A. (1980) Mineralogy and chemistry of a ferromanganese crust from a deep-sea hill, Central Pacific, 'Valdivia' Cruise 13/2. *Marine Geology*, 37, 71–90.

Fryer, P. (1992) Mud volcanoes of the Marianas. *Scientific American*, 266, 46–52.

Fryer, P. (1996) Evolution of the Mariana convergent plate margin system. *Reviews of Geophysics*, 34, 89–125.

Gardner, J.V. and Blakely, R.J. (1984) Shimada seamount: an example of recent mid-plate volcanism. *Geological Society of America Bulletin*, 95, 855–62.

Grigg, R.W. and Epp, D. (1989) Critical depth for the survival of coral islands: effects on the Hawaiian archipelago. *Science*, 243, 638–41.

Grigg, R.W., Malahoff, A., Chave, E.H. and Landahl, J. (1987) Seamount benthic ecology and potential environmental impact from manganese crust mining in Hawaii. In: *Seamounts, Islands, and Atolls* (eds. Keating, B.H., Fryer, P., Batiza, R. and Boehlert, G.W.), pp. 379–90. AGU, Washington, DC.

Haase, K.M. and Devey, C.W. (1994) The petrology and geochemistry of Vesteris seamount Greenland basin – an intraplate alkaline volcano of non-plume origin. *Journal of Petrology*, 35, 295–328.

Harris, R.N., Fisher, A.T. and Chapman, D.S. (2004) Fluid flow through seamounts and implications for global mass fluxes. *Geology*, 32, 725–8.

Hillier, J.K. and Watts, A.B. (2004) 'Plate-like' subsidence of the East Pacific Rise – South Pacific superswell system. *Journal of Geophysical Research*, 109, 1–20.

Holcomb, R.T. and Searle, R.C. (1991) Large landslides from oceanic volcanoes. *Marine Geotechnology*, 10, 19–32.

Janney, P.E., Macdougall, J.D., Natland, J.H. and Lynch, M.A. (2000) Geochemical evidence from the Pukapuka volcanic ridge system for a shallow enriched mantle domain beneath the South Pacific superswell. *Earth and Planetary Science Letters*, 181, 47–60.

Jaroslow, G.E., Smith, D.K. and Tucholke, B.E. (2000) Record of seamount production and off-axis evolution in the western North Atlantic Ocean, 25° 25′–27° 10′N. *Journal of Geophysical Research*, 105, 2721–36.

Jayne, S.R., St. Laurent, L.C. and Gille, S.T. (2004) Connections between ocean bottom topography and Earth's climate. *Oceanography*, 17, 65–74.

Jordan, T.H., Menard, H.W. and Smith, D.K. (1983) Density and size distribution of seamounts in the Eastern Pacific inferred from wide-beam sounding data. *Journal of Geophysical Research*, 88, 10508–18.

Kitchingman, A. and Lai, S. (2004) Inferences on potential seamount locations from mid-resolution bathymetric data. In: *Seamounts: Biodiversity and Fisheries* (eds. Morato, T. and Pitcher, T.J.), pp. 7–12. Fisheries Centre, University of British Columbia, Canada, Vancouver, BC.

Koppers, A.A. and Staudigel, H. (2005) Asynchronous bends in Pacific seamount trails: a case for extensional volcanism? *Science*, 307, 904–7.

Koppers, A.A.P., Staudigel, H., Wijbrans, J.R. and Pringle, M.S. (1998) The Magellan seamount trail: implications for Cretaceous hotspot volcanism and absolute Pacific plate motion. *Earth and Planetary Science Letters*, 163, 53–68.

Koppers, A.A.P., Phipps Morgan, J., Morgan, J.W. and Staudigel, H. (2001) Testing the fixed hotspot hypothesis using 40Ar/39Ar age progressions along seamount trails. *Earth and Planetary Science Letters*, 185, 237–52.

Koppers, A.A.P., Staudigel, H. and Duncan, R.A. (2003) High-resolution 40Ar/39Ar dating of the oldest oceanic basement basalts in the western Pacific basin. *Geochemistry, Geophysics, Geosystems*, 4, doi:10.1029/2003GC000574.

Kroenke, L.W. (1996) Plate tectonic development of the Western and Southwestern Pacific. In: *The Origin and Evolution of Pacific Island Biotas* (eds. Keast, A. and Miller, S.E.), pp. 19–34. SPB Academic Publishing, Amsterdam.

Kroenke, L.W., Wessel, P. and Sterling, A. (2004) Motion of the Ontong Java plateau in the hot-spot frame of reference: 122 Ma–present. In: *Origin and Evolution of the Ontong Java Plateau* (eds. Fitton, J.G., Mahoney, J.J., Wallace, P.J. and Saunders, A.D.), pp. 9–20. Geological Society of London, London.

Kunze, E. and Llewellyn Smith, S.G. (2004) The role of small-scale topography in turbulent mixing of the global ocean. *Oceanography*, 17, 55–64.

Larson, R. and Olson, P. (1991) Mantle plumes control magnetic reversal frequency. *Earth and Planetary Science Letters*, 107, 437–47.

Larson, R.L. (1991) Latest pulse of Earth: evidence for a mid-Cretaceous superplume. *Geology*, 19, 547–50.

Lonsdale, P. (1983) Laccoliths(?) and small volcanoes on the flank of the East Pacific Rise. *Geology*, 11, 706–9.

Ludwig, W.J. and Houtz, R.E. (1979) Isopach map of the sediments in the Pacific Ocean basin, color map with text. American Association of Petroleum Geologists, Tulsa, OK.

Lueck, R.G. and Mudge, T.D. (1997) Topographically induced mixing around a shallow seamount. *Science*, 276, 1831–3.

Macdonald, G.A., Abbott, A.T. and Peterson, F.L. (1986) *Volcanoes in the Sea.* University of Hawaii Press, Honolulu, HI.

McGovern, P.J. and Solomon, S.C. (1993) State of stress, faulting, and eruption characteristics of large volcanoes on Mars. *Journal of Geophysical Research*, 98, 23553–79.

McNutt, M.K., Caress, D.W., Reynolds, J., Jordahl, K.A. and Duncan, R.A. (1997) Failure of plume theory to explain midplate volcanism in the southern Austral Islands. *Nature*, 389, 479–82.

Menard, H.W. (1964) *Marine Geology of the Pacific.* McGraw-Hill, New York.

Mitchell, N.C. (2001) Transition from circular to stellate forms of submarine volcanoes. *Journal of Geophysical Research*, 106, 1987–2003.

Montelli, R., Nolet, G., Dahlen, F.A. *et al.* (2004) Finite-frequency tomography reveals a variety of plumes in the mantle. *Science*, 303, 338–43.

Moore, J.G. (1987) Subsidence of the Hawaiian ridge. *US Geological Survey Professional Paper*, 1350, 85–100.

Moore, J.G., Ingram, B.L., Ludwig, K.R. and Clague, D.A. (1996) Coral ages and island subsidence, Hilo drill hole. *Journal of Geophysical Research*, 101, 11599–605.

Morgan, W.J. (1971) Convection plumes in the lower mantle. *Nature*, 230, 43–4.

Müller, R.D., Roest, W.R., Royer, J.Y., Gahagan, L.M. and Sclater, J.G. (1997) Digital isochrons of the world's ocean floor. *Journal of Geophysical Research*, 102, 3211–14.

Nataf, H.-C. (2000) Seismic imaging of mantle plumes. *Annual Review of Earth and Planetary Sciences*, 28, 391–417.

O'Neill, C., Müller, D. and Steinberger, B. (2005) On the uncertainties in hot spot reconstructions and the significance of moving hot spot reference frames. *Geochemistry, Geophysics, Geosystems*, 6, Q04003, doi: 10.1029/2004GC000784.

Parmentier, E.M. and Haxby, W.F. (1986) Thermal stresses in the oceanic lithosphere: evidence from geoid anomalies at fracture zones. *Journal of Geophysical Research*, 91, 7193–204.

Price, J.P. and Clague, D.A. (2002) How old is the Hawaiian biota? Geology and phylogeny suggest recent divergence. *Proceedings of the Royal Society of London*, 269, 2429–35.

Ribe, N.M. and Watts, A.B. (1982) The distribution of intraplate volcanism in the Pacific Ocean basin: a spectral approach. *Geophysical Journal of the Royal Astronomical Society*, 71, 333–62.

Rogers, A.D. (1994) The biology of seamounts. *Advances in Marine Biology*, 30, 305–50.

Sager, W.W. and Keating, B. (1984) Paleomagnetism of line islands seamounts: evidence for late Cretaceous and early tertiary volcanism. *Journal of Geophysical Research*, 89, 11135–51.

Sandwell, D., Gille, S.T. and Smith, W.H.F. (2002) *Bathymetry from Space: Oceanography, Geophysics, and Climate.* Geoscience Professional Services, Bethesda, MD.

Sandwell, D.T. and Smith, W.H.F. (1997) Marine gravity anomaly from Geosat and ERS-1 satellite altimetry. *Journal of Geophysical Research*, 102, 10039–54.

Sandwell, D.T., Winterer, E.L., Mammerickx, J. *et al.* (1995) Evidence for diffuse extension of the Pacific plate from Pukapuka ridges and cross-grain gravity lineations. *Journal of Geophysical Research*, 100, 15087–99.

Scheirer, D.S., Macdonald, K.C., Forsyth, D.W. and Shen, Y. (1996) Abundant seamounts of the Rano Rahi seamount field near the southern East Pacific Rise, 15S to 19S. *Marine Geophysical Researches*, 18, 13–52.

Schmidt, R. and Schminke, H.-U. (2000) Seamounts and island building. In: *Encyclopedia of Volcanoes* (eds. Sigurdsson, H., Houghton, B., Mcnutt, S.R., Rymer, H. and Stix, J.), pp. 383–402. Academic Press, San Diego, CA.

Searle, R.C., Francheteau, J. and Cornaglia, B. (1995) New observations of mid-plate volcanism and the tectonic history of the Pacific plate, Tahiti to Easter microplate. *Earth and Planetary Science Letters*, 131, 395–421.

Simkin, T. (1972) Origin of some flat-topped volcanoes and guyots. *Geological Society of America Memoir*, 132, 183–93.

Sleep, N.H. (1990) Hotspots and mantle plumes: some phenomenology. *Journal of Geophysical Research*, 95, 6715–36.

Smith, D.K. and Cann, J.R. (1990) Hundreds of small volcanoes on the median valley floor of the Mid-Atlantic Ridge at 24–30 N. *Nature*, 348, 152–5.

Smith, D.K. and Jordan, T.H. (1987) The size distribution of Pacific seamounts. *Geophysical Research Letters*, 14, 1119–22.

Smith, D.K. and Jordan, T.H. (1988) Seamount statistics in the Pacific Ocean. *Journal of Geophysical Research*, 93, 2899–918.

Smith, W.H.F. and Sandwell, D.T. (1997) Global sea floor topography from satellite altimetry and ship depth soundings. *Science*, 277, 1956–62.

Ten Brink, U.S. (1991) Volcano spacing and plate rigidity. *Geology*, 19, 397–400.

Tushingham, A.M. and Peltier, W.R. (1992) Validation of the ICE-3G model of Würm-Wisconsin deglaciation using a global data base of relative sea level histories. *Journal of Geophysical Research*, 97, 3285–304.

Vogt, P.R. (1974) Volcano height and plate thickness. *Earth and Planetary Science Letters*, 23, 337–48.

Vogt, P.R. and Jung, W.-Y. (2000) GOMap: a matchless resolution to start the new millennium. *EOS Transactions of the AGU*, 81, 254–8.

Watts, A.B. (2001) *Isostasy and Flexure of the Lithosphere.* Cambridge University Press, Cambridge.

Watts, A.B. and Ten Brink, U.S. (1989) Crustal structure, flexure, and subsidence history of the Hawaiian islands. *Journal of Geophysical Research*, 94, 10473–500.

Watts, A.B., Sandwell, D.T., Smith, W.H.F., and Wessel, P. (2006) Global gravity, bathymetry, and the distribution of submarine volcanism through space and time. *Journal of Geophysical Research*, 111, doi: 10.1029/2005JB004083.

Wessel, P. (1997) Sizes and ages of seamounts using remote sensing: implications for intraplate volcanism. *Science*, 277, 802–5.

Wessel, P. (2001) Global distribution of seamounts inferred from gridded Geosat/ERS-1 altimetry. *Journal of Geophysical Research*, 106, 19431–41.

Wessel, P. and Kroenke, L.W. (1997) Relocating Pacific hot spots and refining absolute plate motions using a new geometric technique. *Nature*, 387, 365–9.

Wessel, P. and Lyons, S. (1997) Distribution of large Pacific seamounts from Geosat/ERS-1: implications for the history of intraplate volcanism. *Journal of Geophysical Research*, 102, 22459–76.

White, S.M., Macdonald, K.C., Scheirer, D.S. and Cormier, M.-H. (1998) Distribution of isolated seamounts on the flanks of the East Pacific Rise, 15.3 S–20° S. *Journal of Geophysical Research*, 103, 30371–84.

Wilson, J.T. (1963) A possible origin of the Hawaiian islands. *Canadian Journal of Physics*, 41, 863–70.

Winterer, E.L. and Sandwell, D.T. (1987) Evidence from en-echelon cross-grain ridges for tensional cracks in the Pacific plate. *Nature*, 329, 534–7.

Wolfe, C.J., Bjarnason, I.T., Vandecar, J.C. and Solomon, S.C. (1997) Seismic structure of the Iceland mantle plume. *Nature*, 385, 245–7.

Chapter 2

How many seamounts are there and where are they located?

Adrian Kitchingman, Sherman Lai, Telmo Morato and Daniel Pauly

Abstract

Knowledge of seamount distributions is becoming more important in a wide range of research fields. Originally dominated by geophysical research, the identification of seamounts and their distribution now ranges from determining tectonic plate evolution to habitat mapping of seamount species. This, coupled with growing interest from commercial and geopolitical bodies, has further increased the need to geo-locate seamounts and determine the extent of their abundance. Whatever the method used, generally a seafloor peak is identified. Characteristics of the seafloor topography surrounding the peak are then used to determine whether the peak is the top of a seamount. The technologies used to identify seamounts from seafloor topography have advanced in sensitivity and coverage. This has resulted in a variety of methods to identify seamounts and estimate their abundances. The different features of two key techniques for seamount detection, sonar and satellite altimetry, have resulted in two research approaches. Due to coverage limitations, sonar seamount abundance estimates must be based on seamount sub-samples and extrapolated for larger coverage. Satellite altimetry is capable of near-global coverage, but is only able to identify and geo-locate large seamounts (higher than 1 km); it must therefore extrapolate abundance for smaller seamounts. In an attempt to create a global geo-dataset of seamounts, we have developed a relatively simple methodology for identifying and geo-locating potential seamounts. Using bathymetry data derived from altimetry, approximately 14 000 large seamounts greater than 1 km in height were identified. The resulting geo-dataset of seamounts offers an indication of the distribution of seamounts over the seafloor and shows that over 60% of seamounts occur in the Pacific Ocean. It was also determined that 50% of the seamounts occur within the exclusive economic zones (EEZs) of countries. Detailed datasets of seamount locations will help to identify potentially vulnerable regions and improve the understanding of ecological processes involved. This will in turn be an important contribution to the management and conservation of seamounts.

Introduction

The statement 'the surfaces of Mars, Venus, and the Moon are much better mapped than Earth's ocean floors' (Sandwell *et al.*, 2002) becomes clearly evident when one attempts to determine seamount abundance and seafloor distribution. To date, global bathymetric maps largely rely on interpolation to create a complete seafloor topography. In fact, the quality of depth values for some large areas of current global bathymetric data has been likened to the extrapolation of the US topography from its interstate highways (Sandwell *et al.*, 2002). That said, many attempts, mostly geophysical, have been made to determine the extent of seamount seafloor coverage since the mid-twentieth century. The importance of locating and describing seamounts is now reflected in research ranging from the explanation of oceanic crust evolution to habitat mapping of species associated with seamounts, with a growing engagement of commercial and geopolitical interests (see Chapters 1 and 20).

Just as the interest in seamount distributions has increased, so have the forms of data acquisition and analysis broadened over time. Originally relatively limited tracks of bathymetric profiles were used to estimate seamount numbers for single or partial ocean basins. With the accumulation of bathymetric data and advances in a number of bathymetry acquisition techniques, it is now possible to extrapolate seamount populations at a global level.

Bathymetry acquisition and data

The development of active sonar in the early 1900s provided the first means of research into seamount dimensions and their seafloor distribution. At its most basic, active sonar measures depth by timing the return of a sound pulse reflected by the seafloor. A towed or hull-mounted sonar device would measure the bathymetric profile along the path of a research vessel. The early active sonars, wide and single narrow beam, only produced single-dimensional profiles of the seafloor. Wide-beam sonars had little resolution of the actual direction of the sound pulse, so only indicated the shallowest depths within the swath of the sonar track. Single-narrow-beam sonars improved the accuracy of depth measurements by having the ability to focus a sound pulse. Their shortcoming was the limited coverage of the seafloor by a single pulse. The introduction of multi-beam sonar filled the need for accurate depth measurements across a wide track. This system combines the advantages of the narrow- and the wide-beam systems to yield accurate depth profiles across a wide swath. Essentially, in a single pulse, multiple narrow beams simultaneously return more than 100 soundings that measure depth across a wide swath perpendicular to the ship. This allows two- and three-dimensional mapping of the seafloor topography at a high resolution.

Although multi-beam bathymetric readings are still the preferred method for high-resolution bathymetric mapping, their use in global analyses of the seafloor is limited, since only a ship's track is mapped. Trying to obtain a global bathymetric dataset in this fashion would be prohibitively expensive and impractical. Satellite altimetry, a more recent technology, has been used to meet the need for global bathymetric coverage. Satellite altimetry measures the difference between the sea surface and Earth's ellipsoid, which is then used to calculate the Earth's geoid (see Box 1.1).

Many early altimetric seamount studies used US Navy *Seasat* satellite data (e.g., Lararewicz and Schwank, 1982; Dixon *et al.*, 1983; Baudry *et al.*, 1987; Craig and Sandwell, 1988). However, the *Seasat* coverage was sparse, with tracks separated by around 100 km. With such large spacing, any seamounts between tracks would have been missed or incompletely described (Wessel, 2001), affecting any global estimates (e.g., Craig and Sandwell, 1988). More recently, the combined datasets from the US Navy *Geosat* and European Space Agency *ERS-1* satellites have provided a much denser altimetric dataset with only a few kilometres between tracks. Although the resulting gridded map only ranges from 72° north and south, this data has provided the basis for many of the most current global seamount distribution studies (e.g., Wessel, 2001).

While a global bathymetric grid can be inferred from altimetry and sonar bathymetry tracks, such grids are difficult to produce and vary widely in quality and resolution. It is important to remember this when examining any methods that attempt to extract topographic features from these gridded datasets.

There are a number of global bathymetry grids available, mostly composed of various baseline sources, including bathymetric map contours (GEBCO, 2003), echo-sounding tracks, and satellite altimetry. Arguably one of the most frequently used global bathymetry grids is that based on a model pioneered by Dixon *et al.* (1983) from which Smith and Sandwell (1997) interpolated an almost global seafloor bathymetric dataset. They combined *Geosat/ERS-1* altimetry data and bathymetric soundings to produce a bathymetric grid at a resolution of approximately 2 arc minutes. The Smith and Sandwell (1997) dataset has also become the basis for other bathymetry grids that incorporate localised non-altimetric sources (e.g., ETOPO2, 2001; NRL DBDB2, 2003; GINA, 2004) to attempt a complete global coverage. Efforts to combine bathymetric datasets can lead to resolution degradation and in some cases, mistaken registration or mismatches between mapped features and their real-world position (Marks and Smith, 2006). Seafloor topography datasets for global analyses are expected to become increasingly useful with access to higher-resolution satellite altimetry and as military sonar archives are declassified.

Seamount identification

The ability to characterise seamounts from bathymetric data is key to any attempt to estimate seamount abundance. Seamounts are identified using a number of techniques, all of which involve a topographic feature to fit the criteria of a predefined seamount shape. Menard (1959) first suggested a comparative seamount definition and estimated seamount numbers for the Pacific basin. His definition of a seamount as 'an isolated elevation from the seafloor with a circular or elliptical plan, at least 1 km of relief, comparatively steep slopes and a relatively small summit area' (Menard, 1964) has provided a basis for many definition variants throughout the literature. The definition of a seamount in this book (see Preface) is evidence of the way in which different scientific disciplines identify the term 'seamount'.

Despite advances in identification methods, the basic process of isolating seamounts from baseline data has remained the same. This generally involves first identifying topographic maxima or peaks and then comparing the individual surrounding topographic dimensions of each peak with a set of parameter thresholds. Since two-dimensional bathymetric charts or sonar tracks of the seafloor have become available, many studies have relied on closed contours to single out peaked features (e.g., Batiza, 1982; Fornari *et al.*, 1987; Abers *et al.*, 1988; Jaroslow *et al.*, 2000). The introduction of gridded data has introduced automated methods that compare neighbouring grid values to determine local maxima (e.g., Wessel and Lyons, 1997; Wessel, 2001).

With the steady increase in quality and types of data used to identify seamounts, various seamount characteristics have been qualified and incorporated into models. The circular or elliptical nature of a seamount's base perimeter, either visually identified or measured by a length/width ratio, has proved to be a robust parameter and has been used persistently (e.g., Batiza, 1982; Epp and Smoot, 1989; Smith and Cann, 1990; Jaroslow *et al.*, 2000). Batiza (1982) also introduced a seamount slope threshold of >6, broadened to 17° by Smith and Jordan (1988). Height-to-radius ratio has been found to be fairly constant across seamount sizes (Simkin, 1972; Lacey *et al.*, 1981; Jordan *et al.*, 1983), although Abers *et al.* (1988) noted that larger seamounts are more regular in shape than smaller ones. Analysis of wide-beam sonar profiles by Jordan *et al.* (1983) used a seamount flatness threshold, defined by the ratio of the radius at the top to the radius at the bottom, to provide a definition of a seamount as approximating a truncated right circular cone.

Craig and Sandwell (1988) pioneered the use of the *Seasat* satellite altimetry data in seamount distribution research. Possible seamounts were located by visually identifying a characteristic peak and trough on the along-track slope of the geoid or vertical deflection profile (Craig and Sandwell, 1988). By measuring the dimensions of these signatures, researchers could estimate the diameter of a seamount using a modelled Gaussian seamount shape. Wessel and Lyons (1997) used the vertical gravity gradient (VGG) or geoid curvature derived from the denser *Geosat/ERS-1* altimetric data to identify seamounts. Their process involved the automated detection of local peaks in the VGG and the use of a closed VGG contour threshold to isolate possible seamounts. The potential seamount locations were then modelled using a Gaussian seamount shape to estimate seamount radius and height. Their model also incorporated the ability to determine dimensions of overlapping seamounts where more than one maximum occurred within a closed contour. Separating seamounts from peaks along fracture zones and ridges is a persistent problem in the automated identification of local peaks (Craig and Sandwell, 1988; Wessel and Lyons, 1997). Wessel and Lyons (1997) managed to reduce this problem by excluding seamounts within a certain distance of a fracture zone or trough.

The size of seamounts to include is a factor of constant concern within seamount location and abundance research. Height or relief is the most common parameter used to define the size of a seamount. Although originally defined as greater than 1 km (Menard, 1964), seamounts are now characterised at a range of heights from greater than 500 m (Cailleux, 1975) to greater than 50 m (Smith and Cann, 1990). The bathymetric data acquisition technique has furthered the importance of height as a limiting parameter. Modern sonars are able to detect features rising only metres from the seafloor, while altimetry is still restricted to identifying seamounts greater than 1–2 km. This has led to a distinction between large and small seamounts often used when extrapolating seamount abundances.

Seamount count methodologies

The estimation of seamount numbers over the years has been a complicated and difficult task. Most efforts have been restricted to single or partial ocean basins due to limited coverage by shipboard sonar tracks. The incomplete nature of bathymetric data, either through coverage or resolution, has led to the development of statistical techniques to extrapolate seamount abundances. Areal statistics were pioneered by Menard (1959), who treated bathymetric profiles as random samples that yielded an areal/density value of a random distribution of seamounts. Seamount numbers from areal distribution studies have varied widely over time, often using different height ranges, making it difficult to contrast investigations in a standardised manner (Smith, 1991). For the Pacific Ocean, Menard (1959) initially estimated 10 000 large seamounts (>1 km in height). Since then, estimated numbers have ranged widely with small seamounts potentially numbered in millions (see Fig. 2.1).

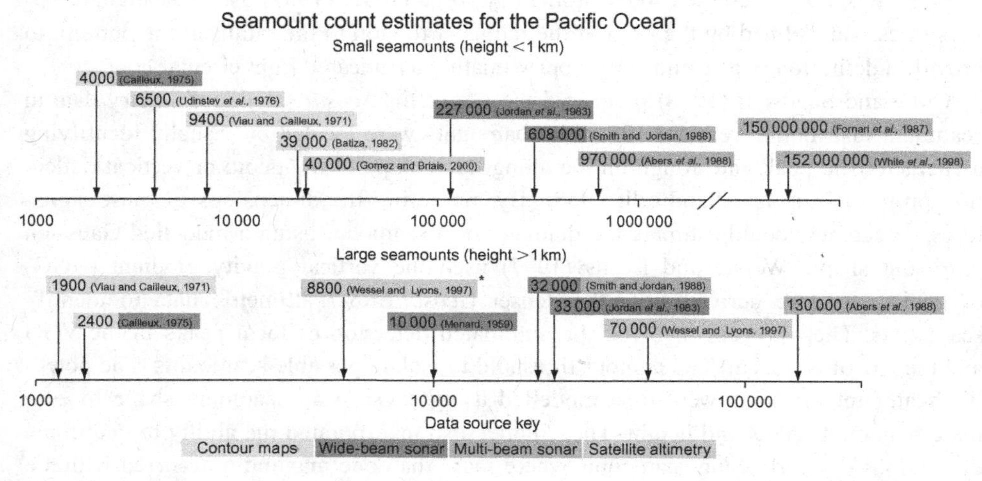

Fig. 2.1 Seamount abundance estimates for the Pacific Ocean from key published literature. The symbol * indicates a count of seamounts directly identified from underlying data, otherwise they will have been statistically derived. Depicted seamount estimates are rounded.

Early on it was noticed that the numbers of Atlantic seamounts decreased as the function of height increased (Litvin and Rudenko, 1971; Udinstev *et al.*, 1976). Batiza (1982) concluded that small seamounts were more abundant than larger ones and suggested that the size–frequency distribution of Pacific seamounts followed a Poisson-like trend. Although the Poisson distribution assumption has been carried throughout subsequent studies, its validity has been debated. Jordan *et al.* (1983) suggested that seamounts tended to form clusters or were strung out in chains (see Chapters 1 and 4).

Jordan *et al.* (1983) used wide-beam sonar profiles to compare the suitability of exponential and power-law models in explaining the size–frequency distribution of seamounts. They found that an exponential size distribution fitted reasonably well for a range of larger seamounts in the Eastern Pacific. Most subsequent studies of seamount size–frequency distributions, using wide- and multi-beam sonar, have focused on those seamounts less than 3 km in height (Smith, 1991) and have used the exponential model. It was noted that for smaller seamounts, within the range of 100–1000 m, the exponential model was only a marginally better fit than the power-law model (Abers *et al.*, 1988).

The introduction of altimetry data provided the first opportunity to estimate seamount numbers at a near-global level and describe the population of larger seamounts adequately

(Smith, 1991). Craig and Sandwell (1988) conducted the first comprehensive examination of global seamount distribution using satellite altimetry. Of the 8556 seamounts they located, one-quarter were previously uncharted (Craig and Sandwell, 1988). This was despite having to use an incomplete dataset due to the early failure of the *Seasat* mission.

Using altimetric data derived from VGG, Wessel and Lyons (1997) investigated the Pacific Ocean seamount population. They then examined the fit of their detected Pacific seamounts (8882) to the exponential and power-law models and concluded that the fit to the model was adequate for seamounts >2 km in height (Fig. 1.8).

Wessel (2001) expanded the examination of the VGG data to a global level and identified 14 639 large seamounts with an extrapolated global population of approximately 100 000 seamounts over 1 km (Fig. 2.2). More than half of the seamounts identified by Wessel (2001) occurred on the Pacific tectonic plate, with numerous seamounts occurring on the African plate (Eastern Atlantic and Western Indian Oceans), Indo-Australian plate (North and Eastern Indian Ocean), and Nazca plate (Central Eastern Pacific Ocean).

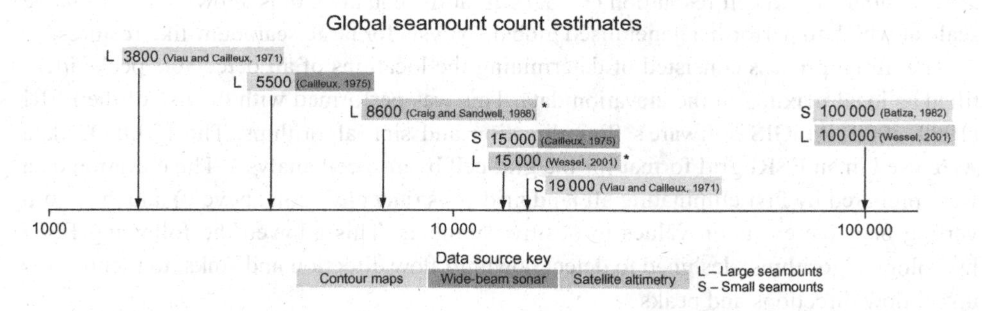

Fig. 2.2 Global seamount abundance estimates for both large and small seamounts from key published literature. The symbol * indicates the number of seamounts directly identified from underlying data, otherwise they will have been statistically derived. Large seamounts are those >1 km in height, while small seamounts are below <1 km. Depicted seamount estimates are rounded.

Despite improved technology and statistical techniques to isolate individual seamount features, there is still a clear difference in the capacity of the various techniques to acquire bathymetry, which leads to varying estimates of seamount numbers. Although bathymetric data derived from altimetry provide a near-global coverage and an opportunity for validation of seamount numbers, currently available data lack resolution to detect seamounts <2 km high (Wessel and Lyons, 1997). Wessel and Lyons (1997) suggested that only a third of the seamounts in the Pacific greater than 1 km in height were visible in current satellite data and that only a third of those were characterised in their study. On the other hand, while the latest sonar technology is able to detect seamounts down to less than 100 m in height, this technology is still restricted in seafloor coverage and must infer numbers through areal density calculations.

Spatial location of seamounts

Seamounts are becoming more important to the scientific community as shown by the breadth of contributions to this book. Good geo-referenced seamount datasets are urgently needed for research and to inform fisheries conservation and management agencies. In an attempt to produce such a dataset, the Sea Around Us Project (SAUP; www.seaaroundus.org) at the University of British Columbia uses a generalised definition to fit seamounts into ecological and management contexts. Unlike more comprehensive global studies by Craig and Sandwell (1988) and Wessel (2001), which use stricter seamount definition criteria in a geophysical context, the SAUP is primarily concerned with the location of seamount-like features regardless of their tectonic characteristics. The SAUP criteria assume that a possible seamount has a rise of 1000 m or more from the seabed and is roughly circular or elliptical in shape. Parameter thresholds used attempt to reflect this most basic of definitions for large seamounts.

The ETOPO2 global elevation dataset is used by SAUP as baseline bathymetric data. ETOPO2 contains the Earth's topographical elevations in a grid format and was created by the US National Oceanic and Atmospheric Administration (NOAA). For the ocean floor, it incorporates a number of bathymetric datasets, though it mainly consists of the Smith and Sandwell (1997) altimetry and sonar track derived bathymetry. The dataset is supplied at a 2 arc minutes grid cell resolution (~13.7 km^2 at the equator), thus allowing a reasonable scale at which to perform a generalised global analysis for large seamount-like features.

The initial process consisted of determining the locations of all detectable peaks identified as local maxima in the elevation data. This was performed with the use of the ESRI (1999–2002) ArcGIS software's flow direction and sink algorithms. The ETOPO2 data were used in an ESRI grid format for the grid cell by grid cell analysis. The elevation data were prepared by first eliminating all land grid cells (any elevation above 0) and then converting negative elevation values to positive numbers. This allowed the following ESRI hydrology algorithms, designed to detect downhill flow direction and sinks, to identify the uphill flow directions and peaks.

The ESRI flow direction algorithm was first used on the bathymetry data. This algorithm produces a grid in which each cell is allocated a flow direction value determined by the steepest descent from the immediate surrounding grid cells. Undefined flow directions occur when no flow direction can be determined for a grid cell. The conditions that cause this are when all surrounding grid cells flow into a single grid cell or when two adjacent grid cells flow into each other. The ESRI sink algorithm is used on the resulting flow direction grid to identify all grid cells that have undefined flow directions. The resulting sink, or its inverse, the seafloor peak grid, can then be overlaid with the ETOPO2 depth grid to indicate all identifiable seafloor peaks.

Once the peaks are detected, two methods are used to identify possible seamounts. The first involves isolating the peaks that have a significant rise from the ocean floor. The second isolates peaks with a circular or an elliptical base in an effort to eliminate small peaks found along steep ridges. The overlapping seamounts found using both these methodologies are used as the SAUP seamount dataset. To determine the overlap, the points of the two datasets have to be within 2 arc minutes of each other.

Method 1

The initial part of the process involved producing a grid of standard deviations of depth to accentuate any depth changes across the ocean floor. The neighbourhood statistics function in ESRI's ArcGIS Spatial Analyst software was used to produce a grid with a depth standard deviation value for each ETOPO2 depth grid cell as compared to its immediate neighbourhood. To enable the identification of possible seamounts, the standard deviation and the seafloor peak grids were overlaid. Using ESRI's ArcGIS Spatial Analyst, each peak grid cell was then compared to a 5 × 5 kernel (localised grid) of its neighbourhood on the standard deviation grid. If any grid cells within the kernel were above a 300-m standard deviation, the focal peak grid cell was considered a possible seamount (see Fig. 2.3).

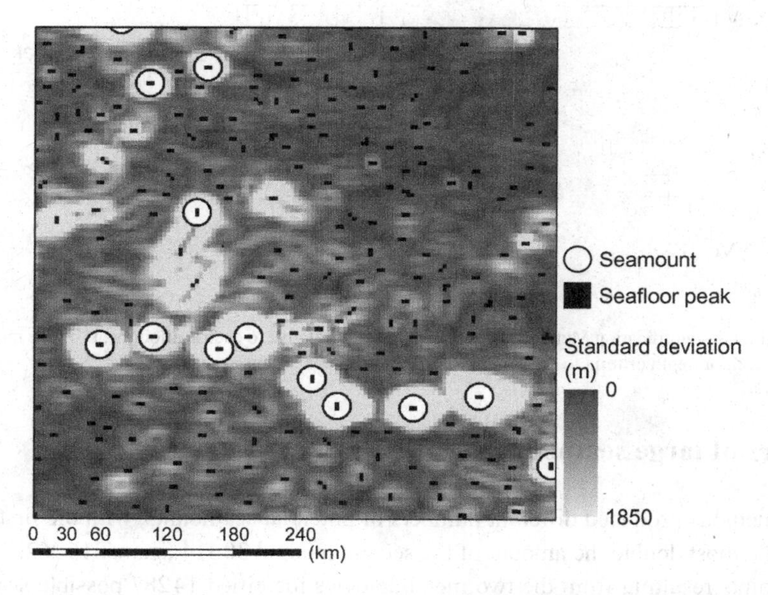

Fig. 2.3 An example of the results of Method 1 for seamount identification. Potential seamount-like features are identified from the standard deviation of depth values surrounding detected seafloor peaks. The peaks marked as seamounts have passed surrounding depth standard deviation threshold criteria.

Method 2

The second method compared the raw peak grid dataset directly with the ETOPO2 data. An algorithm was developed that scanned ETOPO2 depths approximately 90 km around each peak along 8 radii of at 45° intervals (see Fig. 2.4). The lowest and the highest depths over the radii (10 grid cells per radius near the equator, more at higher latitudes) were then recorded. A raw peak was considered to be a seamount when the following three conditions were met:

1. Each of and all the 8 radii included depths differing by at least 300 m. This helped eliminate all peaks of insignificant mounds.
2. If 2 radii included depths between 300 and 1000 m with the shallowest point being closer to the peak than to the deepest point, and if the radii formed an angle of less than 135°. This condition was created to help separate ridges from seamounts.
3. At least 5 of the 8 radii around a peak included depths with a difference of at least 1000 m, with the shallowest point being closer to the peak than to the deepest point.

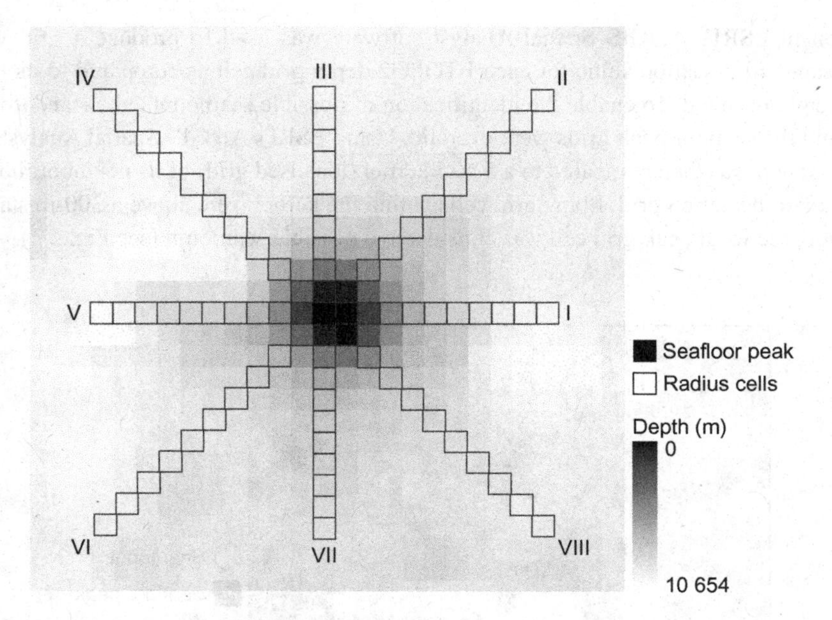

Fig. 2.4 Seamount identification Method 2 uses 8 grid cell radii relative to each seafloor peak. These radii are then used to test seafloor depth criteria to determine whether a peak is a potential seamount.

Numbers of large seamounts

The two methods produced different numbers of potential seamounts, with the first method producing almost double the amount of the second method 30 314 against 15 962. The overlapping points resulting from the two methodologies identified 14 287 possible seamounts. The range of seamount numbers also varied separately for the two methods and their set thresholds. Smaller potential seamounts were identified by Method 1 when the standard deviation threshold was lowered, thus increasing the seamount count (see Table 2.1). Method 2 remained relatively constant, with estimates between 15 000 and 20 000 seamounts, depending on the depth change threshold set between 100 and 500 m. As the threshold was increased for Method 2, the non-linear variation in seamount counts could be attributed to the fact we took into account the proximity to the nearest seafloor rise and the depth of the valley between peaks as well as the change in surrounding depth (see Table 2.1).

Table 2.1 Seamount prediction counts for the two separate methods using various thresholds. Method 1 in the SAUP seamount identification methodology uses standard deviation of depth values surrounding detected seafloor peaks, while Method 2 uses raw depth values surrounding detected seafloor peaks.

Local depth standard deviation or depth threshold (m)	Potential seamount count	
	Method 1	Method 2
100	~142 000	~20 000
300	30 314	15 962
500	~8500	~18 000

The SAUP study has found a relatively simple way of inferring potential seamounts from mid-resolution bathymetric data. Although there is no reference to volcanism in the model parameters, a point dataset of seamount-like features was achieved. However, the production of these potential seamount locations should only be used as a guide for global seamount distribution and is by no means a definitive dataset. Several points that should be noted when making any conclusions with the dataset are as follows:

1. Criteria for the seamount inferences were only sensitive to a broad level, with the definition of seamounts still very generalised. This sensitivity is directly influenced by the depth and depth standard deviation thresholds. The sensitivity of the thresholds is also directly influenced by the resolution of the underlying bathymetry data. Any features smaller than the cell size of the bathymetry data will have had their dimensions blurred with surrounding features, which could bring them outside of the inference criteria bounds. All predicted seamount locations are only indicative of very large seamounts, since the resolution of the underlying bathymetric and altimetric data will only allow for the confident identification of features >2 km high (Wessel and Lyons, 1997).
2. As expected, many of the predicted seamounts occurred along mid-ocean ridges, although Method 2 incorporated criteria that attempted to eliminate peaks along ridges and fracture zones. On the flip side, our attempt to eliminate peaks along ridges could also eliminate actual seamounts. It was decided to keep the results conservative by maintaining relatively restrictive criteria for Method 2 to reduce error.
3. The area around each peak, tested for seamount characteristics, is restrictive in nature and would affect the results to some degree. The kernel used by Method 1 equates to an area of approximately 342 km^2 at the equator. It was hoped that a kernel of this size would allow the detection of large seamounts while eliminating large peaked banks. This kernel size could be further examined to optimise the sensitivity of the analysis. Likewise the lengths of radii in Method 2 could also be optimised. While aiming to be wide enough for the largest of seamounts, the radii could also include some features classed as semi-circular banks or guyots.
4. All points were included, regardless of their proximity to a coastline. As a guide, it was found that around 5% of the predicted seamount locations occurred within 20 km of a coastline. Due to resolution constraints, there is the possibility that the ETOPO2 bathymetry may have registered some small land structures as below sea level, resulting in their inclusion in the dataset.
5. The predicted seamount locations generated by the SAUP model have a much lower resolution than the underlying bathymetric data. One of the results of the ESRI flow direction and sink algorithms is paired points, where two cells flow into each other indicating a sink interpreted in this analysis as a peak. The two methods used were conducted with these paired points present, but the final locations had any paired peaks reduced to a single location at a mid-point between the two, maintaining the shallowest depth value of the pair. The error in real-world location is also enhanced by the mistaken registrations in the underlying ETOPO2 bathymetry dataset (Marks and Smith, 2006).

Ground truthing was performed on a dataset of known seamounts produced from a combination of data recorded in the US Department of Defense (1989) and SeamountsOnline (Stocks, 2003). It was found that approximately 60% of the known seamounts were within 30 arc minutes of predicted seamounts. The 30 arc minutes resolution is used by the SAUP's ecosystem and fisheries models. The total number of large seamount features identified was also notably similar to the number identified by Wessel (2001), although it is evident that the locations of many seamounts differed due to the differences in criteria and the context of the research. As expected, the inferred SAUP seamount count was well below Wessel's extrapolated estimate of 100 000, but it did far exceed numbers identified or estimated (Fig. 2.2) in some previous studies (e.g., Viau and Cailleux, 1971; Cailleux, 1975; Craig and Sandwell, 1988). However, as mentioned above, the dataset used by Craig and Sandwell was incomplete.

Locations of large seamounts

Since many studies are restricted to a particular ocean, an attempt to get an estimate of predicted large seamounts per ocean was performed. The statistical areas of the United Nations' Food and Agriculture Organization (FAO; www.fao.org) were used to identify oceans (see Fig. 2.5). Although these areas do not exactly fit the oceanic boundaries in other studies, a broad comparison was possible. Despite previously noted locational differences of potential seamounts, the number of seamount features for the Pacific Ocean falls within the bounds of the 8882 seamounts identified by Wessel and Lyon (1997). As expected, the modelled numbers fall well below the size–frequency estimates of around 30 000 by Smith and Jordan (1988) and Wessel and Lyon's (1997) own extrapolation of around 70 000 possible seamounts (see Fig. 2.1). The Atlantic contained roughly a fifth of the identified seamount features, while the Indian Ocean held just over a tenth. The counts for the Southern Ocean would have been underestimated by the FAO areas, as the ocean actually reaches the southern coast of Australia. A detailed list of seamount numbers by FAO statistical areas is presented in Table 2.2.

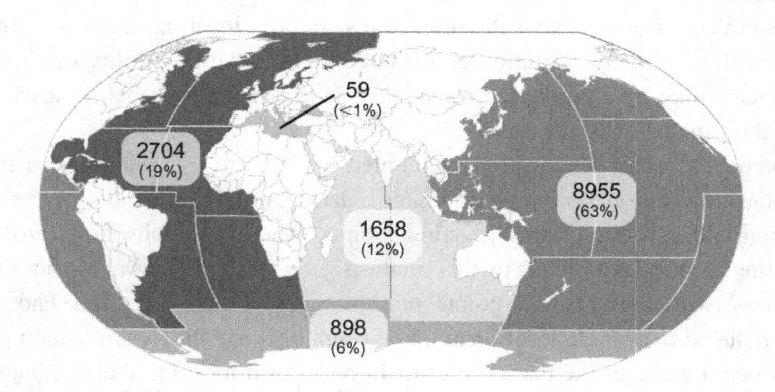

Fig. 2.5 Potential large seamount counts identified by this methodology for oceans defined by the statistical areas of the United Nations' Food and Agriculture Organization.

Table 2.2 Potential seamount counts for oceans defined by FAO statistical areas and those within EEZs.

Ocean areas	FAO area	Potential large seamounts	Potential large seamounts within EEZs
Pacific	All	*8955*	*5415*
Eastern Central	77	2735	1768
Northeast	67	265	89
Northwest	61	1350	720
Southeast	87	939	239
Southwest	81	996	353
Western Central	71	2670	2246
Atlantic	All	*2704*	*745*
Eastern Central	34	536	103
Northeast	27	325	114
Northwest	21	83	6
Southeast	47	639	127
Southwest	41	452	151
Western Central	31	669	244
Indian	All	*1658*	*576*
Eastern	57	588	162
Western	51	1070	414
Mediterranean and Black Sea	37	*59*	*0*
Southern Ocean	All	*898*	*185*
Atlantic, Antarctic	48	498	127
Indian Ocean, Antarctic	58	212	58
Pacific, Antarctic	88	188	0
Arctic	18	*13*	*0*
Total		*14287*	*6921*

EEZ data were licensed from Global Maritime Boundaries Database (2003), General Dynamics Advanced Information Systems (http://www.gd-ais.com/Capabilities/offerings/sr/gmbd.htm).

The occurrence of the SAUP-inferred seamounts within EEZs was tested to indicate the proportion of large seamount that could be individually managed or incorporated into an existing management programme under a country's jurisdiction. It was determined that approximately 50% of potential seamounts occurred within EEZs worldwide (see Table 2.2 and Fig. 2.6). With many possible large seamounts occurring in the high seas, the first attempt of ecosystem management in an international situation may be necessary (Alder and Wood, 2004). Of the seamount features found within country maritime jurisdictions, approximately 60% of the Pacific large seamounts occurred within an EEZ. The highest numbers were in the EEZs under the administration of industrialised countries such as the USA, France, Japan, and New Zealand. Approximately 28% and 35% of seamount features in the Atlantic and Indian oceans, respectively, were within EEZs.

The SAUP methodology has provided a simple way of generating a global seamount dataset directly from elevation data. Although the current output is suitable for a generalised global analysis, it is aimed at providing a possible approach that can be improved and refined with further research and higher-resolution bathymetric datasets.

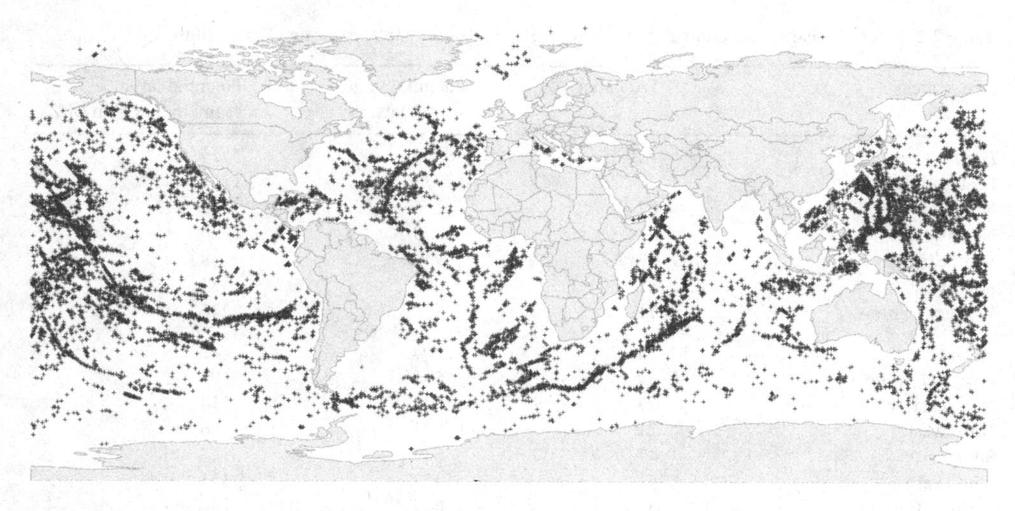

Fig. 2.6 Potential large seamounts points inferred by the Sea Around Us Project analysis.

Importance of mapping seamounts

The availability of detailed datasets of global seamount locations is a major requisite for understanding ecological processes, identifying potentially exploited areas, and effective management proposals. For example, the biogeography and biodiversity of seamounts (Chapter 13) will never be fully understood until detailed maps of seamount locations are available. On the other hand, quantifying seamount fisheries catches (Chapter 18) will only be possible after obtaining a detailed global distribution of seamounts.

As mentioned above, many seamounts occur within the EEZ of maritime countries, with approximately half occurring in international waters. The general perception that most seamounts occur outside of the areas under national jurisdiction is thus incorrect. This has profound implications for the ways in which appropriate levels of seamount management protection are achieved. Beyond areas under national jurisdiction, the responsibility to preserve seamounts as part of the global common heritage belongs to all nations (Alder and Wood, 2004). Clearly, this represents a challenge for the international community and individual countries wishing to conserve the biodiversity of these ecosystems (Chapter 20).

Acknowledgements

The authors would like to acknowledge the support from Pew Charitable Trusts, the staff and students of the Sea Around Us Project and the UBC Fisheries Centre. Thanks also to Irina Fainberg for her feedback on the drafting of the manuscript.

References

Abers, A.A., Parsons, B. and Weissel, J.K. (1988) Seamount abundances and distributions in the southeast Pacific. *Earth and Planetary Science Letters*, 87, 137–51.

Alder, J. and Wood, L. (2004) Managing and protecting seamount ecosystems. In: *Seamounts: Biodiversity and Fisheries* (eds. Morato, T. and Pauly, D.), pp. 67–73. *Fisheries Centre Research Reports*, 12(5), 78.

Baudry, N., Diament, M. and Albouy, Y. (1987) Precise location of unsurveyed seamounts in the Austral archipelago area using Seasat data. *Journal of Geophysical Research Astronomical Science*, 89, 869–88.

Batiza, R. (1982) Abundances, distribution and sizes of volcanoes in the Pacific Ocean and implications for the origin of non-hotspot volcanoes. *Earth and Planetary Science Letters*, 60, 195–206.

Cailleux, A. (1975) Frequence des monts sous-marins dans trois parties des oceans Pacifique et Indien. *Cahiers de geographie du Quebec*, 19, 553.

Craig, C.H. and Sandwell, D.T. (1988) Global distribution of seamounts from Seasat profiles. *Journal of Geophysical Research*, 90(B9), 10408–20.

Dixon, T.H., Naraghi, M., McNutt, M.K. and Smith, S.M. (1983) Bathymetric prediction from SEASAT altimeter data. *Journal of Geophysical Research*, 88(C3), 1563–71.

Epp, D. and Smoot, N.C. (1989) Distribution of seamounts in the North Atlantic. *Nature*, 337, 254–7.

ESRI (1999–2002) *ArcGIS: Release 8.3 [software]*. Environmental Systems Research Institute, Redlands, CA.

ETOPO2 (2001) US Department of Commerce, National Oceanic and Atmospheric Administration, National Geophysical Data Center, 2001. *2-Minute Gridded Global Relief Data*. http://www.ngdc.noaa.gov/mgg/fliers/01mgg04.html

Fornari, D.J., Batiza, R. and Luckman, M.A. (1987) Seamount abundances and distribution near the East Pacific Rise 0°–24° N based on SeaBeam data. In: *Seamounts, Islands, and Atolls* (eds. Keating, B., Fryer, P. and Batiza, R.), pp. 13–21. American Geophysical Union, Washington, DC, 405 pp.

GEBCO (2003) General Bathymetric Chart of the Oceans. http://www.bodc.ac.uk/products/bodc_products/gebco/

GINA (2004) Geographic Information Network of Alaska. http://www.gina.alaska.edu/page.xml?group=data&page=griddata

Gomez, O. and Briais, A. (2000) Near-axis seamount distribution and its relationship with the segmentation of the East Pacific Rise and northern Pacific–Antarctic Ridge, 17° N–56° S. *Earth and Planetary Science Letters*, 175, 233–46.

Jaroslow, G.E., Smith, D.K. and Tucholke, B.E. (2000) Record of seamount production and off-axis evolution in the western North Atlantic Ocean, 25°25′–27°10′N. *Journal of Geophysical Research*, 105, 2721–36.

Jordan, T.H., Menard, W. and Smith, D.K. (1983) Density and size distribution of seamounts in the Eastern Pacific inferred from wide-beam sounding data. *Journal of Geophysical Research*, 88(B12), 10508–18.

Lacey, A., Ockendon, J.R. and Turcotte, D.L. (1981) On the geometrical form of volcanoes. *Earth and Planetary Science Letters*, 54, 139–43.

Lararewicz, A.R. and Schwank, D.C. (1982) Detection of uncharted seamounts using satellite altimetry. *Geophysical Research Letters*, 9, 385–8.

Litvin, V.M. and Rudenko, M.V. (1971) Distribution of seamounts in the Atlantic. USSR Academy of Sciences: Earth Science Sections (English Translation), 213, 944.

Marks, K.M. and Smith, W.H.F. (2006) An evaluation of publicly available global bathymetry grids. *Marine Geophysical Researches*, 27(1), 19–34.

Menard, H.W. (1959) Geology of the Pacific floor. *Experientia*, 15, 205–13.

Menard, H.W. (1964) *Marine Geology of the Pacific*, 271 pp. McGraw-Hill, New York.

NRL DBDB2 (2003) Naval Research Laboratory Digital Bathymetry Data Base 2-Minute Resolution. http://www7320.nrlssc.navy.mil/DBDB2_WWW

Sandwell, D.T., Gille, S.T. and Smith W.H.F. (eds.) (2002) *Bathymetry from Space: Oceanography, Geophysics, and Climate*, June, 24 pp. Geoscience Professional Services, Bethesda, MD. http://www.igpp.ucsd.edu/bathymetry_workshop/

Simkin, T. (1972) Origin of some flat-topped volcanoes and guyots. *Geological Society of America Memoir*, 132, 183–93.

Smith, D.K. (1991) Seamount abundances and size distributions, and their geographic variations. *Reviews in Aquatics Sciences*, 5(3–4), 197–210.

Smith, D.K. and Cann, J.R. (1990) Hundreds of small volcanoes on the median valley floor of the Mid-Atlantic Ridge at 24–30° N. *Nature*, 348, 152–5.

Smith, D.K. and Jordan, T.H. (1988) Seamount statistics in the Pacific Ocean. *Journal of Geophysical Research*, 93(B4), 2899–918.

Smith, W.H.F. and Sandwell, D.T. (1997) Global sea floor topography from satellite altimetry and ship depth soundings. *Science*, 277, 1956–62.

Stocks, K. (2003) SeamountsOnline: an online information system for seamount biology. World Wide Web electronic publication. http://seamounts.sdsc.edu

Udinstev, G.B., Agapova, G.V., Larina, N.I. and Marova, N.A. (1976) Seamounts of the Pacific Ocean, general features of relief on the Pacific Ocean floor. In: *Volcanoes and Tectonosphere*, (eds. Aoki, H. and Lizuka, S.), 370 pp. Tokai University Press, Tokyo.

US Department of Defense (1989) Gazetteer of undersea features. CD-ROM, US Department of Defense, Defense Mapping Agency, USA.

Viau, B. and Cailleux, A. (1971) Fréquence de monts sous-marins dans une parties de l'Océan Indien et du Pacifique. *Zeitschrift für Geomorphologie*, 15, 471–8.

Wessel, P. (2001) Global distribution of seamounts inferred from gridded Geosat/ERS-1 altimetry. *Journal of Geophysical Research*, 106(B9), 19431–41.

Wessel, P. and Lyons, S. (1997) Distribution of large Pacific seamounts from Geosat/ERS-1: implications for the history of intraplate volcanism. *Journal of Geophysical Research*, 102(B10), 22459–75.

White, S.M., Macdonald, K.C., Scheirer, D.S. and Cormier, M.-H. (1998) Distribution of isolated volcanoes on the flanks of the East Pacific Rise, 15.3° S–20° S. *Journal of Geophysical Research*, 103, 30371–84.

Chapter 3
A history of seamount research

Paul E. Brewin, Karen I. Stocks and Gui Menezes

Abstract

Physical and biological phenomena now known to be associated with seamounts have been a source of curiosity since seafarers first encountered seamounts centuries ago. However, only since the technological advancements of the early twentieth century have closer examination of seamounts been possible. Significantly, the development of echo sounding led to an explosion of deep sea exploration and scientific discovery. After early pioneering work with echo sounding in the Gulf of Alaska and Western Pacific, the nature and ubiquity of seamounts throughout the oceans became apparent. Moreover, the characterization and distribution of seamounts led to better understanding of deep ocean ridge development, seafloor spreading, and plate tectonics. Subsequent to this pioneering work, numerous oceanographic and biological characteristics have been described for seamounts, including significant roles in global ocean circulation and biogeography in the deep sea. Despite more than 100 years of study on seamounts, many research challenges remain. With the global expansion of commercial deep sea fishing in the mid-1950s, seamounts have become centres of intensive commercial fishing pressure. The need for their management and conservation has spawned a new era of integrative seamount research, and international collaborative research and conservation initiatives.

Introduction

Seamounts have captured the curiosity of researchers from many disciplines since deep sea exploration began. Charles Darwin may have been the first to describe the process of seamount formation when he proposed a hypothesis for the gradual subsidence of coral reefs and volcanic islands in the South Pacific in his book *The Structure and Distribution of Coral Reefs* (1842). His ideas were largely dismissed by the scientific community of the day and it was a century before his ideas were accepted; it was not until the description of guyot-type deep seamounts (flat-topped, subsided volcanic islands, see Chapter 1) that Darwin's subsidence hypothesis could be tested (see Hamilton, 1956; Menard, 1964 for reviews). If Darwin had possessed the technology aboard the *HMS Beagle* to sound and sample the chain of guyots near the atolls he studied, he would have seen a more complete picture of the fate of volcanic islands in the South Pacific, and would have likely have confirmed many of his theories about subsidence.

The history of seamount discovery dates back to the very earliest ocean explorations. Although seamounts were not specifically known, their increased water movement and higher food production are likely to have been well known and documented

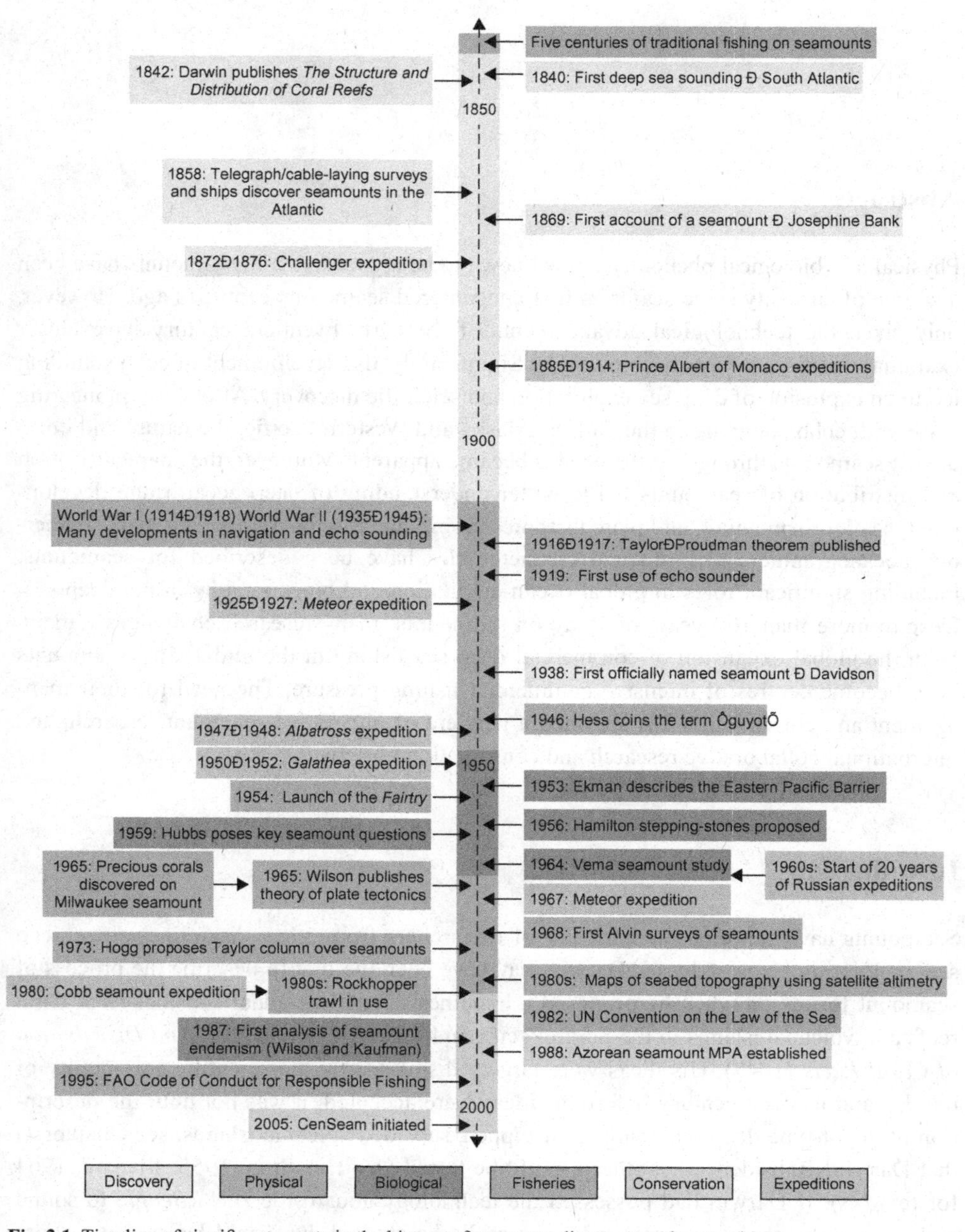

Fig. 3.1 Timeline of significant events in the history of seamount discovery and research.

among seafarers. Early observations of these phenomena were not made by scientists, but fishermen, merchants, settlers, and explorers of the fifteenth century looking for new lands. Explorers were the first 'oceanographers', making meticulous maps and descriptive references of their travels. This chapter outlines the history of seamount discovery and research, highlighting major technical and conceptual milestones, and the evolution of seamount research themes. The early observations of ocean explorers continue to influence present-day research. However, only in relatively recent times has technology allowed closer investigation of seamounts. Technological limitations were overcome post-World War II when sounding technology greatly facilitated deep sea work, as shown by a flurry of deep sea exploration and scientific discovery in the early to mid-1950s (Fig. 3.1). The path from seamount discovery to understanding the relationship between seamounts and large-scale geological oceanographic processes is outlined. Sampling of seamount fauna revealed unexplained high complexity of seamount communities, and prompted the development of large-scale multidisciplinary studies linking pattern, process, and contemporary ecological theory. The exploitation of seamount fisheries is a continuing theme throughout the history of seamount research, and today seamount commercial fisheries are highly valued (Chapter 9). The themes of seamount conservation and management have emerged as primary drivers for future priorities in seamount research.

Early discovery of seamounts

It is thought that the deep ocean was first sounded by Posidonius (C. 135–150 BC), who recorded a depth of 1000 fathoms (approximately 1828 m) in the Sardinian Sea (Deacon, 1971). Fernão de Magalhães (Magellan) of Portugal made several unsuccessful attempts to determine the depth in the Pacific Ocean during the first circumnavigation of the globe in 1521. His soundings between the islands of St Paul and Los Tiburones (Southern Indian Ocean) using sounding lines that were only 100–200 fathoms in length failed to touch the bottom, leading Magellan to conclude that he had reached the deepest part of the ocean (Murray and Hjört, 1965). Sir James Clark Ross, who sounded 2425 fathoms (approximately 4434 m) in the South Atlantic on his 1840 *Erebus* and *Terror* expedition, is often credited for the first accurate deep sea sounding (Deacon, 1971; Schlee, 1973). These latter trials in deep sea soundings started a trend in British and American deep sea exploration, leading to many technical developments, notably the use of a lighter silk or twine line, lighter lead weights, and detachable weight systems.

By the end of the nineteenth century, systematic surveys of the continental shelf and beyond were being carried out, and undersea topography was quickly regarded as being significantly more complex than originally thought. At this time, the ubiquity of seamounts was not fully understood; seamounts were considered to be anomalies in an otherwise flat landscape. Perhaps the first recognized seamount was the Josephine Bank in the North Atlantic west of Gibraltar, which was discovered in 1869 by the Swedish corvette *Josephine* (Ankarcrona, 1969). Numerous benthic dredge samples containing 'very rich findings for the scientists' were taken on the bank. Interestingly, the colour of the water was noted to be lighter than that of the surrounding ocean, and an increase in gull and petrel numbers was observed feeding on the high abundance of fish found in the vicinity of the bank. Evening plankton netting was considered 'very profitable' with an unusually high abundance of life.

Numerous discoveries of deep sea features, as well as major contributions to ocean and atmospheric sciences, were made during the oceanographic expeditions of Prince Albert I of Monaco (1885–1914). On board his yachts *L'Hirondelle*, *Princesse-Alice*, *Princesse-Alice II*, and *L'Hirondelle II*, his explorations ranged from the Mediterranean Sea as far as the Azores in the Atlantic Ocean, where soundings and samples were taken down to depths of more than 4000 m; the *Princesse-Alice* giving its name to the major seamount ('bank') in the Azores region, discovered in 1896 (Minelle, 1986; Carpine-Lancre and Saldanha, 1992).

Further seamount discoveries were made during trans-Atlantic telegraph cable-laying expeditions in the mid to late-1800s. Several seamounts now carry the names of the cable-laying ships from which they were discovered, e.g., the *Dacia* (Box 3.1), *Faraday*, *Cruiser*, and *Seine* seamounts (Murray and Hjört, 1965) (Atlantic Cable History: http://www.atlantic-cable.com/). However, despite the growing knowledge of the existence of distinct 'mountains in the sea', terms such as 'ridge', 'bank', or 'shoal' were still being used to describe seamounts until the early 1900s (Littlehales, 1932). The first official use of the term 'seamount' was by the US Board of Geographic Names in 1938 to describe the Davidson Seamount, discovered by the US Coast and Geodetic Survey ship *Guide* in 1933 (US Board of Geographic Names, 1938).

Seamounts as unique physical features

The advent of the echo sounder revolutionized seafloor mapping and thus facilitated the study of seamounts. It was not until the development of hydrophone and electronic amplifying devices in the early twentieth century that echo sounding came into common use, allowing extensive tracks of seafloor to be mapped continuously without stopping the ship. Although the concept of using sound to 'see' into the ocean had been around since the mid-1800s (NOAA website: http://www.history.noaa.gov); serious work on echo-sounding technology did not expand until French researchers tested the first echo sounder in 1919. This technology became important for submarine detection during World War I and for surveying the seafloor for the laying of deep sea telegraphic cables. It was on the Meteor expedition (1925–1927) that echo sounding was first used extensively (Herdman *et al.*, 1956). Echo soundings of the Atlantic Ocean began to accumulate, and in 1952 Marie Tharp began compiling these soundings to create a bathymetric map of the Atlantic Ocean. Their plots show for the first time the V-shaped valley of the Mid-Atlantic Ridge. In 1959 Heezen, Tharp, and Ewing published *The Floor of the Ocean: The North Atlantic*, which was the first map of its kind, revealing numerous seamounts (Kunzig, 2000).

Echo-sounding technology was quickly adopted by hydrographic survey ships around the world, and highly detailed maps of seabed topography started to be produced. Most notable of the earliest studies taking advantage of this new technology were the seamount surveys of Murray (1941), Hess (1946), and Menard and Dietz (1951), which described extensive soundings of the Gulf of Alaska and the Western Pacific Ocean (Fig. 3.2). These studies

Fig. 3.2 Artist's conception of the Mid-Pacific Mountains from the frontispiece in Hamilton (1956). The original caption reads, 'If the Pacific Ocean were drained away, the mile-deep sunken islands would emerge as truncated volcanic cones. The original painting is by the distinguished scientific illustrator Chesley Bonestall and is based on part of the bathymetric chart of the Mid-Pacific range.'

were particularly novel and important because, for the first time, deep sea exploration was less focused on determining the greatest depths of the ocean than on describing the distribution and formation of topographic features such as seamounts, guyots, and trenches. Hess (1946) mapped and described the chain of flat-topped peaks that paralleled the islands and atolls along the Darwin Rise in the Pacific Ocean, and named these types of seamounts 'guyots' in honour of the late Princeton University Professor of Geology, Arnold Henry Guyot (1807–1884). Hess also recognized that many of those seamounts mapped by Murray (1941) were in fact, guyots. Menard and Dietz (1951) re-evaluated the formation of seamounts in the Gulf of Alaska using previous and new echo sounding data.

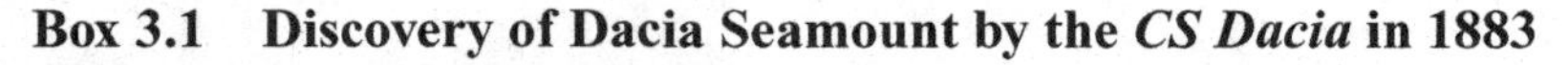

Box 3.1 Discovery of Dacia Seamount by the *CS Dacia* in 1883

The *CS Dacia* in 1883

Several seamounts were discovered and named by the telegraph cable ships that began to operate in the middle of the 19th century. In 1883 the *CS Dacia* discovered several oceanic shoals rising steeply from deep water (one was the Dacia Seamount) while surveying the route for a submarine cable from Cadiz, Spain, to the Canary Islands.

> '... One night we were astonished by the sinker stopping at about one thousand two hundred fathoms, when it ought to have gone nearly twice as deep. It was at once suspected that we were in the neighborhood of a bank. Here, to our surprise and delight, the sinker brought up at sixty-six fathoms! There was immense excitement on board, as it was obvious that we had pitched upon a bank, or rather a mountain, of startling proportions, perhaps the lost island of Atlantis itself. As this submarine mountain lay close to the proposed line of the cable, it was necessary to make a thorough survey, and two days were spent in doing this. A mark-buoy was put down to work by, and numerous soundings were taken in all directions so as to clearly define the limits of the bank. The shoalest water found was forty-nine fathoms, and half a mile distant two hundred and thirty fathoms were obtained, showing a steep slope. When the buoy, which was moored in one hundred seventy-five fathoms, was taken up, the mooring rope was found to be nearly chafed through seventy-five fathoms from the bottom. This showed that the bank must rise almost precipitously, and that there exists a wall of about four hundred and fifty feet in height. A very curious effect observed was a long ripple on the calm sea, apparently caused by the ground-swell breaking on the edge of the bank.'
>
> Extract from Webb (1883).

The accurate mapping and sounding of those seamounts lead researchers to link seamounts to incremental stages of deep ocean ridge development (Schlee, 1973).

Echo sounding was the precursor to more advanced forms of bottom profiling, such as the deep towed and multi-beam sounding instruments of the 1960s after which the discipline of marine geology expanded rapidly. Deep sea camera systems were regularly used at this time, and divers used SCUBA to explore seamounts that reached into shallow depths (e.g., Simpson and Heydorn, 1965; Herlinveaux, 1971). The 1960s and 1970s marked two other key breakthroughs in technology that advanced deep sea research significantly. First, satellites were now commonly used for navigation and position fixing at sea and many carried instrumentation capable of measuring small deviations in sea surface elevation. These anomalies are due to the slight variation in gravity created by underwater features such as seamounts, and thus can be used to map the seafloor remotely (see Box 1.1; Dixon *et al.*, 1983). Smith and Sandwell (1997) produced the first global bathymetry map using satellite altimetry data in which approximately 70% of the true topographic depth was predicted by their model, and delineated for the first time previously undetected whole seamount chains (e.g., the 1400 km long Foundation Seamount chain in the southeastern Pacific). Models built from these maps created the first global estimates of the number of seamounts, and also predicted the locations of ~15 000 seamounts, many of which had not been mapped (see Chapter 2).

The 1960s and 1970s also marked the first comprehensive use of submersible exploration. Most famous was the launch of the deep submersible Alvin, developed at the Woods Hole Oceanographic Institute in 1964. Seamount exploration was one of Alvin's early tasks, diving first on Bear Seamount, and then the Azorean seamounts in 1968 (Alvin dive log: http://www.whoi.edu). Researchers were now able to locate, map, photograph, and observe *in situ* seamounts and other seafloor physical features. Geologists then began to discover and understand processes going on at the global scale, such as changes in sea level (e.g., Schwartz, 1972), seafloor spreading (Hess, 1962), and plate tectonics (Wilson, 1965), which were all strongly supported by the distribution and physical characteristics of seamounts (see Chapter 1).

Physical oceanographers have taken a particular interest in seamounts since their discovery. At the end of the nineteenth century, several attempts had been made to measure the velocity and direction of tidal currents in the ocean, and variations in the latter were often associated with changing undersea topography (Murray and Hjört, 1965). Seamount chains may represent a 'perforated barrier' to impinging oceanic currents (Roden *et al.*, 1982), and create unique patterns of vertical and horizontal flow (see Chapter 4). Therefore the causes of observed variations in current flow, and depth-related changes in temperature and salinity could be better understood with an improved knowledge of seamount morphology (Murray, 1941).

Early studies reported short-lived (in the order of weeks), trapped vortices over seamounts causing flow deflections and accelerations, and doming of isopycnals (see Chapter 4). Hogg (1973) proposed that under certain conditions, these phenomena relate to the theoretical predictions of a Taylor column, where the nature of flow modification depends on seamount orientation and topography, rotation of the Earth, and stratification above the seamount. First theoretically described by Proudman (1916) and Taylor (1917), the term was not widely used until Hide (1961) related the Taylor–Proudman theorem to the formation of the Great Red Spot of Jupiter (Fig. 3.3), which is thought to be caused by an irregularity on the planet's surface (Fig. 3.3; Owens and Hogg, 1980; Roden, 1986). The theoretical formation of Taylor columns around seamounts was also suggested to explain the presence of asymmetric sediment troughs and drifts at the base of seamounts (see Chapter 4; Hogg, 1973; Roberts *et al.*, 1974). Taylor column generation over seamounts was widely inferred initially, but later Owens and Hogg (1980) found direct evidence for Taylor column formation over a seamount. A full discussion of the physics and empirical evidence for Taylor columns and caps is provided in Chapter 4.

Fig. 3.3 The Great Spot of Jupiter: the first described Taylor column? True colour mosaic of images taken from NASA's Cassini spacecraft on 29 December 2000 – Photo credit NASA/JPL/Space Science Institute.

Seamounts as unique biogeographic features

As seamount mapping became more frequent, interest in seamount benthic and pelagic communities grew among taxonomists and biogeographers. Up until the late nineteenth century, there had been considerable interest in studying deep sea communities and determining biogeographic patterns of deep sea fauna; expeditions such as the *Josephine*

(1869) and, most famously, the *Challenger* (1872–1876) drew particular attention to the deep sea realm. However, studies of deep sea biology gave way to oceanographic studies at the turn of the century (Mills, 1983), focusing more on physical and biological properties of the water column and their relationship to sound transmission through water. This shift was possibly due to a change in focus from scientific to military interests around World War I and II. Interest in deep sea biology later re-emerged in the 1950s, as marked by the Swedish *Albatross* cruises (1947–1948), and the Danish *Galathea* expedition (1950–1952). At this time, the field of deep sea biology expanded greatly, and the links between biological and physical patterns were examined. Topical questions of global biogeography in the late nineteenth and early twentieth century related to an explanation of the distribution of marine and terrestrial flora and fauna separated by the large perceived dispersal boundary of the Pacific Ocean. The work of Darwin in the eastern Pacific and Wallace in the western Pacific (see Newman, 1986 for review) proposed that the commonality of east–west patterns of species distributions could be explained simply in terms of chance dispersal upon ocean currents. The discovery of seamounts would soon lead biogeographers to re-think this theory.

The vastness of the Pacific Ocean was regarded by Ekman (1953) as the 'Eastern Pacific Barrier' with respect to marine fauna. He considered that passive dispersal of species with long-lived planktonic larval stages, or the chance transport of larvae or adults on drift material, accounted for similarities and/or discontinuities of east vs west Pacific marine fauna. The discovery of 'mid-Pacific mountains' (i.e., seamounts) prompted Hamilton (1956) to review contemporary theories that might explain the commonality of shallow marine flora and fauna around the Pacific basin. He first assumed that the ocean basins and continents are permanent features, and that dispersal is facilitated along land bridges, for example *via* the Aleutian Islands, or by long-distance oceanic migration. However, the problems of sheer distance, and the unlikelihood of propagules having the wide environmental tolerances needed to endure high or low latitude dispersal pathways (Ekman, 1953), left the matter unresolved. An alternative hypothesis of dispersal by way of an ancient, now sunken, Cenozoic continent that spanned the Pacific Ocean was quickly dismissed by Hamilton (1956), who demonstrated that the existence of ancient Cretaceous seamounts disproved the possibility of a sunken continent. He suggested that seamounts may act as historical or present-day 'stepping-stones' across the Pacific, hence providing the first alternative mechanism for long-distance dispersal of marine flora and fauna, rather than simple passive or 'waif' dispersal as proposed by Ekman (1953) and others.

As seamount taxonomic data accumulated, support for Hamilton's stepping-stone theory grew. New species were being collected, mainly from fisheries reconnaissance, but also as secondary observations to geological investigations (e.g., Pratt, 1963; Budinger, 1967), or assessments of strategic potential (e.g., Herlinveaux, 1971). The distributions of some previously described coastal species were now being extended to seamounts. Hubbs (1959) posed a number of key questions: What species are found on seamounts, where did they come from, and how did they get there? Does the distribution of these species represent past and present oceanic circulation and ocean temperature? Do seamounts act as 'stepping-stones' across the Pacific Ocean? Can deep seamount fauna be used to help interpret seamount subsidence? Has there been speciation due to isolation on seamounts? What physico-chemical or biotic factors drive the apparent high abundance of life over seamounts? With these questions in mind, he examined some of the limited seamount fish faunal data of the day, consisting of a single occurrence of *Embassichthys bathybius* from Pratt Seamount, *Sebastodes ruberrimus* (now *Sebastes ruberrimus*) from Cobb Seamount, *Pterygotrigla picta* from a Chilean seamount, a variety of coastal and pelagic fish and their stomach contents, and anecdotal evidence from fishermen. He suggested that studying seamount fauna may have significant commercial and scientific importance, thereby launching a new era of descriptive biological seamount research.

Seamount researchers today are still challenged by Hubbs' questions (Wilson and Kaufman, 1987; Rogers, 1994). Seamounts are difficult to sample because they are remote, are often deep and steep, have strong currents, and have rocky and highly variable substrates (Rogers, 1994). At least 13 types of sampling gear have been used to sample seamount benthos (Wilson and Kaufman, 1987; Rogers, 1994); perhaps testimony to the variety of habitats and difficulties that sampling seamount benthos presents. Nevertheless, intensive biological surveys of seamounts were carried out, starting in 1964 with the Vema Seamount. The ship commissioned for the study was the diamond prospecting vessel *Emerson K* sponsored by the Marine Diamond Corporation (Simpson and Heydorn, 1965), which may be suggestive of the purpose of the expedition. Although this survey only examined the summit of the seamount within SCUBA-diveable depths, much information was collected on the biological community, oceanography, and geological characteristics. Subsequent publications from that expedition proposed biogeographic links between Vema Seamount and nearby islands (e.g., Penrith, 1967; Berrisford, 1969).

Also in the Atlantic Ocean, the *Meteor* cruise of 1967 made an extensive study of Josephine and Great Meteor Seamounts (Thiel, 1983). During this expedition, the importance of comparative sampling of coastal and deep sea regions surrounding these seamounts was recognized for the first time. The work of Thiel on this cruise was particularly noteworthy: he collected quantitative grab samples of soft sediments, providing information on the variability of species and density within and between seamounts, and among seamount, deep seafloor, and coastal meiofaunas. Interestingly, results of those studies were not well explained by patterns of pelagic productivity, as contemporary theories predicted. Great Meteor Seamount continues to be the subject of extensive biological and physical study, exemplified by publications contained in the special edition of *Archive of Fishery and Marine Research: Oceanography and Ecology of Seamounts – Indications of Unique Ecosystems* (Vol. 51 (1–3), 2004).

The late 1960s marked the beginning of almost 20 years of continuous Russian and Cuban exploration of seamounts on a global scale, with large vessels visiting the north Atlantic, north and south Pacific, western Indian, and Southern oceans (Parin *et al.*, 1997). Significant attention was given to the Nazca and Sala y Gomez Ridge seamounts, which were visited by at least nine research vessels with some making multiple visits. This comprehensive body of work, summarized in Parin *et al.* (1997), describes primarily the bentho-pelagic fish fauna, but also reviews the pelagic and benthic invertebrate fauna, collected on 22 seamounts within the ridge system. They reviewed environmental parameters as they relate to biogeographic patterns observed within the Nazca and Sala y Gomez chain, and throughout the South Pacific. This work gives strong support to much of the current thought on eastward migration of western Pacific fauna, and the notion that seamounts are significant centres of endemism (see below and Chapter 13).

Concentrated effort on particular seamounts continued. For example, the long-term studies of the link between oceanographic and biogeographic patterns at Cobb Seamount between the early 1980s and 1990s, led Parker and Tunnicliffe (1994) to suggest the importance of considering life history when examining species' biogeography. More recently, extensive studies have been done by multinational collaborative expeditions to the Norfolk and Lord Howe Ridges, and Tasmanian seamounts between 1984 and 1998 (Richer de Forges *et al.*, 2000). Over time, along with higher sampling intensity, there has been an increased consideration of sampling design and statistical rigour in recent seamount surveys, providing more potential for quantitative evaluation of ecological patterns (e.g., Richer de Forges *et al.*, 2000; Koslow *et al.*, 2001; Rowden *et al.*, 2002). These studies highlight a clear change in the objectives of seamount biological studies from simple taxonomic collections and resource exploration of the 1950s to addressing wider theoretical concepts. These recent studies emphasize the necessity and usefulness of simultaneous multidisciplinary seamount research.

Although there is much continued interest in intrinsic biological and physical features, it has been realized that seamounts offer unique platforms for studying more generalized aspects of biology and ecology. For example, Levin and Thomas (1989), and Thistle and Levin (1998) used soft sediment habitats at guyots to test some predictions about the effect of flow dynamics on deep sea infaunal assemblages. Considering that most deep sea habitats are soft sediment systems, and that the majority of infaunal or epifaunal inhabitants are detritivores, there has been great interest in testing the role of sediment processes on determining deep sea benthic community structure (Levin *et al.*, 2001). On soft sediment seamount plateaus, one can find or create a variety of flow conditions over a relatively small area while, at the same time, having control of other potentially confounding factors related to depth, region, or differing ecological histories.

With advancing technology in ocean science, and the growing pool of seamount data, studying the spatial and temporal processes associated with generating local endemism became a new theme in seamount research. The occurrence of endemics – species that are possibly unique to individual or groups of seamounts – has been noted since the sampling of seamount assemblages began (see Chapter 13). However, the growing interest in endemism has provided a new conceptual framework for studying seamounts at much larger spatial and temporal scales. Questions of the spatial isolation of seamounts are now being addressed from scales of within-seamount chains, to between-seamount chains separated by oceanic basins, and over geological time scales. Some studies suggest that seamounts may be centres of speciation in the deep sea (e.g., Koslow *et al.*, 2001; Rowden *et al.*, 2002), thereby placing seamounts in a framework of global-scale biodiversity.

Seamounts as a fisheries resource

Seamount-related fisheries represent a significant proportion of the total high seas fish catch (see Chapter 17). Of all the deep sea fisheries, most target species are associated with seamounts (see Chapter 9; FAO, 2004). Historically, seamount research has lagged behind, or at best paralleled, seamount fisheries exploitation, and to some extent, fisheries exploration on seamounts has both financially and intellectually fuelled seamount research since commercial fishing on seamounts began. Possibly the first fisheries associated with seamounts were those traditional handline fisheries of the south Pacific, the Cape Verde islands and the Azores (see Chapter 16). For example, manuscripts of the Azorean historian Gaspar Frutuoso (1522–1599) indicate that fishing on shoals and peaks up to 30 nautical miles (approximately 55 km) offshore of the Azores was important in the fourteenth and fifteenth centuries (Frutuoso, 1963). Most of these traditional fisheries still operate in many oceanic island ecosystems, and are of considerable local importance. Larger-scale commercial fishing on seamounts is thought to have started with the harvesting of precious corals by Japanese fishers. Beds of the commercially valuable pink coral (*Corallium* spp.) were discovered in 1965 on the Milwaukee Seamount (central north Pacific) at a depth of 400 m (Grigg, 1993), and later discovered throughout the Emperor Seamount chain (northwest of Hawaii). The harvesting of precious coral, and exploration for new coral beds continues today for this largely unregulated fishery (Grigg, 1982).

Aggregations of fish over seamounts have been often reported from the onset of seamount research. Hubbs (1959) suggested that new fisheries for rockfish (*Sebatodes*, now *Sebastes* spp.) and deep sea flatfish (e.g., *Embassichthys bathbius*, *Hippoglossus stenolepis*) may develop on seamounts. In the North Atlantic, deep water fisheries were being developed in the 1950s, although fishing in waters deeper than 500 m was considered fruitless as there was a general belief that little life existed at depths greater than the shelf slope (Koslow *et al.*, 2000). However, the discovery of seamounts and their often high densities of fish induced commercial fishers to gear up to go deeper and further offshore for these untapped resources. Shellfish fisheries were also of interest (e.g., Simpson and Heydorn, 1965; Hughes, 1981; Raymore, 1982). The establishment of a commercially valuable rock lobster (*Jasus tristani*) fishery on Vema Seamount (south central Atlantic) also represents the first seamount-related fisheries collapse within only a few years of fishing (Lutjeharms and Heydorn, 1981).

Many technical advances were needed before deep sea and seamount fishing could be viable. Fishing technology made its largest advance in the spring of 1954 with the launch in the United Kingdom of the first deep sea freezer trawler vessel *Fairtry* (Woodard, 2000), a vessel more than four times the size of its largest contemporaries when it began fishing on the Grand Banks off the coast of Newfoundland. The *Fairtry* could stay at sea for many weeks, work around the clock in all weather, and had a large freezer capacity for fish processed at sea. Particularly powerful engines meant that larger otter trawls could be used more effectively and retrieved from very deep water. The *Fairtry* also used the latest echo-sounding technology to find and target schools of fish. By the mid-1970s, large fleets of deep sea fishing freezer trawlers were being deployed from many countries to the oceans of the world; the Soviets had 400 factory trawlers, the Japanese had 125, Spain had 75, West Germany had 50, and France and Britain had 40 ships. A new era in large scale, deep sea commercial fishing had begun (see Chapter 17).

With the new factory ship technology, seamount fisheries were quickly exploited (see Chapter 17). In 1967, during the global ocean explorations of the Soviet Union, large aggregations of pelagic armourhead (*Pseudopentaceros wheeleri*) and alfonsino (*Beryx splendens*) were discovered in the central North Pacific between the Emperor and Hawaiian seamounts (Takahashi and Sasaki, 1977). At the same time, armourhead were being found in the stomachs of Sei whales by Japanese whalers in the vicinity of the Milwaukee Seamount, prompting fishing by Japanese in 1969 (Sasaki, 1974). Armourhead were later discovered by Russian fishers in 1976 at Corning Rise in the Atlantic Ocean (Vinnichenko, 1997). The estimated total catch of armourhead rose to possibly 48 000 t for the period up to 1977, but then rapidly declined to between 5800 and 9900 t in the period up to 1982. The fishery for pelagic armourhead no longer exists (FAO, 2004).

Discovery of new potentially exploitable fish resources came as seamount exploration continued, with at least 77 commercially valuable fish species found on seamounts (Rogers, 1994). The need for management of seamount fisheries was becoming clear, and since the 1970s many countries have implemented fisheries management strategies such as Total Allowable Catch (TAC) and Individual Transferable Quota (ITQ) systems to help manage fisheries within their own Exclusive Economic Zones. However, in the early days of seamount fishing, few seamounts fell within the jurisdiction of a country's fisheries management regulations, and therefore, were left open to uncontrolled fishing in high seas areas. National territorial waters only extended to between 3 and 24 nautical miles (approximately 5 and 44 km) offshore, with only a few countries setting 200 mile territorial limits, primarily to protect their fisheries from long-distance factory ship fishing fleets.

In 1982 the United Nations Convention on the Law of the Sea (UNCLOS) was established, defining a nation's sovereign rights to explore, conserve, and manage natural resources within a 200 nautical mile Exclusive Economic Zone extending from its shore. It became enforceable on 16 November 1994, thus encompassing many seamounts within nationally managed fisheries. UNCLOS also established the first legal high seas regime, laying down conditions that all nations must adhere to when fishing on the high seas, including the adoption of measures for conservation of living resources, a duty to cooperate with other fishing nations in the conservation and management of living resources, and the adoption of sustainable fishing practices. Unfortunately, today seamount fisheries policy makers face ongoing problems of enforcement and compliance with UNCLOS regulations, due to the large number of seamount fisheries in high sea areas (see Chapter 17 and http://www.un.org), and the difficulty in differentiating between seamount- and non-seamount-caught fish from within FAO fisheries management areas (see Chapters 17, 19, and 20). To help address some of these issues, a global Code of Conduct for Responsible Fisheries was established in 1995 (FAO, 1995). This is a non-mandatory instrument, establishing principles and standards applicable to the conservation, management, and development of all fisheries. Although it is voluntary, it reflects regulations set under the UNCLOS and other obligatory legal instruments. The Code is 'directed toward members and non-members of FAO, fishing entities, sub-regional, regional and global organizations, whether governmental or non-governmental, and all persons concerned with the conservation of fishery resources and management and development of fisheries, such as fishers, those engaged in processing and marketing of fish and fishery products and other users of the aquatic environment in relation to fisheries' (FAO, 1995), thereby providing a framework for national and international efforts to ensure sustainable exploitation of aquatic living resources in harmony with the environment.

Seamounts as a conservation focus

As seamount commercial fishing expanded, the need for management became more apparent. Research has shown that seamount fisheries are particularly susceptible to fishing pressure due, in part, to the fact that seamount species are generally K-selected (long lived, late age at maturity, low fecundity). They require large spawning aggregations for successful recruitment, and are concentrated over relatively small and spatially restricted areas. Other contributing factors are common to fisheries in general, such as lack of data for fisheries models. Furthermore, commercial fishers were quick to implement new technological advances such as satellite navigation, satellite altimetry-based maps, high resolution swath bathymetry mapping, electronic fish finding, and video monitoring of nets while fishing. These technological advances enabled fishers to find and exploit new seamount fish stocks more efficiently. The advent of 'rock-hopper' trawls (modified otter trawls) in the mid-1980s gave fishers access to substrata that previously would have resulted in gear damage and loss. They are also particularly destructive of seabed structure, so that natural refugia provided by complex seabed topography typically found on seamounts no longer protected a proportion of fish stocks from fishing.

The action of rock-hopper trawls on the seabed has contributed to significant habitat modification and changes in benthic community structure. Thus, assessing the impact of trawling on seamount benthic habitats created a new conservation focus for managers and scientists (e.g., Probert *et al.*, 1997; Koslow *et al.*, 2001). Trawl nets are highly destructive of the possibly very long-lived and slow-growing sessile suspension feeding organisms that dominate seamount habitats (Jennings and Kaiser, 1998; Watling and Norse, 1998; also see Chapters 7 and 8). Moreover, it has been suggested that organisms such as sponges and corals create further habitat complexity supporting extensive associated sessile and mobile fauna (Sainsbury, 1987; Auster and Langton, 1999). The magnitude of the damage caused by trawling on seamount communities to date, and the long-term effects of habitat destruction on the dynamics of benthic and pelagic interactions within seamounts, are currently unknown (Koslow *et al.*, 2000; Roberts, 2002; Johnston and Santillo, 2004).

Conserving global biodiversity has been an international priority in recent times. Considering that seamounts may be centres for speciation in the deep sea, and are potentially threatened by destructive fishing practices, they may be regarded as 'biodiversity hotspots' and therefore a priority for conservation (Myers *et al.*, 2000). The establishment of Marine Protected Areas (MPAs) has shown to be a useful approach for management and conservation purposes. Roberts *et al.* (2005) cite a growing body of theoretical and empirical studies in support of their statement that, 'marine reserves can simultaneously meet conservation and fishery management objectives'. International acknowledgement that fisheries impact on seamounts is outstripping scientific understanding has prompted a number of countries to establish seamount MPAs for fishery and habitat protection, and scientific investigation of seamount dynamics (see Chapter 20). Countries that have established seamount MPAs include Portugal (the Formigas-Dollabarat Bank, Azores, Marine Reserve since 1988; Santos *et al.*, 1995), the USA (Sitka Pinnacles Marine Reserve, Alaska, 1998; Wing, 2001 – Note in Roberts, 2002), Canada (Bowie seamount, 1998; Fisheries and Oceans Canada, 2006), Australia (Tasmanian seamounts, 1999; Commonwealth of Australia, 2002), and New Zealand (fisheries closures over 19 seamount spread throughout the New Zealand EEZ, 2001; New Zealand Ministry of Fisheries, 2001).

A synthesis of physical and biological patterns of seamounts will be essential for future management and conservation of seamounts. Tools to help both fisheries and conservation biologist are also needed for a holistic approach to studying seamounts. The development of genetic tools has helped determine the nature of biogeographic boundaries in the deep sea (e.g., Bucklin *et al.*, 1987; France and Kocher, 1996). Genetic tools have also aided in determining the boundaries of fisheries management areas (Martin *et al.*, 1992; Hoarau and Borsa, 2000), and improved quantification of endemicity (e.g., Smith *et al.*, 2004). In future, genetic techniques will provide significant contributions to the fields of population ecology and population genetics of seamount communities, leading to the construction of more accurate models of spatial and temporal community dynamics, and ultimately the more informed establishment of MPAs and fisheries management strategies (Kritzer and Sale, 2004).

With new scientific discoveries and approaches to conservation and fisheries management, fundamental concepts for building legal conservation and management frameworks have also evolved. Such concepts include the 'precautionary principle'; now an established principle of environmental governance and policy, one interpretation of which might be 'that absence of definite scientific information should not be an excuse for inaction in curtailing harmful activities' (de Fontaubert, 2001). The 'ecosystem approach' is another principle currently incorporated into international agreements so that nations will take account of the interdependence between target species and associated fauna when developing and adopting new management strategies (de Fontaubert, 2001).

Current and future seamount research

This chapter indicates the extent to which new technology and ideas have advanced our understanding of seamounts over the past 100 years. Much remains to be learned. The hypotheses that Hubbs posed in 1959 are still challenging research frontiers today. Research into the ecology and biology of seamounts has increased dramatically (Fig. 3.4). It is not clear whether this trend will continue, but the volume of seamount research that is currently underway suggests that it may. Several major studies are in progress and have yet to be fully published, and several more have been recently funded and will commence in the future. For example, the European Commission funded a fifth framework programme called OASIS (Oceanic Seamounts: An Integrated Study) that has sponsored a series of expeditions to north Atlantic seamounts. OASIS epitomizes a growing emphasis on interdisciplinary seamount research. Over repeated cruises, geologists, physical oceanographers, taxonomists, ecologists, and conservation scientists collaborated to identify and describe the physical forcing mechanisms affecting the seamounts systems; asses the origin, quality, and dynamics of organic material over and on the seamounts; describe the biodiversity and ecology of the seamounts, their dynamics, and the processes that maintain them; model the trophic web; and use the results to guide management strategies. While OASIS concluded their fieldwork in 2005, a new programme EuroDEEP (under the European Commission initiative called EuroCores) will include seamounts in their study of diversity among deep sea habitats.

Similar interdisciplinary efforts are underway or have been funded by other countries to study other seamount regions. Seamounts off the northeast USA, around Hawaii, and in the Gulf of Alaska are being explored by American researchers through the NOAA Explorations programme; Indian scientists will survey seamounts in the little-studied Indian

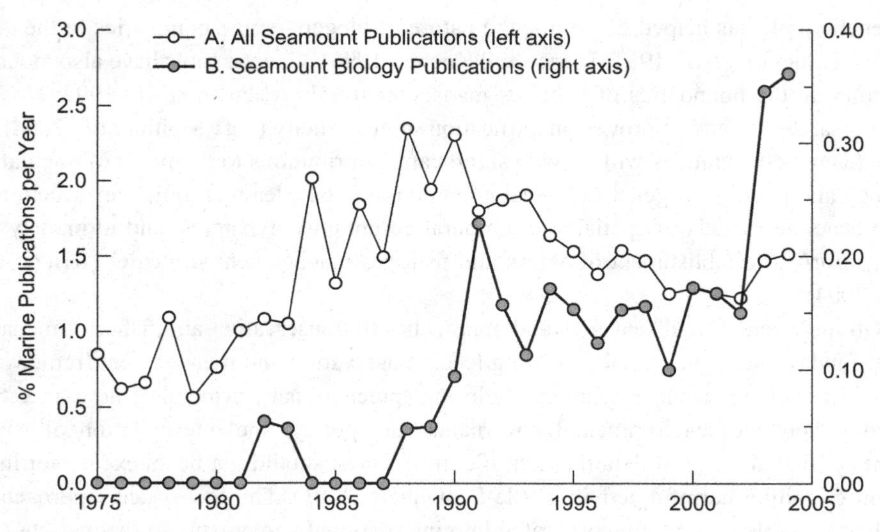

Fig. 3.4 Rate of seamount-related publication as reported in the ISI Web of Science, and standardized as a percentage of total marine-related publications (i.e., those having the words 'ocean' or 'marine'). (a) All publications containing a seamount term (e.g., 'seamount', 'guyot', or 'tablemount'). Publications peaked in the mid- to late-1980s, dropped and levelled off between 1995 and 2005. (b) Publications containing a seamount term and a biological term (e.g., 'biology', 'species', 'ecology', 'community', or 'genetic'). Publications show a strong increase around 1990, and a second large increase in 2003 and 2004. Part of the 2004 peak can be attributed to a special issue *Archive of Fishery and Marine Research* (Vol. 51 (1–3)) on a major expedition to the Great Meteor Seamount, but the increase in 2003 cannot be attributed to any single cruise or research effort. While not all seamount-related publications are indexed and not all papers that have 'seamount' in the abstract can be considered to focus on seamounts, these data suggest that research into the biology is increasing through time.

Ocean Ridge and Central Indian Basin; New Zealand researchers have been funded to study seamounts in the New Zealand region; both New Zealand and the British Antarctic Survey are sampling Antarctic seamounts; a multinational collaboration will revisit the Norfolk Ridge seamounts studied by Richer de Forges *et al.* (2000) (see Chapter 7) to provide physical data to supplement the previous biological collecting; and the Portuguese are continuing active research on the fisheries and ecology of Azorean seamounts.

These studies highlight the growing activity of multidisciplinary, collaborative seamount research. The future of seamount research will involve more highly coordinated research efforts, and large-scale meta-analysis of seamount data. Interdisciplinary and international coordination has been recognized as essential, and a new programme was launched in 2005 to provide a forum for catalysing and coordinating research globally. The Census of Marine Life on Seamounts (CenSeam) is one of several habitat-specific science projects within the Census of Marine Life (2006). CenSeam has several goals: to coordinate and expand existing research through developing standard methods and data reporting; to promote community networking; to foster new seamount research through proposal workshops and mini-grants for planning; to aggregate and synthesize global seamount data through an analysis working group and further develop the SeamountsOnline (2006) as an open-access portal to seamount data; and increase awareness of seamounts through public education and outreach materials.

Conclusion

Since the discovery of seamounts, exploration and research have shown that they are numerous and ubiquitous throughout the deep sea (Chapters 1 and 2). Their special features mean that they are analogous to other deep ocean formations that have received a great deal of recent attention, such as cold seeps and hydrothermal vents. However, seeps and vents have only been discovered and studied in very recent times, while seamounts have been studied for more than 100 years, and have been indirectly known to exist for centuries through their association with increased local biological activity providing a food source. Yet, despite continued great interest in seamounts for their commercial fisheries exploitation, seamounts are poorly understood habitats in terms of the communities they support, the drivers of those communities, and the role that seamounts play in the realm of marine biodiversity. Much of the early discovery and study of seamounts was driven by commercial interests, and this interest persists today, fuelling much of the current oceanographic and biological seamount research. Although many fundamental questions regarding the nature of seamount processes and habitats remain unanswered, it is clear that political interest is increasing as seamount research expands. The establishment of a global network of seamount researchers, and the coordinated study of seamount biological and physical patterns over space and time, will enable more effective decisions regarding seamount fisheries management and protection of seamount habitats, and a better understanding of the ecological role of seamounts in the deep sea.

Acknowledgements

The authors wish to thank Albert 'Skip' Theberge of the National Oceanic and Atmospheric Administration (NOAA) Central Library for useful discussions on early geographic accounts of seamounts. This chapter was improved by the editors and comments of an anonymous reviewer. Thanks to the Census of Marine Life Seamounts Program (CenSeam) for support of a planning workshop for this book. P. Brewin gratefully thanks the Gordon and Betty More Foundation for financial support.

References

Alvin Dive Log (1969) Woods Hole Oceanographic Institute Archives. http://www.whoi.edu (last accessed June 6, 2006).

Ankarcrona, J. (1969) Extract from Commander Johan Ankarcrona's report of August 4 1869. *Tidskrift i Sjöväsendet* (*Journal for the Marine*), 32. Carlscrona.

Atlantic Cable History (2006) http://www.atlantic-cable.com/ for ships names and their history (last accessed June 6, 2006).

Auster, P.J. and Langton, R.W. (1999) The effects of fishing on fish habitat. In: *Fish Habitat: Essential Fish Habitat and Rehabilitation* (ed. Benaka, L.), American Fisheries Society, Bethesda, MD.

Berrisford, C.D. (1969) Biology and zoogeography of the Vema Seamount: a report on the first biological collection made on the summit. *Transactions of the Royal Society of South Africa*, 38, 387–98.

Bucklin, A., Wilson, R.R.J. and Smith, K.L.J. (1987) Genetic differentiation of seamount and basin populations of the deep-sea amphipod *Eurythenes gryllus*. *Deep-Sea Research, Part A: Oceanographic Research Papers*, 34, 1795–810.

Budinger, T.F. (1967) Cobb seamount. *Deep-Sea Research*, 14, 191–201.

Carpine-Lancre, J. and Saldanha, L.V.C. (1992) *Souverains Océanographes*. Fondation Calouste Gulbenkian, Lisboa, 178 pp.

Census of Marine Life (2006) www.coml.org (last accessed June 6, 2006).

Darwin, C. (1842) *The Structure and Distribution of Coral Reefs*, 214 pp. Smith, Elder and Co., London.

de Fontaubert, A.C. (2001) Legal and political considerations. In: *The Status of Natural Resources on the High-Seas* (eds. WWF/IUCN), pp. 69–91. Gland, Switzerland.

Deacon, M. (1971) *Scientists and the sea. 1650–1900 a study of marine science*. Academic Press, London.

Dixon, T.H., Naraghi, M., McNutt, M.K. and Smith, S.M. (1983) Bathymetric predictions from SEASAT altimeter data. *Journal of Geophysical Research*, 88, 1563–71.

Ekman, S. (1953) *Zoogeography of the Sea*. Sidgwick and Jackson, London.

FAO (1995) *Code of Conduct for Responsible Fisheries*, 41 pp. FAO, Rome, Italy.

FAO (2004) *The State of World Fisheries and Aquaculture*. FOA Report, Rome, Italy.

Fisheries and Oceans Canada (2006) http://www.pac.dfo-mpo.gc.ca (last accessed June 6, 2006).

France, S.C. and Kocher, T.D. (1996) Geographic and bathymetric patterns of mitochondrial 16S rRNA sequence divergence among deep-sea amphipods, *Eurythenes gryllus*. *Marine Biology*, 126, 633–43.

Frutuoso, G. (1963) Saudades da Terra, Liv. IV (sobre as ilhas de baixo), Ponta Delgada, Edição do Instituto Cultural de Ponta Delgada.

Grigg, R.W. (1982) Precious corals: an important seamount fisheries resource. *Infofish Marketing Digest*, 2, 8–11.

Grigg, R.W. (1993) Precious coral fisheries of Hawaii and the US Pacific Islands. *Marine Fisheries Review*, 55, 50–60.

Hamilton, E.L. (1956) *Sunken Islands of the Mid-Pacific Mountains*, 97 pp. Geological Society of America Memoir 64, New York.

Herdman, H.F.P., Wiseman, J.D.H. and Ovey, C.D. (1956) Proposed names of features on the deep-sea floor. *Deep-Sea Research*, 3, 253–61.

Herlinveaux, R.H. (1971) Oceanographic features of and biological observations at Bowie Seamount, 14–15 August, 1969. *Technical Report No. 273*, Fisheries Research Board of Canada, 35 pp.

Hess, H.H. (1946) Drowned ancient islands of the Pacific basin. *American Journal of Science*, 244, 772–91.

Hess, H.H. (1962) History of ocean basins. In: *Petrologic Studies: A Volume to the Honour of A.F. Buddington* (eds. Engel, A.E.J., James, H.L. and Leonards, B.F.), pp. 599–620. Geological Society of America, Boulder Colorado.

Hide, R. (1961) Origin of Jupiter's great red spot. *Nature*, 190, 895.

Hoarau, G. and Borsa, P. (2000) Extensive gene flow within sibling species in the deep-sea fish *Beryx splendens*. *Comptes Rendus de l'Academie des Sciences Serie III Sciences de la Vie*, 323, 315–25.

Hogg, N.G. (1973) On the stratified Taylor column. *Journal of Fluid Mechanics*, 58, 517–37.

Hubbs, C.L. (1959) Initial discoveries of fish faunas on seamounts and offshore banks in the eastern Pacific. *Pacific Science*, 13, 311–16.

Hughes, S.E. (1981) Initial US exploration of nine Gulf of Alaska seamounts and their associated fish and shellfish resources. *US National Marine Fisheries Service Fishery Bulletin*, 43, 26–33.

Jennings, S. and Kaiser, M.J. (1998) The effects of fishing on marine ecosystems. *Advances in Marine Biology*, 34, 201–352.

Johnston, P.A. and Santillo, D. (2004) Conservation of seamount ecosystems: application of a marine protected areas. *Archive of Fishery and Marine Research*, 51, 305–19.

Koslow, J.A., Boehlert, G.W., Gordon, J.D.M., Haedrich, R.L., Lorance, P. and Parin, N. (2000) Continental slope and deep-sea fisheries: implications for a fragile ecosystem. *ICES Journal of Marine Science*, 57, 548–57.

Koslow, J.A., Gowlett-Holmes, K., Lowry, J.K., Poore, G.C.B. and Williams, A. (2001) The seamount benthic macrofauna off southern Tasmania: community structure and impacts of trawling. *Marine Ecology Progress Series*, 213, 111–25.

Kritzer, J.P. and Sale, P.F. (2004) Metapopulation ecology in the sea: from Levins' model to marine ecology and fisheries science. *Fish and Fisheries*, 5, 131–40.

Kunzig, R. (2000) *Mapping the Deep. The Extraordinary Story of Ocean Science*, 345 pp. Sorts of Books, London.

Levin, L.A. and Thomas, C.L. (1989) The influence of hydrodynamic regime on infaunal assemblages inhabiting carbonate sediments on central Pacific seamounts. *Deep-Sea Research, Part A: Oceanographic Research Papers*, 36, 1897–916.

Levin, L.A., Etter, R.J., Rex, M.A., Gooday, A.J., Smith, C.R., Pineda, J., Stuart, C.T., Hessler, R.R. and Pawson, D. (2001) Environmental influences on regional deep-sea species diversity. *Annual Review of Ecology and Systematics*, 32, 51–93.

Littlehales, G.W. (1932) The configuration of the oceanic basins. *Physics of the Earth – V Oceanography*, pp. 13–46. The National Academy of Sciences, Washington, DC.

Lutjeharms, J.R.E. and Heydorn, A.E.F. (1981) The rock-lobster *Jasus tristani* on Vema Seamount: drifting buoys suggest a possible recruiting mechanism. *Deep-Sea Research, Part A: Oceanographic Research Papers*, 28, 631–6.

Martin, A.P., Humphreys, R. and Palumbi, S.R. (1992) Population genetic structure of the armorhead *Pseudopentaceros wheeleri* in the North Pacific Ocean. Application of the polymerase chain reaction to fisheries problems. *Canadian Journal of Fisheries and Aquatic Sciences*, 49, 2386–91.

Menard, H.W. (1964) *Marine Geology of the Pacific*. McGraw Hill, New York.

Menard, H.W. and Dietz, R.S. (1951) Submarine geology of the Gulf of Alaska. *Bulletin of the Geological Society of America*, 62, 1263–85.

Mills, E.L. (1983) Problems of deep-sea biology: an historical perspective. In: *Deep-Sea Biology* (ed. Rowe, G.T.), John Willey and Sons, Toronto.

Minelle, P. (1986) Aux Açores – campagne océanographic à bord de la Princesse-Alice (juin, juillet, août 1896). La campagne de la Princese-Alice en 1896. Monaco. Musée océanographic, 87 pp.

Murray, H.W. (1941) Submarine mountains in the Gulf of Alaska. *Bulletin of the Geological Society of America*, 52, 333–62.

Murray, J. and Hjört, J. (1965) *The Depths of the Ocean; A General Account of the Modern Science of Oceanography Based Largely on the Scientific Researches of the Norwegian Steamer Michael Sars in the North Atlantic*. Weinheim, J. Cramer, New York.

Myers, N., Mittermeier, R.A., Mittermeier, C.G., da Fonseca, G.A.B. and Kent, J. (2000) Biodiversity hotspots for conservation priorities. *Nature*, 403, 853–8.

Newman, W.A. (1986) Origin of the Hawaiian marine fauna: dispersal and vicariance as indicated by barnacles and other organisms. *Crustacean Issues*, 4, 21–49.

New Zealand Ministry of Fisheries Press release 10 May 2001. http://www.fish.govt.nz (last accessed June 6, 2006).

NOAA (2006) http://www.history.noaa.gov (last accessed June 6, 2006).

Owens, W.B. and Hogg, N.G. (1980) Oceanic observations of stratified Taylor columns near a bump. *Deep-Sea Research*, 27A, 1029–45.

Parin, N.V., Mironov, A.N. and Nesis, K.N. (1997) Biology of the Nazca and Sala y Gomez submarine ridges, an outpost of the Indo-West Pacific fauna in the Eastern Pacific Ocean: composition and distribution of the fauna, its communities and history. *Advances in Marine Biology*, 32, 145–242.

Parker, T. and Tunnicliffe, V. (1994) Dispersal strategies of the biota on an oceanic seamount: implications for ecology and biogeography. *Biological Bulletin*, 187, 336–45.

Penrith, M.J. (1967) The fishes of Tristan da Cuna, Gough Island and the Vema Seamount. *Annals of the South African Museum*, 48, 523–48.

Pratt, R.M. (1963) Great Meteor Seamount. *Deep-Sea Research*, 10, 17–25.

Probert, P.K., McKnight, D.G. and Grove, S.L. (1997) Benthic invertebrate bycatch from a deep-water trawl fishery, Chatham Rise, New Zealand. *Aquatic Conservation*, 7, 27–40.

Proudman, J. (1916) On the motion of solids in a liquid possessing vorticity. *Proceedings of the Royal Society (A)*, 92, 408–24.

Raymore Jr., P.A. (1982) Photographic investigations on three seamounts in the Gulf of Alaska. *Pacific Science*, 36, 15–34.

Richer de Forges, B., Koslow, J.A. and Poore, G.C.B. (2000) Diversity and endemism of the benthic seamount macrofauna in the southwest Pacific. *Nature*, 405, 944–7.

Roberts, C.M. (2002) Deep impact: the rising toll of fishing in the deep sea. *Trends in Ecology and Evolution*, 17, 242–5.

Roberts, D.G., Hogg, N.G., Bishop, D.G. and Flewellen, C.G. (1974) Sediment distribution around moated seamounts in the Rockall Trough. *Deep-Sea Research*, 21, 175–84.

Roberts, C.M., Hawkins, J.P. and Gell, F.R. (2005) The role of marine reserves in achieving sustainable fisheries. *Philosophical Transactions of the Royal Society B-Biological Sciences*, 360, 123–32.

Roden, G.I. (1986) Aspects of oceanic flow and thermohaline structure in the vicinity of seamounts. *NOAA Technical Report NMFS3-12*.

Roden, G.I., Taft, B.A. and Ebbesmeyer, C.C. (1982) Oceanographic aspects of the Emperor Seamounts region. *Journal of Geophysical Research*, 87, 9537–52.

Rogers, A.D. (1994) The biology of seamounts. *Advances in Marine Biology*, 30, 305–54.

Rowden, A.A., O'Shea, S. and Clark, M.R. (2002) Benthic biodiversity of seamounts on the northwest Chatham Rise. Ministry of Fisheries. *Marine Biodiversity Biosecurity Report No. 2*, 21 pp. Wellington, New Zealand.

Sainsbury, K.J. (1987) Assessment and management of the demersal fishery on the continental shelf of northwestern Australia. In: *Tropical Snappers and Groupers – Biology and Fisheries Management* (eds. Polovina, J.J. and Ralston, S.), pp. 465–503. Westview Press, Boulder, CO.

Santos, R.S., Hawkins, S. Monteiro, L.R., Alves, M. and Isidro, E.J. (1995) Marine research, resources and conservation in the Azores. *Aquatic Conservation: Marine and Freshwater Ecosystems*, 5, 311–54.

Sasaki, T. (1974) The pelagic armorhead, *Pentaceros richardsoni* Smith, in the North Pacific. *Bulletin of the Japanese Society of Fisheries Oceanography*, 24, 156–65.

Schlee, S. (1973) *A History of Oceanography: The Edge of an Unfamiliar World*. Hale, London.

Schwartz, M.L. (1972) Seamounts as sea-level indicators. *Geological Society of America Bulletin*, 83, 2975–80.

SeamountsOnline (2006) http://seamounts.sdsc.edu (last accessed June 6, 2006).

Simpson, E.S.W. and Heydorn, A.E.F. (1965) Vema Seamount. *Nature*, 207, 249–51.

Smith, W.H.F. and Sandwell, D.T. (1997) Global sea floor topography from satellite altimetry and ship depth soundings. *Science*, 277, 1956–62.

Part II

Biophysical Coupling on Seamounts

Chapter 4

Physical processes and seamount productivity

Martin White, Igor Bashmachnikov, Javier Arístegui and Ana Martins

Abstract

A brief review is given of the physical dynamics that occur at seamounts and the implications of these dynamics for seamount productivity highlighted. Several physical seamount characteristics, stratification and oceanic flow conditions interact to provide a number of different local dynamic responses at a seamount. These include Taylor Columns or Cones, doming of density surfaces, enclosed circulation cells and enhanced vertical mixing. Due to oceanic background flow variability, it is likely that the localised seamount dynamics, and resultant bio-physical interaction processes; will also be variable. This makes quantification of an 'idealised' response of a particular seamount to the impinging flow regime difficult. It has been widely accepted that dynamics at seamounts generate conditions such as increased vertical nutrient fluxes and material retention, to promote productivity that fuels higher trophic levels. To date, however, there has been little consistent concrete evidence for this in observations. This is likely due to the non-steady background oceanic forcing which may disrupt the 'idealised' response, such as Taylor Cones and circulation cells generated at the seamount. In addition, the seamount may shed passive tracers such as chlorophyll downstream, providing a source of oceanic bio-physical patchiness in the surrounding ocean. Such variability provides a challenge for the environmental management of seamounts.

Introduction

Seamounts, submerged banks or continental slopes are often characterised by enhanced hydrodynamic activity compared to the flat abyssal ocean. Isolated topographic features of varying shape and size (see Chapters 1 and 2) might therefore be expected to interact with the physical environment to generate a variety of distinct physical flow features. The abrupt nature of seamount topography may disrupt the large scale oceanic flow, generating spatial and temporal variability in the current field (Royer, 1978; Roden, 1994). More localised features include enclosed circulation cells around seamounts (Freeland, 1994), doming of the density surfaces above the seamount (Owens and Hogg, 1980), the amplification and rectification of tidal motions (Brink, 1995) and increased vertical mixing (Kunze and Sanford, 1997; Eriksen, 1998). The most energetic motions are typically concentrated near the top of the seamount (Brink, 1989; Chapman, 1989), and are often hardly distinguishable above the seasonal pycnocline, the depth of strong vertical density gradient (Eriksen, 1991; Brink, 1995; Codiga, 1997b). Depending on distance from the summit to the seasonal pycnocline, a 'seamount effect' on the background flow regime can be restricted to within a hundred metres over the summit of a shallow seamount, or propagate thousands of metres up through the weakly stratified deep ocean over a deep seamount (Roden, 1987).

With the recognition that seamounts were areas of high biomass, biodiversity (Chapter 13), and with the increased interest in seamount biology in the 1980s and 1990s (Chapter 3), increased primary productivity and enhanced chlorophyll levels around some seamounts were reported, especially in oligotrophic regions (Genin and Boehlert, 1985; Dower *et al.*, 1992; Mouriño *et al.*, 2001). A number of arguments relating the distinct physical flow regime at seamounts and the increased biological activity were put forward (Genin and Boehlert, 1985; Dower *et al.*, 1992; Comeau *et al.*, 1995). Evidence for enhanced productivity does exist at some seamounts but is not a consistent feature and a wide range of spatial and temporal variability exists. As will be seen, this is likely due to the complex nature of physical processes that may be at work at different seamounts, which themselves have very different topographic characters.

This chapter identifies the dominant processes at seamounts and describes their basic characteristics using a combination of theoretical concepts and field observations. The theoretical biological response and consequences generated by the interaction of the physical dynamics and the seamount are then presented, and examples given where seamount-generated processes have been observed to influence the chlorophyll distribution or production levels.

Physical processes

A number of physical properties of both the oceanic environment and the seamount itself interact to produce the localised flow and density stratification conditions (Fig. 4.1). These properties include the size and shape of the seamount, such as the linear dimension (L, typically taken as the width of the seamount in metres), and the height of the seamount (h_o) relative to the water depth (H). The magnitude and character of the flow impinging on the seamount is highly important, whether a steady flow or one which varies periodically with a certain (tidal) period. The influence of the Earth's rotation is also fundamental (see Box 4.1), as is the change in the water density (ρ) with depth, i.e., the vertical stratification (N),

$$N = [(-g/\rho) \times (\mathrm{d}\rho/\mathrm{d}z)]^{1/2} \tag{4.1}$$

where g = gravitational acceleration (9.81 m/s^2) and z the vertical coordinate. Both the steady and periodic varying flow impinging on seamount features may generate a multitude of dynamic responses in its vicinity. The determining factors that control whether the local seamount circulation is influenced more by steady or periodic varying flow conditions are illustrated in Fig. 4.1.

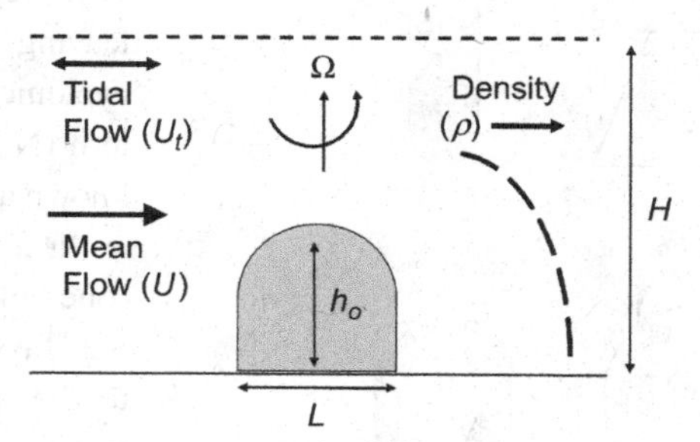

Fig. 4.1 Sketch of the main physical factors that control the localised dynamics at seamounts and other isolated topographic features.

Box 4.1 Vorticity and Taylor Columns

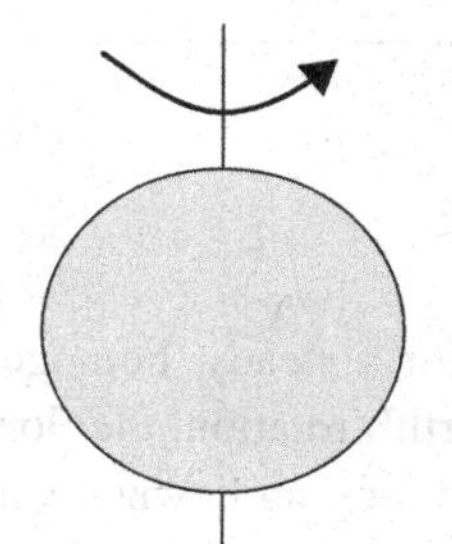

As we stand still on a rotating Earth, then compared to the 'fixed' stars in space, we too are spinning as the Earth spins around on its axis.

The tendency for rotational movement is called ***vorticity***, and the one related to the effect of the Earth's spin is called ***'planetary vorticity'***. It is denoted by the letter '***f***' and increases with latitude from 0 at equator to a maximum at the poles (the axis of rotation).

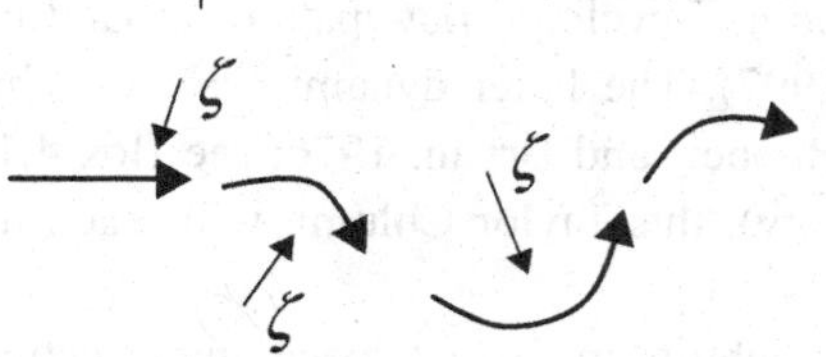

In addition, a body of water itself, as it moves, may have a tendency to spin around in a loop (right) – we often see swirls of water in a river or lake. This is called **'relative vorticity'**, or 'ζ', and is positive when in the same sense as the Earth's rotation.

For a water particle, the sum of the two vorticities should remain constant in the absence of any sources or sinks. For a water column, the total vorticity will also depend on column length, e.g., on water depth. This is similar to the principle used by ice skaters that slow down their spin by stretching out their arms.

So what happens to a column of water as it moves past a seamount?

The water squashes up as it goes over the seamount and stretches again on the other side of the seamount.

The squashing causes the water to acquire $-$ve ζ (see equation above) and stretching $+$ve ζ causing the rotations shown.

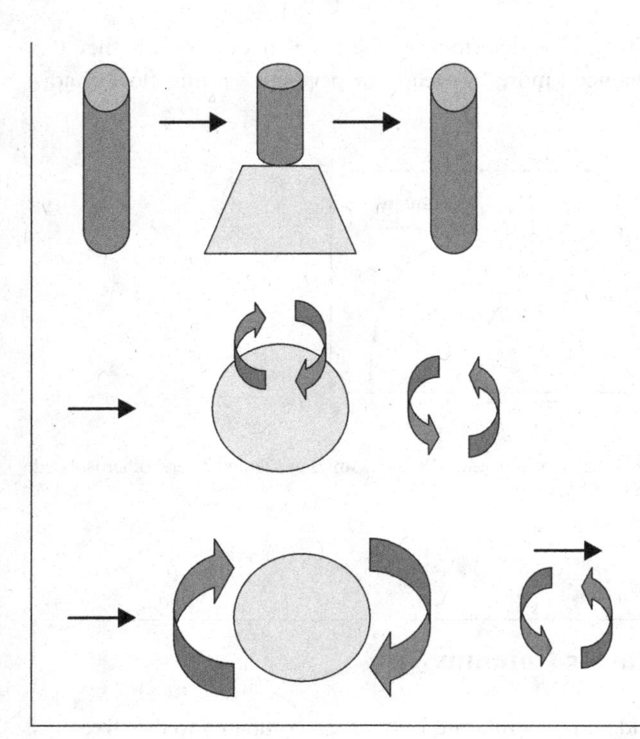

The prevailing current takes the resultant anti-clockwise (cyclonic) swirl away leaving behind just the anti-cyclonic (clockwise) pattern (N Hemisphere). This is known as a 'Taylor Column'.

In a homogeneous ocean (one of constant density), the Taylor Column extends to the ocean surface. If the ocean is stratified, there is likely to be insufficient energy in the water flow to extend the column to the surface. Instead a finite height 'Taylor Cone' is produced above the seamount.

Mean circulation over seamounts

Non-periodic impinging current

Proudman (1916) and Taylor (1917) posed a theoretical concept for a steady, homogeneous flow encountering a seamount. Under the influence of the Earth's rotation, the flow is split by the seamount and is accelerated to one side (the left-hand looking downstream for the northern hemisphere), forming an isolated anti-cyclonic flow pattern around the seamount (Roden, 1987; Codiga and Eriksen, 1997). The latter dynamic structure has been termed a Taylor Column (Huppert, 1975; Huppert and Bryan, 1976; see Box 4.1). For a homogeneous ocean (one of constant density), this Taylor Column will reach the surface, hence the terminology.

In the real ocean environment, stratification, variability in current speed and direction, turbulence and the irregular shape of seamounts disrupt the conditions for ideal Taylor Column flow, leading to more complex patterns. The exact nature of this localised flow pattern is dependent on many factors (Fig. 4.1). In particular, in the presence of stratification the Taylor Column becomes a 'Taylor Cone' of finite height, which may not extend to the surface (Fig. 4.2). A number of basic non-dimensional numbers can be used to help define criteria for the development of the localised seamount induced circulation. These are:

The Rossby number	$Ro = U/f \times L$	(4.2)
The relative height of the seamount to water depth	$\alpha = h_0/H$	(4.3)
The Burger number	$B = N \times H/f \times L$	(4.4)

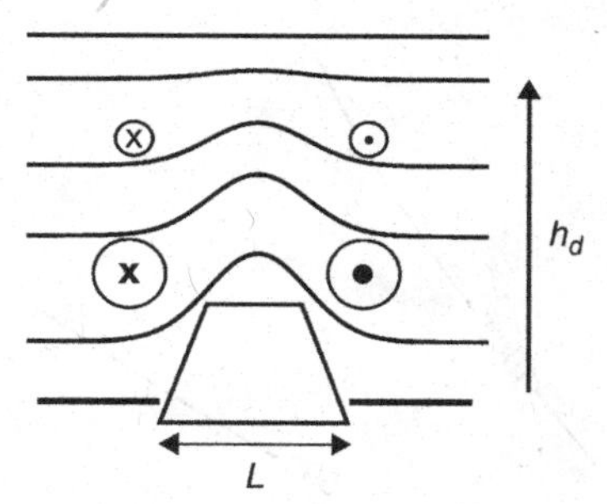

Fig. 4.2 The characteristic residual circulation and isopycnal doming over a seamount found in the presence of a steady flow and vertical stratification due to Taylor Cone formation. Cross and dot shows opposite flow directions.

where $f = 2 \times \Omega \times \sin(\text{latitude})$ is the Coriolis parameter (rads/s), Ω is the Earth's angular velocity (0.0000729 rads/s), U is a typical flow speed for the seamount (m/s), L is the seamount width (m).

The Rossby number describes the importance of the Earth's rotation in determining the governing dynamics over the seamount for any given flow speed and seamount width. A combination of Ro and α gives a blocking parameter $Bl = \alpha/Ro$, which controls the Taylor Column or Cone formation and flow behaviour. For small fractional height Gaussian-shaped seamount and moderate stratification conditions, Taylor Cones (Fig. 4.2) will form for a condition $Bl > \sim 2$ (Chapman and Haidvogel, 1992). Chapman and Haidvogel (1992) also concluded that this threshold value generally increases with fractional height and that there is an upper bound on the Rossby number of $Ro \sim 0.15-0.2$ above which a Taylor Cone would not form. Chapman and Haidvogel (1992) also describe a transitional state possible, whereby Taylor Cones may be generated over the seamount for a short period before being swept downstream. A critical value for Taylor Cone formation of $Bl > \sim 1$ was also found for different parameter configurations by other authors (e.g., Huppert, 1975; Roden, 1987). The Burger number quantifies the relative importance both of the vertical stratification and height scaling to the Coriolis force and the horizontal length scale, over which the Earth's rotation may become important, to determine which factor is most important in controlling the resultant flow characteristics. High B values will indicate strong stratification conditions, and very low B values low stratification conditions more akin to the homogeneous ocean case.

Lifting or doming of the density (isopycnal) surfaces over the seamount is associated with a finite height Taylor Cone (Owens and Hogg, 1980). In this case, both the circulation strength and amplitude of isopycnal doming decreases with height above the seamount. Nearer the seabed, however, bottom friction will reduce the flow magnitude and also drive a down- and off-slope near-seabed current, due to Ekman dynamics. Owens and Hogg (1980) show, from vorticity arguments, that the decay height (H_d) of a Taylor Cone is given by:

$$H_d = f \times L/N \qquad (4.5)$$

The variation of this theoretical doming height with stratification (N) and latitude (which determines f) for an arbitrary seamount of width, $L = 20\,\text{km}$, is shown in Fig. 4.3. Two

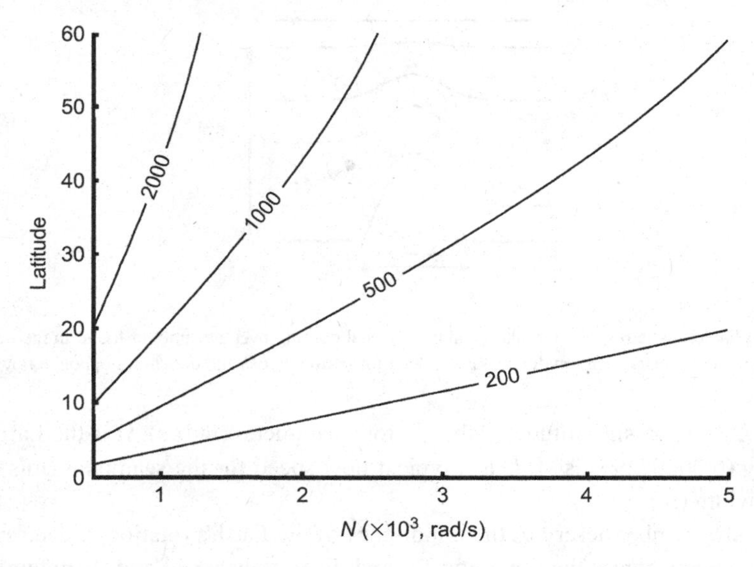

Fig. 4.3 The variation of isopycnal doming decay scale height (in metres) with latitude and vertical stratification for a seamount of typical length scale 25 km, based on the scaling argument of Owens and Hogg (1980).

important results of this scaling are immediately apparent. Firstly for deep seamounts, below the main thermocline where N is small, even a small isolated topographic feature may produce large vertical displacements in the density surfaces. A small seamount may, therefore, cause significant local mesoscale variability in the sub-thermocline layer (Royer, 1978). Secondly, the doming of isopycnals over shallow or intermediate depth seamounts (see seamount definition) may reach the euphotic zone with biological implications that are explored later.

Circulation and density structures measured over several seamounts support the existence of the different features predicted theoretically. Owens and Hogg (1980) presented evidence for a Taylor Column at Corner Rise, a 400 m high seamount in deep NW Atlantic waters. Doming of isopycnals and an anti-cyclonic circulation/vorticity tendency around the seamount was measured from moored instruments and hydrographic surveys. The decay height of the Taylor Cone, ~3000 m, agreed with the scaling described above. Over the relatively shallow Cobb seamount, Freeland (1994) measured strong bottom intensified anti-cyclonic flow from both moored instrumentation and underway acoustic Doppler current profiler (ADCP) measurements. In this case, the measured decay scale of n 40–45 m for the currents above the seamount was half the predicted decay scale height of ~80 m. This discrepancy was likely due to the influence of significant tidal amplification around the seamount.

Ocean currents seldom are steady, being dominated by energetic synoptic scale waves and eddies (Kamenkovich *et al.*, 1986). Thus, a non-steady model of an impinging flow with a gradual change of current velocity and direction may well be more appropriate. As background flow increases, upwelling will occur on the upstream side, and downwelling on the downstream side, of the seamount. A pair of stationary vortices, one anti-cyclonic and one cyclonic one will be formed (Box 4.1). If the flow remains sufficiently weak

($Bl > 10$), the vortices may stay near the seamount for a long time. Stronger impinging flow causes the cyclonic vortex to be shed from the seamount and advected downstream, while the anti-cyclonic vortex remains essentially over the seamount. At a very strong stratification, a cyclonic vortex may occur above an anti-cyclonic one (Roden, 1987). If the impinging flow is quasi-periodic, as well as Taylor Cone formation, a resonant amplification over the seamount summit can occur in the form of seamount-trapped waves (Brink, 1995), analogous to seamount-trapped tidal waves, discussed below.

An array of current meters located around the flanks of Sedlo seamount north of the Azores, just below the summit depth of 780 m, showed persistent daily mean negative values of relative vorticity (Fig. 4.4), which indicated a predominantly anti-cyclonic circulation around the feature. The magnitude of the vorticity varied significantly, however, with some periods of cyclonic vorticity measured. This may, in part, be due to the weak unsteady impinging flow, generally from the western quadrant, inferred from *in situ* data and satellite altimetry. At times the current measurements indicated that flow at the summit depth (800–900 m) might be influenced by westerly flowing Mediterranean Outflow Water (MOW). These results indicated the significance that variability in the impinging flow at a seamount may have in determining the local dynamical response.

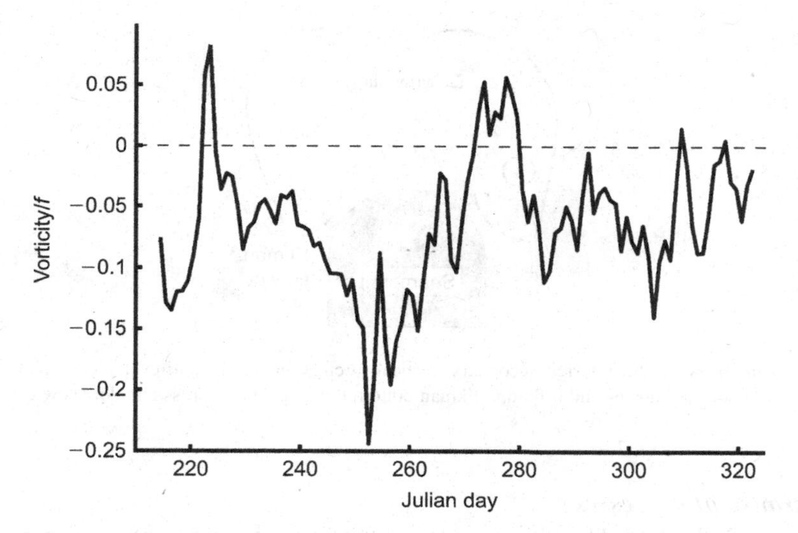

Fig. 4.4 Time-series of daily mean relative vorticity from a triangle of current meter measurements located at 800–900 m depth, close to the summit depth, around the flanks of Sedlo Seamount (40°20′N, 26°40′W, summit depth 780 m), NE Atlantic.

Tidally generated mean flows

Energetic tidal flows also play an important role in the generation of residual anti-cyclonic circulation cells. In many cases (e.g., Fieberling and Cobb seamounts), anti-cyclonic flow around the seamount was mainly due to the tides and, to a lesser extent, by the mean current impinging on the seamount (e.g., Eriksen, 1991; Kunze and Sanford, 1997). This generation of the residual mean currents by the tidal flow over the seamount is commonly referred to as 'tidal rectification'. Tidal rectification is a result of non-linear interaction of a tidal current with steep bathymetry, which, by a combined effect of topographic acceleration and bottom friction, produces asymmetry in tidal transport during a tidal cycle (Eriksen, 1991). This results in the generation of a cold, dense dome over the seamount similar to the uplifted isotherms formed by a Taylor Cone.

Secondary circulation may also be generated over seamounts, consisting of closed vertical cells with downwelling over the summit and radial outflow at and away from the summit plain rim (Eriksen, 1991; Brink, 1995). The circulation cell is essentially comprised of off-seamount flow at the seamount rim, due to Ekman bottom boundary flow, and downwelling above the seamount summit (Fig. 4.5). The cell is closed by upwelling at an outer border of the anti-cyclonic vortex, with inward flow above it (Eriksen, 1991; Haidvogel *et al.*, 1993; Brink, 1995; Mohn and Beckmann, 2002). The secondary circulation velocities are an order of magnitude less than those of the mean anti-cyclonic horizontal flow. This flow results from the complex balance of the mass flow and turbulent mixing processes (Haidvogel *et al.*, 1993; Brink, 1995; Kunze and Sanford, 1997). The downwelling currents can be quite significant. For example, over Fieberling seamount the downwelling was predicted to reach 25 m/day (Brink, 1995), forming a warm water anomaly above the cold summit dome (Haidvogel *et al.*, 1993).

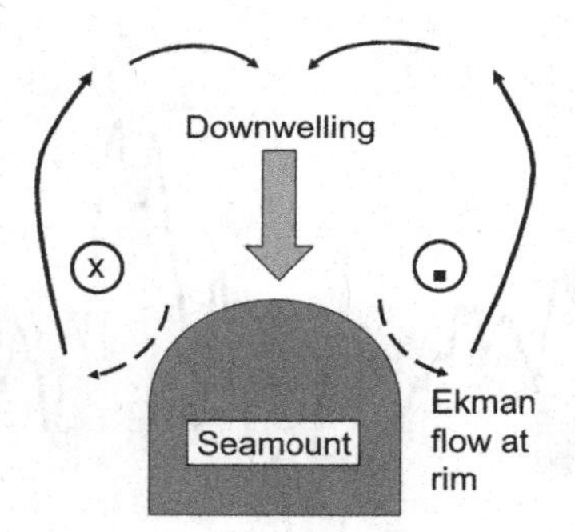

Fig. 4.5 Sketch showing the inferred secondary vertical circulation due to anti-cyclonic currents around a seamount and associated down- and off-slope Ekman bottom boundary flow. Cross and dot show opposite flow directions.

Other transient processes

A collision of a vortex/eddy with a seamount is quite a regular event in many regions of the world ocean. This process may be important for the advection of different water mass and constituent properties to seamounts. North Atlantic seamounts are often impacted by MEDDIES, eddies containing MOW (Bower *et al.*, 1995; Richardson *et al.*, 2000). In the NE Pacific, 'Haida eddies' named after the Aboriginal people of the Haida Gwaii archipelago in British Columbia, Canada, generated at the continental shelf edge; regularly impact offshore seamounts (Miller *et al.*, 2005). Observations and hydrodynamic models suggest that during interaction with a seamount an eddy may either be trapped and disintegrate over the seamount, split into two, or just be rotated around the seamount flank as it passes by. Even in the latter case, when the eddy survives the collision, it may lose up to 30–40% of its initial core transferring water and energy to the seamount system (Richardson *et al.*, 2000; Adduce and Cenedese, 2004).

Tidal motions and mixing over seamounts

Seamount-trapped waves

Observed amplification of tidal flows over seamounts achieves a maximum over the top of the seamount and decreases towards the flanks (Brink, 1995; Codiga and Eriksen, 1997). This amplification is due to general squeezing of the flow when passing over the seamount topography and leads to the formation of wave motions, which are 'trapped' to the seamount (Brink, 1989). In the northern hemisphere, these trapped waves propagate with the shallow water to the right, that is anti-cyclonically around an isolated topographic feature, with an amplitude that quickly decays with distance from the seamount. The waves are composed of one or more pairs of counter rotating circulation cells moving clockwise around the seamount. High and narrow seamounts tend to have the largest response to such tidal forcing (Chapman, 1989).

The background stratification affects these seamount-trapped waves in the same way as the mean flow, producing bottom intensification of the currents (Brink, 1989) and a compression in height above the summit of the response. The vertical trapping scale (H_{dt}) of the currents can be roughly estimated from (Codiga, 1997a):

$$H_{dt} = f \times r/(2\pi \times N) \quad (4.6)$$

where r is the seamount radius.

In the stratified ocean, significant amplification of a tidal flow in the form of seamount-trapped waves may occur for a broad range of frequencies and Burger numbers (Brink, 1990; Haidvogel *et al.*, 1993). The upper frequency limit (σ_{cr}), below which no baroclinic seamount-trapped waves can be excited, can be determined from the expression

$$\sigma_{cr} \leqslant N \sin\alpha \quad (4.7)$$

where α is the angle between bottom slope and the horizontal plain (Codiga, 1997a). Although, in principle, this equation allows waves with semidiurnal period and less to be trapped in subtropical and mid-latitudes for sufficiently steep seamount slopes and strong stratification (N), the seamount-trapped waves generated by semidiurnal tides are generally not reported, though those produced by the tide with the diurnal period are commonly observed. Brink (1995) has presented measurements of a 11-fold amplification of the two principal diurnal tidal components – the O_1 (25.82 h) and K_1 (23.93 h) at Fieberling Guyot. Fortnightly (diurnal spring–neap cycle, 13.66 days) modulation of the mean residual anti-cyclonic flow of about 10 cm/s at the rim of the seamount indicated that the diurnal tide played a leading role in the generation of the residual currents measured.

Internal waves

Internal waves are periodic oscillations of vertical density surfaces within the ocean and are the internal ocean counterparts to surface waves. Internal waves are generally freely propagating and radiate the energy away from the generating source if the period of the wave (T) does not exceed the inertial period ($1/f$) for that particular latitude. For waves of semidiurnal period, this critical latitude is large about 75°. For the diurnal period, however, the critical latitude is approximately 30°. Hence for large portions of the world's oceans baroclinic waves of diurnal period cannot propagate freely. Above the critical latitudes, only trapped internal waves such as seamount-trapped waves can be stable, whereas below these latitudes generation of free internal waves seems to be a better mechanism of tidal energy transfer to baroclinic wave motions.

Freely propagating internal tidal waves may be generated if their period (T) is in the range $f < 1/T < N$, and are an important agent of topography-related mixing (Eriksen, 1982, 1998). Though internal waves of different origin form wide energy spectra in the ocean (Munk, 1981), the most energetic internal waves are typically those of the semidiurnal period, generated by tidal flow over steep topographic features. Near the generation region, free internal tidal waves show ray-like structure, where energy is concentrated in narrow beams and where enhanced mixing is observed. The waves propagate at a particular angle to horizontal (ϕ), which is dependent on the vertical stratification (N), Coriolis parameter (f) and wave frequency (σ), and is given by:

$$\phi = \sqrt{\frac{\sigma^2 - f^2}{N^2 - \sigma^2}} \tag{4.8}$$

For critical seabed slopes, where the seabed slope matches the energy propagation angle (i.e., $\alpha \sim \phi$), the vertical current shear close to the generation region changes significantly. This may lead to wave overturn (breaking) and/or turbulent mixing at the bottom, and along the wave ray path. At Muir seamount, Eriksen (1982) observed spectral enhancement near the frequency at which internal waves would be 'resonant' with the seabed slope. This enhancement was confined to within 100 m of the seabed.

The presence of an anti-cyclonic vortex (Taylor Cone) with an associated negative vorticity anomaly (ζ) may help to promote enhanced mixing at a seamount through internal wave interaction. With a vortex present, the criteria for existence of free internal waves becomes $f + \zeta = f_{eff} < 1/T < N$ (Brink, 1995). If f_{eff} is less than diurnal frequency, free diurnal period internal waves can be generated over the seamount with similar properties to free internal waves otherwise confined to below the critical latitudes (Kunze and Boss, 1998). In this situation, the waves are confined to the anti-cyclonic vortex and dissipate all the energy inside and at the outer border of the vortex (Codiga, 1997b), whereas free internal waves might carry more than 70% of their energy away from the seamount over large distances (Laurent and Garrett, 2002). For this reason 10–100 amplification of vertical mixing at seamounts, as compared with the deep ocean has been measured (Toole *et al.*, 1997; Kunze and Boss, 1998).

Biological response

Idealised processes

Elsewhere in this book it will become clear that seamounts are regions of apparent high biomass, consisting of organisms spanning many trophic levels (e.g., Chapters 6–13) and able to host a diverse range of organisms (Chapters 13 and 14). Here we are concerned with the impact of the physical processes on the lowest trophic level, through seamount enhancement, or retention of, primary productivity. The basic requirement for productivity to occur in the oceans is a combination of sufficient inorganic nutrients and adequate irradiance within the surface waters (euphotic zone). A related condition is the stability of the water column, such that plankton can remain at a water depth where both the above conditions prevail. Generally, either nutrients or light will limit the productivity in any particular region. Before looking in detail at the biological response to the physical dynamics at seamounts, we must first qualify the difference between 'productivity' and 'production' at seamounts. Productivity is equivalent to growth rate and depends on nutrients and light only, whereas production is the result of productivity and accumulation of phytoplankton. The distinction is important, as the physical dynamics may promote productivity but if the dynamics are variable and the productivity is not retained, the resultant production at the seamount will be low as the phytoplankton are advected downstream. This will be highlighted later and in Chapter 5.

The physical processes outlined in the previous section have been thought to produce a biological response over seamounts in the following six ways (Figs. 4.6 and 4.7):

1. Isopycnal doming due to Taylor Cone formation will bring deeper, nutrient-rich waters up to a shallower depth, particularly into the euphotic zone for shallow seamounts (Fig. 4.6; Genin and Boehlert, 1985).
2. Isopycnal doming may generate localised regions of high density stratification over the seamount, which may stabilise the water column, helping to promote productivity (Comeau *et al.*, 1995).
3. Increased vertical mixing due to tidal amplification, flow acceleration, internal wave interaction or deep winter mixing can also mix nutrient-rich deeper water upwards over shallow seamounts (Fig. 4.6).

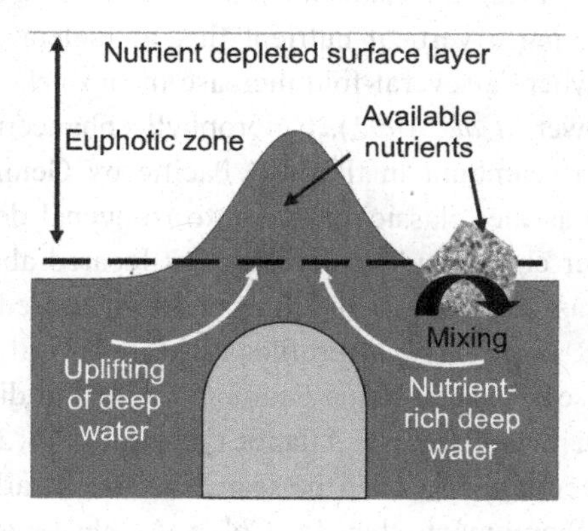

Fig. 4.6 Schematic showing vertical nutrient fluxes associated with dynamical processes acting at a seamount.

4. The enclosed, or semi-enclosed, circulation pattern around the seamount may also be important as a retention mechanism for material produced over, or advected into the vicinity of, the seamount hence increasing seamount production (Fig. 4.7; Goldner and Chapman, 1997; Mohn and Beckmann, 2002).

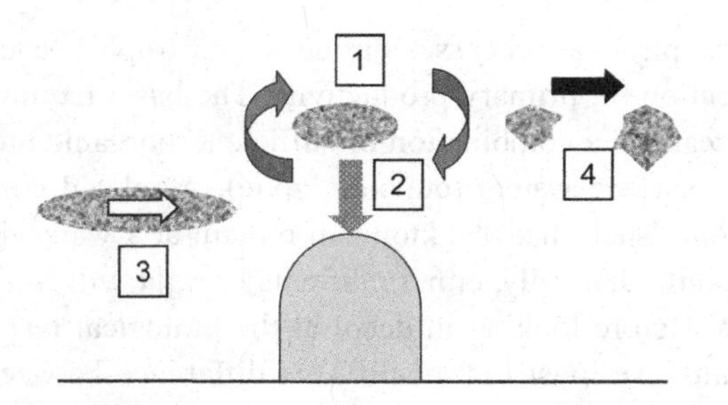

Fig. 4.7 Schematic showing advective processes at a seamount: 1 – organic material and larvae retained by anti-cyclonic and vertical circulation; 2 – downwelling of organic material to benthic communities; 3 – upstream advection and entrainment and 4 – downstream advective loss and 'patchiness' development.

5. Asymmetric flow acceleration at seamount flanks, or summits, may enhance horizontal fluxes of organic material with implications for sediment distribution around the seamount (Roberts *et al.*, 1974; Turnewitsch *et al.*, 2004) and sessile benthic communities (Genin *et al.*, 1986; Chapters 8 and 9).
6. Advection of organic material (phytoplankton, zooplankton, etc.) and nutrients from the far field into the 'sphere of influence' of the seamount can be significant if local dynamic processes do not provide suitable conditions for the enhancement of productivity over the seamount and an external source of nutrients/food is required (Fig. 4.7; Chapters 5, 6 and 15).

Observations of chlorophyll and productivity over seamounts

An increase in chlorophyll a (Chl_a) level has been observed over some seamounts, although Chl_a patches are not always a consistent feature (e.g., Genin and Boehlert, 1985; Mouriño *et al.*, 2001). Chl_a enhancement has been related to uplifting of isotherms around seamounts producing a vertical nutrient flux to fuel productivity, e.g., at Cobb seamount (NE Pacific), where a several-fold increase in subsurface Chl_a was associated with a Taylor Cone (Dower *et al.*, 1992). Chlorophyll enhancement was also observed over the Minami-Kasuga seamount in the NW Pacific by Genin and Boehlert (1985), and has often been cited as the 'classic' response to isopycnal doming over a seamount. Here a sharper and larger chlorophyll maximum was located above the seamount, confined to below 80 m and associated with the limit of the measured isopycnal doming over the seamount. Figure 4.8 shows vertical profiles of March/April chlorophyll and phaeopigments above and off-seamount at Seine Seamount, a 40 km diameter seamount rising from 4000 to ~180 m depth, in the NE Atlantic (33°45′N, 14°20′W). Enhancement of both chlorophyll and phaeopigments over the seamount was clearly visible. The enhancement was greater for phaeopigments than for Chl_a. As phaeopigments may result from the degradation of Chl_a, the results might have indicated the accumulation of passively aggregating plankton. Similar results were not found, however, on two subsequent surveys of the seamount.

Chlorophyll enhancement over seamounts, therefore, is not always the norm. Genin and Boehlert (1985) reported that two further surveys of Minami-Kasuga seamount within a month of the first survey did not show the same chlorophyll enhancement or

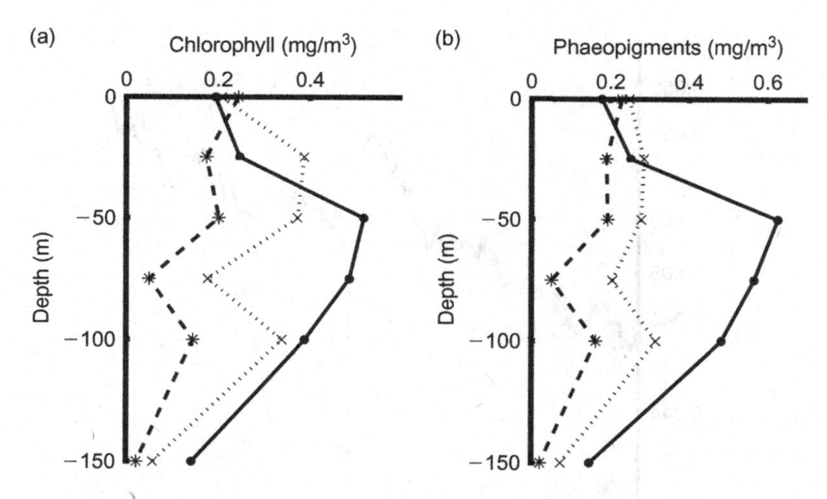

Fig. 4.8 Vertical profiles of (a) chlorophyll and (b) phaeopigments near Seine Seamount, March–April 2003, showing (cross) flank and (star) off-seamount profiles.

the presence of a cold-water dome. Primary productivity estimates over seamounts have also yielded inconclusive evidence for an enhancement response. From several surveys carried out over a 7-year period, Mouriño *et al.* (2001) found significant variability in primary production at Great Meteor Seamount (GMS) at seasonal and shorter time scales. Relative to the adjacent deep water, primary production over the seamount was higher in April 1999 and March 1993 but lower in December 1993.

To illustrate such variability in chlorophyll concentrations over seamounts at different temporal and spatial scales, satellite-based measurements of chlorophyll over the GMS are shown in Fig. 4.9. GMS is a large tablemount with a summit, rising from a surrounding water depth of 4000–5000 m, at approximately 400 m depth, centred at 30° N, 28.5° W. Doming of isopycnals up to the base of the euphotic zone has been both observed and predicted from models (Mohn and Beckmann, 2002). Elevated chlorophyll values have also been observed (e.g., Mouriño *et al.*, 2001; Mohn and Beckmann, 2002). Monthly mean Chl_a data, averaged between the latitude range 29.4–30.4° N, have been retrieved from the Giovanni data base (http://reason.gsfc.nasa.gov/OPS/Giovanni/, last accessed November 2005) along a transect from 35° W to 22° W. Data have a resolution of 0.1° longitude in the west–east direction.

Figure 4.9a shows that the 7-year mean summer Chl_a values were very low, as might be expected for the region, but with a definite trend of increasing values from west (~0.04 mg/m^3) to east (~0.08 mg/m^3). Within the vicinity of GMS (29–28° W) a small peak in Chl_a was apparent relative to mean background values. This would suggest a small but measurable chlorophyll enhancement in the vicinity of GMS. This suggestion is qualified somewhat by the data in Fig. 4.9b, which show that August Chl_a values from individual years and the 7-year August mean. (The mean Chl_a distribution for August paralleled that for the June–August mean.) The individual months indicated both significant interannual and mesoscale variability in the distribution. Peaks were not always associated with GMS and in some years; peaks of similar magnitude were present close to GMS. (Note that smaller scale variability was not represented in the plot as the spatial

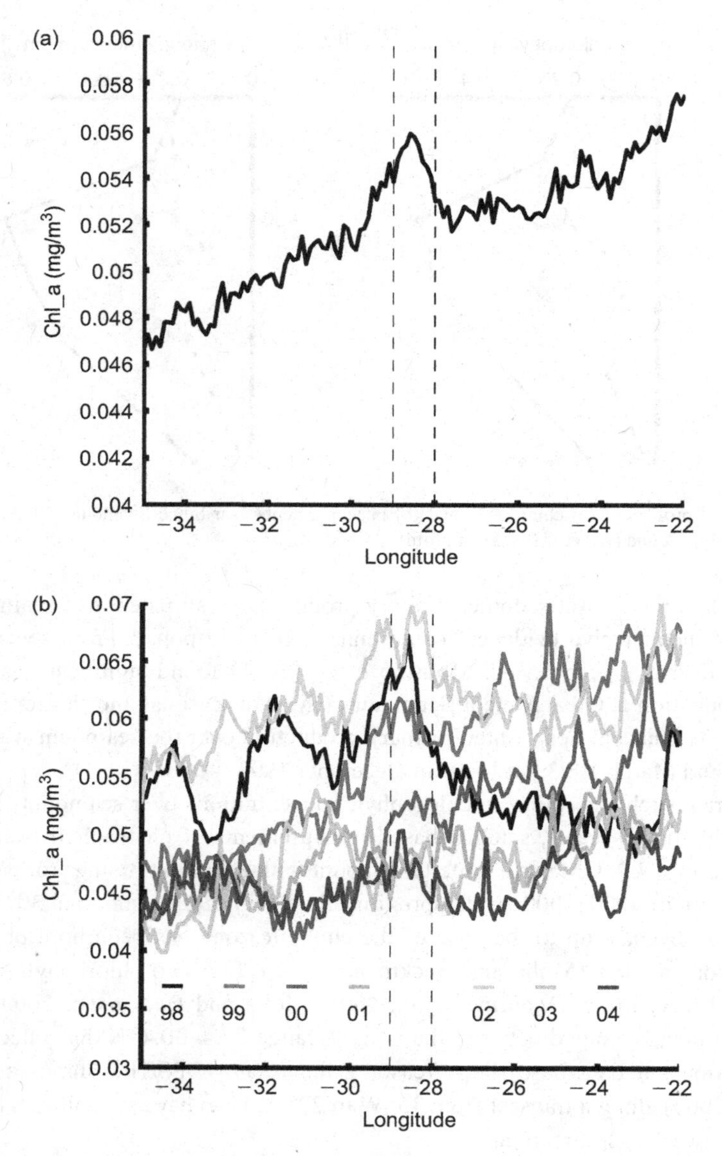

Fig. 4.9 Variation of Chlorophyll from 35° W to 22° W, averaged between 29.4° N and 30.4° N, showing (a) 7-year mean for months June–August and (b) 7 individual years of August Chlorophyll values. The data used in this study were acquired using the GES–DISC Interactive Online Visualisation and Analysis Infrastructure (Giovanni) as part of the NASA's Goddard Earth Sciences (GES) Data and Information Services Center (DISC).

resolution of the observations was ~10 km.) The results indicated that the seamount might be a source of variability in the surface chlorophyll distribution pattern in the surrounding ocean and/or that the localised dynamical processes at GMS themselves were highly variable.

While there is some evidence for a biological response to physical forcing, there is no 'ideal' situation of a regular-shaped seamount in an even flow field, so the response may either not be observed or at the scale one might expect. All seamounts are subject to flow changes ranging from days, through mesoscale up to interannual time scales. Seamounts come in a multitude of shapes and sizes. Variability in the far-field forcing and local density stratification is likely to play a hugely significant role in the response of any particular seamount. This variability has three main consequences:

1. Retention of passive particles reduced (Goldner and Chapman, 1997) with a significant proportion of the particles advected downstream (Fig. 4.7).
2. The seamount becomes a source of both physical and biological patchiness in the surrounding ocean extending from deep sub-thermocline layers (Owens and Hogg, 1980) to the surface layers (Royer, 1978).
3. Biological distributions will vary on similar time and space scales of the physical variability, making quantification of the biological response difficult without extensive measurements (Fig. 4.9).

A further complication is that the horizontal spatial scale of seamounts, typically 10–50 km, matches the typical patchiness in the biological fields in the oceans (Wroblewski *et al.*, 1975). Therefore a biological response at a seamount may be masked by the background variability in the surrounding ocean. This means that observations are likely sampling a pattern within a pattern of a similar spatial or temporal scale. Caution would be advised on drawing definite conclusions from a one-time hydrographic or biological survey of seamount ecosystems.

Circulation and material retention at seamounts

There has been much speculation as to the role of localised anti-cyclonic circulation patterns on organic material retention, thus accumulating plankton and increasing the local biomass at seamounts. Goldner and Chapman (1997) showed from modelling studies that significant retention of passive particles may indeed occur. They used simplified geometry and both steady and periodic forcing flows in the study, but importantly 'ramped up' the far-field forcing from zero to final amplitude (unlike most idealised model studies), which partly represented a real situation of variable flow conditions. As a consequence, particle advection (or retention) was found to be dependent on details of initial location relative to seamount and how quickly or strongly the current forcing was applied. Particle retention was found for both forcing types, but maximum retention was found for a combination of both types of forcing flows, the local response to which were the two superimposed flow patterns. The authors concluded that the retention of particles by the seamount circulation was dependent on the speed of the ramp up of the forcing motion and that particles close to the region of maximum residual current would be retained for longer. In addition, fluid replacing that lost over a seamount came from a relatively narrow band upstream of the seamount so dispersion (or retention) of particles was heavily dependent on the mean flow direction and the time scale of directional changes in the mean flow.

Passive particles over large submerged banks may show long residence times (White *et al.*, 2005), but the situation over the smaller scale seamounts would appear different. Mouriño *et al.* (2001) estimated a residence time of about 3 weeks over the GMS, although this is still larger than a time scale based on flow over the same distance of ocean with no seamount present. Dower *et al.* (1992) reported enhanced chlorophyll levels over Cobb seamount for more than 2 weeks. Variability in the chlorophyll distribution over other seamounts, however, may be high with a range of variability scales. Mouriño *et al.* (2001) observed seasonal and shorter time scale variability in both chlorophyll and productivity estimates, similar to those observed in the satellite-derived Chl_a measurements in Fig. 4.9. Genin and Boehlert (1985) observed variability in the vertical chlorophyll profiles over Pacific seamounts with a time scale of about a week. The retention of passive vs active particles such as vertically migrating plankton is an interesting aspect of the retention problem. Beckmann and Mohn (2002) found a higher retention potential for passive particles, because downward-migrating particles become subject to larger systematic advection in deeper layers, which are decoupled from flows above the thermocline.

Material retention, therefore, is most likely a biologically significant dynamic process at seamounts. Retention may provide suitable conditions to enhance production, through the aggregation of planktonic material, and hence increase the biomass over seamounts (e.g., Figs. 4.8 and 4.9). Over large submerged banks, that have similar circulation patterns to those found at seamounts, the physical dynamics have a major impact on the material retention (White *et al.*, 2005). At seamounts the likely retention time scales have not been adequately quantified. Estimates from days to up to a month have so far been inferred (Genin and Boehlert, 1985; Dower *et al.*, 1992; Mouriño *et al.*, 2001). The importance of retention over seamounts will be discussed further in Chapter 5, particularly in relation to transfer of primary production to higher trophic levels. Retention may be important for other reasons. For example, the vertical secondary circulation pattern implies an inflow towards the seamount. This is often assumed to be at mid-depth above the seamount (Freeland, 1994). Such enclosed circulation cells may well be advantageous for the reproductive strategies of some organisms as their larvae may be retained within the sphere of influence of the seamount (Mullineaux and Mills, 1997; see Chapter 6). Over Cobb seamount for example, Parker and Tunnicliffe (1994) found that most benthic organisms have a short-lived or no planktonic larval stage, presumably as the enclosed circulation measured over the seamount retains these larvae close to their required settling substratum (see also Chapter 13). The downwelling of material over a seamount that results from the circulation pattern is likely important for benthic organisms located near the seamount summit. The flow acceleration measured over seamount summits and flanks will also contribute to organic material fluxes important to suspension feeders such as corals (Genin *et al.*, 1986; Chapters 7 and 8).

Advection from the far field

Seamounts constitute a barrier to flow, and as such may receive material that has been advected from upstream. This, together with the possible retention potential of the local dynamics as highlighted above; means that nutrients, different water masses or organic material such as phyto- and zooplankton, may be brought into a seamount ecosystem. For example, it is known that 'MEDDIES', the lenses of MOW propagating westwards from the Gulf of Cadiz region, impact on subtropical NE Atlantic seamounts (Bower *et al.*, 1995). Elevated levels of chlorophyll over seamounts may also be the result of advection of high Chl_a waters over a seamount. Odate and Furuya (1998) observed high

Chl_a over Komahashi No. 2 seamount which, when measured on different occasions, was attributed to both local production and also advection and accumulation from the surrounding ocean.

'Haida' eddies transport NE Pacific continental shelf waters offshore, influencing the biological functioning of the offshore seamounts (Batten and Crawford, 2005; Mackas *et al.*, 2005). Seamounts located close to the relatively productive continental shelf waters are thus likely to benefit from such allochtonous fluxes. Even isolated deep ocean seamounts may be impacted by residual flows. For example, Lutjeharms and Heydoorn (1981) indicated the importance of advective fluxes for the recruitment of the rock lobster *Jasus tristani* on Vema seamount in the southern Atlantic. Recruitment is thought to be from the advection of developing larvae from the region close to the island of Tristan de Cunha over 1600 km to the SW. Drifting buoy tracks have indicated flow paths from Tristan de Cunha to Vema seamount with a transit time of about 5–7 months, similar to the larval development time of the lobster (9 months) before becoming benthic.

Conclusions

Several decades of observational and modelling research have identified the distinct physical processes that occur at seamounts and demonstrated the main physical forcing mechanisms behind these processes. Theoretical concepts involved have been identified and examples provided of where features akin to those predicted have been observed. Some form of localised dynamics around seamounts, such as amplified near-seabed currents (semi) enclosed circulation patterns, doming of density surfaces and enhanced vertical mixing is likely at any seamount. All these physical dynamics would appear to have the potential to impact significantly on the biological functioning at seamounts. These include: vertical nutrient fluxes to the euphotic zone due to either the uplifting of density surfaces or increased vertical mixing; enhanced organic matter fluxes at localised points of flow acceleration; the retention of organic matter in the vicinity of the seamount and generation of biological patchiness through variability in the physical response at the seamount or other biological effects (discussed further in Chapter 5).

The localised dynamics at seamounts, however, display a wide range of temporal and spatial variability due to the synoptic variability of the impinging flows and different seamount geometry. While the occurrence of these flow features might easily be predicted from assumptions of idealised physical environmental conditions, the strength, persistence and interaction of different individual features has proved harder to quantify precisely due to the complicated interaction of the forcing mechanisms and seamount topographies. This makes it difficult to impossible to assess the bio-physical coupling from a single hydrographic survey. Spatial and temporal physical variability between seamounts makes it difficult to formulate a general seamount management strategy. Even individual seamounts will require long multidisciplinary time series measurements before a management strategy can be formulated, as not enough is known about the physical–biological coupling. While a biological response is likely present at a seamount due to the local dynamics, the variability in the physical forcing makes quantification of the biological response hard, and does not allow for simple measurement of the biological consequences. Both the up- and downstream environments have to be taken into account for the development of management strategies for seamounts.

References

Adduce, C. and Cenedese, C. (2004) An experimental study of a mesoscale vortex colliding with topography of varying geometry in a rotating fluid. *Journal of Marine Research*, 62(5), 611–38.

Batten, S. and Crawford, W. (2005) The influence of coastal origin eddies on oceanic plankton distributions in the eastern Gulf of Alaska. *Deep Sea Research II*, 52(7–8), 991–1009.

Beckmann, A. and Mohn, C. (2002) The upper ocean circulation at Great Meteor Seamount. Part II: Retention potential of the seamount induced circulation. *Ocean Dynamics*, 52, 194–204.

Bower, A.S., Armi, L. and Ambar, I. (1995) Direct evidence for meddy formation off the southwestern coast of Portugal. *Deep Sea Research*, 42, 1621–30.

Brink, K. (1990) On the generation of seamount-trapped waves. *Deep Sea Research*, 37, 1569–82.

Brink, K.H. (1989) The effect of stratification on seamount-trapped waves. *Deep Sea Research*, 36, 825–44.

Brink, K. (1995) Tidal and lower frequency currents above Fieberling Guyot. *Journal of Geophysical Research*, 100(C6), 10817–32.

Chapman, D.C. (1989) Enhanced subinertial diurnal tides over isolated topographic features. *Deep Sea Research*, 36, 815–24.

Chapman, D.C. and Haidvogel, D.B. (1992) Formation of Taylor Caps over a tall isolated seamount in a stratified ocean. *Geophysical and Astrophysical Fluid Dynamics*, 64, 31–65.

Codiga, D.L. (1997a) Physics and observational signatures of free, forced and frictional stratified seamount-trapped waves. *Journal of Geophysical Research*, 102(C10), 23009–24.

Codiga, D.L. (1997b) Trapped-wave modification and critical surface formation by mean flow at a seamount with application at Fieberling Guyot. *Journal of Geophysical Research*, 102(C10), 23025–39.

Codiga, D. and Eriksen C.C. (1997) Observations of low-frequency circulation and amplified subinertial tidal currents at Cobb seamount. *Journal of Geophysical Research*, 102(C10), 22993–3007.

Comeau, L.A., Vezina, A.F., Bourgeois, M. and Juniper, S.K. (1995) Relationship between phytoplankton production and the physical structure of the water column near Cobb seamount, northeast Pacific. *Deep Sea Research*, 42, 993–1005.

Dower, J., Freeland, H. and Juniper, K. (1992) A strong biological response to oceanic flow past Cobb seamount. *Deep Sea Research*, 39, 1139–45.

Eriksen, C.C. (1982) Observations of internal wave reflection off sloping bottoms. *Journal of Geophysical Research*, 87, 525–38.

Eriksen, C.C. (1991) Observations of amplified flows atop a large seamount. *Journal of Geophysical Research*, 96, 15227–36.

Eriksen, C.C. (1998) Internal wave reflection and mixing at Fieberling Guyot. *Journal of Geophysical Research*, 103, 2977–94.

Freeland, H. (1994) Ocean circulation at Cobb seamount. *Deep Sea Research*, 41, 1715–32.

Genin, A. and Boehlert, G.W. (1985) Dynamics of temperature and chlorophyll structures above a seamount. An oceanic experiment. *Journal of Marine Research*, 43, 907–24.

Genin, A., Dayton P.K., Lonsdale, P.F. and Spiess, F.N. (1986) Coral on seamount peaks provide evidence of current acceleration over deep-sea topography. *Nature*, 322, 59–61.

Goldner, D.R. and Chapman, D.C. (1997) Flow and particle motion induced above a tall seamount by steady and tidal background currents. *Deep Sea Research*, 44, 719–44.

Haidvogel, D.B., Beckmann, A., Chapman, D.C. and Lin, R.-Q. (1993) Numerical simulation of flow around a tall isolated seamount. Part II: Resonant generation of trapped waves. *Journal of Physical Oceanography*, 23, 2373–91.

Huppert, H.E. (1975) Some remarks on the initiation of inertial Taylor Columns. *Journal of Fluid Mechanics*, 67, 397–412.

Huppert, H.E. and Bryan, K. (1976) Topographically generated eddies. *Deep Sea Research*, 23, 655–79.

Kamenkovich, V.M., Koshlyakov, M.N. and Monin, A.S. (1986) *Synoptic Eddies in the Ocean*, p. 433. D. Reidel Publishing Company, Dordrecht, The Netherlands.

Kunze, E. and Boss, E. (1998) A model for vortex-trapped internal waves. *Journal of Physical Oceanography*, 28, 2104–115.

Kunze, E. and Sanford, T.B. (1997) Tidally driven vorticity, diurnal shear and turbulence atop Fieberling Seamount. *Journal of Physical Oceanography*, 27, 2663–93.

Laurent, L. St. and Garrett, C. (2002) The role of internal tides in mixing the deep ocean. *Journal of Physical Oceanography*, 32, 2882–99.

Lutjeharms, J.R.E. and Heydoorn, A.E.F. (1981) The rock-lobster *Jasus tristani* on Vema Seamount. Drifting buoys suggest a possible recruiting mechanism. *Deep Sea Research*, 28, 631–6.

Mackas, D.L., Tsurumi, M., Galbraith, M.D. and Yelland, D.R. (2005) Zooplankton distribution and dynamics in a North Pacific Eddy of coastal origin. II. Mechanisms of eddy colonization by and retention of offshore species. *Deep Sea Research*, 52(6–8), 1011–35.

Miller, L.A., Robert, M. and Crawford, W.R. (2005) The large, westward-propagating Haida Eddies of the Pacific eastern boundary. *Deep Sea Research II*, 52(7–8), 845–51.

Mohn, C. and Beckmann, A. (2002) The upper ocean circulation at Great Meteor Seamount. Part I: Structure of density and flow fields. *Ocean Dynamics*, 52, 179–93.

Mouriño, B., Fernadez, E., Serret, P., Harbour, D., Sinha, B. and Pingree, R. (2001) Variability and seasonality of physical and biological fields at the Great Meteor Tablemount (sub tropical NE Atlantic). *Oceanologica Acta*, 24, 1–20.

Mullineaux, L.S. and Mills, S. (1997) A test of the larval retention hypothesis in seamount-generated flows. *Deep Sea Research*, 44, 745–70.

Munk, W.H. (1981) Internal waves and small scale processes. In: *Evolution of Physical Oceanography, Scientific Surveys in Honor of Henry Stommel* (eds. Warren, B. and Wunsch, C.), pp. 264–91. MIT press, Cambridge, MA.

Odate, T. and Furuya, K. (1998) Well-developed chlorophyll maximum near Komahashi No 2 seamount in the summer of 1991. *Deep Sea Research*, 45, 1595–1607.

Owens, W.B. and Hogg, N.G. (1980) Oceanic observations of stratified Taylor Columns near a bump. *Deep Sea Research*, 27, 1029–45.

Parker, T. and Tunnicliffe, V. (1994) Dispersal strategies of the biota of an oceanic seamount. Implications for ecology and biogeography. *Biology Bulletin*, 187, 336–45.

Proudman, J. (1916) On the motion of solids in a liquid processing vorticity. *Proceedings of the Royal Society of London*, A92, 408–24.

Richardson, P.L., Bower, A.S. and Zenk, W. (2000) A census of Meddies tracked by floats. *Progress in Oceanography*, 45(2), 209–50.

Roberts, D.G., Hogg. N., Bishop, D.G. and Flewellen, C.G. (1974) Sediment distribution around moated seamounts in the Rockall Trough. *Deep Sea Research*, 21, 175–84.

Roden, G.I. (1987) Effects of seamounts and seamount chains on ocean circulation and thermohaline structure. In: *Seamounts, Islands, and Atolls, Geophysical Monograph Series*, Vol. XXXXIII (eds. Keating, B.H., Fryer, P., Batisa, R. and Boehlert, G.W.), 335–54. AGU, Washington, DC.

Roden, G.I. (1994) Effects of the Fieberling seamount group upon flow and thermohaline structure in the spring of 1991. *Journal of Geophysical Research*, 99(C5), 9941–61.

Royer, T.C. (1978) Ocean eddies generated by seamounts in the North Pacific. *Science*, 199, 1063–4.

Taylor, G.I. (1917) Motions of solids in fluids when the flow is not irrational. *Proceedings of the Royal Society of London*, A93, 99–113.

Toole, J.M., Schmitt, R.W. and Polzin, K.L. (1997) Near-boundary mixing above the flanks of a mid latitude seamount. *Journal of Geophysical Research*, C102, 947–59.

Turnewitsch, R., Reyss, J-L., Chapman, D.C., Thomson, J. and Lampitt, R.S. (2004) Evidence for a sedimentary fingerprint of an asymmetric flow field surrounding a short seamount. *Earth Planetary Science Letters*, 222(3–4), 1023–36.

White, M., Mohn, C., de Stigter, H. and Mottram, G. (2005) Deep-water coral development as a function of hydrodynamics and surface productivity around the submarine banks of the Rockall Trough, NE Atlantic. In: *Cold-Water Corals and Ecosystems* (eds. Freiwald, A. and Roberts, J.M.), pp. 503–14. Springer-Verlag, Berlin/Heidelberg.

Wroblewski, J.S., O'Brien, J.J. and Platt, T. (1975) On the physical and biological scales of phytoplankton patchiness in the ocean. *Mémoires Société Royale des Sciences de Liège 6e Sèrie*, Tome VII, 43–57.

Chapter 5
Seamount plankton dynamics

Amatzia Genin and John F. Dower

Abstract

The biomass of fishes and other zooplanktivores is unusually high over shallow, intermediate and, sometimes, deep seamounts. This cannot be explained in terms of 'bottom-up control', which is driven by local upwelling and the ensuing augmentation of local primary production. Although upwelling does occasionally occur, it rarely penetrates the photic layer and has not been observed to remain over a seamount long enough to affect the growth of local zooplankton populations. In fact, the biomass of seamount zooplankton is, in many cases, lower above the summit than in the surrounding waters, especially over shallow seamounts. In contrast, the weight of the available evidence indicates that the trophic enrichment over seamounts is due to allochthonous inputs *via* (i) bottom trapping of vertically migrating zooplankton and (ii) greatly enhanced horizontal fluxes of suspended food. Bottom trapping of migrating zooplankton in the early morning is a major mechanism for accumulation and trophic focusing of zooplankton over seamounts at shallow and intermediate depths, the trapped zooplankters being readily consumed by fishes. Enhanced horizontal fluxes of planktonic prey appear to be a major pathway of trophic subsidy over deep seamounts. High fluxes are maintained due to substantial enhancement of currents and amplification of internal waves over the seamount topography. Biophysical interactions are the key mechanisms responsible for the maintenance of unusually high biomass at high trophic levels. Hence, the biological enrichment of seamount communities may be the product of a bottom-up pathway fueled by trophic subsidy to carnivores, rather than by local enhancement of primary production.

Introduction

Seamounts are abrupt, shallow features found over much deeper, nearly flat bottoms (see Chapter 1). As such, seamounts have profound effects on physical and biological characteristics of the surrounding waters and, in turn, on the local dynamics of plankton (see Chapter 4). In this chapter, we consider how bio-physical interactions between seamounts and ocean currents can affect the productivity, distribution and consumption of phytoplankton and zooplankton and their relationships with resident predators. A key goal is to explain the well-documented exceptional production of nektonic populations over many seamounts (see Chapters 6 and 9). We therefore focus on the underlying mechanisms, rather than merely listing planktonic species that have been recorded over seamounts. We also consider this issue from the perspective of both resident seamount species, for which plankton advected over seamounts represents a key energy source, and open-ocean plankton that is occasionally advected over the shallow topography.

We classify seamounts in three depth categories (after Genin, 2004): *shallow* – summit within the photic layer, *intermediate* – summit below the photic layer but shallower than the day-time depth of most of the vertically migrating zooplankton (~400 m) and *deep* (summit below 400 m). Trophic inputs to the seamount community will be termed *autochthonous* if their origin is from local primary production (i.e., over the seamount), or *allochthonous* if they originate in the surrounding open sea.

Plankton dynamics at seamounts

Phytoplankton growth is usually limited by the availability of either light or nutrients. Locally enhanced phytoplankton biomass is often found in association with abrupt topographic features such as headlands, banks and islands due to the injection of inorganic nutrients into the surface waters *via* topographically induced mixing (Mullin, 1993; Mann and Lazier, 1996). It has long been held that the high biomass of nektonic species that often occurs around shallow seamounts is at least in part, due to locally enhanced primary production and the subsequent bottom-up transfer of energy to higher trophic levels in seamount food chains (Uda and Ishino, 1958; Hubbs, 1959; Uchida and Tagami, 1984).

Evidently, current–topography interactions have the potential to enhance primary production over shallow seamounts (see Chapters 4 and 14). Physical oceanographic phenomena such as localized upwelling, enhanced turbulent mixing, the amplification of tidal currents, and Taylor cone formation can all serve to enhance the entrainment of inorganic nutrients to the euphotic zone if the seamount in question penetrates to within a few hundred meters of the surface. In contrast, current–topography interactions at deeper seamounts are unlikely to impact near-surface biological processes, due the damping effects of near-surface stratification (see Chapter 4).

Given that most seamounts occur in offshore, highly oligotrophic waters, away from continental influences, the injection of nutrients into the near-surface layer may represent a very important source of mesoscale variability in the distribution of the inorganic nutrients vital to phytoplankton growth. The vertical uplifting of water over shallow seamounts can also increase average light levels experienced by phytoplankton, further increasing the possibility of locally enhanced primary production. However, the frequency of such topographically induced *in situ* amplification of production above shallow seamount ecosystems is under dispute.

Despite more than 20 years of oceanographic field observations, there have been very few observations of persistently high phytoplankton biomass, usually estimated from chlorophyll concentrations, over seamounts. Whether this is merely a sampling artifact, due to the small number of seamounts that have been studied in detail, or because such effects are rare or highly episodic, remains unclear. The situation is further complicated because the few studies, which have measured phytoplankton stocks over seamounts have only rarely measured subsurface currents at the same time, making it difficult to ascribe any observed biological effects to current–topography phenomena. Here, we review studies that have investigated the effects of seamounts on standing phytoplankton stocks and draw conclusions about the general nature of such interactions and their possible consequences for higher trophic levels in seamount ecosystems.

Local enhancement of standing phytoplankton standing stocks above a seamount was first reported by Genin and Boehlert (1985). While surveying seamounts in the highly oligotrophic western subtropical North Pacific, a region of high subsurface chlorophyll water (~150% of background levels) was observed over Minami-Kasuga seamount (summit depth ~260 m). The patch was associated with a region of colder water caused by localized isothermal doming over the summit, and was therefore clearly a 'seamount effect'. However, a second survey just 2 days later revealed no evidence of the feature, nor did a third survey about 2 weeks later. Thus, although by far the most cited example of the potential for bottom-up forcing to locally enhance primary production at seamounts, the high chlorophyll patch over Minami-Kasuga was rather ephemeral.

A more persistent feature was observed above Cobb Seamount, a shallow NE Pacific seamount that penetrates to within just 25 m of the surface: Dower *et al.* (1992) recorded a 'bulls-eye' of high chlorophyll water centered over the summit in July 1990. Local chlorophyll concentrations within the bulls-eye were almost 10 times higher than background levels for nearly 3 weeks. Contemporaneous current meter observations showed that during that period, Cobb Seamount was capped by an anticyclonic recirculation, consistent with a stratified Taylor cone (Freeland, 1994). Observations from three subsequent cruises in 1991–1992 also revealed locally elevated chlorophyll concentrations over Cobb Seamount (Dower, 1994), suggesting that the feature may be quasi-permanent. Comeau *et al.* (1995) also reported that both primary production and diatom abundances were locally enhanced over Cobb in 1991.

Interestingly, although examination of the Cobb zooplankton community provided evidence of a statistically significant, but biologically subtle, shift in community composition, there was no evidence of any local enhancement of zooplankton biomass (Dower and Mackas, 1996). This suggests that, although current–topography interactions at Cobb Seamount appear to have a measurable effect on the primary producers, this energy is not being transferred to the local secondary producers. If this is a general trend, it may well be that local enhancement of phytoplankton biomass over shallow seamounts is exported downstream, perhaps indicating a role for seamounts in generating mesoscale variability in primary production.

Enhanced subsurface chlorophyll maxima (SCM) layers were also observed during three hydrographic surveys near Komahashi II Seamount (summit depth 289 m), in the highly oligotrophic subtropical gyre of the western North Pacific (Odate and Furuya, 1998). However, at least two factors argue against these features being 'seamount effects' in the classical sense. First, one of the enhanced SCM features was located 60 km downstream, rather than over the seamount summit. Second, the other SCM appeared to result not from localized upwelling but from nutrient injection brought about by mixing, either through shear-driven turbulence or double diffusion, along the interface of two water masses that just happen to meet near Komahashi II Seamount. Thus, although observed in the vicinity of a seamount, the enhanced SCM layers reported by Odate and Furuya (1998) seem more likely to have resulted from simple patchiness and horizontal variations in temperature, nutrients and phytoplankton.

A similar explanation would appear to account for highly variable results from Great Meteor Seamount (summit depth ~300 m) in the Northeast Atlantic (Mourino *et al.*, 2001). Between 1992 and 1999, five surveys collected a suite of physical, chemical and biological oceanographic data, including measurements of currents, hydrographic structure, fluorescence, primary production and nutrient concentrations. During the single survey in which they were deployed, data from a current meter array suggested an anticyclonic flow around the seamount. However, the authors contended that this pattern likely resulted from the amplification and/or rectification of internal tides (e.g., Noble and Mullineaux, 1989; Kunze and Toole, 1997) rather than a Taylor cone recirculation. Although various biological parameters, such as fluorescence, primary production and microzooplankton abundance and composition, were occasionally found correlated with proximity to the seamount, the patterns were largely inconsistent or contradictory from one cruise to the next. Thus, the authors concluded that, even if current–topography interactions were affecting the local phytoplankton populations over Great Meteor Seamount, the signals were impossible to separate from the natural background spatio-temporal variability in hydrographic conditions.

Primary production and chlorophyll concentrations were measured above Fieberling Guyot (eastern North Pacific, summit depth 438 m) in 1989–1991 as part of the TOPO program (Haury *et al.*, 2000). The region is oligotrophic with an average Secchi depth of 34 m. Neither primary production nor chlorophyll was enhanced above the seamount relative to control locations away from the seamount and over its flank (Haury *et al.*, 2000). The thermohaline structure around Fieberling Guyot and its neighboring seamounts Fieberling II and Hoke (summit depths 1050 and 900 m) was studied in 1991 by Roden (1994). While transient, bottom-intensified upwelling cones were encountered on all three seamounts, their persistence at a given location was less than a week and in all cases the water uplift decayed with height above bottom, well below the bottom of the photic layer (Roden, 1994).

An extensive survey designed to examine the occurrence of upwelling-driven enhanced productivity over seamounts was reported by Genin (2004), who surveyed eight different seamounts from 100 to 1440 m in depth across the North Pacific 17 times. Although isothermal doming, indicative of upwelling, was observed in nearly 50% of the visits, only once did the upwelling penetrate the photic layer and result in higher chlorophyll concentrations.

In summary, from the data that are available, some common themes emerge. First, occurrences of enhanced primary productivity and higher phytoplankton biomass over seamounts are uncommon. Second, where such effects have been reported, they all came from seamounts with summits shallower than 300 m and lasted a few days, at most. To date, only the observations from Cobb Seamount are suggestive of the existence of any sort of permanent or recurrent enhancement of primary production over seamounts. Our evidence challenges the widely held view that the immense stocks of seamount nekton are maintained by bottom-up transfer of energy from locally enhanced primary production. It is possible that, far from fostering the growth of higher trophic levels above seamounts, the suite of current–topography interactions that occasionally entrain nutrients to the photic layer are more important to downstream plankton production. This would imply an important role for seamounts in generating mesoscale patchiness in production, particularly in the open oligotrophic ocean.

Zooplankton dynamics over seamounts

Upwelling and secondary production: an unlikely mechanism

Three conditions should be met for topography-generated upwelling to affect the productivity of zooplankton over a seamount: (i) the region should be oligotrophic, that is, primary production is nutrient limited; (ii) the upwelled, nutrient-rich water must penetrate the photic layer and (iii) the 'patch' of upwelled water with augmented productivity should remain above the seamount long enough to affect zooplankton production, that is, at least the duration of a typical zooplankton generation (i.e., weeks to months). The available evidence suggests that seamount-generated upwelling rarely penetrates the photic layer and almost never remains trapped above the seamount more than a few days. Although long residence time is theoretically possible as part of a Taylor cap (see Chapter 4), to the best of our knowledge, no study has yet observed an enduring patch of high zooplankton biomass over a seamount. On the contrary, several studies over seamounts in the eastern Pacific (Genin *et al.*, 1988, 1994) and Atlantic Ocean (Nellen, 1973) reported that the waters overlying the seamounts had lower zooplankton biomass than in the surrounding waters. Our conclusion is that the dynamics of zooplankton above seamounts is likely unrelated to local, seamount-driven changes in the productivity and biomass of phytoplankton.

Trophic focusing: bottom trapping of migrating zooplankton

Ocean wide, numerous zooplankters migrate daily between the deep, aphotic layer and the upper, near-surface waters. The summits of many seamounts are well within the vertical range crossed by those zooplankters in their daily migration. Therefore, during the night, while foraging near the surface, migrating zooplankters are carried with the currents to the area above the seamount summit (Box 5.1). If the summit is shallower than their daytime depth, the animals would be unable to complete their descent in the next morning and thereby become trapped above the summit in shallow, possibly well-lit water, vulnerable to visual predators. This mechanism, first proposed by Isaacs and Schwartzlose (1965), has since been corroborated by a series of studies over several shallow and intermediate seamounts in the Pacific and Atlantic oceans (Genin *et al.*, 1988, 1994; Fock *et al.*, 2002a, b). The food supplied to the seamount community via topographic trapping can be as much as 40 times greater than the local primary productivity (Isaacs and Schwartzlose, 1965).

The depth of the seamount summit is a key parameter in the trapping mechanism. Since most of the migrating zooplankton forage during the night throughout the photic layer, the vertical flux of descending animals in the morning should reach peak values at the bottom of that layer. Where the summit is shallower, some proportion of the migrating animals would be swept around, as opposed to above, the summit during the night and will therefore not add to the biomass accumulated over the summit at dawn. Note, however, that migrating zooplankton swept around the summit may similarly become trapped above the slope.

Since different species of diel migratory zooplankton descend to different depths during the day, the deeper the summit of a given seamount, the less bottom trapping is expected. No significant bottom trapping is expected over seamounts with summits deeper than the deepest reach of the vertically migrating animals. Although some migrating zooplankters can reach 800 m depth (Wiebe *et al.*, 1979), most diel migrations are restricted to the upper few hundred meters (Angel, 1985). Brooks and Mullin (1983) estimated that about half of the zooplankton biomass in the Southern California Bight migrates through the 56 m isobath. Brinton (1967) presented data suggesting that most juvenile and adult euphausiids off California migrate daily through the 100–150 m isobaths. Most migrating micronekton in the waters off Oregon move from a depth of 0–50 m at night to 300–500 m during the day (Pearcy *et al.*, 1977).

Hence, the trophic benefit due to bottom trapping should be most effective for seamounts at the shallow-intermediate depth range (100–250 m). Uchida and Tagami (1984) reviewed historical records of the Soviet and Japanese seamount fisheries in the 1960s–1970s as well as their own observations over several seamounts in the North Pacific. In accordance with our conclusion, high fish catches were reported mostly from seamounts at the intermediate depth range. Estimates based on the average zooplankton biomass and currents suggest that the bottom trapping mechanism together with the horizontal flux of non-migrating zooplankton may provide sufficient food to maintain the commercial fisheries on intermediate seamounts (Tseitlin, 1985).

The formation of dense zooplankton accumulations by the trapping mechanism can greatly augment food capture by zooplanktivorous fishes (Nonacs *et al.*, 1994; Kiflawi and Genin, 1997). Driven by the bottom trapping mechanism, planktivorous rockfish (*Sebastes* spp.) have been shown to aggregate at dawn along the upstream flanks of seamounts where the likelihood of daily prey renewal from the surrounding deep waters is highest (Fock *et al.*, 2002b). The stomachs of rockfishes caught near the upstream flan of Nidever Bank, a seamount in the outer region of the Southern California Bight with a summit at 100–140 m depth, were packed with *Euphausia pacifica*, the dominant migrating species in that region (Genin *et al.*, 1988). Extensive analyses of fish stomachs by Fock *et al.* (2002a) indicated that the topographic trapping mechanism could explain the sustained fish populations, their distribution and diel behavior over the Great Meteor Seamount (subtropical NE Atlantic). Seki and Somerton (1994) observed that feeding by pelagic armourhead (*Pseudopentaceros wheeleri*) at SE Hancock Seamount (265 m, Hawaii seamount chain, central North Pacific) peaked in early morning and that their diet consisted mostly of open water, migrating micronekton that were advected and trapped over the seamount top during the night. That seamount, as well as other intermediate seamounts along the Hawaiian-Emperor ridge, harbored extremely high abundances of micronekton, some of them zooplanktivorous (Boehlert and Seki, 1984; Wilson and Boehlert, 1993, 2004). Additional evidence for locally intensified zooplanktivory over seamounts was reported by Haury *et al.* (1995), who found that the abundance of copepod carcasses was higher above several intermediate depth seamounts in the eastern North Pacific than in the surrounding waters.

In conclusion, bottom trapping appears to be an effective mechanism of trophic focusing through which zooplankters are advected, accumulated and efficiently consumed by seamount fishes and other predators. This mechanism could have been the key trophic pathway in the maintenance of the immense fish populations over intermediate seamounts (Tseitlin, 1985; Fock *et al.*, 2002a, b) prior to the collapse of the seamount fishery due to over fishing (see Part V). As the biomass of vertically migrating zooplankton rapidly declines with depths below several hundred meters, other mechanisms are apparently responsible for the maintenance of rich fishing resources of orange roughy (*Hoplostethus atlanticus*) and oreos (*Allocyttus niger*, *Pseudocyttus maculatus*) over deep (700–1500 m) seamounts (Clark, 2001; see Chapter 17).

Gap formation and zooplankton patchiness

The daily trapping of migrating zooplankton, their consequent consumption by resident predators, and their possible drift off the seamount during the day, result in a situation where only a few migrating zooplankters remain over the seamount by the late afternoon (Genin *et al.*, 1994). During the evening, when migrating zooplankters ascend to the upper water column around the seamount, a gap devoid of migrating zooplankton is created above the seamount (Box 5.1; Fig. 5.1a). Initially, the diameter of the gap matches that of the bathymetric contour of the seamount summit at the depth where these zooplankters reside during the day, for example at 200–300 m for *Euphausia pacifica* (Genin *et al.*, 1994; Haury *et al.*, 2000). Similar occurrences of nocturnal gaps of migrating zooplankton were reported from the Great Meteor and Josephine Seamounts in the NE Atlantic (Hesthagen, 1970; Nellen, 1973). As expected, no gaps are found for non-migrating zooplankters, as those species always remain high above intermediate and deep seamounts (Genin *et al.*, 1988; Fig. 5.1b). Some depletion of non-migrating zooplankton is to be expected above shallow and intermediate seamounts, especially during the day, if resident fish ascend to feed on zooplankton high above the bottom.

During the night, vertically sheared currents degrade the original gap into smaller, narrower gaps that gradually drift off the seamount into the surrounding ocean (Box 5.1). Depending on the strength of the currents and the seamount size, the gap can be found over the neighboring deep waters by dawn, well away from the seamount summit. Furthermore, when the migrating zooplankton around the gap descend at dawn, the gap 'follows' them and can be found in deeper waters. Gaps may thereby be maintained for some time, augmenting zooplankton patchiness around shallow and intermediate depth seamounts. Indeed, using extensive data collected by a Californian survey program (CalCOFI), Genin *et al.* (1988) found that the patchiness of migrating zooplankton (euphausiids and copepods), but not that of non- or weakly migrating animals (chaetognaths), was stronger near seamounts and banks than in the adjacent flat bottom region. The ecological effects of such augmented patchiness on seamount communities have not yet been explored.

Trophic subsidy: enhanced horizontal flux of planktonic prey

The summits and upper slopes of many seamounts are exposed to very strong currents. Several mechanisms amplify flow over seamounts, including the deflection of impinging

Box 5.1 Bottom trapping of vertically migrating zooplankton

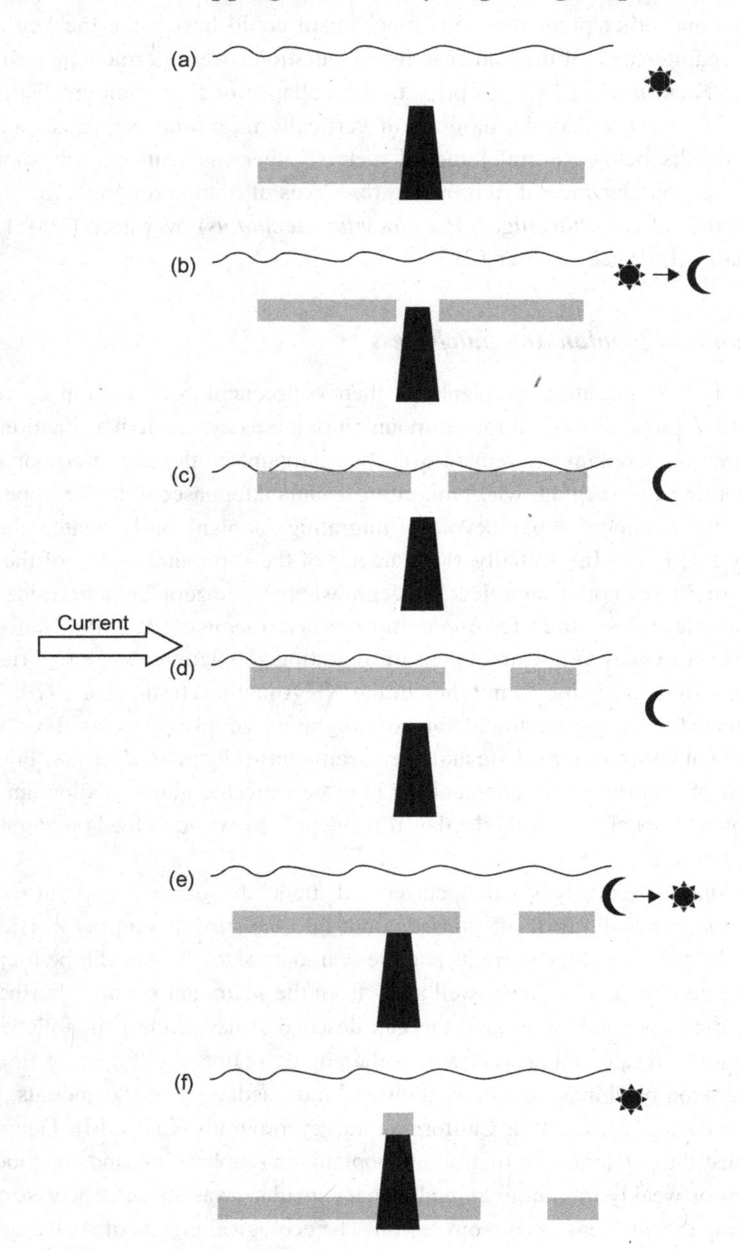

The bottom trapping mechanism over a seamount.

Numerous zooplankters spend the day in deep, aphotic waters, rise after sunset into the food-replete photic layer and return to deep waters early in the morning. The figure describes the bottom trapping mechanism over a seamount where the summit is deeper than the photic layer but shallower than the day-time depth of migrating zooplankters. Dotted areas – a layer of migrating zooplankters; full area – the seamount topography; full wavy line – sea surface; Sun and Moon symbols on the right indicate day and night. For simplicity, the diagram considers the current to flow from left to right throughout the water column. During the day (a) the migrating zooplankton is found in deep waters around the seamount's slopes. After sunset (b) the animals ascend to the photic layer, forming a gap the width of which equals to that of the seamount at the depth where the migrating zooplankters were found during the day. During the night (c, d) the zooplankters drift with the surface current, gradually displacing the gap away from the seamount and bringing 'new' zooplankters from open, upstream region to above the summit. By sunrise (e), when the animals start their descent, a layer of zooplankton is found above the entire summit. Later in the morning (f), the zooplankters found above the seamount become trapped above the shallow summit, unable to complete their descent to the deeper waters. The animals trapped in the illuminated waters above the summit are readily consumed by resident visual predators (e.g., rockfish), so that by late afternoon (a), very few (if any) zooplankters remain above the summit. Based on Fig. 2 in Genin *et al.* (1994) and Fig. 18 in Haury *et al.* (2000).

currents, rectification of internal tides and the formation of bores when internal waves encounter a slope of a certain degree (Eriksen, 1982, 1991; Kunze and Sanford, 1986; Mohn and Beckmann, 2002a, b; see Chapter 4).

Since fluxes of suspended food and planktonic larvae are a linear function of flow speed, the intensification of near-bottom currents over seamounts can substantially augment the growth and survival of benthic and near-bottom planktivores, as well as enhancing the rate of larval supply and recruitment (Genin *et al.*, 1986). The effect is expected to be most pronounced over intermediate and deep seamounts where food is thought to be the major factor limiting secondary productivity.

Benthic trapping of drifting food from the flowing water is a form of trophic subsidy (*sensu* Polis *et al.*, 1997), the import of allochtonous food to a local community. The subsidy de-couples the productivity of the subsidized community, in this case the seamount-resident species, from the regional level of primary production. The magnitude of the subsidy depends on the concentration of plankton and detrital particles, the flow speed and the ability of the benthic community to trap drifting food. We suggest that trophic subsidy is the mechanism responsible for the maintenance of immense biomass of orange roughy and oreos over very deep seamounts (see Chapter 17).

While the benefit of enhanced fluxes of drifting particles is obvious for sessile suspension feeders (Genin *et al.*, 1986; Sebens, 1997), the benefit for mobile animals is more difficult to explain. Instead of remaining stationary and waiting for currents to bring food, mobile animals can simply swim and actively control their search and encounter with prey. Is there a difference in the energy gain between a planktivorous fish, which holds place against the flow and one which actively swims at the same speed relative to water? Obviously, from a physical point of view there is no difference. However, when optimal

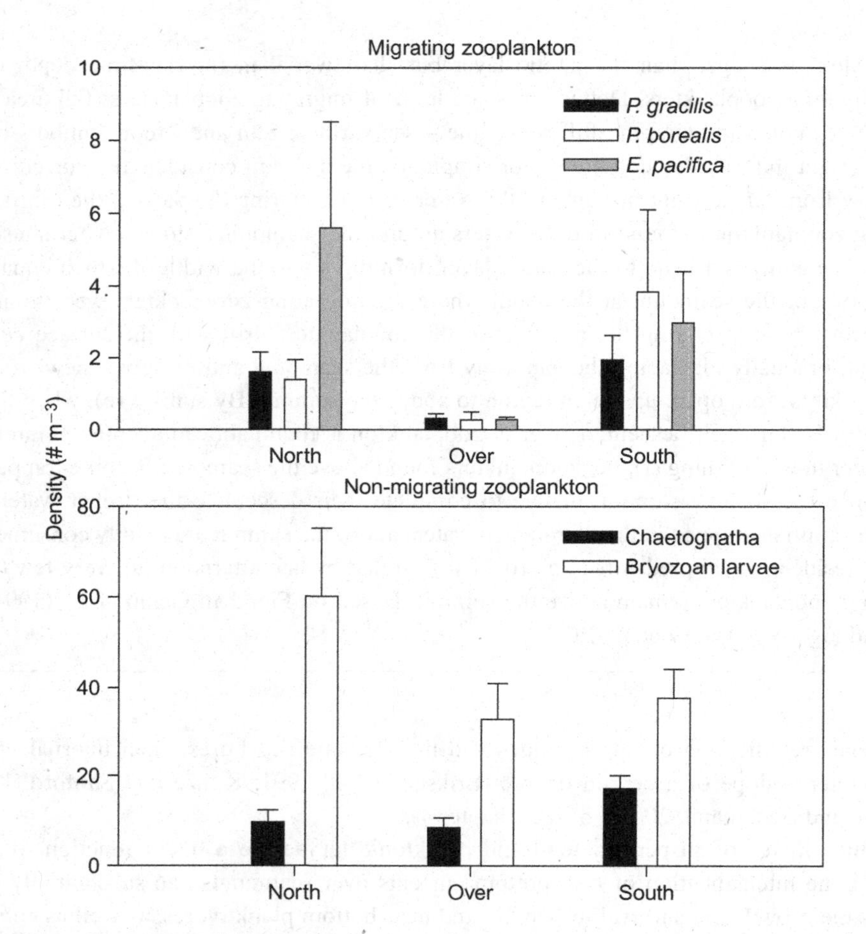

Fig. 5.1 The nocturnal zooplankton gap above an intermediate seamount. The plots present the average (±s.e.) density of migrating zooplankton (*Pleuromamma gracilis*, *P. borealis* and *Euphausia pacifica* – top panel) and non-migrating zooplankton (chaetognaths and bryozoan larvae – bottom panel) sampled during three nights over the shallow (97–150 m) summit of Sixtymile Bank and over the deep (>1000 m) waters 6–15 km north and south of it. The seamount is located in the Southern California Bight, 110 km southwest of San Diego (32°05′N, 118°15′W). Zooplankton was sampled by towing a plankton net (1 m^2 MOCNESS) in a 'yo-yo' pattern between 30 and 55 m below surface. N=9, 10 and 11 samples at the north, over and south sites, respectively. Note the remarkable (6–11 times) decline in the abundance of migrating zooplankters over the summit and the absence of such a difference for non-migrating taxa. *Source*: After Genin *et al.* (1994).

behavior is added, the benefit of being site attached and feeding from the moving water can be substantial.

The suggested mechanism, termed the 'feed-rest' hypothesis (Genin, 2004), is based on the assumption that the fish optimize their energetic gain by (i) limiting their feeding times to periods when prey flux is high and (ii) resting motionless at shelters when food is scarce. The mechanism should benefit zooplanktivorous fish, as the distribution of their prey is ubiquitously patchy (Haury *et al.*, 1978; Mackas *et al.*, 1985). While open water fish must swim to find rich patches of prey, site-attached fish can save energy by remaining stationary in the quiescent benthic boundary layer until rich patches of food drift in.

The rough bottom topography on many seamounts (Fig. 5.2; Genin *et al.*, 1986) provides ample quiescent shelters in close proximity to the strong flow. Such conditions allow fish that rest in the shelters to continuously sense the abundance of prey in the flowing water outside. Site-attached fish are well adapted to feed in the flow (McFarland and Levin, 2002), although their functional response can be complex (Kiflawi and Genin, 1997).

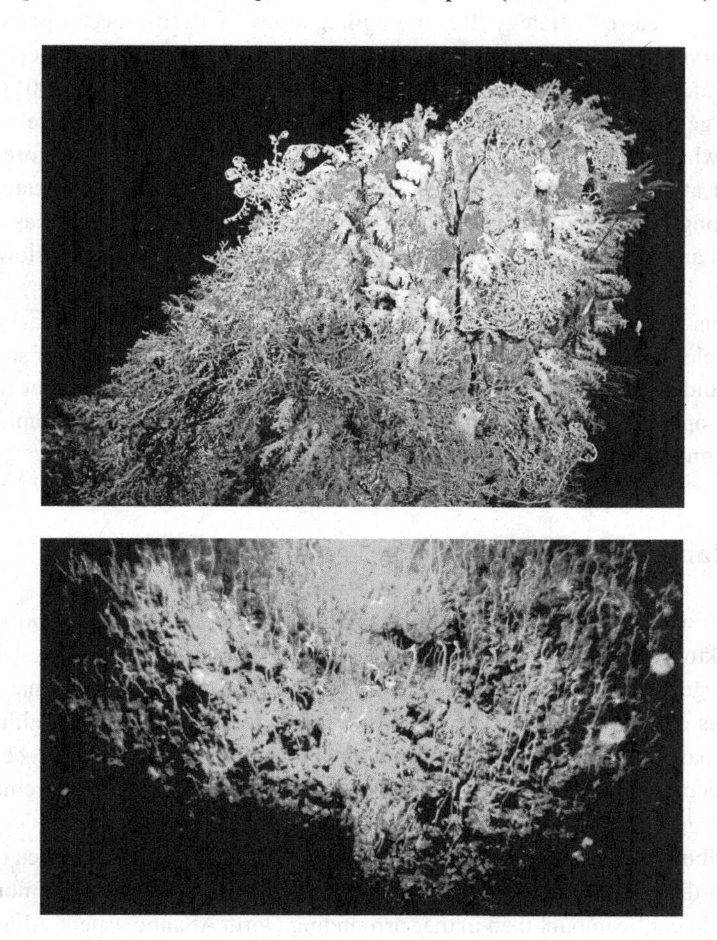

Fig. 5.2 The rich benthic community of suspension feeders on intermediate and deep seamounts. Top panel: basket stars and octocorals on a pinnacle (168 m depth) on East Diamante Seamount in the Mariana Islands. Schools of juvenile fish are visible in the background in the lower left (credit: NOAA Ocean Exploration Program). Lower panel: a 'forest' of black corals, *Stichopathes spiessi* at 450 m depth on the summit of Fieberling Guyot (32°25′N, 127°47′W) (credit: Deep-Tow Group, Marine Physical Lab, Scripps Institution of Oceanography). Note the rough bottom topography and the dense protruding sessile fauna, which together provide quiescence shelters in which planktivorous fish can rest while waiting for drifting prey to arrive with the strong flow outside.

A few behavioral studies of mobile planktivores inhabiting deep, topographically rough habitats support the feed-rest hypothesis. Visual observations made by Lorance *et al.* (2002) on the continental slope in the Bay of Biscay showed dense aggregations of orange roughy in quiescent locations within a canyon. Most of the fish were completely inactive; a few moved slowly, none were feeding. Vigorous currents were encountered around the canyon just outside of the fish shelter. Strong currents are a common characteristic of orange roughy habitats in the deep sea (Clark, 1995). This species, unlike any other deep sea fish, has strong muscle capable of sustained swimming (Koslow, 1997). We speculate that this unique body composition has evolved as an adaptation for effective prey capture under conditions of strong flow and high prey density, thereby maximizing the energetic gain from the feed-rest strategy. Resting aggregations of Pacific ocean perch (*Sebastes alutus*) observed by Brodeur (2001) in Pribilof Canyon, Bering Sea, also seem to use the feed-rest strategy. Using a remotely operated vehicle (ROV), Brodeur (2001) found that dense aggregations of motionless fish form during the night within a 'forest' of benthic sea whips, while during the day the fish were actively feeding above the forest. The fish were absent at sites without sea whips. High densities of sea whips indicate the occurrence of strong currents at that site. Dense forests of sea whips and other passive suspension feeders are found on the summits and slopes of many seamounts shallower than *ca.* 1000 m (Fig. 5.2; Genin *et al.*, 1986).

Seamounts provide optimal conditions for the feed-rest mechanism as the flow is strongly accelerated and quiescent shelters are readily available. The effect of such a trophic subsidy should be most pronounced at deep seamounts where food is a major limiting factor and the trophic contribution by other mechanisms (e.g., upwelling, topographic trapping) cannot be significant.

Ichthyoplankton

The question of whether distinct assemblages of larval fish, different from normal oceanic ichthyoplankton communities, occur at seamounts is of considerable interest in terms of both the biogeographic affinities of seamount fishes and the long-term maintenance of high biomass of seamount fish. The issue is complicated by the fact that, although many fish species have been reported to occur in high abundance over seamounts (see Chapters 6 and 9), the ecology of the early life history stages of most of those species remains largely unknown.

The distribution of seamount ichthyoplankton was first considered by Nellen (1974), who reported that the larvae of two 'topographically associated' fish species were more abundant over Great Meteor Seamount than in the surrounding North Atlantic waters. Although a subsequent study did not find any such pattern (Belyanina, 1984), results from a comprehensive survey in 1998 (Nellen and Ruseler, 2004) re-confirmed the original pattern. It appears that although the overall abundance of ichthyoplankton over Great Meteor is lower than in the surrounding waters, the abundance of neritic larvae is significantly higher. The presence of adults of some of those neritic species prompted the suggestion that their populations are self-sustaining due to the retention of their larvae around the seamount (Nellen, 1974; Nellen and Ruseler, 2004). This claim is supported by the modeling study of Beckmann and Mohn (2002), which indicates the likely existence of a persistent Taylor cone over Great Meteor Seamount. Larval retention in a Taylor cone was also proposed by Mullineaux and Mills (1997) to explain the higher abundance of invertebrate larvae near Feiberling Guyot than away from it. However, the larvae could be more abundant near the guyot also because the seamount was the closest possible source for such larvae in the surrounding deep ocean.

When compared to other seamounts, the results from Great Meteor seem to be the exception, rather than the rule. In a comprehensive review that considered virtually all of the available data, including a series of less widely accessible Russian technical reports, Boehlert and Mundy (1993) found little evidence to conclude that specialized ichthyoplankton assemblages commonly occur around seamounts. Despite this, however, some general trends did emerge. First, ichthyoplankton collections in the vicinity of seamounts are generally dominated by cosmopolitan oceanic taxa. Second, and as is often the case with other zooplankton, the abundance of ichthyoplankton directly over seamounts is usually lower than in the surrounding oceanic waters. Third, and although not a widespread phenomenon, high concentrations of larvae of topographically associated fishes occasionally occur over seamounts, but generally only near shallow seamounts that are reasonably close to continents or large islands (Boehlert and Mundy, 1993). An example of the latter was reported by Dower and Perry (2001) who observed very high abundances of rockfish (*Sebastes* spp.) larvae within a 30 km radius of Cobb Seamount. The fact that the Cobb ichthyofauna is known to be dominated by various rockfish species (Pearson *et al.*, 1993), coupled with the observation that the youngest (i.e., smallest) larvae were collected immediately over the seamount summit, suggest a local origin for the larvae. However, two previous cruises at the same time of year found no such enhancement (Dower, 1994). Similarly variable patterns have been observed at Southeast Hancock Seamount, where the abundance of larvae of two of the dominant seamount fish, *Maurolicus muelleri* and *Pseudopentaceros wheeleri* varied seasonally, being more abundant over the seamount than the surrounding waters in winter, but not in summer (Boehlert, 1988).

References

Angel, M.V. (1985) Vertical migrations in the oceanic realm: possible causes and probable effects. *Contributions in Marine Science*, 27, 45–70.

Beckmann, A. and Mohn, C. (2002) The upper ocean circulation at Great Meteor Seamount. Part II: Retention potential of the seamount induced circulation. *Ocean Dynamics*, 52, 194–204.

Belyanina, T.N. (1984) Observations on the ichthyoplankton in open waters of the Atlantic near the Great-Meteor Seamount. *Journal of Ichthyology*, 24, 127–9.

Boehlert, G.W. (1988) Current–topography interactions at mid-ocean seamounts and the impact on pelagic ecosystems. *GeoJournal*, 16, 45–62.

Boehlert, G.W. and Seki, M.P. (1984) Enhanced micronekton abundance over mid-Pacific seamounts. *EOS, Transactions of the American Geophysical Union*, 65, 298.

Boehlert, G.W. and Mundy, B.C. (1993) Ichthyoplankton assemblages at seamounts and oceanic islands. *Bulletin of Marine Science*, 53, 336–61.

Brinton, E. (1967) Vertical migration and avoidance capability of euphausiids in the California current. *Limnology and Oceanography*, 12, 451–83.

Brodeur, R.D. (2001) Habitat-specific distribution of Pacific ocean perch (*Sebastes alutus*) in Pribilof Canyon, Bering Sea. *Continental Shelf Research*, 21, 207–24.

Brooks, E.R. and Mullin, M.M. (1983) Diel changes in the vertical distribution of biomass and species in the Southern California Bight. *California Cooperative Oceanic Fisheries Investigations Reports*, 24, 210–15.

Clark, M.R. (1995) Experience with management of orange roughy (*Hoplostethus atlanticus*) in New Zealand waters, and the effects of commercial fishing on stocks over the period 1980–1993.

In: *Deepwater Fisheries of the North Atlantic Oceanic Slope* (ed. Hopper, A.G.), pp. 251–66. Kluwer Academic Publishers, Dordrecht, The Netherlands.

Clark, M.R. (2001) Are deepwater fisheries sustainable? – the example of orange roughy (*Hoplostethus atlanticus*) in New Zealand. *Fisheries Research*, 51, 123–35.

Comeau, L.A., Vezina, A.F., Bourgeois, M. and Juniper, S.K. (1995) Relationships between phytoplankton production and the physical structure of the water column near Cobb Seamount, northeast Pacific. *Deep Sea Research*, 42, 993–1005.

Dower, J.F. (1994) Biological consequences of current–topography interactions at Cobb Seamount. PhD Dissertation, University of Victoria, Victoria, Canada, 220 pp.

Dower, J.F. and Mackas, D.L. (1996) "Seamount effects" in the zooplankton community near Cobb Seamount. *Deep-Sea Research*, 43, 837–58.

Dower, J.F. and Perry, R.I. (2001). High abundance of larval rockfish over Cobb Seamount, an isolated seamount in the Northeast Pacific. *Fisheries Oceanography*, 10, 268–74.

Dower, J.F., Freeland, H. and Juniper, K. (1992) A strong biological response to oceanic flow past Cobb Seamount. *Deep-Sea Research*, 39, 1139–45.

Eriksen, C.C. (1982) Observations of internal wave reflection off sloping bottoms. *Journal of Geophysical Research*, 87, 525–38.

Eriksen, C.C. (1991) Observations of amplified flows atop a large seamount. *Journal of Geophysical Research*, 96, 15227–37.

Fock, H.O., Matthiessen, B., Zodowitz, H. and von Westernhagen, H. (2002a) Diel and habitat-dependent resource utilization by deep-sea fishes at the Great Meteor Seamount: niche overlap and support for the sound scattering layer interception hypothesis. *Marine Ecology Progress Series*, 244, 219–33.

Fock, H.O., Uiblein, F., Koster, F. and von Westernhagen, H. (2002b) Biodiversity and species-environment relationships of the demersal fish assemblage at the Great Meteor Seamount (subtropical NE Atlantic), sampled by different trawls. *Marine Biology*, 141, 185–99.

Freeland, H. (1994) Ocean circulation at and near Cobb Seamount. *Deep-Sea Research*, 41, 1715–32.

Genin, A. (2004) Bio-physical coupling in the formation of zooplankton and fish aggregations over abrupt topographies. *Journal of Marine Systems*, 50, 3–20.

Genin, A. and Boehlert, G.W. (1985) Dynamics of temperature and chlorophyll structures above a seamount: an oceanic experiment. *Journal of Marine Research*, 43, 907–24.

Genin, A., Dayton, P.K., Lonsdale, P.F. and Spiess, F.N. (1986) Corals on seamount peaks provide evidence of current acceleration over deep-sea topography. *Nature*, 322, 59–61.

Genin, A., Haury, L.R. and Greenblatt, P. (1988) Interactions of migrating zooplankton with shallow topography: predation by rockfishes and intensification of patchiness. *Deep-Sea Research*, 35, 151–75.

Genin, A., Greene, C., Haury, L.R., Wiebe, P., Gal, G., Kaartvedt, S., Meir, E., Feys, C. and Dawson, J. (1994) Zooplankton patch dynamics: daily gap formation over abrupt topography. *Deep-Sea Research*, 41, 941–51.

Haury, L.R., McGowan, J.A. and Wiebe, P.H. (1978) Patterns and processes in the time–space scales of plankton distributions. In: *Spatial Pattern in Plankton Communities* (ed. Steele, J.H.), pp. 277–327. Plenum Press, New York.

Haury, L.R., Fey, C., Gal, G. and Genin, A. (1995) Copepod carcasses in the ocean. I. Over seamounts. *Marine Ecology Progress Series*, 123, 57–63.

Haury, L.R., Fey, C., Newland, C. and Genin, A. (2000) Zooplankton distribution around four eastern North Pacific seamounts. *Progress in Oceanography*, 45, 69–105.

Hesthagen, I.H. (1970) On the near-bottom plankton and benthic invertebrate fauna of the Josephine Seamount and the Great Meteor Seamount. *Meteo: Forsch Ergeb*, 8, 61–70.

Hubbs, C.L. (1959) Initial discoveries of fish fauna on seamounts and offshore banks in the eastern Pacific. *Pacific Science*, 13, 311–16.

Isaacs, J.D. and Schwartzlose, R.A. (1965) Migrant sound scatterers: interaction with the seafloor. *Science*, 150, 1810–13.

Kiflawi, M. and Genin, A. (1997) Prey flux manipulation and the feeding rates of reef dwelling planktivorous fish. *Ecology*, 78, 1062–77.

Koslow, J.A. (1997) Seamounts and the ecology of deep-sea fisheries. *American Scientist*, 85, 168–76.

Kunze, E. and Sanford, T.B. (1986) Near-inertial wave interactions with mean flow and bottom topography near Caryn Seamount. *Journal of Physical Oceanography*, 16, 109–20.

Kunze, E. and Toole, J.M. (1997) Tidally driven vorticity, diurnal shear, and turbulence atop Fieberling Seamount. *Journal of Physical Oceanography*, 27, 2663–93.

Lorance, P., Uiblein, F. and Latrouite, D. (2002) Habitat, behaviour and colour patterns of orange roughy *Hoplostethus atlanticus* (Pisces: Trachichthyidae) in the Bay of Biscay. *Journal of the Marine Biological Association of UK*, 892, 321–31.

Mackas, D.L., Denman, K.L. and Abott, M.K. (1985) Plankton patchiness: biology in the physical vernacular. *Bulletin of Marine Science*, 37, 652–74.

Mann, K.H. and Lazier, J.R. (1996) *Dynamics of Marine Ecosystems*. Blackwell Science, Oxford.

McFarland, W. and Levin, S.A. (2002) Modelling the effects of current on prey acquisition in planktivorous fishes. *Marine and Freshwater Behaviour and Physiology*, 35, 69–85.

Mohn, C. and Beckmann, A. (2002a) The upper ocean circulation at Great Meteor Seamount. Part I: Structure of density and flow fields. *Ocean Dynamics*, 52, 179–93.

Mohn, C. and Beckmann, A. (2002b) Numerical studies on flow amplification at an isolated shelf-break bank, with application to Porcupine Bank. *Continental Shelf Research*, 22, 1325–38.

Mourino, B., Fernandez, E., Serret, P., Harbour, D., Sinha, B. and Pingree, R. (2001) Variability and seasonality of physical and biological fields at the Great Meteor Tablemount (subtropical NE Atlantic). *Oceanologica Acta*, 24, 1–20.

Mullin, M.M. (1993) *Webs and Scales: Physical and Ecological Processes in Marine Fish Recruitment*. University of Washington Press, Seattle, WA, USA.

Mullineaux, L.S. and Mills, S.W. (1997) A test of the larval retention hypothesis in seamount-generated flows. *Deep Sea Research*, 44, 745–70.

Nellen, W. (1973) Investigations on the distribution of fish larvae and plankton near and above the Great Meteor Seamount. *Meteo: Forsch Ergeb*, 13, 47–69.

Nellen, W. (1974) Investigations on the distribution of fish larvae and plankton near and above the Great Meteor Seamount. In: *The Early Life History of Fish* (ed. Blaxter, J.H.), pp. 213–14. Springer-Verlag, New York.

Nellen, W. and Ruseler, S. (2004) Composition, horizontal and vertical distribution of ichthyoplankton in the Great Meteor Seamount area in September 1998. *Archive of Fishery and Marine Research*, 51, 132–64.

Noble, M. and Mullineaux, L.S. (1989) Internal tidal currents over the summit of Cross Seamount. *Deep Sea Research*, 36, 1791–802.

Nonacs, P., Smith, P.E., Bouskila, A. and Luttbeg, B. (1994) Modeling the behavior of the northern anchovy, *Engraulis mordax*, as a schooling predator exploiting patchy prey. *Deep Sea Research*, 41, 147–69.

Odate, T. and Furuya, K. (1998) Well-developed subsurface chlorophyll maximum near Komahashi No. 2 Seamount in the summer of 1991. *Deep Sea Research*, 45, 1595–607.

Pearcy, W.G., Krygier, E.E., Mesecar, R. and Ramsey, F. (1977) Vertical distribution and migration of oceanic micronekton off Oregon. *Deep Sea Research*, 24, 223–45.

Pearson, D.E., Douglas, D.A. and Barss, B. (1993) Biological observations from the Cobb Seamount rockfish fishery. *Fishery Bulletin*, 91, 573–6.

Polis, G.A., Anderson, W.B. and Holt, R.D. (1997) Toward an integration of landscape and food web ecology: the dynamics of spatially subsidized food webs. *Annual Review of Ecology and Systematics*, 28, 289–316.

Roden, G.I. (1994) Effects of the Fieberling seamount group upon flow and thermohaline structure in the spring of 1991. *Journal of Geophysical Research*, 99(C5), 9941–61.

Sebens, K.P. (1997) Adaptive responses to water flow: morphology, energetics and distribution of reef corals. *Proceedings of the 8th International Coral Reef Symposium*, 2, 1053–8.

Seki, M.P. and Somerton, D.A. (1994) Feeding ecology and daily ration of the pelagic armorhead, *Pseudopentaceros wheeleri*, at Southeast Hancock Seamount. *Environmental Biology of Fishes*, 39, 73–84.

Tseitlin, V.B. (1985) Energetics of fish populations inhabiting underwater rises. *Oceanology*, 2, 308–311.

Uchida, R.N. and Tagami, D.T. (1984) Groundfish fisheries and research in the vicinity of seamounts in the North Pacific Ocean. *Marine Fisheries Review*, 2, 1–17.

Uda, M. and Ishino, M. (1958) Enrichment pattern resulting from eddy systems in relation to fishing grounds. *Journal of the Tokyo University of Fisheries*, 1–2, 105–19.

Wiebe, P.H., Madin, L.P., Haury, L.R., Harbison, G.R. and Philbin, L.M. (1979) Diel vertical migration by *Salpa aspera* and its potential for large-scale particulate organic matter transport to the deep sea. *Marine Biology*, 53, 249–55.

Wilson, C.D. and Boehlert, G.W. (1993) Population biology of *Gnathophausia longispina* (Mysidacea: Lophogastrida) from a central North Pacific seamount. *Marine Biology*, 115, 537–43.

Wilson, C.D. and Boehlert, G.W. (2004) Interaction of ocean currents and resident micronekton at a seamount in the central North Pacific. *Journal of Marine Systems*, 50, 39–60.

Chapter 6
Midwater fish assemblages and seamounts

Filipe M. Porteiro and Tracey Sutton

Abstract

Meso- and bathypelagic fishes are conspicuous components of the 'deep scattering layers' (DSL) of the world oceans. These ichthyofauna interact with resident demersal fauna at seamounts in several ways: (1) horizontal impingement by non- or weak vertical migrators (mainly deep meso- and bathypelagic forms); (2) impingement of mesopelagic vertical migrators during migration, including topographic trapping; (3) adoption of a benthopelagic lifestyle over seamounts by large, adult pelagic fishes; and (4) adoption of a pseudoceanic lifestyle by species from primarily pelagic families, including endemism and active aggregation, the latter being corroborated by acoustic evidence. Bio-physical coupling mechanisms are highly variable and site dependent, but some physical oceanographic features such as Taylor caps may serve to retain pelagic populations over seamounts; the most important physical feature is water depth over the seamount summit. There is evidence that pelagic fauna provide a crucial trophic link in seamount ecosystems and may be responsible for the large biomass of demersal fishes found at these locations. There is little evidence for direct bottom-up enhancement, whereas two behaviour-based explanations are supported by field studies: (1) the 'food-rest' hypothesis that predatory fishes use the seamount to hold station (rest) while taking advantage of the horizontal advection of prey (food); and (2) the 'topographic trapping/interception' hypothesis that predatory fishes rely on seamounts to concentrate the density of vertically migrating pelagic prey. The exact mechanism may vary, but the overall effect is to convert mid-trophic level biomass (pelagic prey) to higher trophic level biomass (seamount-associated fishes) with increased efficiency.

Introduction

Deep-sea midwater fishes are considered ancient 'unevolved' bony fishes (Andriashev, 1953) that belong to ancestral fish orders (Miya *et al.*, 2001; Nelson, 2006). Parin (1984) suggested that most deep pelagics evolved from demersal ancestors at similar depths on the outer continental shelves and slopes. These highly specialized fishes represent about 7.5% of all marine fish species (*ca.* 1200 from 16000; Froese and Pauly, 2006), with the greatest diversity in the mesopelagic layers. Two main groups of deep-sea midwater fishes can be recognized: meso- and bathypelagic fishes. The morphological, physiological and behavioural adaptations (Marshall, 1979; Childress, 1995; Salvanes and Kristoffersen, 2001) developed by each group reflect the environmental conditions and biotic interactions in the disphotic twilight zone (mesopelagial: 200 to 1000 m) and in the aphotic abysses (bathypelagial: 1000 to >3000 m).

In general, mesopelagic fishes are small sized up to 30 cm long, but normally about 10 cm, although some species can reach 1 m. Most migrate vertically into the epipelagic layer at night and have a species-specific ventral array of bioluminescent photophores that are used for camouflage against predators, prey attraction and illumination and intraspecific communication (Herring and Morin, 1978; Young, 1983). They are relatively short-lived, rapid-turnover species with life spans of 1–5 years; cold- and deep-water forms tend to grow larger and live longer than tropical, shallower species (e.g., Childress *et al.*, 1980; Linkowski, 1985; Gartner, 1991). Bathypelagic fishes, on the other hand, are morphologically, physiologically and behaviourally adapted to conserve energy in a food-limited environment. They are larger than mesopelagic fishes and have higher growth rates achieved by higher relative growth efficiencies (between 25% and 50%) (Childress *et al.*, 1980; Childress, 1995). Vertical migration and bioluminescence are uncommon (Young, 1983).

Midwater fishes represent an important link between zooplankton and higher trophic level predators such as seabirds, squids, piscivorous fishes and marine mammals. The primary prey of open-ocean mesopelagic zooplanktivorous fishes are calanoid copepods (Sutton *et al.*, 1998), but gelatinous prey are also important for deep meso- and bathypelagic fishes around mid-ocean ridges and seamount systems (Sutton *et al.*, in press). Micronektivores are also an important guild of midwater fishes, feeding primarily on migrating fishes but also on shrimps and/or cephalopods (Sutton and Hopkins, 1996; Sutton *et al.*, 1998).

Oceanic DSL

The backscattered acoustic signals or the DSL are the most prominent features in the acoustic profiles of ocean water columns. During the day, the DSL is stable and compact, 50–200 m thick, in mid-depths from 400 to 600 m. At dusk, most of the targets ascend into the epipelagic zone, where they split into several vertical sub-layers, while some stay at the daytime depth or displace vertically within the mesopelagial. Before dawn, the main DSL descends again to daytime depths. Apart from the main DSL, other less strong acoustic layers are associated with particular groups of zooplankton or micronekton. Most of the deeper DSL are produced by non-migrant pelagic organisms.

Micronektonic fishes with gas-filled swimbladders, such as vertically migrating myctophids, are the main component of the DSL, while macroplankton with gas chambers, such as some gelatinous siphonophores, also produce strong echoes. Pelagic invertebrates with hard chitinous exoskeletons like euphausiids, decapods, copepods and shelled pteropods are important components. The relative vertical distributions of DSL animals are complex and highly variable in time and space.

Vertically migrating mesopelagic organisms transfer particulate and dissolved organic matter from autotrophic upper waters to the heterotrophic deeper ocean (Angel, 1985; Angel and Pugh, 2000). At depth, vertical migrants produce faecal pellets and excrete metabolic products, such as carbon dioxide and dissolved organic carbon and nitrogen, generating the 'mesopelagic maximum' of organic matter often detected around 500 m, the average daytime depth of most DSL (Walsh *et al.*, 1988). Our understanding of the dynamics and the magnitude of the fluxes in these processes is poor (Angel and Pugh, 2000; Ducklow *et al.*, 2001).

Midwater micronekton–benthopelagic interactions

Typical meso- and bathypelagic organisms enter and accumulate in the benthopelagic layer ('bottom mixed' or 'benthic boundary' layer) where they interact with continental and oceanic island slopes and the summits and slopes of seamounts and mid-oceanic ridges (Parin and Golovan, 1976; Golovan, 1978; Merrett, 1986; Hulley and Lutjeharms, 1989; Mauchline and Gordon, 1991; Reid *et al.*, 1991; Koslow, 1996; Sutton *et al.*, in press).

'Mesopelagic boundary communities' are discrete assemblages of primarily open-ocean, deep-water organisms. They include fishes, decapod crustaceans and cephalopods that otherwise live in the mesopelagic environment and are associated with insular slopes and seamounts between 200 and 800 m in depth (e.g., Hawaiian Islands; Reid *et al.*, 1991; Benoit-Bird *et al.*, 2001; Lammers *et al.*, 2004). They link the oceanic and the neritic realms of oceanic islands and seamounts, providing a predictable food source for epipelagic and demersal micronektivorous organisms. The degree of pelagic–topographic association appears to be site specific; in the Canary Islands, Meteor and Atlantis seamounts, e.g., the midwater assemblages on the slopes interact with the neritic and the benthopelagic domains, but specialized boundary communities were not detected (Bordes *et al.*, 1999; Pusch *et al.*, 2004). A distinction can be made between seamount-associated and continental/insular slope-associated micronekton, as the latter environment is subject to terrestrial influence. However, there are many species and/or genera in common (see Supplementary material).

Deep-pelagic fish–seamount interactions

Most studies of seamount biota have been of plankton (see Genin, 2004) or larger, commercially exploited nekton (see Chapter 9) with relatively few studies of deep-pelagic micronekton. Nevertheless, pelagic micronekton seem to aggregate over seamounts and may play an important role as food (Isaacs and Schwartzlose, 1965; Kashkin, 1977; Koslow, 1997; McClatchie and Dunford, 2003). Knowledge of the distribution patterns of midwater fishes around seamounts is a prerequisite for understanding the trophic interactions of these communities. Froese and Sampang (2004) report about 107 micronekton species that associate with seamounts. Most were sampled at the vicinity of the Bear Seamount (Moore *et al.*, 2003) or at the Nazca and Sala y Gómez submarine ridges (Parin *et al.*, 1997). Midwater fish–seamount interactions can be categorized into four groups:

1. *Non-migrant or weakly migrant midwater fauna that enter the benthopelagic zone around seamounts.* This group includes nearly all bathypelagic fishes as well as several groups of mesopelagic fishes. They may not be able to counter strong currents and are laterally advected towards steep topographies (see Chapter 4). Weakly or non-migrant mesopelagic fishes impinge upon the slopes of shallow, and slopes and summits of mid-depth seamounts. For example, at night the weakly migrant fish *Argyropelecus*

aculeatus is equally abundant over the slope of the Great Meteor and Atlantis Seamounts and at a non-seamount oceanic reference station (Pusch *et al.*, 2004). Higher densities of non-migrant micronekton are found at 300–1000 m depth, and so they are absent from shallow seamount summits. Below 1000 m, fish density decreases with depth, though there is evidence of increased bathypelagic fish biomass associated with mid-ocean ridges (Sutton *et al.*, in press). Bathypelagic fishes are found on the deep slopes of all seamounts and the summits of deeper seamounts, but their overall role in seamount ecosystems is poorly known.

2. *Mesopelagic fauna that migrate to the epipelagic layers at night and interact with seamounts during the migration process.* The interaction and the accumulation of the main migrant DSL organisms with benthopelagic and benthic biotopes of seamounts were first reported by Isaacs and Schwartzlose (1965). Migrant mesopelagic fishes can be laterally advected over shallow topography by surface currents during the night, then can be trapped over the summit of shallow seamounts on their way down to daytime depths (see Box 5.1 and Fig. 4.7). They can also impinge the flanks of shallow seamounts during any phase of the diel cycle and the summits of intermediate seamounts while descending to daytime depth. Seamount interactions therefore depend on local hydrography and the amplitude and intensity of the migration. They may also interact with bottom fauna during active vertical migration or passive horizontal advection (Kashkin, 1984). Maximum interactions range from epipelagic waters at night to about 400–600 m, but vertical migration varies geographically, seasonally and across many space and time scales.

   All vertical migrants in the vicinity of a seamount are subject to interaction with its slopes. For example, a rich diversity of meso- and bathypelagic fishes caught near the flat-topped Bear Seamount (northwest Atlantic; Moore *et al.*, 2003) reflects that of the surrounding environment. The impingement of these organisms with seamounts is accidental with no apparent advantage for the interzonal pelagic migrants that graze in epipelagic layers. Nevertheless, the phenomenon is trophically very important for seamount ecology. Feeding can be so intense that pelagic organisms can be depleted by resident benthopelagic predators, resulting eventually in a 'daily gap formation' over seamounts (see Chapter 5; Genin *et al.*, 1988, 1994; Rogers, 1994; Pusch *et al.*, 2004). Because most of the migrants are confined to the mesopelagic layers, this interaction is expected to decrease over seamounts deeper than 1000 m.

3. *Adults of meso- and bathypelagic micronekton species that dwell in the benthopelagic zones.* A considerable proportion of dragonfish (family Stomiidae) in ichthyological collections worldwide were trawled in seamount and other slope resource surveys (Porteiro, 2005); bottom-trawled specimens were larger than those caught by pelagic gear (160 mm mean standard length vs 79 mm standard length). The stomiids, as well as larger individuals of Gonostomatidae, Myctophidae, Paralepididae and Melamphaidae, dwell over seamounts and other slopes below 700 m, adopting a benthopelagic life strategy (Gushchin and Kukuev, 1981; Kukuev, 1982, 2004; Vinnichenko, 1997a; Melo and Menezes, 2002). These benthopelagic adults of pelagic taxa are not a large part of the seamount community, but they may be important as a concentration of the largest, 'fittest' individuals for reproduction, seeding the surrounding waters with the best of the gene pool. Then carnivores could sit and wait for food to come to them *via* the

mechanism described above, while planktivores benefit from the high concentration of plankton in the near-bottom layers (Lorz *et al.*, 1983), and not have to spend as much energy searching for food.

4. *Members of primarily pelagic families that associate with the benthopelagic layers of seamounts and other abrupt topography.* Some 'pseudoceanic' or 'neritopelagic' species *rsensu* (Hulley and Lutjeharms, 1989; Parin *et al.*, 1997) from primarily pelagic families occur near abrupt topography, but are absent or less abundant in oceanic waters, suggesting a resident boundary community. These species, showing strong ecological interactions with the near-bottom layers and resisting advection off the seamount, fall into two classes: migrators and non-migrators.

Non-migrators include some Sternoptychidae, Phosichthyidae, Platytroctidae and Alepocephalidae. The elongate sternoptychids (subfamily Maurolicinae) *Argyripnus atlanticus*, *A. electronus* and *A. iridescens* live in association with the Great Meteor and other seamounts south of the Azores (Badcock, 1984a; Kukuev, 2004), with the seamounts of the Sala y Gómez Ridge (west of Easter Island; Parin, 1992; Parin *et al.*, 1997) and with the Norfolk Ridge (northwest of New Zealand; Richer de Forges, 2001), respectively. *A. electronus* is endemic to the Sala y Gómez Ridge, while *A. brocki*, a member of the mesopelagic boundary community (Reid *et al.*, 1991), is endemic to the Hawaiian seamount region (Harold and Lancaster, 2003). The highly speciose sternoptychid genus *Polyipnus*, reported as pseudoceanic or benthopelagic, has many representatives that associate with seamounts (Badcock, 1984a; Borets, 1986; Harold, 1994; Parin *et al.*, 1997; Richer de Forges, 2001; Moore *et al.*, 2003). The phosichthyids *Polymetme corythaeola*, *P. thaeocoryla*, *P. andriashevi* and *Yarrella blackfordi* also belong to this group and are known to live preferentially in the continental, insular and seamount slopes (Badcock, 1984b; Shcherbachev *et al.*, 1985; Parin and Borodulina, 1990; Parin *et al.*, 1997). *P. andriashevi* is thought to be endemic to the Sala y Gómez Ridge (Parin *et al.*, 1997). Several bathypelagic platytroctid species are reported to associate with seamounts and other abrupt underwater structures. *Holtbyrnia anomala*, *H. macrops*, *Normichthys operosus*, *Sagamichthys schnakenbecki*, *Maulisia mauli* and *M. microlepis* have been reported in association with many seamounts in the North Atlantic (Kukuev, 1982, 2004). *S. abei* has been found along the Sala y Gómez Ridge (Parin *et al.*, 1997) and over the Northwestern and Hawaiian seamounts (Borets, 1986). Some micronektonic alepocephalids that are considered to be pelagic, among a family of mainly benthopelagic nekton, have also been caught along seamount slopes (e.g., *Bajacalifornia megalops*, *Xenodermichthys copei* and *Photostylus pycnopterus*).

The vertically migrant, seamount-associated mesopelagic fishes include species thought to be endemic to certain seamount chains. These include *Maurolicus rudjakovi*, *Diaphus confusus*, *D. parini* and *Idiolychnus urolampus* from the Nazca and Sala y Gómez Ridges (east central Pacific; Parin *et al.*, 1997) and *D. basileusi* from the Equator Seamount (Indian Ocean; Bekker and Prut'ko, 1984), which migrate vertically off the bottom at night and descend to near-bottom layers during the day. Other species seem to concentrate preferentially on seamounts but occur also elsewhere in the pelagic waters. Parin and Prut'ko (1985) and Parin (1986) reported a myctophid species, *D. suborbitalis*, living in strong association with Equator Seamount: its abundance was much higher over the seamount than off-mount, even 3–5 km away, although it occurs also in other regions. Sassa *et al.* (2002) found similarly greatly enhanced mesopelagic fish numbers and biomass over the Emperor Seamount chain (Hawaii) relative to surrounding waters. This was largely due to dense assemblages of the seamount-endemic sternoptychid *M. imperiatus*. In RSA and Discovery Seamounts in the South Atlantic, Kalinowski and Linkowski (1983) and Linkowski (1983) found aggregations of the congeneric *M. muelleri*.

Boehlert and Seki (1984) reported hydroacoustical 'clouds' of micronekton over Southeast Hancock Seamount (Hawaii); trawling revealed numbers of the sternoptychid *M. muelleri* (later known as *M. imperiatus*; Parin and Kobyliansky, 1996), the lophogastrid mysidacean *Gnathophausia longispina* and the cephalopod *Iridoteuthis iris*. These species remained over the seamount flanks during day and then accumulated over the top to within 100 m of the surface at night to feed (Boehlert, 1988; Wilson, 1992; Boehlert *et al.*, 1994; Wilson and Boehlert, 2004). They were largely absent in waters away from the seamount, suggesting a permanent seamount boundary community. However, the swarms associated with small seamounts may not be self-sustaining and may recruit juveniles from surrounding larger seamounts. Wilson and Boehlert (2004) showed that behaviour was likely responsible for swarms, as the fish resist advection off the seamount by active swimming.

The life cycles of these species appear to be tightly coupled with the hydrography of the seamount to minimize the risk of being lost both during larval and adult phases. They probably benefit either from increased food availability or by a wider range of habitat diversity created by the topography and the hydrography (Wilson and Boehlert, 2004). This pelagic guild is considered the most specialized in terms of seamount interactions, though there are only few localized examples of micronekton that actively aggregate and migrate vertically over seamounts.

Bio-physical coupling mechanisms affecting the interactions of midwater pelagic micronekton with seamounts

Seamounts disrupt seawater flow and affect the hydrography and current patterns relative to adjacent open-ocean regions. Oceanic internal waves and tidal waves can be reflected and amplified, turbulent vertical mixing can be enhanced, localized jets can be produced, and eddy structures, known as Taylor columns, may form around seamounts and generate semi-closed circulations (see Chapter 4).

The physical mechanisms that affect the interaction of different groups of micronekton with seamounts are poorly understood. What is known about bio-physical coupling near abrupt topographies was reviewed by Genin (2004). The meso- and bathypelagic fauna can be horizontally advected by currents and impinge the seamount biotopes as they pass through the benthopelagic layers associated with the seamount (see Chapter 5). These animals probably cannot advect into water bodies over seamounts where a permanent Taylor cap is present (e.g., Great Meteor Seamount; Beckmann and Mohn, 2002; Kaufmann *et al.*, 2002). A substantial degree of isolation of the water above the Great Meteor Seamount is thought to prevent advection of oceanic micronekton organisms towards the plateau of that feature (Diekmann, 2004; Pusch *et al.*, 2004). On the other hand, organisms may become physically aggregated in near-bottom layers where accelerated current (ebb and flow) past seamounts generate a Taylor column and other quasi-permanent eddy structures downstream.

Lateral advection of diel migrant organisms in the epipelagic layer at night over the seamounts summits and slopes is the basic mechanism required for the topographic trapping of micronekton when descending to daytime depth (Isaacs and Schwartzlose, 1965; Dower and Perry, 2001). The scope of this phenomenon is related to the extent of vertical migration, depth of seamounts and intensity of horizontal currents (Chapter 5).

The seamount hydrographic environment significantly influences the behaviour, distribution, recruitment and linkage of seamount-associated micronekton (groups 3 and 4) with the seamount (Wilson and Firing, 1992; Wilson and Boehlert, 2004). To maintain the association, the micronekton must actively keep position by swimming against the prevailing current stream. Wilson and Boehlert (2004) reported that most of the SE Hancock Seamount population of *Maurolicus* stay associated with the seamount even if they face high current intensities. However, the authors observed a horizontal displacement of the *Maurolicus* 'cloud' to the downstream side of the summit by the end of night 'pelagic' phase. These authors suggested that retention of the seamount-associated micronekton organisms by the weakly defined Taylor columns was unlikely. Topographically steered, daytime swimming upflank towards the summit is one of the several mechanisms proposed to explain the active maintenance of those populations over seamounts (Boehlert, 1988; Benoit-Bird *et al*., 2001).

High variability in the topography, physiography (area, depth of the summit), degree of isolation, background regional oceanographic regime, location and seasonality of seamounts renders similarly diverse the ecological processes that determine the structure of local communities (reviewed by Boehlert and Genin, 1987; Rogers, 1994). The water depth of the summit appears to be particularly important, as it determines the composition of the micronekton assemblages that interact with the seamounts. Seamount-associated fauna eating mesopelagic food likely concentrate over the summits and flanks of intermediate seamounts and the flanks of shallow ones. Figure 6.1 summarizes how the different midwater fish assemblages interact with seamounts.

Trophic interactions with seamount predator fishes

Seamounts have higher productivity compared with the surrounding open ocean (see Chapters 4, 5 and 7; Ehrich, 1977; Clark, 1999; Uiblein *et al*., 1999) and may harbour large aggregations of benthopelagic fishes (see Chapter 9; Boehlert and Sasaki, 1988; Rogers, 1994; Koslow, 1996, 1997; Koslow *et al*., 2000) such as orange roughy (*Hoplostethus atlanticus*), pelagic armourhead (*Pseudopentaceros wheeleri*), alfonsinos (*Beryx* spp.) and cardinal fishes (*Epigonus telescopus*) (Morato *et al*., 2006). Three mechanisms, discussed in detail in Chapters 4 and 5 and outlined in Chapter 14, have been suggested to explain how these aggregations are supported trophodynamically (Genin, 2004).

The first proposes that the high biomass of fish results, at least in part, from locally enhanced primary production and subsequent bottom-up transfer of this energy to higher trophic levels in seamount food chains (Uda and Ishino, 1958; Hubbs, 1959; Uchida and Tagami, 1984; Boehlert and Genin, 1987). Upwelling and entrainment processes associated

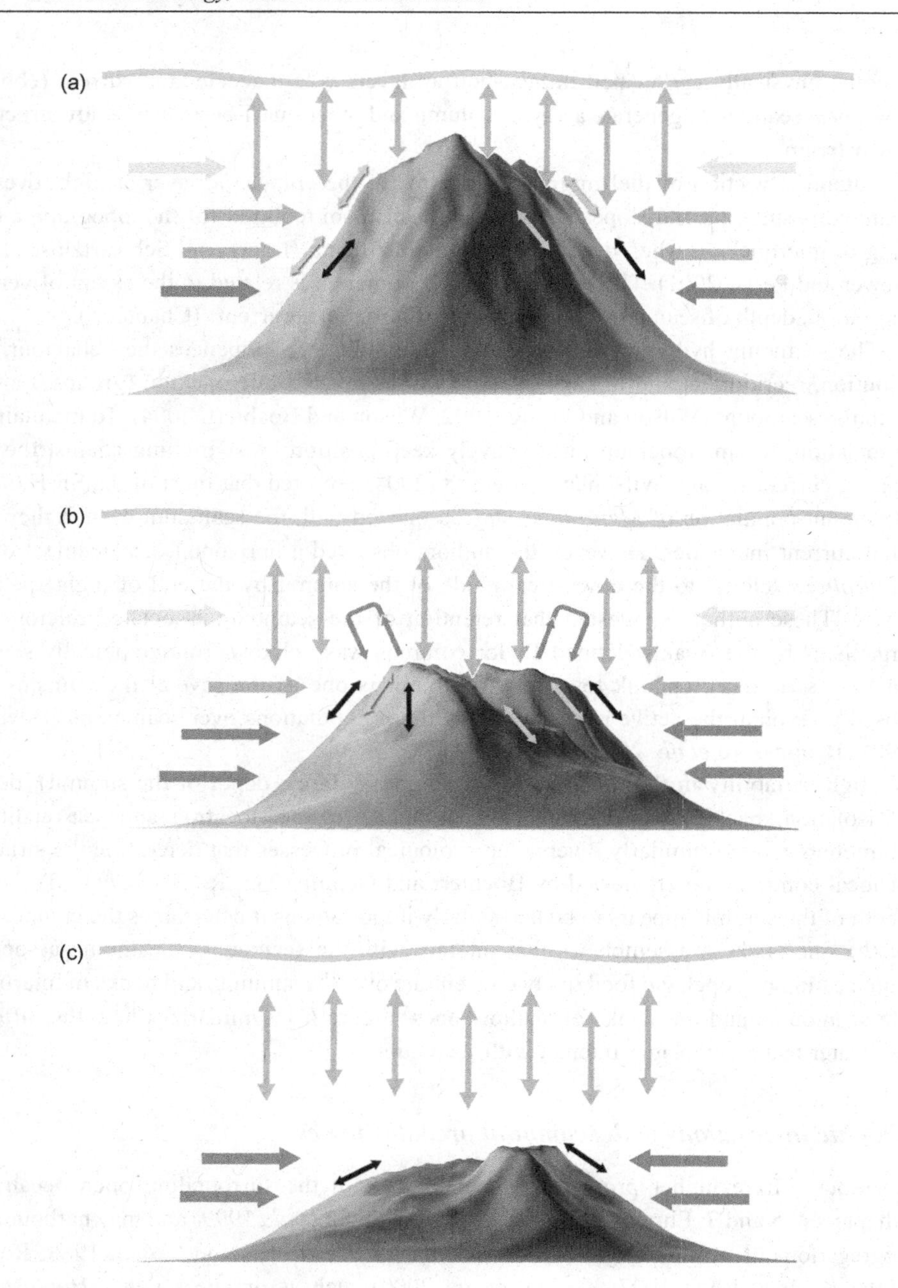

Fig. 6.1 The interactions between the different guilds of deep-pelagic fishes and seamounts of different heights: (a) summit entering the epipelagic layers; (b) summit entering the mesopelagic layers; and (c) summit in the bathypelagic layers. Horizontal arrows: non-migrant or weakly migrant meso- and bathypelagic fishes that are laterally advected to the benthopelagic realm around seamounts; vertical arrows: assemblage of diel vertically migrating fishes that interact with seamounts during the migration process; black arrows near seamount: adults of meso- and bathypelagic micronekton species that dwell in the benthopelagic zones; grey arrows near seamount: non-migrant micronekton fish species that associate preferentially with the benthopelagic layers of seamounts; U-shaped arrows: migrant micronekton fish species associated with the benthopelagic layers of seamounts that perform daily vertical migrations.

with Taylor cap formation would enhance nutrients in epipelagic waters and drive increased primary production, detected over certain well-studied seamounts (Genin and Boehlert, 1985; Dower *et al.*, 1992; Comeau *et al.*, 1995; Odate and Furuya, 1998; Mouriño *et al.*, 2001, 2005). However, it is unlikely that water could be retained around a seamount for a time period needed for production to work its way through the food web to the higher trophic level fishes residing on the seamount. Moreover, most studies have failed to demonstrate persistent high chlorophyll *a* patches over seamounts. Thus, it is not surprising that evidence for enhanced primary production leading to concentrations of fishes over seamounts is sparse (Rogers, 1994).

The second mechanism, termed the 'feed-rest' hypothesis by Genin (2004), proposes that fish aggregations are sustained by the enhanced horizontal flux of pelagic prey organisms past the seamount (Tseytlin, 1985; Dower and Mackas, 1996; Koslow, 1997). This theory suggests that fish-rest motionless in quiescent shelters during non-feeding intervals and when conditions are appropriate they emerge from shelter, feed quickly and then retreat back to rest. The third hypothesis suggests that seamount aggregations are maintained through predation on vertical migrants that are intercepted and trapped during the migration process (Isaacs and Schwartzlose, 1965; Genin *et al.*, 1988, 1994; Williams and Koslow, 1997; Fock *et al.*, 2002); the 'topographic blockage' hypothesis by Genin (2004).

Most studies on the feeding ecology of seamount fishes accord with hypotheses 2 and 3 that imported pelagic food supplies support the large fish aggregations on seamounts. For example, Isaacs and Schwartzlose (1965) estimated that at the 100 m isobath of Banco San Isidro (Baja California), the flux of organic carbon due to topographic trapping of downward-migrating micronekton was about 40 times greater than the primary production at the most productive regions off California. Seki and Somerton (1994) reported that the diet of armourheads over the summit of SE Hancock Seamount (265 m) consisted mostly of open-water, migrating micronekton that were advected and trapped over the seamount summit at night. Again, analyses of fish stomachs by Fock *et al.* (2002) indicated that the hypothesis 3, the topographic blockage mechanism, could explain sustained fish populations, their distribution and diel behaviour over the Great Meteor Seamount. The greatest densities of zooplanktivorous fishes were found at the edge of the seamount plateau, where the likelihood of encounter with demersal predators would be highest (Fock *et al.*, 2002). Several studies corroborate the 'topographic blockage' hypothesis by showing that maximum foraging on vertically migrating micronekton occurs in the early morning, when prey get trapped on their downward migration and become visible in the light, e.g., armourheads (Kitani and Iguchi, 1974; Seki and Somerton, 1994) and rockfish (*Sebastes* spp.; Isaacs and Schwartzlose, 1965; Pereyra *et al.*, 1969; Genin *et al.*, 1988).

Other seamount-associated benthopelagic fishes may have specific feeding strategies. The alfonsino, *Beryx splendens*, which feeds on midwater micronekton, actively tracks the movements of the main DSL instead of waiting passively for prey to be advected to the seamount. During the day, they stay near the bottom, while during night they migrate into the water column up to 250 m above the bottom. This behaviour was observed for *Beryx* populations living at deep seamounts of the northern section of the Mid-Atlantic Ridge (Vinnichenko, 1997b) and the Nazca and Sala y Gómez seamounts in the eastern central Pacific (Dubochkin and Kotlyar, 1989; Parin *et al.*, 1997). The main prey included migrant and non-migrant midwater fishes such as Gonostomatidae, Sternoptychidae, Stomiidae,

Myctophidae; crustaceans such as decapods and euphausids; and other pelagic organisms such as salps (Dubochkin and Kotlyar, 1989; Vinnichenko, 1997b; Dürr and González, 2002). The cardinal fish, *Epigonus* spp., *Zenopsis* spp. and some gempylids show similar feeding behaviour (Parin *et al.*, 1997).

A combination of feeding mechanisms may in fact be present. Tseytlin (1985) observed that the bottom entrapment of descending animals and the continuous horizontal influx of allocthonous micronekton and zooplankton may together provide enough food to maintain the commercial fisheries on intermediate seamounts. In the Azores, Morato *et al.* (2001) reported that blackspot seabream, *Pagellus bogaraveo*, found on seamounts depend both on locally produced benthic food and on a regular supply of midwater fish that drift past.

Because the average daytime depth of the main DSL is around 500 m, aggregations of seamount-associated benthopelagic fishes such as the orange roughy that live below that depth cannot centre their diet on migrant fauna, but rely instead on non-migrant deeper meso- and bathypelagic prey. Indeed, bathypelagic smelts (Bathylagidae), large myctophids and stomiids were found to be the main prey of orange roughy caught between 830 and 1500 m at seamounts off the Azores (Barcelos *et al.*, 2005) and in the Rockall Trough and Porcupine Sea Bight (Mauchline and Gordon, 1984; Gordon and Duncan, 1987). Similarly, orange roughy caught at the seamounts off New Zealand had been feeding on non-migrant pelagic organisms (Rosecchi *et al.*, 1988), Australia (Bulman and Koslow, 1992) and Chile (Labbé and Arana, 2001). Koslow (1997) concluded that seamount aggregators subsist upon on meso- and bathypelagic organisms that drift past seamounts. Most of these studies have not found any clear pattern of daily feeding periodicity, which supports the idea that a constant flux of laterally advected non-migrant pelagic fauna indeed represent the main prey of orange roughy (see Chapter 14).

Other species do seem to profit from the high concentration of daily vertically migrating fishes that associate with the benthopelagic layers of seamounts during the day, instead of open-ocean DSL pelagic organisms. For example, abundance of the myctophid *D. suborbitalis* was substantially higher over Equator Seamount than off mount, as was the abundance of predatory fishes (Parin and Prut'ko, 1985) whose main food was *D. suborbitalis*. In this trophic scenario, micronekton migrate upwards from their daytime depths at the seamount benthopelagic layer into the epipelagic zone at night to feed, are laterally advected over the seamount and then become 'topographically trapped' during their downward morning descent (see Box 5.1). In this manner, organisms that would normally be distributed in three dimensions are concentrated into two dimensions, the horizontal plane over the seamount, thus increasing their density to potential predators. Borets (1986) also reported that the mirror dory, *Zenopsis nebulosa*, mostly eats pearl fish *Maurolicus*, which performs diel vertical migrations and is associated with seamount bottom layers during day.

While the specific mechanisms driving trophic interactions between benthopelagic fauna and water column fauna are not fully understood, it appears that the primary factor is the concentration of pelagic prey that is normally widely distributed. In this manner, greater abundance of higher trophic level predators allows more efficient foraging. Increased trophic efficiency may work simultaneously over several trophic levels. In some cases (e.g., Great Meteor Seamount), higher trophic levels effectively utilize all or most of the vertically migrating prey concentrated by topographic trapping, thus resulting in 'daily gap formation'. In other cases (e.g., SE Hancock Seamount), predation on meso- and bathypelagic fishes by seamount fauna appears to be offset by the benefits of increased prey availability, hence increasing local mesopelagic fish abundance. Whatever the mechanism, it would appear that the trophic subsidy afforded to the demersal seamount fauna in the form of meso- and bathypelagic fishes, cephalopods and crustaceans is an integral part of the ecology of seamount ecosystems.

References

Andriashev, A.P. (1953) Ancient deep-water and secondary deep-water fishes and their importance in a zoogeographical analysis. *Notes on Special Problems in Ichthyology*. Akademi Nauklii SSSR, Moscow.

Angel, M.V. (1985) Vertical migrations in the oceanic realm: possible causes and probable effects. In: *Migration, Mechanisms and Adaptive Significance* (ed. Rankin, M.A.). *Contributions in Marine Science*, Suppl. 27, 45–70.

Angel, M.V. and Pugh, P.R. (2000) Quantification of diel vertical migration by micronektonic taxa in the northeast Atlantic. *Hydrobiology*, 440, 161–79.

Badcock, J. (1984a) Sternoptychidae. In: *Fishes of the North-eastern Atlantic and the Mediterranean* (eds. Whitehead, P.J.P., Bauchot, M.-L., Hureau, J.-C., Nielsen, J. and Tortonese, E.), Vol. I, pp. 302–17. UNESCO, Paris.

Badcock, J. (1984b) Photichthyidae. In: *Fishes of the North-eastern Atlantic and the Mediterranean* (eds. Whitehead, P.J.P., Bauchot, M.-L., Hureau, J.-C., Nielsen, J. and Tortonese, E.), Vol. I, pp. 318–24. UNESCO, Paris.

Barcelos, L.D., Porteiro, F.M., Melo, O., Clarke, M. and Menezes, G. (2005) Feeding ecology of orange roughy, *Hoplostethus atlanticus*, Collett, 1889, in the Azores Archipelago. *40th European Marine Biology Symposium,* Viena Abstracts, 61–2 (poster).

Beckmann, A. and Mohn, C. (2002) The upper circulation at Great Meteor Seamount. Part II: Retention potential of the seamount induced circulation. *Ocean Dynamics*, 52, 194–204.

Bekker, V.E. and Prut'ko, V.G. (1984) A new species of the genus *Diaphus* (Myctophidae) from the Northeastern Indian Ocean. *Journal of Ichthyology*, 24(6), 82–7.

Benoit-Bird, K.J., Au, W.W.L., Brainard, R.E., Lammers, M.O. (2001) Diel horizontal migration of the Hawaiian mesopelagic boundary community observed acoustically. *Marine Ecology Progress Series*, 217, 1–14.

Boehlert, G.W. and Genin, A. (1987) A review of the effects of seamounts on biological processes. In: *Seamounts, Islands and Atolls*. (eds. Keating, B., Fryer, P., Batiza, R. and Boehlert, G.), pp. 319–34. *Geophysical Monographs*, 47. American Geophysical Union, Washington, DC.

Boehlert, G.W. and Sasaki, T. (1988) Pelagic biogeography of the armorhead, *Pseudopentaceros wheeleri*, and recruitment to isolated seamounts in the North Pacific Ocean. *Fishery Bulletin*, 86, 453–66.

Boehlert, G.W. and Seki, M.P. (1984) Enhanced micronekton abundance over mid-Pacific seamounts. *EOS, Transactions of the American Geophysical Union*, 65, 928.

Boehlert, G.W., Wilson, C.D. and Mizuno, K. (1994) Populations of the sternoptychid fish, *Maurolicus muelleri*, on seamounts in the Central North Pacific. *Pacific Science*, 48, 57–69.

Bordes, F., Uiblein, F., Castillo, R., Barrera, A., Castro, J.J., Coca, J., Gomez, J., Hansen, K., Hernandez, V., Merrett, N., Miya, M., Moreno, T., Perez, F., Ramoz, A. and Sutton, T. (1999) Epi- and mesopelagic fishes, acoustic data, and SST images collected off Lanzarote,

Fuerteventura and Gran Canaria, Canary Islands, during cruise 'La Bocaina 04-97'. *Informes Técnicos del Instituto Canario de Ciencias Marinas*, 5, 1–45.

Borets, L.A. (1986) Ichthyofauna of the Northwestern and Hawaiian Ridges. *Journal of Ichthyology*, 26(3), 1–13.

Bulman, C.M. and Koslow, J.A. (1992) Diet and food consumption of a deep-sea fish, orange roughy *Hoplostethus atlanticus* (Pisces: Trachichthyidae), off southeastern Australia. *Marine Ecology Progress Series*, 82, 115–29.

Childress, J.J. (1995) Are there physiological and biochemical adaptations of metabolism in deep-sea animals? *Trends in Ecology and Evolution*, 10(1), 30–35.

Childress, J.J., Taylor, S., Cailliet, G. and Price, M.H. (1980) Patterns of growth reproduction and energy usage in some meso- and bathypelagic fishes. *Marine Biology*, 61, 27–40.

Clark, M. (1999) Fisheries for orange roughy (*Hoplostethus atlanticus*) on seamounts in New Zealand. *Oceanologica Acta*, 22, 593–602.

Comeau, L.A., Vezina, A.F., Bourgeois, M. and Juniper, S.K. (1995) Relationship between phytoplankton production and the physical structure of the water column near Cobb Seamount, northeast Pacific. *Deep-Sea Research I*, 42, 993–1005.

Diekmann, R. (2004) Distribution patterns of oceanic micronekton at seamounts and hydrographic fronts of the subtropical Atlantic Ocean. PhD Dissertation Thesis, Christian-Albrechts-Universität zu Kiel, Germany, 133 + XLIII.

Dower, J. and Mackas, D.L. (1996) 'Seamount effects' in the zooplankton community near Cobb Seamount. *Deep-Sea Research I*, 43, 837–58.

Dower, J. and Perry, R.I. (2001) High abundance of larval rockfish over Cobb Seamount, an isolated seamount in the Northeast Pacific. *Fisheries Oceanography*, 10, 268–74.

Dower, J., Freeland, H. and Juniper, K. (1992) A strong biological response to oceanic flow past Cobb Seamount. *Deep-Sea Research A*, 39, 1139–45.

Dubochkin, A.S. and Kotlyar, A.N. (1989) On the feeding of alfonsino (*Beryx splendens*). *Journal of Ichthyology*, 29(5), 1–8.

Ducklow, H.W., Steinberg, D.K. and Buesseler, K.O. (2001) Upper ocean carbon export and the biological pump. *Oceanography*, 14(4), 50–58.

Dürr, J. and González, J.A. (2002) Feeding habits of *Beryx splendens* and *Beryx decadactylus* (Berycidae) off the Canary Islands. *Fisheries Research*, 54, 363–74.

Ehrich, S. (1977) Die Fishfauna der GroßenMeteorbank. Meteor Forschungsergebrisse, D25, 1–23.

Fock, H., Matthiessen, B., Zidowitz, H. and von Westernhagen, H. (2002) Diel and habitat-dependent resource utilisation by deep sea fishes at the Great Meteor seamount: niche overlap and support for the sound scattering layer interception hypothesis. *Marine Ecology Progress Series*, 244, 219–33.

Froese, R. and Pauly, D. (eds) (2006) Fishbase (www Database) World Wild Web Electronic Publications. http://www.fishbase.org/ (accessed March 2006).

Froese, R. and Sampang, A. (2004) Taxonomy and biology of seamount fishes. In: *Seamounts: Biodiversity and Fisheries* (eds. Morato, T. and D. Pauly). *Fisheries Centre Research Report*, 12(5), 25–31.

Gartner, J.V. (1991) Life histories of three species of lanternfishes (Pisces: Myctophidae) from the eastern Gulf of Mexico. II: Age and growth patterns. *Marine Biology*, 111, 21–7.

Genin, A. (2004) Bio-physical coupling in the formation of zooplankton and fish aggregations over abrupt topographies. *Journal of Marine Systems*, 50, 3–20.

Genin, A. and Boehlert, G.W. (1985) Dynamics of temperature and chlorophyll structures above a seamount: an oceanic experiment. *Journal of Marine Research*, 43, 907–24.

Genin, A., Haury, L. and Greenblatt, P. (1988) Interactions of migrating zooplankton with shallow topography: predation by rockfishes and intensification of patchiness. *Deep-Sea Research, Part A. Oceanographic Research Papers*, 35, 151–75.

Genin, A., Greene, C., Haury, L., Wiebe, P., Gal, G., Kaartvedt, S., Meir, E., Fey, C. and Dawson, J. (1994) Zooplankton patch dynamics: daily gap formation over abrupt topography. *Deep-Sea Research, Part 1. Oceanographic Research Papers*, 41, 941–51.

Golovan, G.A. (1978) Composition and distribution of the ichthyofauna of the continental slope of north-western Africa. *Trudy Instituta Okeanologii*, 111, 195–258.

Gordon, J.D.M. and Duncan, J.A.R. (1987) Aspects of the biology of *Hoplostethus atlanticus* and *H. mediterraneus* (Pisces: Berycomorphi) from the slopes of the Rockall Trough and the Porcupine Sea Bight (North-eastern Atlantic). *Journal of the Marine Biological Association of the United Kingdom*, 67, 119–33.

Gushchin, A.V. and Kukuev, E.I. (1981) On the composition of ichthyofauna of the northern part of the Middle-Atlantic Ridge. In: *Fishes of the Open Ocean* (ed. Parin, N.V.), pp. 36–40. Institute of Oceanology, Academy of Sciences of the USSR (in Russian).

Harold, A.S. (1994) A taxonomic revision of the sternoptychid genus *Polyipnus* (Teleostei: Stomiiformes) with an analysis of phylogenetic relationships. *Bulletin of Marine Science*, 54(2), 428–534.

Harold, A.S. and Lancaster, K. (2003) A new species of the hatchetfish genus *Argyripnus* (Stomiiformes: Sternoptychidae) from the Indo-Pacific. *Proceedings of the Biological Society of Washington*, 116(4), 883–91.

Herring, P.J. and Morin, J.G. (1978) Bioluminescence in fishes. In: *Bioluminescence in Action* (ed. Herring, P.J.), pp. 273–329. Academic Press, New York.

Hubbs, C.L. (1959) Initial discoveries of fish fauna on seamounts and offshore banks in the Eastern Pacific. *Pacific Science*, 13, 311–16.

Hulley, P.A. and Lutjeharms, J.R.E. (1989) Lanternfishes of the southern Benguela region. Part III: The pseudoceanic-oceanic interface. *Annals of South Africa Museum*, 98(19), 409–35.

Isaacs, J.D. and Schwartzlose, R.A. (1965) Migrant sound scatterers: interactions with the sea floor. *Science*, 150, 1810–13.

Kalinowski, J. and Linkowski, T.B. (1983) Hydroacoustic observations on *Maurolicus muelleri* (Sternoptychidae) over R.S.A. and Discovery Seamounts (South Atlantic). *International Council for the Exploration of the Sea*, CM:1983/H: 16, 20 pp.

Kashkin, N.I. (1977) Fauna of Sound-Scattering Layers (SSL). In: *The Biological Structure in the Ocean, Oceanology* (ed. Monin, A.C.), Vol. I, pp. 299–318. Nauka Moscow.

Kashkin, N.I. (1984) Mesopelagic micronekton as a factor of fish productivity on the oceanic banks. In: *Frontal Zones of the Southeastern Pacific Ocean (Biology, Physics, Chemistry)* (eds. Vinogradov, M.E. and Fedorov, K.N.), pp. 285–91. Nauka, Moscow.

Kaufmann, M., von Bröckel, K., Mohn, C. and Beckmann, A. (2002) The influence of the Great Meteor Seamount on the distribution of phytoplankton. *International Council for the Exploration of the Sea*, CM:2002/M: 01.

Kitani, K. and Iguchi, K. (1974) Some observations on the feature of fishing ground on the seamount in the Central North Pacific Ocean as surveyed through the trawl operation. Bulletin of the National Research Institute of Far Seas Fisheries (Shimizu, Japan), 11, 33–6.

Koslow, A. (1996) Energetic and life-history patterns of deep-sea benthic, benthopelagic and seamount-associated fish. *Journal of Fish Biology*, 49(Suppl. A), 54–74.

Koslow, A. (1997) Seamounts and the ecology of deep-sea fisheries. *American Scientist*, 85, 168–76.

Koslow, J.A., Boehlert, G.W., Gordon, J.D.M., Haedrich, R.L., Lorance, P. and Parin, N. (2000) Continental slope and deep-sea fisheries: implications for a fragile ecosystem. *ICES Journal of Marine Science*, 57, 548–57.

Kukuev, E.I. (1982) Ichthyofauna of the Corner Mountains and New England seamounts in the western North Atlantic. In: *Poorly Known Fishes of the Open Ocean* (ed. Parin, N.V.), pp. 92–109. Institute of Oceanology, Academy of Sciences of the USSR, Moscow (in Russian).

Kukuev, E.I. (2004) Ichtyofauna research on underwater mountain within the North Atlantic Ridge and adjacent seas. *Archive of Fishery and Marine Research*, 51(1–3), 215–32.

Labbé, F.J. and Arana, P.M. (2001) Alimentación de orange roughy, *Hoplostethus atlanticus* (Pisces: Trachichthyidae), en el archipiélago de Juan Fernández, Chile. *Revista de Biología Marina y Oceanografía*, 36(1), 75–82.

Lammers, M.O., Brainard, R.E. and Au, W.W.L. (2004) Diurnal trends in the mid-water biomass community of the Northwestern Hawaiian Islands observed acoustically. *Journal of the Acoustical Society of America*, 116(4), 2488–9.

Linkowski, T.B. (1983) Some aspects of the biology of *Maurolicus muelleri* (Sternoptychidae) from the South Atlantic. *International Council for the Exploration of the Sea*, CM:1983/H: 17.

Linkowski, T.B. (1985) Population biology of the myctophid fish *Gymnoscopelus nicholsi* (Gillbert, 1911) from the western South Atlantic. *Journal of Fish Biology*, 27, 683–9.

Lorz, H.V., Pearcy, W.G. and Fraidenberg, M. (1983) Notes on the feeding habits of the yellowtail rockfish, *Sebastes flavidus*, off Washington and in Queen Charlotte sound. *California Fish and Game*, 69, 33–8.

Marshall, N.B. (1979) *Developments in Deep-Sea Biology*, 566 pp. Blandford Press UK.

Mauchline, J. and Gordon, J.D.M. (1984) Occurrence and feeding of berycomorphid and percomorphid teleost fish in the Rockall Trough. *Journal du Conseil International pour l'Exploration de la Mer*, 41, 239–47.

Mauchline, J. and Gordon, J.D.M. (1991) Oceanic prey of benthopelagic fish in the benthic boundary layer of a marginal oceanic region. *Marine Ecology Progress Series*, 74, 109–15.

McClatchie, S. and Dunford, A. (2003) Estimated biomass of vertically migrating mesopelagic fish off New Zealand. *Deep-Sea Research I*, 50, 1263–81.

Melo, O. and Menezes, G. (2002) Exploratory fishing of the orange roughy (*Hoplostethus atlanticus*) in some seamounts of the Azores archipelago. *International Council for the Exploration of the Sea*, CM:2002/M: 26.

Merrett, N.R. (1986) Biogeography and the Oceanic rim: a poorly known zone of the ichthyofaunal interaction. In: *Pelagic Biogeography* (eds. Pierrot-Bults, A.C., van der Spoel, S., Zahuranec, B.J. and Johnson, R.K.), pp. 201–9. *UNESCO Technical Papers in Marine Sciences*, 49.

Miya, M., Kawaguchi, A. and Nishida, M. (2001) Mitogenomic exploration of higher teleostean phylogenies: a case study for moderate-scale evolutionary genomics with 38 newly determined complete mitochondrial DNA sequences. *Molecular Biology and Evolution*, 18(11), 1993–2209.

Moore, J.A., Vecchione, M. Collette, B.B., Gibbons, R., Hartel, K.E., Galbraith, J.K., Turnipseed, M., Southworth, M. and Watkins, E. (2003) Biodiversity of the Bear Seamount, New England Seamount chain: results of exploratory fishing. *Journal of Northwest Atlantic Fishery Sciences*, 31, 363–72.

Morato, T., Sola, E., Grós, M.P. and Menezes, G. (2001) Feeding habits of two congener species of seabreams, *Pagellus bogaraveo* and *Pagellus acarne*, off the Azores (Northeastern Atlantic) during spring of 1996 and 1997. *Bulletin of Marine Science*, 69(3), 1073–87.

Morato, T., Watson, R., Pitcher, T.J. and Pauly, D. (2006) Fishing down the deep. *Fish and Fisheries*, 7, 24–34.

Mouriño, B., Fernández, E., Serret, P., Harbour, D., Sinha, B. and Pingree, R. (2001) Variability and seasonality of physical and biological fields at the Great Meteor Tablemount (subtropical NE Atlantic). *Oceanologica Acta*, 24, 167–85.

Mouriño, B., Fernandez, E., Pingree, R., Sinha, B., Escanez, J. and de Armas, D. (2005) Constraining effect of mesoscale features on carbon budget of photic layer in the NE subtropical Atlantic. *Marine Ecology Progress Series*, 287, 45–52.

Nelson, J.S. (2006) *Fishes of the World*, 4th edition, 601 pp. John Wiley and Sons, Inc., New York.

Odate, T. and Furuya, K. (1998) Well-developed subsurface chlorophyll maximum near Komahashi No. 2 seamount in the summer of 1991. *Deep-Sea Research I*, 45(10), 1595–607.

Parin, N.V. (1984) Oceanic ichthyologeography: an attempt to review the distribution and the origin of pelagic and bottom fishes outside continental shelves and neritic zones. *Archiv für Fischereiwissenschaft*, 35, 5–41.

Parin, N.V. (1986) Distribution of mesobenthopelagic fishes in slope waters and around submarine rises. *UNESCO Technical Papers in Marine Science*, 49, 226–9.

Parin, N.V. (1992) *Argyripnus electronus*, a new sternoptychid fish from Sala y Gomez submarine ridge. *Japanese Journal of Ichthyology*, 39(2), 135–7.

Parin, N.V. and Borodulina, O.D. (1990) Review of the genus *Polymetme* (Phosichthyidae) with the description of two new species. *Journal of Ichthyology*, 30(6), 108–21.

Parin, N.V. and Golovan, G.A. (1976) Pelagic deep-sea fishes of the families characteristic of the open ocean collected over the continental slope off West Africa. *Trudy Instituta Okeanologii*, 104, 250–76 (in Russian).

Parin, N.V. and Kobyliansky, S.G. (1996) Diagnoses and distribution of fifteen species recognized in genus *Maurolicus* Cocco (Sternoptychidae, Stomiiformes) with a key to their identification. *Cybium*, 20(2), 185–95.

Parin, N.V. and Prut'ko, V.G. (1985) The thalassial mesobenthopelagic ichthyocoene above the Equator Seamount in the western tropical Indian Ocean. *Oceanology*, 25, 781–3.

Parin, N.V., Mironov, A.N. and Nesis, K.N. (1997) Biology of the Nazca and Sala y Gómez submarine ridges, an outpost of the Indo-West Pacific Fauna in the Eastern Pacific Ocean: composition of the fauna, its communities and history. *Advances in Marine Biology*, 32, 145–242.

Pereyra, W.T., Pearcy, W.G. and Carvey Jr., F.E. (1969) *Sebastodes flavidus*, a shelf rockfish feeding on mesopelagic fauna, with consideration of the ecological implications. *Journal of the Fisheries Research Board of Canada*, 26, 2211–15.

Porteiro, F.M. (2005) Biogeography and biodiversity of stomiid fishes in the North Atlantic. PhD Dissertations Thesis, University of Liverpool, UK, xix + 397 pp.

Pusch, C., Beckmann, A., Porteiro, F.M. and von Westernhagen, H. (2004) The influence of seamounts on mesopelagic fish communities. *Archive of Fishery and Marine Research*, 51(1–3), 165–86.

Reid, S.B., Hirota, J., Young, R.E. and Hallacher, L.E. (1991) Mesopelagic-boundary community in Hawaii: micronekton at the interface between neritic and oceanic ecosystems. *Marine Biology*, 109, 427–40.

Richer de Forges, B. (2001) *Electronic Database of ORSTOM Sampling on the Norfolk Ridge*. ORSTOM, France.

Rogers, A.D. (1994) The biology of seamounts. *Advances in Marine Biology*, 30, 305–50.

Rosecchi, E., Tracey, D.M. and Webber, W.R. (1988) Diet of orange roughy, *Hoplostethus atlanticus* (Pisces: Trachichthyidae) on the Challenger Plateau, New Zealand. *Marine Biology*, 99, 293–306.

Salvanes, A.G.V. and Kristoffersen, J.B. (2001) Mesopelagic fishes. In: *Encyclopedia of Ocean Sciences* (eds. Steel, J., Thorpe, S. and Turekian, K.), pp. 1711–17. Academic Press, San Diego, CA.

Sassa, C., Kawaguchi, K., Kinoshita, T. and Watanabe, C. (2002) Assemblages of vertical migratory mesopelagic fish in the transitional region of the western North Pacific. *Fisheries Oceanography*, 11, 193–204.

Seki, M.P. and Somerton, D.A. (1994) Feeding ecology and daily ration of the pelagic armorhead, *Pseudopentaceros wheeleri*, at Southeast Hancock Seamount. *Environmental Biology of Fishes*, 39, 73–84.

Shcherbachev, Y.P., Kukuev, E.I. and Shlibanov, V.I. (1985) Bottom and near-bottom ichthyocenes species composition in the underwater mountains of the southern North-Atlantic Ridge. *Voprosy Ichthyologii (Problems of Ichthyology)*, 25(1), 35–50.

Sutton, T.T. and Hopkins, T.L. (1996) Trophic ecology of the stomiid (Pisces: Stomiidae) fish assemblage of the eastern Gulf of Mexico: strategies, selectivity and impact of a top mesopelagic predator group. *Marine Biology*, 127(2), 179–92.

Sutton, T.T., Hopkins, T.L. and Lancraft, T.M. (1998) Trophic diversity of a mesopelagic fish community. *IOC Workshop Report No. 142*, pp. 353–7.

Sutton, T.T., Porteiro, F.M., Heino, M., Byrkjedal, I., Langhelle, G., Anderson, C.I.H., Horne, J., Soiland, H., Falkenhaug, T. and Bergstad, O.A. The structure and biomass of pelagic fish assemblages relative to the Mid-Atlantic Ridge as a function of depth. *Deep-Sea Research II* (in press).

Tseytlin, V.B. (1985) Energetics of fish populations inhabiting seamounts. *Oceanology*, 25, 237–9.

Uchida, R.N. and Tagami, D.T. (1984) Groundfish fisheries and research in the vicinity of seamounts in the North Pacific Ocean. *Marine Fisheries Review*, 46, 1–17.

Uda, M. and Ishino, M. (1958) Enrichment pattern resulting from eddy systems in relation to fishing grounds. *Journal of Tokyo University of Fisheries*, 1–2, 105–19.

Uiblein, F., Geldmacher, A., Köster, F., Nellen, W. and Kraus, G. (1999) Species composition and depth distribution of fish species collected in the area of the Great Meteor seamount, eastern central Atlantic, during cruise M42/3, with seventeen new records. *Instituto Canario de Ciencias Marinas*, Telde (Gran Canaria), 5, 47–85.

Vinnichenko, V.I. (1997a) Russian Investigations and deep water fishery on the Corner Rising Seamount in subarea 6. *NAFO Scientific Council Studies*, 30, 41–9.

Vinnichenko, V.I. (1997b) Vertical diurnal migrations of the slender alfonsino *Beryx splendens* (Berycidae) at the underwater rises of the open North Atlantic. *Journal of Ichthyology*, 37, 438–44.

Walsh, J.J., Dymond, J. and Collier, R. (1988) Rates of recycling of biogenic components of settling particles in the ocean derived from sediment trap experiments. *Deep-Sea Research, Part A. Oceanographic Research Papers*, 35, 43–58.

Williams, A. and Koslow, J.A. (1997) Species composition, biomass and vertical distribution of micronekton over the mid-slope region off southern Tasmania, Australia. *Marine Biology*, 130(2), 1432–793.

Wilson, C.D. (1992) Interactions of ocean currents and diel migrators at a seamount in the Central North Pacific. PhD Dissertations Thesis, University of Hawaii, Honolulu, HI.

Wilson, C.D. and Boehlert, G.W. (2004) Interaction of ocean currents and resident micronekton at a seamount in the Central North Pacific. *Journal of Marine Systems*, 50, 39–60.

Wilson, C.D. and Firing, E. (1992) Sunrise swimmers bias acoustic Doppler current profiles. *Deep-Sea Research I*, 39, 885–92.

Young, R.E. (1983) Oceanic bioluminescence: an overview of general functions. *Bulletin of Marine Science*, 33, 829–45.

Part III

Biology and Ecology of Seamount Organisms

Chapter 7
Seamount benthos

Sarah Samadi, Thomas Schlacher and Bertrand Richer de Forges

Abstract

Seamounts are unique habitats for the deep-sea megabenthos. Several distinctive environmental conditions, such as limited spatial extent, geographic isolation, swift currents, localized upwelling and circulation cells create environments favourable for the establishment of diverse benthic assemblages. Relatively large suspension feeders such as corals, sponges and crinoids, can dominate the biomass of these assemblages and form structural habitat for a diversity of smaller, mobile species. The benthos contains species with apparently limited geographic distributions (endemics) and archaic species thought to have become extinct ('living fossils'). Seamounts rise well above the ocean floor and thus form relatively shallow habitat available for bathyal species above the surrounding abyssal seafloor. Growth of seamount invertebrates can be extraordinarily slow and they often have very long life spans. These life-history traits make the seamount benthos highly vulnerable to destructive bottom trawling. The prominence of suspension feeders suggests a simple trophic web, but in fact, benthic food webs are complex: food-chain lengths and trophic architecture rival other marine ecosystems in both shallow and deep settings. The geographic isolation of seamounts has frequently been likened to oceanic islands, where species differences among seamounts can be very high. Yet, seamount populations may not necessarily be genetically isolated if they produce larvae capable of long-distance dispersal. The fauna of seamounts is poorly documented, and the structure of whole assemblages is known from only a limited number of seamounts worldwide, partly as a consequence of dwindling resources and expertise in taxonomy. Lack of basic ecological knowledge impedes the development of global, integrated structural and functional frameworks concerning seamount benthos.

Introduction

Three factors combine to make a comprehensive global synthesis of seamount benthos difficult: (1) Seamounts occur from the tropics to the poles, resulting in a wide ambit of physico-chemical conditions and differences in rates of primary production and food supply to benthic consumers and since communities respond to such environmental variability, the benthos of seamounts is likely to be geographically diverse. (2) Seamounts cover a broad depth range, including both shallow mounts that extend into the euphotic zone as well as mounts that lie in the bathyal and abyssal zones; as seamount benthos is strongly structured by depth, such wide bathymetric ranges result in a multitude of community types. (3) Seamounts have diverse geological histories and geological ages. Most seamounts are of volcanic origin with basaltic rocks, but guyots (see Chapter 1) can be common in the tropics. Since both substratum type and habitat age influence species composition, the structure of seamount benthos will reflect this diversity.

Theoretically it is possible to envisage faunal surveys that encompass this diversity of geo-morphological types, latitudinal and depth ranges and seamount ages but the reality is starkly different. Of the estimated 100 000 seamounts worldwide (see Chapters 1 and 2), only 232 have been biologically sampled (SeamountsOnline, 2006), and for these very few data on invertebrates are available. In fact, much of our current knowledge on the seamount benthos is a by-product of fisheries studies. A global dearth in taxonomical expertise and resources impedes accurate assessments of benthic diversity, with few surveys identifying the complete spectrum of specimens collected, so that the reported number of seamount species is likely to be a gross underestimate of the total. For example, Wilson and Kaufmann (1987) present a global inventory of the fauna collected on about 100 seamounts that comprises 597 species, many of which are unidentified. Five seamounts account for 72% of the species recorded (Smith and Jordan, 1988; Rogers, 1994), further emphasizing the limited sampling range available for biodiversity assessments of seamounts. Several Russians cruises to the Eastern Pacific (Kuzneksov and Mironov, 1981; Parin *et al.*, 1997) yielded 192 reported species, many unidentified, from 25 seamounts. Koslow and Gowlett-Holmes (1998) list 242 invertebrate species from 14 seamounts off Tasmania. By contrast, 730 species have been described from 18 seamounts in New Caledonia (Richer de Forges *et al.*, 2000, 2005). While this is a substantially higher estimate of the benthic speciosity at seamounts, it is an underestimate of the true seamount species richness because a large part of the catch awaits taxonomic description.

Because of such limitations in published seamount data, this chapter draws mostly on more recent findings from seamount studies in New Caledonia (south west Pacific) to illustrate several key biological aspects of seamount benthos. We highlight six fundamental ecological properties of the seamount benthos which are likely to be conceptually applicable to other seamount systems worldwide; these include: (1) composition of the benthos; (2) species richness and 'new species'; (3) geographic patterns in species composition (beta-diversity); (4) endemism and genetic structures of benthic invertebrate populations; (5) growth rates, longevity and evolutionary ages of benthic invertebrates; and (6) trophic organization of seamount benthos.

Composition of the benthos

Suspension feeders usually dominate the biomass of the megabenthos on seamounts. Currents are amplified around seamounts (see Chapters 4 and 5), and this is thought to be the principal factor that favours suspension feeders. Taxonomic composition varies between seamounts: assemblages may be dominated by sponges and/or corals like stylasterids or gorgonians. These large suspension feeders provide important habitat for smaller, mobile invertebrates, with molluscs, crustaceans and echinoderms being particularly species-rich amongst this vagile fauna (Tables 7.1–7.3). For example, stylasterine corals are well represented in the bathyal zone of New Caledonia. High species diversity has been found on hard bottoms south of New Caledonia island, particularly on Norfolk Ridge seamounts (Richer de Forges *et al.*, 1987). About 49 species, of which 26 are new to science, were collected here (Lindner and Cairns, Personal communication).

Table 7.1 Species richness, and the discovery of species new to science, in expeditions to the bathyal region of New Caledonia (Richer de Forges *et al.* 2005). See supplemental material for expedition details.

Higher taxon	Families	Genera	Species	New species	New %
Porifera	61	135	216	142	65.7
Cnidaria	9	19	73	56	76.7
Brachiopoda	14	20	26	5	19.2
Annelida and Sipuncula	6	11	17	7	41.1
Bryozoa	57	122	206	115	55.8
Mollusca	80	245	710	461	64.9
Pycnogonida	8	23	60	37	61.7
Crustacea	98	298	691	362	52.3
Echinodermata	15	30	36	16	44.4
Tunicata	12	36	66	48	72.7
Vertebrata	109	242	414	73	17.6
Total	469	1181	2515	1322	52.5

Table 7.2 Species richness on the seamounts of Norfolk Ridge and Lord Howe Rise (Richer de Forges *et al.* 2005).

Higher taxon	Families	Genera	Species	New species	New %
Porifera	18	26	34	26	76.4
Cnidaria	3	8	19	15	78.9
Brachiopoda	9	11	13	3	23
Annelida and Sipuncula	2	2	2	2	100
Bryozoa	28	44	56	36	64.2
Mollusca	43	96	201	127	63.1
Pycnogonida	4	9	13	8	61.5
Crustacea	49	118	251	163	64.9
Echinodermata	4	6	8	6	75
Tunicata	7	8	9	6	66.6
Vertebrata	54	94	124	19	15.3
Total	223	423	730	411	56.3

Only broad estimates of the distribution of benthic species richness among trophic guilds can be made from published species lists. Wilson and Kaufmann's (1987) global compilation of benthic seamount species indicates that suspension feeders comprise about 52%. But regional datasets provide much lower figures 15% in the Eastern Pacific from several Russian cruises (Kuznetsov and Mironov, 1981; Parin *et al.*, 1997); about 27% on Tasmanian seamounts (Koslow and Gowlett-Holmes, 1998); about 22% on the Lord Howe and Norfolk Ridge in the SW Pacific (Fig. 7.1; Richer de Forges, Personal

Table 7.3 Species richness of the fauna living on the seamounts only.

Higher taxon	Families	Genera	Species	New species	New %
Porifera	6	8	8	5	62.5
Cnidaria	2	4	4	4	100
Brachiopoda	2	2	2	1	50
Annelida and Sipuncula	0	0	0	0	0
Bryozoa	22	29	35	25	71.4
Mollusca	20	39	55	44	80
Pycnogonida	2	4	5	2	40
Crustacea	22	32	42	30	71.4
Echinodermata	2	3	3	3	100
Tunicata	2	3	3	2	66.6
Vertebrata	24	35	39	9	23
Total	104	159	196	125	63.7

communication). While forming the bulk of the biomass, suspension feeders may not necessarily constitute the largest proportion of species in seamount benthos. Their apparent high biomass has, however, not been reported quantitatively for entire assemblages in the published literature, and their abundance is mostly based on qualitative assessments of catches and underwater imagery.

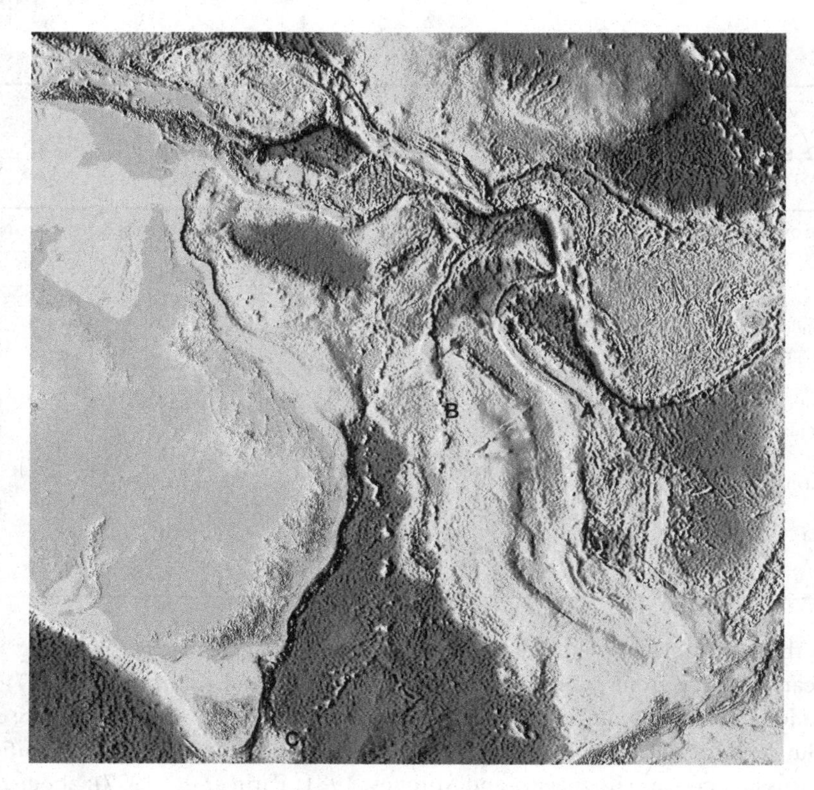

Fig. 7.1 Southwest Pacific map showing ridges, trenches and several seamount lineaments: A: Norfolk Ridge seamounts; B: Lord Howe Rise seamounts; C: Tasmanian seamounts. *Source*: From satellite altimetric mapping NOAA.

Geographic variation in species richness and composition

Several aspects of species richness and geographic variation in community composition are well illustrated by seamount studies in the southwest Pacific (Richer de Forges *et al.*, 2000) based on taxonomically comprehensive collections of megabenthos from three seamount groups: (1) Norfolk Ridge (6 seamounts, 295 samples, 516 species); (2) Lord Howe Ridge (4 seamounts, 35 samples, 108 species) and (3) south of Tasmania (14 seamounts, 34 samples, 297 species).

Estimates of benthic species richness on seamounts can be strongly influenced by collecting effort. Much of the variation in species richness between seamounts in the Coral Sea can be explained by differences in sampling effort (Fig. 7.2). Species richness of

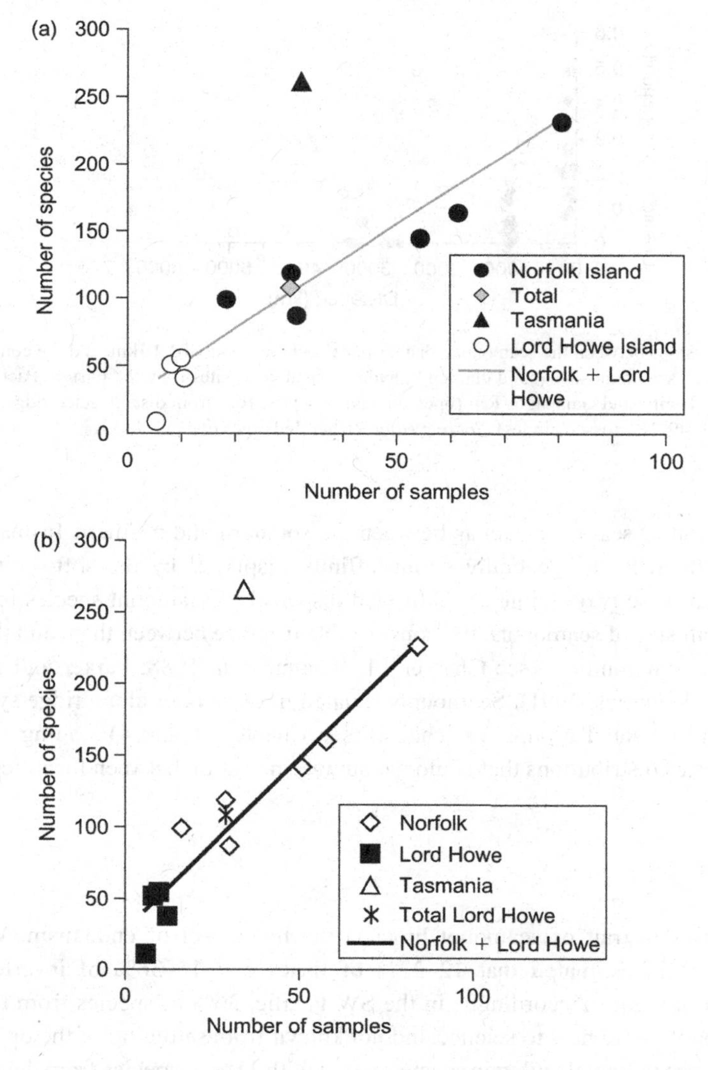

Fig. 7.2 Relationship between number of species and number of samples on the Tasmanian and Coral Sea seamounts. Tasmanian seamounts appear to be richer in species. *Source*: After Richer de Forges *et al.* (2000).

Tasmanian seamounts appears to be higher, but it is unclear whether this is due to more speciose assemblages, is an artefact of aggregating samples over several seamounts, or results from a positive bias from sampling over a greater depth range.

On SW Pacific seamounts, community composition differed significantly between two ridge systems at similar latitudes separated by about 1000 km (Fig. 7.3; Richer de Forges *et al.*, 2000). Over larger spatial scales (~3000 km) corresponding to a 20° difference in latitude, a completely different set of species was observed between deeper Tasmanian seamounts and the shallower seamounts near New Caledonia: indeed, just four benthic invertebrate species from Tasmanian seamounts are known from the seafloor around New Caledonia.

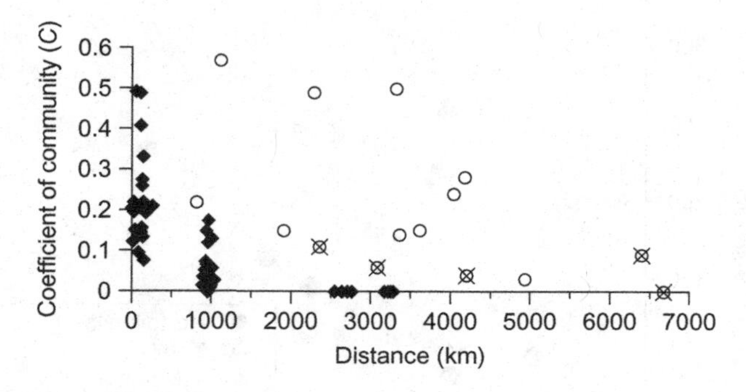

Fig. 7.3 Relationships between the seamounts, distant from each other only by 10 km, and the community coefficient (*C*) between seamount sites (solid diamond), hydrothermal vents sites from the Eastern Pacific Rise in the North and South Pacific and Galapagos Rift (open circles), and vent sites from disconnected ridges in the north-east Pacific at 41–49° N (crossed circles). *Source*: After Richer de Forges *et al.* (2000).

The isolation of seamount faunas between the southern and northern Tasman Sea contrasts markedly with the generally strong affinity displayed by the soft-sediment slope fauna between these two regions. This limited dispersal of seamount species is due to the generally small size of seamounts, the considerable distance between them and their unique oceanographic environment (see Chapter 14; Johannesson, 1988; Parker and Tunnicliffe, 1994; Richer de Forges, 2001). Seamounts situated in clusters or along ridge systems may also function as 'island groups' or 'chains' (see Chapters 1 and 4) leading to observed localized species distributions that could encourage speciation between localities.

Endemism

A frequently cited trait of seamount biota is the high level of endemism. Wilson and Kaufmann (1987) estimated that 12–22% of fishes and 15–36% of invertebrates are endemic to seamounts. Accordingly, in the SW Pacific, 36% of species from the Norfolk Ridge seamounts were new to science and not known from sampling of the open seafloor, and are therefore potential endemic species, along with 31% of species from the Lord Howe seamounts and 16–33% of species from Tasmanian seamounts (Richer de Forges *et al.*,

2000). Similarly, 17 new genera were obtained from the Norfolk Ridge samples, 4 from the Lord Howe Ridge and 7–8 from Tasmanian samples (Richer de Forges *et al.*, 2000). Some species appear to be relicts of groups earlier believed to have become extinct in the Mesozoic (Amèziane-Cominardi *et al.*, 1987; Laurin, 1992; Vacelet *et al.*, 1992).

These figures suggest that the level of endemism may be appreciable on seamounts. However, there are few comprehensive assessments of endemism rates available for other marine habitats, and it is thus difficult to judge. The spatial scale over which endemism is defined and reported is a further complication. Essentially, endemism is defined as 'limited geographic range size' of a species, but what this means depends on the spatial scale over which it is defined. At the smallest scale, it can represent a single site on a seamount, a single seamount, seamounts on a single ridge system and seamounts in sub-ocean basins or whole ocean basins. There are also no unambiguous biological criteria for an ecologically meaningful scale over which endemism is measured, save perhaps for populations showing clear genetic isolation. Hence endemism always needs to be qualified by the spatial scale to which it refers.

The rate at which new species were discovered during the exploration of New Caledonia seamounts was high and constant, so we may have overestimated endemism. In their study of the species richness of molluscs in shallow waters off New Caledonia, Bouchet *et al.* (2002) demonstrate that the true number of species was underestimated in the tropical Indo-Pacific. This conclusion could be extrapolated to other ocean habitats.

Among the organisms found on the seamounts of the Norfolk Ridge, squat lobsters (Galatheidae) are very diverse (e.g., Macpherson and Machordom, 2005). In New Caledonia, many species of galatheids have been described from seamounts, but none is endemic to them or to the local ridge of seamounts (Samadi *et al.*, 2006), and, indeed, all species are also found on nearby island slopes. However, species richness is lower along the continental slope than in any seamount along the Norfolk Ridge (see supplemental information at www.seamountsbook.info) suggesting that the Norfolk seamounts is a diversity hotspot for the family Galatheidae, but not an area of endemism. Other studies have also suggested that seamounts, like other prominent topographic features such as reef islands or shelf breaks, are biodiversity hotspots (see for example Worm *et al.* (2003) for vertebrate predators and Heinz *et al.* (2004) for foraminifera). However, our results suggest that while squat lobsters do not exhibit elevated rates of endemism, seamounts are highly productive zones where many species occur at high abundance. This finding is in accordance with the suggestion that high productivity is a prominent ecological feature of seamounts (see Chapters 5 and 14; Fock *et al.*, 2002; Genin, 2004).

Variability of species richness with depth

On the shallowest seamounts, stalked crinoids (*Metacrinus levii*) are found at a depth of 250–300 m, together with dense assemblages of lithistid sponges and stylasterid corals. On the upper parts of these seamounts species richness and biomass are high. Here, sessile suspension feeders dominate the communities and a species-rich assemblages of mobile invertebrates such as crustaceans, molluscs, echinoderms and sometimes brachiopods are present (Richer de Forges *et al.*, 2005).

On the northern part of the Norfolk Ridge, the seamounts are guyots (Fig. 7.4) with flat summits between 230 and 720 m deep. There are few marine geological studies on these features in the SW Pacific (see Van de Beuque, 1999; Veevers, 2000, 2001), but considering their shape and limestone caps, the summits of these seamounts were historically close to or above the sea surface. The bottom of the deeper (>720 m) seamounts is covered with polymetallic crust, with Fe–Mn layers ranging in thickness from a few millimetres to several centimetres, engulfing other dead objects on the seafloor such as fish otoliths and whale bones (see Chapter 1). The rate of deposition of these crusts is estimated to be 1–5 mm year on Hawaiian seamounts (Verlaan, 1992). Where the seafloor is totally covered by manganese crusts, few invertebrates are able to settle, except for sparse gorgonians, sponges and stalked crinoids (Grigg *et al.*, 1987).

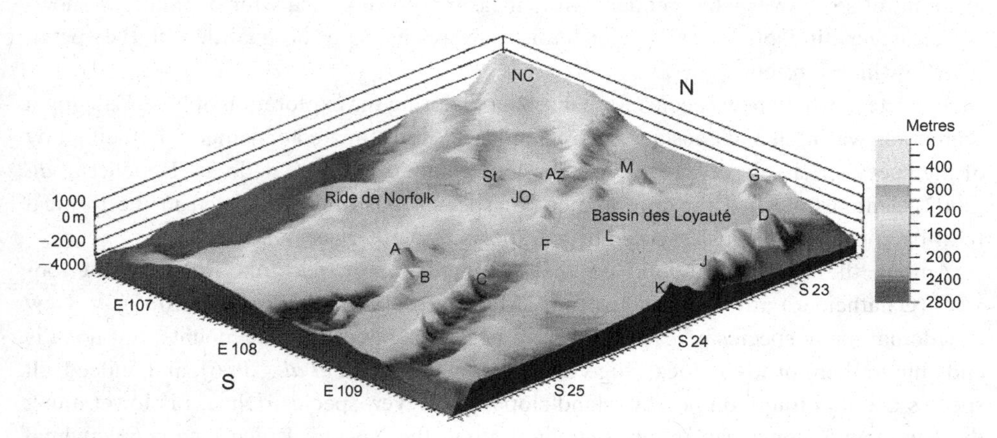

Fig. 7.4 3D view of the south of New Caledonia (NC) showing the relative positions of the main Norfolk Ridge seamounts. M: Munida seamount; Az: Antigonia seamount; St: Stylaster seamount; JO: Jumeau ouest seamount; Je: Jumeau est seamount; A: Kaimon maru seamount; B: Eponge seamount; C: Introuvable seamount.

Seamounts with a summit around 300–600 m host communities of sponges (lithistids and hexactinellids) and stylasterid corals. Species replacement rates in sponges increase with increasing depth separation of sites irrespective of their geographic distance and biological richness declines with depth across several taxonomic levels from species and genera to families; there were between 6 and 47 species of sponges per seamount. Depth appears to be the dominant factor in controlling community composition (Schlacher-Hoenlinger *et al.*, 2005).

'Living fossils': sponges, crinoids, brachiopods

Several archaic species or 'living fossils' have been discovered on seamounts (Lévi, 1991; Vacelet *et al.*, 1992; Kelly, 2000) especially in New Caledonia, where sampling of seamounts is much higher than anywhere else. A new species of Glypheid has recently been described from a seamount of the Coral Sea (Richer de Forges, 2006). Here we briefly review the principal groups for which archaic forms have been described from seamounts.

Sponges

Sponges are especially species-rich on shallow seamounts 200–700 m deep (Bouchet and Metivier, 1982; Lévi, 1991). Sixty-one families with 135 genera and 216 species have been recorded in New Caledonia of which 142 species were new (Table 7.1). This sponge fauna, practically unknown before 1977, was principally studied by Lévi (1991, 1993). Its key attributes include: (a) a high species diversity and a change in species at 700 m depth; (b) lithistids and tetractinellids dominate on the upper parts of seamounts; (c) two-thirds of species were new and thus potential endemics, while several genera were previously unknown from the Pacific Ocean; (d) the high diversity of lithistids is comparable only with the fossils from the Cretaceous in Europe. The sponge fauna has strong affinities with Mesozoic fauna, could be derived from the Tethysian mesogea (Lévi, 1991), and may therefore be considered a 'refuge habitat' for relic fauna such as lithistids that appear to be restricted to caves, continental margins and seamounts (Richer de Forges, 2001). Several relic genera from the Norfolk Ridge seamounts, *Neopelta*, *Aulaxinia*, *Neosiphonia* and *Reidispongia* are also known from the northern part of New Zealand (Kelly, 2000).

Crinoids

Numerous species of crinoids have been described from the bathyal zone of New Caledonia. Stalked crinoids comprise 9 families, 14 genera and 15 species, with 2 new genera and 8 new species. A large proportion of crinoids is considered to be 'living fossils', with affinities to the Jurassic and Cretaceous fauna from the mesogean Tethys Sea (Amèziane-Cominardi *et al.*, 1987, 1990; Bourseau *et al.*, 1991). The description of this crinoid fauna has totally modified the knowledge of this group for the Indo-Pacific (Bourseau *et al.*, 1991). Fourteen genera are represented (*Metacrinus*, *Saracrinus*, *Diplocrinus*, *Proisocrinus*, *Caledonicrinus*, *Porphyrocrinus*, *Naumachocrinus*, *Bathycrinus*, *Gymnocrinus*, *Holopus*, *Proeudesicrinus*, *Thalassocrinus*, *Hyocrinus*, *Guillecrinus*) and eight species were new. The majority of crinoid genera are true 'living fossils' since they appeared in the fossil record before the great cretaceous 'crisis' (Lawton and May, 1995). Spectacular examples of these old forms include two species (*Gymnocrinus richeri* and *Holopus alidis*) from the very archaic Hemicrinidae (Fig. 7.5), which have a short and strongly calcified stalk, an adaptation to the strong currents on seamounts (Hess *et al.*, 1999). Cohen *et al.* (2004) proposed a new molecular phylogeny of this group based mainly on material collected from seamounts.

Brachiopods

Laurin (1997) recorded 14 families of Brachiopods, including 19 genera and 26 species of which 4 were new, from the bathyal zone of New Caledonia, mainly from the seamounts: a new genus and a species, *Neoancistrocrania norfolki*, was similar to Cretaceous species (Cohen *et al.*, 1998; Laurin, 1992).

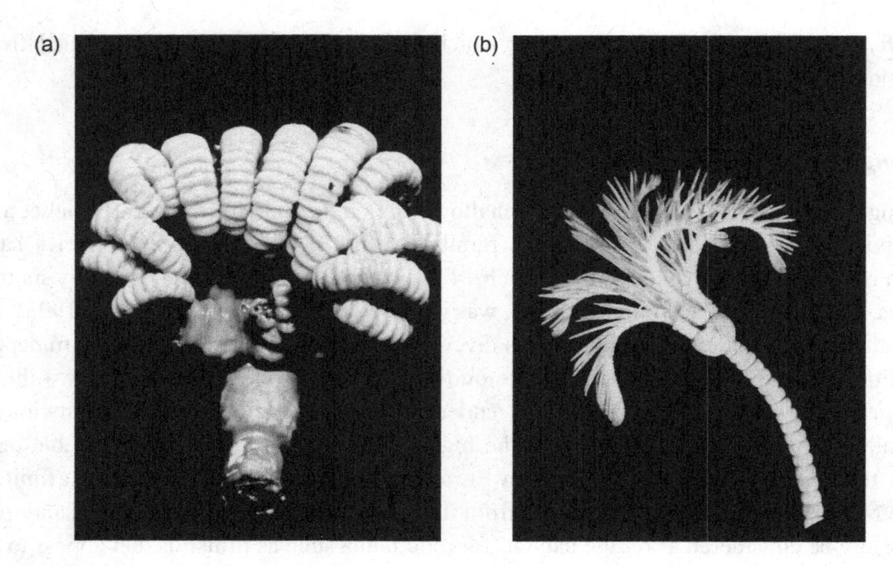

Fig. 7.5 Examples of relic fauna from seamounts: (a) *Gymnocrinus richeri*, from Stylaster seamount, 500 m and (b) *Caledonicrinus vaubani*, from Antigonia seamount, 300 m.

Long-lived animals: sponges, octocorals, crinoids

Deep-sea animals have adapted their growth in response to the environmental conditions of darkness, low temperatures and low food availability. The age of individuals can be determined from skeletal material. Here we discuss estimates of growth and age for some benthic invertebrates from seamounts.

Growth of a calcareous sponge (*Vaceletia* sp.) collected from seamounts was determined at 11 mm per century (Vacelet *et al.*, 1991). Most demosponges contain distinct internal silicious skeletal material, with some species producing extraordinary long glass spicules (Beaulieu, 2001). A specimen of the genus *Monoraphis* collected from the seamounts was estimated to be 440 years old based on an analysis of growth rings in its 3.4 m long spicule (Ellwood and Kelly, 2003). ^{14}C-dating of a 12 cm high stalked crinoid (*Gymnocrinus richeri*) indicates an age of 340 years (Richer de Forges *et al.*, 2004). Gorgonians from the family Isididae, can live over 300 years on New Caledonian seamounts and the age of larger colonies in New Zealand can be 500 years or more. The Isididae are characterized by their axis, which is composed of alternating nodes and internodes, the nodes consisting of horn and the internodes of massive crystalline calcareous substance (Noé and Dullo, 2006). Most species inhabit deep water. The axis of the gorgonians shows rings in cross-section (Fig. 7.6) and it is possible to use these sections to correlate the age and the variation of the rate Mg/Ca depending on water temperature. Surprisingly, in several samples from the Norfolk Ridge collected from depths of 500 m, a drop of temperature of about 2°C over 200 years was observed (Richer de Forges, 2001). In New Zealand, giant bubblegum gorgonians trees (*Paragorgia arborea*) were trawled on seamounts at the beginning of the orange roughy fisheries (see Chapter 19), and their age is estimated at 300–500 years (Tracey *et al.*, 2003). These slow-growing organisms are very long lived, perhaps rivalling some of the well-known forest stands of ancient trees. More recently, preliminary studies using metal isotopes tracers have used the skeletons of deep-sea species as palaeo-environmental indicators (Ellwood and Kelly, 2003).

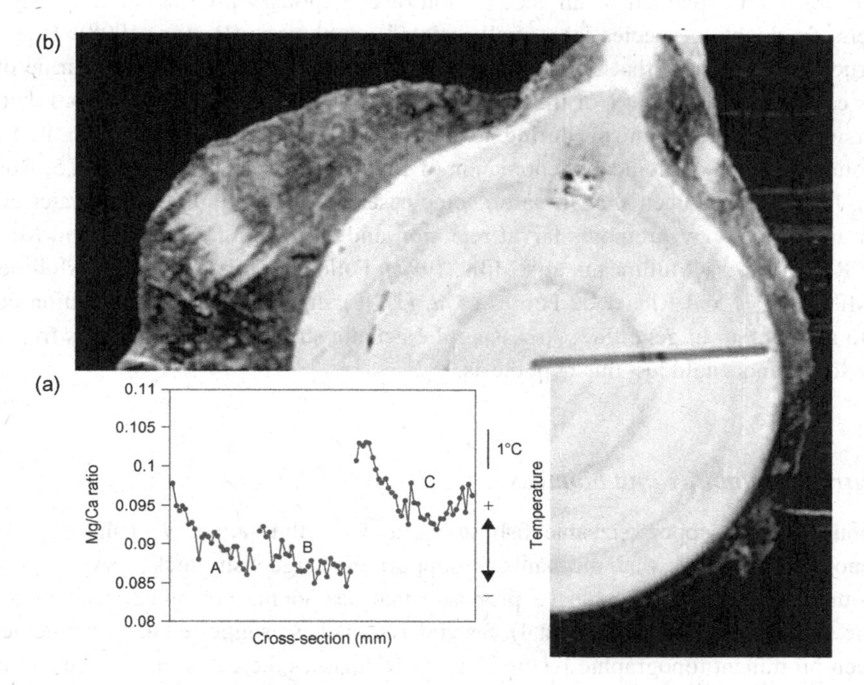

Fig. 7.6 Temperature measurement of three isidid gorgonians samples from Norfolk Ridge seamounts: (a) graphic with, on the left Mg/Ca (molar), on the right water temperature with the scale of 1°C on the graph and (b) cross-section of Isidid gorgonian showing the groth rings and the radial samples series. *Source*: After Richer de Forges (2001), modified by T. Corrège.

Environmental drivers of benthic diversity on seamounts

Data suggest that seamounts can be hotspots of biodiversity, but several of the reported key traits such as high endemism, archaism, slow growth rates and longevity may be an artefact of the greater sampling effort devoted to seamounts compared to other deep-sea habitats. However, the environmental factors that characterize seamounts may influence evolutionary and ecological processes leading to observed patterns of benthic diversity. Here we review these environmental factors and discuss how hypotheses about the origins seamount biodiversity may be tested.

Island effects

Seamounts differ from most other marine habitats by being spatially distinct and, arguably, isolated, topographic features. Most marine environments are relatively homogenous and contiguous over large geographic scales, but seamounts generally occupy smaller and clearly defined areas. This geographic isolation evokes an analogy with terrestrial islands, and so seamounts have been viewed as isolated habitats with populations essentially stranded on them. The endemism and species richness in the terrestrial biota of oceanic islands are often explained by an acceleration of evolutionary processes due to physical barriers, fragmenting species into small isolated populations (Barton, 1998). However, the crucial difference is that seamounts are surrounded by water and most marine organisms can swim at one stage of their life cycle. It has, however, been suggested that the dispersion of benthic organisms during the pelagic stages of their life cycle is limited over seamounts by the hydrological phenomenon of Taylor caps (see Chapters 4 and 5; Roden, 1987). Indeed, it has been shown that in some cases the interaction between water circulation and topography promotes larval retention and aggregation (Boehlert and Mundy, 1993; Rogers, 1994; Mullineaux and Mills, 1997). Following this hypothesis, Mullineaux and Mills (1997) and Richer de Forges *et al.* (2000) suggested that this isolation could lead to a reduction in gene flow. Assessing the genetic structure of populations from different seamounts could test this hypothesis.

Resource availability and biomass

Seamounts often support sizeable fish stocks and are thus attractive fishing grounds. It is not always clear what mechanisms support the large fish stocks associated with seamounts, but some authors have proposed that seamounts are habitats of enhanced trophic subsidy (see Chapters 3 and 4). Several authors have suggested that the interaction between prominent topographic features and water masses increases turbulence and mixing, and enhances local biomass production by moving nutrients up into the euphotic zone (Fock *et al.*, 2002; Worm *et al.*, 2003; Genin, 2004). Enhanced biomass of filter feeders and other consumers should mirror any local concentration of seamount productivity. Indeed, our observations of benthic catches from the New Caledonian seamounts and nearby island slope indicate that the biomass of megabenthos, especially that of filter feeders, appears to be greater on seamounts. These observations of enhanced benthic biomass on seamounts have yet to be put to a quantitative test.

Consequences for life-history traits

Dispersal. If seamounts are isolated by physical barriers preventing larval dispersal, then gene flow should be reduced among populations, and swimming larvae should not confer an adaptive advantage. Thus, if the island hypothesis is correct, populations on seamounts should be genetically isolated and the proportion of species lacking larval dispersal should be greater than in other comparable environments.

Longevity. Some seamount species have particularly great longevity, and it is possible that environmental conditions and food on seamounts are relatively stable over long periods to permit such longevity. It has also been suggested that the longevity of benthic organisms could be an adaptive solution to high variability of reproductive success (Flowers *et al.*, 2002). Neither of these hypotheses are particularly easy to test, as alternative explanations easily suggest themselves.

Gene flow and larval strategies

Genetic studies, of which there are few, permit evaluation of the isolation of seamount populations. Most studies concern fishes (Creasey and Rogers, 1999), many of which indicate isolation of populations at a local scale. For example, Aboim *et al.* (2005) demonstrated that populations of the benthopelagic fish *Helicolenus dactylopterus* were not isolated from one seamount to another, nor from the closest continental slope; genetic isolation among populations was found only at larger oceanic scales. The few genetic studies of benthic organisms from seamounts suggest the same pattern (e.g., Smith *et al.*, 2004). These results are comparable to the genetics of other isolated deep-sea habitats such as hydrothermal vents or cold seeps. Among vent organisms, bivalves from the genus *Bathymodiolus* disperse between very distant sites (Won *et al.*, 2003), and even between hydrothermal and cold-seep sites (Miyazaki *et al.*, 2004). Thus although these environments are markedly fragmented, species associated with these environments can be highly dispersive.

On seamounts, it seems that population fragmentation and restricted gene flow occur only for species with limited larval dispersal. Samadi *et al.* (2006) used benthic organisms from seamounts of the Norfolk Ridge to test whether physical and population fragmentation has encouraged high rates of speciation. If there is no larval retention, we would expect only species with poorly dispersive larvae to exhibit high genetic diversity. This was confirmed by two gastropod species that have contrasting larval dispersal strategies, inferred from the examination of the protoconch, and which allowed us to differentiate between the effects of their dispersal abilities and those of physical fragmentation resulting from hydrological phenomena. The non-planktotrophic species (*Nassaria problematica*) was highly structured, whereas the planktotrophic species *Sassia remensa*, was not. Similarly, five squat lobster species with dispersive larvae were genetically similar among populations on seamounts and the adjacent island slope. These results parallel deep-sea bamboo corals (Smith *et al.*, 2004) among which taxonomists had traditionally suggested a high rate of endemism on seamounts, and low gene flow among distant populations. A genetic survey confirmed that specific diversity is high, but showed that bamboo coral species are not endemic to seamounts and that distant populations are genetically interconnected.

Other studies on Atlantic seamounts revealed the same trends. In the North Atlantic, Gofas (2000) examining Fasciolariidae species, and Dijkstra and Gofas (2004) studying Pectinoidea species, found no seamount-to-seamount endemism, even between seamounts separated as much as 100 km. These results obtained for both planktotrophic and non-planktotrophic species, suggesting that seamounts are not highly isolated patches of habitat. Moreover, when comparing mollusc faunas from North Atlantic and Azores seamounts to the European mainland, planktotrophic development appears overrepresented (Gofas and Beu, 2002).

Overall, the few available studies suggest that, contrary to the island hypothesis, the hydrological phenomena associated with seamounts are not strong physical barriers and that populations of many organisms populations living on seamounts are genetically interconnected. However, this may not apply to all benthic seamount fauna, and should be regarded with caution because there are few data about the larval strategies of many deep-sea species. At Cobb Seamount (Parker and Tunnicliffe, 1994), a large portion of the 117 benthic species do not produce planktonic larva and only 6.8% have a long pelagic larval dispersal, observations supporting the island hypothesis.

Trophic architecture of seamount invertebrate benthos

Seamounts are commonly regarded as habitats where consumers proliferate in the otherwise food-poor environment of the deep ocean (Richer de Forges *et al.*, 2000). Amplified currents over seamounts enhance the growth of resident animals by augmenting the flux of suspended food (Genin, 2004). Hydrological processes that trap material over and around mounts can also locally enhance food resources (see Chapter 5; Rogers, 1994). One consequence of the increased supply of particles to these topographically abrupt habitats is that suspension feeders, such as corals and sponges, dominate the seamount megabenthos (Genin *et al.*, 1986; Wilson and Kaufmann, 1987).

This dominance of suspension feeders on seamounts (Genin *et al.*, 1986) suggests that benthic food webs of seamount may be simple. Because filter feeding is the chief trophic mode, most species would feed at low trophic levels and consume similar resources. The trophic architecture of the consumer guild is thus predicted to have low complexity, resulting in short food chains. However, recent trophic studies on New Caledonian seamounts have tested the hypothesis that food-chain length in seamount benthos is shorter than in other aquatic systems.

Food-chain length was determined with stable nitrogen isotope ratios ($\delta^{15}N$) measured in the tissues of a wide range of benthic invertebrate groups and some small fishes obtained as bycatch collected from several seamounts on the Norfolk Ridge. $\delta^{15}N$ is routinely used to determine the trophic position of consumers in food webs (Post, 2002) because $\delta^{15}N$ values are shifted by 3–4‰ towards more positive values during each trophic transfer (Minagawa and Wada, 1984). Thus, isotopic enrichment from prey to predator can be used to map the trophic level of consumers (Post *et al.*, 2000). Results show that food chains in seamount benthos are not short (Fig. 7.7), and the food web was not compressed to filter-feeding organisms that all feed at a similar trophic position (Fig. 7.7). In fact, the benthic food web on seamounts has a diverse trophic architecture with food-chain length broadly comparable to other aquatic systems, both shallow and deep (Table 7.4). Compared with other aquatic food webs, food-chain lengths in seamount benthos lie towards the upper end of reported values (Table 7.4). For non-seamount deep-sea food webs, measurements of maximum trophic position (MTP) range from 3.5 to 3.8 for the benthos of the Porcupine Abyssal Plain (Iken *et al.*, 2001), 4.6 in the Arctic Canada Basin (Iken *et al.*, 2005), to 4.5 in Astoria Canyon (Bosley *et al.*, 2004). Thus, food chains in the seamount invertebrate benthos – hypothesized to be shorter due to the dominance of suspension feeders – are in fact broadly similar in length to other deep-sea food webs. There

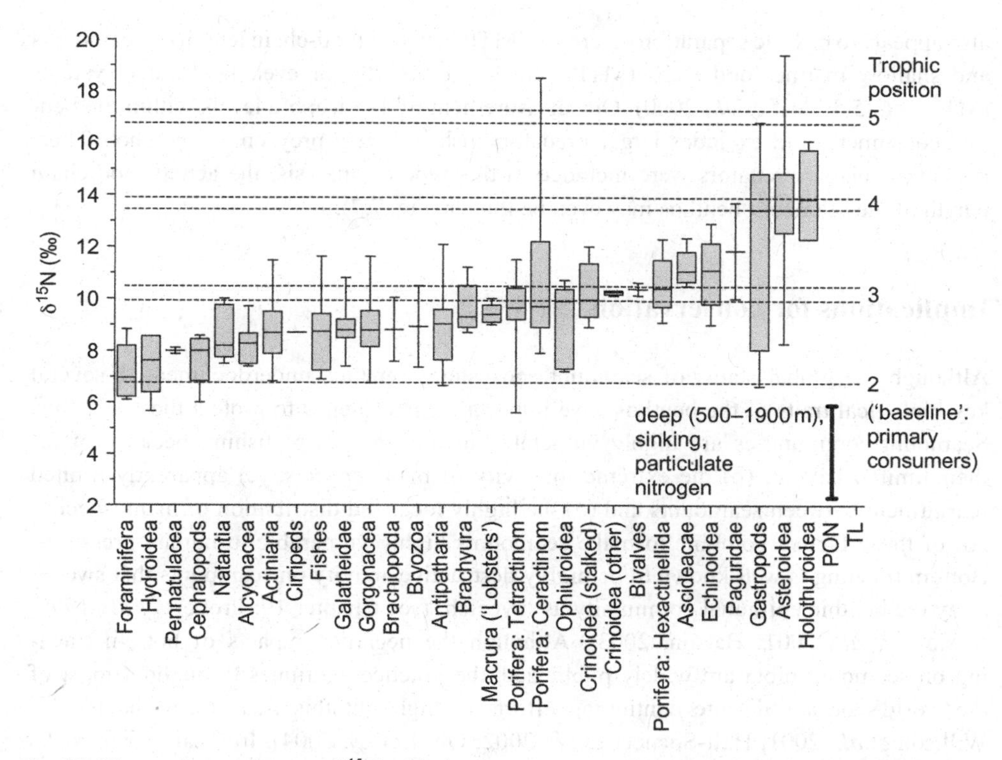

Fig. 7.7 Nitrogen isotope signature (δ^{15}N) of the benthos on seamounts of the Norfolk Ridge. Isotope data from the seamounts are depicted as box and whisker plots (boxes: median plus 1st and 3rd quartile, whiskers: minimum and maximum values). Values for particulate organic nitrogen (PON) in sediment traps are from the literature (Altabet *et al.*, 1991, 1999; Altabet, 2001; Altabet and Francois, 2001; Lourey *et al.*, 2003). Calculation of trophic position is according to Post (2002): $\lambda + (\delta^{15}N_{organism} - \delta^{15}N_{base\ of\ food\ web})/\Delta_n$, where λ is the trophic position of the organism used to estimate the 'baseline' of the food web (e.g., $\Delta = 2$ for primary consumers), and Δ_n is the enrichment per trophic transfer. We used a mean trophic fraction factor of 3.4 (Post, 2002), ±95% confidence limits of 0.26, depicted as dotted lines. We took primary consumers as the 'baseline' and this was calculated as the median of the first decile of all δ^{15}N measurements in the data set (it corresponds very well with an enrichment of 3–4 ppt from deep, particulate nitrogen to primary consumers).

Table 7.4 Comparison of food-chain length (measured as MTP sensu Post (2002) based on stable nitrogen isotopes) in aquatic food webs.

Source	Locality	Depth	MTP
Romanuk and Levings (2005)	Beaches, British Columbia	1 m	~4.2–4.5
Schlacher (unpublished)	Subtropical estuaries (Australia)	<10 m	3.8–4.0
Post *et al.* (2000)	Freshwater lakes ($n = 25$), N-America	<50 m	3.6–5.5
Jennings and Warr (2003)	North- and Irish Seas, English Channel	20–105 m	4.2–4.5
Le Loc'h and Hily (2005)	Bay of Biscay	95–120 m	3.6–4
Davenport and Bax (2002)	Continental Shelf, SE-Australia	20–250 m	3–3.8
Bosley *et al.* (2004)	Astoria Canyon, Oregon (USA)	89–549 m	3.9–4.5
This study	Seamounts Norfolk Ridge	200–900 m	4–5
Polunin *et al.* (2001)	Balearic Island Slope	200–1800 m	4.4
Iken *et al.* (2005)	Arctic, Canada Basin	625–3398 m	4.6
Iken *et al.* (2001)	Porcupine Abyssal Plain	4840 m	3.5–3.8

also appears to be little separation in terms of MTP between food-chain length on seamounts and shallow marine food webs (MTP: 3.8–4.5, Table 7.4) or even freshwater systems (MTP: 3.6–5.5; Post *et al.*, 2000). Our measurement of 4–5 trophic levels within the benthic consumer guild excludes larger predatory fish that may prey on the benthos. Thus, if bentho-pelagic predators were included in this type of analysis, the actual food-chain length of the seamount benthos may even be longer (Fig. 7.7).

Implications for conservation

Although the global status of seamount ecosystems remains underdocumented, several key biological traits of the benthos have important consequence to protect these systems. Seamount communities are highly vulnerable to the impacts of fishing because of (a) their limited habitat; (b) the extreme longevity of many species; (c) apparently limited recruitment between seamounts and (d) the highly localized distribution of many species. All of these factors combine to make seamounts highly vulnerable to human pressures. Bottom trawling is well known to be highly destructive activity on seamounts that sweeps away the benthic epifaunal community as 'bycatch' (see Chapter 19; Probert *et al.*, 1997; Koslow *et al.*, 2001; Batson, 2003). Although the negative impacts of bottom trawling on seamount biota are widely publicized, the practice continues throughout most of the world's oceans despite continuing efforts to highlight this issue (see Chapter 20; Willison *et al.*, 2001; Hall-Spencer *et al.*, 2002; Olu-LeRoy, 2004). Ironically, fish stocks may in part be dependent on the ecological condition of the seamount habitats, and rates of recovery are unknown for the seamount benthos following trawling impacts. Thus, temporary fishing closures may not be efficient in achieving long-term protection of seamounts, which rather requires the establishment and enforcement of permanent protection zones.

If the level of endemism on seamounts is as high as frequently reported or predicted, then the need for conservation of these habitats becomes critical. The highly localized distribution of many benthic seamount species increases the threat of possible extinctions, and as such may require that conservation measures of seamounts are designed and implemented on a local scale. Seamounts are also frequently referred to as highly productive and heterogeneous habitats, similar to rain forests or coral reefs. As such, they are predicted to accommodate large populations of many species and become target areas for protecting highly biodiverse areas (*sensu* Roberts, 2002). In theory, planning and implementation of conservation strategies for seamounts appear more readily achievable, at least from a spatial perspective, due to their geographically clearly defined nature.

Conclusions

Seamounts are unique habitats characterized by abrupt topographies, limited spatial extent, swift currents and hydrodynamic processes that can produce localized upwelling and circulation cells. These environmental conditions are mirrored in a benthic biomass dominated by large suspension feeders that provide structural habitat for a great diversity of the smaller, mobile fauna. These diverse benthic assemblages contain species of limited geographic distribution (endemics) and species with ancient lineages or previously believed to have become extinct ('living fossils'). As a prominent seascape element that frequently rises well above the ocean floor, seamounts provide relatively shallow habitats for bathyal organisms amidst their abyssal surroundings. Invertebrates on seamounts tend to be slow-growing and long-lived, a life-history trait that makes them particularly vulnerable to human impacts such as bottom trawling. Although suspension feeders dominate the megabenthos, food webs are not simple and can rival other shallow and deep ecosystems in food-chain length and trophic architecture. While seamounts have frequently been likened to oceanic islands, their populations may not necessarily be genetically isolated if they produce larva capable of long-distance dispersal. Seamounts can also serve as 'natural laboratories' to test ecological theories like island biogeography in the marine realm.

Given the unique characteristics of seamounts and the diverse assemblages they support, it is surprising that the fauna remains poorly documented on a global scale. Whole assemblages are known only from a few seamounts worldwide, partly as a result of dwindling taxonomic resources. This lack of basic ecological knowledge currently constraints the development of more comprehensive models about the structure and function of the benthos, which would be conceptually applicable worldwide.

References

Aboim, M.A., Menezes, G.M., Schlitt, T. and Rogers, A.D. (2005) Genetic structure and history of populations of the deep-sea fish *Helicolenus dactylopterus* (Delaroche, 1809) inferred from mtDNA sequence analysis. *Molecular Ecology*, 14, 1343–54.

Altabet, M.A. (2001) Nitrogen isotopic evidence for micronutrient control of fractional NO_3– utilization in the equatorial Pacific. *Limnology and Oceanography*, 46, 368–80.

Altabet, M.A. and Francois, R. (2001) Nitrogen isotope biogeochemistry of the antarctic polar frontal zone at 170° W. *Deep-Sea Research, Part II: Tropical Studies Oceanography*, 48, 4247–73.

Altabet, M.A., Deuser, W.E., Honjo, S. and Stienen, C. (1991) Seasonal and depth-related changes in the source of sinking particles in the North Atlantic. *Nature*, 354, 136–9.

Altabet, M.A., Pilskaln, C., Thunell, R., Pride, C., Sigman, D., Chavez, F. and Francois, R. (1999) The nitrogen isotope biogeochemistry of sinking particles from the margin of the Eastern North Pacific. *Deep-Sea Research*, Part I: *Oceanographic Research Paper*, 46, 655–79.

Amèziane-Cominardi, N., Bourseau, J.P. and Roux, M. (1987) Les crinoides pédonculés de Nouvelle-Caledonie (S.W. Pacifique): une faune ancestrale issue de la Mesogée mésozoique. *Comptes rendus hebdomadaires de l'Académie des Sciences Paris*, 304(1), ser. 3, 15–18.

Amèziane-Cominardi, N. (1991) Distribution bathymetrique des pentacrines du Pacifique occidental. Essai de modelisation et d'application aux faunes du Lias. *Documents Laboratories de Geologie, Lyon*, 116, 253 pp, 5pl.

Baba, K. (2005) Deep sea chirostylid and galatheid crustaceans (Decapoda: Anomura) from the Indo-Pacific, with a list of species. *Galathea Report*, 20, 1–317.

Barton, N.H. (1998) Natural selection and random genetic drift as causes of evolution on islands. In: *Evolution on Islands* (ed. Grant, P.R.), pp. 102–23. Oxford University Press, Oxford.

Batson, P. (2003) *Deep New Zealand. Blue Water, Black Abyss*, 240 p. Canterbury University press, Christchurch.

Beaulieu, S.E. (2001) Life on glass houses: sponge stalk communities in the deep sea. *Marine Biology*, 138, 803–17.

Boehlert, G.W. and Mundy, B.C. (1993) Ichthyoplankton assemblages at seamounts and oceanic islands. *Bulletin of Marine Sciences*, 53, 336–61.

Bosley, K.L., Lavelle, J.W., Brodeur, R.D., Wakefield, W.W., Emmett, R.L., Baker, E.T. and Rehmke, K.M. (2004) Biological and physical processes in and around Astoria submarine Canyon, Oregort, USA. *Journal of Marine Systems*, 50, 21–37.

Bouchet, P. and Metivier, B. (1982) Living Pleurotomariidae (Mollusca: Gastropoda) from the South Pacific. *New Zealand Journal of Zoology*, 9, 309–18.

Bouchet, P., Lozouet, P., Maestrati, P. and Heros, V. (2002) Assessing the magnitude of species richness in tropical marine environments: exceptionally high numbers of molluscs at a New Caledonia site. *Biological Journal of the Linnean Society*, 75(4), 421–36.

Bourseau, J.P., Amèziane-Cominardi, N, Avocat, R., and Roux, M., (1991) Echinodermata: Les Crinoides pèdonculès de Nouvelle-Calèdonie. Rèsultats des Campagnes MUSORSTOM 8: 229–333.

Cohen, B.L., Gawthrop, A. and Cavalier-Smith, T. (1998) Molecular phylogeny of brachiopods and phoronids based on nuclear-encoded small subunit ribosomal RNA gene sequences. *Philosophical Transaction: Biological Sciences*, 353, 2039–69.

Cohen, B.L., Améziane, N., Eleaume, M. and Richer de Forges, B. (2004) Crinoid phylogeny: a preliminary analysis (Echinodermata: crinoidea). *Marine Biology*, 144, 605–17.

Creasey, S. and Rogers, A.D. (1999) Population genetics of bathyal and abyssal organisms. *Advances in Marine Biology*, 35, 3–151.

Davenport, S.R. and Bax, N.J. (2002) A trophic study of a marine ecosystem off southeastern Australia using stable isotopes of carbon and nitrogen. *Canadian Journal of Fisheries Aquatic Sciences*, 59, 514–30.

Dijkstra, H., and Gofas, S., (2004). Penctinoidea Bivalvia: Propeamussiidae and Pectinidae from some northeastern Atlantic seamounts. *Sarsia: North Atlantic Marine Science,* 89: 33–78.

Ellwood, M. and Kelly, M. (2003) Sponge 'tree rings'. Marine biodiversity/Palaeoecology. *Water and Atmosphere*, 11(2), 25–7.

Flowers, J.M., Schroeter, S.C. and Burton, R.S. (2002) The recruitment sweepstakes has many winners: genetic evidence from the sea urchin *Strongylocentrotus purpuratus*. *Evolution*, 56, 1445–53.

Fock, H., Uiblein, F., Koester, F. and von Westernhagen, H. (2002) Biodiversity and species–environment relationships of the demersal fish assemblage at the Great Meteor Seamount (sub-tropical NE Atlantic), sampled by different trawls. *Marine Biology*, 141, 185–99.

Genin, A. (2004) Bio-physical coupling in the formation of zooplankton and fish aggregations over abrupt topographies. *Journal of Marine Systems*, 50, 3–20.

Genin, A., Dayton, P.K., Lonsdale, P.F. and Spiess, F.N. (1986) Corals on seamounts provide evidence of current acceleration over deep sea topography. *Nature*, 322, 59–61.

Gofas, S. (2000) Four species of the family Fasciolariidae (Gastropoda) from the North Atlantic seamounts. *Journal of Conchology*, 37, 7–16.

Gofas, S. and Beu, A. (2002) Tonnoidean gastropods of the North Atlantic seamounts and the Azores. *American Malacology Bulletin*, 17, 91–108.

Grigg, R.W., Malahoff, A., Chave, E.H. and Landahl, J. (1987) Seamount benthic ecology and potential environmental impact from manganese crust mining in Hawaii. In: *Seamounts, Islands and Atolls* (eds. Keating, B.H., Fryer, P., Batiza, R. and Boehlert, G.W.). *Geophysical Monographs*, 43, 379–90.

Hall-Spencer, J., Allain, V. and Fossa, J.H. (2002) Trawling damage to Northeast Atlantic ancient coral reefs. *Proceedings of the Royal Society of London B*, 269, 507–11.

Heinz, P., Ruepp, D. and Hemleben, C. (2004) Benthic foraminifera assemblages at Great Meteor Seamount. *Marine Biology*, 144, 985–98.

Hess, H., Ausich, W.I, Brett, C.E. and Simms, M.J. (1999) *Fossil Crinoids*, 275 p. Cambridge university press, Cambridge.

Iken, K., Bluhm, B.A. and Gradinger, R. (2005) Food web structure in the high Arctic Canada Basin: evidence from delta C-13 and delta N-15 analysis. *Polar Biology*, 28, 238–49.

Iken, K., Brey, T., Wand, U., Voigt, J. and Junghans, P. (2001) Food web structure of the benthic community at the Porcupine Abyssal Plain (NE Atlantic): a stable isotope analysis. *Progress in Oceanography*, 50, 383–405.

Jennings, S. and Warr, K.J. (2003) Environmental correlates of large-scale spatial variation in the delta N-15 of marine animals. *Marine Biology*, 142, 1131–40.

Johannesson, K. (1988) The paradox of Rockall: why is a brooding gastropod (*Littorina saxatilis*) more widespread than one having a planktonic larval dispersal stage (*L. littorea*)? *Marine Biology*, 99, 507–13.

Kelly, M. (2000) Description of a new lithistid sponge from northeastern New-Zealand, and consideration of the phylogenetic affinities of families Corallistidae and Neopeltidae. *Zoosystema*, 22(2), 2–18.

Koslow, J.A.and Gowlett-Holmes, K. (1998) The seamount fauna off southern Tasmania: benthic communities, their conservation and impacts of trawling. *Report to the Environmental Australia Fisheries Commission* 95/058.

Koslow, J.A, Gowlett-Holmes, K., Lowry, J., O'Hara, T., Poore, G. and Williams, A. (2001) The seamount benthic macrofauna off southern Tasmania: community structure and impacts of trawling. *Marine Ecology Progress Series*, 213, 111–25.

Kuznetsov, A.P. and Mironov A.N. (eds.) (1981) *Benthos of the Submarine Mountains Marcus-Necker and Adjacent Pacific Regions*. Academy of Sciences of the USSR P.P. Shirshov Institute of Oceanology, Moscow.

Laurin, B. (1992) Decouverte d'un squelette de soutien du lophophore de type 'crura' chez un brachiopode inarticulé: description de *Neoancistrocrania norfolki* gen. sp. nov. (Cranidae). *Comptes rendus hebdomadaires de l'Académie des Sciences, Paris*, 314, ser. 3, 343–50.

Laurin, B. (1997) Brachiopoda: brachiopods récoltés dans les eaux de la Nouvelle-Calédonie et des îles Loyauté, Matthew et Chesterfield. In: Résultats des campagnes MUSORSTOM (ed. Crosnier, A.), Vol. 18. *Mémoires du Muséum national d'Histoire naturelle*, 176, 413–73.

Lawton, J.H. and May, R.M.C. (1995) *Extinction Rates*. Oxford University Press, Oxford.

Le Loc'h, F. and Hily, C. (2005) Stable carbon and nitrogen isotope analysis of *Nephrops norvegicus Merluccius merluccius* fishing grounds in the Bay of Biscay (Northeast Atlantic). *Canadian Journal of Fisheries and Aquatic Sciences*, 62, 123–32.

Lévi, C. (1991) Lithistid sponges from the Norfolk Rise. Recent and Mesozoic Genera. In: *Fossil and Recent Sponges* (eds. Reitner, J. and Keupp, H.), pp. 72–82. Springer-Verlag, Berlin/ Heidelberg.

Lourey, M.J, Trull, T.W. and Sigman, D.M. (2003) Sensitivity of delta N-15 of nitrate, surface suspended and deep sinking particulate nitrogen to seasonal nitrate depletion in the Southern Ocean – Art. No. 1081. *Global Biogeochemical Cycles*, 17, 1081.

Macpherson, E. and Machordom, A. (2005) Description of three sibling new species of the genus *Munida* Leach, 1820 (Decapoda, Galatheidae) from New Caledonia using morphological and molecular data. *Journal of Natural History*, 39, 819–34.

Minagawa, M. and Wada, E. (1984) Stepwise enrichment of ^{15}N along food chains: further evidence and the relation between δ ^{15}N and animal age. *Geochemistica and Cosmochemistica Acta*, 48, 1135–40.

Miyazaki, J.I, Shintaku, M., Kyuno, A., Fujiwara, Y., Hashinoto, J. and Iwasaki, H. (2004) Phylogenetic relationships of deep-sea mussels of the genus *Bathymodiolus* (Bivalvia: Mytilidae). *Marine Biology*, 144, 527–35.

Mullineaux, L.S. and Mills, S.W.A. (1997) A test of the larval retention hypothesis in seamount-generated flows. *Deep-Sea Research Part I: Oceanographic Research Papers*, 44, 745–70.

Noé, S.U. and Dullo, W.C. (2006) Skeletal morphogenesis and growth mode of modern and fossil deep-water isidid gorgonians (Octocorallia) in the West Pacific (New Zealand and Sea of Okhotsk). *Coral Reefs*, DOI 10.1007/s00338-006-0095-8.

Olu-Le Roy P.K. (2004). Deep water corals: biodiversity to be evaluated and preserved. Vertigo 5: 1–10.

Parin N.V., Mironov, A.N. and Nesis, K.N. (1997) Biology of the Nazca and sala y Gomez submarine ridges, an outpost of the Indo-West Pacific fauna in the eastern Pacific Ocean: composition and distribution of the fauna, its communities and history. *Advances in Marine Biology*, 32: 145–242.

Parker, T. and Tunnicliffe, V. (1994) Dispersal strategies of the biota on an oceanic seamount: implications for ecology and biogeography. *Biological Bulletin*, 187, 336–45.

Polunin, N.V.C., Morales-Nin, B., Pawsey, W.E., Cartes, J.E., Pinnegar, J.K. and Moranta, J. (2001) Feeding relationships in Mediterranean bathyal assemblages elucidated by stable nitrogen and carbon isotope data. *Marine Ecology Progress Series*, 220, 13–23.

Post, D.M. (2002) Using stable isotopes to estimate trophic position: models, methods, and assumptions. *Ecology*, 83, 703–18.

Post, D.M., Pace, M.L. and Hairston, N.G. (2000) Ecosystem size determines food-chain length in lakes. *Nature*, 405, 1047–9.

Probert, P.K., McKnight, D.G. and Grove, S.L. (1997) Case studies and reviews. Benthic invertebrate bycatch from a deep-water trawl fishery, Chatham Rise, New Zealand. *Aquatic Conservation: Marine and Freshwater Ecosystems*, 7, 27–40.

Richer de Forges, B. (2001) Les faunes bathyales de l'Ouest Pacifique: diversité et endémisme. Mémoire d'Habilitation à diriger des recherches. Université Pierre et Marie Curie, Paris, Vol. 1, 83 pp.

Richer de Forges B. (2006) Dècouverte en mer de Corail d'une deuxième espèce de glyphèide (Crustacea, Decapoda, Glypheoidea). *Zoosystema*, 28(1), 17–29.

Richer de Forges, B., Koslow, J.A. and Poore, G.C. (2000) Diversity and endemism of the benthic seamount fauna in the southwest Pacific. *Nature*, 405, 944–7.

Richer De Forges, B., Grandperrin, R. and Laboute, P. (1987) La Campagne CHALCAL II sure les guyots de la ride de Norfolk (N.O. CORIOLIS, 26 Octobre-Ler Novembre 1986). Rapports Scientifiques et techniques, sciences de la mer, Rep. 42, Institut Francais de recherche scientifiques pour le devlopmment en cooperation.

Richer de Forges, B., Corrège, T. and Paterne, M. (2004) Estimation de la longévité chez les organismes de profondeur. *Assises de la Recherche française dans le Pacifique, 23–27 août 2004*, Nouméa (poster).

Richer de Forges, B., Hoffschir, C., Chauvin, C. and Berthault, C. (2005) Inventaire des espèces de profondeur de Nouvelle-Calédonie/Census of deep-sea fauna in New Caledonia. *Rapport Scientifique et Technique* II6, volume spécial. IRD Nouméa: 113 p.

Richer de Forges, M. (2001) Endémisme du benthos des monts sous-marins de la ride de Norfolk (Pacifique sud-ouest): eponges et mollusques gastéropodes. *Rapport de Maîtrise de Biologie des populations et écologie, Université Pierre et Marie Curie*, 39 p.

Richer de Forges, S. (2001) Analyse des paléotempératures mesurées sur des squelettes de gorgones Isididae de profondeur. *Rapport de stage de DEUG: IRD/Nouméa* (Maître de stage Thierry Corrège), 11 p.

Roberts, C.M. (2002) Deep impact: the rising toll of fishing in the deep sea. *TREE*, 17, 242–5.

Roden, G.I. (1987) Effects of seamounts and seamount chains on oceanic circulation and thermocline structure. In: *Seamounts, Islands and Atolls* (eds. Keating, B.H. *et al.*), pp. 335–54. *Geophysical Monographs*, ser 43. AGU, Washington, DC.

Rogers, A.D. (1994) The biology of seamounts. *Advances in Marine Biology*, 30, 305–50.

Romanuk, T.N. and Levings, C.D. (2005) Stable isotope analysis of trophic position and terrestrial vs. marine carbon sources for juvenile Pacific salmonids in nearshore marine habitats. *Fisheries Management and Ecology*, 12, 113–21.

Samadi, S., Bottan, L., Macpherson, E., Richer de Forges, B. and Boisselier, M.C. (2006) Seamount endemism questioned by the geographic distribution and population genetic structure of marine invertebrates. *Marine Biology*, DOI 10.1007/s00227-006-0306-4.

Schlacher-Hoenlinger, M.A., Pisera, A. and Hooper, J.N.A. (2005) Deep-sea lithistid assemblages from the norfolk ridge (New Caledonia), with description of seven new species and a new genus (Porifera, Demospongiae). *Zoosystema*, 27(4): p. 649–98.

Smith, D.K. and Jordan, T.H. (1988) Seamount statistics in the Pacific Ocean. *Journal of Geophysical Research*, 93, 2899–919.

Smith, P.J., McVeagh, S.M., Mingoia, J.T. and France, S.C. (2004) Mitochondrial DNA sequence variation in deep-sea bamboo coral (Keratoisidinae) species in the southwest and northwest Pacific Ocean. *Marine Biology*, 144, 253–61.

Stocks, K. (2005) Seamounts Online: an online information system for seamount biology. Version 2005-1. World Wide Web electronic publication. http://seamount.sdsc.edu.

Tracey, D., Neil, H., Gordon, D. and O'Shea, S. (2003) Chronicles of the deep: ageing deep-sea corals in New Zealand waters. *MarineBiodiversity. Water and Atmosphere*, 11(2), 22–4.

Vacelet, J., Cuif, J.P., Gautret, P., Massot, M., Richer de Forges, B. and Zibrowius, H.A. (1992) Colonial sphinctozoan sponge related to triassic reef builders surviving in deep water off New Caledonia. *Comptes rendus hebdomadaires de l'Académie des Sciences, Paris*, 314, 379–85.

Van de Beuque, S. (1999) Evolution géologique du domaine peri-calédonien (Sud ouest Pacifique). *Thèse de doctorat de l'Université de Bretagne occidentale, 19 mars 1999*, 270 p.

Veevers, J.J. (2000) *Billion-Year Earth History of Australia and Neighbours in Gondwanaland*, 388 p. GEMOC Press, Sydney.

Veevers, J.J. (2001) *Atlas of Billion-Year Earth History of Australia and Neighbours in Gondwanaland*, 76 p. GEMOC Press, Sydney.

Verlaan, P.A. (1992) Benthic recruitment and manganese crust formation on seamounts. *Marine Biology*, 113(1): p. 171–4.

Willison, J.H., Hall, J., Gass, S.E., Kenchington, E.L.R., Butler, M. and Doherty, P. (2001) *Proceedings of the First International Symposium on Deep-Sea Corals. Ecology Action Centre.* Nova Scotia Museum, Halifax, 231 p.

Wilson, R.R. and Kaufmann, R.S. (1987) Seamount biota and biogeography. In: *Seamounts, Islands and Atolls* (eds. Keating, B.H., Fryer, P., Batiza, R. and Boehlert, G.W.). *Geophysical Monographs*, 43, 355–77.

Won Y., Hallan, S.J., O'Mullan, G.D. and Vrijenhoek, R.C. (2003) Dispersal barriers and isolation among deep-sea mussel populations (Mytilidae: *Bathymodiolus*) from eastern Pacific hydrothermal vents. *Molecular Ecology*, 12, 3185–90.

Worm, B., Lotze, H.K. and Myers, R.A. (2003) Predators diversity hotspots in the blue ocean. *Proceedings of Natural Academy of Sciences USA*, 100, 9884–8.

Chapter 8
Corals on seamounts

Alex D. Rogers, Amy Baco, Huw Griffiths, Thomas Hart and Jason M. Hall-Spencer

Abstract

Corals are amongst the most conspicuous sessile organisms found on the hard substrates of seamounts. They include a diverse array of groups within the phylum Cnidaria, including the Scleractinia (stony corals), Antipatharia (black corals), Zoanthidea (zoanthids), Octocorallia (gorgonians, sea fans, soft corals) and Stylasteridae (hydrocorals). Estimating global coral diversity on seamounts is complicated by a lack of samples. However, they are an important component of the species diversity found on seamounts, playing a pivotal role in structuring the environment. In particular, some scleractinian corals have the ability to form cold-water reefs that may have a diverse community of associated organisms. Gorgonian corals may form dense stands that also play a structural role in communities of fish and other organisms. Corals found on seamounts show a diverse range of life histories although patterns of recruitment to populations are not understood. Isotopic methods for ageing corals have revealed that individual colonies may be slow growing and can live for hundreds to thousands of years. The distribution of corals on seamounts depends on a variety of physical and biological factors. Relative distribution of the coral groups differs regionally and between oceans, reflecting differences in ocean chemistry, productivity and biogeography. The different coral groups also show significant differences in depth of occurrence resulting from differences in trophic ecology and other aspects of biology. Because of their fragility, conservative life histories and limited geographic distribution, cold-water corals are vulnerable to the impacts of human exploitation through coral harvesting or as bycatch of fishing, especially trawling. Changes in ocean chemistry resulting from climate change are also a significant threat to corals on seamounts and other marine habitats.

Introduction

Corals are conspicuous, long-lived residents of seamounts of great interest to marine biologists and conservationists because of the ability of some species to form cold-water reefs. These reefs may be associated with diverse communities of other species that live in the variety of habitats formed by the living and dead framework-building coral (e.g., Freiwald *et al.*, 2002). However, the number of coral species that form such complex reef structures is relatively small (Roberts *et al.*, 2006). Many species occur as solitary colonies, are themselves associated with other corals, or may enhance habitat complexity in other ways, for example by forming coral meadows, forests or beds (e.g., Brodeur, 2001). This chapter reviews the current knowledge of the ecology and distribution of corals on seamounts, ridges and banks of the world's oceans. This is achieved through review of existing studies on the biology and ecology of corals on seamounts and by the analysis of 3215 records of corals available in scientific literature, museum records and databases. In this analysis, we have used a broad definition of seamounts (see Preface) as there is no 'typical' seamount feature, and they differ markedly in size, depth, elevation, geological associations, origin, distance to the continental slope and oceanographic setting (Rowden *et al.*, 2005; Chapters 1 and 2). Thus, we have included everything from shallow banks, associated with the slope of the continental margins and oceanic islands, to isolated oceanic seamounts. All records used in this analysis are available on Seamounts Online.

Diversity and ecology of corals on seamounts, ridges and banks

Diversity

Corals are all from the phylum Cnidaria. Those that occur on seamounts are mainly from the class Anthozoa and include the hexacorals of the orders Scleractinia (stony corals), Antipatharia (black corals) and Zoanthidea (zoanthids); and the Octocorallia (gorgonians or 'sea fans'). The class Hydrozoa, or hydrocorals, is represented by the family Stylasteridae. Out of the 526 species of corals recorded on seamounts and banks, stony corals are the most diverse and commonly observed group (Table 8.1). This is followed by the octocorals, the stylasterids, the antipatharians and the zoanthids in order of diversity and number of records (Table 8.1). Even when records are considered only from oceanic

Table 8.1 Number of species, genera and families of corals (Scleractinia, Octocorallia, Antipatharia, Zoanthidea, Stylasterida) recorded on seamounts and banks and seamounts only (in parentheses).

Taxonomic category/ taxon	Total number of records	Number of species	Number of genera	Number of families
Scleractinia	1713	249 (165)	85 (61)	20 (14)
Octocorallia	957	161 (110)	68 (49)	21 (17)
Antipatharia	157	34 (24)	22 (17)	6 (6)
Zoanthidea	28	14 (2)	6 (2)	3 (2)
Stylasteridae	372	68 (53)	18 (17)	2 (2)

seamounts, excluding banks connected to the slopes of continents and islands, there are still 354 species recorded. Comparisons with records held in 'Seamounts Online' show that more species of corals (Cnidaria) have been recorded from seamounts than any other group of invertebrates (Stocks, 2004). The species diversity of corals on seamounts and banks is similar to that of fishes found on seamounts, although the latter represent a larger number of families (130 in fish; Froese and Sampang, 2004). The species richness of corals on seamounts shows that the megafauna of these habitats are predominately sessile, filter-feeding organisms (Rogers, 1994; Stocks, 2004). This is because of the nature of seabed and the vigorous hydrodynamic regime on seamounts compared to most other deep-sea habitats (see Chapters 4 and 7).

Approximately 1482 species of scleractinian corals have been described of which more than half (about 800) are associated with shallow tropical reefs (Paulay, 1996), and about 706 are azooxanthellate, which lack symbiotic algae (S.D. Cairns, Smithsonian Institution, Personal communication, 2005). Of the azooxanthellate species, about 615 have been recorded from depths greater than 50 m. Seamounts and banks therefore potentially host a substantial fraction of the global scleractinian fauna and a very large fraction of azooxanthellate species. Many of the zooxanthellate species, genera and even families, common on shallow-water reefs (Paulay, 1996), are eliminated from the seamount fauna if island and continental slope-associated banks are excluded from this dataset (e.g., families Acroporidae, Meandrinidae Mussidae; genus *Agaricia*). This is because many of these topographic features have shallow summits, suitable for formation of warm-water reefs.

Over 2700 species of octocorals have been described (Freiwald *et al.*, 2004) of which about 6% have been recorded from seamounts and banks. This does not reflect the importance of this group either in terms of species diversity or of abundance as components of the seamount fauna. The antipatharians and zoanthids are less well known. The Zoanthidea in particular are poorly represented in this dataset. This is because the taxonomy of these groups is very poorly understood and many zoanthids species are very small and difficult to observe or sample.

Factors determining the distribution of corals on seamounts

Seamounts act as biological hotspots in the oceans and often attract a high abundance and diversity of large predators such as sharks, tuna, billfish, turtles, seabirds and marine mammals (Chapter 12; Worm *et al.*, 2003; Dower and Brodeur, 2004; Yen *et al.*, 2004; Tynan *et al.*, 2005). In the food-stressed environment of oligotrophic oceans, these features may be critical for the survival of many pelagic species. Seamounts can also host diverse and abundant communities of benthic organisms dominated by suspension feeders, including corals, sea anemones, sea pens, hydroids, sponges and feather stars (Rogers, 1994; Stocks, 2004). These communities differ from those found on the sediment-covered continental slopes and abyssal plains that are dominated by deposit feeding organisms such as holothurians (Gage and Tyler, 1991).

Rogers (1994) suggests two main explanations why seamounts host such diverse benthic and pelagic communities. These are increased productivity resulting from upwelling of nutrient-rich deep seawater around seamounts (Chapter 4) or the trapping of layers of diurnally migrating zooplankton, advected over seamount summits at night (Chapter 5). Increased productivity over seamounts and other elevated topography requires that upwelled, nutrient-rich water is resident over the feature for sufficiently long to enhance phytoplankton growth and for this to be converted to increased populations of zooplankton grazers (see Chapter 4).

Evidence of increased primary productivity over seamounts is rare and hard to connect to increased populations of benthic organisms (Genin, 2004). The topographic trapping of descending layers of zooplankton has been observed as providing a source of food for seamount-associated species in several cases (e.g., Genin *et al.*, 1988; Seki and Somerton, 1994; Parin *et al.*, 1997; Haury *et al.*, 2000; Chapter 5). Whether or not this takes place depends on the depth of the seamount summit, with respect to the depths over which the deep scattering layers (DSLs) of plankton migrate, and the intensity of horizontal currents that advect the DSL over the seamount at night (Genin, 2004). This would appear to be an important mechanism of trophic focusing over many seamounts (Rogers, 1994; Genin, 2004). Evidence that seamount corals prey on the DSL is lacking although *Lophelia pertusa* has been observed to prey on planktonic copepods (Freiwald, 1998). Analyses of the ^{15}N and ^{13}C isotope signatures of the tissues of *L. pertusa*, from the Galicia Bank, NE Atlantic, are also consistent with a diet that comprises a significant proportion of zooplankton (Duineveld *et al.*, 2004). Analysis of the distribution of stony corals and stylasterids (see below) shows that the majority of species occur in the depth range 100–1000 m. This depth distribution is consistent with trapping of the DSL and the availability of prey from this food source may be a determining factor in broad-scale patterns of distribution of stony corals on seamounts. Gorgonians and antipatharian corals have a greater tendency to be distributed below 1200 m (see later and also Etnoyer and Morgan, 2005). Octocorals have a relatively low density and diversity of nematocysts (Mariscal and Bigger, 1977) and, unlike stony corals such as *L. pertusa* that can feed on active zooplankton such as copepods, tend to be passive suspension feeders that capture prey items with little or no ability to escape (Ribes *et al.*, 2003). They can also exploit food sources such as particulate organic matter and microbial eukaryotes (Ribes *et al.*, 2003). This may allow them to exploit food sources unavailable to stony corals and stylasterids at abyssal depths. These sources may include micro- or nanozooplankton and dead organic particulate material flowing past or raining down on the seamount and material resuspended from the seabed as a result of accelerated or turbulent current flows. Little is understood about the energetics of deep-sea corals and it may also be the case that food concentration and quality are insufficient at greater depths for many species of scleractinians and stylasterids.

Studies on individual seamounts have demonstrated the strong influence of current strength on the distribution of corals. Current strength may be positively correlated with food supply and negatively correlated with sediment cover on seamounts. Corals require hard substrates for attachment, including bedrock, cobbles, stones and even the skeletons of other corals, the tubes of worms or the shells of marine molluscs. Currents keep feeding structures clear of sediment, remove waste products and disperse gametes or eggs and larvae (Grigg, 1974, 1984). Observations on Pacific seamounts have shown that at large scales, gorgonians and antipatharian corals are abundant near the peaks of seamounts, especially around the rims of summits or near the edges of terraces or basalt dykes where currents are strongest (Genin *et al.*, 1986; Moskalev and Galkin, 1986; Grigg *et al.*, 1987; Matsumoto, 2005). On smaller scales, densities of *Stichopathes* spp. have been observed to increase on rock pinnacles (Genin *et al.*, 1986). Strong negative correlations in coral abundance with sediment cover have also been observed on seamounts (Genin *et al.*, 1986; Grigg *et al.*, 1987).

Studies on Great Meteor Seamount in the Atlantic have shown differences of two orders of magnitude in the abundance of corals on the peaks of the seamount compared to the slopes (Piepenberg and Müller, 2004). This distribution was consistent with strong currents over the seamount. However, even on the seamount summit, the occurrence of corals was very patchy and this was attributable to the presence of sediments and the occurrence of turbulent flow. The effects of turbulence on food supply for corals on seamounts have not been investigated, although the complex tree-like branched form of many species found on seamounts can be viewed as advantageous for feeding in a regime where currents may frequently change direction and strength. Turbulence may also influence patterns of larval settlement over small scales.

Observations of the distribution of the coral *L. pertusa* in non-seamount and seamount habitats also show that this coral is found in areas of strong current flow (Rogers, 1999). On Cobb Seamount, in the NE Pacific, this coral occurs in areas of strong unidirectional flow (Farrow and Durant, 1985), whilst on the Rockall Bank it occurs around the summit 'rim' and on smaller scales, tends to grow on the mounds formed by the edges of old iceberg furrows called levees (Wilson, 1979a). Iceberg furrows have a particularly strong influence on the structure of the large *L. pertusa* reef on the Sula Ridge in the NE Atlantic (Freiwald, 1998).

Reproduction and life histories

There is no specific published information to date on reproduction of corals from seamount habitats, although studies have been carried out on deep-sea corals some of which occur on seamounts and banks (Waller *et al.*, 2002, 2005; Waller, 2005; Waller and Baco-Taylor, 2005; Waller and Tyler, 2005). Corals may reproduce sexually or asexually. Growth of colonial forms takes place by a form of asexual reproduction through budding to form new polyps. Budding may occur by the division of an existing polyp (intratentacular budding) or by the formation of a polyp in the space between two existing polyps (extratentacular budding) (Richmond, 1996). *L. pertusa* and *Solenosmilia variabilis* grow by intratentacular budding, whilst other deep-sea species (e.g., *Enallopsammia* spp.) grow by extratentacular budding (Cairns, 1995). Corals may also reproduce by fragmentation, whereby pieces of a parent colony break off and continue to grow, forming new coral colonies. This process is important in the development of cold-water *L. pertusa* reefs by Wilson (1979b) and reviewed by Rogers (1999). In some corals, pieces of living tissue may regenerate from an otherwise dead coral skeleton, or may leave the coral skeleton, and swim, using cilia covering the epidermis, and settle to form a new colony in a process known as 'polyp bale out' (Sammarco, 1982; Krupp, 1983).

Sexually reproducing corals may have separate sexes (gonochoric) or may be hermaphroditic or may even display both life-history strategies where some individuals are single sex and some have both male and female gonads (Richmond, 1996). About 25% of shallow-water corals are gonochoristic, while the majority are hermaphroditic where reproductive biology is known (Fadlallah, 1983; Richmond and Hunter, 1990). Eight species of deep-sea stony corals e.g., *Lophelia pertusa*, have been found to be gonochoric (Waller *et al.*, 2002; Burgess and Babcock, 2005; Waller and Tyler, 2005), whilst three, e.g., *Caryophyllia ambrosia*, are hermaphrodites (Waller *et al.*, 2005). The octocorals *Corallium secundum*, *C. laauense* and the zoanthidean *Gerardia* spp. from deep-water habitats in Hawaii have also been found to be gonochoric (Waller and Baco-Taylor, 2005). Based on the limited knowledge to date, it appears that deep-sea corals show differences in sexuality to shallow-water hermatypic corals (Roberts *et al.*, 2006). That only one genus displays hermaphroditism probably reflects systematic constraints on life history, although it should be noted that the shallow-water *C. smithii* is gonochoric (Tranter *et al.*, 1982).

Sexually reproducing corals show two modes of larval development, brooding or broadcast spawning. In brooding species, the eggs are fertilised internally and develop into planula larvae prior to being released. The larvae are able to settle and metamorphose immediately after being released (Richmond, 1996). In broadcast spawners, eggs and sperms are released into the water column where fertilisation and subsequent development take place. The eggs have to develop into larvae that must grow for a period of several weeks before being competent to settle. Counter intuitively, the larvae of many brooding, shallow-water tropical corals may have higher dispersal potential than those of broadcast-spawning species. This is because the larvae of brooding species often have zooxanthellae that supplement the energy reserves of the larvae and enable them to survive without settlement and metamorphosis for long periods of time (Richmond, 1996). Only about 15% of the shallow-water coral species studied are brooders, with the majority being broadcast spawners (Richmond, 1996). Timing of reproduction in shallow-water corals is often highly seasonal, taking place over a few months of the year or even just over a few days. In warm-water hermatypic corals where the spawning period is very short, it is often coupled to the lunar cycle (Richmond, 1996).

Twelve species of deep-sea coral, including *L. Pertusa,*, show broadcast spawning (Waller *et al.*, 2002; Waller, 2005; Waller and Baco-Taylor, 2005; Waller and Tyler, 2005). Most of these species produce large-sized oocytes varying between 400 μm maximum diameter in *E. rostrata* and 750 μm in *F. marenzelleri*, suggesting leicithotrophic development. A range of species have smaller-sized eggs but some possess leicithotrophic larvae (Burgess and Babcock, 2005; Waller, 2005; Waller and Baco-Taylor, 2005). It is notable that all the species with smaller eggs are primary reef constructors. The production of leicithotrophic larvae is more common than planktotrophic larvae in deep-sea species (Gage and Tyler, 1991; Young, 1994). Both *Flabellum curvatum* and *F. impensum* off the Antarctic Peninsula are brooding species with very late stage larvae, that are probably competent to settle immediately or shortly after release (Waller *et al.*, in press). Previously, it was thought that harsh ecological conditions at high latitudes favoured brooding of larvae (reviewed by Pearse, 1994). However, many Antarctic species have planktotrophic larvae and only some groups of organisms show a prevalence of brooding in their life histories (Pearse and Lockhart, 2004). The reasons for the prevalence of brooding in some groups are not fully understood, although to some degree it is a result of phylogenetic constraint. It may also be related to the geographic and the oceanographic history of the Antarctic and Sub-Antarctic (Pearse and Lockhart, 2004). *F. impensum* and *F. curvatum* also occur outside of the Southern Ocean in the SW Pacific and SW Atlantic, and *S. variabilis*, a non-brooding species, is found in the Southern Ocean.

The mean fecundity of the deep-sea corals that inhabit seamounts varies from a high of 3146 oocytes per polyp in *L. pertusa* to a low of 10 in *Madrepora oculata* (Waller and Tyler, 2005). The remaining species lie between these two extremes (Waller, 2005). To some extent, this is related to the size of the oocytes and the size of the coral polyps. Solitary corals have large polyps and can develop large number of large eggs, whereas colonial forms can produce large number of small eggs or low number of large eggs (Waller, 2005). The overall fecundity of colonial corals is potentially large.

The timing of reproduction in deep-sea corals that live on seamounts and banks also varies markedly. Some species show seasonal reproduction, such as *L. pertusa* in the NE Atlantic, which probably spawns in winter (January/February; Waller and Tyler, 2005) and *Gerardia* spp. in the North Pacific (Waller and Baco-Taylor, 2005). Others such as *M. oculata* in the NE Atlantic show periodic reproduction that is not seasonal (Waller and Tyler, 2005). Still others such as *F. marenzelleri* from the NE Atlantic, *C. lauuense* and *C. secundum* from the North Pacific, are quasi-continuous spawners, where reproduction is continuous but possibly with some variation in spawning intensity through the year (Waller *et al.*, 2002; Waller and Baco-Taylor, 2005). The hermaphroditic *Caryophyllia* spp. that have been studied show a cyclicity of development of male and female gametes within individuals so that self-fertilisation is not possible (Waller *et al.*, 2002). The timing of spawning in species of corals recorded as living on seamounts therefore varies markedly and is likely to be controlled by a range of environmental and genetic (phylogenetic) factors.

Age and growth of deep-sea corals on seamounts

Deep-water scleractinians can form ancient reefs through accumulation and repeated settlement on older coral skeletons over time. Schröder-Ritzrau *et al.* (2005) used U/Th dating to determine the ages of Atlantic deep-water corals exposed at the seabed. They found that seamounts off NW Africa, the low latitude Mid-Atlantic Ridge and the Azores had conditions suitable for coral growth during glacial times, as well as the interglacial periods, with coral samples spanning the past 50 000 years. In contrast, deep-water reefs from northern parts of the Atlantic were merely thousands of years old. All were of Holocene age, indicating that glacial periods were unfavourable for coral growth at high latitudes. Few geological cores have yet been analysed from deep-water reefs, but the current record for the oldest living reef comes from giant coral-topped mounds off Ireland that have been building up intermittently since the Pleistocene (De Mol *et al.*, 2005).

Deep-water reefs are generating strong interest in climate change research because zooxanthellate scleractinians have been shown to provide important archives of seasonal variations in temperature, salinity, and productivity in shallow waters of the tropics (Tudhope *et al.*, 2001; Cohen *et al.*, 2004; Roberts *et al.*, 2006). It is hoped that analyses of deep-water corals and the remains of their associated biota will provide a detailed understanding of subsurface oceanic circulation patterns as this is key to accurate predictions of future climate variability (Smith *et al.*, 1997; Adkins *et al.*, 1998; Thresher *et al.*, 2004; López-Correa, 2005; Risk *et al.*, 2005). The deep-water scleractinians examined for paleoclimate signals to date have complex internal banding patterns that makes extracting time series of environmental change difficult (Sinclair *et al.*, 2005). However, the skeletons of deep-water gorgonians, antipatharians, and zoanthids can show a much clearer banding similar to tree-rings and this can be used to estimate age (Sherwood *et al.*, 2005). Growth band studies (sclerochronology) of this sort require validation using radiometric analyses whereby naturally occurring radioisotopes are used to determine an independent estimate of age or growth rate. Such studies have revealed that the oldest known deep-water coral to date is *Gerardia* spp., a zoanthid from 620 m off Florida, carbon dated to 1800 years old (Druffel *et al.*, 1995). Gorgonians can also be slow growing and long lived, the oldest known being fossil specimens of *Primnoa resedaeformis* from Georges

Bank dated as 320 years (Risk *et al.*, 2002) and 700 ± 100 years (Sherwood *et al.*, 2006). Andrews *et al.* (2002) dated a 112 years old *P. resedaeformis* colony in the Gulf of Alaska and noted that larger colonies in the vicinity were probably older.

Evidence shows that deep-sea corals have the potential to live for thousands of years and are typically slow growing (Andrews *et al.*, 2005a), yet where food supply and water conditions are optimal some can grow quickly. For example, the scleractinian *L. pertusa* has rapidly colonised oil rigs and exhibited growth rates of up to 33 mm/year (Gass and Roberts, 2006). Recent carbon dating work on isidid corals by Roark *et al.* (2005) revealed ages of 75–126 years confirming the longevity of certain deep-water corals, although lead-210 dating of an isidid (*Lepidisis* spp.) from 690 to 800 m off New Zealand by Tracy *et al.* (2005) showed that it was 38–48 years old, indicating a linear growth rate of *ca.* 30 mm/year. Andrews *et al.* (2005b) used the same methods on another isidid (*Keratoisis* sp.), collected at 1425 m on Davidson Seamount off California. This colony was 97–197 years old and had grown about three times more slowly than the shallower New Zealand isidid, suggesting that food supply may limit growth at depth.

Role of corals in structuring seamount communities

Some of the corals that occur on seamounts and banks are capable of forming reefs. However, most of our knowledge on the ecology of deep-sea or cold-water coral reefs relates to frameworks constructed by *L. pertusa*. This species has a wide distribution, but to date, large complex reefs formed by *L. pertusa* have only been found in fjords, on the continental slope and on slope or near-slope banks rather than oceanic seamounts*. *S. variabilis*, one of the main reef-forming corals, builds extensive frameworks on seamounts, along with several secondary reef-constructing species such as *M. oculata* and *Desmophyllum dianthus* (e.g., Koslow *et al.*, 2001; Clark and O'Driscoll, 2003). Reefs formed by this species have been observed off New Zealand and Australia. There is also evidence of reefs on the Pacific–Antarctic Ridge and on a small seamount in Drake Passage in the Southern Ocean. Most corals occur on seamounts as thickets or isolated colonies. Although these may influence the diversity of the seamount habitat by structuring the environment, they are not reefs.

Cold-water coral reefs, like shallow-water tropical corals reefs, host a distinct community of associated species that contrast with the surrounding deep seabed in terms of taxonomic composition and biomass (Rogers, 1999; Freiwald *et al.*, 2002). This is because the corals secrete calcium carbonate to form a complex three-dimensional skeleton that alters the local hydrodynamic and sedimentary regime (Freiwald *et al.*, 2002). Such coral species are defined as being hermatypic and may be classed as 'ecosystem engineers' defined by Jones *et al.* (1994) as a species that alters the environment by its physical presence. The coral frameworks present a variety of sub-habitats, including coral rubble, sediment clogged coral framework, exposed dead coral framework and the living coral (Jensen and Frederiksen, 1992; Mortensen *et al.*, 1995; Rogers, 1999; Freiwald *et al.*, 2002).

The presence of a variety of substrata and habitats that are not present on surrounding deep-sea sediments explains the high diversity and contrasting faunal composition of

*See Supplementary Material for map.

deep-sea coral reefs compared with the background communities (Rogers, 1999; Freiwald *et al*., 2002). The number of macrofaunal and megafaunal species and the density of organisms tend to be high on cold-water coral reefs compared to the surrounding habitats (Jensen and Frederiksen, 1992; Koslow *et al*., 2001; Jonsson *et al*., 2004; Thiem *et al*., 2006). For many groups of organisms, the diversity found in the few studies of deep-water coral reefs is comparable to those on shallow-water reefs, although data are very limited for both (Rogers, 1999). There are notable exceptions to this, and the diversity of reef-forming corals, fish and molluscs tends to be much lower on cold water than tropical coral reefs (Rogers, 1999). Many biophysical processes on deep-water and shallow-water, tropical coral reefs are similar, including processes of reef accretion and destruction (Rogers, 1999; Freiwald *et al*., 2002).

One of the main questions is how many species are endemic, or obligate associates of living or dead deep-sea coral communities. More than 1300 species have been described as associated with *L. pertusa* reefs in the NE Atlantic, although the vast majority of these occur in other habitats (Rogers, 1999; Roberts and Gage, 2003; Roberts *et al*., 2006). For example, in the NE Atlantic, fish species such as redfish (*Sebastes marinus*), tusk (*Brosme brosme*) and ling (*Molva molva*) occur both on and off *L. pertusa* reefs, although they may be more numerous and larger in reef habitats (Husebø *et al*., 2002). Between 24% and 43% of the 262 invertebrate species collected from *S. variabilis* reefs on the seamounts south of Tasmania were undescribed and about 16–43% thought to be endemic (Richer de Forges *et al*., 2000; Koslow *et al*., 2001). If seamounts conform to expectations of island biogeography, it is likely that they will recruit species from the regional species pool, which in the SW Pacific may be huge (see later and Chapter 13). In general, islands have a lower species diversity than the nearest mainland fauna (Vermeij, 2004), although the ratio of endemic to surrounding species varies and may be as high as 42%.

It is now recognised that other types of coral can form distinct habitats with associated communities of animals. In particular, large colonies of gorgonians can form dense stands (Auster *et al*., 2005). These have been located in the North Pacific, throughout Hawaii, along the Aleutian Island chain, and also in the Bering Sea and Gulf of Alaska (e.g., Stone, 2006). These habitats have been found to be rich in redfish (*Sebastes* spp.), shrimp, galatheid lobsters and other crustaceans (e.g. Stone 2006). They also host other attached suspension feeders such as crinoids, basket stars, sponges and other corals (e.g., Parrish and Baco, in press). NE Atlantic gorgonians *Acanthogorgia* spp. have been found to host the amphipod *Pleusymtes comitari*, thought to be an obligate commensal species (Myers and Hall-Spencer, 2004). The legs of the amphipod have special hooks to grip on to the coral. Predators that specialise on feeding on the polyps of the octocorals have also been identified, including starfish (*Hippasteria heathii*) and a nudibranch (*Tritonia exulsans*) (Krieger and Wing, 2002). Gorgonians and other corals also form dense populations in areas such as canyons and may also have a highly diverse associated fauna. *Paragorgia arborea* has been found to host up to 16 species of crustaceans and *Primnoa resedaeformis* 7 (Buhl-Mortensen and Mortensen, 2004).

Predatory fish may take advantage of the energetic environment of seamounts and other deep-sea habitats by feeding at certain times of day in the current and then resting in cryptic habitat, such as behind rocks, or amongst corals. Hundreds of the rockfish, *S. alutus*, have been observed resting amongst forests of sea whip corals (*Halipteris willemoesi*) at night and then swimming above the corals during the day to prey on passing food items (Brodeur, 2001). Such behaviour may explain the association of many species of seamount fish with biogenic habitats, although there is evidence that any physical structure creating shelter from current may serve this function (Auster *et al.*, 2005; ICES, 2005).

Studies in the Hawaiian Archipelago on associations between black corals (*Antipathes* spp.) in shallow water and fish have indicated that many fish may routinely pass through the branches of coral colonies treating them as general habitat. A few species regularly used the coral for protection from perceived threats and only one species of fish was restricted to the branches of coral trees (Boland and Parrish, 2005). The fish communities of deeper slopes in Hawaii also use corals (*Gerardia* and *Corallium* spp.) as shelter interchangeably with non-biotic habitat (Parrish and Baco, in press). Taller coral colonies (*Gerardia* spp.) are more attractive to fish than other corals. In some cases, observations suggest that fish and corals may have similar habitat requirements on seamounts and banks (e.g., exposure to currents or areas with a high supply of planktonic food; Mundy and Parrish, 2004; Parrish and Baco, in press). Resolution of the question of importance of stands of gorgonians to fish species will require further detailed *in situ* observational studies and habitat classification (Auster *et al.*, 2005). The rich communities of animals associated with coral reefs, or forests of gorgonians or sponges may in themselves provide food for predators. Evidence from electronic tagging and submersible observations on banks off Hawaii indicate that Hawaiian monk seals (*Monachus schauinslandi*) may preferentially forage for fish amongst beds of deep-sea octocorals and antipatharians (Parrish *et al.*, 2002). Observations from submersibles have also suggested that some fish that occur on seamounts may feed on invertebrates associated with coral colonies or on the coral polyps themselves (Auster *et al.*, 2005). There may be a direct link between commercial fish species and benthic habitat-forming organisms (see also the 'feed-rest' hypothesis in Chapter 6). This means that fishing activities on seamounts may not only have the potential to directly remove target fish species but may destroy or severely reduce habitat essential for the existence of fish populations in these areas and negatively impact on the predators that feed on them.

Biogeography

Geographic and taxonomic coverage of sampling for corals

A total of 3215 species of corals have been recorded on seamounts, making them one of the best studied groups of animals in these habitats. The question arises as to whether such a dataset is informative in terms of patterns of biodiversity on seamounts. Previous studies have recorded species from only about 200 seamounts (Stocks, 2004). Estimates from satellite gravity maps indicate that there are probably 100 000 or more large seamounts (1000 m + elevation; Wessel, 2001; see Chapters 1 and 2), i.e., the number sampled is a small proportion of the total. In the present study, records of corals and stylasterids were obtained for 271 seamounts, banks and ridges, reducing to a total of 184 if shelf or continental slope-associated banks are removed from the dataset. This number of seamounts and banks is larger than the number from which cnidarians have previously been recorded

(84 in Stocks, 2004), but is still tiny compared to the overall number of seamounts in the world's oceans. The geographic distribution of records of corals from seamounts, banks and ridges makes it apparent that whilst some areas have been well sampled, such as around New Zealand, Hawaii, off western N America and in the NE and NW Atlantic, vast areas of the ocean remain unsampled (see Fig. 8.1). The present dataset contains only five records of corals from the entire Indian Ocean and all these are associated with

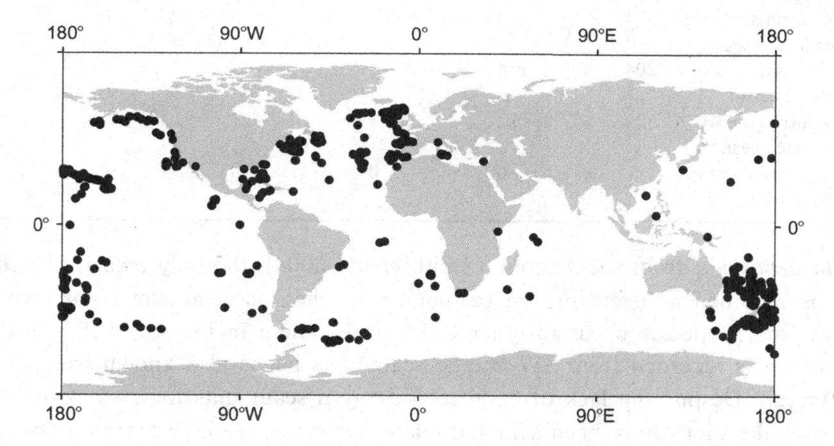

Fig. 8.1 Octocorallia global map of the distribution of the samples of corals on seamounts (dark circles) included in analyses of distribution and biogeography.

shallow banks. Large areas of the NE, W, equatorial and S central Pacific and S Atlantic oceans have not been sampled. Some of these areas, especially the Pacific Ocean, contain the largest concentrations of seamounts in the world (Wessel, 2001; Stone *et al.*, 2004).

The global deficiency of scientific expertise in coral taxonomy is another significant barrier to understanding the diversity of corals and many other organisms on seamounts. De Vogelaere *et al.* (2005) report that Davidson Seamount, off California, hosts at least 20 coral species, with *Paragorgia* spp. being abundant and forming large and dense patches. However, they were only able to identify three other coral species to genus level and none to species. Many records obtained in the present study only identified corals to genus or family level, greatly decreasing the usefulness of the data in a global analysis of coral distribution and in the establishment of programmes to manage and minimise human impacts.

Global distribution

One of the most striking features of the occurrence of corals on seamounts and banks is that the majority of species have been observed only within one region of one ocean (Table 8.2). This is even considering land-locked seas, such as the Mediterranean and Caribbean as separate regions from the major oceans to which they are connected. This regionalisation partially reflects the poor sampling of seamounts in many parts of the world. Difficulties in identification of corals also mean that the true geographic distribution of many species remains unresolved. Recent analyses of mitochondrial DNA sequences from corals in the New Zealand region, for example, have identified haplotypes identical to

Table 8.2 Distribution patterns of corals (Scleractinia, Octocorallia, Antipatharia, Zoanthidea, Stylasterida) across oceans and regions within oceans.

Taxon/geographic scale	Scleractinia	Octocorallia	Antipatharia	Stylasterida	Zoanthidea	Total
Species occurring in one ocean or inland sea	218	156	27	64	14	479
Species occurring in two oceans or inland seas	30	5	7	4	0	46
Species occurring in three oceans or inland seas	4	0	0	0	0	4
Species occurring in one region in one ocean	204	134	26	63	13	440
Species occurring in two regions in one ocean	22	22	2	1	1	48
Species occurring in three regions in one ocean	5	0	0	0	0	5

Acanella arbuscula from the Atlantic (Smith *et al.*, 2004), the only area in which this species is recorded as occurring on seamounts. Studies on seamount coral taxonomy show that many species occur in other habitats elsewhere in the world. For example, *F. lowekeyesi* is recorded from SW Pacific seamounts but is also known from the SW Indian Ocean. Despite the lack of coral records from seamounts for many parts of the globe, several regions have been sampled relatively well, especially the SW Pacific, the NE Pacific and the N Atlantic. If the distribution of corals on seamounts were widespread then this should be obvious, even within the limitations of the present dataset. Some coral species are endemic to specific regions and are commonly recorded on seamounts and banks.

The limitations of the dataset make it difficult and complex to explain the apparent regionalisation of the seamount coral fauna. A great variety of methods of sampling corals from seamounts and banks have been used from dredging to point sampling with submersibles. Sampling methodology is known to affect estimates of community diversity. The depth to the seamount or bank summit and the physical environment provided by different seamounts must influence coral distribution. The most obvious examples of this are zooxanthellate species, such as those belonging to the genus *Acropora*, which require sunlight and warm waters and will therefore only occur in regions where seamounts and banks with shallow summits occur in the tropics or sub-tropics such as the Caribbean and around Hawaii. Seamounts and banks also represent island-like patches of habitat that contrast strongly in their physical characteristics with the surrounding seabed that is at greater depth, may lie in a different oceanographic regime and is often composed of sediment. As such, seamounts may be expected to conform to the predictions of island biogeographic theory in terms of community assembly (MacArthur and Wilson, 1967) where distance to source populations, which are usually on the continental shelf or slope, the age of the seamount, the proximity to other seamounts and the size of the seamount play a critical role in determination of rates of species immigration, colonisation and extinction (see also Chapter 13).

Specific habitat availability on seamounts and island biogeography community dynamics may be sufficient to explain why the distribution of corals on seamounts does not reflect global occurrence of many species even within the sampling limitations of this study. It may also explain why the few detailed investigations of seamount biology have revealed relatively little similarity in the species composition of communities, even in adjacent seamounts within chains or between seamounts located on different chains (e.g., Richer de Forges *et al.*, 2000). Island biogeography predicts that the fauna of 'islands' will be a subset of that of the nearest source regions and this is consistent with at least some studies of seamount faunas. An example would be the megafauna of the Great Meteor Seamount (Piepenberg and Müller, 2004; reviewed by Wilson and Kaufmann, 1987). This contrasts with the high levels of diversity described for seamount communities in the SW Pacific, which include the South Australian Seamounts, those near New Caledonia and on Lord Howe Rise (Richer de Forges *et al.*, 2000). However, it must be considered that the SW Pacific studies have taken place on the periphery of the most biologically diverse marine region in the world, the tropical Indo-West Pacific (e.g., Briggs, 2003). The fauna of the deep shelf, continental slopes and abyssal plains of this region are largely unexplored. Seamounts probably recruit species from the regional species pool, and diversity is likely to reflect regional species diversity (Cornell and Lawton, 1992; Karlson *et al.*, 2004; Witman *et al.*, 2004; Chapter 13) that in this region is likely to be very high. This also may mean that point comparisons with small areas of seabed on the immediately adjacent continental slope may not be meaningful.

Another obvious aspect of the coral dataset is that very few species of corals occur across a very wide geographic area (Table 8.2). Only four species, *D. dianthus*, *S. variabilis*, *Stenocyathus vermiformis* and *M. oculata*, occur on seamounts in three oceans (see Supplementary Material for maps*). All the widespread corals are scleractinians, probably because these corals have a good ability to disperse by planktonic larvae, although this is unproven (see earlier). Three out of four of these species are also primary or secondary framework constructors in cold-water coral reefs. Other widespread species of scleractinians are also reef-building species and at least one of these, *L. pertusa*, has been observed to colonise remote oil platforms presumably by dispersal of pelagic planula larvae (Gass and Roberts, 2006). Why reef-building species should have such wide geographic distributions is unknown. Distributions may result from vicariance, which has left fragmented populations from a previously wider distribution. There is evidence that cold-water coral reefs have been eliminated from many areas as a result of climate change (e.g., Rogers, 1999; Remia and Taviani, 2005). It is also possible that the population dynamics of reef-forming species, or the environment in which they occur, selects for planktotrophic larval development that confers high dispersal ability for these species. The detailed systematics of such widely distributed species of coral also remains uninvestigated using molecular techniques and they may represent geographically separated sibling species.

Analysis of the species richness of corals on seamounts on a 10° latitude and 10° longitude grid shows that the SW Pacific has the highest species diversity irrespective of whether seamounts and banks are considered or just seamounts alone (Figs. 8.2 and 8.3). Other areas of high diversity include the NE Pacific and the NE Atlantic. However, plotting the number of species for each 10° × 10° box against the number of samples (Fig. 8.4) shows a strong relationship between sampling intensity and species richness. To try and correct for sampling intensity, the number of species for each 10° × 10° box was divided by the number of samples and re-plotted on the world map (Fig. 8.5). The results are in many ways the reciprocal of the plot of species richness, as for many areas

*See Supplementary Material on-line at www.seamounts book.info

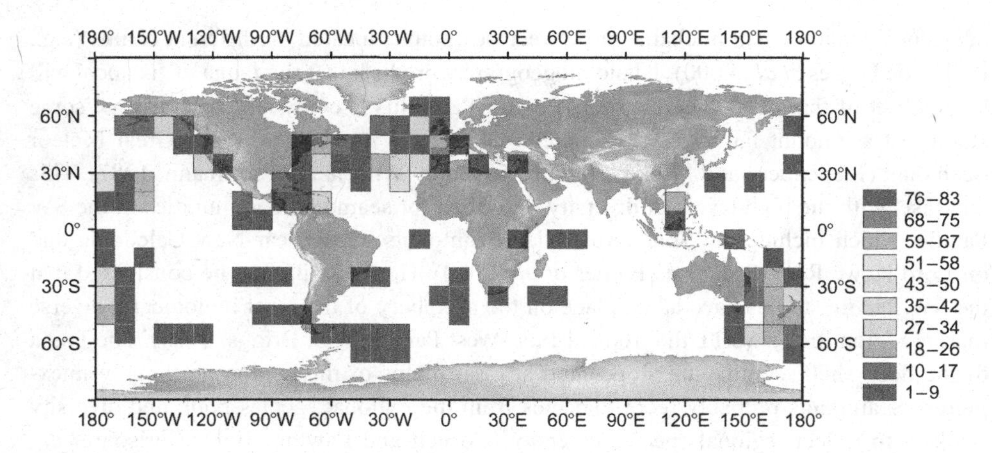

Fig. 8.2 Global species richness of corals on seamounts, banks, ridges and plateaus analysed on 10° × 10° grid of latitude and longitude. *Key*: Number of species; red: high diversity; blue: low diversity.

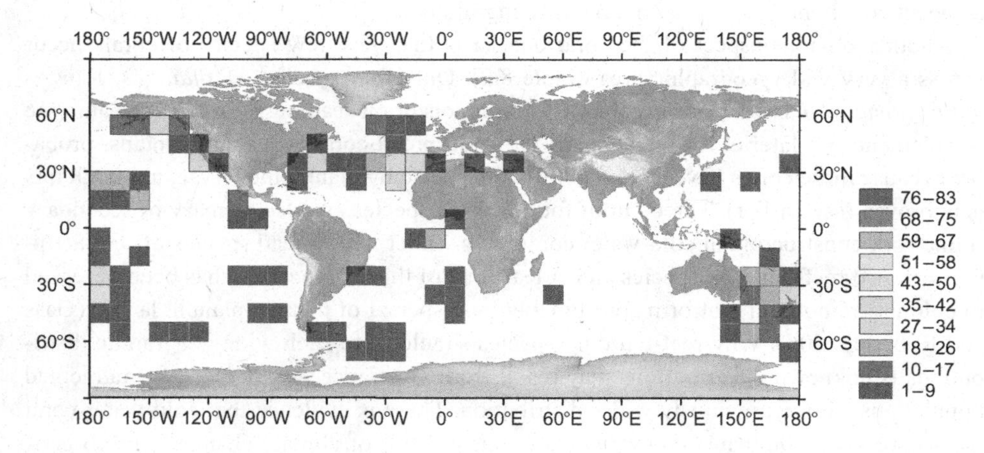

Fig. 8.3 Global species richness of corals on seamounts only, analysed on 10° × 10° grid of latitude and longitude. *Key*: Number of species; red: high diversity; blue: low diversity.

of low sampling effort, a relatively high number of species per sample are identified. This is expected from species-accumulation curves, as the number of species detected in a few consecutive samples rises very steeply and only levels off when the regional fauna has been well sampled. It is notable that the SW and NE Pacific areas still have quite a high species richness to sampling ratio. The species-rich boxes in the NE Atlantic (Figs. 8.2 and 8.3), however, have a low species richness to sampling ratio in Fig. 8.5. This indicates that high sampling effort has probably over-represented the importance of parts of this area in terms of species richness. This is confirmed by detailed examination of the values that fall well above and well below the trend line in Fig. 8.4. Grid boxes from the NE and SW Pacific show a high number of species per sample. Those from the N Atlantic show low number of species per sample, as well as one off SE New Zealand that probably represents a well-sampled area or an area where samples are dominated by a few abundant species. The northern N Atlantic region not only shows low species richness, but also a

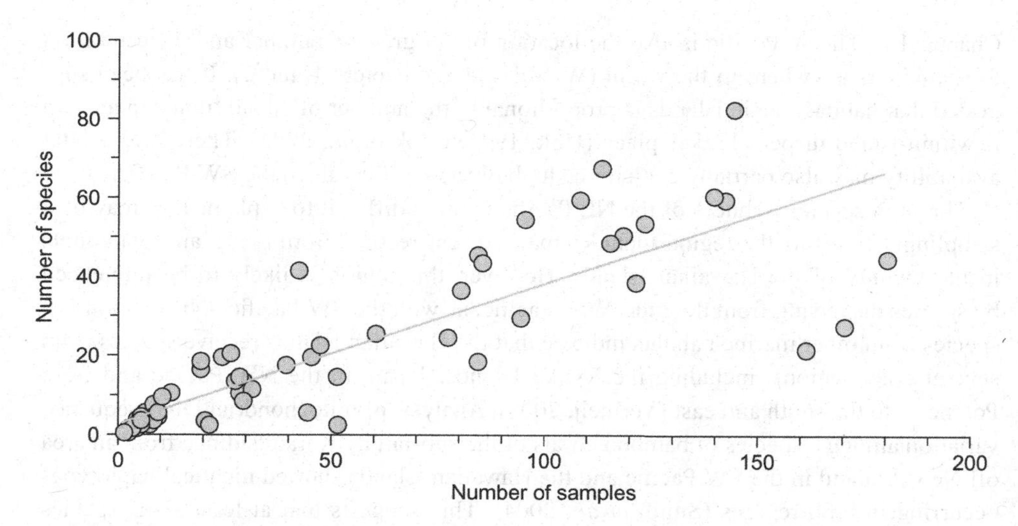

Fig. 8.4 Graph of number of samples vs number of species found on seamounts and banks in each 10° × 10° box of a global grid. Line of best fit shown.

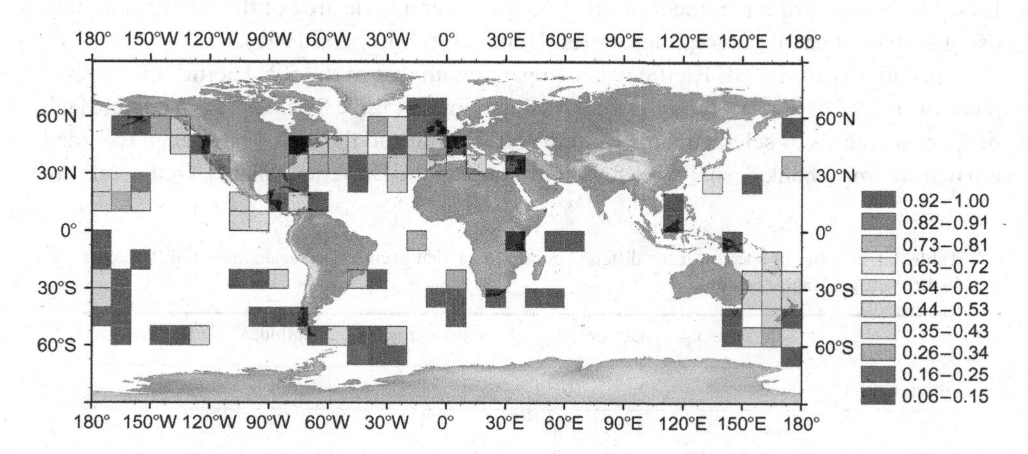

Fig. 8.5 Ratio of number of samples over number of species of corals observed on seamounts and banks in each 10° × 10° latitudinal and longitudinal box of a global grid. Key at lower right: Red: high number of species per sample; blue: low number of species per sample.

low species richness to sample ratio, possibly indicating that this region has a genuinely low diversity. The Arctic is known to have an impoverished fauna as a result of cycles of glaciation (Gage and Tyler, 1991).

The SW Pacific region is on the periphery of the Indo-West Pacific biodiversity hotspot, the most species-rich area of the world's oceans, including the 'coral triangle', the greatest concentration of shallow coral reef biodiversity (e.g., Bechtel *et al.*, 2004; Wilkinson, 2004; Briggs, 2005; Heads, 2005). This region, which had its origins in the Tethys Sea and was established by the early Miocene, has acted as a centre for speciation in shallow water and subsequent species dispersal across the SW Pacific (Briggs, 2000, 2003, 2005). Onshore–offshore patterns of speciation likely had a positive influence on the diversity of the deep seas in this region including the SW Pacific (Zezina, 1997; Briggs, 2003;

Chapter 13). The W Pacific is also the location of the greatest number and concentration of seamounts anywhere in the world (Wessel, 2001; Chapters 1 and 2). It has been suggested that habitat size in islands is proportional to the number of island 'units' in an area if within-island dispersal takes place (Holt, 1992; Cook *et al.*, 2002). Therefore, habitat availability may also partially explain the high diversity of corals in the SW Pacific.

The high species richness of the NE Pacific is more difficult to explain. This may be a sampling artefact as the region includes many recent records from banks and seamounts in the vicinity of the Hawaiian Islands. However, this region is likely to be influenced by species dispersing from the Indo-West Pacific, as with the SW Pacific. Observations on species from other marine habitats indicate that the Hawaiian Islands receive species from several other regions, including the Ryuku Islands, Japan, in the NW Pacific and from Polynesia to the south and east (Vermeij, 2004). Analysis of mitochondrial DNA sequence variation amongst species of bamboo corals of the sub-family Keratoisidinae from an area off New Zealand in the SW Pacific and the Hawaiian Islands showed identical haplotypes occurring in both regions (Smith *et al.*, 2004). This suggests that at least some species have the capacity to disperse even to the distant seamount chains in the Pacific Ocean. It is possible, therefore, that corals have colonised the seamounts and banks around the Hawaiian Islands from a number of other regions over a wide area of the Pacific, possibly using seamount chains as stepping stones (Hubbs, 1959; Rogers, 1994).

A peculiarity of the NE Pacific data compared with that of the SW Pacific is the prevalence of species of octocorals compared to scleractinians (Table 8.3). There is a notable lack of species records of scleractinians from the NE Pacific and those that have been recorded are mainly from 'banks', shallower features mainly in the Hawaiian Islands. Octocorals are

Table 8.3 Number of species of the different coral groups that occur on (a) seamounts and (b) banks in different geographic regions.

	Scleractinia		Octocorallia		Antipatharia		Zoanthidea		Stylasterida	
Group/region	a	b	a	b	a	b	a	b	a	b
NEA	48	24	27	3	8	1	–	1	7	4
NWA	9	9	7	17	2	–	1	2	–	–
SEA	10	–	1	–	–	–	–	4	–	–
SWA	5	5	1	–	–	–	–	–	–	–
Med	2	7	–	–	–	–	–	–	–	–
Car	–	31	–	21	–	8	–	3	–	3
NEP	15	29	54	25	13	9	1	1	1	–
NWP	3	5	3	19	3	–	–	1	–	2
SEP	3	–	–	–	–	–	–	–	–	–
SWP	108	17	20	1	–	–	–	–	46	19
SCS	–	18	–	–	–	–	–	1	–	–
Cel	–	–	–	1	–	–	–	–	–	–
WIO	–	3	–	–	–	–	–	1	–	–
EIO	–	1	–	–	–	–	–	–	–	–
SO	8	–	4	–	1	–	–	–	4	–

NEA: North East Atlantic; NWA: North West Atlantic; SEA: South East Atlantic; SWA: South West Atlantic; Med: Mediterranean; Car: Caribbean; NEP: North East Pacific; NWP: North West Pacific; SEP: South East Pacific; SWP: South West Pacific; Cel: Celebes Sea; SCS: South China Sea; WIO: Western Indian Ocean; EIO: Eastern Indian Ocean; SO: Southern Ocean.

more diverse in the NE Pacific and occur mainly on seamounts, although in the SW Pacific there are relatively few species. It is tempting to ascribe these differences to bias in collecting and identifying animals, but submersible observations in the Hawaiian Islands confirm the relative abundance of octocorals compared to scleractinians (Baco, Personal observation). It is likely that the changes in relative species diversity of different coral groups between these regions reflect differences in the physical characteristics of the seamounts and the biological and the physical oceanographic settings in which they occur (e.g., see Rowden *et al.*, 2005). In particular, the depth of the aragonite saturation horizon (ASH) is shallow in the N Pacific compared to other ocean regions such as the N Atlantic. Higher dissolution rates of aragonite in the N Pacific are likely to make this region less favourable for scleractinian corals and for formation of cold-water coral reefs (Guinotte *et al.*, 2006; Roberts *et al.*, 2006). Octocorals are less affected by a shallow ASH as their skeletons are formed by calcite. The extreme geographic isolation of the Hawaiian Island chain may also select for colonisation by octocorals from distant sources of species because of the life-history characteristics of these species. The differences in the biology of these groups may also be related to the occurrence of octocorals in deeper water than scleractinians (see later).

Sampling intensity has probably led to an overestimate of relative species richness of the N Atlantic in this study. The coral fauna of seamounts, however, is well developed and the presence of many banks located on the continental slopes of this region, beneath bands of high productivity surface waters associated with the shelf break (e.g., Pingree and Mardell, 1981; Holligan and Groom, 1986), may have provided large areas of suitable habitat for many coral species. Tropical regions, such as the W Atlantic, may have acted as evolutionary centres of origin for the Atlantic deep sea (Briggs, 2003). This hypothesis cannot yet be tested as little is known about tropical and sub-tropical W Atlantic seamounts and other deep-sea habitats with the exception of hydrocarbon seeps. A further source of species may have been the Great Trans-Arctic exchange between the N Pacific and Atlantic oceans (Briggs, 1995). Several cold-water scleractinians and octocorals found in the North Atlantic also occur in the North Pacific. Dispersal *via* this route may have occurred although this is unproven.

Latitudinal patterns of diversity

Examination of the latitudinal distribution of species of corals on seamounts suggests that species richness peaks at mid-latitudes (20–40° N/S), with low values at the poles and very low values at equatorial latitudes (Fig. 8.6). This is not consistent with reports of parabolic patterns of species richness with latitude in marine species, where peak biodiversity is recorded at low latitudes and decreases progressively towards the poles (e.g., Hillebrand, 2004; Witman *et al.*, 2004). Note that the latitudinal diversity in shallow benthic communities reported by Witman *et al.* (2004) shows high diversity values at mid-latitudes and moderate-to-low values at the equator despite reporting parabolic patterns of species richness with latitude (Fig. 3 of Witman *et al.*, 2004). Seamount coral diversity would appear to be congruent with latitudinal biodiversity patterns in large, oceanic predators (Worm *et al.*, 2003). However, examination of the latitudinal pattern of distribution of sampled seamounts shows the same broad pattern as species richness in corals. A hypothesis that

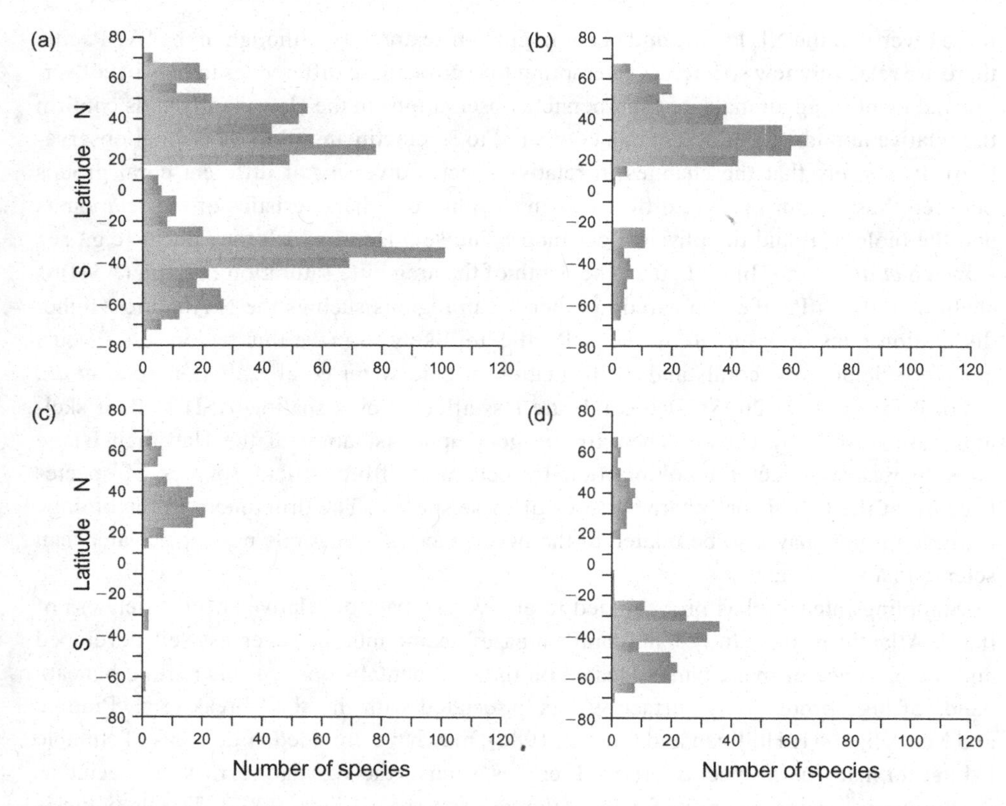

Fig. 8.6 Distribution of number of species at different latitudes: (a) Scleractinia, (b) Octocorallia, (c) Antipatharia and (d) Stylasterida.

the latitudinal pattern of species richness reflects sampling intensity of seamounts cannot be rejected. To some extent, this may reflect the actual distribution of seamounts (e.g., see maps in Stone *et al*., 2004), although it is likely that it simply reflects lack of scientific sampling effort on equatorial seamounts.

Comparison of the latitudinal species richness of corals (Fig. 8.6) with the number of seamounts (Fig. 8.7) does reveal some differences. The species richness of corals at 20–30° N and 25–35° S does seem to be higher than would be expected simply from the number of seamounts and banks sampled. These latitudinal bands correspond to the SW Pacific and NE Pacific biodiversity hotspots; 45–55° N and 05–15° N appear to have low diversities given the number of seamounts and banks. However, the number of sites sampled in equatorial regions and high latitudes is so low that it is not possible to place any confidence in estimates of species richness.

Depth distribution

An analysis of variance using the general linear model (GLM) was used to compare the depth distribution of the different groups of corals. This was done by estimating the maximum and mean depths for all species and testing the hypothesis that the coral group (scleractinian, octocoral, antipatharian, stylasterid) was a predictor of the depth of distribution

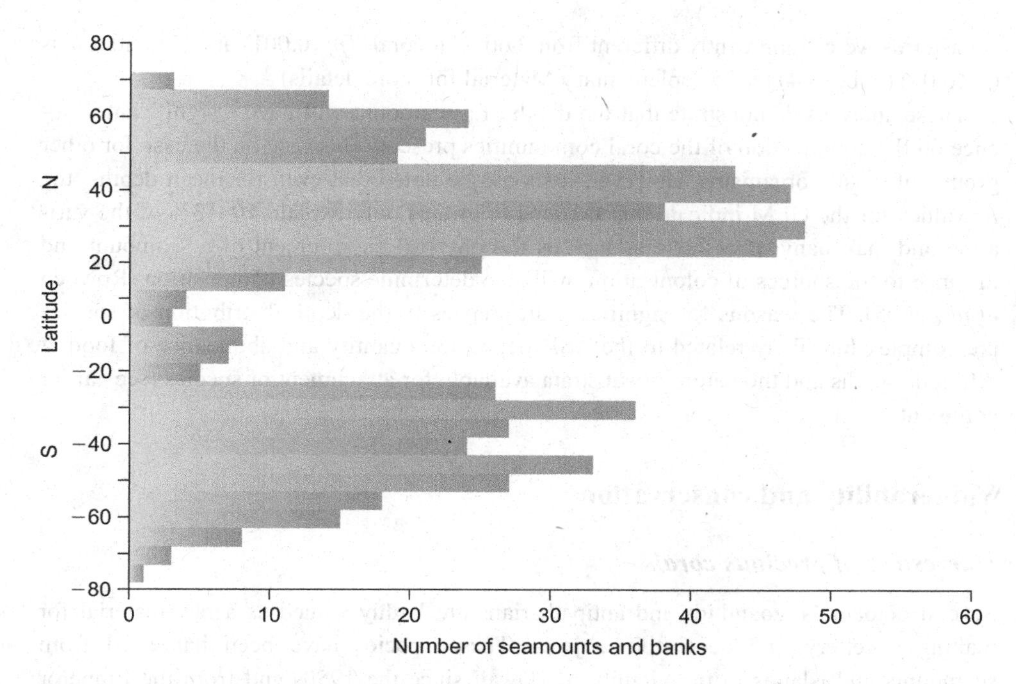

Fig. 8.7 Number of seamounts sampled at different latitudes.

of species. Data were root transformed to normalise the residuals. When this analysis was carried out on the entire dataset, it showed that the depth distributions were significantly different between the four coral groups ($p < 0.001$; Table 8.4). Pairwise analyses showed that the depth distribution of scleractinian and stylasterid corals was not significantly different. Likewise, depth distributions of octocorals and antipatharians were not significantly different, although this analysis was weak given the poor sample size for antipatharians. The GLM analyses showed that depth distributions of scleractinians and

Table 8.4 Analysis of variance using GLM of the maximum and mean depths of distribution of species of Scleractinia, Octocorallia, Antipatharia and Stylasterida.

		F	p	R^2 (adj.) (%)
All data	Maximum	8.98	<0.001	5.67
	Mean	18.21	<0.001	11.48
Scleractinia vs Octocorallia	Maximum	18.44	<0.001	5.76
	Mean	46.78	<0.001	13.13
Scleractinia vs Antipatharia	Maximum	5.04	0.026	1.89
	Mean	6.52	0.011	2.56
Scleractinia vs Stylasterida	Maximum	0.24	0.624	0
	Mean	0.01	0.921	0
Octocorallia vs Antipatharia	Maximum	0.05	0.823	0
	Mean	1.51	0.220	0.35
Octocorallia vs Stylasterida	Maximum	16.25	<0.001	7.54
	Mean	25.35	<0.001	11.52
Stylasterida vs Antipatharia	Maximum	5.44	0.022	4.51
	Mean	5.05	0.027	4.14

stylasterids were significantly different from both octocorals ($p<0.001$) and antipatharians ($p<0.05$) (Table 8.4) (see Supplementary Material for more details).

These analyses demonstrate that the depth of a seamount will have a significant influence on the composition of the coral communities present. This may be the case for other groups of sessile organisms. However, it should be noted that even for mean depths, the R^2 values for the GLM indicate that taxonomic groups only explain 10–13% of the variation and that many other factors, such as the physical environment of a seamount and distance to the sources of colonisation, will also determine species composition (Rowden *et al.*, 2005). The reasons for significant differences in the depth distribution of species are complex but likely related to the ASH depth, the quantity and abundance of food at different depths and the nature of substrata available for attachment of species (see earlier comments).

Vulnerability and conservation

Harvesting of precious corals

Several octocorals, zoanthids and antipatharians are highly valued as a raw material for making jewellery and decorative objects. These species have been harvested from seamounts and islands in the vicinity of Hawaii since the 1950s and from the Emperor Seamount chain since the 1960s. These corals are all slow-growing, long-lived species that have low levels of natural mortality and recruitment (see above; Grigg, 1984). As a result, populations can only sustain a very low level of harvesting. The history of precious coral fisheries is one of frequent depletion (Grigg, 1984, 1986). Apart from the direct impacts of such fisheries on the target species, tangle-net dredges cause significant damage to other benthic sessile megafauna. The use of coral beds as foraging areas for the Hawaiian monk seals (see above) also means that such fisheries may impact on this extremely rare and endangered species.

Fishing

The South Australian Seamounts provide striking evidence for the impact of trawling on deep-water coral reefs. These seamounts have been subject to intensive trawling for orange roughy (*Hoplostethus atlanticus*) and oreo (*Pseudocyttus* spp., *Allocyttus* spp., *Neocyttus* spp.). On the most intensively trawled seamounts, the deep-water coral reefs, formed mainly by *S. variabilis*, have been totally removed or reduced to rubble (Koslow and Gowlett-Holmes, 1998; Koslow *et al.*, 2001). Deeper unfished seamounts hosted a rich deep-water coral reef community with a high proportion of undescribed and potentially endemic species. Photographic surveys on seamounts off New Zealand also identified a negative correlation between trawling intensity and the coverage of coral frameworks formed by *S. variabilis* (Clark and O'Driscoll, 2003). Trawling intensity, especially on small seamounts, has been found to be extremely intense with up to 17400 km of tows per km^2 of seamount summit, explaining the large-scale destruction of coral communities on some seamounts (O'Driscoll and Clark, 2005). In many other parts of the world where trawling has been observed to coincide with the occurrence of cold-water corals, large bycatches of corals have been observed in the fisheries, and there has been photographic evidence of widespread destruction of benthic communities (reviewed by Freiwald *et al.*, 2004).

As discussed, deep-sea corals take thousands of years to develop. Data on the recruitment of the larvae of deep-sea corals are very scarce, although at least one study on *Primnoa* spp. in the Gulf of Alaska has suggested that this may be sporadic (Krieger, 2001). Genetic and reproductive studies suggest that trawling can reduce deep-water coral colonies to a size where sexual reproduction is no longer viable (Le Goff *et al.*, 2004). Given these factors, recovery of deep-water corals reefs from significant trawling impacts is likely to be extremely slow and where the habitat is altered may never happen (Hall-Spencer *et al.*, 2002). The environmental impacts of trawling have a high risk of causing species extinction on seamounts and other deep-sea areas if they possess endemic species. In addition, whilst the links between the abundance of commercially valuable species of fish and the presence of coral communities are unproven the possibility still remains.

Climate change

The interaction between the background current field, the pelagic ecosystem and the communities living on or around seamounts is particularly strong (Rogers, 1994; Genin, 2004). In the past, natural climate change has dramatically affected the distribution of deep-sea corals (Rogers, 1999). It is likely that present climate change, driven by the burning of fossil fuels, will also impact the distribution of deep-sea corals through local and regional changes in primary productivity, organic carbon flux, the position and strength of major ocean currents and as a result of ocean acidification (Glover and Smith, 2003; Guinotte *et al.*, 2006). Ocean acidification is of particular concern as over the next 100 years there may be marked changes in the depth of the ASH of the oceans, especially in the Southern Ocean, Sub-Arctic Pacific and northern N Atlantic (Orr *et al.*, 2005). The effects of under-saturation of carbonates on corals are not well understood, but will at least decrease the rate of calcification of cold-water corals and will become corrosive to dead coral skeletons that are composed of aragonite, that form most of the habitat of deep-sea reefs. Overall, there may be large-scale changes in the faunal composition of seamount communities, especially where corals play a role in structuring the environment and providing habitats for other species.

Conclusions

Corals are important components of seamount ecosystems in terms of their diversity and their role in providing habitat for other species. It is likely that trophic focusing plays a significant role in maintenance of seamount coral communities. Other physical factors at scales from metres to thousands of kilometres also play significant roles in determining the distribution of coral on seamounts. The different coral groups respond to these factors in contrasting ways, leading to differences in regional and vertical distribution. Knowledge of the life history of corals on seamounts is limited, although patterns of sexuality, larval development and timing of reproduction vary amongst the species that have been studied. It is now understood that many species of deep-sea corals are slow growing and long lived. The conservative life history of corals and their fragility means that they are vulnerable to impacts from deep-sea fishing, especially trawling. However, corals are also sensitive to changes in the physical environment and the effects of global climate change represent a significant threat to corals on seamounts and other habitats globally.

Acknowledgements

Valuable assistance was given by Linda Ward and Dr Stephen Cairns, Smithsonian Institution, Washington, DC, in locating coral records at the museum. Dr Alex Rogers and Huw Griffiths would like to thank Prof. Christopher Rapley, Prof. Paul Rodhouse and Dr Alistair Crame for providing the facilities required for execution of this chapter. Finally, Dr Alex Rogers would like to dedicate this work to the memory of Prof. John Gage, who encouraged his first foray into the biology of seamounts and will be missed by the global community of deep-sea scientists.

References

Adkins, J.F., Cheng, H., Boyle, E.A., Druffel, E.R.M. and Edwards, R.L. (1998) Deep-Sea coral evidence for rapid change in ventilation of the deep North Atlantic rapid change in ventilation of the deep North Atlantic 15,400 years ago. *Science*, 280, 725–8.

Andrews, A.H., Cordes, E.E., Mahoney, M.M., Munk, K., Coale, K.H., Cailliet, G.M. and Heifelz J. (2002) Age growth and radiometric age validation of a deep-sea habitat-forming gorgorian (*Primnoa resedaeformis*) from the behalf Alaska. *Hydrobiologia*, 471, 101–10.

Andrews, A.H., Cailliet, G.M., Kerr, L.A., Coale, K.H., Lundstrom, C. and De Vogelaere, A.P. (2005a) Investigations of age and growth for three deep-sea corals from the Davidson Seamount off central California. In: *Cold-Water Corals and Ecosystems* (eds. Freiwald, A. and Roberts, J.M.), pp. 1021–38. Springer-Verlag, Berlin/Heidelberg.

Andrews, A.H., Tracey, D.M., Neil, H., Cailliet, G.M. and Brooks, C.M. (2005b) *Proceedings of the Third International Symposium on Deep-Sea Corals*, November 28–December 2, 2005. University of Miami, p. 79. Miami, Florida, USA.

Auster, P.J., Moore, J., Heinonen, K.B. and Watling, L. (2005) A habitat classification scheme for seamount landscapes: assessing the functional role of deep-water corals as fish habitat. In: *Cold-Water Corals and Ecosystems* (eds. Freiwald, A. and Roberts, J.M.), pp. 761–9. Springer-Verlag, Berlin/Heidelberg.

Bechtel, J., Werner, T.B., Llewellyn, G., Salm, R.V. and Allen, G.R. (2004) Coral triangle. In: *Defying Ocean's End: An Agenda for Action* (eds. Glover, L.K. and Earle, S.), pp. 89–103. Island Press, Washington, DC.

Boland, R.C. and Parrish, F.A. (2005) A description of fish assemblages in the black coral beds of Lahaina, Maui, Hawai'i. *Pacific Science*, 59, 411–20.

Briggs, J.C. (1995) *Global Biogeography*. Elsevier, Amsterdam.

Briggs, J.C. (2000) Centrifugal speciation and centres of origin. *Journal of Biogeography*, 27, 1183–8.

Briggs, J.C. (2003) Marine centres of origin as evolutionary engines. *Journal of Biogeography*, 30, 1–18.

Briggs, J.C. (2005) The marine East Indies: diversity and speciation. *Journal of Biogeography*, 32, 1517–22.

Brodeur, R.D. (2001) Habitat-specific distribution of Pacific Ocean perch (*Sebastes alutus*) in Pribilof Canyon, Bering Sea. *Continental Shelf Research*, 21, 207–24.

Buhl-Mortensen, L. and Mortensen, P.B. (2004) Crustaceans associated with the deep-water gorgonian corals *Paragorgia arborea* (L., 1758) and *Primnoa resedaeformis* (Gunn., 1763). *Journal of Natural History*, 38, 1233–47.

Burgess, S. and Babcock, R.C. (2005) Reproductive ecology of the three reef-forming, deep-sea corals in the New Zealand region. In: *Cold-Water Corals and Ecosystems* (eds. Freiwald, A. and Roberts, J.M.), pp. 701–13. Springer-Verlag, Berlin/Heidelberg.

Cairns, S.D. (1995) The marine fauna of *New Zealand*: Soleractinia (Cnidaria: Anthozoa). *New Zealand* Oceanoguaphic Institute Memoris, 103, 210 pp.

Clark, M. and O'Driscoll, R. (2003) Deepwater fisheries and aspects of their impact on seamount habitat in New Zealand. *Journal of Northwestern Atlantic Fisheries Science*, 31, 441–58.

Cohen, A.L., Smith S.R., McCartney, M.S. and VanEtten, J. (2004) How brain corals record dimate: an integration of skeletal structure, growth and chemistry of *Diploria labyrinthiformis* from Bermuda. *Marine ecology Progress Series,* 271, 147–58.

Cook, W.M., Lane, K.T., Foster, B.L. and Holt, R.D. (2002) Island theory, matrix effects and species richness patterns in habitat fragments. *Ecology Letters*, 5, 619–23.

Cornell, H.V. and Lawton, J.H. (1992) Species interactions, local and regional processes, and limits to the richness of ecological communities: a theoretical perspective. *Journal of Animal Ecology*, 61, 1–12.

De Mol, B. and Shipboard Party of IODP LEG 307 (2005) IODP Expedition 307 unravelled the deep secrets of the cold-water coral banks in the Porcupine Seabight. *Proceedings of the Third International Symposium on Deep-Sea Corals Science and Management*, Miami, FL. University of Florida, IFAS, pp. 19–20.

De Vogelaere, A.P., Burton, E.J., Trejo, T., King, C.E., Clague, D.A., Tamburri, M.N., Cailliet, G.M., Kochevar, R.E. and Douros, W.J. (2005) Deep-sea corals and resource protection at the Davidson Seamount, California, USA. In: *Cold-Water Corals and Ecosystems* (eds. Freiwald, A. and Roberts, J.M.), pp 1189–98. Springer-Verlag, Berlin/Heidelberg.

Dower, J.F. and Brodeur, R.D. (2004) The role of biophysical coupling in concentrating marine organisms around shallow topographies. *Journal of Marine Systems*, 50, 1–2.

Druffel, E.R.M., Griffin, S., Witter, A., Nelson, E., Southan J., Kashgarian, M. and Vogel, J. (1995) *Gerardia*: brisklecone pine of the deep-sea? *Geochimica et Casmochimica Acta*, 59, 5031–6.

Duineveld, G.C.A., Lavalaye, M.S.S. and Berghuis, E.M. (2004) Particle flux and food supply to a seamount cold-water coral community (Galicia Bank, NW Spain). *Marine Ecology Progress Series*, 277, 13–23.

Etnoyer, P. and Morgan, L.E. (2005) Habitat-forming deep-sea corals in the Northeast Pacific Ocean. In: *Cold-Water Corals and Ecosystems* (eds. Freiwald, A. and Roberts, J.M.), pp. 331–43. Springer-Verlag, Berlin/Heidelberg.

Fadlallah, Y.H. (1983) Sexual reproduction, development and larval biology in scleractinian corals: a review. *Coral Reefs*, 2, 129–50.

Farrow, G.E. and Durant, G.P. (1985) Carbonate-basaltic sediments from Cobb Seamount, Northeast Pacific: zonation, bioerosion and petrology. *Marine Geology*, 65, 73–102.

Freiwald, A. (1998) *Geobiology of Lophelia pertusa (Scleractinia) reefs in the North Atlantic.* Habilitationsschrift zur Erlangung der venia legendi am Fachbereich Geowissenschaften der Universität Bremen, 116 pp.

Freiwald, A., Hühnerbach, V., Lindberg, B., Wilson, J.B. and Campbell, J. (2002) The Sula reef complex, Norwegian shelf. *Facies*, 47, 179–200.

Freiwald, A., Fosså, J.H., Grehan, A., Koslow, T. and Roberts, J.M. (2004) *Cold-Water Corals: Out of Sight – No Longer Out of Mind*, 84 pp. UNEP World Conservation Monitoring Centre, Cambridge, UK.

Froese, R. and Sampang, A. (2004) Taxonomy and biology of seamount fishes. In: *Seamounts: Biodiversity and Fisheries* (eds. Morato, T. and Pauly, D.). *Fisheries Centre Research Reports*, 12(5). Fisheries Centre, University of British Columbia, Vancouver, BC, Canada. ISSN 1198-6727, pp. 25–31.

Gage, J.D. and Tyler, P.A. (1991) *Deep-Sea Biology: A Natural History of the Deep-Sea Floor*, 504 pp. Cambridge University Press, Cambridge, UK.

Gass, S.E. and Roberts, J.M. (2006) The occurrence of the cold-water coral *Lophelia pertusa* (Scleractinia) on oil and gas platforms in the North Sea: Colony growth, recruitment and environmental controls on distribution. *Marine Pollution Bulletin*, 52, 549–59.

Genin, A. (2004) Bio-physical coupling in the formation of zooplankton and fish aggregations over abrupt topographies. *Journal of Marine Systems*, 50, 3–20.

Genin, A., Dayton, P.K., Lonsdale, P.F. and Speiss, F.N. (1986) Corals on seamount peaks provide evidence of current acceleration over deep topography. *Nature*, 322, 59–61.

Genin, A., Haury, L. and Greenblatt, P. (1988) Interactions of migrating zooplankton with shallow topography: predation by rockfishes and intensification of patchiness. *Deep-Sea Research*, 35, 151–75.

Glover, A.G. and Smith, C.R. (2003) The deep-sea floor ecosystem: current status and prospects of anthropogenic change by the year 2025. *Environmental Conservation*, 30, 219–41.

Grigg, R.W. (1974) Distribution and abundance or precious corals in Hawaii. In: *Proceedings of the Second International Coral Reef Symposium* (eds. Cameron, A.M., Campbell, B.M., Cribb, A.R., Endean, R., Jell, J.S., Jones, O.A., Mather, P. and Talbot, F.H.), Vol. 2, 235–40. The Great Barrier Reef Committee, Brisbane, Australia.

Grigg, R.W. (1984) Resource management of precious corals: a review and application to shallow-water reef building corals. *Marine Ecology*, 5, 57–74.

Grigg, R.W. (1986) Precious corals: an important seamount fishery resource. In: *The Environment and Resources of Seamounts in the North Pacific* (eds. Uchida, R.N., Hayasi, S. and Boehlert, G.W.). *Proceedings of the Workshop on the Environment and Resources of Seamounts in the North Pacific*. US Department of Commerce, *NOAA Technical Report*, NMFS 43, pp. 43–44.

Grigg, R.W., Malahoff, A., Chave, E.H. and Landahl, J. (1987) Seamount benthic ecology and potential environmental impact from manganese crust mining. In: *Seamounts, Islands and Atolls* (eds. Keating, B.H., Fryer, P., Batiza, R. and Boehlert, G.W.). *Geophysical Monograph*, 43. American Geophysical Union, Washington, DC, pp. 379–90.

Guinotte, J.M., Orr, J., Cairns, S., Freiwald, A., Morgan, L. and George, R. (2006) Will human-induced changes in seawater chemistry alter the distribution of deep-sea scleractinian corals? *Frontiers in Ecology and the Environment*, 1, 141–6.

Hall-Spencer, J.M., Allain, V. and Fosså, J.H. (2002) Trawling damage to Northeast Atlantic ancient coral reefs. *Proceedings of the Royal Society of London B*, 269, 507–11.

Haury, L., Fey, C., Newland, C. and Genin, A. (2000) Zooplankton distribution around four eastern North Pacific seamounts. *Progress in Oceanography*, 45, 69–105.

Heads, M. (2005) Towards a panbiogeography of the seas. *Biological Journal of the Linnean Society*, 84, 675–723.

Hillebrand, H. (2004) Strength, slope and variability of marine latitudinal gradients. *Marine Ecology Progress Series*, 273, 251–67.

Holligan, P.M. and Groom, S.B. (1986) Phytoplankton distributions along the shelf break. *Proceedings of the Royal Society of Edinburgh B*, 88, 230–63.

Holt, R.D. (1992) A neglected facet of island biogeography: the role of internal spatial dynamics in area effects. *Theoretical Population Biology*, 41, 354–71.

Hubbs, C.L. (1959) Initial discoveries of fish faunas on seamounts and offshore banks in the eastern Pacific. *Pacific Science*, 13, 311–16.

Husebø, A., Nøttestad, L., Fosså, J.H., Furevik, D.M. and Jørgensen, S.B. (2002) Distribution and abundance of fish in deep-sea coral habitats. *Hydrobiologia*, 471, 19–28.

ICES (2005) *Report of the Working Group on Deep-Water Ecology (WGDEC)*, 8–11 March 2005, ICES Headquarters, Copenhagen. ICES CN 2005/ACE:02, 76 pp.

Jensen, A. and Frederiksen, R. (1992) The fauna associated with the bank-forming deepwater coral *Lophelia pertusa* (Scleractinia) on the Faroe Shelf. *Sarsia*, 77, 53–69.

Jones, C.G., Lawton, J.H. and Shachek, M. (1994) Organisms as ecosystem engineers. *Oikos*, 69, 373–86.

Jonsson, L.G., Nilsson, P.G., Floruta, F. and Lundälv, T. (2004) Distributional patterns of macro- and megafauna associated with a reef of the cold-water coral *Lophelia pertusa* on the Swedish west coast. *Marine Ecology Progress Series*, 284, 163–71.

Karlson, R.H., Cornell, H.V. and Hughes, T.P. (2004) Coral communities are regionally enriched along an oceanic biodiversity. *Nature*, 429, 867–70.

Koslow, J.A. and Gowlett-Holmes, K. (1998) The seamount fauna off southern Tasmania: benthic communities, their conservation and impacts of trawling. Final Report to Environment Australia and the Fisheries Research Development Corporation. FRDC Project 95/058, CSIRO, Australia, 104 pp.

Koslow, J.A., Gowlett-Holmes, K., Lowry, J.K., O'Hara, T., Poore, G.C.B. and Williams, A. (2001) Seamount benthic macrofauna off southern Tasmania: community structure and impacts of trawling. *Marine Ecology Progress Series*, 213, 111–25.

Krieger, K.J. (2001) Coral (*Primnoa*) impacted by fishing gear in the Gulf of Alaska. In: *Proceedings of the First International Symposium on Deep-Sea Corals* (eds. Wilson, J.H.M., Hall, J., Gass, S.E., Kenchington, E.L.R., Butler, M. and Doherty, P.). Ecology Action Centre, Halifax, Canada, pp. 106–16.

Krieger, K.J. and Wing, B.L. (2002) Megafauna associations with deep-water corals (*Primnoa* spp.) in the Gulf of Alaska. *Hydrobiologia*, 471, 83–90.

Krupp, D.A. (1983) Sexual reproduction and early development of the solitary coral *Fungia scutaria* (Anthozoa: Scleractinia). *Coral Reefs*, 2, 159–64.

Le Goff-Vitry, M.C., Pybus, O.G. and Rogers, A.D. (2004) Genetic structure of the deep-sea coral *Lophelia pertusa* in the North East Atlantic revealed by microsatellites and ITS sequences. *Molecular Ecology*, 13, 537–49.

Lopez-Correa, M., Freiwald, A., Hall-Spencea, J.M. and Taviani, M. (2005) Distribution and habitats of *Acesta excavata* (Bivalvia: Limidae) with newdata on its shell ultranstructure. In: *Cold-water corals and Ecasystems* (eds. Freiwald, A. and Roberts J.M.), pp. 173–205. Springer-Verlag, Berlin/Heidelberg.

MacArthur, R.H. and Wilson, E.O. (1967) *The Theory of Island Biogeography. Monographs in Population Biology*, 1. Princeton University Press, Princeton, NJ.

Mariscal, R.N. and Bigger, C.H. (1977) Possible ecological significance of octocoral epithelial ultrastructure. In: *Proceedings of the Third International Coral Reef Symposium* (ed. Taylor, D.H.), Vol. 1. University of Miami, Rosentiel School of Marine and Atmospheric Science, Miami, FL, pp. 127–34.

Matsumoto, A.K. (2005) Recent observations on the distribution of deep-sea coral communities on the Shiribeshi Seamount, Sea of Japan. In: *Cold-Water Corals and Ecosystems* (eds. Freiwald, A. and Roberts, J.M.), pp. 345–56. Springer-Verlag, Berlin/Heidelberg.

Mortensen, P.B., Hovland, M., Brattegard, T. and Farestveit, R. (1995) Deep-water bioherms of the scleractinian coral *Lophelia pertusa* (L.) at 64 N on the Norwegian shelf: structure and associated megafauna. *Sarsia*, 80, 145–58.

Moskalev, L.I. and Galkin, S.V. (1986) Investigations of the fauna of submarine upheavals during the 9th trip of the research vessel 'Academic Mstislav Keldysh'. *Zoologicheskii Zhurnal*, 65, 1716–20.

Mundy, B.C. and Parrish, F.A. (2004) New records of the fish genus *Grammatonotus* (Teleostei: Perciformes: Percoidei: Callanthiidae) from the Central Pacific, including a spectacular species in the northwestern Hawaiian Islands. *Pacific Science*, 58, 403–17.

Myers, A.A. and Hall-Spencer, J. (2004) A new species of amphipod crustacean, *Pleusymtes comitari* sp. nov., associated with gorgonians on deep-water coral reefs off Ireland. *Journal of the Marine Biological Association of the United Kingdom*, 84, 1029–1032.

O'Driscoll, R.L. and Clark, M.R. (2005) Quantifying the relative intensity of fishing on New Zealand seamounts. *New Zealand Journal of Marine and Freshwater Research*, 39, 839–50.

Orr, J.C., Fabry, V.J., Aumont, O., Bopp, L., Doney, S.C., Feely, R.A., Gnanadesikan, A., Gruber, N., Ishida, A., Joos, F., Key, R.M., Lindsay, K., Maier-Reimer, E., Matear, R., Monfray, P., Mouchet, A., Najjar, R.G., Plattner, G.K., Rodgers, K.B., Sabine, C.L., Sarmiento, J.L., Schlitzer, R., Slater, R.D., Totterdell, I.J., Weirig, M.F., Yamanaka, Y. and Yool, A. (2005) Anthropogenic ocean acidification over the twenty-first century and its impacts on calcifying organisms. *Nature*, 437, 681–6.

Parin, N.V., Mironov, A.N. and Nesis, K.N. (1997) Biology of the Nazca and Sala Y Gómez submarine ridges, an outpost of the Indo-West Pacific fauna in the eastern Pacific Ocean: composition, and distribution of the fauna, its communities and history. *Advances in Marine Biology*, 32, 145–242.

Parrish, F.A., Abernathy, K., Marshall, G.J. and Buhleier, B.M. (2002) Hawaiian monk seals (*Monachus schauinslandi*) foraging in deep-water coral beds. *Marine Mammal Science*, 18(1), 244–58.

Parrish, F.A. and Baco, A.R. State of the US Deep Coral Ecosystems in the Hawaiian Archipelago and the United States Pacific Islands Region (Chapter 8). NOAA Technical Memorandum NMFS-OPR-29 (in press).

Paulay, G. (1996) Diversity and distribution of reef organisms. In: *Life and Death of Coral Reefs* (ed. Birkeland), pp. 298–353. Chapman and Hall, New York.

Pearse, J.S. (1994) Cold-water echinoderms break 'Thorson's rule'. In: *Reproduction, Larval Biology and Recruitment in the Deep-Sea Benthos* (eds. Eckelbarger, K.J. and Young, C.M.), pp. 26–39. Columbia University Press, New York.

Pearse, J.S. and Lockhart, S.J. (2004) Reproduction in cold water: paradigm changes in the 20th century and a role for cidaroid sea urchins. *Deep-Sea Research II*, 51, 1533–49.

Piepenberg, D. and Müller, B. (2004) Distribution of epibenthic communities on the Great Meteor Seamount (North-East Atlantic) mirrors pelagic processes. *Archive of Fishery and Marine Research*, 51, 55–70.

Pingree, R.D. and Mardell, G.T. (1981) Slope turbulence, internal waves and phytoplankton growth at the Celtic Sea shelf break. *Philosophical Transactions of the Royal Society of London A*, 302, 663–82.

Remia, A. and Taviani, M. (2005) Shallow-buried Pleistocene *Madrepora*-dominated coral mounds on a muddy continental slope, Tuscan Archipelago, NE Tyrrhenian Sea. *Facies*, 50, 419–25.

Ribes, M., Coma, R. and Rossi, S. (2003) Natural feeding of the temperate asymbiotic octocoral-gorgonian *Leptogorgia sarmentosa* (Cnidaria: Octocorallia). *Marine Ecology Progress Series*, 254, 141–50.

Richer de Forges, B., Koslow, J.A. and Poore, G.C.B. (2000) Diversity and endemism of the benthic seamount fauna in the southwest Pacific. *Nature*, 405, 944–7.

Richmond, R.H. (1996) Reproduction and recruitment in corals: critical links in the persistence of reefs. In: *Life and Death of Coral Reefs* (ed. Birkeland, C.), pp. 175–97. Chapman and Hall, New York.

Richmond, R.H. and Hunter, C.L. (1990) Reproduction and recruitment of corals: comparisons among the Caribbean, the Tropical Pacific and the Red Sea. *Marine Ecology Progress Series*, 60, 185–203.

Risk M.J., Hall-Spencer, J.M. and Williams, B. (2005) Climate records from the Faroe-Scotland Channelusing *Lophelia pertusa* (Linneaus 1758). In: *Cold-WaterCorals and Ecosystems* (eds. Friewald, A. and Roberts, J.M.), pp. 1097–1108. Springea-Verlag, Berlin/Heidelberg.

Roark, E.B., Fallon, S., Guilderson, T.P., Dunbar, R.B., McCulloch, M. and Ingram, B.L. (2005) Development of radiocarbon, trace element and stable isotopic records from a deep sea coral: Isididae sp. *Proceedings of the Third International Symposium on Deep-Sea Corals Science and Management*, Miami, FL. University of Florida, IFAS, p. 64.

Roberts, J.M. and Gage, J.D. (2003) Scottish Association of Marine Science. Workpackage 3 of ACES Project: 'To describe the deep-water coral ecosystem, its dynamics and functioning; investigate coral biology and behaviour and assess coral sensitivity to natural and anthropogenic stressors'. Final Report to the Atlantic Coral Ecosystem Study. Internal Report, SAMS, Oban, Scotland, 95 pp.

Roberts J.M., Wheeler, A.J. and Freiwald, A. (2006) Reefs of the deep: the biology and geology of cold-water coral ecosystems. *Science*, 312, 543–7.

Rogers, A.D. (1994) The biology of seamounts. *Advances in Marine Biology*, 30, 305–50.

Rogers, A.D. (1999) The biology of *Lophelia pertusa* (Linnaeus 1758) and other deep-water reef-forming corals and impacts from human activities. *International Review of Hydrobiology*, 84, 315–406.

Rowden, A.A., Clark, M.R. and Wright, I.C. (2005) Physical characterisation and a biologically focused classification of 'seamounts' in the New Zealand region. *New Zealand Journal of Marine and Freshwater Research*, 39, 1039–59.

Sammarco, P.W. (1982) Polyp bail out: an escape response to environmental stress and a new means of reproduction in corals. *Marine Ecology Progress Series*, 10, 57–8.

Schröder-Ritzrau, A., Freiwald, A. and Mangini, A. (2005) U/Th-dating of deep-water corals from the eastern North Atlantic and western Mediterranean. In: *Cold-Water Corals and Ecosystems* (eds. Freiwald, A. and Roberts, J.M.), pp. 157–72. Springer-Verlag, Berlin/Heidelberg.

SeamountsOnline. http://www.seamounts.sdsc.edu/ (last accessed August 9, 2006).

Seki, M.P. and Somerton, D.A. (1994) Feeding ecology and daily ration of the pelagic armourhead *Pseudopentaceros wheeleri* at Southeast Hancock Seamount. *Environmental Biology of Fishes*, 39, 73–84.

Sherwood, O.A., Scott, D.B., Risk M.J. and Guilderson, T.P. (2005) Radiocarbon evidence for annual growth rings in the deep-sea octocoral *Primnoa resedaeformis. Marine Ecology Progress Series*, 301, 129–34.

Sherwood, O.A., Scott D.B. and Risk, M.J. (2006) Late Holocene radiocarbon and aspartic acid racemization dating of deep-sea octocorals. *Geochimica et Cosmochimica Acta*, 70, 2806–14.

Sinclair D.J., Sherwood, O.A., Risk M.J., Hillaire-Marcel, C., Tubrett, M., Sylvester, P., McCulloch, M. and Kinsley, L. (2005) Testing the reproducibility of Mg/Ca profiles in the deep-water coral *Primnoa resedaeformis*: putting the proxy through its paces. In: Cold-Water Corals and Ecosystems (eds. Freiwald, A. and Roberts J.M.), pp. 1039–62.

Smith, J.E., Risk M.J., Schwarz, H.P. and McConnaughey, T.A. (1997) Rapid climate change in the North Atlantic during the younger Dryas recorded by deep-sea corals. *Nature*, 386, 818–20.

Smith, P.J., McVeagh, S.M., Mingoia, J.T. and France, S.C. (2004) Mitochondrial DNA sequence variation in deep-sea bamboo coral (Keratoisidinae) species in the southwest and northwest Pacific Ocean. *Marine Biology*, 144, 253–61.

Stocks, K. (2004) Seamount invertebrates: composition and vulnerability to fishing. In: *Seamounts: Biodiversity and Fisheries* (eds. Morato, T. and Pauly, D.). *Fisheries Centre Research Reports*,

12(5). Fisheries Centre, University of British Columbia, Vancouver, BC, Canada. ISSN 1198-6727, pp. 17–24 + Appendices.

Stone, G.S., Madin, L.P., Stocks, K., Hovermale, G., Hoagland, P., Schumacher, M., Etnoyer, P., Sotka, C. and Tausig, H. (2004) Seamount biodiversity, exploitation and conservation. In: *Defying Ocean's End: An Agenda for Action* (eds. Glover, L.K. and Earle, S.), pp. 43–70. Island Press, Washington, DC.

Stone R.P. (2006) Coral habitat in the Aleutian Islands of Alaska: depth distribution, fine-scale species associations, and fisheries interactions. *Coral reefs*, 25, 229–38.

Thiem, Ø., Ravagnan, E., Fosså, J.H. and Berntsen, J. (2006) Food supply mechanisms for cold-water corals along a continental shelf edge. *Journal of Marine Systems*, 60, 207–19.

Thresher, R., Rintoul, S.R., Koslow, J.A., Weidman, C., Adkins, J. and Proctor C. (2004) Oceanic evidence of climate change in southern Australia over the last three centuries. *Geophysical Research Letters*, 31, L07212.

Tracy, D.M., Sanchez, J.A., Neil, H., Marriott, P., Andrews, A.H. and Cailliet, G.M. (2005) Age and growth, and age validation of deep-sea coral family Isididae. *Proceedings of the Third International Symposium on Deep-Sea Corals Science and Management*, Miami, FL. University of Florida, IFAS, p. 80.

Tranter, P.R.G., Nicholson, D.N. and Kinchington, D. (1982) A description of spawning and post-gastrula development of the cool temperate coral *Caryophyllia smithii* (Stokes and Broderip). *Journal of the Marine Biological Association of the UK*, 62, 845–54.

Tudhope, A.W., Chilcott, C.P., McCulloch, M.T., Cook E.R., Chappell. J., Ellam, R.M., Lea, D.W., Lough, J.M. and Shimmield, G.B. (2001) Variability in El Nino-Southern Oscillation through a glacial-interglacial cycle. *Science*, 291, 1511–17.

Tynan, C.T., Ainley, D.G., Barth, J.A., Cowles, T.J., Pierce, S.D. and Spear, L.B. (2005) Cetacean distributions relative to ocean processes in the northern Californian Current System. *Deep-Sea Research II*, 52, 145–67.

Vermeij, G.J. (2004) Island life: a view from the sea. In: *Frontiers of Biogeography: New Directions in the Geography of Nature* (eds. Lomolino, M.V. and Heaney, L.R.), pp. 239–54. Sinauer Associates, Inc., Sunderland, MA.

Waller, R.G. (2005) Deep-water Scleractinia (Cnidaria: Anthozoa): current knowledge of reproductive processes. In: *Cold-Water Corals and Ecosystems* (eds. Freiwald, A. and Roberts, J.M.), pp. 691–700. Springer-Verlag, Berlin/Heidelberg.

Waller, R.G. and Baco-Taylor, A. (2005) Some findings on the reproduction of Hawaiian precious corals. *Proceedings of the Third International Symposium on Deep-Sea Corals Science and Management*, Miami, FL. University of Florida, IFAS, p. 205.

Waller, R.G. and Tyler, P.A. (2005) The reproductive biology of two deep-water, reef-building scleractinians from the NE Atlantic Ocean. *Coral Reefs*, 24, 514–22.

Waller, R.G., Tyler, P.A. and Gage, J.D. (2002) Reproductive ecology of the deep-sea scleractinian coral *Fungiacyathus marenzelleri* (Vaughan, 1906) in the northeastern Atlantic Ocean. *Coral Reefs*, 21, 325–31.

Waller, R.G., Tyler, P.A. and Gage, J.D. (2005) Sexual reproduction in three hermaphroditic deep-sea *Caryophyllia* species (Anthozoa: Scleractinia) from the NE Atlantic Ocean. *Coral Reefs*, 24, 594–602.

Waller, R.G., Tyler, P.A. and Smith, C.R. Fecundity and embryo development of three Antarctic deep-water scleractinians: *Flabellum thouarsii*, *F. curvatum* and *F. impensum*. *Deep-Sea Research II* (in press).

Wessel, P. (2001) Global distribution of seamounts inferred from gridded Geosat/ERS-1 altimetry. *Journal of Geophysical Research*, 106(B9), 19431–41.

Wilkinson, C. (2004) Executive summary. In: *Status of Coral Reefs of the World 2004* (ed. Wilkinson, C.), Vol. 1, pp. 7–66. Australian Institute of Marine Science, Townsville, Queensland, Australia.

Wilson, J.B. (1979a) The distribution of the coral *Lophelia pertusa* (L.) [*L. prolifera* (Pallas)] in the North East Atlantic. *Journal of the Marine Biological Association of the UK*, 59, 149–64.

Wilson, J.B. (1979b) 'Patch' development of the deep-water coral *Lophelia pertusa* (L.) on Rockall Bank. *Journal of the Marine Biological Association of the UK*, 59, 165–77.

Wilson, R.R. and Kaufmann, R.S. (1987) Seamount biota and biogeography. In: *Seamounts, Islands and Atolls* (eds. Keating, B.H., Fryer, P., Batiza, R. and Boehlert, G.W.). *Geophysical Monograph*, 43. American Geophysical Union, Washington, DC, pp. 319–34.

Witman, J.D., Etter, R.J. and Smith, F. (2004) The relationship between regional and local species diversity in marine benthic communities: a global perspective. *Proceedings of the National Academy of Sciences of the USA*, 101, 15664–9.

Worm, B., Lotze, H.K. and Myers, R.A. (2003) Predator diversity hotspots in the blue ocean. *Proceedings of the National Academy of Sciences of the USA*, 100, 9884–8.

Yen, P.P.W., Sydeman, W.J. and Hyrenbach, K.D. (2004) Marine bird and cetacean associations with bathymetric habitats and shallow-water topographies: implications for trophic transfer and conservation. *Journal of Marine Systems*, 50, 79–99.

Young, C.M. (1994) The tale of two dogmas: the early history of deep-sea reproductive biology. In: *Reproduction, Larval Biology and Recruitment of the Deep-Sea Benthos* (eds. Young, C. and Eckelbarger, K.J.), pp. 1–25. Columbia University Press, New York.

Zezina, O.N. (1997) Biogeography of the bathyal zone. *Advances in Marine Biology*, 32, 389–426.

Chapter 9

Seamount fishes: ecology and life histories

Telmo Morato and Malcolm R. Clark

Abstract

Seamounts are biologically distinctive open ocean habitats exhibiting a number of unique features. They have received much attention mainly because of the presence of substantial aggregations of mid- and deep-water fish, which have became the target of highly developed commercial fisheries. Based on life history and ecological characteristics, several authors have placed 'seamount fishes' at the extreme end of the vulnerability spectrum. Although the terms 'seamount' and 'seamount-aggregating' species are in general use, no rigorous criteria yet exist to identify these taxa. Some of the best-known representatives of 'seamount-aggregating fishes' include deep-water fishes like orange roughy, alfonsinos, Patagonian toothfish, oreos, and pelagic armourhead. Many other fish species also occur on seamounts or congregate over their summits to feed. This may be the case for some sharks, tunas, and other large pelagic predators. Other fish species aggregate around shallow seamounts mainly for spawning, such as reef-associated fish like serranids and jacks. This chapter describes seamount fish species, evaluates their adaptations to ecological conditions on seamounts, and describes life history characteristics of frequently occurring seamount species.

Can a 'seamount fishes' category be defined?

What are 'seamount fishes'? This is a simple question, yet the answer remains elusive. The designation of 'seamount fishes' or seamount species has been widely employed (Koslow, 1996; Probert *et al.*, 1997; Probert, 1999; Koslow *et al.*, 2000; Fock *et al.*, 2002; Tracey *et al.*, 2004) however the criteria used in identifying those taxa are rarely defined. Pioneer work (see Chapter 3) focused on the question of what species inhabit individual banks and seamounts? Since then, a large number of studies have described the fish fauna inhabiting these features. The results of early studies have been summarized in a number of reviews (Wilson and Kaufmann, 1987; Rogers, 1994; Froese and Sampang, 2004; Morato *et al.*, 2004), and so exhaustive lists are not given here.

Most fish species appear to occupy a range of habitats. The issue is how to distinguish obligatory seamount dwellers from those more typical of other habitats, or species that span both. Many fish species occur on seamounts or congregate over their summits to feed due to enhanced levels of planktonic production, hydrographic retention mechanisms such as eddies, or because they are able to remain close to the bottom yet reach shallower depths (see Chapters 4, 5, 10, 12, and 14). This may be the case for some commercially important species of deep-water fish, such as orange roughy, *Hoplostethus atlanticus*, pelagic armourhead, *Pseudopentaceros wheeleri*, oreosmatids, such as *Allocyttus niger* and *Pseudocyttus maculatus*, and alfonsinos, *Beryx* spp. and some sharks (see Chapter 10B; Klimley *et al.*, 1988; Hazin *et al.*, 1998), tunas (see Chapter 10A; Holland *et al.*, 1999; Itano and Holland, 2000; Sibert *et al.*, 2000), and other large pelagic predators (see Chapter 12; Ward *et al.*, 2000; Sedberry and Loefer, 2001).

A range of fish species periodically aggregate around shallow seamounts for spawning; for instance, reef-associated fish like serranids (*Mycteroperca rosacea*, *Paranthias colonus*) and jacks (*Caranx sexfasciatus*, *Seriola lalandi*) (Sala *et al.*, 2003). Recently, Tsukamoto *et al.* (2003) discovered the spawning site of the Japanese eel (*Anguilla japonica*) in the northwest Pacific near to three seamounts, 2000–3000 km away from their freshwater habitats. Further examples are the deep-bodied species of the orders Zeiformes (mainly the genera *Antigonia*, *Capros*, *Zenopsis*, and *Cyttopsis*) and Syngnathiformes (in particular the genus *Macroramphosus*), which are the dominant fishes in depths greater than 500 m at the Great Meteor Seamount in the central eastern Atlantic (Fock *et al.*, 2002). These fish are also the main prey of large demersal predators inhabiting the slopes of the Azores islands and seamounts (Morato *et al.*, 1999, 2000, 2001, 2003). However, as well as occurring on seamount features, they are among the most abundant fishes from some adjacent continental shelves.

Coral reef scientists faced exactly the same problem when trying to provide a definition of 'reef fishes' (see Choat and Bellwood, 1991; Bellwood, 1996, 1998; Robertson, 1998). They first tried to find potential taxonomic and ecological characteristics that could distinguish coral fish assemblages from other fish assemblages (Choat and Bellwood, 1991). They also proposed a consensus list of fish families that would better describe, not define, a coral reef assemblage (Bellwood, 1996). They concluded from this list that most reef fishes are characteristic of, but not restricted to, coral reefs (Bellwood, 1996). Coral reef scientists have yet to progress beyond the tautological definition of reef fishes as those that live on coral reefs.

The definition of 'seamount fishes' may be similar and involve the same redundancy with trying to define a functional type of label that applies only in part to the ecology of the species: seamount fishes are those individual fishes that live on seamounts.

Several species found over and around seamounts have been intensely exploitated since the late 1970s because of their high abundance and good flesh quality (see Chapters 16–18), changing the notion that deep-sea fish were too scarce and/or unpalatable to support commercial fisheries. Koslow (1996, 1997) found that 'some fishes occurring on seamounts' differ markedly from other deep-water species in their relatively high levels of food consumption and energy expenditure, low growth and productivity, a robust body composition, and a body plan suited for swimming in strong currents. By addressing this problem in an energetic perspective (see below), he concluded that they appear to form a distinct guild of 'seamount-associated fishes' or 'seamount-aggregating fishes' (Koslow, 1996; Koslow *et al.*, 2000).

In this chapter, we will consider 'seamount fishes' as those species that live on seamounts, 'seamount-aggregating fishes' as those that form large aggregations around these features and that comprise the main target for fisheries on seamounts.

What fishes species occur on seamounts?

Seamount fishes

Numerous studies have described the species richness and diversity of fish fauna on seamounts (see Chapter 13). In their global review of seamount biota, Wilson and Kaufmann (1987) reported some 450 fishes collected from more than 60 seamounts. Rogers (1994) provided a list of 77 commercial species fished on seamounts. Since then, more detailed studies of certain seamounts and seamount chains provide more comprehensive species lists, especially with an increase in exploratory fishing in the last two decades (e.g., Parin *et al.*, 1997; Koslow and Gowlett-Holmes, 1998; Grandperrin *et al.*, 1999; Fock *et al.*, 2002; Clark and Roberts, 2003; Moore *et al.*, 2003; Tracey *et al.*, 2004). Froese and Sampang (2004) found 535 fish species recognized as seamount fishes. Morato *et al.* (2004) augmented Froese and Sampang's (2004) list from additional sources and compiled a total of 798 species of seamount fishes.

These last two studies present the most comprehensive checklist of seamount fishes, even if incomplete. Although the number of known seamount fishes is comparatively small (2.8% of known species), they encompass a third of fish families (165 of 515 known), and about half of the orders. They thus represent a relatively large and unique portion of fish biodiversity (Froese and Sampang, 2004). They have a range of different habitat preferences (associations), as defined in 'Fishbase' (an online fish database: www.fishbase.org). Forty-three species are pelagic, 94 reef associated, 118 demersal, 68 benthopelagic, 223 bathypelagic, and 252 bathydemersal. A large proportion of the seamount fish community is composed of deep-sea fishes, but many shallow-water species are also known to occur on these features (Fig. 9.1). According to Froese and Sampang (2004), only six seamount fishes are included in the 2000 IUCN (World Conservation Union) Red List (Hilton-Taylor, 2000): *Sebastes paucispinis* is listed as 'critically endangered', *Sphoeroides pachygaster* and *Hexanchus griseus* are listed as 'vulnerable', and *Squalus acanthias*, *Dalatias licha*, and *Prionace glauca* are listed as 'lower risk, near threatened'. Other seamount fishes have not been evaluated so far.

Seamount-aggregating fish

Morato *et al.* (2006) compiled a list of 23 fish species that could fall into the 'seamount-aggregating' category (Table 9.1). They acknowledge that this list is preliminary and its accuracy will improve as we gain more knowledge about the ecology of seamount and deep-water fish species. Some of the most well-known representatives of this group include the deep-water fishes: orange roughy, alfonsinos (*Beryx splendens* and *B. decadactylus*), Patagonian toothfish (*Dissostichus eleginoides*), oreos, pelagic armourhead, several species of rockfishes (*Sebastes* spp.) (Koslow, 1996; Koslow *et al.*, 2000) and the roundnose grenadier (*Coryphaenoides rupestris*) (Vinnichenko, 2002a). These species are the main targets of large-scale fisheries that occur on top and around seamounts (see Chapter 17).

Studies of fish composition on seamounts have often reported high levels of endemism, exceeding 40% in one case (e.g., 12%, Wilson and Kaufmann, 1987; 44%, Parin *et al.*,

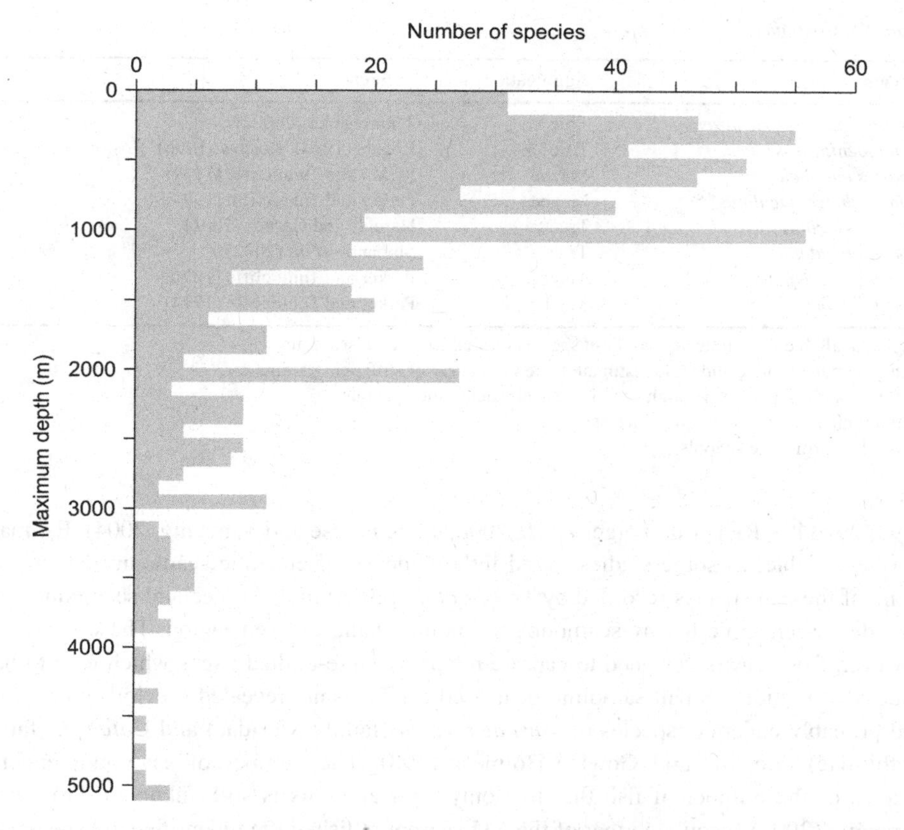

Fig. 9.1 Number of seamount fish species with known depth distribution (n = 694) by lower end of their depth range (adapted from Froese and Sampang, 2004; Morato *et al.*, 2004).

Table 9.1 'Seamount-aggregating' fish species (*sensu* Koslow 1996).

Species	Aggregation	Reference
Alepocephalus bairdii	Maybe	Piñeiro *et al.* (2001), Allain *et al.* (2003)
Allocyttus niger[a]	True	Koslow (1996), Koslow *et al.* (2000)
Allocyttus verrucosus[a]	Maybe	Froese and Pauly (2003)
Aphanopus carbo[b]	True	Vinnichenko (2002b)
Beryx decadactylus	True	Koslow *et al.* (2000), Vinnichenko (2002a)
Beryx splendens	True	Koslow (1996), Koslow *et al.* (2000), Ramos *et al.* (2001), Vinnichenko (2002a)
Coryphaenoides rupestris	True	Hareide and Garnes (2001), Shibanov *et al.* (2002)
Dissostichus eleginoides	True	Koslow *et al.* (2000)
*Epigonus telescopus**	True	Hareide and Garnes (2001), Vinnichenko (2002b)
Hoplostethus atlanticus	True	Koslow (1996), Koslow *et al.* (2000), Hareide and Garnes (2001), Shibanov *et al.* (2002)
Hoplostethus mediterraneus	Maybe	Piñeiro *et al.* (2001)
*Lepidion eques**	Maybe	Piñeiro *et al.* (2001)
Mora moro	Maybe	Piñeiro *et al.* (2001)
*Neocyttus rhomboidalis**,[a]	Maybe	Allain *et al.* (2003)
Pseudocyttus maculatus[a]	True	Koslow (1996), Koslow *et al.* (2000)

(*Continued*)

Table 9.1 *Continued*

Species	Aggregation	Reference
Pseudopentaceros richardsoni	True	Vinnichenko (2002a)
*Pseudopentaceros wheeleri**	True	Rogers (1994), Koslow (1996), Koslow *et al.* (2000)
*Sebastes entomelas**[,c]	Maybe	Parker and Tunnicliffe (1994)
*Sebastes helvomaculatus**[,c]	Maybe	Parker and Tunnicliffe (1994)
Sebastes marinus	True	Hareide and Garnes (2001)
Sebastes mentella	True	Shibanov *et al.* (2002)
Sebastes paucispinis[c]	Maybe	Parker and Tunnicliffe (1994)
Sebastes ruberrimus[c]	Maybe	Parker and Tunnicliffe (1994)

Maybe: available data indicate that these species may aggregate on seamount.
*Intrinsic vulnerability index not estimated due to the lack of sufficient parameters.
[a]Forming large shoals over rough ground near pinnacles and canyons.
[b]Not a typical seamount-aggregating fish.
[c]Juveniles form large schools.

1997; 29–34%, Richer de Forges *et al.*, 2000; 12%, Froese and Sampang, 2004). Estimates can be variable, as some studies found little evidence of endemic seamount fish species. None of the fish species recorded by Tracey *et al.* (2004) on New Zealand seamounts were regarded as endemic to any seamount, seamount chain, or even region. These data were, however, from trawls designed to capture relatively large-bodied fishes which tend to have wider distributions. Small sampling gear used off Tasmania revealed previously unknown, and probably endemic, species of *Paralaemonena* (family Moridae) and *Cataetyx* (family Bythitidae) (Koslow and Gowlett-Holmes, 1998). The number of seamount endemic species or the number of fish that live only on seamounts is still unknown. Froese and Sampang (2004) speculated that of the 535 seamount fished they identified, 62 species are reported from only one seamount, suggesting a high rate of endemism (see Chapter 8). Generally, studies focus on the samples collected from the seamount features, and have not considered how specific the species composition is to seamounts (see Chapter 13).

Ecological conditions existing on seamounts that require certain adaptations by fish

There is some evidence of localized upwelling and eddies around seamounts (Genin *et al.*, 1986; Boehlert, 1988; Lueck and Mudge, 1997). This may enhance primary productivity near the surface (Genin and Boehlert, 1985), but it is unlikely that water could be retained around a seamount long enough for productivity to work its way through the food web to the higher trophic level fish residing on the seamount itself. Evidence for enhanced primary production leading to concentrations of fish over seamounts is sparse (see Chapters 4 and 5).

Several studies support the hypothesis that seamounts sustain large fish communities by trapping migrating, vertically migrating, or advected prey (see Chapter 5) rather than by any increase of primary productivity resulting from current topography interactions, which might increase levels of nutrients in oligotrophic waters found over and around seamounts (e.g., Genin *et al.*, 1988). In addition to the trapping of migrating prey (Rogers, 1994), the regular flow of organisms over and past seamounts may support fish populations

(Tseytlin, 1985; Koslow, 1997). Aggregations of seamount fishes do however appear to subsist very near the margin of energetic sustainability (see Chapters 14 and 15; Koslow, 1997).

Seamounts may also act as pockets of suitable bathymetry where demersal fishes are able to remain close to the bottom at lesser depths than the surrounding seafloor. Seamounts with peaks that may interrupt the bottom of the deep scattering layer effectively bring food to the fish rather than the fish having to ascend into midwater to feed on their zooplankton prey (see Chapter 14). The prey of many commercial demersal fishes, such as orange roughy (e.g., Rosecchi *et al*., 1988) and alfonsino (e.g., Horn and Massey, 1989) is benthopelagic.

How seamount specific are 'seamount fishes'?

There is no question that some species aggregate over or around seamounts, whether for spawning or feeding purposes (e.g., Rogers, 1994), but how does this relate to areas of flatter terrain on the adjacent continental shelf or slope? Tracey *et al*. (2004) examined the species composition and abundance of fish from trawl surveys on seamounts and surrounding slopes in deep-water areas around New Zealand. They found a consistent pattern of lower total species richness on seamounts than on the adjacent slope, and suggested this may be due in part to seamounts having less soft sediment than the slope (and therefore restricting some fish species which require soft sediment habitat). They observed that even though some species could be classed as characteristic of seamounts, for example orange roughy were found on all 10 seamounts surveyed, they were not obligate seamount dwellers were, but found widely on the open slope as well. Only one species (black oreo *A. niger*) occurred more frequently in seamount than slope tows.

Koslow (1997) described several characteristics of seamount-aggregating species around Australia. He noted attributes of high levels of food consumption, relatively high-energy expenditure, low growth rates, robust body form and composition designed for strong swimming. Orange roughy, oreosomatid fishes, pelagic armourhead, and redfish were recognized as having several of these characteristics. As these species are also found on slope areas, it is unclear how typical these findings are, especially those of food consumption and energetic rates. Orange roughy distribution on seamounts can also vary with times of the year, with certain seamounts acting as spawning sites (and during spawning the fish do not feed). Fish can migrate to and from the spawning seamounts. There may also be resident and migratory components of an orange roughy population, which might explain how the fisheries often contract from occurring throughout much of the year initially, to being restricted to the spawning period (Clark and Tracey, 1994). The distribution of orange roughy outside the spawning season appears to be a combination of dispersed fish over a wide area of the slope, and feeding aggregations on seamounts. Energetic adaptations may differ between these 'types' but no data were found.

What can influence seamount fish abundance?

There is a considerable literature dealing with the composition of fish assemblages, and their association with various environmental factors. Aspects of bottom depth, latitude, longitude, sediment type, bottom temperature, and oceanographic water masses are frequently recorded as important in determining fish species composition and abundance (e.g., Haedrich and Merrett, 1990; Koslow, 1993; Koslow *et al.*, 1994; Francis *et al.*, 2002). Seamounts affect local ocean circulation. Current flow can be interrupted by tidal mixing and localized upwellings and eddies are common around these features. The dynamics of these interactions and their effect on biological processes are by no means fully understood (see Chapter 4; Owens and Hogg, 1980; Robinson, 1981; Roden *et al.*, 1982; Roden, 1987; Genin *et al.*, 1989; Eriksen, 1991).

Other studies have addressed the role of environmental factors in explaining species diversity, dominance, or rarity, of seamount fauna; but most have carried out more general examinations of benthic fauna as well as fish. Rogers (1994) concluded that there was morphological and genetic evidence that populations of some seamounts organisms are distinct from populations located on other seamounts, the abyssal plain and continental shelf. Some of the major studies were further summarized by Stone *et al.* (2004), which indicated mixed results, although they also concluded that high levels of endemism were hard to explain if seamount faunas were not isolated to an extent (see Chapters 8 and 13). Wilson and Kaufmann (1987) concluded that deep seamount biota were dominated by widespread or cosmopolitan species as opposed to shallow seamounts that comprised equal amounts of regional and widespread species. Boehlert *et al.* (1994) described seamount populations as 'dependent' due to different larvae settling on different seamounts.

Clark *et al.* (2001) examined the relationships between physical variables and the size of orange roughy populations on New Zealand seamounts. Multiple regression procedures were used to model the effects of the physical variables. Seamount location, depth of the peak, slope of the seamount flanks, and geological 'association' (continental/oceanic) were significant factors in determining stock size in various analyses. These are credible explanations for variability in species composition and abundance, but there was a wide scatter in the data, and results were somewhat inconsistent between seamounts. This reflects the numerous observations that the benthic and fish fauna on nearby seamounts can vary considerably (e.g., Wilson and Kaufmann, 1987; Koslow *et al.*, 2001; Richer de Forges *et al.*, 2000).

Life history characteristics of seamount fishes and seamount-aggregating fishes

Studies on the life history of seamount fishes have focused on a few of the main commercial species, such as orange roughy (e.g., Gordon and Duncan, 1987; Koslow *et al.*, 1995; Horn *et al.*, 1998; Branch, 2001; Lorance *et al.*, 2002); oreos (e.g., Clark *et al.*, 1989; Conroy and Pankhurst, 1989; Lyle and Smith, 1997); redfish (e.g., Cailliet *et al.*, 2001); roundnose grenadier (e.g., Bergstad, 1990; Atkinson, 1995; Elekseyev, 1995; Kelly *et al.*, 1996, 1997; Allain, 2001; Lorance *et al.*, 2001); Patagonian toothfish (Horn, 2002); and alfonsinos (Lehodey *et al.*, 1997; Rico *et al.*, 2001; Gonzalez *et al.*, 2003). These show a wide range of life history strategies and biological parameter values. Based on the specific life history and ecological characteristics, several authors have placed seamount fishes (mainly seamount-aggregating fish) at the extreme end of the vulnerability spectrum (Koslow, 1997; Boyer *et al.*, 2001; Branch, 2001; Clark, 2001). This probably relates mainly to the characteristics exhibited by the deep species like orange roughy and oreos that typically have slow growth, high longevity, low fecundity, and low productivity (in a fisheries sense).

Comparing basic life history characteristics

Is there anything about the life history of seamount-associated fish that might make them more vulnerable to over-fishing than other deep-sea species? Morato *et al.* (2006) compiled life history characteristics of 14 000 'non-seamount fishes', 'seamount fishes', and 'seamount-aggregating fishes'. Data on longevity (T_{Max}), age at maturity (T_m), asymptotic length (L_∞), total fecundity (F_T), von Bertalanffy growth parameter (K), and natural mortality rate (M), showed that seamount fishes and, in particular, seamount-aggregating fishes, have life history traits (Fig. 9.2) that would make them more vulnerable than non-seamount fishes. They found that (Fig. 9.2a) seamount-aggregating fish have a longer lifespan (62.7 years) than seamount fishes (41.5 years), and non-seamount fish (16.9 years). They also suggested that sexual maturation (Fig. 9.2b) occurs later for those species that aggregate on seamounts (12.8 years) than for the other two categories. Comparisons of natural mortality rate (Fig. 9.2c) and the von Bertalanffy growth parameter (K) (Fig. 9.2d) among the three categories of fishes show similar, but reciprocal, trends in longevity and age at maturity. Seamount-aggregating fishes have the lowest natural mortality and von Bertalanffy growth parameter (mean $M = 0.16$ and mean $K = 0.10$), while non-seamount fishes have the highest among the three fish categories (mean $M = 1.05$ and mean $K = 0.57$). A vulnerability index (Morato *et al.*, 2006) indicated

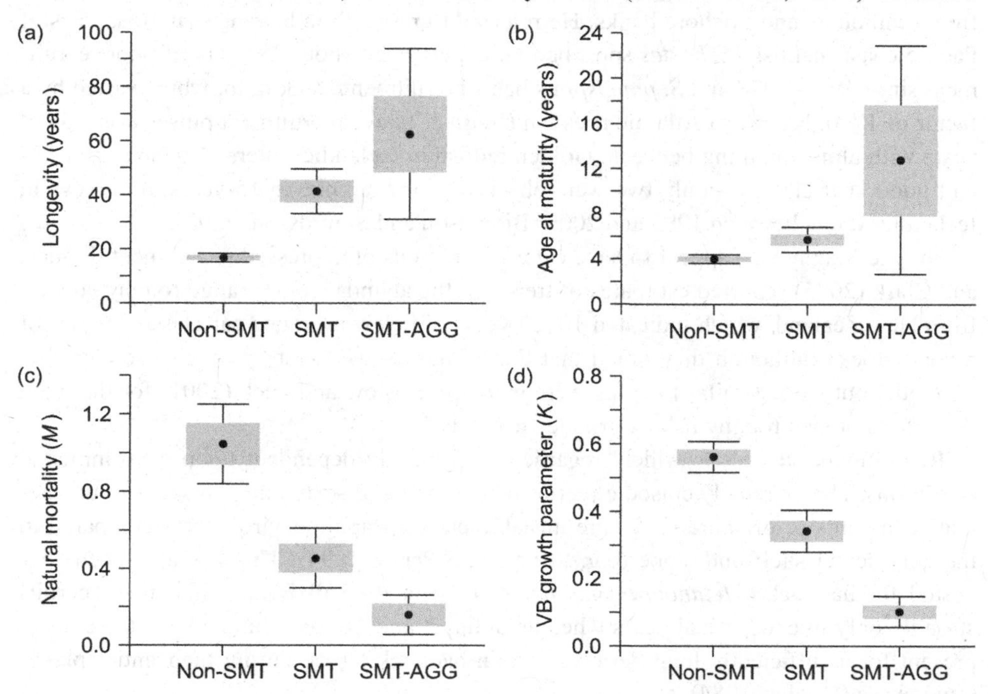

Fig. 9.2 Comparison of some life history characteristics of 'non-seamount fish' species (non-SMT), fish occurring on seamounts (SMT), and 'seamount-aggregating' species (SMT-AGG): (a) longevity (T_{Max}), (b) age at maturity (T_m), (c) natural mortality (M), and (d) von Bertalanffy (VB) growth parameter (K). In the graphs the middle point is the mean, the box is the mean ± SE, and the whisker is the mean ± 95% CL.

seamount species were more at risk than species that did not occur on seamounts, and that aggregating species were even more vulnerable (see later section).

In general, these findings accord with the life history qualities for 'seamount fishes' proposed by Koslow (1996, 1997). It must however be stressed that such generalizations need to be applied carefully. There is considerable variability in life history parameters between seamount species. The overall productivity of alfonsino, for example, is much greater than for orange roughy. The number of species included may also influence the results (see Morato *et al.*, 2006). As we gain more knowledge of what lives on seamounts and what aggregates there, we will be able to further evaluate these findings.

Recruitment

Recruitment level is a key element in the sustainability of any fishery and its consistency over time. As discussed earlier, the life history characteristics of species associated with seamounts are highly variable; hence one cannot generalize about what patterns of any single biological factor may apply. Many of the deeper seamount species exhibit relatively high longevity, with important implications for recruitment.

A number of studies comparing species with differing life histories found a positive relationship between longevity and recruitment variability: the longer the lifespan of a species, the more time it has to breed successfully, and therefore it may have longer period between good year classes (Leaman and Beamish, 1984; Spencer and Collie, 1997; Longhurst, 2002). Longhurst (2002) reviewed information on a number of species from seamounts and offshore banks. He referred to research indicating that 10 species of Pacific coast rockfish (*Sebastes* spp.) had all experienced about 25 years of poor recruitment since 1977–1978 and *S. paucispinis* has a recruitment/stock ratio, which varied by a factor of 1790. Northeast Atlantic rockfish ('redfish') have recruitment pulses every 5–10 years with almost nothing between. Golden redfish in Icelandic waters also have intermittent good year classes – only two were observed, for example, in 15 years of surveys in Icelandic waters between 1985 and 2000 (Bjornsson and Sigurdsson, 2003).

Orange roughy also appear to have extensive periods of depressed recruitment. Francis and Clark (2005) reported estimates of trends in the abundance of orange roughy cohorts from New Zealand, which indicated 10–20 years of below average year class strength for several stocks (although they noted that these findings were not based on age data, but on model outputs). Similar findings were made by Koslow and Tuck (2001) for the main Australian orange roughy fishery around Tasmania.

Recruitment success is widely regarded as primarily dependent upon environmental conditions. The extremely episodic recruitment of some of seamount fish species suggests that seamounts or offshore banks are unstable oceanographic environments compared to the continental shelf and slope (e.g., Myers and Pepin, 1994). This has also been suggested for haddock (*Melanogrammus aeglefinus*) on Rockall Bank. This stock recruits strongly only in exceptional years when an anticyclonic Taylor column above the bank is present for a sufficiently long period to retain haddock larvae during their entire planktonic phase (Dooley, 1984).

The evidence seems overwhelming that the life history traits of each species are tuned to match the problems posed by its natural habitat (Longhurst, 2002). This must be taken into account if such fisheries are to be managed in a sustainable manner.

Residency

The degree of residency of seamount species is an important issue in defining the degree to which a species is regarded as 'seamount' associated. However, little is known about individual fish movements around and between seamounts.

Pelagic armourhead on Hawaiian seamounts have a peculiar life history. Juveniles spend the first 1–2 years of their lives in near surface waters where they build up large fat reserves. The fish then become resident on a seamount for the remaining 3–4 years of life during which time the fat reserves are depleted (Humphreys *et al.*, 1989).

Extensive migrations may occur with alfonsino. Alekseev *et al.* (1986) suggest that alfonsino stocks may have differing distributions at various stages of their life history. Oceanic eddy systems disperse eggs and larvae, and juveniles are separated from adults. Once sexually mature, they migrate back to the parent spawning stock. Tagging studies in Japan however indicate that fish do not migrate over large distances in periods of a year or less (Masuzawa *et al.*, 1975). Regular annual spawning migrations do not appear common. Short-term migrations are unlikely in New Zealand waters, where the size structure remains relatively consistent between fishing grounds sampled over a 12-month period. There is evidence from New Zealand studies that some age-specific migration may occur between seamounts (Horn and Massey, 1989). Alfonsino off New Caledonia are pelagic as larvae and juveniles for several months, and then become more demersal on shallow seamounts (Lehodey and Grandperrin, 1996), later moving to deeper features as they become larger.

Seamount aggregations may vary in composition (mixed species), density, and abundance. Localized aggregations often occur in predictable locations in predictable seasons, making them particularly vulnerable to overexploitation. They can also be dynamic, moving around various flanks of the seamounts, and also into the midwater zone, where they are relatively inaccessible to bottom trawling. Heavy fishing pressure on orange roughy is thought to cause smaller-sized aggregations to break up and disperse (Clark and Tracey, 1991), which can provide some protection from extreme depletion, although such disruption to spawning aggregations may have longer-term implications for spawning success and sustainability of the population.

Vulnerability of 'seamount fishes' to fishing

Responses of a fish species to exploitation may be partly determined by life history and ecological characteristics (Cheung *et al.*, 2005). Fish that mature late and have low growth and mortality rates, likely have higher vulnerability to fishing. Species that display social aggregation behaviours such as shoaling, schooling (Pitcher and Parrish, 1993), or shoal spawning may also have higher vulnerability because of increased catchability, leading to hyperstability of catch rates (Hilborn and Walters, 1992; Pitcher, 1995, 1997; Walters, 2003), and the possible disruption of spawning behaviour by fishing.

Morato *et al.* (2006) used a fuzzy expert system developed by Cheung *et al.* (2005) to predict intrinsic vulnerability to fishing. Cheung *et al.* (2005) defined intrinsic vulnerability as the inherent capacity to respond to fishing that relates to the fish's maximum rate of population growth and strength of density dependence: a fuzzy expert system classifies fishes into different levels of vulnerability based on basic life history and ecological characteristics. The input variables include maximum length, age at first maturity, longevity, von Bertalanffy growth parameter (K), natural mortality rate, fecundity, strength of spatial behaviour, and geographic range. Heuristic rules were incorporated to describe the relationships between these biological traits and fish's intrinsic vulnerability, through which the latter can be predicted. Intrinsic vulnerability was expressed on an arbitrary scale from 1 to 100, with 100 being the most vulnerable. Comparisons with empirical data showed that the fuzzy expert system successfully predicted the intrinsic vulnerability of fishes to fishing.

Morato *et al.* (2006) predicted the intrinsic vulnerability for 1600 species of fish for comparison between 'non-seamount fishes', 'seamount fishes', and 'seamount-aggregating fishes'. They showed that seamount-aggregating fishes have a higher intrinsic vulnerability to exploitation than other fishes (Fig. 9.3). Median intrinsic vulnerabilities of 45.0, 51.8, and 68.2 were estimated for 'non-seamount fishes', 'seamount fishes', and 'seamount-aggregating fishes'. The results confirmed that seamount-aggregating fishes are at the extreme end of the vulnerability spectrum (Koslow, 1997; Boyer *et al*., 2001; Branch, 2001; Clark, 2001). However, as fish vulnerability was strongly related to depth range, seamount association, as in the 'seamount fishes' group, may not be the proximal factor. Higher vulnerability of fish found at seamounts may be compounded because more deep-water species, which are more vulnerable, are included in this category and deep-sea species aggregate on seamounts.

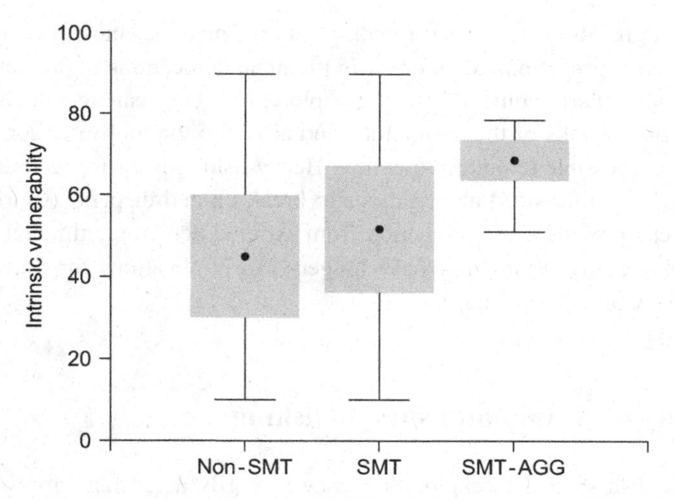

Fig. 9.3 Intrinsic vulnerability (V_I) index for fish species not occurring on seamounts (non-SMT), occurring on seamounts (SMT), and 'seamount-aggregating' species (SMT-AGG). In the graphs the middle point is the median, the box the 25–75% percentiles, and the whisker is the range.

Morato *et al.* (2006) have also shown that the intrinsic vulnerabilities estimated from the fuzzy system were significantly related to population declines of marine fish groups caused by fishing. Groups of species with higher vulnerabilities had larger biomass declines than species with lower vulnerabilities. Even at modest levels of fishing, seamount-aggregating species were depleted, not sustained. The high vulnerability raises serious conservation concerns about the exploitation of seamount fishes (see Chapter 20).

Management implications of 'seamount fishes'

This chapter casts serious doubt on the long-term sustainability of seamount fisheries. Simulation work by Morato *et al.* (2006) suggested that exploitation rates of more than 5% are not sustainable (see Chapter 15). Examples from all over the world have shown the 'boom and bust' pattern of seamount trawl fisheries (see Chapters 17, 19, and 20), with rapid stock reduction and serial depletion of successively exploited new seamounts (Fig. 9.4a–f). The case of the orange roughy, a 'seamount-aggregating fish', is well known.

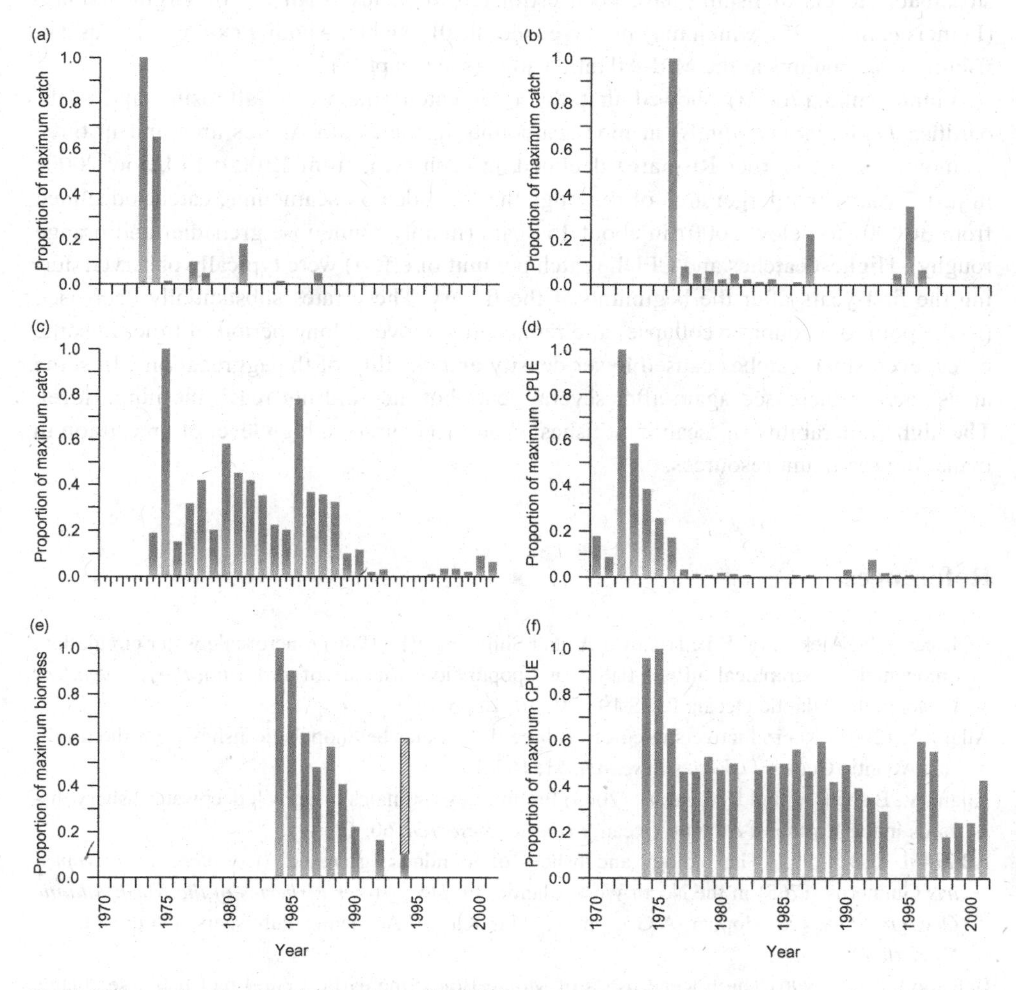

Fig. 9.4 Time series data (catch, biomass, or CPUE) for several seamount fisheries as proportion of the maximum value observed for the time period covered: (a) catches from the South Azores Area by the Soviet/Russian fisheries (9 seamounts; targeting alfonsino, *B. splendens* and scabbardfish, *L. caudatus*); (b) catches from the Corner Rise by the Soviet/Russian fisheries (3 seamounts; targeting alfonsino); (c) catches from the Mid-Atlantic Ridge by the Soviet/Russian fisheries (34 seamounts; targeting roundnose grenadier, *C. rupestris*); (d) CPUE by Japanese trawlers fishing the seamounts in international waters beyond the Hancock Seamounts (targeting Pelagic armourhead, *P. richardsoni*); and (e) estimated trawl survey abundance indices for the northeast Chatham Rise orange roughy stock (in 1994 shows two alternative estimates). (a)–(c) and (f) adapted from Vinnichenko (2002a); (d) adapted from Humphreys and Moffitt (1999); (e) adapted from Clark *et al.* (2000); (f) CPUE by Soviet/Russian fisheries in the Mid Atlantic Ridge.

In Namibian waters, orange roughy were fished down to 10% of virgin biomass in 6 years (Branch, 2001), while in Australia biomass levels dropped to 7–13% in about 15 years (Lack *et al.*, 2003). The orange roughy stocks in New Zealand have been fished down to varying degrees (Clark, 1995; Francis and Clark, 2005), with most below the target level of 30%, and many below 20%. Serial depletion occurred with a number of seamount-based fisheries (Clark, 1999). The main stock on the Chatham Rise was depleted to levels of 15–20%, although strong management action has reduced catch levels, and stock assessment models indicate the population may now be rebuilding (Clark, 2001). Annual sustainable levels of fishing have been estimated to be less than 2% of virgin biomass (Francis *et al.*, 1995), which may not be economically viable. Another example is Russian fishing on seamounts at the Mid-Atlantic Ridge (see Chapter 17).

Vinnichenko (2002a) showed that the total catch (mainly of alfonsino and scabbardfish *Lepidopus caudatus*) at nine seamounts in the South Azores area and in three seamounts at the Corner Rise area declined, in each area, from 12000t to below 2000t in just 2 years. In a larger area of the ridge that included 34 seamounts, catches declined from 30000t to below 2000t in about 15 years (mainly roundnose grenadier and orange roughy). Highest catches and CPUE (catch per unit of effort) were typically observed during the first years after the beginning of the fishery. These rates substantially decreased (to the point of economic collapse) and remained low over a long period of time. In some cases, even small catches caused lower density and stability of the aggregations. In some areas catches increased again after several years but they did not reach the initial level. The high vulnerability of 'seamount fishes' should encourage a high level of precaution in managing seamount resources.

References

Alekseev, F.E., Alekseeva, E.I., Trunov, I.A. and Shibanov, V.I. (1986) Macroscale water circulation, ontogenetic geographical differentiation and population structure of alfoncino, *Beryx splendens* Lowe, in the Atlantic Ocean. ICES 1986/C: 10, 16 pp.

Allain, V. (2001) Reproductive strategies of three deep-water benthopelagic fishes from the northeast Atlantic Ocean. *Fisheries Research*, 51, 165–76.

Allain, V., Biseau, A. and Kergoat, B. (2003) Preliminary estimates of French deepwater fishery discards in the Northeast Atlantic Ocean. *Fisheries Research*, 60, 185–92.

Atkinson, D.B. (1995) The biology and fishery of roundnose grenadier (*Coryphaenoides rupestris* Gunnerus, 1765) in the North West Atlantic. In: *Deep-Water Fisheries of the North Atlantic Oceanic Slope* (ed. Hopper, A.G.), pp. 51–111. Kluwer Academic Publishers, Dordrecht, The Netherlands.

Bellwood, D.R. (1996) The Eocene fishes of Monte Bolca: the earliest coral reef fish assemblage. *Coral Reefs*, 15, 11–19.

Bellwood, D.R. (1998) What are reef fishes? Comment on the report by D.R. Robertson: Do coral-reef fish faunas have a distinctive taxonomic structure? *Coral Reefs*, 17, 187–9.

Bergstad, O.A. (1990) Distribution, population structure, growth and reproduction of the roundnose grenadier *Coryphaenoides rupestris* (Pisces: Macrouridae) in the deep waters of the Skagerrak. *Marine Biology*, 107, 25–39.

Bjornsson, H. and Sigurdsson, T. (2003) Assessment of golden redfish (*Sebastes marinus* L.) in Icelandic waters. *Scientia Marina*, 67, 301–14.

Boehlert, G.W. (1988) Current-topography interactions at mid-ocean seamounts and the impact on pelagic ecosystems. *GeoJournal*, 16, 45–52.

Boehlert, G.W., Wilson, C.D. and Mizuno, K. (1994) Populations of the sternoptychid fish *Maurolicus muelleri* on seamounts in the Central North Pacific. *Pacific Science*, 48, 57–69.

Boyer, D.C., Kirchner, C.H., McAllister, M.K., Staby, A. and Staalesen, B.I. (2001) The orange roughy fishery of Namibia: lessons to be learned about managing a developing fishery. *South African Journal of Marine Science*, 23, 205–21.

Branch, T.A. (2001) A review of orange roughy *Hoplostethus atlanticus* fisheries, estimation methods, biology and stock structure. *South African Journal of Marine Science*, 23, 181–203.

Cailliet, G.M., Andrews, A.H., Burton, E.J., Watters, D.L., Kline, D.E. and Ferry-Graham, L.A. (2001) Age determination and validation studies of marine fishes: Do deep-dwellers live longer? *Experimental Gerontology*, 36, 739–64.

Cheung, W.W.L., Pitcher, T.J. and Pauly, D. (2005) A fuzzy logic expert system to estimate intrinsic extinction vulnerabilities of marine fishes to fishing. *Biological Conservation*, 124, 97–111.

Choat, J.H. and Bellwood, D.R. (1991) Reef fishes: their history and evolution. In: *The Ecology of Fishes on Coral Reefs* (ed. Sale, P.F.), pp. 39–66. Academic Press, New York.

Clark, M.R. (1995) Experience with the management of orange roughy (*Hoplostethus atlanticus*) in New Zealand, and the effects of commercial fishing on stocks over the period 1980–1993. In: *Deep-Water Fisheries of the North Atlantic Oceanic Slope* (ed. Hopper, A.G.), pp. 251–66. Kluwer Academic Publishers, Dordrecht, The Netherlands.

Clark, M.R. (1999) Fisheries for orange roughy (*Hoplostethus atlanticus*) on seamounts in New Zealand. *Oceanologica Acta*, 22(6), 593–602.

Clark, M.R. (2001) Are deepwater fisheries sustainable? The example of orange roughy (*Hoplostethus atlanticus*) in New Zealand. *Fisheries Research*, 51, 123–35.

Clark, M.R. and Roberts, C.D. (2003) NORFANZ marine biodiversity survey uncovers mysteries of the deep. *Aquatic Biodiversity and Biosecurity*, 5, 1.

Clark, M.R. and Tracey, D.M. (1991) Trawl survey of orange roughy on the Challenger Plateau July 1990. *New Zealand Fisheries Technical Report No. 26*, 20 pp.

Clark, M.R. and Tracey, D.M. (1994) Changes in a population of orange roughy (*Hoplostethus atlanticus*) with commercial exploitation on the Challenger Plateau, New Zealand. *Fishery Bulletin*, 92, 236–53.

Clark, M.R., King, K.J. and McMillan, P.J. (1989) The food and feeding relationships of black oreo, *Allocyttus niger*, smooth oreo, *Pseudocyttus maculatus*, and eight other fish species from the continental slope of the South-West Chatham Rise, New Zealand. *Journal of Fish Biology*, 35, 465–84.

Clark, M.R., Anderson, O.F., Francis, R.I.C.C. and Tracey, D.M. (2000) The effects of commercial exploitation on orange roughy (*Hoplostethus atlanticus*) from the continental slope of the Chatham Rise, New Zealand, from 1979 to 1997. *Fisheries Research*, 45, 217–38.

Clark, M.R., Bull, B. and Tracey, D.M. (2001) The estimation of catch levels for new orange roughy fisheries on seamounts: a meta-analysis of seamount data. *New Zealand Fisheries Assessment Report 2001/75*, 40 pp.

Conroy, A.M. and Pankhurst, N.W. (1989) Size–fecundity relationships in the smooth oreo, *Pseudocyttus maculatus*, and the black oreo, *Allocyttus niger* (Pisces: Oreosomatidae). *New Zealand Journal of Marine and Freshwater Research*, 23, 525–7.

Dooley, H.D. (1984) Aspects of oceanographic variability on Scottish fishing grounds. PhD Thesis, University of Aberdeen, Aberdeen, 154 pp.

Elekseyev, F.Y. (1995) Reproductive cycle of the roundnose grenadier, *Coryphaenoides rupestris* (Macrouridae), from the North Atlantic. *Journal of Ichthyology*, 35, 123–34.

Eriksen, C.C. (1991) Observations of amplified flows atop a large seamount. *Journal of Geophysical Research*, 96, 15227–36.

Fock, H., Uiblein, F., Köster, F. and Westernhagen, H. (2002) Biodiversity and species-environment relationships of the demersal fish assemblage at the Great Meteor Seamount (subtropical NE Atlantic), sampled by different trawls. *Marine Biology*, 141, 185–99.

Francis, R.I.C.C. and Clark, M.R. (2005) Sustainability issues for orange roughy fisheries. *Bulletin of Marine Science*, 76(2), 337–51.

Francis, R.I.C.C., Clark, M.R., Coburn, R.P., Field, K.D. and Grimes, P.J. (1995) Assessment of the ORH 3B orange roughy fishery for the 1994–1995 fishing year. *New Zealand Fisheries Assessment Research Documents* 95/4, NIWA.

Francis, M.P., Hurst, R.J., McArdle, B.H., Bagley, N.W. and Anderson, O.F. (2002) New Zealand demersal fish assemblages. *Environmental Biology of Fishes*, 65, 215–34.

Froese, R. and Pauly, D. (eds.) (2003) Fishbase. World Wide Web electronic publication. www.fishbase.org (last accessed February 16, 2004).

Froese, R. and Sampang, A. (2004) Taxonomy and biology of seamount fishes. In: *Seamounts: Biodiversity and Fisheries* (eds. Morato, T. and Pauly, D.), pp. 25–31. *Fisheries Centre Research Report No. 12(5)*, 78 pp.

Genin, A. and Boehlert, G.W. (1985) Dynamics of temperature and chlorophyll structures above a seamount: an oceanic experiment. *Journal of Marine Research*, 43, 907–24.

Genin, A., Dayton, P.K., Lonsdale, P.F. and Spiess, F.N. (1986) Corals on seamount peaks provide evidence of current acceleration over deep-sea topography. *Nature*, 322(6074), 59–61.

Genin, A., Haury, L. and Greenblatt, P. (1988) Interactions of migrating zooplankton with shallow topography: predation by rockfishes and intensification of patchiness. *Deep-Sea Research*, 35, 151–75.

Genin, A., Noble, M. and Lonsdale, P.F. (1989) Tidal currents and anticyclonic motions on two North Pacific seamounts. *Deep-Sea Research*, 36, 1803–15.

Gonzalez, J.A., Rico, V., Lorenzo, J.M., Reis, S., Pajuelo, J.G., Dias, M.A., Mendonça, A., Krug, H.M. and Pinho, M.R. (2003) Sex and reproduction of the alfonsino *Beryx splendens* (Pisces: Berycidae) from the Macaronesian archipelagos. *Journal of Applied Ichthyology*, 19, 104–8.

Gordon, J.D.M. and Duncan, J.A.R. (1987) Aspects of the biology of *Hoplostethus atlanticus* and *H. mediterraneus* (Pisces: Berycomorphi) from the slopes of the Rockall Trough and the Porcupine Sea Bight (North-Eastern Atlantic). *Journal of the Marine Biological Association of the United Kingdom*, 67, 119–33.

Grandperrin, R., Auzende, J.-M., Henin, C., LaFoy, Y., Richer de Forges, B., Seret, B., van de Beuque, S. and Virly, S. (1999) Swath-mapping and related deep-sea trawling in the southeastern part of the economic zone of New Caledonia. In: *Proceedings of the Fifth Indo-Pacific Fish Conference*, Noumea (eds. Seret, B. and Sire, J.-Y.), pp. 459–68. Société Française d'Ichtyologie, Paris.

Haedrich, R.L. and Merrett, N.R. (1990) Little evidence for faunal zonation or communities in the deep sea demersal fish faunas. *Progress in Oceanography*, 24, 239–50.

Hareide, N.-R. and Garnes, G. (2001) The distribution and catch rates of deep water fish along the Mid-Atlantic Ridge from 43 to 61°N. *Fisheries Research*, 519, 297–310.

Hazin, F.H.V., Zagaglia, J.R., Broadhurst, M.K., Travassos, P.E.P. and Bezerra, T.R.Q. (1998) Review of a small-scale pelagic longline fishery off northeastern Brazil. *Marine Fisheries Review*, 60, 1–8.

Hilborn, R. and Walters, C.J. (1992) *Quantitative Fisheries Stock Assessment: Choice, Dynamics and Uncertainty*. Chapman and Hall, New York.

Hilton-Taylor, C. (2000) *2000 IUCN Red List of Threatened Species*. IUCN, Gland, Switzerland and Cambridge, UK, xviii + 61 pp.

Holland, K.N., Kleiber, P. and Kajiura, S.M. (1999) Different residence times of yellowfin tuna, *Thunnus albacares*, and bigeye tuna, *T. obesus*, found in mixed aggregations over a seamount. *Fishery Bulletin*, 97, 392–5.

Horn, P.L. (2002) Age and growth of Patagonian toothfish (*Dissostichus eleginoides*) and Antarctic toothfish (*D. mawsoni*) in waters from the New Zealand subantarctic to the Ross Sea, Antarctica. *Fisheries Research*, 56(3), 275–87.

Horn, P.L. and Massey, B.R. (1989) Biology and abundance of alfonsino and bluenose off the lower east coast North Island, New Zealand. *New Zealand Fisheries Technical Report No. 15*, 32 pp.

Horn, P.L., Tracey, D.M. and Clark, M.R. (1998) Between-area differences in age and length at first maturity of the orange roughy *Hoplostethus atlanticus*. *Marine Biology*, 132, 187–94.

Humphreys, R. and Moffitt, R. (1999) Western Pacific bottomfish and armourhead fisheries. Unit 17. In: *Our Living Oceans. Report on the Status of US Living Marine Resources, 1999* (ed. NMFS). US Department of Commerce, NOAA Technical Memorandum NMFS-F/SPO-41, on-line version, http://spo.nwr.noaa.gov/unit17.pdf

Humphreys Jr., R.L. Winans, G.A. and Tagami, D.T. (1989) Synonymy and life history of the North Pacific pelagic armourhead, *Pseudopentaceros wheeleri* Hardy (Pisces: Pentacerotidae). *Copeia*, 1, 142–53.

Itano, D.G. and Holland, K.N. (2000) Movement and vulnerability of bigeye (*Thunnus obesus*) and yellowfin tuna (*Thunnus albacares*) in relation to FADs and natural aggregation points. *Aquatic Living Resources*, 13, 213–23.

Kelly, C.J., Connolly, P.L. and Bracken, J.J. (1996) Maturity, oocyte dynamics and fecundity of the roundnose grenadier *Coryphaenoides rupestris* (Gunnerus, 1765) from the Rockall Trough. *Journal of Fish Biology*, 49(Suppl. A), 5–17.

Kelly, C.J., Connolly, P.L. and Bracken, J.J. (1997) Age estimation, growth, maturity and distribution of the roundnose grenadier from the Rockall Trough. *Journal of Fish Biology*, 50, 1–17.

Klimley, A.P., Butler, S.B., Nelson, D.R. and Stull, A.T. (1988) Diel movements of scalloped hammerhead sharks, *Sphyrna lewini* Griffith and Smith, to and from a seamount in the Gulf of California (Mexico). *Journal of Fish Biology*, 33, 751–62.

Koslow, J.A. (1993) Community structure in North Atlantic deep-sea fishes. *Progress in Oceanography*, 31, 321–38.

Koslow, J.A. (1996) Energetic and life-history patterns of deep-sea benthic, benthopelagic and seamount-associated fish. *Journal of Fish Biology*, 49(Suppl. A), 54–74.

Koslow, J.A. (1997) Seamounts and the ecology of deep-sea fisheries. *American Scientist*, 85, 168–76.

Koslow, J.A. and Gowlett-Holmes, K. (1998) The seamount fauna off Southern Tasmania: benthic communities, their conservation and impacts of trawling. Final Report to Environment Australia and the Fisheries Research and Development Corporation. Report FRDC Project 95/058, CSIRO, Hobart, Tasmania, Australia, p. 104.

Koslow, J.A. and Tuck, G. (2001) The boom and bust of deepwater fisheries: Why haven't we done better? NAFO SCR Document 141, Series 4535, 10 pp.

Koslow, J.A., Bulman, C.M. and Lyle, J.M. (1994) The mid-slope demersal fish community off southeastern Australia. *Deep-Sea Research I*, 41, 113–41.

Koslow, J.A., Bell, J., Virtue, P. and Smith, D.C. (1995) Fecundity and its variability in orange roughy: effects of population density, condition, egg size, and senescence. *Journal of Fish Biology*, 47, 1063–80.

Koslow, J.A., Boehlert, G.W., Gordon, J.D.M., Haedrich, R.L., Lorance, P. and Parin, N. (2000) Continental slope and deep-sea fisheries: implications for a fragile ecosystem. *ICES Journal of Marine Science*, 57, 548–57.

Koslow, J.A., Gowlett-Holmes, K., Lowry, J.K., O'Hara, T., Poore, G.C.B. and Williams, A. (2001) Seamount benthic macrofauna off southern Tasmania: community structure and impacts of trawling. *Marine Ecology Progress Series*, 213, 111–25.

Lack, M., Short, K. and Willock, A. (2003) Managing risk and uncertainty in deep-sea fisheries: lessons from orange roughy. Traffic Oceania and WWF Endangered Seas Programme.

Leaman, B.M. and Beamish, R.J. (1984) Ecological and management implications of longevity in some northeast Pacific groundfishes. *Bulletin of the International North Pacific Fisheries Commission*, 42, 85–97.

Lehodey, P. and Grandperrin, R. (1996) Age and growth of the alfonsino *Beryx splendens* over the seamounts off New Caledonia. *Marine Biology*, 125, 249–58.

Lehodey, P., Grandperrin, R. and Marchal, P. (1997) Reproductive biology and ecology of a deep-demersal fish, alfonsino *Beryx splendens*, over the seamounts off New Caledonia. *Marine Biology*, 128, 17–27.

Longhurst, A. (2002) Murphy's law revisited: longevity as a factor in recruitment to fish populations. *Fisheries Research*, 56, 125–31.

Lorance, P., Garren, F. and Vigneau, J. (2001) Age estimation of the roundnose grenadier (*Coryphaenoides rupestris*), effects of uncertainties on ages. Scientific Council Research Document 01/123. Northwest Atlantic Fisheries Organization, Dartmouth, NS, 15 pp.

Lorance, P., Uiblein, F. and Latrouite, D. (2002) Habitat, behaviour and colour patterns of orange roughy *Hoplostethus atlanticus* (Pisces: Trachichthyidae) in the Bay of Biscay. *Journal of the Marine Biological Association of the United Kingdom*, 82, 321–31.

Lueck, R.G. and Mudge, T.D. (1997) Topographically induced mixing around a shallow seamount. *Science*, 276, 1831–33.

Lyle, J.M. and Smith, D.C. (1997) Abundance and biology of warty oreo (*Allocyttus verrucosus*) and spiky oreo (*Neocyttus rhomboidalis*) (Oreosomatidae) off south-eastern Australia. *Marine and Freshwater Research*, 48, 91–102.

Masuzawa, T., Kurata, Y. and Onishi, K. (1975) Results of group study on a population of demersal fishes in waters from Sagami Bay to southern Izu Islands: population ecology of Japanese alfonsin and other demersal fishes. *Japan Aquatic Resources Conservation Association Fishery Research Paper 28*, 105 pp (in Japanese).

Moore, J.A., Vecchione, M., Collette, B.B., Gibbons, R., Hartel, K.E., Galbraith, J.K., Turnipseed, M., Southworth, M. and Watkins, E. (2003) Biodiversity of Bear Seamount, New England seamount chain: results of exploratory trawling. *Journal of Northwest Atlantic Fishery Science*, 31, 363–72.

Morato, T., Cheung, W.W.L. and Pitcher, T.J. (2004) Additions to Froese and Sampang's checklist of seamount fishes. In: *Seamounts: Biodiversity and Fisheries* (eds. Morato, T. and Pauly, D.). Fisheries Centre Research Reports, 12(5), Appendix 1: 1–6. Fisheries Centre, University of British Columbia, Canada.

Morato, T., Solà, E., Grós, M.P. and Menezes, G. (1999) Diets of forkbeard (*Phycis phycis*) and conger eel (*Conger conger*) off the Azores during spring of 1996 and 1997. *Arquipélago: Life and Marine Sciences*, 17A, 51–64.

Morato, T., Santos, R.S. and Andrade, P. (2000) Feeding habits, seasonal and ontogenetic diet shift of blacktail comber, *Serranus atricauda* (Pisces: Serranidae), from the Azores, Northeastern Atlantic. *Fisheries Research*, 49(1), 51–60.

Morato, T., Solà, E., Grós, M.P. and Menezes, G. (2001) Feeding habits of two congener species of seabreams, *Pagellus bogaraveo* and *Pagellus acarne*, off the Azores (northeastern Atlantic) during spring of 1996 and 1997. *Bulletin of Marine Science*, 69(3), 1073–87.

Morato, T., Solà, E., Grós, M.P. and Menezes, G. (2003) Diet of the two most common elasmobranchs in the bottom longline fishery of the Azores, northeastern Atlantic: thornback ray *Raja cf. clavata* and tope shark, *Galeorhinus galeus*. *Fishery Bulletin*, 101(3), 590–602.

Morato, T., Cheung, W.W.L. and Pitcher, T.J. (2006) Vulnerability of seamount fish to fishing: fuzzy analysis of life history attributes. *Journal of Fish Biology*, 68, 209–21.

Myers, R.A. and Pepin, P. (1994) Recruitment variability and oceanographic stability. *Fisheries Oceanography*, 3, 246–55.

Owens, W.B. and Hogg, N.G. (1980) Oceanic observations of stratified Taylor columns near a bump. *Deep-Sea Research*, 27A, 1029–45.

Parin, N.V., Mironov, A.N. and Nesis, K.N. (1997) Biology of the Nazca and Sala y Gomez submarine ridges, an outpost of the Indo-West Pacific fauna in the eastern pacific ocean: composition and distribution of the fauna, its communities and history. *Advances in Marine Biology*, 32, 145–242.

Parker, T. and Tunnicliffe, V. (1994) Dispersal strategies of the biota on an oceanic seamount: implications for ecology and biogeography. *Biological Bulletin*, 187(3), 336–45.

Piñeiro, C.G., Casas, M. and Araujo, H. (2001) Results of exploratory deep-sea fishing survey in the Galician Bank: biological aspects on some of seamount-associated fish (ICES Division IXb). Scientific Council Research Document 01/146. Northwest Atlantic Fisheries Organization, Dartmouth, NS.

Pitcher, T.J. (1995) The impact of pelagic fish behaviour on fisheries. *Scientia Marina*, 59, 295–306.

Pitcher, T.J. (1997) Fish shoaling behaviour as a key factor in the resilience of fisheries: shoaling behaviour alone can generate range collapse in fisheries. In: *Developing and Sustaining World Fisheries Resources: The State of Science and Management* (eds. Hancock, D.A., Smith, D.C., Grant, A. and Beumer, J.P.), pp. 143–8. CSIRO, Collingwood, Australia.

Pitcher, T.J. and Parrish, J. (1993) The functions of shoaling behaviour. In: *The Behaviour of Teleost Fishes* (ed. Pitcher, T.J.), pp. 363–439. Chapman and Hall, London.

Probert, P.K. (1999) Seamounts, sanctuaries and sustainability: moving towards deep-sea conservation. *Aquatic Conservation: Marine and Freshwater Ecosystems*, 9, 601–5.

Probert, P.K., McKnight, D.G. and Grove, S.L. (1997) Benthic invertebrate bycatch from a deep-water trawl fishery, Chatham Rise, New Zealand. *Aquatic Conservation: Marine and Freshwater Ecosystems*, 7, 27–40.

Ramos, A., Moya, F., Salmerón, F., García, P., Carroceda, A., Fernández, L., González, J.F., Tello, O., Sánz, J.L. and Ballesteros, M. (2001) Demersal fauna on deep seamounts of Sierra Leone rise (Gulf of Guinea, Africa). Scientific Council Research Document 01/149. Northwest Atlantic Fisheries Organization, Dartmouth, NS.

Richer de Forges, B., Koslow, J.A. and Poore, G.C.B. (2000) Diversity and endemism of the benthic seamount macrofauna in the southwest Pacific. *Nature*, 405, 944–7.

Rico, V., Lorenzo, J.M., Gonzalez, J.A., Krug, H.M., Mendonça, A., Gouveia, E. and Dias, M.A. (2001) Age and growth of the alfonsino *Beryx splendens* Lowe, 1834 from the Macaronesian archipelagos. *Fisheries Research*, 49, 233–40.

Robertson, D.R. (1998) Do coral-reef fish faunas have a distinctive taxonomic structure? *Coral Reefs*, 17, 179–86.

Robinson, J.S. (1981) Tidal vorticity and residual circulation. *Deep-Sea Research*, 28A, 195–212.

Roden, G.I. (1987) Effect of seamounts and seamount chains on ocean circulation and thermohaline structure. In: *Seamounts, Islands, and Atolls* (eds. Keating, B.H., Fryer, P., Batiza, R. and Boehlert, G.W.), pp. 335–54. *Geophysical Monograph*, 43.

Roden, G.I., Taft, B.A. and Ebbesmeyer, C.C. (1982) Oceanographic aspects of the Emperor seamounts region. *Journal of Geophysical Research*, 87, 9537–52.

Rogers, A.D. (1994) The biology of seamounts. *Advances in Marine Biology*, 30, 305–50.

Rosecchi, E., Tracey, D.M. and Webber, W.R. (1988) Diet of orange roughy, *Hoplostethus atlanticus* (Pisces: Trachichthyidae) on the Challenger Plateau, New Zealand. *Marine Biology*, 99, 293–306.

Sala, E., Aburto-Oropeza, O., Paredes, G. and Thompson, G. (2003) Spawning aggregations and reproductive behavior of reef fishes in the Gulf of California. *Bulletin of Marine Science*, 72, 103–21.

Sedberry, G.R. and Loefer, J.K. (2001) Satellite telemetry tracking of swordfish, *Xiphias gladius*, off the eastern United States. *Marine Biology*, 139, 355–60.

Shibanov, V.N., Vinnichenko, V.I. and Pedchenko, A.P. (2002) Prospects of fisheries on seamounts. Russian investigation and fishing in the northern part of the Mid-Atlantic Ridge. ICES CM:2002/L: 35 (poster).

Sibert, J., Holland, K. and Itano, D. (2000) Exchange rates of yellowfin and bigeye tunas and fishery interaction between Cross seamount and near-shore fads in Hawaii. *Aquatic Living Resources*, 13, 225–32.

Spencer, P.D. and Collie, J.S. (1997) Patterns of population variability in marine fish stocks. *Fisheries Oceanography*, 6, 188–204.

Stone, G.S., Madin, L.P., Stocks, K., Hovermale, G., Hoagland, P., Schumacher, M., Etnoyer, P., Sotka, C. and Tausig, H. (2004) Seamount biodiversity, exploitation and conservation. In: *Defying Ocean's End* (eds. Glover, L.K. and Earle, S.A.), pp. 45–70. Island Press, Washington, DC.

Tracey, D.M., Bull, B., Clark, M.R. and Mackay, K.A. (2004) Fish species composition on seamounts and adjacent slope in New Zealand waters. *New Zealand Journal of Marine and Freshwater Research*, 38, 163–82.

Tseytlin, V.B. (1985) Energetics of fish populations inhabiting seamounts. *Oceanology*, 25(2), 237–9.

Tsukamoto, K., Otake, T., Mochioka, N., Lee, T.-W., Fricke, H., Inagaki, T., Aoyama, J., Ishikawa, S., Kimura, S., Miller, M.J., Hasumoto, H., Oya, M. and Suzuki, Y. (2003) Seamounts, new moon and eel spawning: the search for the spawning site of the Japanese eel. *Environmental Biology of Fishes*, 66, 221–9.

Vinnichenko, V.I. (2002a) Prospects of fisheries on seamounts. ICES CM2002/M32 (poster).

Vinnichenko, V.I. (2002b) Russian investigations and fishery on seamounts in the Azores area. In: *Relatório das XVIII e XIX Semana das Pescas dos Açores* (ed. Anonymous), pp. 115–29. Secretaria Regional da Agricultura e Pescas, Horta.

Walters, C. (2003) Folly and fantasy in the analysis of spatial catch rate data. *Canadian Journal of Fisheries and Aquatic Science*, 60, 1433–6.

Ward, P., Porter, J.M. and Elscot, S. (2000) Broadbill swordfish: status of established fisheries and lessons for developing fisheries. *Fish and Fisheries*, 1, 317–36.

Wilson, R.R. and Kaufmann, R.S. (1987) Seamount biota and biogeography. In: *Seamounts, Islands and Atolls* (eds. Keating, B.H., Fryer, P., Batiza, R. and Boehlert, G.W.), pp. 319–34. *Geophysical Monograph*, 43.

Chapter 10A

Fish visitors to seamounts: Tunas and billfish at seamounts

Kim N. Holland and R. Dean Grubbs

Abstract

A growing body of data from a variety of sources indicates that seamounts exert a strong influence on the behavior and distribution of tunas and other large, highly vagile pelagic fishes. Fishing fleets target seamounts and available catch data indicate that catch per unit effort (CPUE) is higher around seamounts than in adjacent areas of ocean (Fonteneau, 1991; Campbell and Hobday, 2003). Two general hypotheses have been proposed to explain the high densities of large pelagic fishes associated with seamounts. First, there may be enhanced food availability at seamounts relative to the surrounding areas. Second, seamounts may play a role as navigation aids in fish movements. These movements might be localized wherein the fish leave the seamount for short periods of time in search of nearby food or the seamount may act as a waypoint in larger, pan-oceanic migrations. Seamounts frequently have distinctive geo-magnetic signatures (Klimley, 1993), which may make them particularly important as navigation landmarks. The two general hypotheses are not mutually exclusive – seamounts used as navigation 'waypoints' may also provide enhanced foraging opportunities. This chapter summarizes the available information on the interaction between pelagic fish species and seamounts and describes a suite of experiments aimed at understanding the behavior, dispersion and trophic ecology of tuna aggregations associated with the Cross Seamount near the Hawaiian Islands. Although focused on a specific seamount, the results of this case study give an indication of what factors might be at work at other seamounts around the world.

Introduction

Although most fishers and fisheries scientists believe that seamounts to be sites of increased densities of tunas and other pelagic species, empirical data to support this contention are few. However, the data base is growing and comes from a variety of sources. These include analysis of fishing boat log books, tagging experiments using traditional tag-and-recapture methods and various types of electronic tags and gut content analyses of fish captured seamounts.

Logbook data from the international tuna fleet (both pole-and-line and purse seiners) fishing in the tropical eastern Atlantic showed enhanced tuna biomass associated with certain seamounts in the region (Fonteneau, 1991). The most enhanced CPUE was associated with seamounts furthest from the continental shelf and at seamounts apparently embedded within adjacent areas of poor productivity. The catch comprised fairly equal proportions of sub-adult yellowfin tuna (*Thunnus albacares*), bigeye tuna (*T. obesus*) and small adult skipjack tuna (*Katsuwonus pelamis*). In Hawaii, the Cross Seamount produces very high catch rates relative to adjacent areas with approximately a 3:1 ratio of sub-adult bigeye to yellowfin tuna (Itano and Holland, 2000; Adam *et al.*, 2003). An exhaustive analysis of swordfish (*Xiphias gladius*) catches from the eastern Australian fishery indicated that seamounts in the Tasmantid and Lord Howe chains supported enhanced densities of swordfish that were subsequently eliminated as the fishery matured (Campbell and Hobday, 2003). The authors suggested that the unfished population might have contained a sub-population of seamount-associated swordfish that could not replenish itself once the overall population was reduced by fishing activity. Enhanced densities of tuna at seamounts probably also support enhanced densities of tuna predators such as billfishes and mako (*Isurus oxyrhincus*), thresher (*Alopias* spp.) and cookie cutter sharks (*Isistius brasiliensis*). These species are all reported as occurring in seamount communities (Rogers, 1994).

Data from a variety of tagging studies also indicate that seamounts exert great influence over tuna movements. One of the more extensively studied seamounts is Espiritu Santo in the Gulf of California. This seamount, which comes to within 18 m of the surface, supports a broad assemblage of pelagic species including yellowfin tuna and hammerhead sharks (*Sphyrna lewini*) (Klimley and Butler, 1988). By placing individually coded sonic transmitters in yellowfin tuna and placing acoustic data loggers ('listening stations') on the seamount, Klimley and colleagues (2003) observed residence patterns ranging from less than a day to over a year. Regardless of their overall length of stay, individual tunas were constantly moving in and out of range of the receivers suggesting they were using this seamount as a base of operations for short-range foraging excursions. Similar seamount-based excursions were observed in a skipjack tuna tracked using active acoustic telemetry around a seamount in the eastern Atlantic (Fonteneau, 1991). Hammerhead sharks associated with the Espiritu Santo seamount also used it as a base from which to launch nighttime foraging excursions (Klimley *et al.*, 1988).

The fact that some of the yellowfin tunas tagged at the Espiritu Santo seamount were only detected for a short time suggests that a portion of the fish were transient (Klimley *et al.*, 2003). This pattern of behavior has been observed in Pacific bluefin tuna (*T. orientalis*) equipped with internally implanted data loggers. Two individuals interrupted what were otherwise very direct trans-Pacific westerly migrations to spend several days associated with the Emperor seamounts located west of the Hawaiian archipelago before proceeding to the bluefin spawning grounds near Japan (C. Farwell, Personal communication). Results from tag-and-recapture experiments at the Cross Seamount (see below) also suggest that for yellowfin and bigeye tuna in the central Pacific seamounts may be part of larger movement patterns rather than a place of permanent habitation.

Two general hypotheses have been presented to explain the higher densities of tunas and other pelagic species found at certain seamounts. One is that seamounts may represent navigational waypoints in larger movement patterns (Klimley, 1993; Holland *et al.*, 1999; Klimley *et al.*, 2003). The second holds that seamounts provide enhanced foraging opportunities for apex predators. Of course, these two hypotheses are not mutually exclusive. Seamounts often have distinct geo-magnetic signatures (Klimley, 1993) making the navigation hypothesis particularly applicable to species such as tunas and sharks that are known to be able to detect magnetic fields (Walker, 1984; Meyer *et al.*, 2005). Intriguingly, in the Australian swordfish fishery, there was a significant negative relationship between the mean CPUE and the median magnetic anomaly associated with each seamount ($R^2 = 0.25$, $p < 0.016$, $n = 23$) raising the possibility that 'swordfish are attracted to seamounts with strong negative magnetic anomalies' (Campbell and Hobday, 2003).

The second hypothesis is that seamounts support enhanced forage communities for higher trophic level predators such as tunas and swordfish. Many possible mechanisms exist as discussed by Genin (2004). One commonly cited explanation is that seamounts may cause upwelling events that bring nutrient-rich waters into the photic zone resulting in enhanced primary productivity and resultant enrichment of higher trophic levels (Uda and Ishino, 1958; Boehlert and Genin, 1987; Genin, 2004). For this mechanism to influence tuna distribution the enhancement phenomenon must persist long enough to result in enhancement of the nekton that are consumed by tunas. A period of 90 days may be required to turn an upwelling event in the tropical Pacific into forage suitable for skipjack tuna (Lehody *et al.*, 1998). Such persistent upwelling may occur in locations such as the edges of continental shelves (Genin, 2004). An example can be found on the coast of western North America where certain topographic features are associated with seasonally high catches of albacore tuna (*T. alalunga*; Dotson, 1980). This mechanism does not seem to hold true for most mid-ocean locations where high tuna densities are found. Though possible in theory, the available data suggest that upwelling events associated with seamounts rarely reach the photic zone, and those that do are generally too short lived to significantly enhance zooplankton production (Genin, 2004; Chapters 4 and 5).

Another mechanism whereby seamounts may provide enhanced forage is by increasing access to the mesopelagic boundary fauna (MBF). The MBF is found where the ocean floor rises shallower than 1200 m. It is typically patchy in distribution but is denser than the deeper bathypelagic assemblage that it replaces in the shallower depths (Reid *et al.*, 1991). These organisms form part of the deep scattering layer (DSL) that makes nocturnal vertical excursions toward the surface (Chapter 6). These mesopelagic prey may simply get swept over the seamount by prevailing currents, thereby increasing local availability to predators. Additionally, it has been demonstrated that some DSL organisms were capable of holding their position over the Hancock Seamount (an intermediate seamount in the northwestern part of the Hawaiian Ridge), and were not advected from the seamount even when faced with strong currents (Wilson, 1992; Wilson and Boehler, 2004). This and the fact that all life stages were present for the two dominant DSL species (one shrimp and one fish) suggest the populations of these particular species are resident (Wilson and Boehlert, 1993; Boehlert *et al.*, 1994; Genin, 2004; Wilson and Boehlert, 2004).

Two additional plausible explanations for the enhanced availability MBF prey were proposed by Genin (2004); the 'feed-rest' and 'topographic blockage' hypotheses. The complex microtopography of a seamount provides calm shelter and may result in decreased current speeds at the benthic boundary layer, thereby increasing suitable habitat for some

MBF organisms. Genin (2004) proposed the 'feed-rest' hypothesis whereby micronekton may conserve energy by taking advantage of this calm habitat during the day, only swimming into the water column during their nightly vertical feeding migrations. This may impart a significant energetic advantage to the MBF associated with seamounts compared to those that are in open ocean and must swim constantly.

Another promising explanation is the 'topographic blockage' hypothesis whereby the pre-dawn migratory descent of mesopelagic organisms that are carried over the seamount during the night is temporarily halted by the seamount topography (Isaacs and Schwartzlose, 1965; Genin, 2004; Chapter 5). Such a mechanism could trap mesopelagic prey in illuminated waters at the foraging depth of tunas and other visual predators (Genin, 2004). Fock *et al.* (2002) proposed this mechanism to explain the dominance of mesopelagic prey in benthopelagic predators on the Great Meteor seamount. As the biomass of trapped migrators is depth-dependent, the greatest concentration of trapped organisms should occur at seamounts of intermediate depth with plateaus between the bottom of the photic zone and 400 m (Genin, 2004; Chapter 5). Indeed the highest catches in groundfish fisheries often occur on intermediate seamounts (Uchida and Tagami, 1984). We initiated a series of tag-and-recapture experiments and stomach analyses at Cross Seamount to determine if enhanced forage availability was responsible for the high densities of tuna over seamounts and to determine the residence times of these tuna populations.

The Cross Seamount: a case study

The Cross Seamount is a prominent bathymetric feature located at 18°40′N, 158°10′W; ~280 km South of Honolulu. It rises from depths of more than 4000 m to a minimum charted depth of 330 m. There are many other seamounts in the region, but the shallowest of these is the Pensacola Seamount at 624 m and most are deeper than 1000 m. These deeper seamounts are fished by the pelagic longline fleet (Chapter 17) but none hold tuna aggregations comparable to the Cross Seamount (Itano and Holland, 2000). In recent years, the Cross Seamount has supported a significant tuna fishery conducted by small vessels less than 15 m in length that use a combination of trolling and jigging techniques (Chapter 16). Landings have been relatively stable since the late 1980s at around 700 t/year (Hawaii Department of Aquatic Resources), although significant underreporting may exist. Although originally described as a yellowfin tuna fishery, the catch is actually dominated (75%) by juvenile (sub-adult) bigeye tuna with juvenile yellowfin composing the rest of the catch.

Residence time and dispersal patterns

Tag-and-recapture experiments were conducted on board small commercial fishing vessels targeting tuna schools associated with the Cross Seamount. Bigeye and yellowfin tuna were tagged and released at the Cross Seamount as well as at other areas around the Hawaiian archipelago. Residence time at the seamount was determined by plotting an 'attrition curve' for the time at liberty of tunas recaptured at their point of release (i.e., the seamount). The results indicate that the seamount residence times of these two closely related species are significantly different and, for both species, residence time is quite brief (Holland *et al.*, 1999). The 'half life' for yellowfin was about 18 days as opposed to ~32 days for bigeye (Fig. 10A.1). Subsequent analysis of a slightly larger dataset indicated a somewhat longer 'half life' for bigeye tuna of 97.6 ± 18.5 days (Sibert *et al.*, 2000). Larger bigeye tended to remain at the seamount longer than smaller individuals (Adam *et al.*, 2003). Of course, tag-and-recapture experiments are not able to distinguish between continuous residency and repeated revisitation to the seamount and the recapture of tagged fish at nearby locations (offshore weather buoys) indicates that a certain amount of 'shuttling' might occur. Nevertheless, it appears that bigeye tuna are associated with this feature significantly longer than yellowfin.

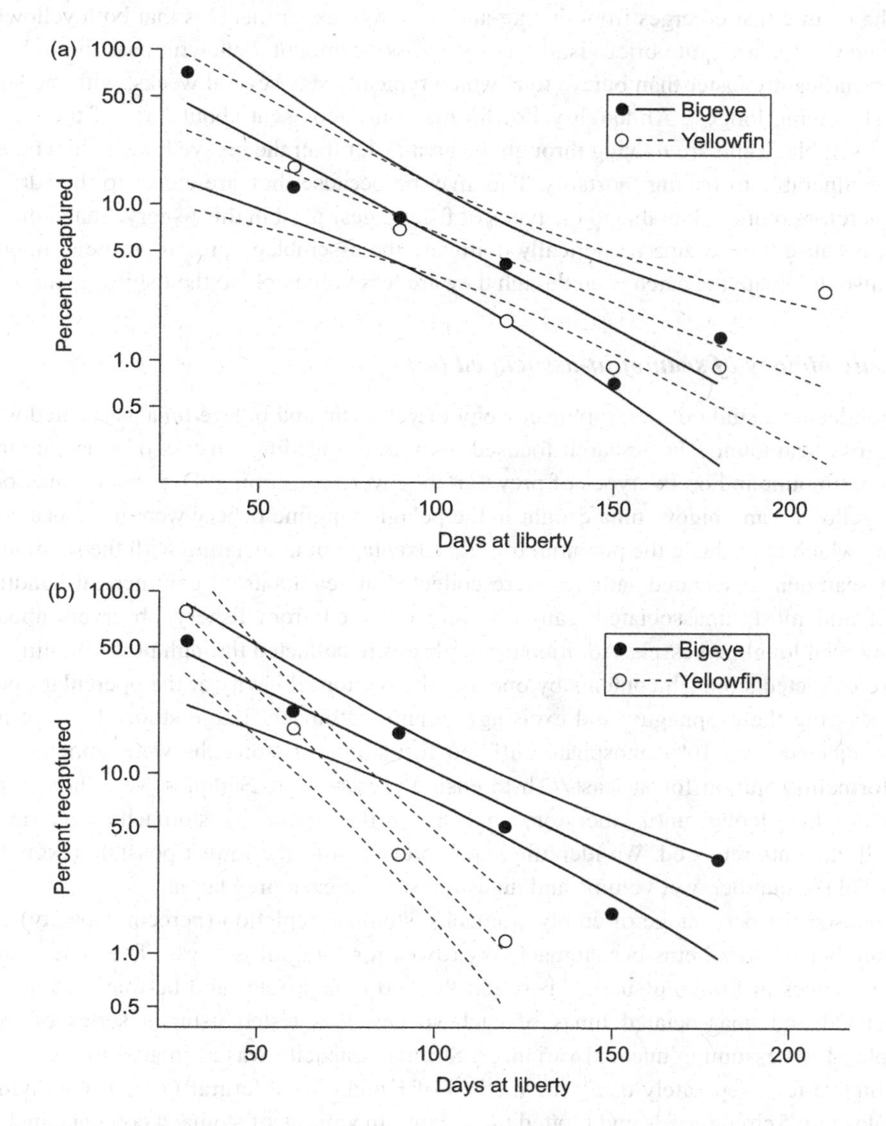

Fig. 10A.1 Tag recapture attrition curves (semi-log plots with 95% confidence contours) for bigeye and yellowfin tuna tagged and released at Cross Seamount: (a) all recaptures from all locations throughout the Hawaii region; (b) recaptured at Cross Seamount (from Holland *et al.* 1999).

For both species, the gross emigration rates from the seamount are around 70%. This means that most fish leave the seamount rather than being caught, but vulnerability to capture is higher for yellowfin than for bigeye. Estimates of fishing mortality for bigeye ranged from 5% to 12% vs 7% to 30% for yellowfin, depending on the size class (Sibert *et al.*, 2000; Adam *et al.*, 2003). Patterns of recaptures away from the seamount indicated quite low recruitment into both the nearshore fisheries and the open ocean longline fisheries. Recruitment into the longline fishery only occurred after sufficient time for the subadult fish tagged at the Cross Seamount to reach the larger sizes typically captured by longline techniques (Adam *et al.*, 2003).

The picture that emerges from the tag-and-recapture experiments is that both yellowfin and bigeye tuna are quite brief visitors to the Cross Seamount. Yellowfin pass through the area significantly faster than bigeye tuna which typically stay several weeks, with the largest fish staying longest. Although yellowfin may only represent about 25% of the mixed tuna assemblage and are moving through the area faster than the bigeye tuna, yellowfin are more vulnerable to fishing mortality. This may be because they are closer to the surface and therefore more vulnerable to the types of fishing gear used in this fishery. Stated differently, because bigeye tuna numerically dominate the assemblage present at the seamount they also dominate the catch even though they are less vulnerable to the fishing gear.

Trophic biology of seamount-associated tuna

We conducted a study of the trophic ecology of yellowfin and bigeye tuna associated with the Cross Seamount. The research focused both on the feeding success of tuna captured at the seamount and on the types of prey that they were consuming. Data from 'unassociated' yellowfin and bigeye tuna caught in the pelagic longline fishery were used as a control by which to evaluate the potential trophic advantage of associating with the seamount. Most seamount-associated samples were collected at sea aboard a commercial handline vessel and most 'unassociated' samples were obtained from fishery observers aboard commercial longline vessels. Additional samples were collected through port sampling.

We collected stomach contents by opening the peritoneal cavity at the opercular opening, severing the esophagus, and excising the entire stomach. Large stomachs were initially injected with 10% phosphate-buffered formalin. All stomachs were immersed in the formalin solution for at least 72 h to ensure preservation. Samples were then stored in 50% ethyl alcohol until laboratory analysis. In the laboratory, stomachs were rinsed and all contents removed. We identified each prey item to the lowest possible taxon and recorded the number, wet volume and digestion state of each prey taxon.

We used the percentage of empty stomachs, stomach repletion (percent capacity) and the number of prey items per stomach as proxies for foraging success. The significance of differences in foraging success between the two tuna species and between seamount-associated and unassociated tunas of each species was tested using a series of two-sample t-tests assuming unequal variances. Stomach capacity was estimated for yellowfin and bigeye tuna separately using the method of Knight and Margraf (1982). We divided samples into 5 cm intervals and plotted the maximum volume of stomach contents, including bait, vs fork length for each interval. These data were used to derive the following regression equations that predict stomach capacity (C) as a function of fork length (FL).

$$\text{Yellowfin tuna: } C = 0.0078\ (\text{FL}^{2.4382}) \quad \text{Bigeye tuna: } C = 0.0164\ (\text{FL}^{2.2481}) \qquad (10\text{A}.1)$$

We collected stomachs from 630 seamount-associated tunas (474 bigeye, 176 yellowfin) and 214 unassociated tunas (138 bigeye, 76 yellowfin). Analysis of foraging success indicated that for bigeye tuna there is a significant trophic advantage to associating with the Cross Seamount but this is not the case for yellowfin tuna. The proportion of empty stomachs in seamount-associated yellowfin tunas (27.8%) was much higher than for bigeye tunas (15.9%). For those yellowfin tuna containing prey, stomach repletion and the number of prey per stomach were slightly higher when associated with the seamount as opposed to being in 'open water' but these differences were insignificant (Table 10A.1, Figs. 10A.2 and 10A.3). By contrast, for bigeye tuna containing prey, stomach repletion when associated with the Cross Seamount was more than three times that of unassociated bigeye (Fig. 10A.2) and the number of prey per stomach in seamount-associated bigeye was more than double that of unassociated bigeye (Fig. 10A.3). On the Cross Seamount stomach repletion for bigeye tuna was more than four times that of yellowfin tuna (Fig. 10A.2). These differences were highly significant (Table 10A.1) and held for all size classes (Fig. 10A.4). By contrast, when caught at locations away from the Cross Seamount, the feeding success of the two species was similar and there were no statistical differences in prey numbers per stomach or stomach repletion (Table 10A.1, Figs. 10A.2 and 10A.3).

Table 10A.1 Statistical comparison (*t*-tests) of foraging success by unassociated and seamount-associated yellowfin tuna and bigeye tuna in Hawaiian waters. Stomach fullness and the number of prey present in each stomach were used as proxies for foraging success.

		t-statistic	Probability
Seamount vs unassociated yellowfin	Prey #/stomach	0.097	0.9226
Seamount vs unassociated bigeye	Prey #/stomach	5.467	<0.0001
Seamount vs unassociated yellowfin	Mean repletion	0.887	0.3802
Seamount vs unassociated bigeye	Mean repletion	6.722	<0.0001
Yellowfin vs bigeye – unassociated	Prey #/stomach	0.817	0.4164
Yellowfin vs bigeye – seamount associated	Prey #/stomach	5.183	<0.0001
Yellowfin vs bigeye – unassociated	Mean repletion	0.873	0.3873
Yellowfin vs bigeye – seamount associated	Mean repletion	8.011	<0.0001

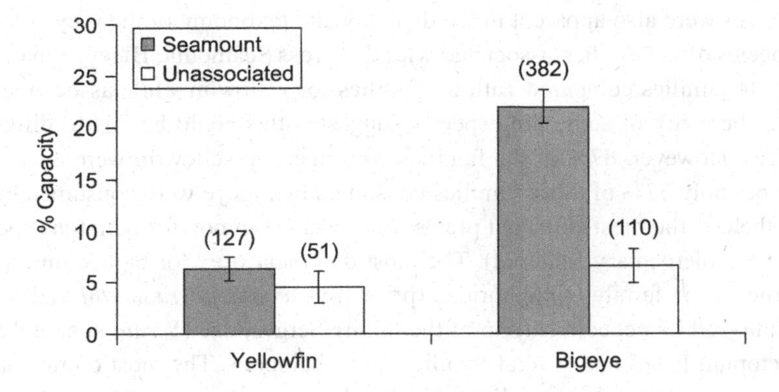

Fig. 10A.2 Mean stomach repletion (percent capacity) for unassociated and seamount-associated yellowfin and bigeye tunas in Hawaii. Sample sizes are reported parenthetically and only include non-empty stomachs (error bars = standard error or mean, SE).

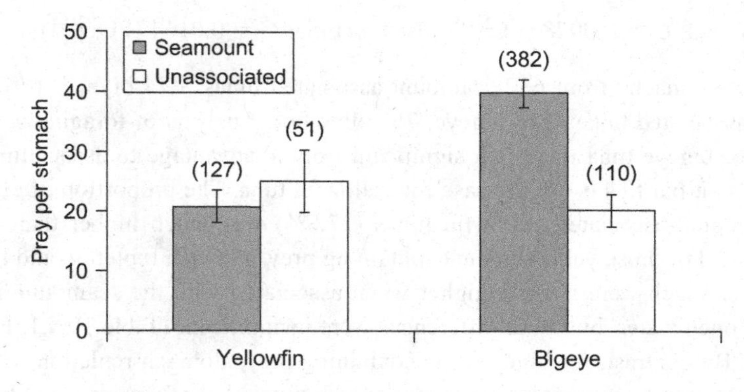

Fig. 10A.3 Mean number of prey per stomach for unassociated and seamount-associated yellowfin and bigeye tunas in Hawaii. Sample sizes are reported parenthetically and only include non-empty stomachs (error bars = SE).

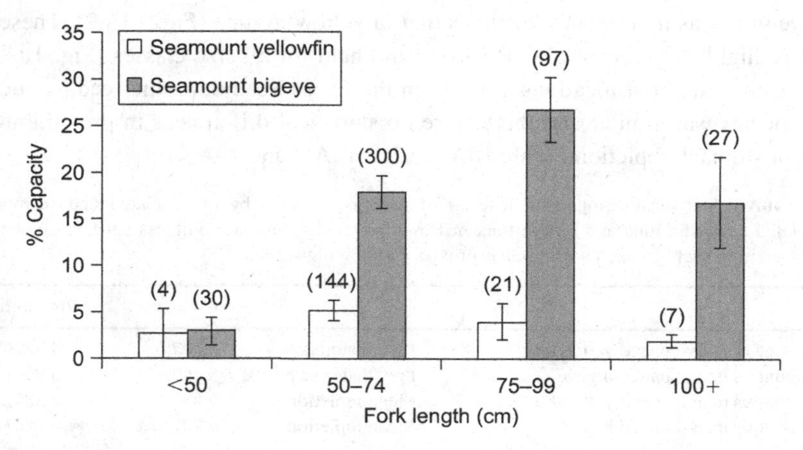

Fig. 10A.4 Mean stomach repletion (percent capacity) including empty stomachs for four size classes of yellowfin and bigeye tunas associated with the Cross Seamount in Hawaii. Sample sizes are reported parenthetically (error bars = SE).

Differences were also apparent in the diversity and taxonomy of the prey consumed by the two species of tunas when associated with the Cross Seamount. Bigeye tuna consumed prey from 94 families compared with 67 families for yellowfin while associated with the seamount. The rarity of some prey species suggested this could be due to differences in sample sizes. However, 87% of the families consumed by yellowfin were also consumed by bigeye but only 57% of those families consumed by bigeye were consumed by yellowfin. Nevertheless, the most abundant prey species were common for both tuna species (see figure 1 in Supplementary Material). The most dominant prey for bigeye tuna were caridean shrimp of the family Oplophoridae (primarily *Acanthephyra smithi* and *Systelaspis debilis*), followed by peneoid shrimp of the family Sergestidae (*Sergia* spp. and *Sergestes* spp.), myctophid fishes and several families of cephalopods. The most dominant prey for yellowfin were sergestid shrimp followed by brachyuran crab megalopae, cephalopods and oplophorid shrimps. However, closer inspection indicated that although the two tuna species were feeding on the same prey families, the species and life history stages of the prey were different for the two tuna species. Unlike bigeye, yellowfin tuna primarily consumed species of oplophorid shrimp and life stages of sergestid shrimp, which have vertical distributions that extend up into the surface mixed layer (Fig. 10A.5) whereas the bigeye consumed species and life history stages found in deeper sections of the water column. In summary, bigeye were eating a wider range of prey than yellowfin tuna and apparently feeding throughout a deeper and more extensive portion of the water column.

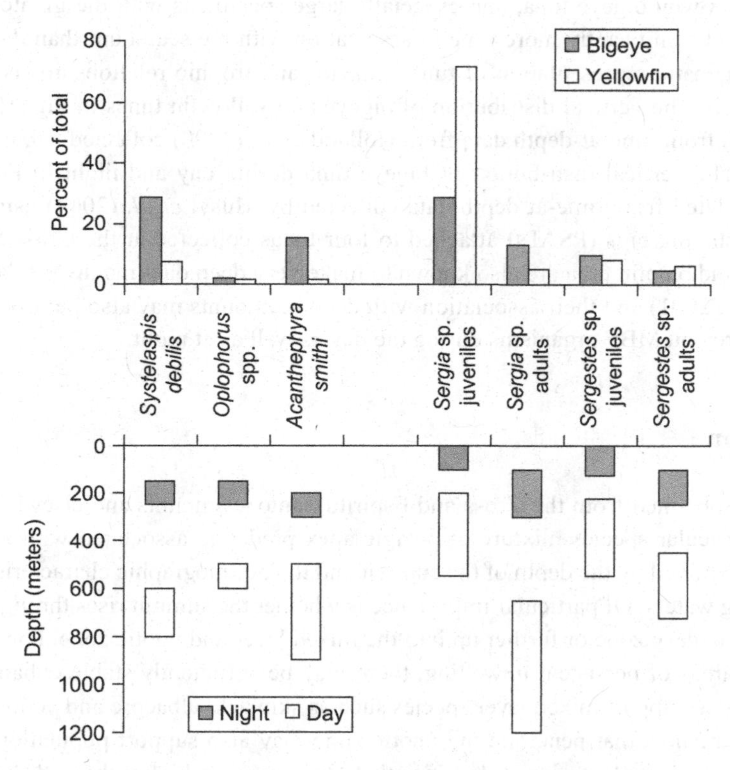

Fig. 10A.5 Relative percentage of oplophorid and sergestid shrimp in yellowfin and bigeye tunas associated with the Cross Seamount. Diel depth distributions for prey are also shown.

Bigeye tuna appear to gain a significant feeding advantage by associating with the Cross Seamount where they prey on mesopelagic organisms from the boundary fauna that are found where the ocean floor rises above 1200 m. By contrast, yellowfin tuna feed on a different suite of organisms that are primarily found in the mixed layer and do not appear to switch their feeding strategy to take advantage of deeper seamount-associated organisms.

These gut content analyses are supported by what is known about the vertical distribution of these two tuna species. When moving through open water, away from seamounts and floating debris or man-made fish aggregating devices, yellowfin tuna are predominantly found in the mixed surface layer and the thermocline that defines its lower boundary. By contrast, bigeye tuna occupying the same geographic areas select much deeper, colder water and can be found in depths in excess of 500 m (Holland *et al.*, 1990; Dagorn *et al.*, 2000). Using internally implanted tags, Musyl and colleagues demonstrated that bigeye tuna retained this deep diving pattern when associated with the Cross Seamount

(Musyl *et al.*, 2003). In fact, even the shallowest daytime depths were deeper at the seamount than when in the bigeye were in open water.

These behavioral data strongly suggest that the deeper diving exhibited by bigeye tuna enables them to access the enhanced MBF forage associated with the seamount both during the day and at night. This would explain the dramatic difference in feeding success between sympatric bigeye and yellowfin tuna. This disparity in feeding success also probably explains why bigeye tuna, and especially large specimens with the greatest thermal inertia, spend significantly more time in association with the seamount than do yellowfin tuna. A schematic representation of tuna behavior and trophic relationships is presented in Fig. 10A.6. The vertical distribution of bigeye and yellowfin tunas in Fig. 10A.6a was interpolated from time-at-depth data from Holland *et al.* (1990) collected using ultrasonic telemetry. The vertical distribution of bigeye tuna during day and night in Fig. 10A.6b was interpolated from time-at-depth data collected by Musyl *et al.* (2003) using pop-off satellite archiving tags (PSATs) attached to four tunas collected at the Cross Seamount. Swordfish and bluefin tuna are also known to make very deep daytime dives (Carey, 1990; Block *et al.*, 2001) and their association with deep seamounts may also be driven by their ability to prey on MBF organisms during the day as well as at night.

Discussion

The results obtained from the Cross and Espiritu Santo seamounts and elsewhere suggest that the particular species mixture of pelagic apex predators associated with a seamount will be determined by the depth of the summit and the oceanographic characteristics of the surrounding waters. Of particular importance is whether the summit rises through the mesopelagic boundary zone or further up into the mixed layer and photic zone. For seamounts located in areas of persistent upwelling, there may be sufficiently stable enhancement of productivity to support mixed layer species such as skipjack, albacore and yellowfin tunas. Isolated seamounts that penetrate the photic zone may also support populations of these more surface-oriented species at densities that are somewhat higher than adjacent waters. Deeper seamounts can support enhanced densities of deeper diving tunas such as bigeye and bluefin and swordfish by providing circumstances that make organisms of the mesopelagic boundary zone more accessible to predation during both day and night.

Seamounts may also act as navigational 'waypoints' in the larger movement patterns of these species – such a role is supported by tracks of bluefin and bigeye tuna and of the quite brief residence times of yellowfin tuna at the Cross and Espiritu Santo seamounts. Even bigeye tuna that seem to fare well at seamounts appear not to stay there for more than a few weeks and the overall biomass of tunas at seamounts is variable (Fonteneau, 1991; Holland, Personal observation). This suggests that either the available forage resource becomes depleted or is swept away or that, regardless of feeding success, the tuna have innate tendencies to migrate that override the importance of localized feeding opportunities.

From a management perspective, there is little doubt that seamounts are very important in the life history strategies of tunas and in shaping their distribution patterns. Certainly, much of the human activity around seamounts is related to fishing fleets targeting tunas and

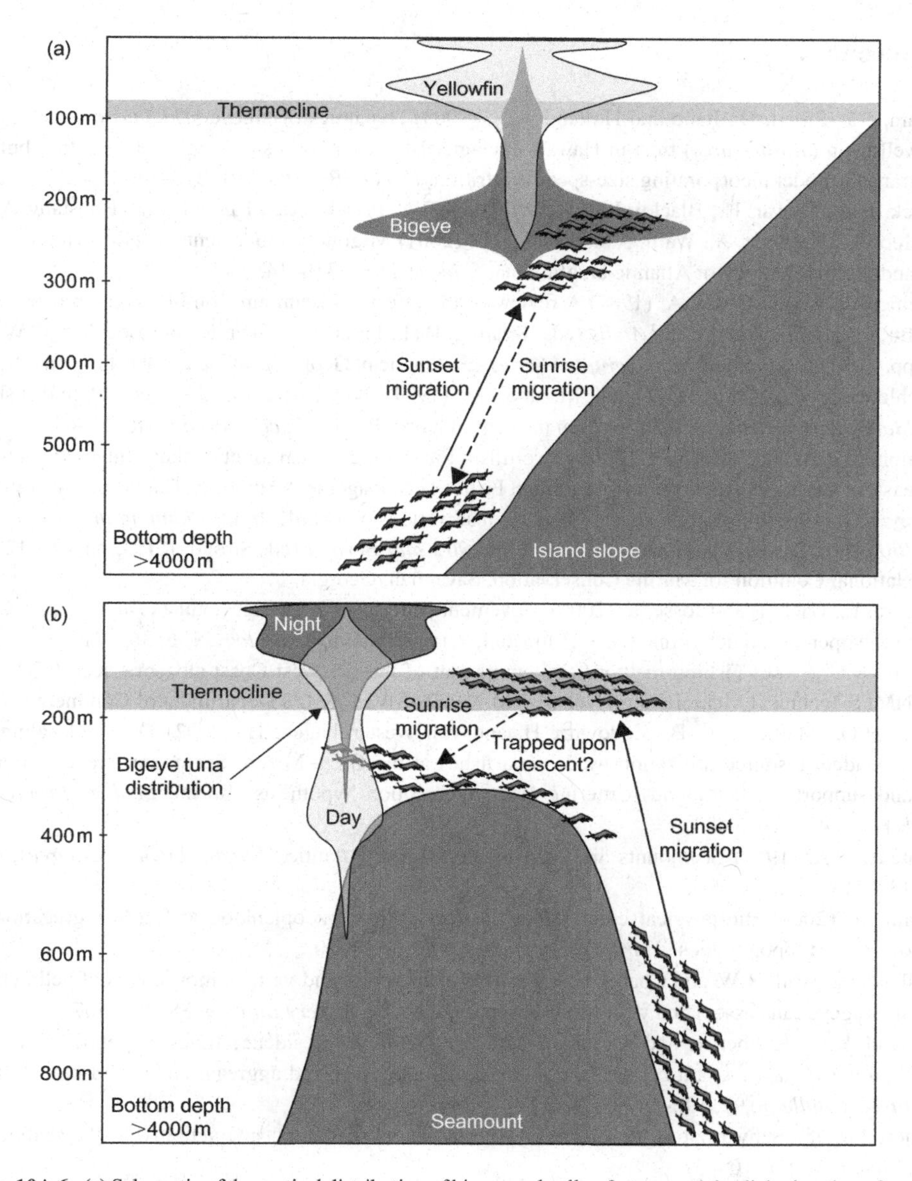

Fig. 10A.6 (a) Schematic of the vertical distribution of bigeye and yellowfin tuna and the diel migration of mesopelagic shrimp in nearshore waters. Tuna distribution is based on sonic telemetry data from Holland *et al.* (1990). Mesopelagic migration is based on data for oplophorid shrimp from Ziemann (1975). (b) Schematic of the vertical distribution of bigeye tuna during day and night while associated with the Cross Seamount relative to the diel migration of mesopelagic oplophorid shrimp. Tuna distribution is based on archival telemetry data from Musyl *et al.* (2003). Mesopelagic migration is based on data for oplophorid shrimp from Ziemann (1975).

associated species. Because their locations are known, seamounts increase the vulnerability of these species to harvesting. Since, for unknown reasons, seamounts seem to attract disproportionate amounts of sub-adults, catches from seamounts should be particularly closely monitored for species that are being harvested at or beyond sustainable levels.

References

Adam, S.J., Sibert, D., Itano and Holland, K.N. (2003) Dynamics of bigeye (*Thunnus obesus*) and yellowfin (*T. albacares*) tuna in Hawaii's pelagic fisheries: analysis of tagging data with a bulk transfer model incorporating size-specific attrition. *Fishery Bulletin*, 101, 215–18.

Block, B.A., Dewar, H., Blackwell, S.B., Williams, T.D., Prince, E.D., Farwell, C.J., Boustany, A., Teo, S.L.H., Seitz, A., Walli, A. and Fudge, D. (2001) Migratory movements, depth preferences, and thermal biology of Atlantic bluefin tuna. *Science*, 293, 1310–14.

Boehlert, G.W. and Genin, A. (1987) A review of the effects of seamounts on biological processes. In: *Seamounts, Islands and Atolls* (eds. Keating, B.H., Fryer, P., Batiza, R. and Boehlert, G.W.), pp. 319–34. *Geophysics Monographic Series* 43. American Geophysical Union. Washington, DC.

Boehlert, G.W., Wilson, C.D. and Mizuno, K. (1994) Populations of the sternoptichid fish, *Maurolicusmuelleri*, on seamounts in the central North Pacific. *Pacific Science*, 48, 57–69.

Campbell, R.A. and Hobday, A. (2003) Swordfish–Environment–Seamount–Fishery Interactions off eastern Australia. Report to the Australian Fisheries Management Authority, Canberra, Australia.

Carey, F.G. (1990) Further acoustic telemetry observations of swordfish. In: *Planning the Future of Billfishes; Research and Management in the 90's and Beyond* (ed. Stroud, R.H.), pp 103–122. National Coalition for Marine Conservation, Savannah, Georgia.

Dagorn, L., Bach, P. and Josse, E. (2000) Movement patterns of large bigeye tuna (*Thunnus obesus*) in the open ocean determined using ultrasonic telemetry. *Marine Biology*, 136, 361–71.

Dotson R.C. (1980) Fishing methods and equipment of the US West Coast albacore fleet. NOAA-NMFS Technical Memorandum NOAA-TM-NMFS-SWFC-8. US Department of Commerce.

Fock, H.O., Matthiessen, B., Zodowitz, H. and von Westernhagen, H. (2002) Diel and habitat-dependent resource utilization by deepsea fishes at the Great Meteor Seamount: niche overlap and support for the sound scattering layer interception hypothesis. *Marine Ecology Progress Series*, 244, 219–33.

Fonteneau, A. (1991) Seamounts and tuna in the tropical Atlantic. *Aquatic Living Resources*, 4, 13–25.

Genin, A. (2004) Bio-physical coupling in the formation of zooplankton and fish aggregations overabrupt topographies. *Journal of Marine Systems*, 50, 3–20.

Holland, K., Brill, R.W. and Chang, R.K.C. (1990) Horizontal and vertical movements of yellowfin and bigeye tuna associated with fish aggregating devices. *Fishery Bulletin*, 88, 493–507.

Holland, K.N., Kleiber, P. and Kajiura, S.M. (1999) Different residence times of yellowfin tuna, *Thunnus albacares*, and bigeye tuna, *T. obesus*, found in mixed aggregations over a seamount. *Fishery Bulletin*, 97, 392–5.

Isaacs, J.D. and Schwartzlose, R.A. (1965) Migrant sound scatterers: interaction with the seafloor. *Science*, 150, 1810–13.

Itano, D. and Holland, K. (2000) Movement and vulnerability of bigeye (*Thunnu sobesus*) and yellowfin tuna (*T. albacares*) in relation to FADs and natural aggregation points. *Aquatic Living Resources*, 13, 213–23.

Klimley, A.P. (1993) Highly directional swimming in scalloped hammerhead sharks, *Sphyrna lewini*, and subsurface irradiance, temperature, bathymetry and geomagnetic field. *Marine Biology*, 117, 1–22.

Klimley, A.P. and Butler, S.B. (1988) Immigration and emigration of a pelagic fish assemblage to seamounts in the Gulf of California related to water mass movements using satellite imagery. *Marine Ecology Progress Series*, 49, 11–20.

Klimley, A.P., Butler, S.B., Nelson, D.R. and Stull, A.T. (1988) Diel movements of hammerhead sharks *Sphyrna lewini*, Griffith and Smith, to and from a seamount in the Gulf of California. *Journal of Fishery Biology*, 33, 752–61.

Klimley, A.P., Jorgensen, S.J., Muhlia-Melo, A. and Beavers, S.C. (2003) The occurrence of yellowfin tuna (*Thunnus albacares*) at Espiritu Santo Seamount in the Gulf of California. *Fishery Bulletin*, 101, 684–92.

Knight, R.L. and Margraf, F.J. (1982) Estimating stomach fullness fishes. *North American Journal of Fisheries Management*, 2, 413–14.

Lehodey, P., Andre J.-M., Bertingnac, M., Hampton, J., Stoens, A., Menkes, C., Memery, L. and Grima, N. (1998) Predicting skipjack forage distributions in the equatorial Pacific using a coupled dynamical bio-geochemical model. *Fisheries Oceanography*, 7(3/4), 317–25.

Meyer, C.G., Holland, K.N. and Papastamatiou, Y.P. (2005) Sharks can detect changes in the geomagnetic field. *Journal of the Royal Society Interface*, 2, 129–30.

Musyl, M.K., Brill, R.W., Boggs, C.H., Curran, D.S., Kazama, T.K. and Seki, M.P. (2003) Vertical movements of bigeye tuna (*Thunnus obesus*) associated with islands, buoys, and seamounts near the main Hawaiian Islands from archival tagging data. *Fisheries Oceanography*, 12(2), 1–18.

Reid, S.B., Hirota, J., Young, R.E. and Hallacher, L.E. (1991) Mesopelagic-boundary community in Hawaii: micronekton at the interface between neritic and oceanic systems.

Rogers, A.D. (1994) The biology of seamounts. *Advances in Marine Biology*, 30, 305–50.

Sibert, J.R., Holland, K. and Itano, D. (2000) Exchange rates of yellowfin and bigeye tunas and fishery interaction between Cross Seamount and near-shore FADs in Hawaii. *Aquatic Living Resources*, 13, 225–32.

Uchida, R.N. and Tagami, D.T. (1984) Groundfish fisheries and research in the vicinity of seamounts in the north Pacific ocean. *Marine Fisheries Review*, 46(2), 1–17.

Uda, M. and Ishino, M. (1958) Enrichment pattern resulting from eddy systems in relation to fishing grounds. *Journal of Tokyo University Fisheries*, 1–2, 105–19.

Walker, M.M. (1984) Learned magnetic field discrimination in yellowfin tuna (*Thunnus albacares*). *Journal of Comparative Physiology*, 155, 673–9.

Wilson, C.D. and Boehlert, G.W. (1993) Population biology of *Gnathophausia longispina* (Mysidacea: Lophogastrida) from a central North Pacific seamount. *Marine Biology*, 115, 537–43.

Wilson, C.D. and Boehlert, G.W. (2004) Interaction of ocean currents and resident micronekton at a seamount in the central North Pacific. *Journal of Marine Systems*, 50, 39–60.

Wilson, D.W. (1992) Interactions of ocean currents and diel migrators at a seamount in the central North Pacific Ocean. PhD Dissertation, University of Hawaii, USA.

Ziemann, D.A. (1975) Patterns of vertical distribution, vertical migration and reproduction in the Hawaiian mesopelagic shrimp of the family Oplophoridae. PhD Dissertation, University of Hawaii, 112 pp.

Chapter 10B

Fish visitors to seamounts: Aggregations of large pelagic sharks above seamounts

Feodor Litvinov

Abstract

Field observations and sampling from widely scattered locations around the globe suggest that pelagic sharks may also be attracted to seamounts. Russian surveys in the 1970s and 1980s noted large aggregations of blue and other pelagic sharks near seamounts, up to 20 times as dense as that of the open ocean. Seamounts may facilitate shark navigation and social mating behavior, which may render some sharks exceptionally vulnerable to overfishing. Hence sharks at seamounts need protection and management.

Introduction

Tagging studies show that some seamounts host transient populations of tuna species, consequently some fisheries have taken advantage of these aggregations to increase their yields (Yasui, 1986; Fonteneau, 1991; Adam *et al.*, 2003). Other pelagic fishes such as billfishes (Ward *et al.*, 2000; Sedberry and Loefer, 2001) appear also to be attracted to complex high relief bottom structures where they may be subject to local depletion. Pelagic shark aggregations over seamounts are poorly understood and underreported in the literature. For example, in the Gulf of California, hammerhead sharks (*Sphyrna lewini*) congregate round El Bajo Espiritu Santo seamount during the day, but move separately into the surrounding pelagic environment at night (see Chapter 10A; Klimley *et al.*, 1988). This pattern was observed over a 10-day period. Catches of gray sharks (*Carcharhinus* spp.) off northeast Brazil were significantly higher around seamounts, particularly those with relatively deep summits (300 m) and gentle slopes (Hazin *et al.*, 1998). Queiroz *et al.* (2005) tagged 168 blue sharks (*Prionace glauca*), along the Portuguese coast of which 34 were recaptured. From these, 32 sharks were recaptured in the vicinity of areas with high bottom relief, such as seamounts, canyons or the continental shelf slope. Based on these observations it has been hypothesized that pelagic sharks may also be attracted to seamounts for feeding or orientation. However, this hypothesis has been poorly tested.

A worldwide case study

Several pelagic longline surveys (290 drifts and 76931 hooks in total) in the eastern Atlantic and eastern Pacific from 1978 to 1987 estimated catch per unit effort (CPUE) as the number of sharks caught per 100 hooks (Litvinov, 1989, 2004; Table 10B.1). Litvinov (1989) found that blue sharks dominated shark communities in oceanic waters. He also found the abundance of pelagic sharks to be low, being about 1.3 for blue shark and less than 0.2 for other elasmobranchs (Fig. 10B.1). However, dense aggregations of sharks were identified (up to 20 times more abundant comparing adjacent waters). Such aggregations were found in the NE Atlantic over Meteor, Yer, Erving and Atlantis seamounts and

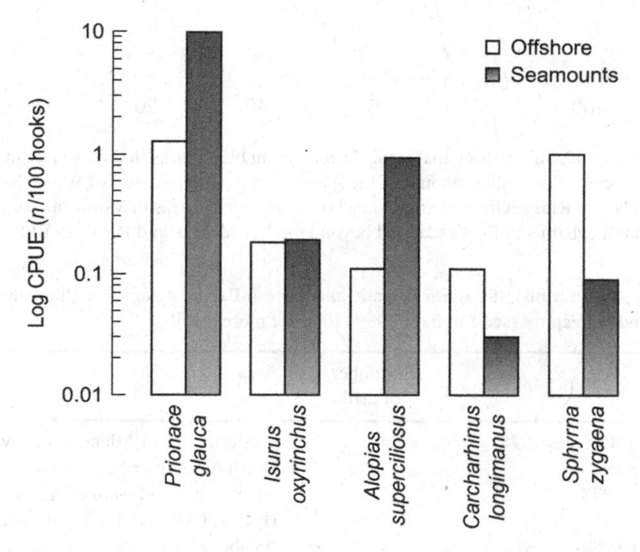

Fig. 10B.1 Average shark catch *per* 100 hooks in the open waters of the east central Atlantic in July 1980 (white bars). Gray bars show December 1980 catch over Great Meteor, Atlantis and Irving seamounts south of the Azores.

in the Southeast Atlantic over the Whale Ridge (Fig. 10B.2). In the eastern Pacific, such aggregations were observed over Nazca Ridge and around the Galapagos Islands. These aggregations were mainly formed by large (200–300 cm TL) adult male blue sharks but higher CPUE values for bigeye thresher shark (*Alopias superciliosus*) were also observed over seamounts.

Two general, not mutually exclusive, hypotheses have been presented to explain the higher densities of pelagic species at certain seamounts. One is that seamounts may represent navigational waypoints in larger movement patterns (Klimley, 1993; Holland *et al.*, 1999; see Chapter 12 for turtles and Chapter 10A for tunas). The second holds that seamounts provide enhanced foraging opportunities for apex predators (see Chapter 12C for seabirds). Seamounts often have distinct geo-magnetic signatures (Klimley, 1993), making it possible for blue sharks to detect those fields and thus, use seamounts as meeting points for copulation aggregations.

Male blue sharks form aggregations or 'male clubs', most probably for first copulation with subadult females. Blue sharks are oceanic, but spawn in shelf and neritic waters. Young

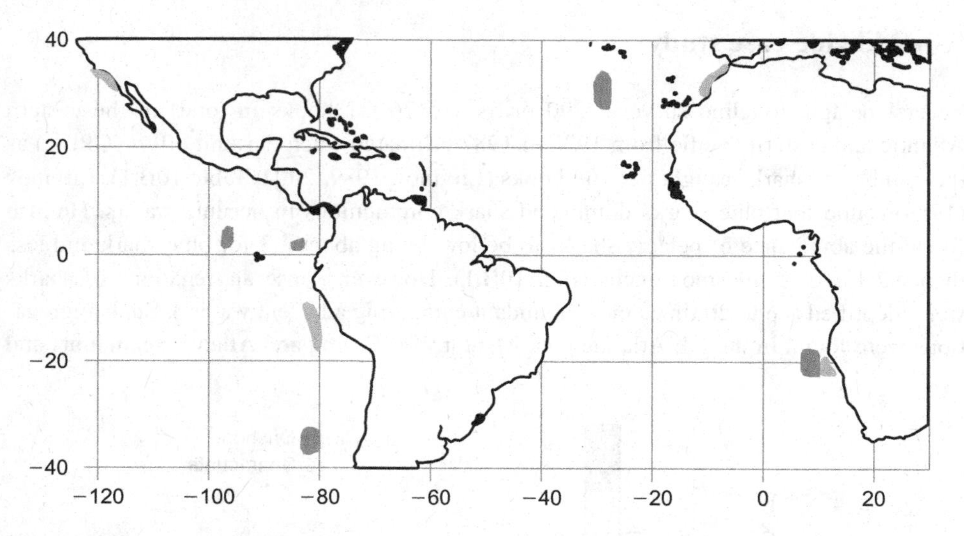

Fig. 10B.2 Aggregations of adult males ('male clubs') and young blue sharks in the east Atlantic and east Pacific. Red shading: aggregations of blue sharks south of the Azores, on the oceanic part of Whale Ridge, west and east of Galapagos and on Nazca Ridge. Green shading: newborn and young sharks in Moroccan waters, the neritic area of Whale Ridge, Santa Barbara and Ensenada, and beyond the Ecuadorean and Peruvian EEZs.

Table 10B.1 Date and locations of eastern Atlantic and eastern Pacific pelagic longline surveys from 1978 to 1987 by Litivinov and colleagues (see Litvinov, 1989, 2006 for more details).

No.	Period	Number of drifts	Area
1.	September–October 1978	24	Western Central Atlantic open waters
2.	November 1978	4	North Atlantic open waters
3.	November 1978	7	Seamounts southward of Azores: Great Meteor, Hyeres, Irving, Plato, Flamingo, Atlantis
4.	December 1980	4	As above
5.	July–September1980	43	Open Atlantic
6.	September 1980	9	Whale Ridge, oceanic waters
7.	September1980	5	Whale Ridge, neritic waters
8.	November 1979–January 1980	14	Moroccan waters
9.	January–February1980	12	Sierra-Leone waters
10.	June–September 1982	27	Guinea Bay
11.	June–September 1982	23	San-Tome and Principe waters
12.	May–July 1982	24	Nicaraguan waters, Atlantic coast
13.	November 1985–March 1986	21	Nicaraguan waters, Atlantic coast
14.	May–July 1986	26	Cuban waters
15.	May–June 1981	10	Waters of Grenada Island
16.	April–June 1984	7	East Pacific, East and West of Galapagos Islands
17.	June 1984	37	East Pacific, off Peru
18.	April 1987	2	Nazca and Sala y Gomez seamounts
19.	September 1992 (purse seine)	1	Ensenada, Mexico

females leave 'kindergartens' and go for the first copulation to 'male clubs', which form in oceanic waters. Females leave immediately after copulation and may come back after fertilization, at the various stages of pregnancy, when signs stimulating mating behavior in males are absent (Litvinov, 2006). Such male aggregations are situated above peaks of seamounts, but may occur in other places such as features of water structure or gradients.

The second and further copulations of adult females may occur in other areas. Undoubtedly, dense male aggregations of the blue shark are crucial points for the species survival. That also make blue shark (and other species) even more vulnerable to overfishing. The extremely high demand for shark fins suggests that sharks may be even more vulnerable to overfishing than whales were in the past.

There are no year-round observations to confirm the permanent or seasonal nature of such aggregations. However, the occurrence of the very different stages of pregnancy at the same place, from newly fertilized eggs to 41 cm length embryos in uteri (Litvinov, 2006), suggests a long residence period and probably all-the-year-round spawning and copulation. It is thus clear that shark aggregations are very important for the species survival and that these phenomena need serious protection. On the other hand, the long period aggregations of large pelagic sharks may seriously influence seamount populations through predation on a wide variety of fish, squids and crustaceans.

Conclusions

Blue sharks are among the most abundant, widespread, fecund and faster growing of the elasmobranchs, and a pelagic species that is widely distributed throughout the world's oceans. They are also the most heavily fished sharks in the world (Baum *et al.*, 2003). Annual fisheries mortality estimated at 10–20 million individuals, is likely having an impact on the world population, but monitoring data are inadequate to assess the scale of any population decline. There is concern over the removal of such large numbers of this likely keystone predator from the oceanic ecosystem (Stevens, 2000). It is to be noted, that according to studies of shark fossil teeth in bottom sediments, the blue shark occupied a dominant position very recently, between 302 and 3940 years ago, having substituted representatives of the genus *Isurus* that already dominated during the Holocene (Litvinov, in press). The reason for such substitution is not understood, but it is clear, that the oceanic ecosystem in quite fragile, and its keystone species are subject to some rather slight influence. Thus, if dense shark aggregations on seamounts are indeed the rule, protective measures need to be undertaken immediately on these ecosystems and on other shark aggregation spots.

Acknowledgment

The author thanks all the captains and crew members for assistance and great patience during data collection in the field. I am very much obliged to the editors for helping with the revision.

References

Adam, M.S., Sibert, J., Itano, D. and Holland, K. (2003) Dynamics of bigeye (*Thunnus obesus*) and yellowfin (*T. albacares*) tuna in Hawaii's pelagic fisheries: analysis of tagging data with a bulk transfer model incorporating size-specific attrition. *Fishery Bulletin*, 101, 215–28.

Baum, J.K., Myers, R.A., Kehler, D.G., Worm, B., Harley, S.J. and Doherty, P.A. (2003) Collapse and conservation of shark populations in the Northwest Atlantic. *Science*, 299, 389–92.

Fonteneau, A. (1991) Monts sous-marins et thons dans l'Atlantique tropical est. *Aquatic Living Resources*, 4, 13–25.

Hazin, F.H.V., Zagaglia, J.R., Broadhurst, M.K., Travassos, P.E.P. and Bezerra, T.R.Q. (1998) Review of a small-scale pelagic longline fishery off northeastern Brazil. *Marine Fisheries Review*, 60(3), 1–8.

Holland, K.N., Kleiber, P. and Kajiura, S.M. (1999) Different residence times of yellowfin tuna, *Thunnus albacares*, and bigeye tuna, *T. obesus*, found in mixed aggregations over a seamount. *Fishery Bulletin*, 97, 392–5.

Klimley, A.P. (1993) Highly directional swimming by scalloped hammerhead sharks, *Sphyrna lewini*, and subsurface irradiance, temperature, bathymetry, and geomagnetic field. *Marine Biology*, 117, 1–20.

Klimley, A.P., Butler, S.B., Nelson, D.R. and Stull, A.T. (1988) Diel movements of scalloped hammerhead sharks, *Sphyrna lewini* Griffith and Smith, to and from a seamount in the Gulf of California (Mexico). *Journal of Fish Biology*, 33(5), 751–62.

Litvinov, F.F. (1989) Structure of epipelagic elasmobranch communities in the Atlantic and Pacific oceans and their change in recent geological time. *Journal of Ichthyology*, 29(8), 75–87.

Litvinov, F.F. (2004) The dense male aggregation over submarine mounts as an integral part of species range in the blue shark *Prionace glauca*. ICES CM:2004/Session K: 11, 6 pp.

Litvinov, F.F. (2006) On the role of dense aggregations of males and juveniles in the functional structure of the range of the blue shark *Prionace glauca*. *Journal of Ichthyology*, 46(8), 613–24.

Litvinov, F.F. (2007) Foundations of the method to dating global fluctuations of the abundance of oceanic sharks on the basis of the amount of fossil teeth in bottom sediments. *Journal of Ichthyology*, 47(4), 267–70.

Queiroz, N., Lima, F.P., Maia, A., Ribeiro, P.A., Correia, J.P. and Santos, A.A. (2005) Movement of blue shark, *Prionace glauca*, in the north–east Atlantic based on mark – recapture data. *Journal of the Marine Biological Association of the United Kingdom*, 85(5), 1107–12.

Sedberry, G.R. and Loefer, J.K. (2001) Satellite telemetry tracking of swordfish, *Xiphias gladius*, off the eastern United States. *Marine Biology*, 139, 355–60.

Stevens, J. (2000) Prionace glauca. In: *IUCN 2003. 2003 IUCN Red List of Threatened Species.* <www.redlist.org>

Ward, P., Porter, J.M. and Elscot, S. (2000) Broadbill swordfish: status of established fisheries and lessons for developing fisheries. *Fish and Fisheries*, 1, 317–36.

Yasui, M. (1986) Albacore *Thunnus alalunga* pole-and-line fishery around the Emperor Seamounts. *NOAA Technical Report NMFS*, 43, 37–40.

Chapter 11
Seamounts and cephalopods

Malcolm Clarke

Abstract

The relationship between cephalopods and seamounts has been inadequately studied. Here, the work on seamounts is supplemented by collections made around oceanic islands with similar cephalopod communities and compared with sampling over the abyssal plain. Cephalopods differ from fish in several respects that make their populations resilient and their study particularly difficult. Nets sample only small numbers of juveniles or small species. Most adult cephalopods are adequately sampled only by their predators, and their presence is deduced from their predators' stomach contents. Fortunately, digestive processes leave their chitinous jaws ('beaks') intact. These can often be identified and body mass estimated from beak lengths. Some net collections of pelagic cephalopods from the NE Atlantic are compared, together with samples from predators' stomachs. The large majority of species which live clear of the bottom do not show any association with, or retention by, seamounts: only *Neorossia caroli* and possibly *Liguriella* appear restricted to seamounts. The results support six of the ecological groups of cephalopods considered by Nesis to occur on and around seamounts. Bottom octopods have been sampled globally, but speciation related to seamounts has not been detected. The importance of size and other structural features to cephalopods as predators and as prey is related to biomass considerations at seamounts. While there are formidable problems of investigation, it is likely that cephalopods play a major role in supplying energy to seamount ecosystems. By swimming or drifting to seamounts for spawning, or possibly feeding, cephalopods may provide accumulations of great attraction to passing fish, birds, seals and cetaceans. Seamounts may have a wide oceanic influence by fuelling such migrators.

Introduction

Up to the turn of the twentieth century, nearly all sampling of pelagic cephalopods over and around seamounts was focused on deducing geographic distribution rather than contributing to seamount ecology. This is useful for drawing general conclusions, but detailed analysis is hampered by the variation in sampling methods, geographic area and time. Nesis (1993) pointed out that 'data on seamount cephalopods is scanty and fragmentary', and divided cephalopods on and around seamounts into seven ecological groups, listing species for each group.

This chapter compares pelagic cephalopods, sampled with a very diverse collection of nets and by various predators, in various localities in the NE Atlantic between about 30ºN and 40ºN. They include samples from above seamounts, over the abyssal plain, over island slopes and close to island shores (Clarke, 2006). By this means, any special features of the seamount cephalopod fauna should become evident.

Nearly all the net caught samples of pelagic species in these collections are of small cephalopods (<6 cm body length), either larvae or juveniles of large species, or adults of very small species. To sample the adults of the majority of species our only recourse is to find them in the stomachs of predators such as cetaceans, seals, fish and birds. Fortunately, the chitinous beaks (jaws) of cephalopods are not digested and often accumulate in the digestive tract. In 17 sperm whales (*Physeter catodon*) sampled in the Azores there were an average of 1692 lower beaks. By their identification and their size, it is possible (Clarke, 1986b) to find the contribution of a species by number and mass to the diet, and sometimes also growth rate (Clarke, 1980, 1987, 1993). It is also often possible to examine intact specimens from stomach contents to determine their sexual maturity, fecundity and spawning season (Clarke, 1980).

North Atlantic data are used to determine which cephalopod species are associated exclusively or otherwise with seamounts and to attempt to illustrate the interactive biomass they may represent. Data on most large cephalopods do not come directly from seamounts because the predators which give us all the information have to be examined on shores where they strand (cetaceans); haul out for breeding (seals) or nest (birds), hence the positions where they ate the cephalopods are not known. In the last century, commercial whaling for sperm whales provided much data on cephalopods, including from the Antarctic (Clarke, 1980) and some considered here from Madeira and the Azores. Only fish are obtained directly from seamounts. The widespread distributions of many oceanic cephalopod genera and species make conclusions from the North Atlantic generally applicable.

Differences between cephalopods and fish

Several differences between cephalopods and fish are relevant to understanding the cephalopod contribution to, and effect on, seamount communities and the problems presented to the investigator.

Number of species

There are far fewer species of cephalopods than fish. Nesis (1987) lists ~650 species now described, with 786 in the online database CephBase (www.cephbase.utmb.edu) compared with 28 500 finfish species (www.fishbase.org). The 200–230 oceanic species have very wide distributions. The very lack of speciation suggests few barriers to distribution and genetic mixing which a close adhesion to seamount communities would imply.

Evidence for spawning

Spawn is known for very few species, possibly because of its extreme delicacy. In some species (e.g., *Illex illecebrosus*, O'Dor *et al.*, 1982), each tiny egg is held within a relatively enormous bag of thin jelly, which is readily destroyed by any net, be it a midwater plankton net or bottom commercial trawl. Many individuals of the species eaten by sperm whales are in spawning condition (Clarke, 1980).

Very young larvae are sampled at shallow depths and many squids descend during growth so that adults, particularly at spawning, often live deeper than 1000 m (Clarke and Lu, 1974). If they do need a substratum for attachment of the spawn, it is very likely that they use seamounts because the abyssal plain, several 1000 m deeper, is, as far as we know, well beyond their vertical range.

Life cycle

Cephalopods that have been investigated, mostly neritic species, are short lived, usually less than 2 years, fast growing and die after spawning once (Boyle and Rodhouse, 2005). We know that at least some of the oceanic ommastrephid squids are short lived. If this holds for most of the oceanic species; it implies that any biomass input into the energy of the seamount ecosystem could be large and variable from year to year, relative to the stocks of their slow-growing, long-lived predators. Heavy predation of spawning adults or their spawn would be rapidly replaced given the right conditions. However, there is little evidence about growth for oceanic species except for Ommastrephidae and Pholidoteuthidae, and it is quite possible some of the most important oceanic groups, including Histioteuthidae, Octopoteuthidae and Architeuthidae, have lifespans of several years, which would make their removal more important to the ecological balance (Clarke, 1993).

Environmental variability

Cephalopods show great resilience to environmental change by having short lives, high turnover of generations, plasticity in their seasons of reproduction, variable growth rates and migration (Boyle and Boletsky, 1996). The populations of shelf species can fluctuate enormously from year to year and from season to season, and they survive to flourish in later years. This is true also for the oceanic *I. illecebrosus*, although the species of the deep sea must have to cope with smaller environmental changes unless human commercial activities reduce their food or predators.

Predator/prey interaction

While fish collect prey with their mouths in one consuming or debilitating bite, cephalopods use the spread of their eight short tentacles ('arms') as their effective mouth size, so that they can catch and consume, at leisure, fish as large as, or larger than, themselves (Fig. 11.1 and Table 11.1; Rodhouse and Nigmatullin, 1996). They hold their prey and bite off pieces small enough to pass along the oesophagus, where its diameter is restricted by the encircling brain and cranial connective tissue. Many midwater species, such as octopoteuthids and architeuthids, grow larger than most midwater fish such as cyclothonids, myctophids, gonostomatids (Clarke, 1996b), and some even larger than big bottom fish such as orange roughy (e.g., *Hoplostethus atlanticus*) and rattails (e.g., *Coryphaenoides rupestris*). This, coupled with their effective 'mouth' size, makes them predators of many midwater fish and probably some larger bottom fish.

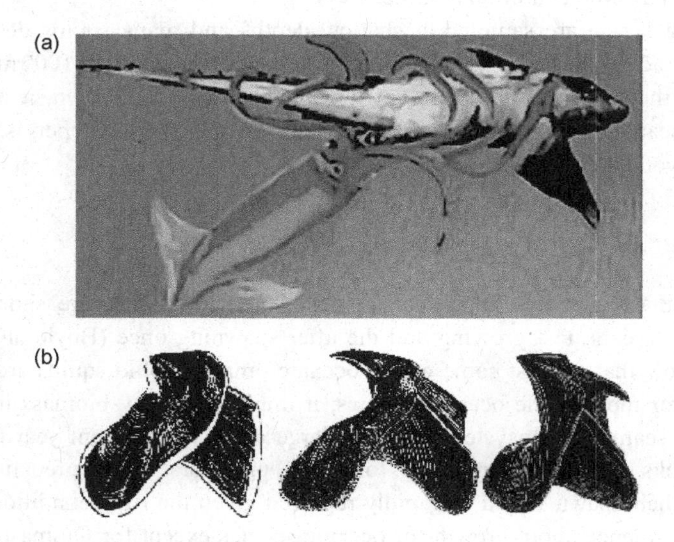

Fig. 11.1 (a) Method of attack by a squid showing how it uses the spread of its arms to hold a fish larger than itself. (b) Three lower beaks to show differences in leverage for hard scaled (left), tough skinned (centre) and soft muscled prey. See text for further explanation.

Sampling differences

Cephalopods are not sampled as well as fish and rarely represent more than a small proportion of the nekton caught in research nets; in one seamount comparison there was only 1 cephalopod to 18 fish (Diekmann, 2004); but it is often much less than this. This makes statistical comparisons less reliable compared to those for fish and tempts one to assume that conclusions based on fish can be accepted for cephalopods. In fact, the rarity of oceanic cephalopods in research and commercial nets and their importance in the diet of many cetacean, fish, bird and seal species indicates their ability to avoid capture rather than their scarcity. Many squid species in the diets of larger predators are almost unknown from commercial or research net catches of cephalopods. This may be because fish tend to be herded and overrun by nets, while squids turn and shoot through the meshes.

Evidence of cephalopod distribution

Diekmann (2004) identified 47 species in a collection of 1346 cephalopods from four seamounts in the eastern North Atlantic, the Great Meteor Seamount, 'Atlantis',

Table 11.1 Features which influence cephalopods as potential food and predators.

Typical species	Size		Numbers		Squid as food		Squid as predator			Notes
	Maximum adult stand L. cm	Arm spread	Relative commonness	Estimated mass g	Food value	Known predators	Beak ratio	Food quality	Probable food	
Heteroteuthis dispar	5	5.5	5		m	f	3 s	Hard	s, f, c	b
Architeuthis dux	600	670	4		m-g	sp	4 s	Hard	v, l, f, q	
Ancistrocheirus lesueurii	40	36	2		m-g	sp	2–3 w	Fairly soft	m, f	h
Chiroteuthis veranyi	57	184	3		g	sp, d, f,	2–3 s	Fairly soft	m, f	l
Liguriella	24	23				sp				
Liocranchia sp.	27	12	4		g	sp, f	3–4	Fairly hard	f	
Megalocranchia sp. (*Phasmotopsis*)	115	50	4		g	sp	1	Soft	f	l
Taonius pavo	53	10	3		g	sp, b, f	1	Soft	f	h
Teuthowenia megalops	32	16	3		g	sp, w, b	1–2	Soft	f	
(*Mesonychoteuthis hamiltoni*)			1		m-g	sp	1–2 vs	Hard	l, f	h
Discoteuthis spp.					m-g	sp	1–2 s	Hard	med, f	
Cycloteuthis sirventi	96	70	4	1800	m-g	sp	1	Soft	l, f	
Abraliopsis pfefferi	10.6	9.1	1		m	d, f	1.5–2	Soft	s, m, f, c	l, h
Gonatus steenstrupi	50	32	1		m	sp, w, s, f, b	1	Hard	med, f	h
Histioteuthis arcturi	66	90	1	950	m-g	sp	1	Soft	med, f	
Histioteuthis bonnellii	31	40	1	250	m-g	sp, f, b	2	Hard	s, f	l, w
Lepidoteuthis grimaldii	162	125	5		m-g	sp	1	Soft	l, f, q	s
Selenoteuthis scintillans	6	5.4	5		m		1	Soft	s, m, c	l
Idioteuthis c.f. hjorti	3.8	8.5	5		m-g	f	1	Soft	s, f	
Taningia danae	80	45	1	4100	m-g	sp, f	1	Soft	l, f, q	l, h
Octopoteuthis rugosa	46	43	3	383	m-g	sp, w, f	1	Soft	f	l, h
Octopoteuthis 'giant'	100	90	5		m-g	sp	1	Soft	l, f, q	l, h
Ommastrephes bartrami	136	98	1		m	f, b	2	Hard	l, f, q	
Sthenoteuthis c.f. pteropus	61	43	1		m	f, b	2	Hard	l, f, q	

(*Continued*)

Table 11.1 *Continued*

Typical species	Size		Numbers		Squid as food		Squid as predator			Notes
	Maximum adult stand L. cm	Arm spread	Relative commonness	Estimated mass g	Food value	Known predators	Beak ratio	Food quality	Probable food	
Todarodes sagittatus	125	100	1		m	sp, d, f	2	Hard	l, f, q	
Onychoteuthis banksii	22	14	4		m	sp, w, f, b	2	Hard	med, f	h
Moroteuthis sp.			2		m	sp	2	Hard	l, f	s, h
Pholidoteuthis	76	49	2	4648	m	sp, f	3	Hard point	t, f	s
Pterygioteuthis spp.		2.7	3		m	d		Hard	s, f, c	h
Pyroteuthis margaritifera		8.8	3		m	d, f		Hard	s, f, c	h
Thysanoteuthis rhombus	106	55	5		m		4	Hard	l, f	
Alloposus mollis	200	290	5		g	sp, f	7	Hard	l, f, q	w
Argonauta argo	120	138	5		m	f, b	<10	Soft	v, s	wc
Eledonella pygmaea	35	38	5		g	f	–	Soft	s	w
Japetella diaphana	28	24	5		g	f, c	–	Soft	s	w

Common to both: adult length (cm), relative commonness (one very common to five scarce). *As food*: estimated mass, food value (m: muscular and high calorific value; m-g: moderate value; g: gelatinous and about half value). *As a predator*: arm spread (cm), beak ratio (lower rostral length divided into the wing length). Some known predators are listed (sp: sperm whale; w: other toothed whales; d: dolphins; s: seals; f: fish; b: birds). A list of probable food based on beak shape and hardness very small (vs), small (s), medium sized (med), large (l), very large (vl), fish (f), crustaceans (c), squids (q) and midwater (m). 'Notes' show special qualities of the squids, e.g., possession of light for luring prey (l) or blinding prey (b), hooks to hold slippery food (h), webs to smother food (w) or collect it (wc) and scales to prevent abrasion (s).

'The Twins' and one near the Azores (Fig. 11.2). They were sampled by a Young Fish Trawl and, more intensively, at the Great Meteor Seamount, with a 1 m² 'BIOMOC' net; in total 172 hauls were made.

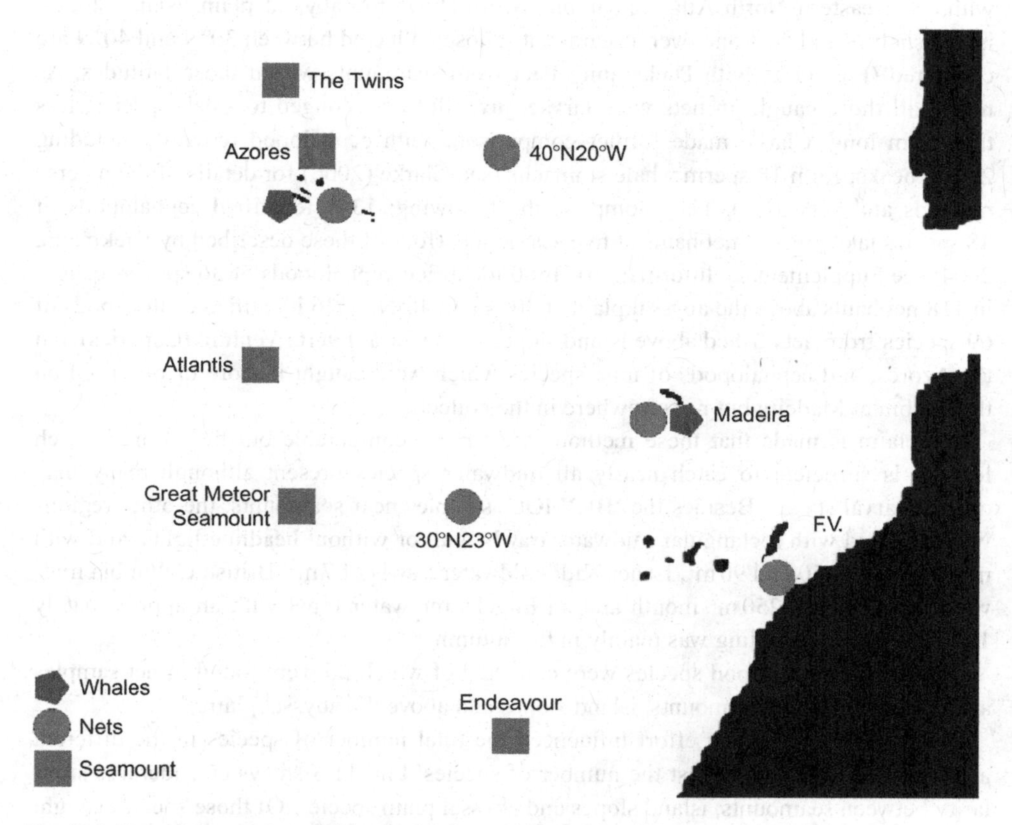

Fig. 11.2 Map of the east central Atlantic showing positions of the collections of cephalopods compared by Clarke (2006) examined by Diekmann (2004) and discussed here.

Diekmann showed, by non-metric multidimensional scaling and related techniques, no difference between the open ocean cephalopods and the community in the immediate vicinity of the seamount. However, horizontal distribution patterns of both fish and cephalopods corresponded well to the structure of a closed circulation cell (Taylor cap formation: see Chapter 4) detected above the upper slope and plateau area. There was a significant difference between fish assemblages within the 1500 m contour and the outer seamount regime, but this difference was not established for cephalopods. Diekmann concluded that flow at the seamount might retain early stages nearby and sustain local populations. However, while the fish larvae included many bottom living species, only two near-bottom cephalopods (*Heteroteuthis dispar*) and an octopod (possibly *Scaegus unicirrhus*) were represented in the larvae, and it may be that few cephalopod species, if any, have populations retained on seamounts. He concluded, from abundance, that the Great Meteor Seamount was not an area of high spawning activity for cephalopods during the sampling period (September 1998), although his nets did catch larvae and small juveniles and he gave no comparative numbers from other areas to make a sound judgement on the degree of spawning activity in an oceanic situation.

Clarke (2006) described the distribution of oceanic midwater cephalopod species within the eastern North Atlantic sampled from above the abyssal plain, island slopes, island shelves and, just one over a seamount. Those collected between 30°N and 40°N are compared (Fig. 11.2) with Diekmann's data from seamounts within those latitudes. As nearly all those caught in nets were larvae, juveniles or belonged to small species, less than 6 cm long. Clarke made further comparisons with cephalopod remains, including 29 226 beaks; from 18 sperm whale stomachs (see Clarke (2006) for details of the diverse methods and nets used). They comprise the following: 1346 identified cephalopods of 48 species taken in 172 net hauls at five seamounts (four of these described by Diekmann, 2004; see Supplementary Information); 1560 identified cephalopods of 36 species caught in 118 net hauls above the abyssal plain at 30°N and 40°N; 3336 identified cephalopods of 69 species from nets fished above island slopes at Madeira, Fuerta Ventura (Canaries) and the Azores; and cephalopods of nine species which were caught inshore or observed on the bottom at Madeira but not elsewhere in the collection.

No claim is made that these methods are strictly comparable but the effort in each locality is sufficient to catch nearly all midwater species present although many may only be larval stages. Besides the 'BIOMOC' samples near seamounts, the other regions were sampled with rectangular midwater trawls with or without headline lights, and with mouths of 1, 8, 50 and 90 m^2, Isaacs Kidd midwater trawls of 7 m^2, British Columbia midwater trawl with a 250 m^2 mouth and an Engel's midwater trawl with an approximately 1600 m^2 mouth. Sampling was mainly in the autumn.

In total, 78 cephalopod species were collected of which 20 were found in net samples at all three localities, seamounts, island slopes and above the abyssal plain.

Very variable sampling effort influenced the total number of species in the different areas but, even so, taking just the number of species, Fig. 11.3 shows considerable similarity between seamounts, island slopes and abyssal plain species. Of those species caught over the abyss, 78% were also caught over seamounts and 83% over island slopes; 94% of species caught on the seamounts were also present on island slopes. Thus, the majority of species living over seamounts are part of the community living over the abyssal plains, and the presence of their larvae show that they spawn over seamounts but not necessarily only there.

Clarke (2006) considered that only the sepiolid *N. caroli* was restricted to seamounts. *H. dispar*, which was caught over seamounts, was also caught over the abyssal plain supporting Nesis' (1993) contention that is it not rigidly linked to seamounts as had often previously been supposed.

Species regularly caught commercially close to Madeira are *Sepia officinalis*, *Loligo forbesi*, *Ommastrephes bartrami*, *Sthenoteuthis pteropus*, *Todarodes sagittatus*, several octopodids and, occasionally, *Thysanoteuthis rhombus*. Of these, *O. bartrami*, *S. pteropus*, *T. sagittatus* and *T. rhombus* have been caught by lines across much of the area studied here and *Octopus vulgaris*, *Sepia* and *Loligo* are normally regarded as continental shelf genera. Of all these only *S. pteropus*, *O. bartrami* and *Octopus* sp. were sampled on the seamounts. None of the species mentioned in this paragraph, although of the right size, were eaten by sperm whales sampled here.

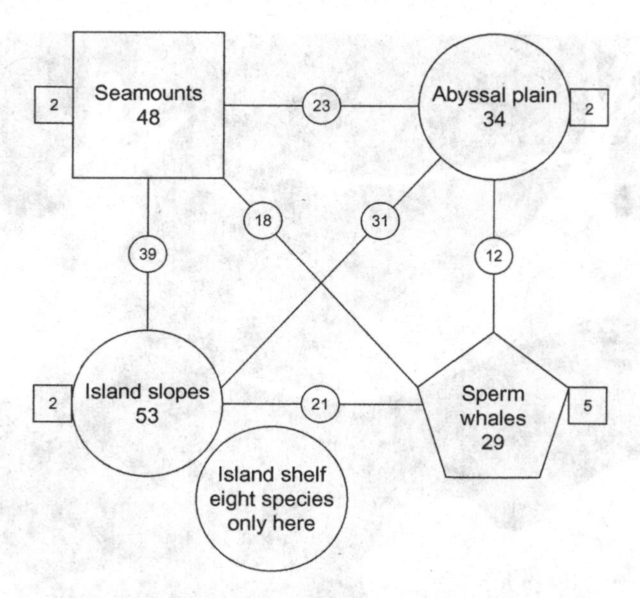

Fig. 11.3 The number of species in each source of the collections listed in the text in large numbers and shapes. Small figures in small circles show how many species were found in both the shapes connected and small figures in small squares show the number of species found in the collection to which it is attached but not in other collections (from Clarke, 2006).

Adult cephalopods in the sperm whale diet

Beaks representing all the species eaten by whales are of a size and advanced pigmentation to indicate they are from adult cephalopods; 63% of those in whales were sampled by nets over seamounts and 72% over island slopes (Fig. 11.3). These are high proportions, considering that sampling over seamounts was restricted in season and caught only larvae and small animals.

On the seamount, there are larvae of 16 species having adult lengths of over 20 cm. Of these, 11 species, *Ancistrocheirus lesueuri*, *Megalocranchia* sp., *Taonius pavo*, *Discoteuthis* sp., *Histioteuthis bonnellii*, *H. arcturi*, *H. celetaria*, *H. meleagroteuthis*, *H. reversa*, *Taningia danae* and *Onychoteuthis banksii*, are eaten by sperm whales, and also are found on island slopes. The other five genera, *O. bartrami*, *S. pteropus*, *T. rhombus*, *Tremoctopus violaceus* and *Octopus* sp. are not normally eaten by sperm whales.

Six species are important as whales' food but were not caught in any nets. These are *Architeuthis dux*, *Gonatus* sp., *Lepidoteuthis grimaldii*, *Octopoteuthis rugosa*, *Octopoteuthis* '*giant*' and *Pholidoteuthis boschmai*. *Gonatus* does not occur in the nets in this area since it lives further North (Lu and Clarke, 1975) and the whales probably ate them during a southward migration. Larvae are generally not recognised or are very rare for the other five species. Considering the great importance in the diet of *H. bonnelli* (Fig. 11.4), it is remarkable that only four were caught in nets and those over seamounts (Clarke, 2006). However, good specimens of this species (and the octopod *Haliphron atlanticus*) are regularly picked up at the surface during whale watching off the Azores, after being vomited by, or having escaped from, cetaceans and must have been eaten close inshore, probably in

Fig. 11.4 Four juveniles of cephalopods important in diets of predators in the Azores region: (a) *Histioteuthis* sp., (b) *Taningia danae*, (c) octopod *Haliphron atlanticus* and (d) cranchiid *Teuthowenia*.

canyons where any net fishing is impossible. This also possibly applies to the other four species. Absence of larvae may be due to their hatching and growing where the adults live, in canyons. The squid that contributes most weight to the diet, *T. danae*, occurs in nets in the three types of locality, but rarely.

While these comparisons provide little evidence that sperm whales use seamounts to feed as well as island slopes, feeding in midwater over the abyssal plain is likely to be more difficult for sperm whales than if the prey are concentrated against a canyon slope. The scarcity of island slopes, compared to seamounts reaching to within 500–1000 m of the surface (to which depths sperm whales regularly dive) strongly argues for their feeding on seamounts as well as island slopes. All these species are large and must be numerous to cater for any sperm whales present (see below) so are probably an important contribution to seamount as well as to island slope communities.

A collection from whales during a passage of a Japanese factory ship through the Tasman Sea (Clarke and MacLeod, 1974) showed the importance of Histioteuthidae and Octopoteuthidae by number and Octopoteuthidae and Architeuthidae by mass, although the whales were caught far from land. It is likely that these whales obtained the squids from seamounts accessible to whales (coming to within 1000 m of the surface, as many of them do in the region) since much of the sea is very deep. Even most of Lord Howe rise is deeper than 1000 m. As representatives of these families probably live deeper than 500 m around seamounts and islands, it seems likely that sperm whales use seamounts which have tops shallower than 500–1000 m to act as convenient 'restaurants' as they cross the abyssal plains (see Chapter 14).

Azores region

Cephalopods have been recorded, mainly as beaks, from the stomachs of three cetacean species (the sperm whale, a pygmy sperm whale, *Kogia breviceps*, a dolphin, *Delphinus delphis*) and four fish (swordfish, *Xiphius gladius*; blue shark, *Prionace glauca*; orange roughy; and opah, *Lampris guttatus*) caught near Pico Island in the Azores (see Supplementary Information). Only the orange roughy and most of the other fish were caught on seamounts. The rest are all likely to have been feeding either above the island slope or over seamounts near Pico. Only one swordfish (of 110 sampled) contained an island species (*Loligo forbesi*). All other cephalopod species were oceanic and pelagic. Cephalopods contributed almost 100% of the food organisms in the sperm whale and pygmy sperm whale, but less than 6% in swordfish, 15% in blue shark and 39% in orange roughy. Of the families of cephalopods, Chiroteuthidae, Mastigoteuthidae and Histioteuthidae are in more than one predator and, of these, the Histioteuthidae (Fig. 11.4) is by far the biggest percentage. In the sperm whale it comprises 71% by number (32% by mass) and in the pygmy sperm whale 67%, in the blue shark 23% and in the *Lampris* 20%. *H. bonnellii* is by far the most numerous, but rarely appears in any nets. The 60% represented by Sepiolidae in *Lampris* were *Heteroteuthis dispar* and these strongly suggest the fish was feeding on a seamount. Octopoteuthidae, mainly the large *T. danae*, are the most important by mass, in the diet of sperm whales, and also rarely occurs in nets. Of the four most numerous families in nets from seamounts, Onychoteuthidae, Lycoteuthidae, Ancistrocheiridae and Ommastrephidae, none are important in the food of the predators sampled. Similarly, none of the predators' diets link them to any specific topographical groups sampled; although four fish species sampled *Argonauta*, which was sampled only by nets on the seamounts (see Supplemental Information for data behind this analysis).

Figure 11.5, which was largely obtained from sperm whales caught during commercial activities and therefore not repeatable, shows the worldwide importance of the nine most numerous families in sperm whale diet. *Architeuthis dux*, *Pholidoteuthis boschmai*, *Lepidoteuthis grimaldii*, *Histioteuthis bonnellii*, *Taonius pavo* and *Ancistrocheirus alessandrini* (as Enoploteuthidae in Fig. 11.5) are worldwide in temperate and tropical regions. Histioteuthids are numerically important in the diet and sometimes provide a considerable proportion of the mass. Gonatids are numerous in the North Pacific and North Atlantic, and the giant cranchiid *Mesonychoteuthis hamiltoni* and an onychoteuthid *Kondakovia longimana* are the main food species of sperm whales in the Antarctic.

While the association with seamounts of the cephalopods eaten by sperm whales is not proven, it seems highly probable from the data analysed above.

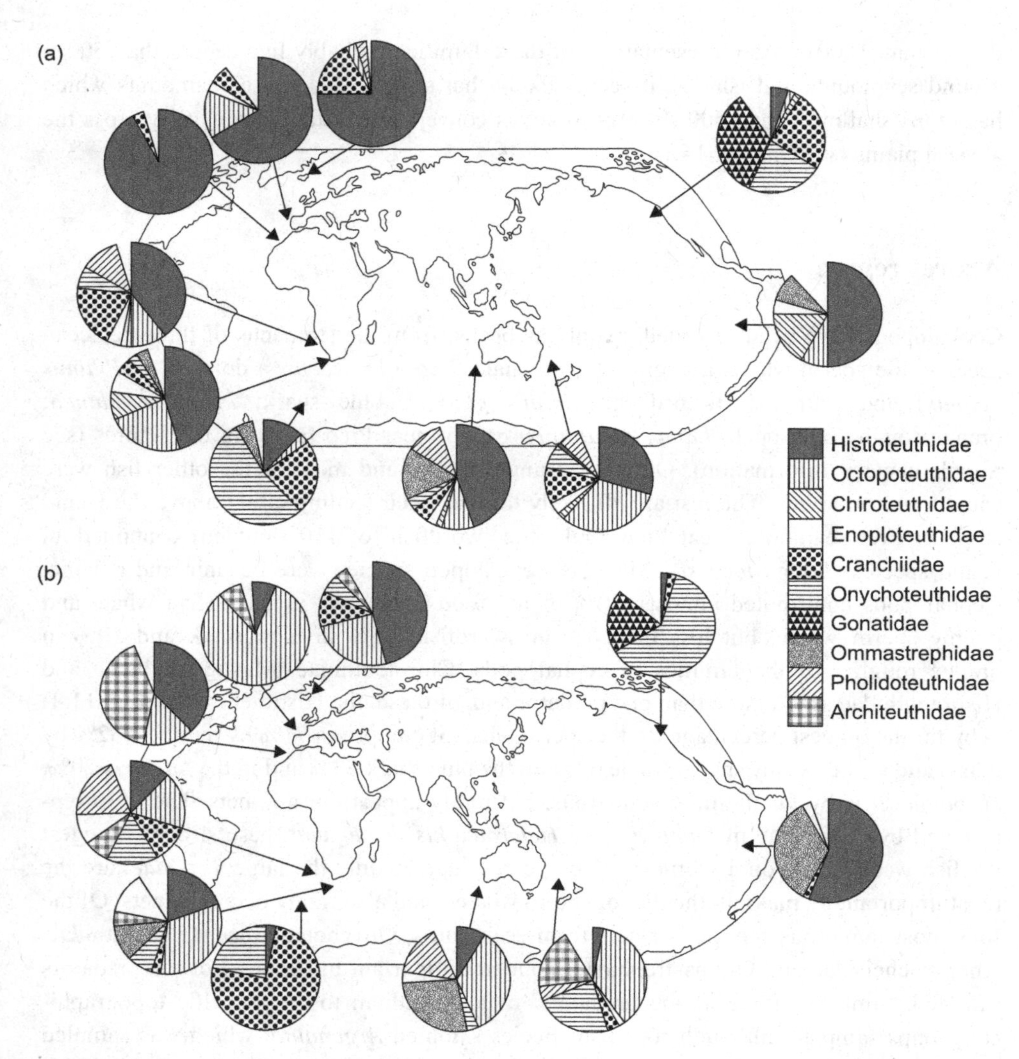

Fig. 11.5 Pie charts showing diet (see central key) of nine squid families found in the diet of sperm whales caught in various regions of the world. (a) Composition (%) based on the number of lower beaks. (b) Composition by weight (%) estimated from the numbers and sizes of beaks (see Clarke, 1986a).

Discussion of Nesis' (1993) classification of cephalopods associated with seamounts

From his own participation in three major expeditions, two to the western Indian Ocean and one to the SE Pacific using a diversity of nets, traps, longlines and a remotely operated vehicle (ROV), Nesis (1993) examined the relationship between cephalopods and seamounts. His conclusions are based on these and many USSR cruises throughout the world, but many are published in Russian and 'almost inaccessible to colleagues abroad'. His survey included 17 banks, seamounts and ridges in the 3 world oceans. Here, I assess the validity of Nesis' groups and his assignment of species to each (Fig. 11.6).

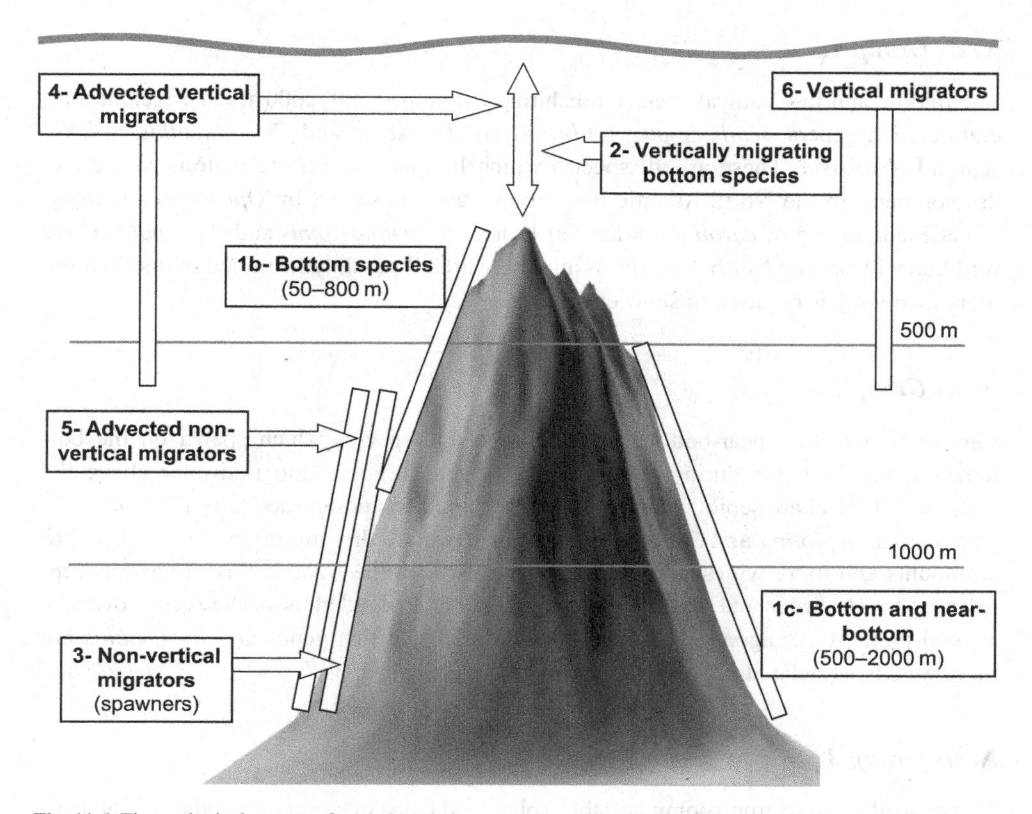

Fig. 11.6 The ecological groups of cephalopods in association with seamounts described by Nesis.

Nesis Group 1a

This group includes shallow-water species inhabiting the coastal zone and the shelf, mainly from the Saya de Malha Bank in the Indian Ocean. We exclude these on the basis they are not true seamounts. The cephalopods listed, *Loligo*, *Sepioteuthis*, *Sepia* and *Octopus*, are inshore, shelf genera, but may also inhabit islands even as far from continental shores as Madeira and the Azores. The true seamount species belong to the following six groups.

Nesis Group 1b

Bottom and near-bottom species which live permanently on seamounts. Lower littoral and upper bathyal species mainly on the tops of seamounts at 50–800 m. *Scaergus* has a pelagic stage and is very widespread. Three other species in the group, *Austrorossia mastigophora*, *Pteroctopus tetracirrhus* and *Danoctopus hoylei*, do not have a pelagic stage, but are also widespread.

In the material considered by Clarke (2006), *Scaergus unicirrhus* was found off Madeira (at 150 m depth) and Fuertaventura, and was assumed to be present over seamounts by Diekmann on the basis of octopod larvae, too small to identify positively. *P. tetracirrhus* was present at Madeira (at 791 m).

Nesis Group 1c

The middle and low bathyal species inhabiting the slopes 500–2000 m deep include several octopod genera *Benthoctopus*, *Bathypolypus*, *Froekenia* and *Grimpoteuthis*, and the sepiolid *Neorossia*. These are all species which live near or on the bottom with direct development. In the North Atlantic this group was represented by *Opisthoteuthis agassizi* at Madeira and *N. caroli* at Endeavour Bank, but *Benthoctopus* and *Bathypolypus* are well known from the North Atlantic. While these are all possibly restricted in distribution, more taxonomy is required to show endemism.

Nesis Group 2

This group includes near-bottom and benthopelagic species which spawn on the bottom, but regularly (or during some stage of their life) rise into midwater above the seamount. Typical are sepiolids in the Heteroteuthinae, including species of *Heteroteuthis*, *Iridoteuthis*, *Sepiolina* and *Stoloteuthis*. The first two are diel migrators, less attached to seamounts and more widespread than the other two. In the Azores, this group was represented by *H. dispar* and was on both seamounts and island slopes. The record over the abyssal plain at 40° agrees with Nesis' observation that this genus and *Iridoteuthis* are more weakly associated with seamounts and are widespread.

Nesis Group 3

This bathyal–pelagic group dominates the cephalopod fauna of seamounts and nearly all have widespread distributions. They comprise nerito-oceanic species living near or over the slopes such as species of *Enoploteuthis, Abralia, Ancistrocheirus, Moroteuthis, Pholidoteuthis, Todarodes, Martialia, Nototodarus, Ornithoteuthis, Lycoteuthis, Nematolampis, Taningia, Histioteuthis spp., Lepidoteuthis, and Architeuthis*. Their larvae, juveniles and (in some species) subadults inhabit midwater in the epipelagic and mesopelagic zones but spawn on or over the bottom or near the surface over the tops or slope (Fig. 11.6).

Lepidoteuthis grimaldii, which is only known from very few juveniles (Clarke, 1992), and *Architeuthis* which, from evidence from commercial fishing for orange roughy, are likely resident on the deep slopes of seamounts and island slopes, should be added to this group. A few larvae have been identified from near surface sampling nets, but in regions where there are seamounts (Personal observations). *Moroteuthis*, *Pholidoteuthis* and *Lepidoteuthis* all have scales covering their bodies, and in the latter two species these are leathery and tough, which might suggest that adults spend some time in contact with the bottom.

Nesis Group 4

This group comprises interzonal diel vertical migrators advecting by currents over the seamounts at night and descending to the bottom in daylight if it is shallower than the depth of their usual daily vertical distribution (300–800 m). Nesis included 'many species' of Enoploteuthidae, Histioteuthidae, Pyroteuthidae, Octopoteuthidae and Ctenopterigidae and Diekmann contributed evidence for this 'gap formation' process and reasonably assumed that oceanically distributed species were more vulnerable to bottom predation above seamounts than elsewhere.

Evidence for vertical migration of the Octopoteuthidae is poor and these should probably be considered as belonging to the next group. *Gonatus steenstrupi* and *Teuthowenia megalops* execute a reversed migration, being deeper at night than in the day (Lu and Clarke, 1975) and could provide food for bottom animals at night, but these are from higher latitudes (north of 50°N).

Nesis Group 5

Non-migrating, mostly bathypelagic species advected over the tops or slopes of seamounts by currents and, on deep seamounts, to the tops throughout the 24 h (becoming food for bottom species). These include many species of Chiroteuthidae, Mastigoteuthidae, Cranchiidae, Vampyroteuthidae and Bolitaenidae.

Nesis Group 6

Group 6 includes pelagic, nektonic squids such as Sthenoteuthis and Ommastrephes which, as juveniles and adults, avoid areas over the summits of shallow seamounts but are advected over seamounts as larvae.

Discussion

Nesis believed that squids in Groups 4 and 5 are so fragile they would lie passively on the substrate when they came in contact with it and would fall easy victims to predators. He considered these groups provide a very important trophic link between pelagic and benthic realms. The author agrees that Group 5 species would be passive and vulnerable, but this seems unlikely for most of the species in Group 4, particularly the Enoploteuthidae, Pyroteuthidae and Ctenopteridae. All species associated with seamounts, as defined here, are widely distributed and many are circumpolar or ocean wide. This is supported by work on diets of predators.

Size

Neither Nesis (1993) nor Diekmann (2004) considered size of the cephalopods, and this is of critical importance if we are to understand their importance in the dynamics of seamount ecology. Cephalopod adults have a standard length (including the body and arms but not the long tentacles) from 2 cm to over 500 cm according to the species.

Food of cephalopods

As with any animal, the sizes of squids influence the sizes of the species they consume but, because they have arms and tentacles, they can capture and take bites from much larger prey than if they caught them with a mouth (Fig. 11.1; Rodhouse and Nigmatullin, 1996). Table 11.1 shows the standard sizes of a range of squids playing a part in the seamount ecosystem, with sizes of the fish which may be caught by them and their known predators. The food of squids is not known for most oceanic species; juveniles eat small crustaceans and small fish and turn to larger fish and other cephalopods as they grow (Rodhouse and Nigmatullin, 1996). Myctophids are known to be important food in some oceanic species. The soft bodied *H. arcturi*, *Megalocranchia* sp. and *Idioteuthis hjorti* have been identified from the stomachs of the muscular squid *O. bartrami*.

A few suggestions on likely food can be made based on the spread of the arms and the shape of the lower beak (Fig. 11.1). Beaks with short rostral (cutting) edges and long wings for muscle attachments to give high leverage (Architeuthidae) cut hard surfaces like skin with scales such as those on many bottom fish; if beaks also have sharp points (Pholidoteuthidae), penetration between fish scales or through very tough skin such as bottom sharks is suggested. Beaks with long cutting edges and short wings probably are for cutting softer tissues and skin similar to midwater fishes such as myctophids. A range of species have beaks with intermediate features, and some have thicker or thinner parts for modifying these basic properties.

If larger squid go to the seamounts for spawning they are probably not actively feeding and their appearance there may also be seasonal.

Cephalopods as food

There are different predators equipped to eat the complete range of adults as well as the range of sizes from larvae and subadults each species produces. The enormous wastage, much of it due to predation, is indicated by the large quantity of eggs laid, hundreds of thousands in some species (Boyle and Rodhouse, 2005), and the short life cycles of 1 year with terminal spawning.

The species of squid eaten by each type of predator partly depends on size, behaviour and speed of travel relative to the predator. The calorific value may also influence the predator, gelatinous drifters being easier to catch but much less energetically rewarding than muscular squids (Clarke *et al.*, 1985). Slower, low-energy predators, such as many oceanic bottom fish, might be expected to choose low-energy squids such as Mastigoteuthidae and Chiroteuthidae, while tuna and dolphins would catch muscular Ommastrephids and Enoploteuthids. Our observations tend to support this.

Squids figure in the diets of most species of cetaceans (Clarke, 1996a) and, in oceanic waters, they all take a variety of families. The deep diving beaked whales (Ziphiidae) favour the softer ammoniacal squids. Bottlenose dolphins (Globiocephalidae) and Risso's dolphin (*Grampus griseus*) vary the diet between the muscular species (such as Ommastrephidae) and the softer species (such as Histioteuthidae and Octopoteuthidae). The shallower divers in the Delphinidae and Stenidae take more fish with the cephalopods and consume more of the muscular squids such as Ommastrephidae, Onychoteuthidae, Lycoteuthidae and Enoploteuthidae (Clarke, 1996a).

All seals which inhabit oceanic water include some cephalopods in their diet but few regularly eat appreciable quantities except, possibly, seasonally (Klages, 1996). Most seals favour muscular or oily species (Gonatidae), although the deep diving elephant seals (North and South) and Ross seal (*Ommatatophoca rossi*) also include soft, ammoniacal species. The Southern Elephant seal (*Mirounga angustirostris*), which dives to several 100 m, is the major pinnipeds consumer of cephalopods. Examinations of their stomach contents in the Southern Ocean are always on oceanic islands, but they likely take advantage of any accumulation of pelagic squids near the bottom on shallow seamounts during their wide foraging. The food includes some bottom octopods such as *Pareledone charcoti* and 16–17 taxa, and favours the onychoteuthid *Moroteuthis knipovitchi*, *Psychroteuthis glacialis* and *H. eltaninae*.

Very few fish feed exclusively on cephalopods but many include them in their diet during growth, seasonally and opportunistically (Smale, 1996). Some bottom sharks such as *Centroscymnus coelolepis*, *Sphyraena*, *Isurus*, *Galeocarda* and *Dalatius*, include Histioteuthidae and Ommastrephidae in their diet.

It is very likely that the dynamic marine life conditions around the tops of shallow seamounts attract seabirds (see Chapter 12C). Although squids figure in the diet of most seabirds, only in some species of albatrosses and petrels are they consistently more important than fish and crustaceans (Croxall and Prince, 1996). Most is known about southern polar and temperate seabird diets during the nesting season when sampling is easiest. Squids also probably play a key role in the diet of tropical seabirds and in the non-breeding season of temperate and polar seabirds. The Ommastrephidae, Onychoteuthidae, Gonatidae and Histioteuthidae make the largest contribution to the diet. Albatrosses and the larger petrels mainly eat living juveniles 5–30 cm long and probably scavenge the larger squid in their diet. Some of the deep-water squids such as Histioteuthidae, Octopoteuthidae and Chiroteuthidae may be obtained by birds scavenging on vomit of cetaceans or by the cetaceans actively herding the squids to the sea surface.

Biomass considerations

In considering the food web of a seamount, we know that the birds take small species and larvae up to perhaps a standard length of 30 cm, fish take from small up to perhaps 60 cm while cetaceans, particularly sperm whales, take up to the largest cephalopods known, with a standard length of 500 cm. These differences are supported by the calculation of standard lengths from beaks in a few of each type of predator. Sperm and pygmy sperm whales eat the same species of squid but larger sperm whales eat relatively more of the larger species and the larger individuals of the species (unpublished data).

Biomass of the commoner species removed by each predator

The biomass provided to resident or visiting species at a seamount can be roughly estimated only in a few species, because the basis of calculations depend on multiplying several parameters of the predator, such as population and average mass, which are rarely, even approximately, known.

If seamounts are 'restaurants' for sperm whales during their continuous movement about the oceans one might expect many more visits to more isolated seamounts such as Great Meteor. Visits by sperm whales provide an insight into the importance of cephalopods in the food web of the Azores. A calculation for sperm whales visiting the Azores can be made from past commercial catches making a few assumptions. More than the total caught (5600) over 20 years were visiting the area, since numbers showed no obvious decline and whales take more than 20 years to reach the mean size caught. The whales averaged over 20t in weight and each ate 3.25% of body weight each day of low calorie squids (the low calorific values requiring twice the mass of the estimate based on muscular food). If whales each stayed 10 days in the area, annual consumption would be $5600 \times 20 \times 0.0325 \times 2 \times 10 = 72\,800$t. This would suggest that an average of 154 whales would be in the Azores area each day. Taking the smoothed length of the coastline of the seven islands in the central and Eastern groups, one can estimate that a whale watcher's lookout ('*vigia*') observes over one twenty-fifth of the total length. If the estimate of the number of whales is correct and they spend most of the time inshore in the feeding grounds, one would expect an observer to see up to an average of six whales a day. That is, close to the value through the 6 summer months of whale-watching south of Pico Island (notes of a Vigia, Sidónio Gonçalves relayed by Manuel Fernandes).

The total mass of food would include annual consumption of approximately 9000t of *Architeuthis dux*, 28 000t of *Taningia danae* and 20 500t of *Histioteuthis bonnellii*! This compares with the maximum annual landings at the Azores of about 10 000t of pelagic fish and 5000t of demersal fish (see other chapters). Extending a similar calculation to a seamount, one average adult staying 1 day would consume approximately 1.3t of cephalopods.

Other toothed whales are very much smaller as the following adult weights show: *Tursiops truncatus* 209kg, *D. delphis* 60kg, *Stenella longirostris* 66kg, *Lagenorhynchus obliquidens* 91kg, *G. griseus* 328kg and *Globicephala melas* 943kg (Perrin *et al.*, 2002). Their food requirements are proportionately smaller and it would take about 20 *Globicephalus* to eat the same amount as one average sperm male.

Boyd *et al.* (1994) estimated the cephalopod consumption of the South Georgia population of elephant seals at 2.8 million tonnes.

Adult swordfish average only 500kg and their diet includes 50% by weight of largely muscular squids, so it would probably take about 20 fish to consume the same amount of squid as one 20t sperm whale.

Although their consumption per weight is far higher, because of their small size (perhaps 40 000 shearwaters to the average male sperm whale), seabirds eat far less weight of squids than cetaceans. For example, in the NE Atlantic seabirds are estimated to take only 40 000t/year (Furness, 1994) – perhaps about half of the estimate for sperm whales off the Azores.

Energy transfer

Seamounts, certainly ones coming to within less than 500m of the surface, are probably centres of high biological activity in ocean areas of general low activity. The cephalopods comprise a few species living deep on the slopes (octopodids) and numerous pelagic squid species, which are carried to the seamount by currents and by swimming, from the ocean around, many of them for spawning. They include migrators and non-migrators and some may linger there for food or die after spawning.

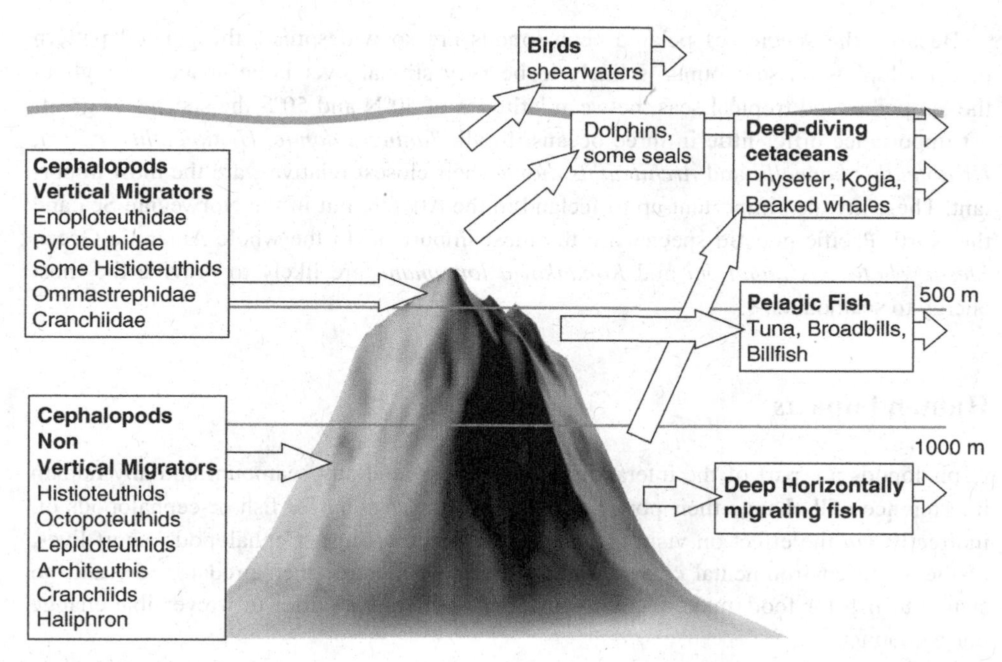

Fig. 11.7 The main sources and removal of energy derived from cephalopods around seamounts which come to within 1000 m of the sea surface. See text for explanation.

The energy they introduce to the seamount environs is partially absorbed by benthos, but far more is probably removed by passing visitors, cetaceans, possibly seals particularly in the Antarctic, large fish and birds (Fig. 11.7).

Birds of greatest importance are shearwaters in the North Atlantic, albatrosses and possibly penguins in the Antarctic and parts of the Pacific. These mainly take epipelagic species coming to within a few metres of the sea surface although, surprisingly, their diet also includes species of squid which we normally catch much deeper such as Chiroteuthidae, Octopoteuthidae and Histioteuthidae. How they are picked up by birds at the surface has been the cause of much discussion (Croxall and Prince, 1996), but they are possibly vomited, or chased to the surface by dolphins or large fish such as tuna. Most dolphins probably normally feed at less than 500 m or so, and a few baleen whales such as Sei whales are, at times, heavy predators of squids at these shallow depths. Deeper diving cetaceans including all the beaked whales, *Kogia*, *Hyperoodon* and *Physeter* probably eat any pelagic squids collecting near seamounts from 300 m to over 1000 m, and possibly over 2000 m. Large epipelagic fish such as tunas, billfish and broadbills often have considerable vertical ranges to over 1000 m but favour the muscular cephalopods while the large bathypelagic migratory fish take slower gelatinous species. In general, of all visitors, sperm whales probably remove most energy and nutrient from seamounts and seabirds probably the least although groups of other cetaceans (including Sei whales), seals and large fish, may have considerable impact on seamounts in particular regions at particular times. Energy from seamounts will be dispersed throughout the oceans according to migration patterns of the visitors, but it will be ocean wide and even extend onto land as guano and *via* fisheries.

Because the species of pelagic cephalopods are so widespread, the general pattern of cephalopods on seamounts is likely to be very similar over large areas. Throughout the temperate and tropical seas, between latitudes of 50°N and 50°S the species of greatest importance differ little in three oceans. In all, *Taningia danae*, *Histioteuthis arcturi*, *Histioteuthis bonnellii* and *Architeuthis dux* or their closest relatives, are the most important. These are also important up to Iceland in the Atlantic but in the Norwegian Sea and the North Pacific gonatid species are the most important. In the whole Antarctic Ocean *Mesonychoteuthis hamiltoni* and *Kondakovia longimana* are likely to provide the most energy to seamounts.

Human impacts

Cephalopods are part of the interacting biosphere around any seamount and any human interference will change their populations, directly by fishing for fish or cephalopods or, indirectly, *via* the effect on visiting species or climate change. Cephalopods' short lives, resilience to environmental change and the fact that most of their predators are able to switch to fish for food, make them a considerable energetic buffer to irreversible change on seamounts.

Climate change is likely to have the greatest effect on seamount ecology through rises in sea level since this would reduce the number of seamounts available to cetacean and seal visitors. It could change cetacean and seal migratory routes and reduce predation of cephalopods on the seamounts. Minor changes in water temperature are less likely to adversely affect cephalopods directly since they have such broad distributions and tolerance to temperature change.

Because of the long migrations of so many visitors there is little doubt that any drastic changes to seamount ecosystem stability would have very wide oceanic effects. Understanding the complexity of the involvement of cephalopods in the seamount ecosystem can only be achieved through greater knowledge of the diets of their predators, life cycles of the larger oceanic species and by counting the cetacean, seal and bird visitors. It is much less clear how one can then obtain typical population estimates for the visiting and resident fish predators of cephalopods, without disturbing the balance by fishing. Where fishing does take place, the opportunity for diet analysis should not be missed.

To preserve the biological content of some seamounts for the future may be desirable, but it is probably impossible since their ecosystems are directly tied to the wider ocean by their visitors (see Chapter 21). While we might prevent direct human interference on selected seamounts by stopping fishing and 'squidding', we would also need to preserve changes in visitors by preserving nesting and haul out sites on more than just adjacent islands and prevent killing Odontocetes (almost achieved but under pressure).

The knowledge necessary to conserve and manage seamounts in general can probably only be gained by selection of specific 'typical' untouched seamounts and then by applying a large effort with many specialists to study as many aspects as possible. Regular research samples with midwater trawls over seamounts would show presence of particular cephalopod species (not absence) and seasonal or yearly changes; but will not be useful in estimating their biomass. Capture of appreciable numbers of predators to establish the presence of adults of larger species would possibly require periods of recovery.

Little can be learned about cephalopods without widespread identification of stomach contents of their predators. This might involve taking opportunistic advantage of strandings on nearby island shores, unless the seamount was selected where such knowledge exists from bygone whaling or sealing activities. Otherwise, general knowledge of diets, gleaned from elsewhere, might be sufficient, provided that good estimates of the number of migrant fish, bird, seal and cetacean visitors can be made. Where conservation measures exist, such as preventing direct removal of cephalopods from seamounts by fishing and the moratorium on sperm whaling, any changes should be resisted.

Summary

1. Considerable differences between cephalopods and fish make cephalopods both important consumers and important prey on seamounts. Their size, variable and often rapid growth, high fecundity, plasticity in their seasons of reproduction, short lives and migratory habits provide high resilience to environmental change, including population variation of their predators. They can survive enormous fluctuations in population and recuperate in very few years. Their effective mouth size enables them to eat larger fish than themselves and possibly the largest near the bottom on seamounts.
2. They provide a trophic link between the pelagic and benthic realms.
3. The ecological groups of cephalopods identified by Nesis (1993) on seamounts have been checked against diverse collections from the North Atlantic. Group 1a has been rejected as comprising shallow water species. With the aid of diverse collections in the North Atlantic a few important species have been added.
4. Comparisons in the North Atlantic show that pelagic seamount species are almost the same as those found around islands and over the abyssal plain.
5. Only very few species, and those all benthic without pelagic stages, can be considered as special to seamounts but even these have wide distributions or poor taxonomic data.
6. Many species probably go to seamounts to spawn and subsequently die. They probably do not eat there and, with their spawn, provide considerable nourishment for cetacean, seal and bird visitors to the seamounts as well as to visiting and resident fish.
7. Calculations show that energy removed from seamounts by visiting cetaceans is probably more than that removed by any other visiting group and that sperm whales extract the most of any species. These consume more of each of the squids *Histioteuthis, bonnellii*, *Taningia danae* and *Architeuthis dux* than humans catch of either pelagic or demersal fish in the Azores waters.
8. The wide distributions of most oceanic cephalopod species suggest that conclusions made on the basis of the material considered here will be of general application.
9. The ecological importance of seamounts is that they are probably centres where midwater cephalopods collect for spawning, feeding or by accidental drifting. Cephalopod presence near the bottom at less than 1000 m from the surface probably makes their

capture by cetaceans, seals and some fish more energy cost effective than over deeper water.

References

Boyd, I.L., Arnbom, T.A. and Fedak, M.A. (1994) Biomass and energy consumption of the South Georgia population of southern elephant seals. In: *Elephant Seals: Population Ecology, Behaviour and Physiology* (eds. Le Boef, B.J. and Laws, R.M.), pp. 98–117. University of California Press, Berkley.

Boyle, P.R. and Boletsky, S.V. (1996) Cephalopod populations, definition and dynamics. *Philosophical Transactions of the Royal Society of London B*, 351, 985–1002.

Boyle, P.R. and Rodhouse, P. (2005) *Cephalopods Ecology and Fisheries*, Blackwell Science, pp. 1–452. Oxford.

Clarke, A., Clarke, M.R., Holmes, L.I. and Waters, T.D. (1985) Calorific values and elemental analysis of eleven species of oceanic squids (Mollusca: Cephalopoda). *Journal of the Marine Biological Society of the United Kingdom*, 65, 983–6.

Clarke, M.R. (1980) Cephalopoda in the diet of sperm whales of the southern hemisphere and their bearing on sperm whale biology. *Discovery Reports*, 37, 1–324.

Clarke, M.R. (1986a) Cephalopods in the diet of odontocetes. In: *Research on Dolphins* (eds. Brydon, M.M. and Harrison, R.), pp. 281–321. Oxford University Press, Oxford.

Clarke, M.R. (1986b) *A Handbook for the Identification of Cephalopod Beaks*, 273 pp. Clarendon Press, Oxford.

Clarke, M.R. (1987) Cephalopod biomass – estimation from predation. In: *Cephalopod Life Cycles. Vol. 2. Comparative Reviews* (ed. Boyle, P.R.), pp. 221–37. Academic Press, London.

Clarke, M.R. (1992) Lepidoteuthidae. In: *"Larval" and Juvenile Cephalopods: A Manual for their Identification* (eds. Sweeney, M. Roper, C.F.E., Mangold, K. Clarke, M.R. and Boletzky, S.V.) *Smithosinan Contributions to Zoology*, 513, 282 pp.

Clarke, M.R. (1993) Age determination and common sense. In: *Recent Advances in Cephalopod Fisheries Biology* (eds. Okutani, T., O'Dor, R.K. and Kuborera, T.), pp. 670–8. Tokai University Press, Tokyo.

Clarke, M.R. (1996a) Cephalopods as prey. III. Cetaceans. In: *The Role of Cephalopods in the World's Oceans* (ed. Clarke, M.R.). *Philosophical Transactions of the Royal Society of London B*, 351, 1053–65.

Clarke, M.R. (1996b) General conclusions and the future. *Philosophical Transactions of the Royal Society of London B*, 351, 1112.

Clarke, M.R. (2006) Cephalopod distribution and topographical features in the Eastern North Atlantic. *Arquipélago. Life and Marine Sciences*, 23A, 27–46.

Clarke, M.R. and Lu, C.C. (1974) Vertical distribution of cephalopods at 30°N 23°W in the North Atlantic. *Journal of the Marine Biological Association of the United Kingdom*, 54, 969–84.

Clarke, M.R. and MacLeod, N. (1974) Cephalopod remains from the stomachs of sperm whales caught in the Tasman Sea. *Memoirs of the Victoria Museum*, 43, 25–42.

Croxall, J.P. and Prince, P.A. (1996) Cephalopods as prey. I. Seabirds. In: *The Role of Cephalopods in the World's Oceans* (ed. Clarke, M.R.). *Philosophical Transactions of the Royal Society of London B*, 351, 1023–43.

Diekmann, R. (2004) Distribution patterns of oceanic micronekton at seamounts and hydrographic fronts of the subtropical Atlantic Ocean. Dissertation zur Erlangung des Doktorgrades der

Mathematisch-Naturwissenschaftlichen Fakultät der Christian-Albrechts-Universität zu Kiel, Kiel, 183 pp. + XLIII.

Furness, R.W. (1994) An estimate of the quantity of squid consumed by seabirds of the eastern North Atlantic and adjoining seas. *Fisheries Research*, 21, 165–77.

Klages, N.T.W. (1996) Cephalopods as prey. II. Seals. In: *The Role of Cephalopods in the World's Oceans* (ed. Clarke, M.R.). *Philosophical Transactions of the Royal Society of London B*, 351, 1045–52.

Lu, C.C. and Clarke, M.R. (1975) Vertical distribution of Cephalopods at 40°N, 53°N and 60°N at 20°W in the North Atlantic. *Journal of the Marine Biological Association of the United Kingdom*, 55, 143–63.

Nesis, K. (1987) *Cephalopods of the World*. TNP Publications, New York.

Nesis, K. (1993) Cephalopods of seamount and submarine ridges. In: *Recent Advances in Cephalopod Fisheries Biology* (eds. Okutani, T., O'Dor, R.K. and Kuborera, T.), pp. 365–74. Tokai University Press, Tokyo.

O'Dor, R.K., Balch, N. and Amaratunga, T. (1982) Laboratory observations of midwater spawning by *Illex illecebrosus*. North Atlantic Fisheries Organization, SCR Doc. No. 82/vi/5, 8 pp.

Perrin, W.F., Wursig, B. and Thewissen, J.G.M. (eds.) (2002) *Encyclopaedia of Marine Mammals*. Academic Press, London.

Rodhouse, P.G. and Nigmatullin, C.M. (1996) Role as consumers. *Philosophical Transactions of the Royal Society of London B*, 351, 1003–22.

Smale, M.J. (1996) Cephalopods as prey. IV. Fishes. In: *The Role of Cephalopods in the World's Oceans* (ed. Clarke, M.R.). *Philosophical Transactions of the Royal Society of London B*, 351, 1067–81.

Chapter 12A
Air-breathing visitors to seamounts: Marine mammals

Kristin Kaschner

Abstract

Pinnipeds and cetaceans are widely reported as more abundant in the vicinity of abrupt topographies like seamounts, yet there are few rigorous quantitative analyses. Some species such as the Hawaiian monk and Galapagos fur seals, Commersons and spinner dolphins, with slope or mesopelagic foraging habits, are characteristic of island arcs with many seamounts. Elephant and crabeater seals and beaked whales also tend to forage at seamounts, while Common dolphins, Bottlenose whales, Risso's dolphin, Dall's porpoise and pilot whales appear to be associated with strong bathymetric gradients, including seamounts. Here, a global habitat suitability model is used to forecast local abundances of marine mammal species by species. Results show a significant association of marine mammal abundance with seamount-rich locations.

Introduction

Whatever the exact mechanism, it seems intuitive that the flow-driven accumulation of zooplankton, large invertebrates and fish around seamounts discussed in the previous chapters (see Chapters 5, 6, 8, 10 and 11) attract what Genin (2004) called a 'transient accumulation' of larger and highly mobile animals. This includes marine mammals, which are known to actively seek out predictable regions of enhanced prey densities (Jaquet and Whitehead, 1999; Bonadonna *et al.*, 2000; Acevedo-Gutierrez *et al.*, 2002; Benoit-Bird and Au, 2003). The importance of seamounts, among other hydrographic features with steep bathymetric gradients, as foraging posts or migration stops for marine mammals has therefore repeatedly been referred to in the literature dealing with management and conservation strategies for seamount habitats (Hyrenbach *et al.*, 2000; Canessa *et al.*, 2003). Most existing information about seamount association of marine mammal species appears however to be anecdotal, supported by very little available quantitative evidence. Although the occurrence of a diversity of cetaceans and pinnipeds around seamounts has been well documented throughout the world (Baba *et al.*, 2000; Moore *et al.*, 2002;

Seabra *et al.*, 2005); this alone provides little evidence for an actual preference of any species for seamounts (Table 12A.1).

Table 12A.1 Documented association of marine mammal species with seamounts.

Species	Geographic region/ seamount	Strength of evidence	Source
Baird's beaked whale	North Pacific	Qualitative observation	Ohsumi (1983), Balcomb (1989), Reeves and Mitchell (1993)
Dall's porpoise	California Current System/Cordell Bank	Quantitative investigation of environment correlates	Yen *et al.* (2004), Tynan *et al.* (2005)
Humpback whale	California Current System/Hecate Bank	Quantitative investigation of environment correlates	Tynan *et al.* (2005)
Short-finned pilot whale	California Current System	Quantitative investigation of environment correlates	Hui (1985)
Common dolphins	California Current System	Quantitative investigation of environment correlates	Hui (1985)
Spinner dolphins	Hawaii	Indirect evidence based on feeding and diurnal tracking of mesopelagic boundary layer	Benoit-Bird and Au (2003)
Southern elephant seals (juveniles)	South Pacific/ De Gerlache Seamounts	Qualitative observation based on tagging data	Bornemann *et al.* (2000)
Crabeater seals	Antarctic waters near Queen Maud Land	Qualitative observation based on tagging data	Nordøy *et al.* (1995)

Dedicated marine mammal surveys are very costly, particularly the relatively large-scale surveys needed to investigate and possibly detect gradients of increasing density of species around seamounts in comparison to the surrounding ocean areas. Most of the surveys conducted are restricted to continental shelf areas (Barlow, 1997; Hammond *et al.*, 2002). The few oceanic surveys that exist have focused on the estimation of species' abundances rather than habitat usage (Sigurjónsson *et al.*, 1989; IWC, 2001). Despite the dearth of quantitative data, available information about marine mammal life histories, dietary preferences and general habitat usages, summarized below, may provide some indirect indication for the importance of seamounts for pinnipeds and cetaceans. The few studies of marine mammal occurrences in areas of steep bottom topography were mainly conducted in shelf and slope waters. The results nevertheless allow some conclusions about why abrupt hydrographic features such as offshore seamounts may attract marine mammals.

Anecdotal information or indirect evidence

For the most part, pinnipeds and cetaceans have wide, even panglobal distributions and roam widely during their regular migrations (Jefferson *et al.*, 1993). Consequently, we would not expect to see the degree of endemism that distinguishes seamount communities of other trophic levels (see Chapters 5 and 13; de Forges *et al.*, 2000). It is noteworthy, however, that distributions of the few marine mammal species with very restricted ranges often centre around offshore volcanic islands or island chains in areas of high seamount density. These species include the Hawaiian monk seal (*Monachus schauinslandi*), the Galapagos fur seal (*Arctocephalus galapagoensis*) and a recognized subspecies of the Commerson dolphin (*Cephalorhynchus commersonii*) occurring around Kerguelen Island. All are relatively shallow divers, rarely venturing beyond 200 m (Reijnders *et al.*, 1993; Dawson, 2002; Parrish *et al.*, 2002). Although most feed on a wide array of pelagic prey types, benthic and mesopelagic organisms are also part of their diets (Pauly *et al.*, 1998). Hence, islands slopes and nearby shallow seamount plateaus may represent important foraging grounds (Stewart and Yochem, 2004).

The Hawaiian spinner dolphin (*Stenella longirostris*) is an example of a less restricted marine mammal species, which may utilize seamount areas as feeding posts. This species has been shown to primarily feed on prey taken from the mesopelagic boundary layer (Norris *et al.*, 1994). The regular inshore–offshore movement of spinner dolphins around Hawaii also matches the diurnal vertical and horizontal migrations of the organisms in this layer (Benoit-Bird and Au, 2003). The mechanisms resulting in the aggregation of prey around the slopes of islands (Benoit-Bird *et al.*, 2001) are comparable to those described for seamounts (see Chapters 4–6). Although this remains to be investigated; it seems likely that spinner dolphins would similarly exploit elevated prey densities associated with the surrounding seamounts.

Seamount slopes and plateaus may be equally of importance for many of the most elusive marine mammal species, the beaked whales. To date, very little is known about the majority of these primarily oceanic and deep-diving toothed whales. From what we do know, beaked whales seem to be most common in slope waters and around offshore volcanic islands and are known to feed primarily on mesopelagic and deep sea benthic fish and squid species (Pauly *et al.*, 1998; Kasuya, 2002; Pitman, 2002; Kaschner *et al.*, 2006). Again, there has been no quantitative assessment of seamount habitat usage by any beaked whale species, even though (Ohsumi, 1983), for instance, states that Baird's beaked whales (*Berardius bairdii*) are most commonly encountered in regions with submarine escarpments and seamounts. Despite the lack of quantitative data, dietary preferences and information about habitat usage would suggest that seamounts might be important feeding grounds for beaked whales.

Bornemann *et al.* (2000) hypothesized that southern elephant seals (*Mirounga leonina*) also rely on the tropic focusing abilities of seamounts and similar features. This was based on the analysis of tagging data of several juvenile males, which spent several weeks foraging near the De Gerlache Seamounts in the South Pacific (Bornemann *et al.*, 2000). Interestingly, the exploitation of seamount-associated prey patches may represent an exclusively male foraging strategy (Bornemann *et al.*, 2000), possibly only used during the austral winter and throughout the northern parts of their range, where bathypelagic myctophids such as *Gymnoscopelus nicholsi* represent an important part of the diet of southern sea elephants (Daneri and Carlini, 2002).

The crabeater seal (*Lobodon carcinophagus*) feeds at a lower trophic level and on smaller prey – namely krill, and so are even more dependent on regions of reliable high prey availability than elephant seals (Burns *et al.*, 2004). This species is most commonly observed in shallow waters thickly covered by ice and associated with high prey concentration (Burns *et al.*, 2004). However, Nordøy *et al.* (1995) found that crabeater seals near Queen Maud Land also spent considerable time in very deep waters along the shelf break and near seamounts where krill may be aggregated by the local physical and hydrographic conditions.

Quantitative investigations of seamount habitat usage

Numerous studies of marine mammal presence have found strong correlations between bathymetry and patterns in inter- or intraspecific occurrences for both cetaceans and pinnipeds in different regions and ocean basins (Payne and Heinemann, 1993; Moore *et al.*, 2000; Baumgartner *et al.*, 2001; Hamazaki, 2002). Many species have been shown to occur in higher densities in areas marked by strong bathymetric gradients that define topographic features such as shallow water banks and ridges (Common dolphins (*Delphinus delphis*) around Georges Bank, NW Atlantic; Griffin, 1997), submarine canyons (e.g., Northern bottlenose whale (*Hyperoodon ampullatus*) in the Gully, NW Atlantic; Hooker *et al.*, 1999) or the continental slope itself (e.g., Risso's dolphin (*Grampus griseus*) in the Mediterranean Sea; Cañadas *et al.*, 2002). However, the few existing quantitative investigations of marine mammal habitat usage that may shed some light on the relative importance of seamounts in particular have all been conducted in the California Current System (CCS) over the past 20 years.

Hui (1985) made a particular reference to underwater escarpments in his study of common dolphins and short-finned pilot whale (*Globicephalus macrorhynchus*) occurrence in the CCS and noted that both species appeared to be concentrated in areas of steep bottom topography. However, pilot whales are known to feed primarily on oceanic and neritic squids of the family Ommastrephidae, which has been thought to be an indication that this cetacean may mainly forage over the upper continental slope and shelves (see Chapter 11; Clarke, 1996). This would suggest that pilot whales might only exploit prey aggregations over shallower seamounts. Clarke also suggests that sperm whales may exploit squid species characteristic of canyons and steep seamount slopes (see Chapter 11). Similarly, the diet of common dolphins – a mixed array of myctophids and small pelagics – would suggest a preference for waters surrounding shallower escarpments where, for example, anchovies (*Engraulidae*) are known to aggregate rather than the utilization of 'true' seamounts.

Tynan *et al.* (2005) found that Heceta Bank, a submarine bank in the CCS, and the associated flow – topography interactions greatly influenced the distribution of cetaceans in the area. Specifically they report seasonally higher densities of a number of species including humpback whales (*Megaptera novaeangliae*), Pacific white-sided dolphins (*Lagenorhynchus obliquidens*), Dall's porpoise (*Phocoenoides dalli*) and even the more coastal harbour porpoise (*Phocoena phocoena*). The results of a similar analysis conducted by Yen *et al.* (2004), who also studied cetacean habitat usage in the CCS, are somewhat contradictory in terms of the association of marine mammals with seamount-like features. Although, again, the occurrence of Dall's porpoises was significantly and positively correlated with high contour index areas, no direct link could be found between this species presence and the distance to the Cordell Bank, another slope 'seamount' (Yen *et al.*, 2004). This suggests that usage of steep bottom topography areas by Dall's porpoise is not restricted to habitat found around underwater escarpments or seamount-like features.

Although Heceta and Cordell Bank are not seamounts in the strict sense, the prey aggregating mechanisms and trophic dynamics around these escarpments are probably quite similar to those occurring around seamounts. Similar trophic focusing does however occur along the shelf edge, submarine canyons and other upwelling areas. Consequently, the extent to which the association of cetaceans with areas of high bottom topography and underwater escarpments or banks is indicative of a close link between the species and seamount habitat in offshore areas remains to be investigated. Such investigations are greatly hampered by the difficulties of determining marine mammal habitat usage on larger geographic and temporal scales, particularly in oceanic areas as mentioned above.

Results from habitat suitability modelling

The lack of reliable marine mammal distribution data led Kaschner *et al.* (2006) to develop a global habitat suitability model for predicting heterogeneous patterns of occurrence of 115 marine mammal species. Validation of species-specific predictions showed that observed high species encounter rates strongly and positively correlate with predicted highly suitable habitat for almost all species tested (Kaschner *et al.*, 2006). Areas of predicted high relative environmental suitability (RES) thus likely represent important core habitat. A preliminary analysis of the possible importance of seamount habitat for marine mammals at very large scales was conducted using RES predictions for all species (Kaschner, unpublished data). Figure 12A.1 shows mapped hotspots of marine mammal

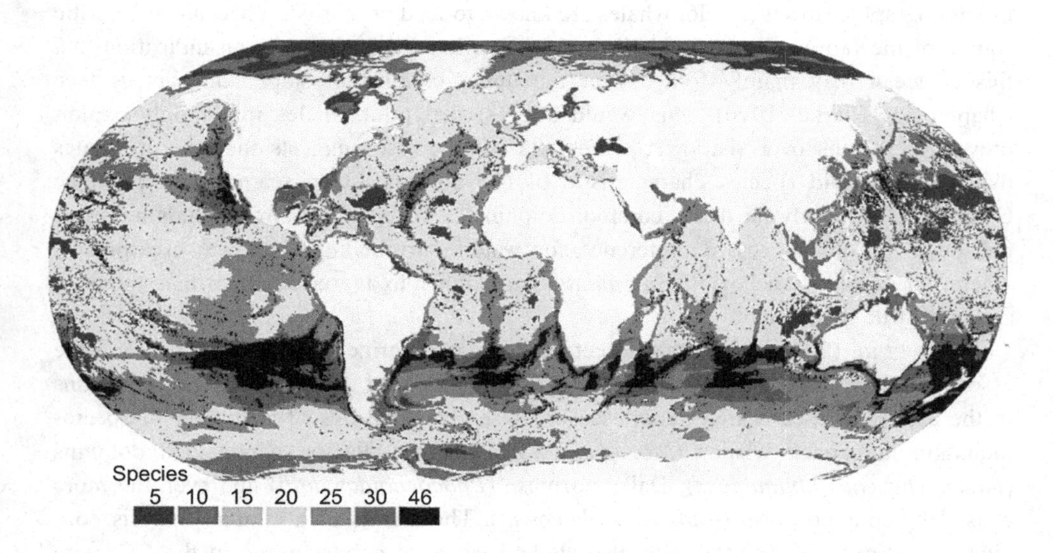

Fig. 12A.1 Map of predicted global marine mammal species richness (115 species; core areas only) and superimposed seamount locations (black dots). Species numbers shown in key are the maximum in each colour category.

species richness (generated by summing predicted core areas across all species) and superimposed geographic locations of the world's seamounts, as defined by Kitchingman *et al.* (see Chapter 2; Kaschner, unpublished data). Statistical comparison showed a strong positive, linear and highly significant relationship between mean seamount density per 0.5° latitude by 0.5° longitude grid cell and predicted marine mammal species richness (Spearman's rho = 0.76, $p < 0.0001$) (Fig. 12A.2). Interestingly, a linear relationship between marine mammal species richness and average cell depth (Spearman's rho = 0.15, $p = 0.34$) or between bottom depth and mean seamount density (Spearman's rho = 0.41, $p = 0.21$) was not detected. These preliminary findings provide some evidence that areas of high seamount density may indeed deserve some special attention in the context of preserving marine mammal biodiversity.

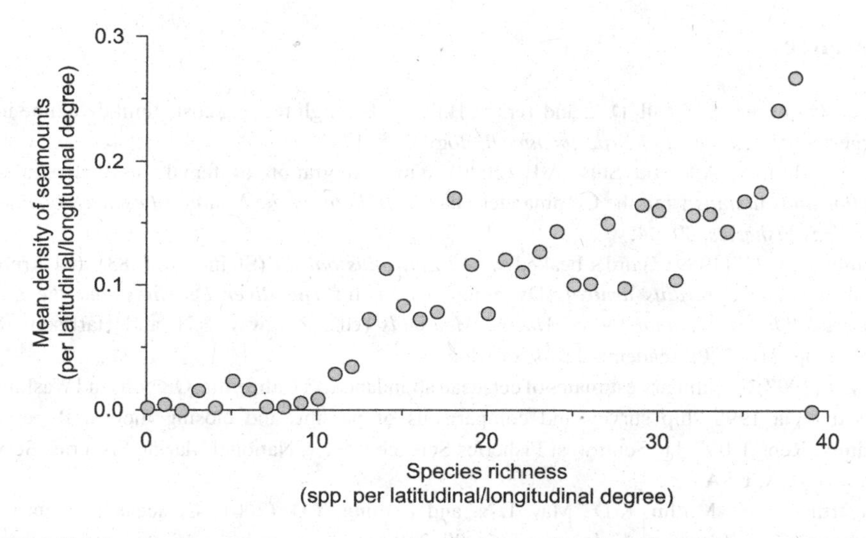

Fig. 12A.2 Relationship between mean seamount density across all cells falling within a species richness category (Spearman's rho = 0.76, $p < 0.0001$).

Conclusions

The notion that marine mammals are an important if transient component of seamount communities is currently supported by very little actual data. Most available information is based on anecdotal or qualitative descriptions, but there are nevertheless a few studies – mostly conducted in the CCS – that could demonstrate a general association of a number of marine mammal species with areas of high contour indexes. This is a common characteristic of seamounts and underwater escarpments, but steep bathymetric gradients are also a distinguishing mark of areas around banks, canyons and the continental slope. These features are equally known to represent important habitat for many marine mammal species; most likely because of the associated elevated prey densities. Available evidence suggests, that, unlike many other members of seamount communities, the vast majority of marine mammal species are probably only loosely associated with particular seamounts. Relying on the dependable trophic focusing abilities of seamounts and other hydrographic features, oceanic pinnipeds and cetaceans may nevertheless visit seamounts within their range on a regular basis on foraging trips. This feeding post-hypothesis is frequently mentioned in the context of management plans involving the protection of seamounts (Hyrenbach *et al.*, 2000; Canessa *et al.*, 2003). The preliminary investigation of the association of predicted marine mammal species richness and seamount density conducted here provides some support for the relative importance of seamount habitat for many of these predators on larger scales. More quantitative studies of pinniped and cetacean foraging strategies and resulting underwater topography preferences are needed to assess the actual reliance of individual species on true seamounts and identify the underlying mechanisms that may be responsible for drawing these warm-blooded visitors to seamounts.

References

Acevedo-Gutierrez, A., Croll, D.A. and Tershy, B.R. (2002) High feeding costs limit dive time in the largest whales. *Journal of Experimental Biology*, 205, 1747–53.

Baba, N., Boltnev, A.I. and Stus, A.I. (2000) Winter migration of female northern fur seals *Callorhinus ursinus* from the Commander Islands. *Bulletin of the National Research Institute of Far Seas Fisheries*, 39–44.

Balcomb III, K.C. (1989) Baird's beaked whale *Berardius bairdii* (Stejneger, 1883) and Arnoux's beaked whale *Berardius arnuxii* (Dyvernoy, 1851). In: *The River Dolphins and the Larger Toothed Whales – Handbook of Marine Mammals* (eds. Ridgway, S.H. and Harrison, R.H.) Vol. 4, pp. 216–320. Academic Press, London.

Barlow, J. (1997) Preliminary estimates of cetacean abundance off California, Oregon, and Washington based on a 1996 ship survey and comparisons of passing and closing modes. Report No. Admin. Rept. LJ-97-11, Southwest Fisheries Science Center, National Marine Fisheries Service, La Jolla, CA, USA.

Baumgartner, M.F., Mullin, K.D., May, L.N. and Leming, T.D. (2001) Cetacean habitats in the Northern Gulf of Mexico. *Fishery Bulletin*, 99, 219–39.

Benoit-Bird, K.J. and Au, W.W.L. (2003) Prey dynamics affect foraging by a pelagic predator (*Stenella longirostris*) over a range of spatial and temporal scales. *Behavioral Ecology and Sociobiology*, 53, 364–73.

Benoit-Bird, K.J., Au, W.W.L., Brainard, R.E. and Lammers, M.O. (2001) Diel horizontal migration of the Hawaiian mesopelagic boundary community observed acoustically. *Marine Ecology Progress Series*, 217, 1–14.

Bonadonna, F., Lea, M.-A. and Guinet, C. (2000) Foraging routes of Antarctic fur seals (*Arctocephalus gazella*) investigated by the concurrent use of satellite tracking and time-depth recorders. *Polar Biology*, 23, 149–59.

Bornemann, H., Kreyscher, M., Ramdohr, S., Martin, T., Carlini, A., Sellmann, L. and Ploetz, J. (2000) Southern elephant seal movements and Antarctic sea ice. *Antarctic Science*, 12, 3–15.

Burns, J.M., Costa, D.P., Fedak, M.A., Hindell, M.A., Bradshaw, C.J.A., Gale, N.J., McDonald, B., Trumble, S.J. and Crocker, D.E. (2004) Winter habitat use and foraging behavior of crabeater seals along the Western Antarctic Peninsula. Deep Sea Research (Part II). *Topical Studies in Oceanography*, 51, 2279–303.

Cañadas, A., Sagarminaga, R. and García-Tiscar, S. (2002) Cetacean distribution related with depth and slope in the Mediterranean waters off southern Spain. Deep Sea Research (Part I). *Oceanographic Research Papers*, 49, 2053–73.

Canessa, R., Conley, K., Smiley, B. and Gueret, D. (2003) Building an ecosystem-based management plan for the proposed Bowie seamount marine protected area 5th International SAMPAA conference: making ecosystem based management work, Victoria, British Columbia, Canada.

Clarke, M.D. (1996) Cephalopods as prey III. Cetaceans. *Philosophical Transactions of the Royal Society of London (Series B)*, 351, 1053–65.

Daneri, G.A. and Carlini, A.R. (2002) Fish prey of Southern elephant seals, *Mirounga leonina*, at King George Island. *Polar Biology*, 25, 739–43.

Dawson, S. (2002) *Cephalorhynchus* dolphins – *C. heavisidii, C. eutropia, C. hectori, and C. commersonii*. In: *Encyclopedia of Marine Mammals* (eds. Perrin, W.F., Würsig, B. and Thewissen, J.G.M.) pp. 200–203. Academic Press, San Diego, California.

de Forges, B.R., Koslow, J.A. and Poore, G.C.B. (2000) Diversity and endemism of the benthic seamount fauna in the southwest Pacific. *Nature*, 45, 944–7.

Genin, A. (2004) Bio-physical coupling in the formation of zooplankton and fish aggregations over abrupt topographies. *Journal of Marine Systems*, 50, 3–20.

Griffin, R.B. (1997) Relationships between odontocete distributions and zooplankton community structure along the southern edge of Georges bank. *Journal of Northwest Atlantic Fishery Science*, 22, 27–36.

Hamazaki, T. (2002) Spatiotemporal prediction models of cetacean habitats in the mid-western North Atlantic ocean (from Cape Hatteras, North Carolina, U.S.A. to Nova Scotia, Canada). *Marine Mammal Science*, 18, 920–39.

Hammond, P.S., Berggren, P., Benke, H., Borchers, D.L., Collet, A., Heide-Jorgensen, M.P., Heimlich, S., Hiby, A.R., Leopold, M.F. and Øien, N. (2002) Abundance of harbour porpoise and other cetaceans in the North Sea and adjacent waters. *Journal of Applied Ecology*, 39, 361–76.

Hooker, S.K., Whitehead, H. and Gowans, S. (1999) Marine protected area design and the spatial and temporal distribution of cetaceans in a submarine canyon. *Conservation Biology*, 13, 592–602.

Hui, C.A. (1985) Undersea topography and the comparative distributions of two pelagic cetaceans. *Fishery Bulletin*, 83, 472–5.

Hyrenbach, K.D., Forney, K.A. and Dayton, P.K. (2000) Viewpoint: marine protected areas and ocean basin management. *Aquatic Conservation: Marine and Freshwater Ecosystems*, 10, 437–58.

IWC (2001) IDCR-DESS SOWER Survey data set (1978–2001). International Whaling Commission.

Jaquet, N. and Whitehead, H. (1999) Movements, distribution and feeding success of sperm whales in the Pacific Ocean, over scales of days and tens of kilometers. *Aquatic Mammals*, 25, 1–13.

Jefferson, T.A., Leatherwood, S. and Webber, M.A. (1993) *Marine Mammals of the World*, FAO Species Identification Guide, FAO, Rome.

Kaschner, K., Watson, R., Trites, A.W. and Pauly, D. (2006) Mapping worldwide distributions of marine mammals using a relative environmental suitability (RES) model. *Marine Ecology Progress Series*, 316, 285–310.

Kasuya, T. (2002) Giant beaked whales – *Berardius bairdii* and *B. arnuxii*. In: *Encyclopedia of Marine Mammals* (eds. Perrin, W.F., Würsig, B. and Thewissen, J.G.M.), pp. 519–22. Academic Press, San Diego, California.

Moore, S., Watkins, W.A., Daher, M.A., Davies, J.R. and Dahlheim, M. (2002) Blue whale habitat associations in the Northwest Pacific: analysis of remotely-sensed data using a Geographic Information System. *Oceanography*, 15, 20–25.

Moore, S.E., DeMaster, D.P. and Dayton, P.K. (2000) Cetacean habitat selection in the Alaskan Arctic during summer and autumn. *Arctic*, 53, 432–47.

Nordøy, E.S., Folkow, L. and Blix, A.S. (1995) Distribution and diving behaviour of crabeater seals (*Lobodon carcinophagus*) off Queen Maud Land. *Polar Biology*, 15, 261–8.

Norris, K.S., Würsig, B., Wells, R.S. and Würsig, M. (1994) *The Hawaiian Spinner Dolphin*, Vol. University of California Press, Berkeley, CA, 408 pp.

Ohsumi, S. (1983) Population assessment of Baird's beaked whales in the waters adjacent to Japan. *Reports of the International Whaling Commission*, 33, 633–41.

Parrish, F.A., Abernathy, K., Marshall, G.J. and Buhleier, B.M. (2002) Hawaiian monk seals (*Monachus schauinslandi*) foraging in deep-water coral beds. *Marine Mammal Science*, 18, 244–58.

Pauly, D., Trites, A.W., Capuli, E. and Christensen, V. (1998) Diet composition and trophic levels of marine mammals. *ICES Journal of Marine Science*, 55, 467–81.

Payne, P.M. and Heinemann, D.W. (1993) The distribution of pilot whales (*Globicephala spp.*) in shelf/shelf-edge and slope waters of the Northeastern United States, 1978–1988. In: *Biology of Northern Hemisphere Pilot Whales – Reports of the International Whaling Commission (Special Issue 14)* (eds. Donovan, G.P., Lockyer, C.H. and Martin, A.R.), pp. 51–68. IWC, Cambridge, UK.

Pitman, R.L. (2002) Mesoplodont whales – *Mesoplodon spp.* In: *Encyclopedia of Marine Mammals* (eds. Perrin, W.F., Würsig, B. and Thewissen, J.G.M.). pp. 738–42, Academic Press.

Reeves, R.R. and Mitchell, E. (1993) Status of Baird's Beaked Whale, *Berardius bairdii*. *Canadian Field Naturalist*, 107, 509–23.

Reijnders, P., Brasseur, S., van der Toorn, J., van der Wolf, R., Boyd, I.L., Harwood, J., Lavigne, D.M. and Lowry, L. (1993) Seals, fur seals, sea lions, and walrus. Status survey and conservation action plan, IUCN/SSC Specialist Group, International Union for the Conservation of Nature and Natural Resources, Gland, Switzerland.

Seabra, M.I., Silva, M., Magalhães, S., Prieto, R., August, P., Vigness-Raposa, K., Lafon, V. and Santos, R.S. (2005) Distribution and habitat preferences of bottlenose dolphins (*Tursiops truncatus*) and Sperm Whales (*Physeter macrocephalus*) with respect to physiographic and oceanographic factors in the waters around the Azores (Portugal). In: *Proceedings of the 19th Annual Conference of the European Cetacean Society* (ed. Evans, P.G.H.). ECS, La Rochelle.

Sigurjónsson, J., Gunnlaugsson, T. and Payne, M. (1989) NASS-87: Shipboard sightings surveys in Icelandic and adjacent waters June–July 1987. In: *Reports of the International Whaling Commission*, Vol. 39. IWC, Cambridge, UK, pp. 395–409.

Stewart, B.S. and Yochem, P.K. (2004) Use of marine habitats by Hawaiian monk seals (*Monachus schauinslandi*) from Laysan island: satellite-linked monitoring in 2001–2002. Report No. Administrative Report H-04-02C, Pacific Islands Fisheries Science Center – NMFS.

Tynan, C.T., Ainley, D.G., Barth, J.A., Cowles, T.J., Pierce, S.D. and Spear, L.B. (2005) Cetacean distributions relative to ocean processes in the northern California Current System. Deep Sea Research (Part II). *Topical Studies in Oceanography*, 52, 145–67.

Yen, P.P.W., Sydeman, W.J. and Hyrenbach, K.D. (2004) Marine bird and cetacean associations with bathymetric habitats and shallow-water topographies: implications for trophic transfer and conservation. *Journal of Marine Systems*, 50, 79–99.

Chapter 12B
Air-breathing visitors to seamounts: Sea turtles

Marco A. Santos, Alan B. Bolten, Helen R. Martins, Brian Riewald and Karen A. Bjorndal

Abstract

Marine turtles have a complex life cycle with ontogenetic shifts that involve major changes in their ecology, behaviour and distribution. In oceanic habitats, oceanographic and topographic features may generate eddies that can provide prime habitats where sea turtles can find food and shelter. Evidence from the pelagic longline swordfish fishery in the Azorean Exclusive Economic Zone (EEZ) and from satellite telemetry suggests that seamounts may affect loggerhead turtle distribution. Seamounts appear to be important habitats for juvenile oceanic loggerhead turtles.

Introduction

Sea turtles have complex life cycles, with ontogenetic shifts that involve major changes in their ecology, behaviour and distribution (Bolten, 2003a). In oceanic habitats, oceanographic and topographic features, such as seamounts, can generate eddies and convergent zones (see Chapter 4) that provide prime habitats where sea turtles find food and shelter. Our knowledge of the association of sea turtles with seamounts is based primarily on our understanding of the life history of the North Atlantic loggerhead (*Caretta caretta*) populations. This chapter will therefore focus on this species during its oceanic juvenile stage.

Most loggerheads occurring in the Eastern North Atlantic deposit their eggs on beaches in the SE United States. After hatching, young turtles of about 5 cm carapace length swim offshore where the Gulf Stream/Azores Current carries them to the eastern Atlantic, including the areas around the Azores, Madeira, and Canary Islands (Carr, 1986; Bolten *et al.*, 1998). Based on length frequency analyses and skeletochronology, the duration of the oceanic juvenile stage has been estimated to be from 6.5 to 11.5 years, depending on the size/age at which the turtles leave the oceanic habitat (Bjorndal *et al.*, 2000, 2003). Juvenile turtles return to coastal waters in the western Atlantic to continue development and, after an additional 20 years, reach sexual maturity at approximately 30 years of age (Bolten, 2003b). In Azorean waters, most loggerheads are between 10 and 65 cm curved carapace length (Fig. 12B.1) and are primarily epipelagic, spending 75% of their time in

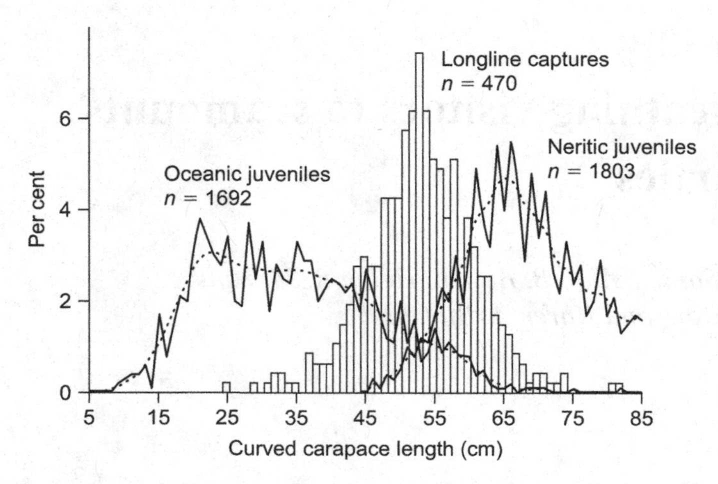

Fig. 12B.1 Size distributions of loggerhead turtles: left curves are oceanic loggerheads in Azorean waters; histogram shows loggerheads caught in longline fisheries in Azorean waters; right curves are neritic loggerheads in western Atlantic along east coast of USA (modified from Bolten, 2003b).

the top 5 m of the water column, but occasionally diving to over 200 m (Bolten, 2003b). The movements of juvenile loggerheads are both active and passive relative to winds and surface and subsurface oceanic currents (Bolten, 2003b).

Evidence for associations of sea turtles with seamounts

We evaluated the associations of sea turtles with seamounts using two approaches. First, we evaluated the distribution patterns of oceanic juvenile loggerheads with respect to seabed slope. We used data from 343 longline sets collected during experiments conducted in 2000–2004 within the EEZ of the Azores archipelago. The total distance covered by the gear line was 26 774 km. The study region was divided into 1 min squares. In each square that contained a longline set, loggerhead catch per unit effort (CPUE) was calculated as the number of turtles caught per 100 km of gear: $100 \times (n/L)$ where n is number of loggerheads captured and L is the number of kilometres of fishing effort.

Bathymetry data were obtained from GEBCO (General Bathymetric Chart of the Oceans) Digital Atlas (1 min resolution). Seabed slope was computed from the bathymetry lattice using ArcGIS (v. 8.3, ESRI). Mean slope was calculated for each 1 min square and determined for each loggerhead position. The results support the hypothesis that distribution patterns of loggerheads are related to local topographic characteristics. Mean slope showed a significant positive association with loggerhead CPUE (Fig. 12B.2; Spearman correlation, $r = 0.068$, $n = 13\,117$, $p < 0.0001$). Chi-square analysis also showed a greater use of steep slopes (Fig. 12B.2; 12°–18°, chi square $= 20.42$, df $= 8$, $p < 0.0001$). Slope may exert its influence through increased food abundance for loggerheads; steep slopes, characteristic of seamounts, are associated with water features that can generate higher productivity through several different mechanisms (see Chapters 4 and 5). Seamounts may therefore be hotspots for sea turtles.

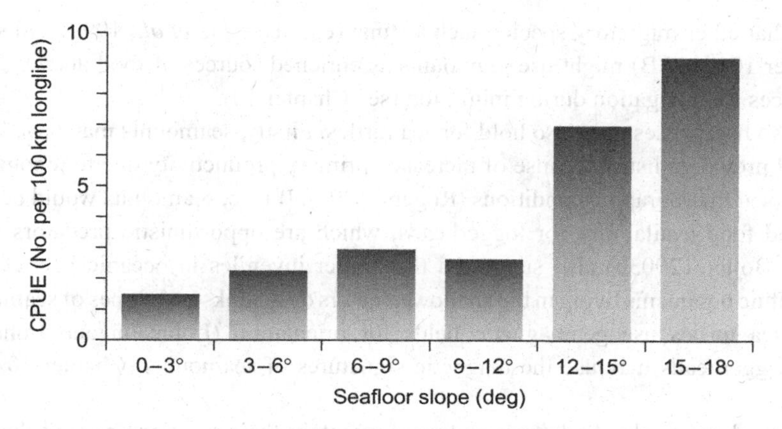

Fig. 12B.2 Relationship of longline CPUE of loggerheads (number of loggerheads captured per 100 km of longline) in Azorean waters relative to mean slope of sea bottom (see text).

Our second approach to evaluate the association of sea turtles with seamounts employs biotelemetry. Polovina *et al.* (2004) reported an association between loggerheads in the Pacific and local oceanographic anomalies, such as fronts and currents. Satellite tracking data obtained by Bolten and Riewald (unpublished data) may be used to evaluate whether sea turtles visit North Atlantic seamounts during their oceanic juvenile stage. A qualitative assessment of turtle tracks indicates that loggerheads range widely throughout the Azores Archipelago. Visual inspection of the tracks shows that loggerheads visit seamounts within the Azorean EEZ (e.g., Princess Alice seamount) and seamounts outside the EEZ (e.g., Altair, Great Meteor, Atlantis). Some of these seamounts are located hundreds of kilometres from the archipelago. Dellinger (2005) reported similar behaviour for loggerheads tracked near seamounts in the Madeira archipelago. Tracked turtles appear to move towards seamounts and to increase residence time in these areas. Santos *et al.* (2006) analysed some movement parameters and verified that mean distance and daily speed were always lower and the number of bearings higher in the vicinity of seamounts. Track length and residence time were also higher close to Altair seamount. These biotelemetry results demonstrate that the turtles exhibit different movement behaviours near seamounts. This is a further evidence that these topographic features might be hotspots for juvenile loggerheads.

Why sea turtles visit and how they find seamounts

Oceanic regions are often oligotrophic, characterized by low productivity and low standing stocks of plankton and nekton. Seamounts provide topographic features that can result in environments with higher productivities (see Chapters 4, 5 and 7). Seamounts play an important role in the distribution of various pelagic fish species (see Chapter 9), as well as marine mammals (see Chapter 12A). The importance of seamounts for the biology of oceanic juvenile loggerheads is poorly understood. Fréon and Dagorn (2000) suggested that other migratory species such as tuna (e.g., Holland *et al.*, 1999) and sharks (see Chapter 10A and B) might use seamounts as enriched sources of food and/or as spatial references for navigation during migration (see Chapter 10).

These two hypotheses may also hold for sea turtles. Firstly, seamounts may have higher densities of prey organisms because of increased primary productivity due to topographic effects on local hydrographic conditions (Rogers, 1994). If true, seamounts would be areas of increased food availability for loggerheads, which are opportunistic predators in the epipelagic. Bolten (2003b) also suggested that larger juveniles in oceanic habitats may feed on benthic organisms living in the shallower waters over banks and slopes of seamounts. Secondly, sea turtles use geomagnetic fields for orientation (Lohmann and Lohmann, 2003), so loggerheads may use the magnetic signatures of seamounts (Chapter 10A) for navigation.

How loggerhead turtles find seamounts and maintain their positions around them are important questions. Magnetic cues may be important because juvenile loggerheads are able to detect the magnetic inclination angle and magnetic field intensity of the Earth to assess global position (Lohmann and Lohmann, 2003). Chemical cues may also be used to locate seamounts. Sea turtles may use odour gradients to navigate towards a distant odour plume source (Lohmann *et al.*, 1997; Bartol and Musick, 2003). Chemical compounds, such as dimethyl sulphide (DMS), are associated with areas of high productivity and areas where planktivores feed (Hay and Kubanek, 2002). Nevitt *et al.* (1995) reported an association between DMS and migratory sea birds in Antarctic waters.

Conservation implications

Seamounts and their related oceanographic features are important habitats for commercial fish species (see Chapters 9, 10 and 18) and so are targeted by fishing fleets (Fonteneau, 1991). The association of loggerheads with seamounts makes them vulnerable to bycatch in these fisheries. The most significant anthropogenic threat to the survival of juvenile loggerhead turtles during the oceanic stage is the risk of incidental capture in commercial fisheries, mainly as bycatch of pelagic longline fleets. The impact of longline fisheries on sea turtles has recently received attention by the scientific community (Long and Schroeder, 2004). In Azorean waters, the size classes of loggerheads being impacted by the swordfish (*Xiphias gladius*) fishery correspond to the largest size classes of loggerheads occurring in the area (Fig. 12B.1; Ferreira *et al.*, 2001; Bolten, 2003b). Crouse *et al.* (1987) reported that these size classes are the most important for the recovery of North Atlantic loggerhead populations.

The discovery of the importance of seamounts for sea turtles raises the possibility of protecting these animals by establishing marine protected areas around seamounts. Other fishery management options (e.g., gear modifications, line retrieval times, time/area closures) in these critical areas should be considered by regional fishery management organizations to reduce incidental capture of turtles. In conclusion, seamounts appear to be important habitats for juvenile oceanic loggerhead sea turtles. More research is needed on the role that sea turtles play in the trophic structure of seamount communities and how sea turtles locate these habitats.

Acknowledgements

We would like to thank the editors for the invitation to participate in this book. This study was supported by a PhD grant by FCT/MCT – Portuguese Ministry of Science (SFRH/BD/10960/2002), US National Marine Fisheries Service, and Disney Wildlife Conservation Fund. This chapter is dedicated to the memory of Brian Riewald, our colleague and friend, who was one of the pioneers of satellite tracking studies of oceanic juvenile loggerhead turtles.

References

Bartol, S.M. and Musick, J.A. (2003) Sensory biology of sea turtles. In: *The Biology of Sea Turtles* (eds. Lutz, P.L., Musick, J. and Wyneken, J.), Vol. II, pp. 79–102. CRC Press, Boca Raton.

Bjorndal, K.A., Bolten, A.B. and Martins, H.R. (2000) Somatic growth model of juvenile loggerhead sea turtles: duration of the pelagic stage. *Marine Ecology Progress Series*, 202, 265–72.

Bjorndal, K.A., Bolten, A.B., Dellinger, T., Delgado, C. and Martins, H.R. (2003) Compensatory growth in oceanic loggerhead sea turtles: response to stochastic environment. *Ecology*, 84, 1237–49.

Bolten, A.B. (2003a) Variation in sea turtle life history patterns: neritic vs. oceanic developmental stages. In: *The Biology of Sea Turtles* (eds. Lutz, P.L., Musick, J. and Wyneken, J.), Vol. II, pp. 243–57. CRC Press, Boca Raton.

Bolten, A.B. (2003b) Active swimmers passive drifters: the oceanic juvenile stage of loggerheads in the Atlantic system. In: *Loggerhead Sea Turtles* (eds. Bolten, A.B. and Witherington, B.E.), pp. 63–78. Smithsonian Institution Press, Washington, DC.

Bolten, A.B., Bjorndal, K.A., Martins, H.R., Dellinger, T., Biscoito, M.J., Encalada, S.E. and Bowen, B.W. (1998) Transatlantic developmental migrations of loggerhead sea turtles demonstrated by mtDNA sequence analysis. *Ecological Applications*, 8, 1–7.

Carr, A.F. (1986) Rips, fads, and little loggerheads. *BioScience*, 36, 92–100.

Crouse, D.T., Crowder, L.B. and Caswell, H. (1987) A stage-based population model for loggerhead sea turtles and implications for conservation. *Ecology*, 68, 1412–23.

Dellinger, T. (2005) The importance of seamounts for turtles. In: *Seamounts and Fisheries – Conservation and Sustainable Use* (eds. Neumann, C., Christiansen, S. and Christiansen, B.), pp. 14–16. First OASIS Stakeholder Workshop. *OASIS* Report. Horta, Azores April 1–2, 2004.

Ferreira, R.L., Martins, H.R., Bolten, A.B. and Silva, A.A. (2001) Impact of swordfish fisheries on sea turtles in the Azores. *Arquipélago. Life and Marine Sciences*, 18A, 75–9.

Fonteneau, A. (1991) Seamounts and tuna in the tropical Atlantic. *Aquatic Living Resources*, 4, 13–25.

Fréon, P. and Dagorn, L. (2000) Review of fish associative behaviour: toward a generalisation of the meeting point hypothesis. *Reviews in Fish Biology and Fisheries*, 10, 183–207.

Hay, M.E. and Kubanek, J. (2002) Community and ecosystem level consequences of chemical cues in the plankton. *Journal of Chemical Ecology*, 28, 2001–16.

Holland, K.N., Kleiber, P. and Kajiura, S.M. (1999) Different residence time of yellowfin tuna, *Thunnus albacares*, and bigeye tuna, *T. obesus*, found in mixed aggregations over a seamount. *Fishery Bulletin*, 97, 392–6.

Lohmann, K.J. and Lohmann, C.M.F. (2003) Orientation mechanisms of hatchling loggerheads. In: *Loggerhead Sea Turtles* (eds. Bolten, A.B. and Witherington, B.E.), pp. 44–62. Smithsonian Institution Press, Washington, DC.

Lohmann, K.J., Witherington, B.E., Lohmann, C.M.F. and Salmon, M. (1997) Orientation, navigation and natal beach homing in sea turtles. In: *The Biology of Sea Turtles* (eds. Lutz, P.L. and Musick, J.), pp. 107–35. CRC Press, Boca Raton.

Long, K.J. and Schroeder, B.A. (eds.) (2004) *Proceedings of the International Technical Expert Workshop on Marine Turtle Bycatch in Longline Fisheries*. US Department of Commerce, NOAA Technical Memorandum NMFS-F/OPR/26, 189 pp.

Nevitt, G., Veit, R.R. and Kareiva, P. (1995) Dimethyl sulfide as a foraging cue for Antarctic Procellariiform seabirds. *Nature*, 376, 680–2.

Polovina, J., Balazs, G., Howell, E., Parker, D., Seki, M. and Dutton, P. (2004) Forage and migration habitat of loggerhead (*Caretta caretta*) and olive ridley (*Lepidochelys olivacea*) sea turtles in the central North Pacific Ocean. *Fisheries Oceanography*, 13, 36–51.

Rogers, A.D. (1994) The biology of seamounts. *Advances in Marine Biology*, 30, 305–50.

Santos, M.R., Bolten, A.B., Martins, H.R., Riewald, B., Bjorndal, K., Ferreira, R. and Gonçalves, J. (2006) Distribution of oceanic stage North Atlantic loggerheads: are seamounts important hotspots (Poster). In: *Book of Abstracts. 26th Annual Symposium on Sea turtle Biology and Conservation* (compilers Frick, M., Panagopoulou, A., Rees, A. and Williams, K.), pp. 110–11. International Sea turtle Society, Athens, Greece.

Chapter 12C

Air-breathing visitors to seamounts: Importance of seamounts to seabirds

David R. Thompson

Abstract

A wide range of seabird species appear to make use of resources associated with seamounts. Knowledge of the relationships between seabirds and seamounts is based either on ship-based observational studies, or studies that employ seabirds as data-gatherers. The majority of studies fall into the 'observational' category, and many of these report 'one-off' counts of seabirds at a particular feature. A small number of studies extend this approach to incorporate a temporal component to seabird observations. To date, relatively little research has made use of electronic data-gathering devices, attached to seabirds, as a means to elucidate their seamount resource-use. Continued refinement of these technologies, particularly the use of miniaturized global positioning system (GPS) technology; holds much promise for better understanding of how seabirds exploit marine resources associated with seamounts.

Introduction

Seabirds are conspicuous predators that are able to exploit marine resources over relatively large spatial scales. Although constrained to return to land at regular, usually annual, intervals to raise offspring, pelagic seabirds, particularly members of the order Procellariiformes (petrels, shearwaters and albatrosses), can undertake foraging trips of many thousands of kilometres, often over relatively deep ocean waters far beyond the continental margins (Weimerskirch *et al.*, 1999; Klomp and Schultz, 2000; Fernandez *et al.*, 2001). Most seabirds are restricted to obtaining food at the ocean surface. Others, notably species from the 'diving' groups like penguins and auks, can exploit prey at considerable depth (Charrassin and Bost, 2001; Falk *et al.*, 2002). Seabirds take a wide range of prey, from small zooplankton through to fish and squid, taken both during the day and night, and can exploit vertically migrating prey that move into surface waters (see chapters in Croxall, 1987).

Seabirds are well able to exploit marine resources associated with seamounts of continental shelf slopes and beyond. Early researchers such as Darwin noted seabirds associated with seamounts (see Chapter 3). While there is a general perception that seamounts are often sites of seabird aggregation (Ballance *et al.*, 2001), it is highly likely that seabird prey species composition and abundance varies between seamounts due to factors such as seamount location, topography, summit depth and how the seamount interacts with flows, currents and tides (see Chapter 4). There is considerable inter-annual, seasonal and even daily (diurnal vs nocturnal) variation in prey availability associated with seamounts (see Chapters 5 and 6). In short, although some seabirds apparently associate with some seamounts, detailed analyses of these associations over clearly defined spatio-temporal scales, and studies that have explicitly examined links between seabirds and seamounts are relatively few.

This paucity of information reflects the fact that most ship-based research, whether specifically focused on seamounts or coincidentally incorporating seamounts as part of a more general itinerary, seldom includes collection of seabird data, either general observations, *ad hoc* counts or quantitative measurements of seabird numbers and densities. Despite the difficulties of identifying and counting seabirds from a moving platform, current knowledge of seabirds and seamounts is based almost entirely on ship-based observations. Here, I summarize the available information regarding the importance of seamounts to seabirds, and suggest how this field might develop in the future. Seabird nomenclature follows that of BirdLife International (2004a).

Ship-based observations

Ship-based observations fall into two general categories: single observations or 'snapshots' of seabird–seamount characteristics at a particular site and time; and more frequent observations, possibly with an inter-seasonal or inter-annual component. Repeated observations enable researchers to be somewhat more firm in their conclusions, although the logistic and financial constraints of repeat visits to an oceanic seamount make it difficult to adhere to a scientifically robust sampling protocol.

Dealing with single observation studies initially, Powell *et al.* (1952) were among the first workers to record the presence of a range of seabirds specifically at a seamount. They noted black-footed albatross *Phoebastria nigripes*, sooty shearwater *Puffinus griseus*, fork-tailed storm-petrel *Oceanodroma furcata* and 'Beal's petrel' (=Leach's storm-petrel) *O. leucorhoa* at Cobb Seamount off southern Washington, USA in August, 1950 (Powell *et al.*, 1952). At the same seamount, Sanger (1970) cites unpublished University of Washington records, of western gull *Larus occidentalis* (although herring gull *L. argentatus* could not be ruled out due to identification problems) in March, 1960. In another early report, Grindley (1967) noted a wide range of species over the Vema Seamount, off the west coast of South Africa, during November, 1966, dominated numerically by Cape petrel *Daption capense* and European storm-petrel *Hydrobates pelagicus*. It seems likely, however, that bird observations were confounded by boats actively fishing over the seamount, providing food in the form of offal and discards (Grindley, 1967). Also in the South Atlantic Ocean, Bourne (1992) recorded seabird species and numbers over the Republic of South Africa Seamount in February, 1985. Great shearwater *Puffinus gravis* and white-bellied storm-petrel *Fregetta grallaria* were particularly numerous over the seamount compared to other locations along the cruise track. In contrast, Bourne (1992) also noted that when crossing St Joseph Seamount in the NE Atlantic in April 1960, the only seabird present was a single Cory's shearwater *Calonectris diomedea*. Seabirds were counted over the Emperor Seamounts, north Pacific, during July 1978 (Anon, 1979), and at stations elsewhere away from seamounts, during 10 min periods in a 300 m zone to one side of the ship. There were no clear patterns in seabird density at stations over or away from seamounts, but several species of gadfly petrel, including scaled (=mottled) *Pterodroma inexpectata*, Solander's (=providence) *P. solandri*, Bonin *P. hypoleuca*, Stejneger's *P. longirostris*, Cook's *P. cookii*, and unidentified *Pterodroma* sp., were generally only recorded at stations over seamounts. Numbers of fork-tailed storm-petrel were higher at stations over seamounts compared to numbers at other stations (Anon, 1979). B. Herbert (Personal communication cited in Monteiro *et al.*, 1996), concluded that seamounts were the preferred summer foraging areas for three seabird species from the Azores archipelago: Cory's shearwater, band-rumped storm-petrel *Oceanodroma castro*, with aggregations of up to a few dozen between July and September, and yellow-legged gull *Larus cachinnans*, particularly juveniles in flocks of up to 100 birds during July and August.

Vermeer *et al.* (1985) recorded the distribution and numbers of ancient murrelet *Synthliboramphus antiquus* and Cassin's auklet *Ptychoramphus aleuticus* NW of the Queen Charlotte Islands, Canada, during May–July, 1981, and found that whilst both species tended to forage over the continental shelf break, Cassin's auklet also fed over seamounts. Vermeer *et al.* (1985) also correlated higher densities of Cassin's auklet over seamounts with densities of the calanoid copepod *Neocalanus cristatus* their preferred prey.

In a study of seabird distribution and numbers during April, 1985, Blaber (1986) followed a transect SE from Tasmania to the Soela Seamount, and recorded birds using 10 min, 360° counts of all birds within about 500 m of the ship. Some species were notably more abundant around and over the seamount, compared to stations along the transect to the seamount (e.g., wandering albatross *Diomedea exulans* and probably *D. antipodensis*, black-browed albatross (when identified = Campbell albatross) *Thalassarche impavida*, white-chinned petrel *Procellaria aequinoctialis* and black-bellied storm-petrel *Fregetta tropica*). Three species, shy albatross *Thalassarche cauta*, white-chinned petrel and black-bellied storm-petrel, were more abundant over than around the seamount. One species, Antarctic prion *Pachyptila desolata*, was more abundant around the seamount compared to over the seamount (Blaber, 1986). Haney *et al.* (1995) recorded the numbers and distribution of seabirds over and away from Fieberling Guyot, a seamount in the eastern North Pacific Ocean, during June 1991. Seabird density within a 30 km radius was 2.4 times higher compared to adjacent waters, and dominated by black-footed albatross and Cook's petrel. Haney *et al.* (1995) also sampled potential prey over the seamount, later in the year during September, but results were generally inconclusive, perhaps due, in part, to the different sampling dates: however, some prey groups sampled were more abundant over the seamount.

Guzman and Myres (1983) reported the seasonal occurrence (April to October) of six species of shearwater *Puffinus* spp. off the coast of British Columbia, Canada, based on counts conducted from 1975 to 1978. Of the six species recorded, only sooty shearwater, during June and July, and Buller's shearwater *Puffinus bulleri*, during June, showed an association with seamounts, specifically Giacomini, Surveyor and Welker seamounts in the central Gulf of Alaska (Guzman and Myres, 1983). Haney (1987) investigated the abundance and distribution of black-capped petrels *Pterodroma hasitata* off the coast of southeastern USA over 173 days from 1982 to 1985. Although overall seasonal abundance estimates were produced, Haney (1987) was able to conclude that black-capped petrels were observed only over seamounts and other submarine features when to the east of the western Gulf Stream frontal boundary. Briggs *et al.* (1987), recorded species and abundance of seabirds off California, USA from April 1975 to March 1978 through ship and plane-based observations. They recorded flesh-footed shearwater *Puffinus carneipes* associated with the Cordell Bank and Guide seamounts between May and October; fork-tailed storm-petrel associated with San Juan Seamount between May and January; ashy storm-petrel *Oceanodroma homochroa* in most months at Guide Seamount, and horned puffin *Fratercula corniculata* at San Juan Seamount during April (Briggs *et al.*, 1987). Interestingly, Briggs *et al.* (1987) reported no association between Cassin's auklet and seamounts, in contrast with the findings of Vermeer *et al.* (1985).

Reid *et al.* (2002) compiled ship-based observational data from SE Australian waters, spanning 1975 to 1993, to produce an 'atlas' of seabird distributions, with seasonal resolution. Although cruise tracks were not the same for each 3-month seasonal period, and seamounts within the area covered were not necessarily targeted for seabird observations, Reid *et al.* (2002) concluded that black-browed albatross *Thalassarche melanophris*, Campbell albatross, sooty albatross *Phoebetria fusca*, southern royal albatross *Diomedea epomophora* and Cape petrel were especially numerous over seamounts. For both southern royal albatross and Cape petrel, however, Reid *et al.* (2002) noted that relatively high numbers of birds over St Helen's Rise were associated with trawler activity and in the case of Cape petrel, sperm whales *Physeter catodon*.

In a comprehensive study of seabird and marine mammal associations with bathymetric features off central California, including Cordell Bank Seamount, between 1996 and 2002, Yen *et al.* (2004) reported that, of the four seabirds investigated, only Cassin's auklet exhibited a significant relationship with the seamount. This species was found in greater densities near Cordell Bank with numbers declining rapidly with distance away from the feature, although there was a suggestion that aggregations of Cassin's auklet at Cordell Bank were increasing through the period of the study (Yen *et al.*, 2004). In a similar study across the eastern Gulf of Alaska, Yen *et al.* (2005) did not report any associations between seabirds and seamounts, even though the transect along which observations were made crossed over several seamounts.

Birds as data-gatherers

It has only been in the last 20 years or so that tracking technology has become reliable and small enough to be deployed on seabirds as they forage and migrate over the ocean. Equipping birds with data-gathering or transmitting devices has some advantages compared to platform-based observations of seabird distributions at sea. The provenance and, in many cases, the status (sex, breeding, non-breeding) of the birds is known, distributions are not biased by being attracted to boats, as is the case for some species (Hyrenbach, 2001), and device costs tend to be relatively low compared to ship, or other platform, costs.

Tracking methods fall into four general categories, each with advantages and disadvantages: radio VHF tags (Anderson and Ricklefs, 1987), satellite telemetry (Jouventin and Weimerskirch, 1990), light-based geolocation (Shaffer *et al.*, 2005) and, most recently, GPS tags (Weimerskirch *et al.*, 2002, 2005). Radio tags operate over relatively short distances, which is of limited utility when tracking pelagic seabirds, but Anderson and Ricklefs (1987) employed this approach to identify two seamount foraging areas favoured by masked *Sula dactylatra* and blue-footed *S. nebouxii* boobies at the Galapagos archipelago. Location information derived from light-based archival tags tends to be too imprecise (Phillips *et al.*, 2004; Shaffer *et al.*, 2005) to link seabirds and specific seamounts with any certainty. Position information using satellite telemetry is more accurate than that from geolocation, usually within a few kilometres of the true location, but is relatively expensive to acquire and would require repeated and extensive deployment of transmitting devices to determine seamount use by seabirds. Nevertheless, Waugh and Weimerskirch (2003) and Birdlife International (2004b) were able to conclude that wandering albatross breeding at the Crozet Islands frequented the seamounts to the west, based on satellite telemetry.

GPS technology perhaps offers the most potential in quantifying the use of seamounts by seabirds: location information is accurate to within a few metres, and can be generated frequently throughout a foraging trip so that time spent over seamounts, compared to elsewhere, can be quantified. Awkerman *et al.* (2005) employed GPS dataloggers to identify two areas close to Banco Ruso Seamount within the Galapagos archipelago as being favoured foraging locations for waved albatross *Phoebastria irrorata* during chick brooding in 2003. Weimerskirch *et al.* (2005), also employed GPS technology and noted that red-footed booby *Sula sula* did not preferentially forage over, or in the wake of, Bassas da India and Jaguar seamounts in the Mozambique Channel during August and September, 2003, although these seamounts were at the limit of the foraging range for this species (Weimerskirch *et al.*, 2005).

Conclusions

A wide range of seabird species exploit marine resources associated with seamounts. Apart from the few integrated and comprehensive studies, our knowledge of this association is often based on little more than a single observation of a named species at or near a specified seamount. These mere 'presence' data cannot resolve apparent contradictions where seabird–seamount associations reported by one study are not substantiated by a second study, separated in space and time. In this situation, it is impossible to conclude whether differences in seamount characteristics, prey availability or sampling time, or a combination of these, produced the apparent discrepancy.

Advances in bird-borne tracking technology, in particular GPS dataloggers, represent an exciting opportunity to examine the relationships between seabirds and seamounts in a more detailed and ultimately useful manner. Guzman and Myres (1983) wondered, 'whether the birds that are found near the seamounts have arrived by chance while wandering over the open ocean and stay for a while, are attracted to a source of food there in some way, or return after previous chance visits'. Clarke (see Chapter 11), Livinov (see Chapter 10B), Kaschner (see Chapter 12A) and Santos *et al.* (see Chapter 12B) make a strong case for seamounts being 'way stations' or 'cafes' for sperm whales, sharks, some marine mammals and marine turtles. Given what we know about the navigational abilities of birds, it would seem likely that they have the ability to 'map' and return to seamounts. GPS technology offers the chance to answer some of Guzman and Myres' (1983) questions directly, particularly when allied to complementary research that addresses productivity and prey availability over seamounts. Indeed, without a fully integrated approach, unequivocal associations between seabirds and seamounts, and hence the functional role seabirds play within seamount-based marine ecosystems, will remain elusive.

A much more complete understanding of spatio-temporal variation in seabird abundance is essential to understanding their functional role in seamount ecosystems. Our ability to conserve these charismatic marine predators and protect them from the direct and indirect impacts of fishing (see Chapter 20) requires a fully integrated approach.

Acknowledgements

Ashley Rowden and an anonymous reviewer provided constructive comments that improved an earlier version of this work, and the New Zealand Foundation for Research, Science and Technology (contract number: CO1X0508) provided financial support.

References

Anderson, D.J. and Ricklefs, R.E. (1987) Radio-tracking masked and blue-footed boobies (*Sula* spp.) in the Galapagos Islands. *National Geographic Research*, 3, 152–63.

Anon (1979) The Hokusei Maru cruise 6 (1–3) to the sea area off Hokkaido and the sea area of the Emperor Seamount, northwest Pacific Ocean in June–August 1978. *Data Record of Oceanographic Observations and Exploratory Fishing Hokkaido University*, 22, 72–186.

Awkerman, J.A., Fukuda, A., Higuchi, H. and Anderson, D.J. (2005) Foraging activity and submesoscale habitat use of waved albatrosses *Phoebastria irrorata* during the chick-brooding period. *Marine Ecology Progress Series*, 291, 289–300.

Ballance, L.T., Ainley, D.G. and Hunt, G.L. (2001) Seabird foraging ecology. In: *Encyclopedia of Ocean Sciences* (eds. Steele, J.H., Thorpe, S.A. and Turekian, K.K.), Vol. 5, pp. 2636–44. Academic Press, London, UK.

BirdLife International (2004a) *Threatened Birds of the World 2004*. CD-ROM. BirdLife International, Cambridge, UK.

BirdLife International (2004b) *Tracking Ocean Wanderers: The Global Distribution of Albatrosses and Petrels*. Results from the Global Procellariiform Tracking Workshop, September 1–5, 2003, Gordon's Bay, South Africa. Birdlife International, Cambridge, UK.

Blaber, S.J.M. (1986) The distribution and abundance of seabirds south-east of Tasmania and over the Soela seamount during April 1985. *Emu*, 86, 239–44.

Bourne, W.R.P. (1992) A concentration of great shearwaters and white-bellied storm-petrels over the RSA seamount in the south Atlantic east of Gough Island. *Sea Swallow*, 41, 51–3.

Briggs, K.T., Tyler, W.B., Lewis, D.B. and Carlson, D.R. (1987) Bird communities at sea off California: 1975 to 1983. *Studies in Avian Biology*, 11, 1–74.

Charrassin, J.-B. and Bost, C.-A. (2001) Utilisation of the oceanic habitat by king penguins over the annual cycle. *Marine Ecology Progress Series*, 221, 285–98.

Croxall, J.P. (ed.) (1987) *Seabirds: Feeding Ecology and Role in Marine Ecosystems*. Cambridge University Press, Cambridge, UK.

Falk, K., Benvenuti, S., Dall'Antonia, L., Gilchrist, G. and Kampp, K. (2002) Foraging behaviour of thick-billed murres breeding in different sectors of the north water polynya: an inter-colony comparison. *Marine Ecology Progress Series*, 231, 293–302.

Fernandez, P., Anderson, D.J., Sievert, P.R. and Huyvaert, K.P. (2001) Foraging destinations of three low-latitude albatrosses (*Phoebastria*) species. *Journal of Zoology, London*, 254, 391–404.

Grindley, J.R. (1967) Birds at the Vema Seamount and off the west coast of South Africa. *Ostrich*, 38, 281–2.

Guzman, J.R. and Myres, M.T. (1983) The occurrence of shearwaters (*Puffinus* spp.) off the west coast of Canada. *Canadian Journal of Zoology*, 60, 2064–77.

Haney, J.C. (1987) Aspects of the pelagic ecology and behavior of the black-capped petrel (*Pterodroma hasitata*). *Wilson Bulletin*, 99, 153–68.

Haney, J.C., Haury, L.R., Mullineaux, L.S. and Fey, C.L. (1995) Sea-bird aggregation at a deep north Pacific seamount. *Marine Biology*, 123, 1–9.

Hyrenbach, K.D. (2001) Albatross response to survey vessels: implications for studies of the distribution, abundance, and prey consumption of seabird populations. *Marine Ecology Progress Series*, 212, 283–95.

Jouventin, P. and Weimerskirch, H. (1990) Satellite tracking of wandering albatrosses. *Nature*, 343, 746–8.

Klomp, N.I. and Schultz, M.A. (2000) Short-tailed shearwaters breeding in Australia forage in Antarctic waters. *Marine Ecology Progress Series*, 194, 307–10.

Monteiro, L.R., Ramos, J.A., Furness, R.W. and Del Nevo, A.J. (1996) Movements, morphology, breeding, molt, diet and feeding of seabirds in the Azores. *Colonial Waterbirds*, 19, 82–97.

Phillips, R.A., Silk, J.R.D., Croxall, J.P., Afanasyev, V. and Briggs, D.R. (2004) Accuracy of geolocation estimates for flying seabirds. *Marine Ecology Progress Series*, 266, 265–72.

Powell, D.E., Alverson, D.L. and Livingstone, R. (1952) North Pacific albacore tuna exploration – 1950. *US Fish and Wildlife Service, Fishery Leaflet*, 402, 56 pp.

Reid, T.A., Hindell, M.A., Eades, D.W. and Newman, M. (2002) *Seabird Atlas of South-Eastern Australian Waters*. Birds Australia Monograph 4. Birds Australia, Melbourne. p. 146.

Sanger, G.A. (1970) The seasonal distribution of some seabirds off Washington and Oregon, with notes on their ecology and behavior. *Condor*, 72, 339–57.

Shaffer, S.A., Tremblay, Y., Awkerman, J.A., Henry, R.W., Teo, S.L.H., Anderson, D.J., Croll, D.A., Block, B.A. and Costa, D.P. (2005) Comparison of light- and SST-based geolocation with satellite telemetry in free-ranging albatrosses. *Marine Biology*, 147, 833–43.

Vermeer, K., Fulton, J.D. and Sealy, S.G. (1985) Differential use of zooplankton prey by ancient murrelets and Cassin's auklets in the Queen Charlotte Islands. *Journal of Plankton Research*, 7, 443–59.

Waugh, S.M. and Weimerskirch, H. (2003) Environmental heterogeneity and the evolution of foraging behaviour in long ranging greater albatrosses. *Oikos*, 103, 374–84.

Weimerskirch, H., Catard, A., Prince, P.A., Cherel, Y. and Croxall, J.P. (1999) Foraging white-chinned petrels *Procellaria aequinoctialis* at risk: from the tropics to Antarctica. *Biological Conservation*, 87, 273–5.

Weimerskirch, H., Bonadonna, F., Bailleul, F., Mabille, G., Dell'Omo, G. and Lipp, H.-P. (2002) GPS tracking of foraging albatrosses. *Science*, 295, 1259.

Weimerskirch, H., Le Corre, M., Jaquemet, S. and Marsac, F. (2005) Foraging strategy of a tropical seabird, the red-footed booby, in a dynamic marine environment. *Marine Ecology Progress Series*, 288, 251–61.

Yen, P.P.W., Sydeman, W.J. and Hyrenbach, K.D. (2004) Marine bird and cetacean associations with bathymetric habitats and shallow-water topographies: implications for trophic transfer and conservation. *Journal of Marine Systems*, 50, 79–99.

Yen, P.P.W., Sydeman, W.J., Morgan, K.H. and Whitney, F.A. (2005) Top predator distribution and abundance across the eastern Gulf of Alaska: temporal variability and ocean habitat associations. *Deep-Sea Research II*, 52, 799–822.

Part IV

Synoptic Views of Seamounts

Chapter 13

Biogeography and biodiversity of seamounts

Karen I. Stocks and Paul J.B. Hart

Abstract

This chapter evaluates the emergent community properties of biodiversity, endemism, and biogeography at seamounts. As with all deep-sea environments, the paucity of sampling, varying, and incompatible methods create problems, but certain patterns emerge, which differ between the pelagic and benthic components of the community. Pelagic fishes and plankton found over seamounts do not seem to be strongly differentiated from nearby oceanic pelagic communities. Abundances may vary, with certain species concentrated or depleted over a seamount, but the same suite of species is often found, and endemics (species restricted to one seamount or seamount chain) are not commonly reported. This indicates that the management of pelagic fisheries over seamounts should be within the context of a regional or stock-wide management plan. It also raises the potential for bioregionalization schemes, such as Longhurst's pelagic provinces, to be used to categorize seamount communities globally, for use in management and to prioritize future research. The benthic community of fishes and invertebrates, in contrast, can differ more strongly from either the surrounding deep seafloor or nearby continental margins of similar depth. Rates of endemism between 10% and 50% have been reported in medium and large-scale studies, though these numbers will likely change with more deep-sea sampling. Overall benthic abundances are often elevated on seamounts compared to soft-sediment deep-sea communities, though there is no clear trend towards seamounts supporting either an elevated or a depressed number of species (species richness) compared to the deep sea or an continental margins. Biogeographically, when endemic species are not considered, the seamounts appear to support a subset of the regional species pool. However, the rates of endemism are high enough, and the types of communities supported are distinct enough, that categorizing the benthic communities on seamounts by the larger biogeographic region they fall within may not be a useful approach. It also indicates that seamounts need to be recognized and managed as distinct habitat types, and not simply as part of a larger region.

Introduction

One of the reasons why seamounts have been the focus of substantial research efforts is that to some degree they are thought to represent unique communities, supporting assemblages of species that are distinct from the surrounding seafloor. Other chapters in this volume describe the characteristics of specific components of the seamount community, such as the benthic fauna (Chapter 7), fishes (Chapters 9 and 10), plankton (Chapter 5), cephalopods (Chapter 11), and visitors (Chapter 12). Here, we review the higher-order properties of biodiversity and biogeographic affinities.

One of the unifying questions for seamount research is to what degree seamounts represent unique, isolated habitats. At one extreme, seamounts could be distinct biotopes, supporting species found nowhere else in the oceans and evolving independently. At the other extreme, they could host only species also found in other habitats, accompanied by a high degree of migration and genetic exchange with surrounding environments. Understanding where seamounts lie on this spectrum has clear management implications, for example determining whether marine protected areas on seamounts are needed and how they should be placed (see Chapter 20). Describing and comparing seamount community structure also has theoretical implications: once patterns are determined, they can be compared to predictions from biogeographic theory to assess their usefulness for extrapolating from the handful of studied seamounts to the tens of thousands of unknown ones. And, where theory does not match reality, seamount research can contribute to improvement of our framework for understanding the patterns of life in the oceans.

This chapter explores how different characteristics support or contradict the concept of seamounts as isolated habitats. It begins with a brief overview of the seamount environment, showing how seamounts differ from other marine habitats in ways likely to influence community structure. The second section examines patterns of biodiversity, asking how the patterns of species richness on seamounts compare to non-seamount areas. Then endemism, a specific aspect of diversity that has attracted much attention on seamounts, is considered by reviewing the prevalence of endemic species on different seamounts. The fourth section focuses on genetic and life history evidence for and against population isolation on seamounts, while the fifth section covers the problems attendant upon establishing a biogeography of the open ocean. Finally, we cover the biogeographic affinities of seamounts, asking in particular whether seamount faunas are most closely affiliated with the nearest continental margin of similar depth, and whether emerging biogeographic regionalizations, such as Longhurst's (1998) provinces, provide a useful framework for examining and predicting species biogeography on seamounts.

We may expect that patterns of community structure, connectivity, and biogeography vary among nektonic, planktonic, and benthic components of the community. These components are considered separately because their mobility differences are likely to change the dispersal potential of the species and its characteristic spatial and temporal scales and patterns: nekton are larger organisms that can swim against currents, plankton are smaller organisms that float passively in the prevailing currents (thought they often can adjust their vertical position), and sedentary benthic invertebrates have limited mobility outside of a restricted larval stage.

In this chapter, we do not follow a strict definition of seamounts as being features over 1000 m high (see Preface). Studies from low seamounts are included and not differentiated.

The seamount environment

Seamounts are small 'islands' of shallower habitat usually surrounded by deeper ocean (see Chapter 1). Depth is known to be an important factor for marine communities, and different communities are found in different depth zones (Tyler, 1995). This is likely due to a suite of environmental factors correlated with depth: light, amount of sinking organic matter, the presence of dense migrating zooplankton layers, pressure, oxygen, nutrients, and physical disturbance. Benthic communities found near the summits of seamounts are therefore unlikely to flourish in the deep sea as well, and, unlike the continental margin, are isolated from habitats of similar depth (see Chapter 7).

Several other factors differentiate the seamount benthic environment from the deep sea. The majority of the deep sea is covered with fine, unconsolidated sediments; hard substrates are rare (Tyler, 1995). In contrast, the steep slopes and accelerated currents over seamounts keep sediment from depositing and make bare rock surfaces common (Rogers, 1994). While unconsolidated sediments are found on seamounts, many of the emergent epifauna common on seamounts, such as corals, sponges, and crinoids, that need a hard bottom cannot live in the typical unconsolidated deep seafloor.

Finally, some seamounts generate hydrographic features like Taylor cones or recirculating eddies (see Chapter 4). These may serve to retain water over the seamount and may restrict the dispersal of larvae and plankton, though how persistent and important these features are is an area of some debate (see Chapter 4). While the pelagic environment over a seamount is less clearly distinguished from the surrounding waters, there are still environmental differences. As discussed in Chapters 4 and 5, the movement of water over and by seamounts can concentrate plankton and retain it over the seamount, migrating zooplankton can impinge on the seamount summits, and upwellings can change nutrient and productivity characteristics of the waters.

Biodiversity

Biodiversity, a concept central to both the management and science of seamounts, refers to the diversity of all aspects of life in an area, from genetic and molecular diversity through to the diversity of major phyla and body plans. In practice, biodiversity on seamounts has most commonly been measured by looking at the diversity of species. Theory might predict that seamounts would have comparatively low diversity. In many terrestrial and marine habitats, a general empirical relationship between habitat size and the number of species has been found (May, 1975; Rosenzweig, 1995). All else being equal, seamounts would be predicted to have fewer total species than, for example, large continental margins of similar depth. The theory of Island Biogeography similarly predicts that small, isolated patch habitats like seamounts should have fewer species than the source population, which is presumably a larger region like continental margins or the deep sea (MacArthur and Wilson, 1967). This assumes that colonization to seamounts comes from some external 'mainland' source, and that immigration decreases with distance (or isolation in a more general sense) from that source, two assumptions that have not been rigorously examined on seamounts and which may not hold as well in the marine environment as for terrestrial islands. An alternative hypothesis argues that if the environment of deep seamounts is more stable over evolutionary time than continental margins (Vöros, 2005), they may have accumulated more species.

Good comparative biodiversity data are, unfortunately, sparse on seamounts. Estimates of diversity are influenced by both sampling effort and gear type. Comparing the species richness (the total number of species found) on a seamount with that on a nearby continental margin is often biased because the margins generally have received much more sampling, and gear is often more effective on smoother shelves and slopes than steep, rocky seamounts (see Chapter 7; Rowden *et al.*, 2004).

Richer de Forges *et al.* (2000) plotted the number of species collected vs the number of samples for seamounts off New Caledonia in the southwest Pacific. They found that the relationship was linear: even for seamounts that had received ~80 trawls or dredges, new species continued to be found at the same rate in each additional sample. This makes estimating the total species diversity of the seamount nearly impossible. Tittensor and Myers (unpublished data, 2004) expanded this approach to look at all of the seamounts in the SeamountsOnline database (Stocks, 2005; http://seamounts.sdsc.edu), which have species records and for which the sampling effort is known. Of the 180 seamounts, the species accumulation curve had not begun to approach a horizontal asymptote. This dataset does not yet include recent intense research on, for example, the Great Meteor Seamount in the NE Atlantic and around New Zealand, but it is likely that the total species diversity is not known for any seamount.

Nevertheless, some comparisons of diversity on seamounts have been made, as measured by species richness in most cases (Table 13.1). The most complete data are for benthic or near-bottom invertebrates. Of the six studies that compared seamount fauna to a nearby continental margin of similar depth, two found higher diversity, two found lower, and two found similar diversity (see table for references). Of the four studies comparing benthic invertebrates on seamounts to nearby deep seafloor, two found higher diversity on seamounts, one found lower, and one was similar. Data for fishes and pelagic invertebrates are too scattered to evaluate methodically, but no overall trend of elevated or depressed diversity emerges across groups.

These studies should be interpreted with caution, as most do not consider the difference in sampling effort between the seamount and the margin or deep-sea habitats, and they do not necessarily match the depth of sampling precisely between the seamount and margin. Contrary to the theoretical predictions, there is no indication from this small sample of a trend towards either lowered or elevated species richness on seamounts. This may be because the theories do not hold well on seamounts. Or it may be that a simple comparison does not account for other factors that can influence diversity (Rosenzweig, 1995), such as a possible increased abundance per unit area on seamounts or differential habitat variability or patchiness on seamounts. It will be interesting to examine how diversity may relate to latitude, depth, ocean basin, and distance from the continental margin as further data are collected.

The work reviewed here focuses on species richness. There are indications that species evenness, the other component of species diversity may be low on seamounts. Species evenness refers to the relative abundance of the species: a community where one or two

Table 13.1 Diversity of bottom-associated communities on seamounts compared to other habitats.

Diversity	Comparison	Effort consid.	Taxa	Seamount(s)	Reference
Bottom-associated fishes					
Lower (at habitat level)	Margin	Yes	Fish	New Zealand: 10 groups of seamounts (hills)	Tracey *et al.* (2004)
Bottom-associated invertebrates					
High	Margin	No	Invertebrates	Eratosthenes seamount	Galil and Zibrowius (1998)
High	None	No	Bamboo coral	New Zealand seamounts	Smith *et al.* (2004)
High	None	No	Epsilonematidae) (Nematoda)	Great Meteor	Gad (2004a)
Higher	Deep sea	Yes	Loricifera	Great Meteor	Gad (2004b)
Higher (measured by 'taxonomic richness')	Margin and deep sea	Yes	Bryozoa	New Zealand region	Rowden *et al.* (2004)
Low	Margin	No	Invertebrates	Cobb	Parker and Tunnicliffe (1994)
Lower	Margin	Yes	Ophiuroidea	Great Meteor and Josephine	Bartsch (1991)
Lower	Deep sea	Yes	Primarily invertebrate megafauna (some fish)	Cross	
Similar	Deep sea	Yes	Benthic forams	Great Meteor	Heinz *et al.* (2004)
Similar	Margin	No?	Gorgonaria and Antipatharia	Great Meteor and Josephine	Grasshoff (1985)
Similar or higher	Margin	Yes	Azooxanthellate Scleractinia	Norfolk Ridge seamounts: Gascyone, Taupo, Derwent Hunter, Elizabeth Reef, Britannia, Gifford, Argo, Nova	Cairns (2004)
Pelagic fishes					
Lower	Deep sea	Yes	Fishes (mesopelagic)	Atlantis and Great Meteor	Pusch *et al.* (2004)
Higher	Deep sea	Yes	Ichthyoplankton	Great Meteor	Nellen and Ruseler (2004)
Pelagic invertebrates					
Lower – but see notes	Deep sea	Yes	Euphausiids	Great Meteor	Weigmann (1974)
Fishes – depth zone unknown or mixed					
Low	None	No	Fishes (shallow)	Walters Shoals	Collette and Parin (1991)
Lower	Margin	No	Fishes	Vema seamount	Penrith (1967)
Invertebrate – depth zone unknown or mixed					
High	None	No	Caridea (Decapoda)	Mid-Pacific Mountains: Horizon, Hess, Hamilton, Agassiz, Unnamed guyot (=SIO), Darwin, unnamed seamount	Allen and Butler (1994)
Lower	Margin	No	Cephalopods	Walters Shoals	Nesis (1994)
Lower	Margin	Yes	Shrimp	Walters Shoal	Burukovskiy (1992)

Comparison indicates whether the comparison is to the surrounding deep sea, to the nearest continental margin, or if a general statement is given about the diversity (e.g., 'unusually high diversity') without an explicit comparison. *Effort consid.* indicates whether differences in sampling effort and method between the seamount and the comparison habitat were considered. *Yes* indicates some consideration or discussion, though it may not be a statistically rigorous treatment. Data are categorized by whether the sampled taxa are predominately pelagic or not. We use the term *benthic associated* in the widest interpretation: it includes demersal, near-bottom, and 'benthopelagic' groups. Diversity is generally measured as species richness (compiled mid-2005).

species are highly abundant and all the others are rare is considered less even and thus less diverse than one with the same number of species, but roughly even abundances of most of the species. Evenness on seamounts has not been rigorously examined, but many examples exist of seamounts dominated by one or a few very abundant species (e.g., Parin *et al.*, 1997; Gad, 2004b; Tracey *et al.*, 2004), a factor well known by commercial fisherman targeting dense schools of commercial fishes over seamounts.

Endemism

Endemics are species found in only a limited geographic region. With respect to seamounts, endemics are defined here as species that are found on one seamount or one chain of seamounts and nowhere else in the oceans. In discussions of endemism it is important to remember that the true rates of endemism can never be known: just because a species has been found only on one seamount to date, one cannot be certain it is not present elsewhere and simply has not been sampled. Thus, published rates of endemism are best considered as imperfect estimates of endemism; values may decrease as species ranges are more fully described, or may increase as cryptic species are differentiated. The uncertainty over levels of endemism is greatest for those taxa that have been least studied globally.

One of the drivers of seamount research has been the finding of highly endemic communities on some seamounts. Wilson and Kaufmann (1987) reviewed species lists for seamounts worldwide to create the first overview of global rates of endemism on seamounts. They concluded that 11.6% of fishes and 15.4% of invertebrates reported from seamounts were endemic. These data covered 449 species of fish and 596 species of invertebrates from 100 seamounts, although, because of uneven sampling, 72% of the data came from just 5 seamounts.

Since Wilson and Kaufmann, two major studies have raised the potential for far higher rates of endemism. Richer de Forges *et al.* (2000) describe intense sampling efforts on three groups of seamounts in the SW Pacific: the Norfolk Ridge seamounts, the Lord Howe seamounts, and the Tasmanian seamounts (see Chapter 5). They found that 29–34% of all fish and invertebrates captured were new to science and potential endemics. A total of 516 species were collected, which represents an almost 50% increase in known seamount species over Wilson and Kaufmann's (1987) review. In a less-cited work, Parin *et al.* (1997) summarized multiple Soviet expeditions to 22 seamounts on the Nazca and Sala-y-Gómez chains in the southeast Pacific. They found endemism rates of 51% for bottom/near-bottom invertebrates and 44% for fishes. The number of species this represents is substantial: 192 species of invertebrates and 171 species of fishes.

A later global summary of seamount fish data, however, found lower endemism. Froese and Sampang (2004) collected data on fishes from 60 seamounts globally and found that 12% were endemic. They defined endemics as species recorded from a single seamount; this is more restrictive than the more common definition that allows endemic species to be found on several seamounts of a single seamount chain or group, so it may not contradict the higher rates reported for chain-level endemism. Note also that the data in Froese and Sampang (2004) partially overlap with the data in Wilson and Kaufmann (1987).

Since Wilson and Kaufmann (1987), there have been a number of published rates of endemism for particular seamounts or restricted taxonomic groups. As shown in Table 13.2,

Table 13.2 Published levels of seamount endemism (compiled mid-2005).

% Endemism	Taxa	Seamount(s)	Reference
Invertebrates – bottom associated			
0% of 15 spp.	Brachiopods	NE/North Central Atlantic: 6 seamounts	Gaspard (2003)
0% of 8 spp.	Brachiopods	NE/North Central Atlantic: 7 seamounts	Logan (1998)
0% of 20 spp.	Forams (living)	NE/North Central Atlantic: Great Meteor	Heinz *et al.* (2004)
0% of 15 spp.	Gorgonaria and Antipatharia	NE/North Central Atlantic: Great Meteor and Josephine	Grasshoff (1985)
0% of 3 spp.	Hydroids	NW Atlantic: Argus, Bowditch, Challenger	Calder (2000)
0% of 18 spp.	Madreporaria coral	Atlantic: Great Meteor, Atlantis, Rockaway, Reykjanes ridge	Keller (1985)
0% of 31 spp.	Molluscs	NE/North Central Atlantic: Ormonde	Avila and Malaquias (2003)
0% of 16 spp.	Tonnoidea (Gastropoda)	NE/North Central Atlantic: 10 seamounts	Gofas and Beu (2002)
Low or none of 95 spp.	Invertebrates	NE Pacific: Cobb	Parker and Tunnicliffe (1994)
2% of 57 spp.	Azooxanthellate Scleractinia	SW Pacific: 8 Norfolk Ridge seamounts	Cairns (2004)
6% of 53 spp.	Polychaetes	NE/North Central Atlantic: Atlantis, Hyeres, Josephine, Meteor	Gillet and Dauvin (2000)
6–75% of 84 spp.	Polychaetes	NE/North Central Atlantic: Plato, Irving, Meteor, Hyeres, Atlantis	Gillet and Dauvin (2003)
7% of 13 spp.	Tunicates	NE/North Central Atlantic: 5 seamounts	Monniot and Monniot (1992)
9% of 11 spp.	Hydrozoans	NE/North Central Atlantic: 6 seamounts	Ramil *et al.* (1998)
10% of 10 spp.	Pycnogonids	NE/North Central Atlantic: Gorringe, Josephine, Seine, Ampere, Galice	Stock (1991)
11% of 9 spp.	Calanoid copepods	NE/North Central Atlantic: Great Meteor	Markhaseva and Schnack-Schiel (2003)
15%+ of 414 spp.	Benthic megafauna	SW Pacific: New Zealand hills	Rowden *et al.* (2002)
25% of 4 spp.	Fasciolariidae (Gastropoda)	NE/North Central Atlantic: 11 seamounts	Gofas (2000)
43% of 7 spp.	Ascidians	SE Atlantic: Vema	Primo and Vazquez (2004)
51% of 171 spp.	Invertebrates	SE Pacific: 22 seamounts, Nazca and Sala-y-Gómez chains	Parin *et al.* (1997)
59%+ of 17 spp.	Epsilonematidae (Nematoda)	NE/North Central Atlantic: Great Meteor	Gad (2004a)
62% of 8 spp.	Madreporaria coral	Pacific: Marcus-Necker ridge, Hawaii ridge	Keller (1985)
63% of 27 spp.	Pyramidellidae (Gastropoda)	NE/North Central Atlantic: Great Meteor, Hyeres, Irving, Atlantis	Penas and Rolan (1999)
94% of 33 spp.	Harpacticoid copepods	NE/North Central Atlantic: Great Meteor	George and Schminke (2002)
100% of 12 spp.	Loricifera	NE/North Central Atlantic: Great Meteor	Gad (2004b)
100% of 10 spp.	Trituba (Gastropoda)	NE/North Central Atlantic: 5 seamounts	Gofas (2002)
'Highly endemic' of ? spp.	Meiofauna	NE/North Central Atlantic: Great Meteor	Gad and Schminke (2004)

(*Continued*)

Table 13.2 *Continued*

% Endemism	Taxa	Seamount(s)	Reference
Invertebrates – mixed benthic and pelagic, or bottom-association unknown			
0–5% of 52 or 53 spp.	Cephalopods	Indian: Walters Shoal	Nesis (1994)
2% of 45 spp.	Cephalopods	Indian: Fred, Error, Equator, Farquhar	Nesis (1986)
6% of 33 spp.	Caridea (Decapoda)	Central/NW Pacific: 8 seamounts	Allen and Butler (1994)
7% of 29 spp.	Shrimp	Indian: Walters Shoal	Burukovskiy (1992)
15% of 596 spp.	Invertebrates	Global: 100 seamounts worldwide	Wilson and Kaufmann (1987)
36% of 11 spp.	Anomuran and Brachyuran decapods	Central/NW Pacific: Nintoku, Jingu, Kinmei	Sakai (1980)
0% of 31 spp. (but not all identified to species)	Cephalopods	NE/North Central Atlantic: Great Meteor	Diekmann and Piatkowski (2004)
3% of 39 spp.	Shrimp: Penaeidea and Caridea	NE Pacific: Erben, Fieberling, 350 seamounts, Dowd	Hanamura (1983)
Fishes – bottom associated			
0% of 15 spp.	Fishes (deep water)	Central/NW Pacific: Cross	Chave and Mundy (1994)
21% of 28 spp.	Fishes	Pacific: multiple seamounts	Fujii (1986)
44% of 192 spp.	Fishes	SE Pacific: 22 seamounts on Nazca and Sala-y-Gómez	Parin *et al.* (1997)
Fishes – mixed benthic and pelagic, or bottom-association unknown			
0% of 9 spp. (only 'unusual' species considered)	Fishes	SE Atlantic: Bonaparte	Edwards (1993)
10–20% of 20 spp.	Fishes (shallow water)	Indian: Walters Shoal	Collette and Parin (1991)
12% of 449 spp.	Fishes	Global: 100 seamounts worldwide	Wilson and Kaufmann (1987)
12% of 535 spp. – single seamount endemism	Fishes	Global: 60 seamounts	Froese and Sampang (2004)
Fishes and invertebrates – bottom associated			
11% of 19+ spp.	Fishes and invertebrates	Central/NW Pacific: Horizon Magelland Rise	Kaufmann *et al.* (1989)
Fishes and invertebrates – mixed benthic and pelagic, or bottom-association unknown			
29–34% of 516 spp.	Fish and mega-invertebrates	SW Pacific: 6 Norfolk Ridge, 4 Lord Howe, and 14 Tasmanian seamounts	Richer de Forges *et al.* (2000)

Only papers not in Wilson and Kaufmann (1987) are included. Data are categorized by whether the sampled taxa are predominately pelagic or not. We use the term 'benthic associated' in the widest interpretation: it includes demersal, near-bottom, and 'benthopelagic' groups.

the reported rates of endemism vary between 0% and 100%, though it should be noted that the extreme values are often for very small studies of just a few species, or for very poorly known groups such as the Loricifera. The studies also vary in how they counted endemism. Together, these studies give an average endemism of just under 20% when studies are weighted by the total number of species.

These rates of endemism need to be considered in comparison to other marine habitats. Though there is much variability, seamounts have a range of endemism on a par with marine communities on isolated islands: e.g., 3–28% for shorefishes of the central Atlantic and eastern Pacific Islands (Robertson, 2001) and 18–24% for molluscs on isolated tropical Pacific Islands and archipelagos (Kay, 1979; Rehder, 1980; Brook, 1998). Seamount rates appear lower than those for hydrothermal vent fauna, another distinct and isolated deep-sea community: of 443 vent species report in Tunnicliffe *et al.*'s (1998) review, 61% have been recorded only from vents in one geographic region, with an additional 6% where the distribution is unclear. This count did not include 32 species categorized as 'vagrants' or 'visitors' observed at vents.

It is important to remember that these data represent a small and biased sampling of seamounts. Taxonomically, fishes are covered best, followed by molluscs, and then crustaceans. Representation of other groups is scattered and sparse. One pattern that does emerge to date is the suggestion of lower endemism in fishes than benthic invertebrates. Wilson and Kaufmann (1987) found this in their review, as did Parin *et al.* (1997) in their extensive SE Pacific work. None of the fish studies in Table 13.2 found rates of endemism over 50%, while seven of the invertebrate studies had rates this high. Whether this reflects a real pattern in endemism, or merely our more advanced taxonomic and biogeographic knowledge of the fishes, is not clear. It is not a surprising pattern, though, considering that adult fishes are more mobile than adult benthic invertebrates, and thus may have more genetic mixing with off-seamount populations.

Geographically, the most comprehensive data come from the NE Atlantic, the SW Pacific (including work on the Norfolk, Lord Howe, Tasmanian, and New Zealand seamounts), and the Nazca and Sala-y-Gómez chains in the SE Pacific. Limited data are available for the SE Atlantic (primarily from Vema seamount), the central Pacific (Hawaiian, Emperor, and Mid-Pacific Mountains), a couple of seamounts from the Indian Ocean, and a few isolated reports. Major gaps exist in the low-latitude seamounts of all oceans, the Arctic and Antarctic, the western Atlantic, most of the Indian, the Indonesian–Philippine region, and the south central Pacific, though expeditions underway at the time of writing will bring new information for the NW Atlantic, Alaskan Seamounts, Hawaiian chain, and Antarctic region. Overall, the existing data are too patchy for geographic patterns such as relationships to latitude, longitude, or water mass in endemism to be examined.

Despite these gaps, two interesting points emerge that are relevant to assessing the applicability of existing marine biogeographic theory to seamounts. First, the Nazca and Sala-y-Gómez seamounts and the Hawaiian/Emperor seamounts are both near the outer limits of the Indo-West Pacific (IWP) region, but vary in their endemism. The Nazca and Sala-y-Gómez area has high endemism (44–52%; Parin *et al.*, 1997), but the Hawaiian/ Emperor seamounts are relatively low, as least with respect to the fish fauna (5–12%; Stocks, 2004). Mora *et al.* (2003) found that shorefish endemism in the IWP region increased with distance from the centre of diversity in Indonesia and the Philippines and predicted this would be a general biogeographic pattern. While it is not wise to draw firm conclusions from a sample of two seamount chains, the existing seamount data do not support this conclusion.

Second, of the seamount areas that have received community surveys, the two with the highest endemism are the Nazca and Sala-y-Gómez seamounts and the Norfolk Ridge chain. The Nazca and Sala-y-Gómez seamounts are a long chain of tall seamounts that vary greatly in depth, are isolated from the nearest continental margin by a deep trench, and are at the furthest extreme of the IWP (Parin *et al.*, 1997). The Norfolk seamounts are a few small seamounts and hills in the SW Pacific near New Caledonia, quite close to nearby islands and closer to the centre of IWP diversity. In short, these two areas vary in the primary characteristics thought to influence endemism. It looks likely, therefore, that endemism on seamounts will not be predictable from any single, simple, factor, but will need more complex models.

In conclusion, seamounts span a spectrum of endemicity. While improvements in taxonomy and sampling in the deep sea in general might substantially change our current views of the magnitude and patterns in seamount endemicity, there is evidence that at least some seamounts or seamount chains may support unique species. This argues that seamounts host, to some degree, isolated and distinct communities. At present, we are not able to generalize the patterns in endemism to predict from sampled seamounts to the approximately 100 000 unsampled seamounts. Some progress in this area could be made with further analysis of existing data. The references cited in Table 13.2 are restricted to studies where the authors reported endemism levels; many more studies have reported lists of species found on seamounts. A better perspective on endemism could be developed by examining the known distribution of each of the species reported from a seamount to determine which could be endemic. The highest potential for future progress is held by sampling using standardized methods including both genetic and morphological identification in different regions and along gradients of depth, distance, and other variables.

Population isolation

Another approach to considering the level of isolation on seamounts is to ask whether species that are not endemic to seamounts have genetic exchange with conspecifics (individuals of the same species) living off seamounts, or whether the seamounts support distinct populations. There is ample evidence that large, mobile, pelagic species such as tuna, turtles, marine mammals, and seabirds actively move between seamount and non-seamount areas (see Chapter 10; Hui, 1985; Blaber, 1986; Haney *et al.*, 1995; Yen *et al.*, 2004). Beyond these taxa, the picture is less clear. This can be evaluated through several methods: tracking individual organisms to see if they move between seamount and non-seamount areas, looking for genetic differentiation between seamount and non-seamount conspecifics, and examining the life history characteristics of seamount species. Morphology has also been used to look for differentiation between seamount and non-seamount conspecifics (see Rogers, 1994). However, it can be difficult or impossible to determine with morphology alone whether or to what degree any differentiation seen is due to genetic isolation or simply to different phenotypes emerging in response to environmental differences.

Estimating population isolation on seamounts is a question of degree. Widely distributed species are found in the oceans, including on seamounts, indicating that the continuous fluid environment of the ocean can transport species globally. One example is fishes of the family Liparidae (snailfishes) which are thought to have originated in the North Pacific (Andriashev, 1986) and dispersed south to the Antarctic where this family now forms 31% of the fish fauna. Conversely, there is also ample evidence that the oceans are not fully mixed with respect to dispersal: biogoegraphic boundaries, species invasions, hotspots, and genetic isolation do exist (Ekman, 1953; Creasey and Rogers, 1999; Hellberg *et al.*, 2002; Briggs, 2003).

Genetic evidence

Genetic studies can provide compelling evidence regarding population isolation but, to date, few have been published from seamounts. Three studies of fishes found no genetic differentiation between seamount and nearby non-seamount populations: the trevalla *Hyperoglyphe antarctica* around Australia (Bolch *et al.*, 1994), a boarfish species (Pentacerotidae) on the Hawaiian chain (Borets, 1980), and the alfonsino *Beryx splendens* in the southwest Pacific (Hoarau and Borsa, 2000). Two other studies looked at the genetics of seamount populations and found no differentiation, but could not be certain that the sequences used had sufficient mutation rates for recent speciation to be found: France and Hoover (2002) for octocorals of the subclass Octocorallia, and Smith *et al.* (2004) for the bamboo coral *Keratoisis*, also an octocoral. Bucklin *et al.* (1987) compared the amphipod *Eurythenes gryllus* on the top and base of Horizon seamount to individuals collected at widely separated deep-sea sites and found that all the deep-sea specimens were similar, but the seamount specimens were distinct. However, France and Kocher's (1996) work on a larger set of samples from this species suggests the alternative interpretation that this species is divergent across depth zones, but homogeneous within depth zones even over large geographic areas. There have also been genetic studies on orange roughy (Smith *et al.*, 1986, 2002), spiny lobster (Gopal *et al.*, 2006), Patagonian toothfish (Rogers *et al.*, 2006), and the deep-sea fish *Helicolenus* (Aboim *et al.*, 2005), none of which have supported endemism.

None of these studies have found convincing evidence of genetic isolation on seamount populations; indeed seamount endemism has been questioned by the population genetic structure of marine invertebrates (Samadi *et al.*, 2006). It is difficult to reconcile the lack of genetic differentiation with the common occurrence of endemism on seamounts. If one assumes that endemicity is the result of local speciation and not relict populations, then genetic isolation should be present. It is possible that these few studies, which are biased towards more mobile fishes, are not representative of seamount communities in general.

Dispersal characteristics

Another way to look at population isolation on seamounts is through examining life history characteristics. Models for the recruitment of seamount species are suggested by

Rogers (1994; Fig. 4). He proposes four mechanisms for recruitment, each of which has some evidence supporting it:

1. *Immigration from non-seamount areas.* The lack of genetic differentiation found in the studies reported above indicates that some external recruitment must occur in some species. However, the isolation of many seamounts from areas of similar depth and habitat type argues that entirely external recruitment is unlikely; some young produced locally on a seamount are likely to settle on that seamount.
2. *Eggs and larvae are transported away from the seamount but juveniles recruit back to it.* The armourhead *Pseudopentaceros wheeleri* is a species known to use this strategy, with larvae dispersing widely but juveniles recruiting back to the seamount after about 2 years (Boehlert and Sasaki, 1988).
3. *Larvae are retained over the seamount by hydrographic conditions and remain on their seamount of origin.* As discussed in Chapter 4, recirculating Taylor cones or columns have been recorded over seamounts and could serve to retain larvae over seamounts. While there is doubt whether these features are persistent enough through time to facilitate local resettlement of larvae, there is evidence of larval trapping and resettlement over seamounts. Dower and Perry (2001) found elevated concentrations of fish larvae from seamount species over, compared to away from, a seamount that had a Taylor column, implying that retention can occur. Mullineaux and Mills (1997) also found local concentrations of planktonic larvae, and additionally documented that settlement plates placed *in situ* on around Fieberling Guyot for 3–6 months showed a larval settlement pattern consistent with larval retention in the circulation cell observed over the seamount.
4. *Larval behaviour retains larvae over their seamount of origin.* Little work has been done on the larval behaviour of seamount species in particular. In other habitats, larvae have been found to move vertically in the water column to affect where they settled and how they disperse (Young, 1995), or post-larvae can attach to floating material like kelp to assist dispersal (Havenhand, 1995; and references therein). Multiple examples have been found of small-scale genetic isolation in species with larvae capable of long-distance dispersal (Taylor and Hellberg, 2003; references in Palumbi and Warner, 2003), or species with no free larvae having wide distributions (Johannesson, 1988): presumably, behaviour must modify the dispersal potential of these species.

Researchers considering whether seamount populations have external or local recruitment have also examined the kind of larvae seamount species produce. The hypothesis is that larvae having no planktonic phase or a very short one are unlikely to be carried away from their seamount of origin. Parker and Tunnicliffe (1994) catalogued the presence and length of the pelagic larval phase in all of the species collected from Cobb Seamount in the NE Pacific and compared them to coastal communities at a similar depth. They found that species with direct development (no free larval stage for dispersal) and species with short-lived larvae were more common on the seamount than those with medium or long-lived larvae, supporting the hypothesis. Calder (2000) found a similar result for hydroids: 71% of the species found on seamounts near Bermuda have fixed sporosacs (i.e., no free medusa stage), compared with only 17% with free or short-lived medusa. In contrast, two other studies (Gillet and Dauvin, 2000, for polychaetes, and Avila and Malaquias, 2003, for molluscs) found no pattern with respect to direct vs planktotrophic development.

Recent studies on the origin of biodiversity in the ocean present an entirely different perspective on how life history characteristics influence seamount communities. They propose that certain areas in the ocean are speciation hotspots which act as a source of species that are distributed widely (Briggs, 1995, 2003; Mora *et al.*, 2003). A study by Mora *et al.* (2003) examined the large-scale patterns in the distributions of reef fishes in the Indian and Pacific oceans. Although reefs are not seamounts, these findings can infer what might be expected on seamounts if similar processes are at work. The first part of the study established that the Indonesian and Philippine region is likely to be a centre for the origin of new species of reef fishes. The study then examined data on the larval duration of Labrid and Pomacentrid fishes and found that it supported the hypothesis that species found further from the hotspot are more likely to have longer-lived larvae. If seamounts prove to be isolated communities, then one might expect a higher, not a lower, proportion of dispersal-adapted species.

Gofas and Beu (2002) found support for this trend in their study of tonnoidean gastropods on a suite of NE Atlantic seamounts. They found that species with teleplanic (extremely long-lived) larvae were much more common on these seamounts than in continental margin populations, even though some of the species appear to be endemic or have isolated populations. They further show that some of the now-isolated populations had much wider distributions in the fossil record, indicating their capacity for long-distance dispersal. They hypothesize that the observed isolation of seamount populations results from narrow environmental requirements and not the inability to disperse.

As mentioned above, behaviour can also influence the actual dispersal of a population. In light of this, and the lack of a consistent correlation between larval duration and dispersal, larval type is perhaps best viewed as just one of the several factors influencing how open or closed a population is (Havenhand, 1995).

It is also important to remember that the debate over whether seamount populations are self-recruiting or open to external recruits depends on whether one has an ecological or evolutionary/biogeographic perspective. A seamount population that receives only a few external recruits per year can be said to be self-recruiting from an ecological perspective: clearly the population can maintain itself. But a relatively few recruits might be sufficient to ensure genetic mixing with the external members of the community, and also allow a distant site to serve as a source of immigrating species over time.

Establishing a biogeography of seamounts

The biogeography of the ocean as we sample it will be a snapshot of a dynamic process in which species are evolving adaptations to local conditions (Briggs, 2003). The most desirable approach to estimating the present distributions of the ocean's flora and fauna is to obtain detailed data on the distributions of the major species of plants and animals, and to analyse these distributions to discern their spatial patterns and the variation in community structure. In terrestrial systems this has produced well-defined biomes that are separated from each other by clear discontinuities. The ocean, however, presents unique challenges.

Terrestrial biogeography is primarily based on the distribution of sessile plant communities, which are correlated with global climatic patterns (MacArthur, 1972). In contrast, most primary producers in the open ocean are microscopic in size, have short generation times and are much more temporally variable. The major vegetation types on land can be remotely sensed, but this is not true in the oceans. Obtaining synoptic data with wide spatial coverage on marine communities requires expensive *in situ* sampling, and difficult taxonomic analysis needing a significant investment in the training of personnel. The pelagic components of the ocean bring the further challenge that they are not rooted in place, but rather live in a fluid environment where boundaries are less distinct and more variable than on land.

Despite the challenges, biogeographic regionalizations have been proposed in the marine realm. Multiple hotspots of diversity have been identified in shallow-water areas (Briggs, 2003). Mora *et al.* (2003) used data from the IWP region to demonstrate that diversity decreased with distance away from the IPR centre of this region, and that this pattern was more consistent with the IPR being a centre of speciation than with the competing hypotheses that it has high diversity because it is an area where several other regions overlap, or that hydrographically it acts as a passive accumulator of species evolved elsewhere. In the pelagic realm, various biome schemes have been developed, which are discussed more fully below. For the deep seafloor, biogeographic boundaries have been proposed (Briggs, 1995; Vinogradova, 1997), but evaluations are still in the preliminary stages (Levin *et al.*, 2001).

If biogeographic regions have any value, they should be able to support predictions about the types of communities to be found on seamounts. Because different models and states of knowledge exist for benthic and pelagic communities, they are discussed separately below.

Biogeographic affinities of benthic invertebrate communities on seamounts

Wilson and Kaufmann (1987) concluded that the majority of seamount communities are most closely affiliated with the nearest continental margin. Later studies (Table 13.3) do not obviously contradict this finding. When endemic species are not considered, seamount benthic invertebrate communities seem similar to the nearby margins or, for isolated seamounts, to nearby islands. Of 24 studies reviewed in Table 13.3, just two invertebrate surveys and one macroalgal study (Cecere and Perrone, 1988; Markhaseva and Schnack-Schiel, 2003; Primo and Vazquez, 2004) found communities that were not typical of the general biogeographic region. Rao and Newman (1972) and Parin *et al.* (1997) also noted that finding IWP faunas on mid- and SE Pacific seamounts suggest that these areas properly belong in the IWP region (Randall, 1998; Mora *et al.*, 2003).

The remainder of the studies fall into two groups. Some explicitly reported that the seamount faunas found were similar to nearby non-seamount regions. Others that the seamounts supported a mix of cosmopolitan species and those limited to the nearby margin. Without direct comparison, it is possible that this latter group may have different proportions of local vs widespread species, but as reported they do not contradict an affinity to the nearby regional species pool. Though several range extensions or exceptions were noted (e.g., Kensley, 1981; Nesis, 1994; Gofas and Beu, 2002), they did not change the

Table 13.3 Biogeographic affinities for non-endemic seamount algae and invertebrates (compiled mid-2005).

Region	Seamount(s)	Taxa	Reference
Algae – bottom associated			
Mediterranean	• Amendolara (Ionian Sea) Fauna: Allied with Atlantic-Boreal and Mediterranean, not West Mediterranean province as expected	Macroalgae (benthic)	Cecere and Perrone (1988)
Invertebrates – bottom associated			
NE Pacific	• Gulf of Alaska seamounts: Dickins, Welker, Quinn, Giacomini, Patton, Durgin, Pratt, Applequist, Surveyor Fauna: Typical of nearby slopes	Commercial fish and crabs	Alton (1986)
	• Cobb Fauna: Nearly all species also found on nearby pacific coast	Invertebrates	Parker and Tunnicliffe (1994)
SW Pacific	• Norfolk Ridge seamounts: Gascyone, Taupo, Derwent Hunter, Elizabeth Reef, Britannia, Gifford, Argo, Nova Fauna: Similar to nearby margin	Azooxanthellate Scleractinia	Cairns (2004)
NE/N Central Atlantic	• Ormonde Fauna: Similar to nearby margins and Mediterranean	Molluscs (mainly benthic)	Avila and Malaquias (2003)
	• Great Meteor Fauna: Mainly similar to nearby margins	Meiofauna	Gad and Schminke (2004)
	• Gorringe, Galicia, Josephine, Ampere, Lion, Seine Fauna: Most species found in nearby non-seamount areas or are widespread	Brachiopods	Gaspard (2003)
	• Atlantis, Hyeres, Josephine, Meteor Fauna: Mainly widespread or cosmopolitan species	Polychaetes	Gillet and Dauvin (2000)
	• Plato, Irving, Meteor, Hyeres, Atlantis Fauna: Similar to nearby margins and islands	Polychaetes (benthic)	Gillet and Dauvin (2003)
	• Gorringe, Josephine, Seine, Ampere, Meteor, Hyeres, Irving, Plato, Atlantis, Tyro Fauna: 25% spp. unknown or rare on nearby margin; rest were on nearby margin	Tonnoidea (gastropods)	Gofas and Beu (2002)
	• Great Meteor and Josephine Fauna: Typical of eastern Atlantic	Gorgonaria and Antipatharia	Grasshoff (1985)
	• Great Meteor Fauna: Most species on summit were also found in deep sea	Living forams	Heinz *et al.* (2004)
	• Meteor, Hyeres, Cruiser (=Irving), Plato, Atlantis, Tyro, Antialtair Fauna: Typical of nearby continental margin	Brachiopods	Logan (1998)

(*Continued*)

Table 13.3 *Continued*

Region	Seamount(s)	Taxa	Reference
	• Great Meteor Fauna: Strong affinity with Norway (4 of 9 taxa)	Calanoid copepods	Markhaseva and Schnack-Schiel (2003)
	• Gorringe, Josephine, Seine, Ampere, Galice Fauna: 10 of 13 spp. known from nearby margin	Tunicates	Monniot and Monniot (1992)
	• Gorringe, Josephine, Seine, Ampere, Galice Fauna: 9 of 10 spp. found in 'neighbouring areas'	Pycnogonids	Stock (1991)
NW Atlantic	• Argus, Bowditch, Challenger Fauna: Most species found elsewhere in western Atlantic tropical region	Hydroids	Calder (2000)
SE Atlantic	• Vema Fauna: Very distinct from South African coast and nearby islands	Ascidians	Primo and Vazquez (2004)
Indian	• Walters Shoals Fauna: Most species known from region, but some exceptions	Cephalopods	Nesis (1994)
Invertebrates – depth zone unknown or multiple			
Indian	• Walters Shoal Fauna: Mix of widespread and nearby coastal species	Shrimp	Burukovskiy (1992)
	• Fred, Error, Equator, Farquhar Fauna: All species found on nearby shelves and slopes	Cephalopods	Nesis (1986)
Fish and invertebrates – bottom associated			
Central/NW Pacific	• Horizon Magellan Rise Fauna: Most species widespread	Fish and invertebrates	Kaufmann *et al.* (1989)
	• Great Meteor Fauna: Most species widespread and also in North Atlantic and/or Mediterranean	Megafauna (mainly invertebrate)	Piepenburg and Muller (2004)
SE Pacific	• 22 seamounts, Nazca and Sala-y-Gómez chains Fauna: Most species closely affiliated with IWP	Invertebrates and fishes	Parin *et al.* (1997)

Data are categorized by whether the sampled taxa are predominately pelagic or not. We use the term benthic associated in the widest interpretation: it includes demersal, near-bottom, and 'benthopelagic' groups.

overall affiliation of the fauna. If it holds with further study, this affinity of seamount and local region fauna suggests that theory developed for non-seamount habitats might also apply to seamount biogeography. That said, the prevalence of endemic species on seamounts, and the variability in species from seamount to seamount within a region based on seamount-specific factors like depth and substrate type mean that bioregionalizations have limited ability to predict the benthic community on any particular seamount.

Influence of pelagic biogeography on planktonic biota over seamounts

A biogeography of the pelagic realm of the ocean must use different criteria to those used for terrestrial systems and perhaps also for benthic marine organisms (Briggs, 1995; Longhurst, 1998). Pelagic biogeography has to be related more to the three-dimensional structure of the upper ocean and distributions need to be defined in terms of current systems which may well move in different directions at different depths.

The only dataset that comes close to achieving a synoptic view of the distribution of organisms in the pelagic zone is that gathered by the Continuous Plankton Recorder (CPR) Survey, which has been running since 1947 in the North Atlantic (Barnard *et al.*, 2004; Beaugrand, 2004; Edwards, 2004). These datasets have been used to define communities with spatial limits and characteristic sets of plants and animals (Colebrook, 1979; Beaugrand *et al.*, 2002). The latter study defines ecotones (boundaries between ecological regions) based on the distribution of calanoid copepods over a 40-year period. Other attempts at an ocean-wide biogeography have been tried by using a restricted set of species, e.g., euphausiids (Brinton, 1962; Mauchline and Fisher, 1969; Gibbons, 1997) and calanoid copepods (Mauchline, 1998; Woodd-Walker *et al.*, 2002).

Biological oceanographers have also tried to define the limits of pelagic biomes using physical oceanographic data. Variables such as temperature, salinity, and pressure are easier to measure with electronic devices and the coming of satellite data has ramped up the level of spatial and temporal coverage. In particular, it is now possible to obtain near synoptic data on surface colour which can be used to estimate chlorophyll pigment concentration and therefore phytoplankton biomass.

Several such biome schemes exist. One is the large marine ecosystem (LME) approach of Sherman (1991, 1992) where LMEs for coastal areas are based on bathymetry, hydrography, productivity, and trophically dependent populations. Another is Briggs' (1995) classification of the ocean into five broad biogeographic regions: cold or polar, boreal, warm, temperate, and tropical. A third, and more detailed, is Longhurst's (1998) delineation and mapping of pelagic ecosystems based on productivity regime, developed using extensive satellite data and 26 000 profiles of primary production. The key variables in Longhurst's scheme are shown in Table 13.4 and his conceptual approach is outlined in Fig. 13.1. Longhurst's scheme has been supported, and refined through, work by Beaugrand *et al.* (2002) on calanoid copepod biogeography, by Woodd-Walker *et al.*'s (2002) study of copepod distributions along the North–South length of the Atlantic, and by study of the presence/absence distribution of euphausiids in the Southern Ocean (Gibbons, 1997).

The Longhurst biogeography (see Longhurst, 1998, endpapers, for maps) predicts that the planktonic fauna and flora should reflect that of the biome and province in which the seamount is situated. So, the patterns of the main zooplankton groups shown in

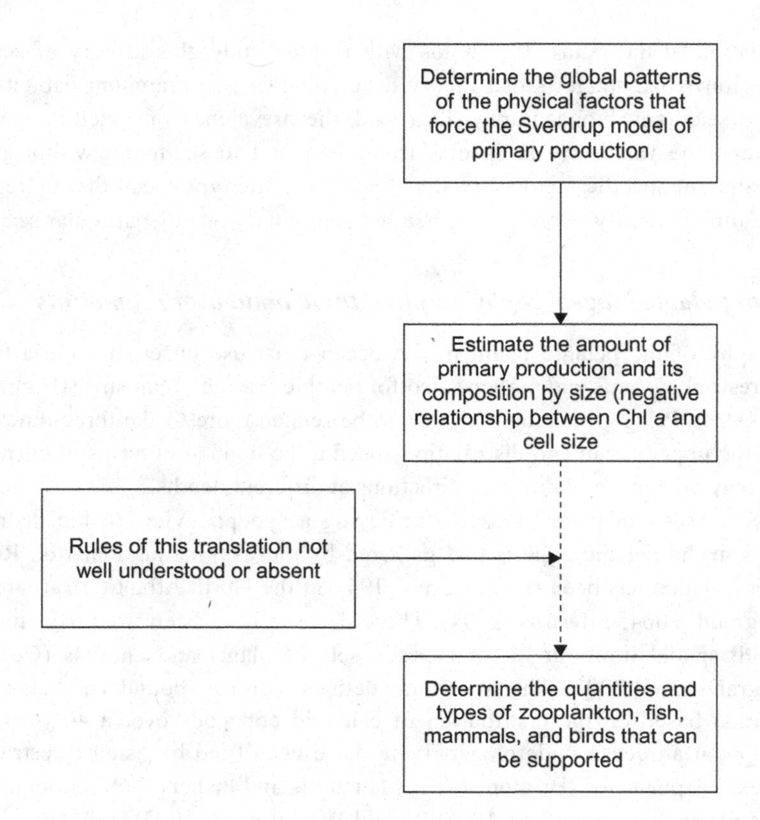

Fig. 13.1 The hypothesis behind Longhurst's (1998) delineation of pelagic biomes.

Table 13.4 The environmental variables required to provide a complete picture of how the Sverdrup model of primary production will change with global position.

Variable	Comment
Latitude	Determines if wind stress induces mixing or momentum
Depth of water	Stratification may be overturned in shallow water by tidal stress
Proximity of coastline	Effects of river effluents and coastal upwelling
Season irradiance cycle	Forces photosynthesis, stratification, and ice
Local wind regime	Forces mixing or vertical Ekman transport of nutrients
Local precipitation regime	May induce near-surface stratification
Distant forcing of pycnocline depth	Modifies effects of local wind mixing
Nutrients in intermediate water mass	Modifies upward nutrient flux
External sources of Fe	Lack of which may constrain nutrient uptake

From Longhurst (1998, p. 43).

Fig. 13.2 should be found reflected in the waters overlying the seamounts in the respective biomes. As the evidence indicates that the biomass of phytoplankton and zooplankton can be higher over seamounts or downstream of them (Rogers, 1994; Parin *et al.*, 1997), we might expect a concentration of species in the areas of seamounts but the species compositions will not necessarily be different (Genin, 2004; Valle-Levinson *et al.*, 2004). It has been shown that seamount topography can locally focus prey availability for fish and

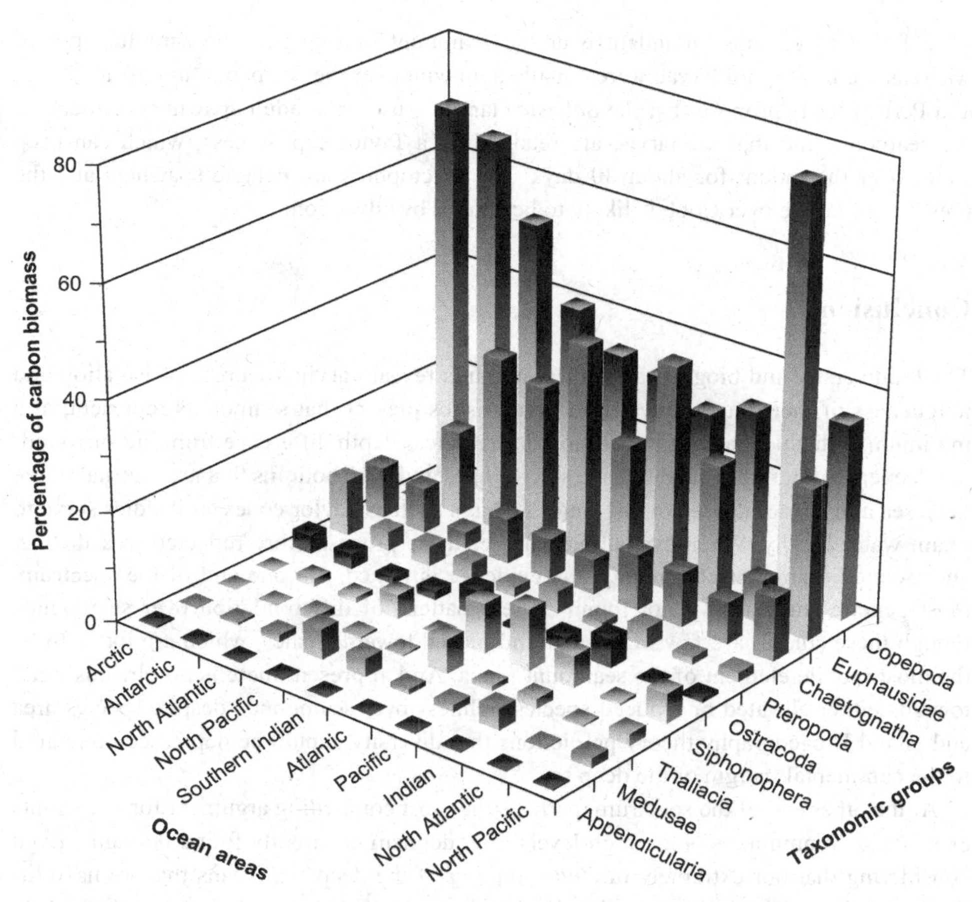

Fig. 13.2 The proportion of zooplankton carbon biomass in each taxonomic group in each ocean area: Arctic and Antarctic – polar biome; North Atlantic, North Pacific, Southern Indian – westerlies biome; Atlantic, Pacific, Indian – trades biome; North Atlantic, North Pacific – coastal biome. From Table 4.1 in Longhurst (1998). Data from 4166 stations with nets worked from 250 m to the surface and collated in the Smithsonian Institute's archive.

marine mammals, and more mobile fauna can be attracted to seamounts (see Chapters 10–12). There is however no suggestion that the focusing process can create a long-term community structure over a seamount that is different from the surrounding pelagic community. As discussed in the second part of this book, retention times of water over seamounts through a Taylor Proudman process (i.e., Taylor cone or column) may not be permanent enough to ensure a unique planktonic flora or fauna to develop over a seamount (Freeland, 1994).

An interesting study of fish over Cobb Seamount by Dower and Perry (2001) illustrates how the fish fauna can be a combination of local species and species that are representative of the biogeographic area in which the seamount is located. They found that the community of fish larvae in the waters above the seamount was divided into larvae of *Sebastes* spp. (rockfish) and larvae of representatives of the mesopelagic fish fauna of the NE Pacific. These were species belonging to the family Myctophidae. *Sebastes* larvae were found to be most abundant over the seamount during the June sampling period whereas the myctophid larvae were equally abundant over the seamount and off it. Dower and Perry (2001) propose that the *Sebastes* larvae arose from adult spawners resident on the seamount and that the larvae are retained by a Taylor cap process, which can keep water over the mount for about 30 days. The myctophids are pelagic spawners and the presence of larvae over Cobb is likely to be caused by advection.

Conclusions

The biodiversity and biogeography of seamounts reveal varying degrees of isolation and uniqueness of their fauna. Physical characteristics predict that seamounts represent, at a minimum, a distinct habitat: in addition to their clear depth difference from the surrounding deeper seafloor, they tend to have steep slopes and hard bottoms that are unusual in the deep sea and, on some seamounts, water features such as Taylor cones and eddies serve to retain water locally. When evaluating whether these processes are reflected in a distinct and isolated seamount community, the evidence is mixed. On one end of the spectrum, most genetic studies have not found a clear pattern of differentiation over seamounts, though these studies are few in number and biased towards fishes, which are likely to be the most mobile element of the seamount fauna. And at present there is no obvious trend towards either elevated or reduced species richness over seamounts, despite species-area and Island Biogeography theory predictions that diversity should be depressed compared to the continental margin or the deep sea.

At the other end of the spectrum, perhaps the most compelling argument for seamounts as isolated communities is the high levels of endemism apparently found on many. Even considering that our extremely limited sampling of the deep sea means that we have little knowledge of the true rates of endemism in most, if not all, species, it is difficult to dismiss as pure artefacts the rates of 10–50% that have been found in large-scale studies of benthic and near-bottom fauna. Examining reproductive characteristics of species from seamounts offers an intriguing method for linking endemism to processes of dispersal, though the topic is in its infancy, both on seamount and in the deep sea in general.

Findings of distinct seamount assemblages and substantial levels of endemism are largely restricted to the benthic and near-bottom organisms. What we know of pelagic biogeography leads us to conclude that the pelagic fauna of a seamount should reflect that of the region in which it is located. There might be some ephemeral concentration of species over the short term, but in general the long-term hydrography around seamounts is too unstable to retain a distinct pelagic fauna or flora. This remains to be rigorously tested empirically, but if this is the case then schemes such as Longhurst's (1998) will satisfactorily predict the composition of pelagic communities over seamounts.

The differences between the pelagic and benthic realms are important for framing seamount management and research. If, as the data suggest, the pelagic component of the seamount community is not highly distinct in composition, though abundances may vary, from surrounding waters, management scenarios should consider seamount fisheries as part of a larger regional management plan. The benthic community, however, is more likely to support species that are either not found (endemic) or rare in the larger deep sea or continental margins, and which may have some degree of local recruitment. In this case, recognizing seamounts as a distinct habitat to manage, though one not entirely disconnected from its surroundings, seems wise (see Chapter 20). With respect to research, bioregionalization schemes may be useful for grouping seamounts and identifying gaps in sampling globally for the pelagic community. In contrast, benthic communities may reflect more local conditions like depth and substrate type, and may have endemic species. Developing classification schemes specifically for seamounts, as Rowden *et al.* (2005) have done, and then testing them may be more informative than using regional classifications.

Not surprisingly for a habitat that is deep, isolated, and difficult to sample, the clearest conclusion supported by this survey is how much research remains to be done. The rise in seamount research during the 1990s and early 2000s discussed in Chapter 3 is encouraging. Perhaps in the near future we will be able to write a seamount book that includes predictive models and a process-based understanding, instead of description.

Acknowledgements

The authors thank an anonymous reviewer, Nigel Haggan, and the participants of the *Seamounts: Ecology, Fisheries, and Conservation* workshop for comments and suggestions on an earlier draft. Karen Stocks would like to thank the Census of Marine Life on Seamounts and NSF Grant OCE-0340839 for supporting her work on seamount benthic communities.

References

Aboim, M.A., Menezes, G.M., Schlitt, T. and Rogers, A.D. (2005) Genetic structure and history of populations of the deep-sea fish *Helicolenus dactylopterus* (Delaroche, 1809) inferred from mtDNA sequence analysis. *Molecular Ecology*, 14(5), 1343–54.

Allen, J.A. and Butler, T.H. (1994) The Caridea (Decapoda) collected by the Mid-Pacific mountains expedition, 1968. *Pacific Science*, 48, 410–45.

Alton, M.S. (1986) Fish and crab populations of Gulf of Alaska seamounts. In: *Environment and Resources of Seamounts in the North Pacific* (eds. Uchida, R.N., Hayasi, S. and Boehlert, G.W.). US Department of Commerce, *NOAA Technical Report NMFS*, 43, 45–51.

Andriashev, A.P. (1986) *Review of the Snailfish Genus Paraliparis (Scorpaeniformes: Liparididae) of the Southern Ocean*. Theses Zoologicae 7, Koeltz Scientific, Koenigstein.

Avila, S.P. and Malaquias, M.A.E. (2003) Biogeographical relationships of the molluscan fauna of the Ormonde Seamount (Gorringe Bank, Northeast Atlantic Ocean). *Journal of Molluscan Studies*, 69, 145–50.

Barnard, R., Batten, S.D., Beaugrand, G., Buckland, C., Conway, D.V.P., Edwards, M., Finlayson, J., Gregory, L.W., Halliday, N.C., John, A.W.G., Johns, D.G., Johnson, A.D., Jonas, T.D., Lindley, J.A. and Nyman, J. (2004) Continuous plankton records: plankton atlas of the North Atlantic Ocean (1958–1999). 2. Biogeographical charts. *Marine Ecology Progress Series*, Suppl., 11–75.

Bartsch, I. (1991) Brittle stars (Ophiuroidea) in the northeast Atlantic, in the Ibero-Moroccan-Mauritanian region: comparison of the faunas with taxonomic remarks. *Spixiana*, 14, 95–112.

Beaugrand, G. (2004) Continuous plankton records: plankton atlas of the North Atlantic Ocean (1958–1999). I. Introduction and methodology. *Marine Ecology Progress Series*, Suppl., 3–10.

Beaugrand, G., Ibañez, F., Lindley, J.A. and Reid, P.C. (2002) Diversity of calanoid copepods in the North Atlantic and adjacent seas: species associations and biogeography. *Marine Ecology Progress Series*, 232, 179–95.

Blaber, S.J.M. (1986) The distribution and abundance of seabirds south-east of Tasmania and over the Soela Seamount during April 1985. *Emu*, 86, 239–44.

Boehlert, G.W. and Sasaki, T. (1988) Pelagic biogeography of the armorhead, *Pseudopentaceros wheeleri*, and recruitment to isolated seamounts in the North Pacific Ocean. *Fishery Bulletin*, 86(3), 453–65.

Bolch, C.J.S., Ward, R.D. and Last, P.R. (1994) Biochemical systematics of the marine fish family Centrolophidae (Teleostei: Stromateoidei) from Australian waters. *Australian Journal of Marine and Freshwater Research*, 45(7), 1157–72.

Borets, L.A. (1980) The population structure of the boarfish *Pentaceros richardsoni* (Smith) from the Emperor Seamounts and the Hawaiian Ridge. *Journal of Ichthyology*, 19, 15–20.

Briggs, J.C. (1995) *Global Biogeography*. Elsevier, Amsterdam.

Briggs, J.C. (2003) Marine centres of origin as evolutionary engines. *Journal of Biogeography*, 30, 1–18.

Brinton, E. (1962) The distribution of Pacific euphausiids. *Bulletin of the Scripps Institution of Oceanography*, 8, 51–270.

Brook, F.J. (1998) The coastal molluscan fauna of the northern Kermadec Islands, southwest Pacific Ocean. *Journal of the Royal Society of New Zealand*, 28(2), 185–233.

Bucklin, A., Wilson Jr., R.R. and Smith Jr., K.L. (1987) Genetic differentiation of seamount and basin populations of the deep-sea amphipod *Eurythenes gryllus*. *Deep-Sea Research, Part A. Oceanographic Research Papers*, 34(11), 1795–1810.

Burukovskiy, R.N. (1992) Shrimp at Walters Bank (Indian Ocean). *Oceanology*, 32(3), 328–31.

Cairns, S.D. (2004) The azooxanthellate Scleractinia (Coelenterata: Anthozoa) of Australia. *Records of the Australian Museum*, 56(3), 259–329.

Calder, D.R. (2000) Assemblages of hydroids (Cnidaria) from three seamounts near Bermuda in the western North Atlantic. *Deep-Sea Research, Part I. Oceanographic Research Papers*, 47(6), 1125–39.

Cecere, E. and Perrone, C. (1988) First contribution to the knowledge of macrobenthic flora of the Amendolara sea-mount (Ionian Sea). *Oebalia*, 14, 43–67.

Chave, E.H. and Mundy, B.C. (1994) Deep-sea benthic fish of the Hawaiian archipelago, Cross Seamount, and Johnston Atoll. *Pacific Science*, 48(4), 367–409.

Colebrook, J.M. (1979) Continuous plankton records: seasonal cycles of phytoplankton and copepods in the North Atlantic Ocean and the North Sea. *Marine Biology*, 51(1), 23–32.

Collette, B.B. and Parin, N.V. (1991) Shallow-water fishes of Walters Shoals, Madagascar Ridge. *Bulletin of Marine Science*, 48(1), 1–22.

Creasey, S.S. and Rogers, A.D. (1999) Population genetics of bathyal and abyssal organisms. *Advances in Marine Biology*, 35, 1–151.

Diekmann, R. and Piatkowski, U. (2004) Species composition and distribution patterns of early life stages of cephalopods at Great Meteor Seamount (subtropical North-East Atlantic). *Archive of Fishery and Marine Research*, 51(1–3), 115–31.

Dower, J.F. and Perry, R.I. (2001) High abundance of larval rockfish over Cobb Seamount, an isolated seamount in the Northeast Pacific. *Fisheries Oceanography*, 10(4), 268–74.

Edwards, A.J. (1993) New records of fishes from the Bonaparte Seamount and Saint Helena Island, South Atlantic. *Journal of Natural History*, 27(2), 493–503.

Edwards, M. (2004) Preface. *Marine Ecology Progress Series*, Suppl., 1–2.

Ekman, S. (1953) *Zoogeography of the Sea*. Sidgwick and Jackson, London.

France, S.C. and Hoover, L.L. (2002) DNA sequences of the mitochondrial COI gene have low levels of divergence among deep-sea octocorals (Cnidaria: Anthozoa). *Hydrobiologia*, 471(1–3), 149–55.

France, S.C. and Kocher, T.D. (1996) Geographic and bathymetric patterns of mitochondrial 16S rRNA sequence divergence among deep-sea amphipods, *Eurythenes gryllus*. *Marine Biology*, 126(4), 633–43.

Freeland, H. (1994) Ocean circulation at and near Cobb Seamount. *Deep-Sea Research, Part I. Oceanographic Research Papers*, 41, 1715–32.

Froese, R. and Sampang, A. (2004) Taxonomy and biology of seamount fishes. In: *Seamounts: Biodiversity and Fisheries* (eds. Morato, T. and Pauly, D.). *Fisheries Centre Research Reports*, 12(5), 25–32.

Fujii, E. (1986) Zoogeographical features of fishes in the vicinity of seamounts. *NOAA Technical Report NMFS*, 43, pp. 67–9.

Gad, G. (2004a) Diversity and assumed origin of the Epsilonematidae (Nematoda) of the plateau of the Great Meteor Seamount. *Archive of Fishery and Marine Research*, 51(1–3), 30–42.

Gad, G. (2004b) The Loricifera fauna of the plateau of the Great Meteor Seamount. *Archive of Fishery and Marine Research*, 51(1–3), 9–29.

Gad, G. and Schminke, H.K. (2004) How important are seamounts for the dispersal of interstitial meiofauna? *Archive of Fishery and Marine Research*, 51(1–3), 43–54.

Galil, B. and Zibrowius, H. (1998) First benthos samples from Eratoshenes Seamount, Eastern Mediterranean. *Senckenbergiana Maritima*, 28, 111–21.

Gaspard, D. (2003) Recent brachiopods collected during the 'Seamount 1' cruise off Portugal and the Ibero-Moroccan Gulf (Northeastern Atlantic) in 1987. *Geobios*, 36, 285–304.

Genin, A. (2004) Bio-physical coupling in the formation of zooplankton and fish aggregations over abrupt topographies. *Journal of Marine Systems*, 50, 3–20.

George, K.H. and Schminke, H.K. (2002) Harpacticoida (Crustacea, Copepoda) of the Great Meteor Seamount, with first conclusions as to the origin of the plateau fauna. *Marine Biology*, 144, 887–95.

Gibbons, M.J. (1997) Pelagic biogeography of the South Atlantic Ocean. *Marine Biology*, 129, 757–68.

Gillet, P. and Dauvin, J.C. (2000) Polychaetes from the Atlantic seamounts of the southern Azores: biogeographical distribution and reproductive patterns. *Journal of the Marine Biological Association of the United Kingdom*, 80, 1019–29.

Gillet, P. and Dauvin, J.C. (2003) Polychaetes from the Irving, Meteor and Plato seamounts, North Atlantic Ocean: origin and geographical relationships. *Journal of the Marine Biological Association of the United Kingdom*, 83, 49–53.

Gofas, S. (2000) Four species of the family Fasciolariidae (Gastropoda) from the North Atlantic Seamounts. *Journal of Conchology*, 37, 7–16.

Gofas, S. (2002) An endemic radiation of Trituba (Mollusca, Gastropoda) on the North Atlantic Seamounts. *American Malacological Bulletin*, 17, 45–63.

Gofas, S. and Beu, A. (2002) Tonnoidean gastropods of the North Atlantic Seamounts and the Azores. *American Malacological Bulletin*, 17(1–2), 91–108.

Gopal, K., Tolley, K.A., Groeneveld, J.C. and Matthee, C.A. (2006) Mitochondrial DNA variation in spiny lobster *Palinurus delagoae* suggests genetically structured populations in the southwestern Indian Ocean. *Marine Ecology Progress Series*, 319, 191–8.

Grasshoff, M. (1985) Die Gorgonaria und Antipatharia der Grossen Meteor-Bank und der Josephine-Bank. (Cnidaria: Anthozoa). *Senckenbergiana Maritima*, 17, 65–87.

Hanamura, Y. (1983) Pelagic shrimps (Penaeidea and Caridea) from Baja California and its adjacent region with description of a new species. *Bulletin of the Biogeographical Society of Japan*, 38(8), 51–85.

Haney, J.C., Haury, L.R., Mullineaux, L.S. and Fey, C.L. (1995) Sea-bird aggregation at a deep North Pacific seamount. *Marine Biology*, 123, 1–9.

Havenhand, J.N. (1995). Evolutionary ecology of larval types. In: *Ecology of Marine Invertebrate Larvae* (ed. McEdward, L.), pp. 79–122. CRC Press, New York.

Heinz, P., Ruepp, D. and Hemleben, C. (2004) Benthic foraminifera assemblages at Great Meteor Seamount. *Marine Biology*, 144, 985–98.

Hellberg, M.E., Burton, R.S., Neigel, J.E. and Palumbi, S.R. (2002). Genetic assessment of connectivity among marine populations. *Bulletin of Marine Science*, 70(Suppl. 1), 273–90.

Hoarau, G. and Borsa, P. (2000) Extensive gene flow within sibling species in the deep-sea fish *Beryx splendens*. *Comptes Rendus de l'Academie des Sciences Serie III Sciences de la Vie*, 323, 315–25.

Hui, C.A. (1985) Undersea topography and the comparative distributions of two pelagic cetaceans. *Fisheries Bulletin*, 83, 472–5.

Johannesson, K. (1988) The paradox of Rockall: Why is a brooding gastropod (*Littorina saxatilis*) more widespread than one having a planktonic larval dispersal stage (*L. littorea*)? *Marine Biology*, 99, 507–13.

Kaufmann, R.S., Wakefield, W.W. and Genin, A. (1989) Distribution of epibenthic megafauna and lebensspuren on two Central North Pacific Seamounts. *Deep-Sea Research, Part A. Oceanographic Research Papers*, 36, 1863–96.

Kay, E.A. (1979) *Hawaiian Marine Shells*. Bernice P. Bishop Museum Special Publication 64. Bishop Museum Press, Honolulu, HI.

Keller, N.B. (1985) Coral populations of underwater ridges in the North Pacific and Atlantic Oceans. *Oceanology*, 25, 784–6.

Kensley, B. (1981) On the zoogeography of Southern African decapod crustacea, with a distributional checklist. *Smithsonian Contributions to Zoology*, 338, 1–64.

Levin, L.A., Etter, R.J., Rex, M.A., Gooday, A.J., Smith, C.R., Pineda, J., Stuart, C.T., Hessler, R.R. and Pawson, D. (2001) Environmental influences on regional deep-sea species diversity. *Annual Review of Ecology and Systematics*, 32, 51–93.

Logan, A. (1998) Recent Brachiopoda from the oceanographic expedition SEAMOUNT 2 to the north-eastern Atlantic in 1993. *Zoosystema*, 20(4), 549–62.

Longhurst, A.R. (1998) *Ecological Geography of the Sea*. Elsevier, New York.

MacArthur, R.H. (1972) *Geographical Ecology: Patterns in the Distribution of Species*. Harper and Row, New York.

MacArthur, R.H. and Wilson, E.O. (1967) *The Theory of Island Biogeography*. Princeton University Press, Princeton, NJ.

Markhaseva, E.L. and Schnack-Schiel, S.B. (2003) New and rare calanoid copepods from the Great Meteor Seamount, North Eastern Atlantic. *Ophelia*, 57(2), 107–23.

Mauchline, J. (1998) The biology of calanoid copepods. *Advances in Marine Biology*, 33, 1–710.

Mauchline, J. and Fisher, L.R. (1969) The biology of euphausiids. *Advances in Marine Biology*, 7, 1–454.

May, R.M. (1975) Patterns of species abundance and diversity. In: *Ecology and Evolution of Communities* (eds. Cody, M.L. and Diamond, J.M.). Belknap Press of Harvard University Press, Cambridge.

Monniot, C. and Monniot, F. (1992) Ascidians from the Lusitanian seamounts (Seamount I cruise). *Bulletin du Museum National d'Histoire Naturelle Section A Zoologie*, 14(3–4), 591–603.

Mora, C., Chittaro, P.M., Sale, P.F., Kritzer, J.P. and Ludsin, S.A. (2003) Patterns and processes in reef fish diversity. *Nature*, 421, 933–6.

Mullineaux, L.S. and Mills, S.W. (1997) A test of the larval retention hypothesis in seamount-generated flows. *Deep-Sea Research I: Oceanographic Research Papers*, 44(5), 745–70.

Nellen, W. and Ruseler, S. (2004) Composition, horizontal and vertical distribution of ichthyoplankton in the Great Meteor Seamount area in September 1998. *Archive of Fishery and Marine Research*, 51(1–3), 132–64.

Nesis, K.N. (1986) Cephalopods of seamounts in the western Indian Ocean. *Oceanology*, 26(1), 91–6.

Nesis, K.N. (1994) Teuthofauna of Walters Shoals, a seamount in the Southwestern Indian Ocean. *Ruthenica*, 4(1), 67–77.

Palumbi, S.R. and Warner, R.R. (2003) Why gobies are like hobbits. *Science*, 299, 51–2.

Parin, N.V., Mironov, A.N. and Nesis, K.N. (1997) Biology of the Nazca and Sala y Gómez submarine ridges, an outpost of the Indo-West Pacific fauna in the eastern Pacific Ocean: composition and distribution of the fauna, its communities and history. *Advances in Marine Biology*, 32, 145–242.

Parker, T. and Tunnicliffe, V. (1994) Dispersal strategies of the biota on an oceanic seamount: implications for ecology and biogeography. *Biological Bulletin*, 187(3), 336–45.

Penas, A. and Rolan, E. (1999) Pyramidellidae (Gastropoda, Heterostropha) from oceanographic mission 'Seamount 2'. *Iberus*, 5(Suppl.), 151–99.

Penrith, M.J. (1967) The fishes of Tristan da Cuna, Gough Island and the Vema Seamount. *Annals of the South African Museum*, 48(22), 523–48.

Piepenburg, D. and Muller, B. (2004) Distribution of epibenthic communities on the Great Meteor Seamount (North-East Atlantic) mirrors pelagic processes. *Archive of Fishery and Marine Research*, 51, 55–70.

Primo, C. and Vazquez, E. (2004) Zoogeography of the southern African ascidian fauna. *Journal of Biogeography*, 31(12), 1987–2009.

Pusch, C., Beckmann, A., Porteiro, F.M. and von Westernhagen, H. (2004) The influence of seamounts on mesopelagic fish communities. *Archive of Fishery and Marine Research*, 51, 165–86.

Ramil, F., Vervoort, W. and Ansin, J.A. (1998) Report on the Haleciidae and Plumularioidea (Cnidaria, Hydrozoa) collected by the French SEAMOUNT 1 Expedition. *Zoologische Verhandelingen*, 322, 1–42.

Randall, J.E. (1998) Zoogeography of shorefishes of the Indo-Pacific region. *Zoological Studies*, 37(4), 227–68.

Rao, M.V. and Newman, W.A. (1972) Thoracic Cirripedia from guyots of the Mid-Pacific Mountains. *Transactions of the San Diego Society of Natural History*, 17, 69–94.

Rehder, H.A. (1980) The marine mollusks of Easter Island (Isla de Pascua) and Sala-y-Gómez. *Smithsonian Contributions to Zoology*, 289, 1–167.

Richer de Forges, B., Koslow, J.A. and Poore, G.C.B. (2000) Diversity and endemism of the benthic seamount fauna in the southwest Pacific. *Nature*, 405, 944–7.

Robertson, D.R. (2001) Population maintenance among tropical reef fishes: inferences from small-island endemics. *Proceedings of the National Academy of Sciences of the USA*, 98(10), 5667–70.

Rogers, A.D. (1994) The biology of seamounts. *Advances in Marine Biology*, 30, 305–50.

Rogers, A.D., Morley, S., Fitzcharles, E., Jarvis, K. and Belchier, M. (2006) Genetic structure of Patagonian toothfish (*Dissostichus eleginoides*) populations on the Patagonian Shelf and Atlantic and western Indian Ocean Sectors of the Southern Ocean. *Marine Biology*, 149(4), 915–24.

Rosenzweig, M.L. (1995) *Species Diversity in Space and Time.* Cambridge University Press, Cambridge.

Rowden, A.A., O'Shea, S. and Clark, M.R. (2002) Benthic biodiversity of seamounts on the northwest Chatham Rise. Ministry of Fisheries, *Marine Biodiversity Biosecurity Report No. 2*, Wellington.

Rowden, A.A., Warwick, R. and Gordon, D. (2004) Bryozoan biodiversity in the New Zealand region and implications for marine conservation. *Biodiversity and Conservation*, 13(14), 2695–721.

Rowden, A.A., Clark, M. and Wright, I.C. (2005) Physical characterisation and a biologically focused classification of 'seamounts' in the New Zealand region. *New Zealand Journal of Marine and Freshwater Research*, 39, 1039–59.

Sakai, T. (1980) On new or rare crabs taken from Japanese and central Pacific waters. *Researches of Crustacea*, 10, 73–84.

Samadi, S., Bottan, L., Macpherson, E., De Forges, B.R. and Boisselier, M.C. (2006) Seamount endemism questioned by the geographic distribution and population genetic structure of marine invertebrates. *Marine Biology*, 149(6), 1463–75.

Sherman, K. (1991) The large marine ecosystem concept: research and management strategy for living resources. *Ecological Applications*, 1(4), 349–60.

Sherman, K. (1992) Productivity, perturbations and options for biomass yields in large marine ecosystems. In: *Large Marine Ecosystems: Patterns, Processes and Yields* (eds. Sherman, K., Alexander, L.M. and Gold, B.D.), pp. 206–19. AAAS Press, Washington, DC.

Smith, P.J. (1986) Genetic similarity between samples of the orange roughy *Hoplostethus atlanticus* from the Tasman Sea, South-West Pacific Ocean and North-East Atlantic Ocean. *Marine Biology*, 91, 173–80.

Smith, P.J., Robertson, S.G., Horn, P.L., Bull, B., Anderson, O.F., Stanton, B.R. and Oke, C.S. (2002) Multiple techniques for determining stock relationships between orange roughy, *Hoplostethus atlanticus*, fisheries in the eastern Tasman Sea. *Fisheries Research*, 58, 119–40.

Smith, P.J., McVeagh, S.M., Mingoia, J.T. and France, S.C. (2004) Mitochondrial DNA sequence variation in deep-sea bamboo coral (Keratoisidinae) species in the southwest and northwest Pacific Ocean. *Marine Biology*, 144(2), 253–61.

Stock, J.H. (1991) Pycnogonides de la campagne Seamount 1 au large de la peninsule iberique et dans le golfe ibero-morocain. *Bulletin du Museum National d'Histoire Naturelle Section A Zoologie*, 13, 135–42.

Stocks, K.I. (2004) SeamountsOnline: an online information system for seamount biology. In: *Proceedings of the Colour of Ocean Data Symposium*, Brussels, 25–27 November (eds. Vanden Berghe, E., Brown, M., Costello, M.J., Heip, C., Levitus, S. and Pissierssens, P.), pp. 77–89. *IOC Workshop Report No. 188*, VLIZ Special Publication 16.

Stocks, K.I. (2005) SeamountsOnline: an online information system for seamount biology. Version 2005–1. World Wide Web electronic publication. http://seamounts.sdsc.edu (last accessed October 1, 2005).

Taylor, M.S. and Hellberg, M.E. (2003) Genetic evidence for local retention of pelagic larvae in a Caribbean reef fish. *Science*, 299, 107–9.

Tracey, D.M., Bull, B., Clark, M.R. and Mackay, K.A. (2004) Fish species composition on seamounts and adjacent slope in New Zealand waters. *New Zealand Journal of Marine and Freshwater Research*, 38(1), 163–82.

Tunnicliffe, V., McArthur, A.G. and McHugh, D. (1998) A biogeographical perspective of the deep-sea hydrothermal vent fauna. *Advances in Marine Biology*, 34, 353–442.

Tyler, P.A. (1995) Conditions for the existence of life at the deep-sea floor: an update. *Oceanography and Marine Biology: An Annual Review*, 33, 221–4.

Valle-Levinson, A., Trasvina Castro, A., Gutiérrez de Velasco, G. and Gonzales Armas, R. (2004) Diurnal vertical motions over a seamount of the southern Gulf of California. *Journal of Marine Systems*, 50, 61–77.

Vinogradova, N.G. (1997) Zoogeography of the abyssal and hadal zones. *Advances in Marine Biology*, 32, 325–88.

Vörös, A. (2005) The smooth brachiopods of the Mediterranean and Jurassic: refugees or invaders? *Palaeogeography, Palaeoclimatology, Palaeoecology*, 223, 222–42.

Weigmann, R. (1974) Investigations on the distribution of the euphausiids [Crustacea] near and above the Great Meteor Seamount. *Biologische Anstalt Helgoland*, 17, 17–32.

Wilson, R.R. and Kaufmann, R.S. (1987) Seamount biota and biogeography. *American Geophysical Union Geophysical Monographs*, 43, 355–78.

Woodd-Walker, R.S., Ward, P. and Clarke, A. (2002) Large-scale patterns in diversity and community structure of surface water copepods from the Atlantic Ocean. *Marine Ecology Progress Series*, 236, 189–203.

Yen, P.P.W., Sydeman, W.J. and Hyrenbach, K.D. (2004) Marine bird and cetacean associations with bathymetric habitats and shallow-water topographies: implications for trophic transfer and conservation. *Journal of Marine Systems*, 50, 79–99.

Young, C.M. (1995) Behavior and locomotion during the dispersal phase of larval life. In: *Ecology of Marine Invertebrate Larvae* (ed. McEdward, L.), pp. 249–77. CRC Press, New York.

Chapter 14

Raiding the larder: a quantitative evaluation framework and trophic signature for seamount food webs

Tony J. Pitcher and Cathy Bulman

Abstract

A logical framework for analysing food webs on seamounts is proposed in which each functional component is driven by physical processes that derive from ocean currents and geomorphology, or by the presence of other food web organisms. Not every functional component may be present on every seamount. Seamount food webs are based upon retention or local enhancement of primary and secondary plankton, driven by upwelling and other nutrient enrichments. In some cases, sessile filter feeders may encourage both detrital food chains and foraging visitors, but many seamounts attract large fish, seabirds and marine mammals. Seamounts seem to act as a food store ('larders') and may display a characteristic food web signature. We explore the linkages between the trophic ecology of seamount organisms, their life history strategies, and the complexity and resilience of seamount food webs. Inevitable gaps in field data might be plugged by meta-analysis and seamount food web modelling.

Introduction

In a key, classic paper on seamount ecology, Koslow (1997) mentions 'dining at the seamount café'; the analysis of food webs in this chapter suggests support for this idea that seamounts are like living larders or oases for visiting marine animals that also live and feed elsewhere in the ocean. Organisms resident at seamounts are raided by many animals important for conservation that are attracted to feed. Human fisheries for residents and visitors are opportune, but have proved hard to control and have tended to 'empty the larder'. Here we suggest an additional mechanism contributing to the larder; a factor that may have also encouraged the evolution of characteristic seamount life history traits such as longevity and the ability to withstand intermittent food supplies.

Seamount food webs are structured by the habitat template created by ocean geomorphology and physics (Chapters 1, 2, 4 and 5), and in particular by the various mechanisms that may retain and locally enhance plankton production. The basis of all seamount food webs is driven by these physical processes, while other functional components of the ecosystem also rely upon them, or upon the arrival and residence of other food web organisms. On account of variation in the physical processes, not every functional component in the food web may be present on every seamount. Most seamount food webs are based upon retention or local enhancement of primary and secondary plankton, driven by upwelling and other nutrient enrichments. In some cases sessile filter feeders may encourage the establishment of detrital food chains and encourage foraging visitors, but most seamounts attract aggregations of fish, and their marine mammal, large fish and cephalopod predators (Chapters 10 and 11). Relatively shallow seamounts attract seabirds, tuna and possibly marine turtles, as will be shown later in this chapter (see also Chapter 12B).

There is considerable information about the diets of seamount residents and visitors, but little quantitative data upon which to base management and conservation interventions. Moreover, seamount organisms vary greatly in their robustness to local extirpation by the vagaries of ocean climate or by the depredations of human fishing. Based on the functional groups of organisms, we present a logical framework for analysing food webs on seamounts, which can serve as a way of evaluating the likely impacts of human and climatic factors.

A logical framework for analysing food webs on seamounts

The aim is to describe a framework of measurable attributes of seamount food webs that shows how the different elements interact to produce the range of seamount ecosystems. Although geologically young, on biological time scales, most seamounts have existed for a very long periods of time (Chapter 1). Seamounts are very varied, and it is hard to find 'typical' attributes that all may posses. For example, in a survey of 800 seamounts in the New Zealand EEZ, Rowden *et al.* (2005) found that they, 'span a wide range of sizes, depths, elevation, geological associations and origins, and occur over the latitudinal range of the region, lying in different water masses of varying productivity, and both near shore and off shore'. Hence, our emphasis is on dynamic processes that generate the seamount food 'larder' and we attempt to avoid problems that arise from adopting rigid definitions. The scheme is explained using a series of diagrams presented in Fig. 14.1.

In the first panel, Fig. 14.1a, the depth of the seamount peak determines whether macrophytes may grow or not, although currents and turbidity may also inhibit possible plant growth. Generally, only very shallow seamounts have kelp, such as Bowie seamount in British Columbia (see Chapter 20; McDaniel *et al.*, 2003; WWF, 2003), and the seaweed may also act as a refuge for fish and may foster detrital retention. And, as noted by Darwin (see Chapters 3 and 21) coral reefs grow on very shallow flat-topped eroded seamounts termed guyots (Chapter 1). Even without coral or algal growth, shallow seamounts that have their summit in the photic layer are classified separately because of the implications for summit-dwelling organisms (Genin, 2004). Seamounts vary between 100 and 10 000 m summit-to-base (Wessel, 2007; Chapter 1), and over geological time scales, isostatic effects affect the height. Most are at least 2.5 km tall, matching the typical mid-ocean ridge depth, and therefore only large seamounts have their summit in shallow water (Chapter 1).

Secondly, Fig. 14.1b shows that the depth of the peak and the height of the mount itself have important physical consequences (Chapter 4) that may include a connection of the seamount to deep ocean processes, including upwelling of nutrients. Intermediate

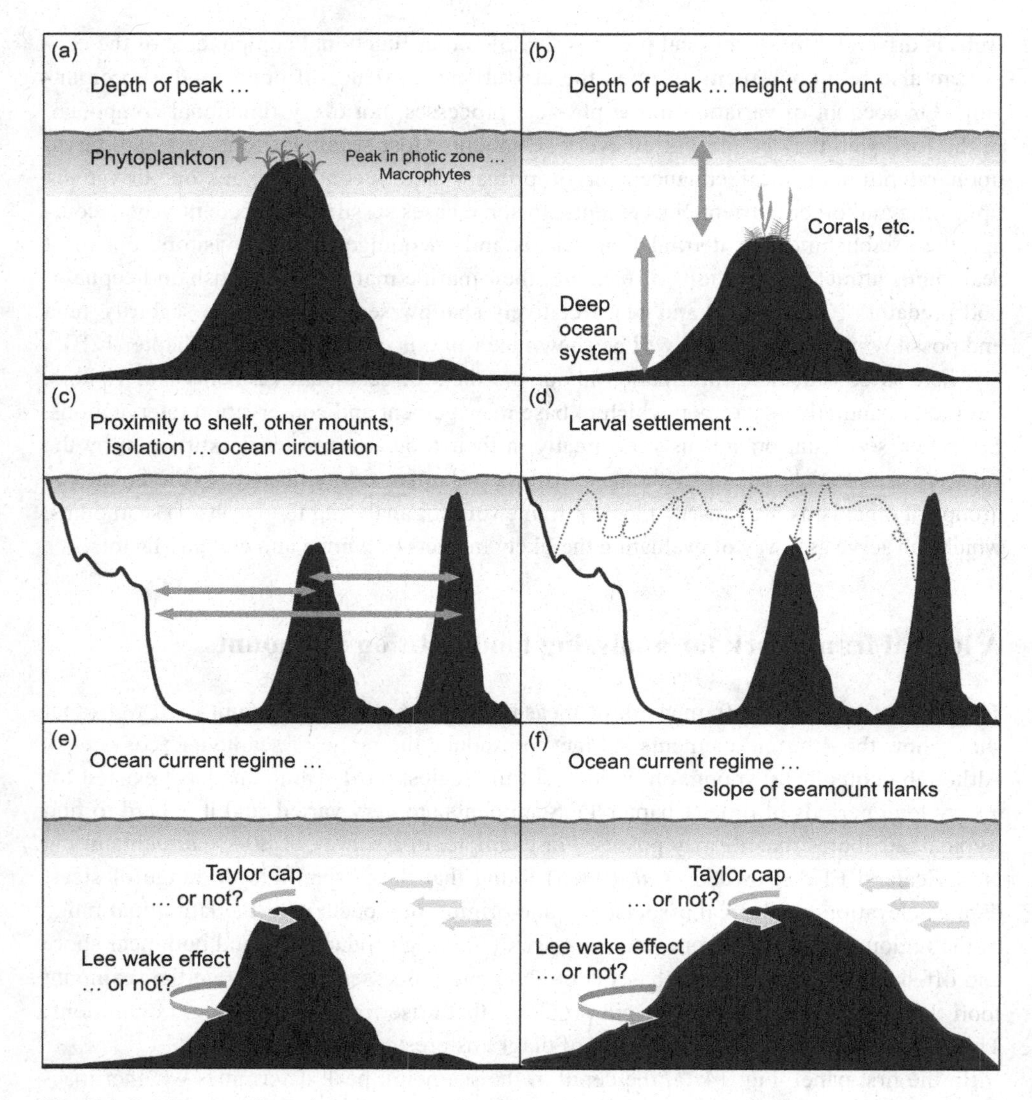

Fig. 14.1 Panels a to r: The various components that may make up a seamount food web and used as the basis of the Ecosystem Evaluation Framework (EEF) described in Chapter 21. For further explanation see text.

seamounts are those with the summit below the photic layer but shallower than the daytime depth of most of the vertically migrating zooplankton (~400 m), and deep seamounts have their summit below 400 m (Genin, 2004).

The third panel (Fig. 14.1c) shows that proximity to coastlines and other seamounts affects the local current regime on a seamount. Many seamounts are aligned in linear, sub-parallel chains in the direction of tectonic plate movements, at hotspots, at plate boundaries and at plate subduction zones (Montelli *et al.*, 2004). As the distribution of seamounts is approximated by a power law, the result of this geomorphology is that many seamounts have near neighbours that may be aligned with ocean currents (e.g., Wessel, 2001). Large-scale ocean eddies can be split by seamounts and give up over half of their water (White *et al.*, 2007; Chapter 4).

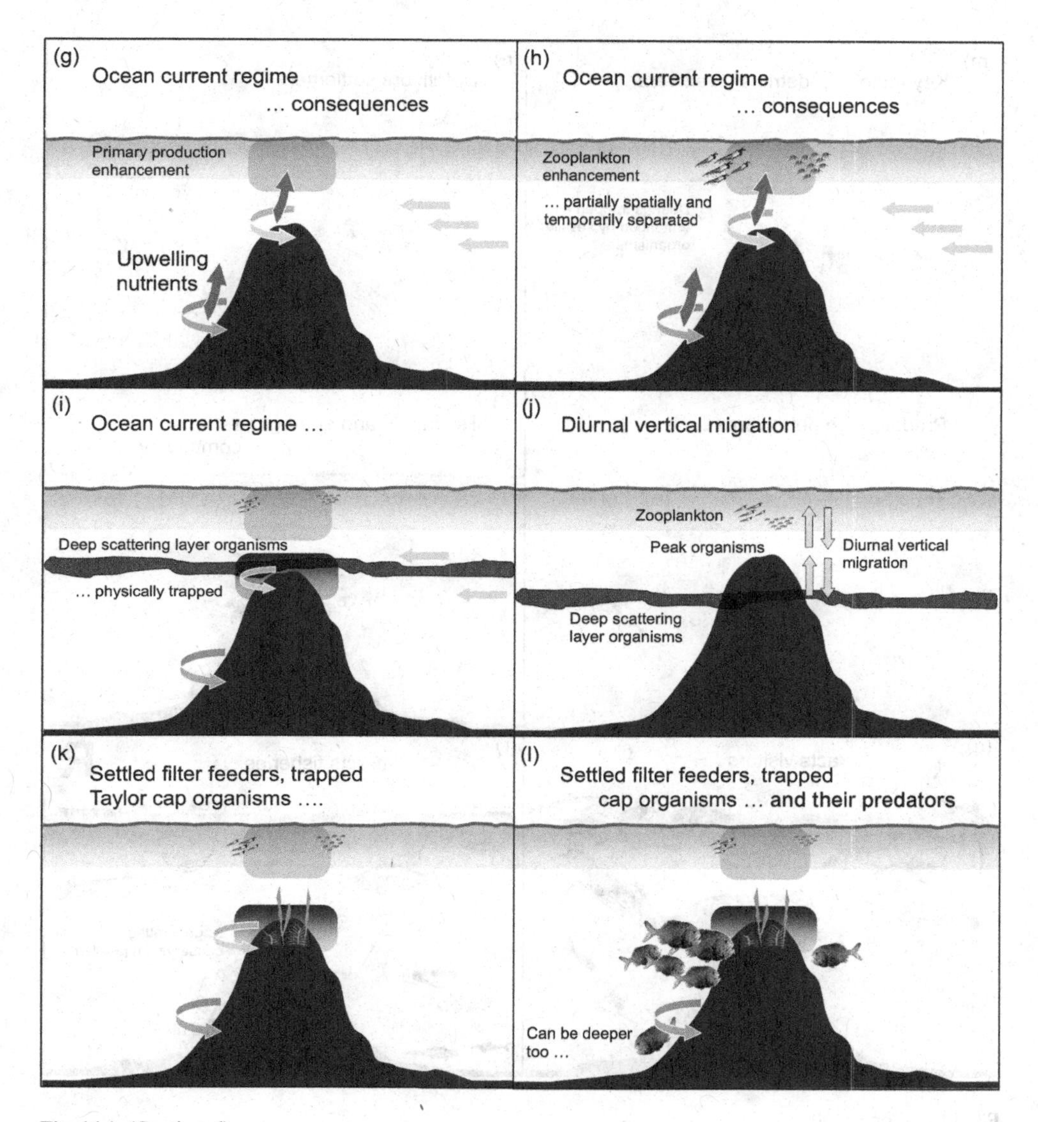

Fig. 14.1 (Continued).

Figure 14.1d illustrates the settlement of planktonic larvae of plants, benthic filter feeding organisms such as corals, gorgonids, sponges and benthic detritivores as for example crabs, lobsters and shrimps (Samadi *et al.* 2007; Chapter 7). Settlement patterns for these organisms depend on ocean currents or sometimes the water trapped in gyres.

Figure 14.1e covers one of the most often mentioned 'seamount effects', the formation of Taylor caps and lee wake effects that may trap water masses and create upwelling currents. White *et al.* (2007) (Chapter 4) show that a Taylor Column is formed by the effect of the Earth's rotation on directional current flow split by a seamount (Huppert, 1975), and the stratification of the water mass creates a Taylor cone (or cap). Taylor caps form, or not, over a particular seamount depending on width, current, height and the Coriolis force, which varies by latitude with the critical parameters being the Rossby and Burger numbers

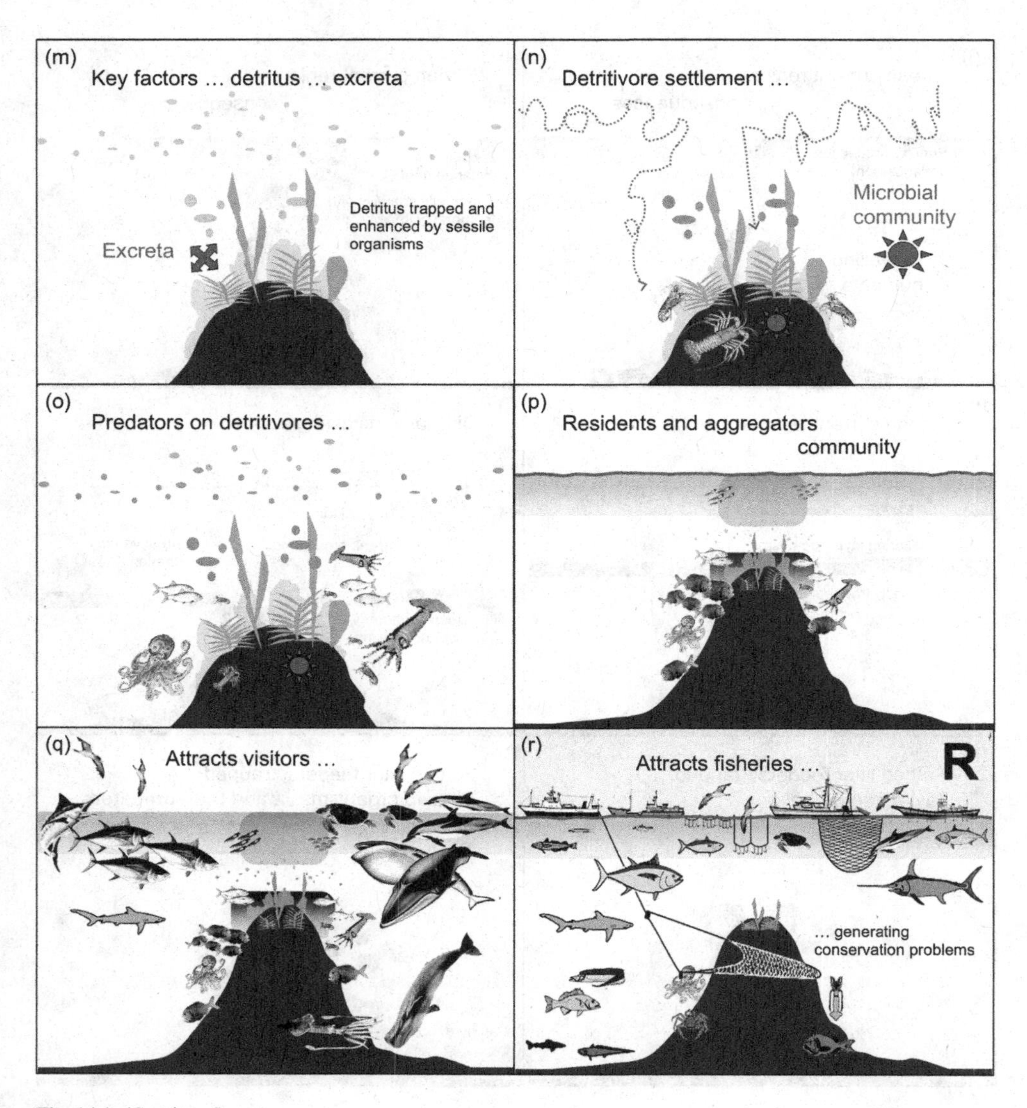

Fig. 14.1 (Continued).

(Chapter 4). Some seamounts always produce Taylor cones, others not. Intermittent, or spun off Taylor cones may result as currents vary above and below a threshold. The cone can encourage a doming of water density layers, creating two physical effects, each with biological implications. Firstly, below the thermocline, isolated topographic features may produce large vertical displacements in the density gradient so that small deep seamounts can cause significant local current structures (Royer, 1978). Secondly, density domes over shallow seamounts may reach the euphotic zone and have powerful effects on local plankton abundance including increased vertical mixing (Eriksen, 1998). In addition, tides can generate similar density cones and wave spin-offs over seamounts (Chapter 4), and eddies like this can be formed by several different mechanisms (e.g., Huppert and Bryan, 1976). Chapters 4 and 5 outline considerable field evidence for all of these processes, although no one seamount may exhibit all of them, and some seamounts may show none.

Figure 14.1f illustrates how the slope and shape of seamount flanks can influence the formation of upwelling currents. Pressure and cooling by seawater affects the shape of a seamount as it forms, allowing steeper sides than terrestrial volcanoes (Macdonald *et al.*, 1986). Also the magma chamber shape, rift zones, erosion and landslides modify the basic conical volcano shape into stellate forms (e.g., Mitchell, 2001).

In Fig. 14.1, panels g to l illustrate the biological consequences of the formation of Taylor caps and spin-offs; turbulent mixing at seamounts helps to distribute nutrients in the global ocean (Kunze and Llewellyn Smith, 2004). Some seamounts have been measured as enhancing primary production (PP) by 5–10% (Dower *et al.*, 1992; Odate and Furuya, 1998; Mouriño *et al.*, 2001), whilst at very shallow seamounts it may be as much as 60%. As shown in Fig. 14.1g, at some seamounts there is evidence for enhanced PP, but it is not a consistent feature and there is wide spatial and temporal variability (Chapter 5). Enhancement, or retention, of PP appears to depend on six processes:

1. Density-layered doming from a Taylor cap bringing deeper, nutrient-rich waters up, into the euphotic zone for shallow seamounts (Genin and Boehlert, 1985).
2. Domes may generate localised high density stratification which may stabilise the water column, promoting PP (Comeau *et al.*, 1995).
3. Increased vertical mixing, due to tidal amplification, flow acceleration, internal wave interaction or deep winter mixing can mix nutrient rich deeper water upwards, especially over shallow seamounts.
4. Enclosed or semi-enclosed circulation pattern around seamounts may be a retention mechanism for local or advected PP (Mohn and Beckmann, 2002).
5. Asymmetric flow acceleration at seamount flanks or summits may enhance horizontal fluxes of organic material affecting sediment distribution around the seamount (Turnewitsch *et al.*, 2004) and its sessile benthic communities (Genin *et al.*, 1986).
6. Advection of allochthonous nutrients and organic material into the local area of the seamount (Bulman *et al.*, 2002).

Seamount modelling (see Chapter 15) suggests that enhancement by local ocean dynamics cannot fully provide for the observed increase of PP at some seamounts, and an external source of nutrients and food is required (Morato and Pitcher, 2005). A model of the seamounts off southern Tasmania, which support very large abundances of orange roughy (*Hoplostethus atlanticus*) and oreostomatids, demonstrated that advection of mesopelagic fishes and crustaceans was not only necessary but possible given the current regime in the area (Bulman *et al.*, 2002).

Figure 14.1h shows that a Taylor cap and circular flow around the top of the mount can physically trap zooplanktonic organisms, which, as described above, may in some cases have more phytoplankton food. So some seamounts will have more zooplankton and jellyfish. But field evidence for finding more zooplankton over seamounts is conflicting. Huskin *et al.* (2001) found a 2-fold increase while Fedosova (1974) reported an 8-fold increase. Sime-Ngando *et al.* (1992) report an increase of micro-zooplankton (ciliates). Dower and Mackas (1996) found no increase in zooplankton, but at the same seamount (Cobb Seamount: see Fig. 16.11) there was a persistent stratified Taylor cap and abundant larval rockfish (Dower and Perry, 2001). It has been suggested that low observed zooplankton biomass might result from heavy grazing (Genin *et al.*, 1994; Haury *et al.*, 2000), and that secondary production could therefore be high.

Figure 14.1i and j shows how the seamount food web may be enhanced with small vertically migrating organisms. Small drifting mesopelagic organisms such as crustacea and myctophid fish in the deep scattering layer, may also be partially retained and feed in the Taylor cap of seamounts of suitable depth. Furthermore, the example here is of a seamount whose top, upper flank and Taylor cap may intercept the mesopelagic diel vertical migration, as organisms rise from their deeper water daytime refuge from predators to feed on zooplankton during the night. These organisms of trophic level 2 and above may be up to 10 times their density in the open ocean, so that this mechanism helps some seamount food webs to support a much larger biomass. Downward vertical migration of zooplankton during the day may also bring these organisms within range of a seamount summit.

Seasonally, it has been estimated that aestivating plankton takes about 0.1% of surface PP down to seamount food webs; on a daily time scale vertical migration takes about 0.5% down to a seamount food web of suitable depth (Koslow, 1997) of 1 gC/m^2 transported by aestivating zooplankton of 200 gC/m^2 net PP, ultimately 0.1 gC reaches third trophic level fish such as orange roughy. Williams and Koslow (1997) state that day-time depths of vertically migrating micronektonic species off the southern Tasmania mid-slope are between 400 and 500 m, but nycto-epipelagic species are in the upper 200 m; another peak of biomass occurred at 775–900 m. These peaks of biomass are coincident with the dominant water masses in the region. It is suggested that advection of mesopelagics sustain orange roughy at these depths on the southern slope seamounts.

Enhanced currents, steep slopes and volcanic rocks favour the growth of suspension feeders in benthic seamount communities (e.g., Genin *et al.*, 1986; Rogers, 1994). In Fig. 14.1k we show the sessile benthic filter feeding organisms, such as sponges, gorgonians and other corals, bivalves and tunicates benefiting from the seamount water retention effects described above, and possibly enhanced by locally generated detritus at some seamounts (Chapter 7). Filter feeding organisms may especially colonise the peak of the seamount where current fluxes are high even if the peak is in deep water (see above). Records from deep seamounts suggest that some are covered with tall forests of giant long-lived cold-water corals, some of which are extremely long-lived (100+ years, Grigg, 1993; 2–500 years, Chapter 8).

Fish predators that can feed on these organisms are shown in Fig. 14.1l. As they are hunting for sparse food, these fish may move large distances, or may only aggregate on seamounts at certain times. The figure implies that aggregators such as orange roughy are therefore feeding on filter feeders, but in fact, for roughy this is not the case. Hareide and Garnes (1998) state that along the Mid-Atlantic Ridge schools of *Beryx splendens* and *Sebastes marinus* (south to north, respectively) feed on mesopelagic fishes on seamounts (Morato *et al.*, 1998; Vinnichenko, 1998) while *Corphaenoides rupestris* feeds on pelagic prey in the water column (Haedrich, 1974). Other macrourids migrate into the water column to feed on pelagic species, although none of these species might need to go far if advected organisms are brought to them.

A more controversial suggestion that may enhance some seamount food webs is shown in Fig. 14.1m. Even in the open ocean, some falling detritus such as moulted exoskeletons, corpses and faecal pellets can support pelagic detritivores, mainly crustaceans. Australian work (Bulman, Personal communication) suggests open ocean detritus rain is on average about 7% of surface PP, but of lower nutritional value. Observations suggest that, at a seamount, this detrital rain may be increased by the water retention mechanisms, and support a community of sessile filter feeders. Hence, on a peak more heavily colonised by filter feeders, more detritus may accumulate from breakages, mortalities and excreta, trapped by the fine skeletal structures of growing sessile organisms, which would include macrophyte plants on very shallow seamounts. Such detritus layers would attract decomposing microorganisms that recycle nutrients and energy. This microbial community and its products in turn can attract larger detritivores, such as polychaete worms, crabs, lobsters and shrimps, shown in Fig. 14.1n. This happens only if suitable planktonic larvae are able to reach the seamount *via* ocean currents and gyres, although the density of polychaetes on most seamounts is reported to be low (e.g., Gillet and Dauvin, 2000, 2003). We have seen above that the general geomorphology of seamounts will often mean that there are close neighbours from where such larval colonisation may take place. Establishment of a microbial detritivore community might then be the critical factor for a seamount to have a maximum enhancement effect on ocean productivity and biodiversity. Physical detritus trapping and the refuge effects of the structural habitats created by sessile organisms on top of seamounts are components that are physically destroyed by trawling. Hence trawling can have disastrous effect on localised ocean biodiversity by depleting the 'seamount larders'.

The presence of resident detritivores and scavengers may soon lead to colonisation by larger predatory invertebrates and small fish (Fig. 14.1o), which in turn, may lead to resident members of the food web of trophic level 3 and above, provided suitable opportunities to colonise occur. And once there are residents, then visitors and aggregators arrive (Fig. 14.1p). These are analogous to 'house guests' who are shorter-term residents or visitors. As a consequence of the limited food supply stored on seamounts, seamount fish tend to be long-lived, slow-growing, late maturing and have low reproductive potential (Koslow, 1997). However, if these fish are residents, as are orange roughy, oreos and even the scavenging squalids, they are largely relying on pelagic rather than benthic food resources.

Figure 14.1q illustrates the vast range of temporary visitors who come to take advantage of, and hence contribute to, the seamount food web, and are described in detail in Chapters 9–12. Short-term visitors – who come to 'raid the larder' – include turtles (Chapter 12), sharks (Klimley *et al.*, 1988; Chapter 10B), billfish and tuna (Itano and Holland, 2000; Sibert *et al.*, 2000; Chapter 10A), swordfish (Sedberry and Loefer, 2001), some cephalopods (Nesis, 1986; Chapter 11), some cetaceans (Baird's beaked whale, Reeves and Mitchell, 1993; Chapter 12A) and some seabirds (Haney *et al.*, 1995; Monteiro *et al.*, 1996; Chapter 12C).

Human fisheries therefore often focus on seamounts (Fig. 14.1r), and thereby come the problems of sustainable use, depletion, biodiversity conservation and management that are the subjects of Chapters 15–20.

Most studies support the hypotheses that imported and retained food supplies support the observed large fish aggregations on seamounts, but the additional detrital chain mechanism proposed here may operate on some seamounts. Both of these hypotheses show why seamount 'larders' are so attractive to a wide range of foraging visitors, but production levels are quite low compared to coastal waters and major upwellings; hence resident organisms often exhibit long-lived, slow-growing life histories, making these communities quite fragile to exploitation.

A scoresheet of the scheme of measurable or scorable seamount attributes described here is set out in Table 14.1; the intention is that this scheme could be used for any seamount.

Table 14.1 Attributes scheme that may be objectively scored to provide an evaluation of seamount food webs, ecology and exploitation status, an Ecosystem Evaluation Framework (EEF).

	Scores	Notes
Oceanographic factors		
Depth of peak		
Depth of surrounding ocean		
Height of peak		
Slope of seamount		
Proximity to shelf		
Proximity to neighbour seamounts		
Ocean currents link to shelf		
Ocean currents to neighbour seamounts		
Taylor cap forms		
Total oceanographic status		
Ecological factors		
Macrophytes present		
Corals present		
Larval settlement regime		
Nutrient upwelling occurs		
Phytoplankton enhancement		
Zooplankton enhancement		
Deep scattering layer organisms entrapped		
Settled filter feeders		
Zooplankton migrates in feeding range		
Predators/grazers present		
Detritus build-up present		
Detritivores present		
Small resident invertebrate predators		
Small resident fish predators		
Resident cephalopods		
Aggregating deep sea fish		
Visiting fish predators		
Visiting elasmobranch predators		
Visiting marine turtles		
Visiting mammal predators		
Visiting seabird predators		
Total ecological status		
Fisheries		
Trawl fishery		Presence and status
Longline fishery		
Handline fishery		
Purse seine fishery		
Others		
Total fisheries status		
Conservation concern issues		
Corals and benthos damage		
Turtle by-catch issues		
Shark by-catch issues		
Dolphin by-catch issues		
Whale by-catch issues		
Seabird by-catch issues		
Others		
Total conservation concern status		

A characteristic trophic signature for seamounts

If the seamount-as-a-larder may have the attributes set out in Table 14.1, can we detect and measure a trophic signature for the seamount food web? We used a generic spatial seamount ecosystem simulation model (taken from Morato and Pitcher, 2005;

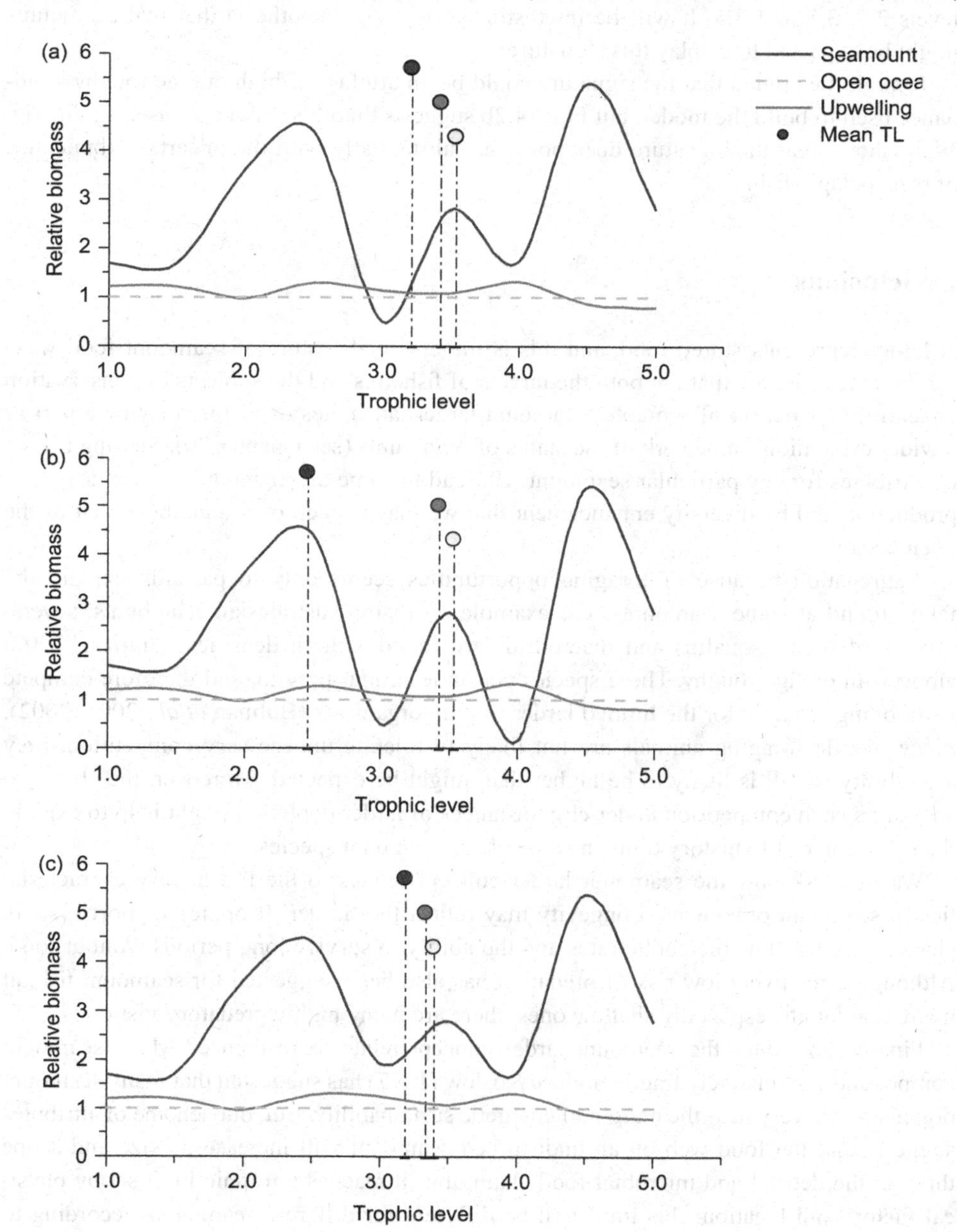

Fig. 14.2 (a) Trophic signature for seamount ecosystem obtained from a generic seamount ecosystem model. Biomass is plotted by trophic level relative to open ocean for 39 functional groups. (b) As top, but with orange roughy biomass the same in all habitats. (c) As top, but with mesopelagic abundances 30 times greater.

Chapter 15) to output biomasses and trophic levels of 30 functional groups in the system. Biomasses, partitioned by three locations (at the seamount, in upwelling regions and in the open ocean) were grouped in 0.5 trophic level categories. Figure 14.2a shows the abundances relative to open ocean values plotted by trophic level, leading to a clear 'seamount signature' from the model with peaks of biomass at approximately trophic levels 2.5, 3.5 and 4.5. It will be interesting to test our hypothesis that real seamounts might be expected to display this signature.

It might be argued that the signature could be an artefact of high orange roughy abundance used to build the model, but Fig. 14.2b suggests that this is not the case. Again, Fig. 14.2c shows that the signature does not alter substantially with the uncertain abundance of mesopelagic fish.

Conclusions

A larder represents stored food, and this is the essential feature of seamount food webs for the many visitors that are both the targets of fisheries and the subjects of conservation concerns. Our matrix of scorable 'seamount larder' attributes or factors may be a part of a wider evaluation framework of the status of seamounts (see Chapter 20). Scoring the set of attributes for any particular seamount will lead to some understanding of the degree of production and biodiversity enhancement that we may expect, over and above that of the open ocean.

Aggregation because of foraging opportunities seems only to partially explain the fauna found at some seamounts. For example, in many Australasian seamounts, several species of oreos, squalids and macrourids are found at high densities (Bulman, 2002) along with orange roughy. These species consume similar prey to, and therefore compete with, orange roughy for the limited larder of prey organisms (Bulman *et al.*, 2001, 2002). Since mobile foraging animals are not likely to tolerate unnecessary competition, prey availability for all is likely to be higher than might be expected. Moreover, the ability to withstand such competition under circumstances of larder depletion might help to explain the evolution of life history traits in some of the seamount species.

We may ask how the seamount larder concept relates to the life history characteristics of seamount organisms. Longevity may reflect the 'larder' (trophic) support system, characterised by low metabolic rates and the ability to survive long periods without food. Although a relatively low risk of predation has also been suggested for seamount fish, at many seamounts, especially shallow ones, there are many highly predatory visitors.

Finally, how does the seamount larder concept relate to resilience? Many seamount components are relatively fragile; indeed Koslow (1997) has suggested that many seamount organisms are very near the margin of energetic sustainability. But, our scheme of attributes suggests that the food web on an undisturbed seamount will increase its size and scope through the detrital and microbial food chain until it reaches a trophic limit set by physical factors and location: this limit will be different for different seamounts according to the factors. So perhaps the seamount food web is not as fragile after all; perhaps it is the 'extra' components that are labile. For example, foraging visitors can easily go somewhere else if the food supply on a seamount is poor. Currents, depth and proximity factors are fixed and will lead to the development and maintenance of some of the food web if there is enough time. But the detritus 'larder' probably takes a very long time to develop its maximum potential. Again, this system endows seamounts with resilience to natural climate fluctuations, but as it relies on sessile organisms, it is easily destroyed by trawling. Inevitable gaps in field data might be remediated in part by using the hypotheses we raise here, and by meta-analysis and seamount food web modelling.

References

Bulman, C.M. (2002) Trophic ecology and food web modelling of midslope demersal fishes off southern Tasmania, Australia. PhD Thesis, University of Tasmania, Hobart, Australia.

Bulman, C., Althaus, F., He, X., Bax, N. and Williams, A. (2001) Diets and trophic guilds of demersal fishes of the south-eastern Australian shelf. *Marine and Freshwater Research*, 52, 537–48.

Bulman, C.M., He, X. and Koslow, J.A. (2002) Trophic ecology of the mid-slope demersal fish community off southern Tasmania, Australia. *Marine and Freshwater Research*, 53, 59–72.

Comeau, L.A., Vezina, A.F., Bourgeois, M. and Juniper, S.K. (1995) Relationship between phytoplankton production and the physical structure of the water column near Cobb Seamount, northeast Pacific. *Deep-Sea Research I*, 42, 993–1005.

Dower, J.F. and Mackas, D.L. (1996) 'Seamount effects' in the zooplankton community near Cobb Seamount. *Deep-Sea Research I*, 43, 837–58.

Dower, J.F. and Perry, R.I. (2001) High abundance of larval rockfish over Cobb Seamount, an isolated seamount in the Northeast Pacific. *Fisheries Oceanography*, 10(3), 268–74.

Dower, J., Freeland, H. and Juniper, K. (1992) A strong biological response to oceanic flow past Cobb Seamount. *Deep-Sea Research*, 39, 1139–45.

Eriksen, C.C. (1998) Internal wave reflection and mixing at Fieberling Guyot. *Journal of Geophysical Research*, 103, 2977–94.

Fedosova, R.A. (1974) Distribution of some copepods species in the vicinity of an underwater Hawaiian Ridge. *Oceanology*, 14, 724–7.

Genin, A. (2004) Bio-physical coupling in the formation of zooplankton and fish aggregations over abrupt topographies. *Journal of Marine Systems*, 50, 3–20.

Genin, A. and Boehlert, G.W. (1985) Dynamics of temperature and chlorophyll structures above a seamount: an oceanic experiment. *Journal of Marine Research*, 43, 907–24.

Genin, A., Dayton, P.K., Lonsdale, P.F. and Spiess, F.N. (1986) Corals on seamount peaks provide evidence of current acceleration over deep-sea topography. *Nature*, 322, 59–61.

Genin, A., Greene, C., Haury, L., Wiebe, P., Gal, G., Kaartvedt, S., Meir, E., Fey, C. and Dawson, J. (1994) Zooplankton patch dynamics: daily gap formation over abrupt topography. *Deep-Sea Research*, 41, 941–51.

Gillet, P. and Dauvin, J.-C. (2000) Polychaetes from the Atlantic seamounts of the southern Azores: biogeographical distribution and reproductive patters. *Journal of Marine Biology Assessment* UK, 80(6), 1019–29.

Gillet, P. and Dauvin, J.-C. (2003) Polychaetes from the irving, meteor and plato seamounts, North Atlantic ocean: origin and geographical relationships. *Journal of Marine Biology Assessment* UK, 83(1), 49–53.

Grigg, R.W. (1993) Precious coral fisheries of Hawaii and US Pacific islands. *Marine Fisheries Review*, 55, 50–60.

Haedrich, R.L. (1974) Pelagic capture of the epibenthic rattail *Coryphaenoides rupestris*. *Deep-Sea Research*, 21, 977–9.

Haney, J.C., Haury, L.R., Mullineaux, L.S. and Fey, C.L. (1995) Sea-bird aggregation at a deep North Pacific seamount. *Marine Biology*, 123, 1–9.

Hareide, N.-R. and Garnes, G. (1998) The distribution and abundance of deep water fish along the mid-Atlantic Ridge from 43 N to 61N. ICES CM: 1998/O: 39.

Haury, L., Fey, C., Newland, C. and Genin, A. (2000) Zooplankton distribution around four eastern North Pacific seamounts. *Progress in Oceanography*, 45, 69–105.

Huppert, H.E. (1975) Some remarks on the initiation of inertial Taylor Columns. *Journal of Fluid Mechanics*, 67, 397–412.

Huppert, H.E. and Bryan, K. (1976) Topographically generated eddies. *Deep-Sea Research*, 23, 655–79.

Huskin, I., Anadon, R., Medina, G., Head, R.N. and Harris, R.P. (2001) Mesozooplankton distribution and copepod grazing in the subtropical Atlantic near the Azores: influence of mesoscale structures. *Journal of Plankton Research*, 23, 671–91.

Itano, D.G. and Holland, K.N. (2000) Movement and vulnerability of bigeye (*Thunnus obesus*) and yellowfin tuna (*Thunnus albacares*) in relation to FADs and natural aggregation points. *Aquatic Living Resources*, 13(4), 213–23.

Klimley, A.P., Butler, S.B., Nelson, D.R. and Stull, A.T. (1988) Diel movements of scalloped hammerhead sharks, *Sphyrna lewini* Griffith and Smith, to and from a seamount in the Gulf of California (Mexico). *Journal of Fish Biology*, 33(5), 751–62.

Koslow, J.A. (1997) Seamounts and the ecology of deep-sea fisheries. *American Scientist*, 85, 168–76.

Kunze, E. and Llewellyn Smith, S.G. (2004) The role of small-scale topography in turbulent mixing of the global ocean. *Oceanography*, 17, 55–64.

Macdonald, G.A., Abbott, A.T. and Peterson, F.L. (1986) *Volcanoes in the Sea.* University of Hawaii Press, USA, Honolulu, HI.

McDaniel, N., Swanston, D., Haight, R., Reid, D. and Grant, G. (2003) Biological observations at bowie seamount August 3–5, 2003. Report to Fisheries and Oceans Canada, 25 pp.

Mitchell, N.C. (2001) Transition from circular to stellate forms of submarine volcanoes. *Journal of Geophysics Research*, 106, 1987–2003.

Mohn, C. and Beckmann, A. (2002) The upper ocean circulation at Great Meteor Seamount. Part I: Structure of density and flow fields. *Ocean Dynamics*, 52, 179–93.

Monteiro, L.R., Ramos, J.R., Furness, R.W. and del Nevo, A.J. (1996) Movements, morphology, breeding, molt diet and feeding of seabirds in the Azores. *Colonial Waterbirds*, 19, 82–97.

Montelli, R., Nolet, G., Dahlen, F.A. *et al.* (2004) Finite-frequency tomography reveals a variety of plumes in the mantle. *Science*, 303, 338–43.

Morato, T. and Pitcher, T.J. (2005) Ecosystem simulations in support of management of data-limited seamount fisheries. In: *Fisheries Assessment and Management in Data-Limited Situations* (eds. Kruse, G.H., Gallucci, V.F., Hay, D.E., Perry, R.I., Peterman, R.M., Shirley, T.C., Spencer, P.D., Wilson, B. and Woodby, D.), pp. 467–86. University of Alaska Fairbanks, Alaska Sea Grant.

Mouriño, B., Fernández, E., Serret, P., Harbour, D., Sinha, B. and Pingree, R. (2001) Variability and seasonality of physical and biological fields at the Great Meteor Tablemount (subtropical NE Atlantic). *Oceanologica Acta*, 24, 167–85.

Nesis, K.N. (1986) Cephalopods of seamounts in the western Indian Ocean. *Oceanology*, 26, 91–6.

Odate, T. and Furuya, K. (1998) Well-developed subsurface chlorophyll maximum near Komahashi No. 2 seamount in the summer of 1991. *Deep-Sea Research*, 45, 1595–607.

Reeves, R.R. and Mitchell, E. (1993) Status of Baird's beaked whale, *Berardius bairdii*. *Canadian Field-Naturalist*, 107(4), 509–523.

Rogers, A.D. (1994) The biology of seamounts. *Advances in Marine Biology*, 30, 305–50.

Rowden, A.A., Clark, M.R. and Wright, I.C. (2005) Physical characterisation and a biologically focused classification of 'seamounts' in the New Zealand region. *New Zealand Journal of Marine and Freshwater Research*, 39, 1039–59.

Royer, T.C. (1978) Ocean eddies generated by seamounts in the North Pacific. *Science*, 199, 1063–4.

Sedberry, G.R. and Loefer, J.K. (2001) Satellite telemetry tracking of swordfish, *Xiphias gladius*, off the eastern United States. *Marine Biology*, 139, 355–60.

Sibert, J., Holland, K. and Itano, D. (2000) Exchange rates of yellowfin and bigeye tunas and fishery interaction between cross seamount and near-shore fads in Hawaii. *Aquatic Living Resources*, 13(4), 225–32.

Sime-Ngando, T., Juniper, K. and Vezina, A. (1992) Ciliated protozoan communities over Cobb Seamount: increase in biomass and spatial patchiness. *Marine Ecology Progress Series*, 89(1), 37–51.

Turnewitsch, R., Reyss, J.-L., Chapman, D.C., Thomson, J. and Lampitt, R.S. (2004) Evidence for a sedimentary fingerprint of an asymmetric flow field surrounding a short seamount. *Earth Planetary Science Letters*, 222(3–4), 1023–36.

Vinnichenko, V.I. (1998) Alfonsino (*Beryx splendens*) biology and fishery on the seamounts in the open North Atlantic. ICES CM: 1998/O: 13.

Wessel, P. (2001) Global distribution of seamounts inferred from gridded Geosat/ERS-1 altimetry, *J. Geophysics Research*, 106(19), 431–41.

Williams, A. and Koslow, J.A. (1997) Species composition, biomass and vertical distribution of micronekton over the mid-slope region off southern Tasmania, Australia. *Marine Biology*, 130, 259–76.

WWF (2003) Management direction for the Bowie Seamount MPA: links between conservation, research, and fishing. World Wildlife Fund, Canada, 69 pp.

Chapter 15
Modelling seamount ecosystems and their fisheries

Beth Fulton, Telmo Morato and Tony J. Pitcher

Abstract

Ecosystem models are one way of implementing an integrated approach to the recovery of heavily depleted systems like those on and over seamounts. We review the three main types of ecosystem models that have been applied in understanding seamounts, their ecology and fisheries; we describe *Ecopath-with-Ecosim* food web simulation models, agent-based models, and a more complex biogeochemical spatially explicit model (*Atlantis*). We discuss their capabilities, limitations, uncertainty, advantages and disadvantages, and present some examples. Ecosystem models of seamounts enable us to evaluate true fisheries sustainability in an ecosystem context and analyse alternative management scenarios for seamount ecosystems.

Introduction: seamounts and ecosystem models

The world's continental shelf fisheries resources have been seriously overfished, creating new pressures on alternative fishing grounds in a process of serial depletion. Like the deep sea more generally (Morato *et al.*, 2006a), seamounts are among those newly targeted ecosystems that have been intensively fished since the second half of the twentieth century (Chapters 3, 16, and 17). Targeted seamount fish communities tend to form localized aggregations and are highly vulnerable to overfishing (Morato *et al.*, 2006b; Chapters 9 and 19). The experience of the past few decades has been that fish of a certain seamounts area are typically depleted within 5–10 years of exploitation (Chapter 17). Once depleted, seamount populations will likely require decades to recover. Side effects caused by overfishing or extensive trawling on seamounts are also raising serious concerns: for example, damage to benthic communities (Chapter 19) typically dominated by corals (Chapter 8) and other suspension feeders (Chapter 7), and impacts on transient migratory species whose life history relies on seamount food webs (Chapters 10 and 12). Because of the complex trophic links on seamounts (Chapter 14), ecosystem-based modelling approaches can help in understanding ecosystem functioning and in planning effective management of these fragile ecosystems

(Chapter 20). However, there have been only a few attempts to model seamount biota as a part of marine ecosystems.

Sustainable resource management has been an issue of prominent, public, political, and scientific concern for at least 20 years (Holling, 1978; Sainsbury, 1988; IWC, 1992; McLain and Lee, 1996; Butterworth and Punt, 1999; Sainsbury *et al.*, 2000; Pitcher, 2001). Within the last decade ecosystem approaches to fisheries management have become more widely accepted, and popular political objectives (Pitcher, 2000). This is evident from the wide range of ecosystem-based local, national, and international declarations, guidelines and legislation. Examples are Australia's National Strategy for ESD 1992 and Oceans Policy 1998, Canada's Oceans Act 1997, the Reykjavik Declaration 2001, the World Summit on Sustainable Development 2002, and the European Union's Common Fisheries Policy 2003. Despite the reservations of some scientists about the uncertainty associated with bringing ecosystem considerations into fisheries management (e.g., Rice, 2000), exploratory tools such as ecosystem models have much to offer strategically (Christensen and Pauly, 2004; Walters and Martell, 2004). It is clear from a wide variety of systems from around the globe that single species-by-species management cannot account for, or cope with, multispecies and habitat impacts (Sainsbury and Sumaila, 2003; Hall and Mainprize, 2005). These types of impacts can reverberate through an ecosystem and ultimately lead to an unanticipated decline in the target species or the loss of other vulnerable species (Pauly *et al.*, 2002). Ecosystem models are one way to achieve a more integrated consideration of the topics and explore recovery and rebuilding scenarios for heavily depleted systems like seamounts (Pitcher, 2005). Answering many of the big questions concerning ecosystem functioning, marine resource management and prediction of future trends requires the integration of a wide range of scientific knowledge, encompassing physical, chemical, and biological processes. Ecosystem modelling is therefore a key technique for investigating how these processes interact.

In this chapter, we discuss why modelling approaches are necessary for seamounts; describe different approaches that have been used, discussing the capabilities, limitations, uncertainty, advantages, and disadvantages of each of them. We also present some examples of ecosystem models for a generic seamount, discuss fisheries sustainability in an ecosystem context, and analyse alternate management scenarios for seamount ecosystems.

Ecosystem modelling techniques

Ecosystem models developed over the past decade can be classified into three forms:

(i) Trophodynamic (or aggregate) ecosystem models based on food webs and energy budgets.
(ii) Biogeochemical models based on nutrient dynamics in addition to trophic and non-trophic interactions.
(iii) Individual- (or agent-) based (IBM or ABM) models that may employ a range of agents from individuals to populations, and typically use a mix of decision trees and differential equations to focus on system processes. ABMs can focus easily on the different scales (biological, spatial, or temporal) required, and can help to elucidate emergent properties not intrinsic to the agents themselves.

Each technique has its own strengths and weaknesses, is characterized by different internal structure and is better suited to answering particular questions. They can all aid our understanding of systems and their response to pressure and change by capturing the multidirectional flows inherent to system dynamics. Their particular strength is in simulating how human (or other) impacts can potentially have unanticipated effects across the system (Hollowed *et al.*, 2000; Mace, 2001; Fulton *et al.*, 2003). The three approaches are compared in Table 15.1.

Another common feature of ecosystem models is their complexity. While this can give them the potential to analyse complicated mathematical interactions, it means they are typically presented as time-dependent simulation models. The degree of detail in the models also means they may require more data; and outputs can be uncertain and difficult to summarize and interpret (Silvert, 1981; Jørgensen, 1994; Duplisea, 2000; Mace, 2001). Fulton *et al.* (2003) consider that this is not a serious constraint if the models are being used for strategic consideration of broad issues rather than finely focused tactical management questions. For instance, ecosystem models are useful in the consideration of possible impacts of fishing or implications of alternative management policies. The results can be treated with more confidence if they are tested across a range of plausible alternative model structures and assumptions (Reichert and Omlin, 1997; Duplisea, 2000; Fulton *et al.*, 2003). This type of modelling involves tradeoffs between simplicity and parameter availability, and the utility of being able to capture real ecological processes with reasonable accuracy: recent advances in fitting models to times series data improve confidence in their use.

Trophodynamic ecosystem models

Trophodynamic ecosystem models are typically constrained by, or initialized with, mass balance and hence they can more easily make use of existing information and meta-analyses. They are useful for the consideration of production, multispecies and fisheries effects on marine ecosystems, comparison of alternative fishing regulations (e.g., *Ecopath with Ecosim, EwE*; Walters *et al.*, 1997; Christensen and Walters, 2004a), comparison among practical rebuilding policies (Pitcher, 2005; Ainsworth and Pitcher, 2005; Pitcher *et al.*, 2005a), and spatially explicit versions are useful for evaluating spatial dynamics and the utility of closed areas (e.g., Sayer *et al.*, 2005). Among the many tools meant to assist in the time consuming, data exigent, multidisciplinary, and uncertain task of producing ecosystem models of marine systems, the *EwE* modelling approach and software has proved to be one of the most successful. It has yielded some helpful insights (Robinson and Frid, 2003) and is a powerful tool that has recently seen widespread use (Whipple *et al.*, 2000).

EwE has three main components (Walters *et al.*, 1997, 1999, 2000; Christensen and Walters, 2004a): *Ecopath*, which creates static, mass-balanced snapshots of the biomass pools in the ecosystem; *Ecosim*, which adds a time dynamic component for simulating policy exploration; and *Ecospace*, a spatial and a temporal dynamic simulation designed for exploring the impact and placement of fisheries management zones and protected areas. These tools are meant to address ecological questions, evaluate ecosystem effects of fishing, explore management policy options, analyse impact and placement of marine protected areas (MPAs), and model the effect of environmental changes (Christensen and Walters, 2004a; Fulton *et al.*, 2005; Pitcher *et al.*, 2005b; Gray *et al.*, 2006).

Table 15.1 Comparison of three model types used for seamount ecosystems.

Feature	Model *EwE*	*Atlantis*	*InVitro*
General features			
Biomass units	t/km^2 (wet weight)	mg N/m^3 (dry weight)	kg (wet weight) for mass and m for spatial dimensions
Input forcing	Yes (primary producers and recruitment), interannual	Yes (nutrient point sources and physical transport), interannual, seasonal, tidal	Yes (nutrient and environmental fields such as winds, rainfall, and tides), interannual, seasonal, tidal
Level of group detail	Variable (age group of species up to entire trophic levels)	Variable (species through to functional group)	Variable (individual of species, through schools to entire patches of habitat or assemblages/species complexes)
Software availability	User-friendly software is freely available and partially documented; users report many bugs in the software	Software in C is freely available but is restricted to expert use; documentation thorough, but are the responsibility of the use	No generally available software for non-expert use
Formulation related			
Consumption formulation	Forage arena	Mixed (types II and III typically)	Explicit interactions for IBM through to mixed (types II and III) for classical model components
Formulation detail	Simple (expansion of *Ecopath* master equation)	General (growth, mortality, and excretion explicit)	General through to fine details and decision trees (assimilation, basal/activity/stress respiration, defecation, excretion, ingestion, and mortality all explicit)
Nutrient limitation	No	Yes	Yes (abstracted in the lower trophic level field representations)
Light limitation	No	Yes	Yes
Oxygen limitation	No	Yes	Yes
Sediment chemistry	No	Yes (dynamic, with sediment bacteria)	No (not explicit, assumed to be effective)
Spatial structure	Not spatially explicit unless *Ecospace* model included	Explicit (coarse 3D, with polygons and slab vertical layers)	Explicit (multiple resolutions from continuous 3D for IBM to gridded or polygonal for habitat or fields)
Temperature dependency	Possible via forcing functions	Yes	Potentially
Age-structured vertebrates	Yes (juvenile + adult, stanzas)	Yes (multiple age classes)	Yes (from traditional age-structured population models to continuous time aged IBM)
Stock–recruit relationship	Dynamic	Multiple options available	Dependent on agent type (Beverton Holt through to explicit agent spawning and tracked development)

(*Continued*)

Table 15.1 *Continued*

Feature	Model *EwE*	*Atlantis*	*InVitro*
Stock structure	Self-seeding (closed stock)	Multiple stocks and external supply options possible	Multiple stocks (even multiple phenotypes) and external supply options possible
Non-trophic habitat dependency	Possible via facilitation relationships	Explicit dependency available	Explicit preferences (and threat matrix)
Movement	Gradient based on habitat preferences and seasonal migrations	Multiple options (e.g., habitat, forage, and density dependent) with weighting for seasonal migrations	Short-scale random walk wandering, with condition dependent directed habitat/forage searches and seasonal migration options
Industry related			
Sectors	Fisheries (explicit)	Fisheries (explicit), pollution, climate change, and habitat degradation (scenarios)	Fisheries, oil and gas, tourism, shipping, ports, mining, conservation (explicit); and climate change (scenario)
Management	F changes	Simple magnitude changes for all but fisheries where all major levers (input and catch controls) are considered	Sector management levers all explicit, plus options for regional scale integrated management
Economics	Yes (drivers)	Yes (drivers, trading, investment and disinvestment, and compliance)	Yes (drivers, compliance, and regional development)
Strengths (for modelling seamounts)	Easily make use of existing information and meta-analyses; inbuilt policy search optimizations; explicit biomass fitting available; Monte Carlo based viability analysis tools directly available*	Primary productivity and depth-based environmental forcing explicit; polygonal maps focus spatial attention effectively; migration in/out of the system explicitly handled; modular construction and multiple options for all processes (flexible mix-n-match construction)	Flexible hybrid form and asynchronous temporal handling allows for optimal representation of all physical and ecological system components (even transients) and supply
Potential weaknesses	Uncertainty associated with input parameters; forage arena dynamics; structural sensitivity; seasonal processes can be difficult to capture	Process and parameter intensive; large uncertainties; moderate-long model development time	Restricted food web representation (parts of webs rather than complete multi-part webs typical); can be computationally expensive (depending on agents selected can have multi-day run times); moderate-long model development time; large uncertainties; parameter intensive

*It is possible to perform Monte Carlo considerations for the other model types, indeed it is necessary in the stochastic *InVitro* model, but for *Atlantis* and *InVitro* the supporting tools to analyse and synthesis the various model outputs over the various runs must be carried out using additional software, whereas these tools are built into *EwE*.

The parameterization of an *Ecopath* model is based on satisfying two 'master' equations. The first describing how production for each group can be divided up, and the second is based on the principle of conservation of matter (Christensen and Pauly, 1992; Walters *et al.*, 1997; Pauly *et al.*, 2000). As a trophic mass-balanced model, *Ecopath* assumes that, for each functional group i in an ecosystem, mass balance should occur over a given time period (Christensen and Walters, 2004a). In the first 'master' equation, biomass production of a compartment (P_i) is balanced by catches (Y_i), predation mortality ($B_i \cdot M2_i$, where B_i is the biomass of the group and $M2_i$ is the total predation rate for the group), biomass accumulation (BA_i), net migration (E_i, emigration–immigration), and other mortality (MO_i), such that:

$$P_i = Y_i + B_i \cdot M2_i + E_i + BA_i + MO_i \tag{15.1}$$

Production is usually estimated from the production/biomass ratio (P/B) and the average annual biomass (B) and can be expressed as ($P_i = B_i \cdot (P/B)_i$). Predation mortality can be expressed as the sum of consumption by all predators (j) preying upon group (i), i.e.:

$$B_i \cdot M2_i = \sum_{j=1}^{n} B_j \cdot (Q/\mathrm{B})_j \cdot DC_{ji} \tag{15.2}$$

where $(Q/B)_j$ is the consumption/biomass ratio of the predator (j) and DC_{ji} is the fraction of the prey (i) in the average diet of the predator (j). The other mortality can also be expressed as:

$$MO_i = P_i \cdot (1 - EE_i) \tag{15.3}$$

where EE is the ecotrophic efficiency, or the proportion of the production that is not utilized in the system.

Substituting (15.2) and (15.3) into (15.1) means it can be re-expressed as:

$$B_i \cdot (P/B)_i \cdot EE_i = Y_i + \sum_{j=1}^{n} B_j \cdot (Q/B)_j \cdot DC_{ji} + E_i + BA_i \tag{15.4}$$

The second 'master' equation is:

$$Q_i = P_i + R_i + U_i \tag{15.5}$$

where R_i is respiration and U_i is unassimilated food.

Ecosim uses the *Ecopath* model to estimate its initial parameters. It then uses a system of differential equations of the form given in (15.6) to calculate the biomass fluxes between pools through time:

$$\frac{\mathrm{d}B_i}{\mathrm{d}t} = g_i \sum_j Q_{ji} - \sum_j Q_{ij} + I_i - (M_i + F_i + e_i) \cdot B_i \tag{15.6}$$

where the Q terms refer to predation on and consumption by group i, and are calculated using the 'foraging arena' concept (Christensen and Walters, 2004a). For further details regarding the equations and their solutions, see Walters *et al.* (1997), Walters *et al.* (2000), and Christensen and Walters (2004a).

Three features of *Ecosim* are especially useful. Firstly, time series of biomass estimates from abundance surveys of fish, bird, and marine mammals, or fish stock assessments may be used in a fitting process that adjusts predator–prey and diet parameters to minimize the differences between these data and simulated biomass trajectories (e.g., for 35 functional groups over 50 years in northern British Columbia; Ainsworth and Pitcher, 2005).

Secondly, climate factors derived from time series including tree rings, ice cores, upwelling indices, or ocean dynamics indices such as ENSO; can be used to drive, or partially drive, primary production and other parameters (such as recruitments) in the simulation. Climate-driven versions may be used in the fitting process described above, and residuals from the fitting can be examined to look for any further external factors.

Thirdly, Monte Carlo simulation of the major input parameters has begun to tackle the effect of parameter uncertainties upon forecast policy outcomes, as required in modern fishery analysis (Patterson *et al.*, 2001). The four parameters describing each biomass pool are selected at random from predetermined probability distributions: each selection of parameters (e.g., 156 for the 39-pool *EwE* seamount model given here) is challenged with mass balance, and a new random selection made if this is not achieved. Hence, many simulations of a chosen fishing policy are run, each with initialization parameters that meet the mass-balance criterion. In this way, an ecosystem-based viability analysis can be applied to alternative fishing policies, and the risks to policy (Francis and Shotton, 1997) from parameter uncertainty and climate made explicit (Pitcher *et al.*, 2005b). This promising technique needs to be further developed to include predator–prey parameters and model specification uncertainty.

Ecospace takes this assessment capability one step further by implementing the *Ecosim* dynamic simulation of biomass pools across a grid of cells linked by movement and migration parameters (Walters *et al.*, 1999). This allows for spatial dispersal, variation in productivity, different habitats and spatial fleet dynamics and management. *Ecospace* has been employed for evaluating cost and benefits of alternative size and placement of no-take MPAs (Walters, 2000; Cheung and Pitcher, 2005); the optimal size and location of MPAs (Beattie *et al.*, 2002); and the design, size and placement of protected fishing zones around human-made reefs (Pitcher *et al.*, 2002; Sayer *et al.*, 2005).

The *EwE* approach, its methods, capabilities, and pitfalls are described in detail by Christensen and Walters (2004a). *EwE* is a 'powerful and sophisticated tool' that, as with any modelling tool, needs to be used carefully, as without cautious, thoughtful and active use of the software, there is a risk of misleading results (Cochrane, 2002). This caveat applies to all of the complicated and complex modelling approaches discussed here. Unfortunately, limitations of the approach have lead to some criticisms; specific points of contention are the uncertainty associated with the input parameters, the assumption of foraging arena dynamics for predator–prey interactions (Plagányi and Butterworth, 2004), and the emergence of different results from different structural arrangements of the model biomass pools ('model specification uncertainty'; Pinnegar *et al.*, 2005).

In response, Christensen and Walters (2004a) recently reiterated that this approach was not intended to replace traditional single species assessment methods. Instead, it was developed to conduct fisheries policy analyses that cannot be addressed with traditional assessment methods. In this context, 'closed' and 'open' loop policy exploration routines have been developed and included in the software to facilitate the examination of ecological, social, and fishing industry aspects of fisheries and ecosystem-based management (Walters *et al.*, 2002).

An example *EwE* model structure that attempts to capture the ecology of a generic seamount is given in Fig. 15.1 (presented in greater detail later in this chapter). It serves to illustrate the general form of an ecosystem model and the steps involved in its construction. While *Ecospace* models are not seen as commonly as *Ecopath* and *Ecosim* models, other ecosystem models are usually spatial, so for comparative purposes an example *Ecospace* map is given in Fig. 15.2.

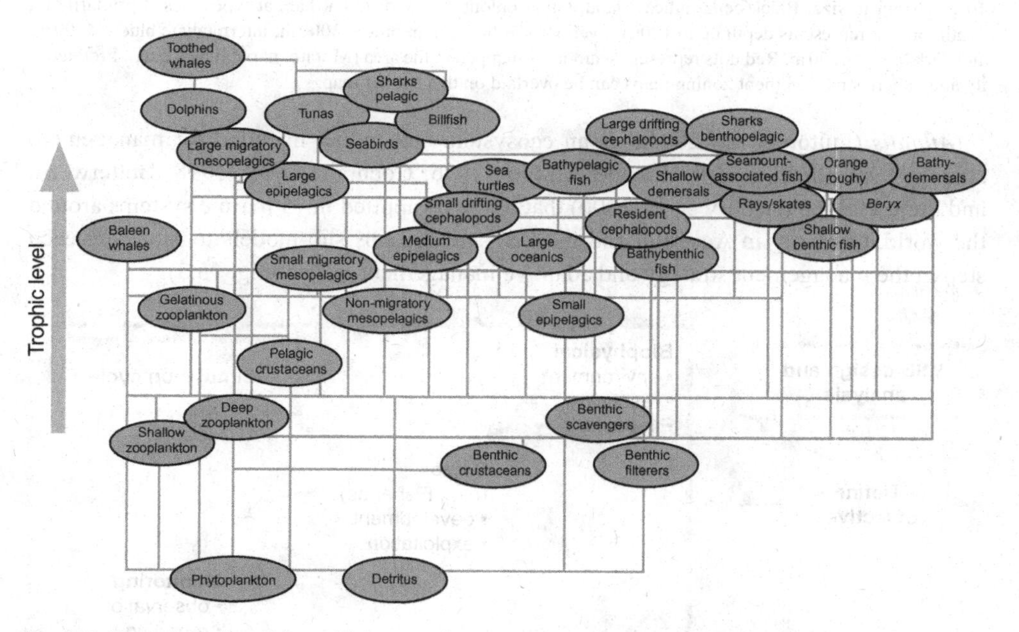

Fig. 15.1 Example *EwE* trophic model structure for a seamount ecosystem, showing functional groups modelled explicitly (ellipses) and major flows shown by thick blue lines. Vertical position shows trophic level.

Biogeochemical models

Biogeochemical ecosystem models follow the cycling of nutrients throughout the trophic web. They incorporate trophic and non-trophic interactions, and are often embedded in circulation models or their derivatives. These dynamic models grew from a background in water quality monitoring, e.g., the nutrient–phytoplankton–zooplankton (NPZ) physico-chemical models of Fransz *et al.* (1991) and Fasham (1993), but have been extended to consider all sections of marine ecosystems and the industries that impact upon them (e.g., ERSEM II, Baretta-Bekker and Baretta, 1997; *Atlantis*, Fulton *et al.*, 2004a).

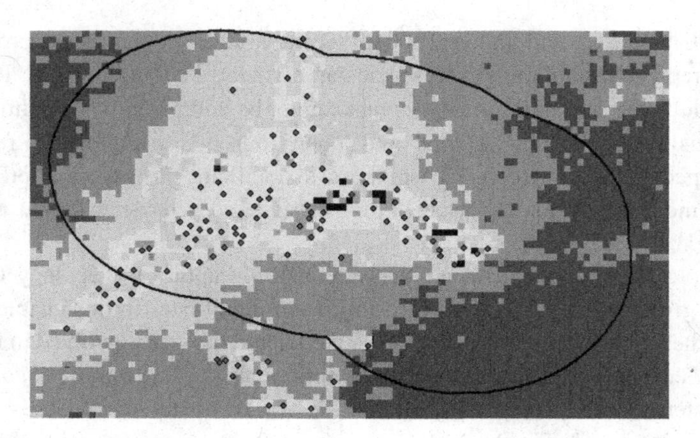

Fig. 15.2 Example *Ecospace* habitat map based on the bathymetry of the Azores EEZ (black line). Each cell is 10 × 10 min in size. Black cells indicate land. Other colours show different habitat types, based primarily on depth: orange represents depth up to 1000 m; yellow <2000 m; light blue <3000 m; intermediate blue <4000 m; and dark blue >4000 m. Red dots represent seamounts mapped in the area (Morato, unpublished data). Productivity and fisheries management zoning maps can be overlaid on this habitat map.

Atlantis (Fulton *et al.*, 2004a) is an ecosystem box model intended for management strategy evaluation (described in de la Mare, 1996; Cochrane *et al.*, 1998; Butterworth and Punt, 1999; Sainsbury *et al.*, 2000) that has been applied in 13 marine systems around the world, primarily in Australia and the USA. It contains sub-models to represent each step in the management strategy and adaptive management cycles (Fig. 15.3).

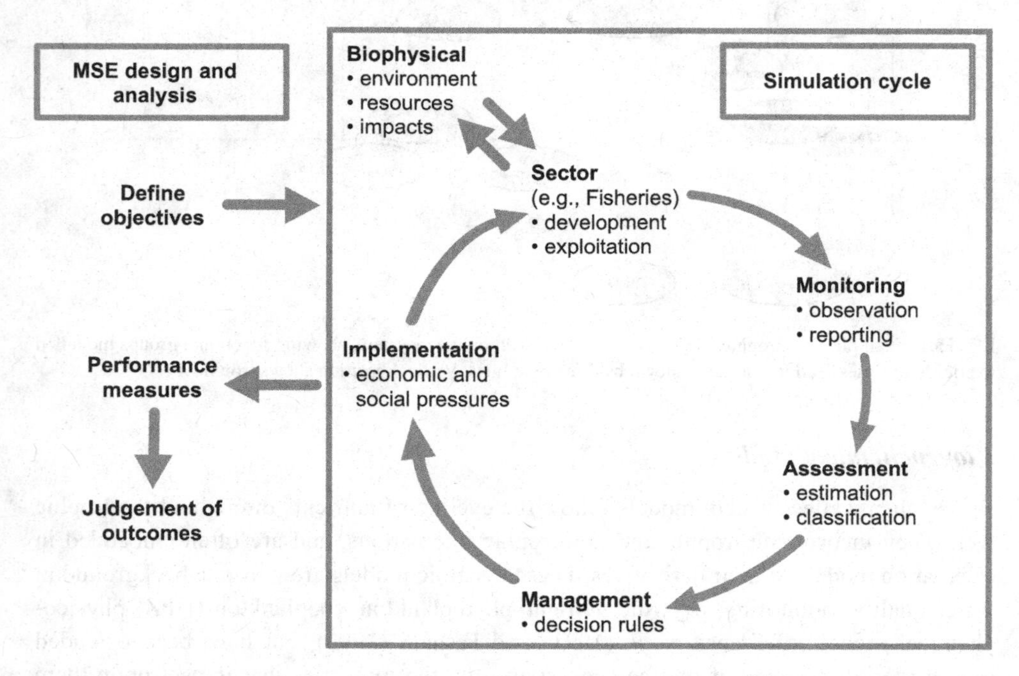

Fig. 15.3 Management strategy evaluation (MSE) cycle. The steps in the simulation cycle match those in the adaptive management cycle; within an MSE model (such as *Atlantis* or *InVitro*), each of these steps is represented by a sub-model.

At the core of *Atlantis* is a deterministic biophysical sub-model, coarsely spatially resolved in three dimensions, which tracks nutrient flows, usually expressed as nitrogen and silica, through the main biological groups in the system. The primary ecological processes modelled are consumption, production, waste production, migration, predation, recruitment, habitat dependency, and mortality. The trophic resolution is more evenly spread than in typical models constructed using *EwE* but, nonetheless, *Atlantis* aggregates lower trophic levels represented by invertebrates more heavily than higher trophic levels, representing them as biomass pools, while vertebrates are represented using an explicit age-structured formulation. The physical environment is represented explicitly, *via* a set of polygons matched to the major geographical and bioregional features of the simulated marine system. Biological model components are replicated in each depth layer of each of these polygons. Movement between the polygons is by advective transfer or by directed movements depending on the variable in question. Figure 15.4 shows an example *Atlantis* polygon map of a seamount for comparison with the *Ecospace* model structure in Fig. 15.2. Similarly, the trophic structure of *Atlantis* in Fig. 15.5 can be compared with the *EwE* structure in Fig. 15.1. For ease of comparison, the groups and maps in each example have been kept as similar as possible, while being typical of each model type.

Atlantis also includes a detailed exploitation sub-model. This model deals not only with the impact of pollution, coastal development, and broad-scale environmental change, but

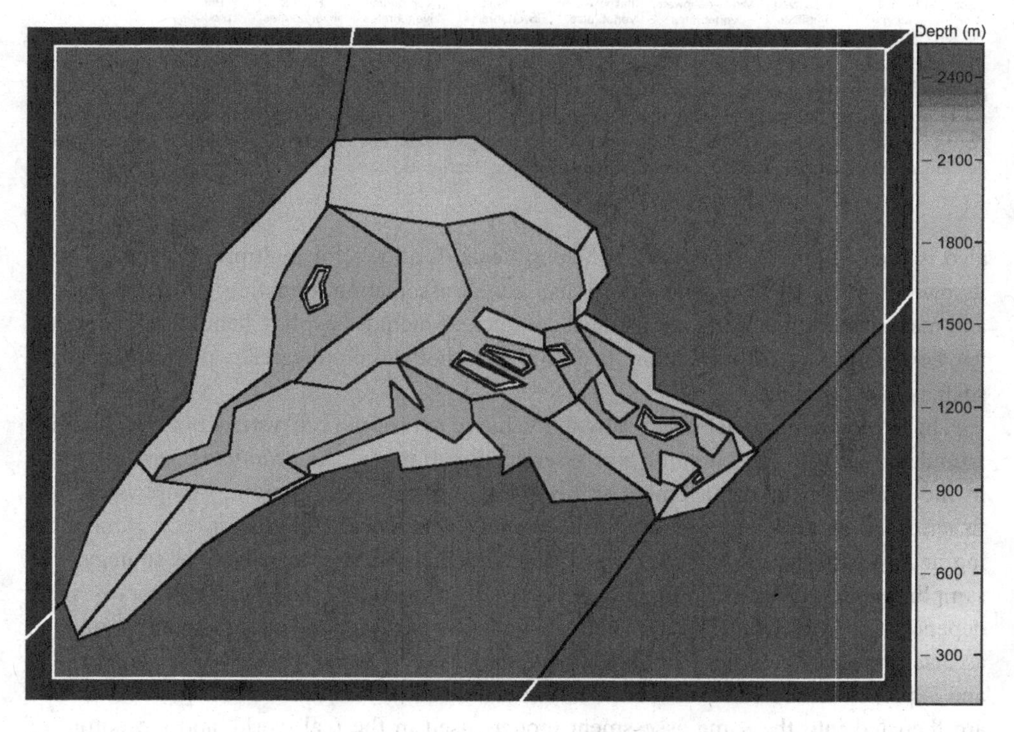

Fig. 15.4 Polygon map of example spatial structure for an *Atlantis* seamount model (based on the bathymetry of the Azores). The dark blue polygons represent boundary boxes, and the brown polygons are landmasses. Fisheries management zones and monitoring patterns can be overlaid on this basic map.

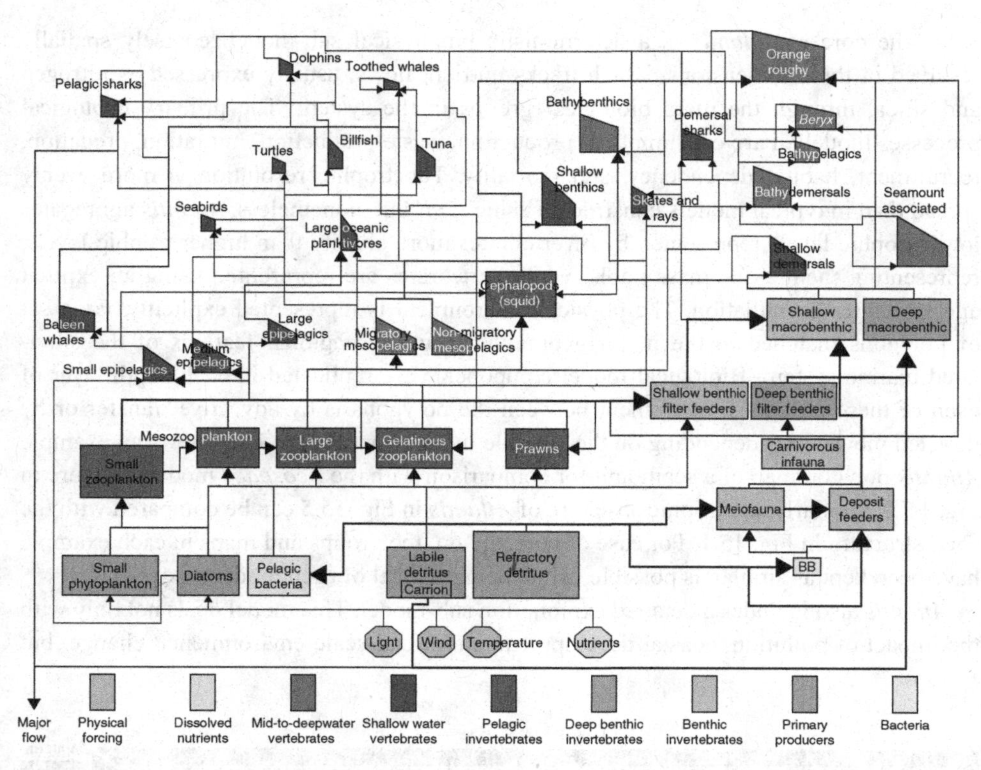

Fig. 15.5 Example *Atlantis* model structure. Boxes represent biological groups, rectangles represent biomass pools, and trapezoids are for age-structured vertebrate groups (with the shape proportional to the size structure in the population – left side showing contribution of small fish and right side the large fish in each population). Arrows show flows; arrow size is proportional to path strength. Octagons represent physical and chemical components of the biophysical model. Note that BB stands for Benthic Bacteria.

also is focused on the dynamics of fishing fleets. It allows for multiple fleets, each with its own characteristics of gear selectivity, habitat association, targeting, effort allocation, and management structures. At its most complex, it includes explicit handling of economics, compliance decisions, exploratory fishing, and other complicated real-world concerns such as quota trading.

The exploitation model interacts with the biotic part of the ecosystem, but also supplies 'simulated data' to the sampling and assessment sub-model. As *Atlantis* is primarily used in the evaluation of management strategies, it includes both operating biophysical and fisheries sub-models and assessment sub-models, which deal with the efficacy of monitoring and assessment models considered hand in hand with any management strategy. The sampling and assessment sub-models in *Atlantis* are therefore designed to generate sector dependent and independent data with realistic levels of measurement uncertainty evaluated as bias and variance. These simulated data are based on the outputs from the biophysical and exploitation sub-models, with a user-specified monitoring scheme as given. The data are then fed into the same assessment models used in the real world, and the output of these is input to a management sub-model, which is typically a set of decision rules and management actions, which are currently only detailed for the fisheries sector. The model framework includes an extensive list of fishery management instruments which includes gear restrictions, days at sea, quotas, spatial and temporal zoning, discarding restrictions, size limits, bycatch mitigation, and biomass reference points.

One of *Atlantis* greatest strengths is its modular construction. A wide range of alternative assumptions and sub-model implementations are provided so that a user has freedom to set complexity at any desired level, from a few groups with simple trophic interactions and a simple catch equation, to complex models with complicated stock structure, multiple fleets, detailed economics, and multiple management options. Another of its strengths is its ability to capture age–size dependencies and temporal and spatial variation, producing realistic ecosystem dynamics (Fulton *et al.*, 2004b). Unfortunately, this flexibility and mechanistic basis also makes the model process- and parameter intensive, which can make validation difficult and lead to large uncertainties. As with all ecosystem models, the output of biogeochemical models must be regarded with care.

Biogeochemical models are attractive for seamounts because they can incorporate explicitly the primary production and depth-based environmental forcing that makes these systems more productive than the surrounding waters. Models that use polygonal maps can more effectively focus spatial attention where needed, capturing critical dynamics while still achieving computational efficiency (Nihoul and Djenidi, 1998; Fulton *et al.*, 2004c).

Agent-based models

ABM (also termed IBMs) is a relatively new technique for ecosystem modellers, although it has long been used in evolutionary theory, archaeology, economics, and the social sciences. ABMs have great potential for increasing the breadth of resource management questions that may be considered. Early decision-based ABMs were often tightly focused on only a small subset of the entire system, for example a single fish in DeAngelis *et al.* (1993), or a small part of the food web, as in van Nes *et al.* (2002). Recent hybrid models incorporate a wide variety of ecosystem components and tie classical dynamic models, using differential equations, with decision-based agents. In this way, they allow for the best means of representing each ecosystem component to be used; for example, classical meta-population models can model habitats while ABMs may represent higher trophic levels or species of conservation concern, like whales.

InVitro (Gray *et al.*, 2006) is an example of just such a hybrid model. It has been built around the basic steps of management strategy evaluation, with a sub-model for each major step in the cycle. It is much broader than the typical fisheries focused ecosystem models discussed above and has the potential to simulate the effects of a wide range of human activities on a regional ecosystem, including fisheries, tourism, oil and gas, or salt production, and to forecast system response to different management regimes.

In practice, model construction depends upon tradeoffs between model complexity, available data or knowledge and computing resources. While *InVitro* can consider multiple uses of the marine environment it does so by constraining the list of ecological components of an ecosystem under consideration. In contrast to the complete trophic web included in models such as *EwE* or *Atlantis*, *InVitro* deals with a subset of the web that incorporates the dominant system components. These usually include commercially valuable fish and crustaceans, top predators, species of special interest such as vulnerable species like turtles, benthic communities, or forage communities if in the pelagic system, and primary producers. The behaviour and representation of the ABM agents is specific to their type, such that mobile agents may be represented as individuals, e.g., turtles and sharks, or small groups such as schools or sub-populations of finfish and prawns. In contrast, sedentary habitat-defining agents represent entire patches (e.g., entire seagrass beds, mangrove forests and reefs). This ecological structure is set within a continuous, three-dimensional environment model, which dictates the appropriate agent actions based on the bathymetry, currents, temperature, light intensity, chemical concentrations, habitat type, and resident communities. To complete the set of examples of seamount models, *InVitro* spatial maps are given in Fig. 15.6, and an example ecological model structure is presented in Fig. 15.7.

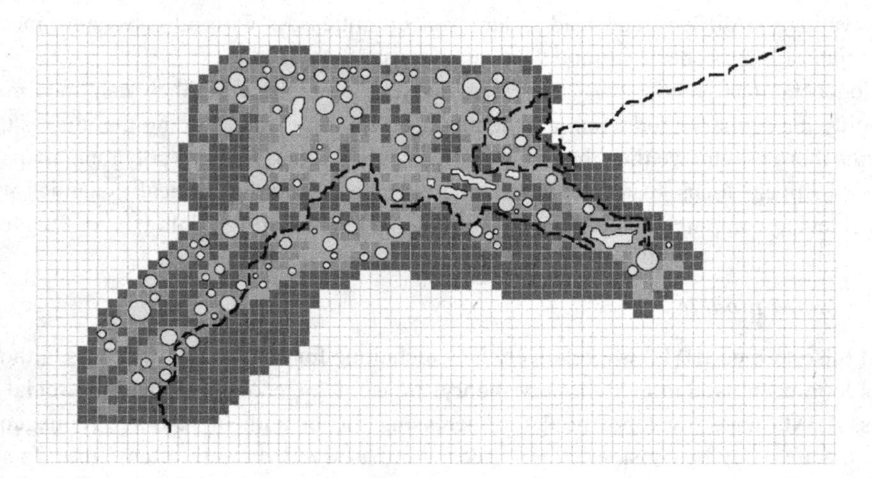

Fig. 15.6 Examples of the spatial maps (based on the Azores) that could be used in a seamount version of *InVitro*. The light grey grid (or a finer one) would contain bathymetry data (the islands are the irregular shapes marked in very pale green). The habitat quality model is represented by the red and green squares of different hue – it uses a grid (or set of polygons), which need not match the bathymetry grid as it does here. The habitat grid could use a different scale and cover a smaller area. Mobile animals (either individuals or small aggregations) are represented here by circles of various size and colour, with yellow representing a species preferring extensive biogenic habitat (green cells) and blue a species that prefers open ground (red and brownish cells); they are not restricted to the grid-scale resolution of space and actually operate in continuous three-dimensional space. The dotted line represents an example track that a transient (like a tuna) may have followed while visiting the seamount. As with *Atlantis* and *Ecospace*, fisheries management zones and monitoring patterns (and other anthropogenic features, such as vessel traffic routes) can be overlaid on these maps.

'Industry' sub-models in *InVitro* employ its hybrid nature to make the best use of existing representations of the different sectors. In practice, this has meant that sectors other than fishing are represented *via* simple empirical models that mimic the impact and basic action of these industries. They include nutrient pollution, salt extraction, shipping, tourism, coastal development, conservation, aquaculture, and the oil and gas sector. Fisheries sub-models are much more detailed, using economic and mechanical drivers for a true IBM, with Kalman filters to represent the process of learning and updating of individual fisher's knowledge (Gray *et al.*, 2006). A simple management model is in place for each industrial sector, though again the fisheries model is more structured, allowing for management of gear, target species, bycatch, discards, catch rates, quotas, and spatial closures. *InVitro*'s

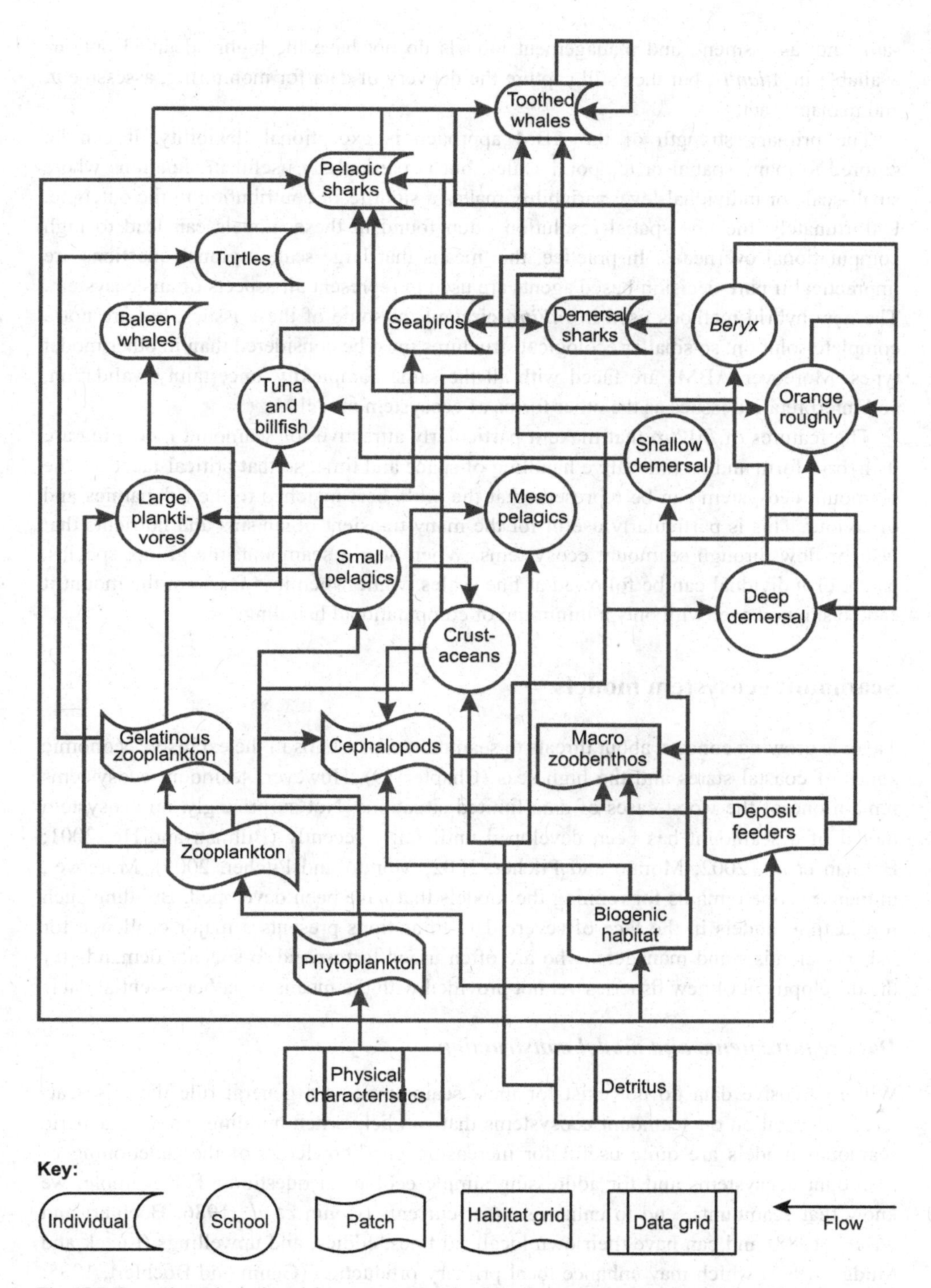

Fig. 15.7 Example ecological model structure and major flows for a seamount version of the ABM *InVitro*. The shape of the components indicates the agent type (see key). Note that in comparison with the *EwE* and *Atlantis* models there are fewer ecological components in the *InVitro* biophysical model structure.

sampling, assessment, and management models do not have the highly detailed options available in *Atlantis*, but they still capture the delivery of data for monitoring, assessment, and management.

The primary strength of the ABM approach is exceptional flexibility. It can be tailored to many spatial or temporal scales, but is particularly useful in situations where small-scale or individual-level variability makes a significant contribution to the outcome. Unfortunately, the fine spatial resolution often found in these models can lead to high computational overheads. In practice, this means that large-scale regional questions are impractical if pure decision-based agents are used to represent all aspects of an ecosystem. The new hybrid methods used in *InVitro* circumvent some of these issues, but are not a complete solution, so smaller ecological structures must be considered than in other model types. Moreover, ABMs are faced with all the same complexity, uncertainty, validation, and interpretation issues as the other forms of ecosystem model.

The features of *InVitro* that make it particularly attractive for seamount modelling are its hybrid form and its innovative handling of space and time, so that critical facets of the seamount ecosystem can be represented at the scale best matched to their dynamics and behaviour. This is particularly useful for the many transient organisms and nutrients that visit or flow through seamount ecosystems. When near a seamount, the group, species, patch, or individual can be followed at fine scales, while when it is far from the mount it can be skipped over with only a minimum of computational handling.

Seamount ecosystem models

There is growing concern about threats to seamount ecosystems in the exclusive economic zones of coastal states and the high seas (Chapter 19). However, seamount ecosystems remain one of the worst cases of data-limited situations. Not surprisingly, no ecosystem model of a seamount has been developed until fairly recently (Bulman and He, 2001; Bulman *et al.*, 2002; Morato and Pitcher, 2002; Morato and Pitcher, 2005). Moreover, immense scope remains for refining the models that have been developed. Building such forecasting models in the face of severe data limitations presents a major challenge for fishery scientists and managers, who are often urged to respond to societal demands for the development of new fisheries yet not provided with the means to gather essential data.

Data requirements and model construction

While extensive data do not exist for most seamounts, as a general rule there is scattered information on seamount ecosystems that can help when building models. Generic seamount models are quite useful for increasing our knowledge of the functioning of seamount ecosystems and for addressing simple ecological questions. For example, we know that seamounts tend to enhance water currents (Genin *et al.*, 1986; Boehlert and Sasaki, 1988) and can have their own localized tides, eddies, and upwellings (Lueck and Mudge, 1997) which may enhance local primary production (Genin and Boehlert, 1985; Dower *et al.*, 1992; Odate and Furuya, 1998; Mouriño *et al.*, 2001). Zooplankton biomass is often high over seamounts (Fedosova, 1974; Dower and Perry, 2001; Huskin *et al.*, 2001), though evidence for this is equivocal since some studies detected reduced abundance of zooplankton above seamounts due to grazing (Genin *et al.*, 1994; Haury *et al.*,

2000), while other studies reported no differences in zooplankton biomass compared to the open sea (Voronina and Timonin, 1986; Dower and Mackas, 1996; see Chapter 5).

It is also known that seamounts are characterized by the presence of substantial aggregations of deep-bodied fishes in the water column (Boehlert and Sasaki, 1988; Koslow, 1996, 1997; Koslow *et al.*, 2000). These aggregations are supported by the enhanced flux of prey organisms past the seamounts and the interception and trapping of vertical migrators by the uplifted topography (Tseytlin, 1985; Genin *et al.*, 1988; Koslow, 1997). Sharks (Klimley *et al.*, 1988; Hazin *et al.*, 1998), tuna (Fonteneau, 1991; Holland *et al.*, 1999; Itano and Holland, 2000; Sibert *et al.*, 2000), billfishes (Sedberry and Loefer, 2001), cephalopods (Nesis, 1986), and marine mammals (Reeves and Mitchell, 1993) also congregate over seamounts to feed, and even seabirds (Haney *et al.*, 1995; Monteiro *et al.*, 1996) have been shown to be more abundant in the vicinity of shallow seamounts. This productive nature extends to the benthos; enhanced currents and steep slopes expose the volcanic rocks and favour the growth of suspension feeders in these benthic seamount communities (Genin *et al.*, 1986; Grigg *et al.*, 1987; Wilson and Kaufmann, 1987; Rogers, 1994). Consequently, seamounts often support rich communities dominated by suspension feeders like gorgonians and corals (Richer de Forges *et al.*, 2000; Koslow *et al.*, 2001; Ohkushi and Natori, 2001), features that may be particularly susceptible and sensitive to disturbance by trawling (Probert *et al.*, 1997; Koslow *et al.*, 2001).

Information of this nature can be used to construct an ecosystem model for a general seamount. To start this process, it is informative to begin by building a sketch of a 'typical' seamount system (Fig. 15.8). As noted above, ecosystem models that span large proportions of the entire system usually have differential treatment of the various trophic levels. Lower trophic levels are nearly always represented using aggregate functional

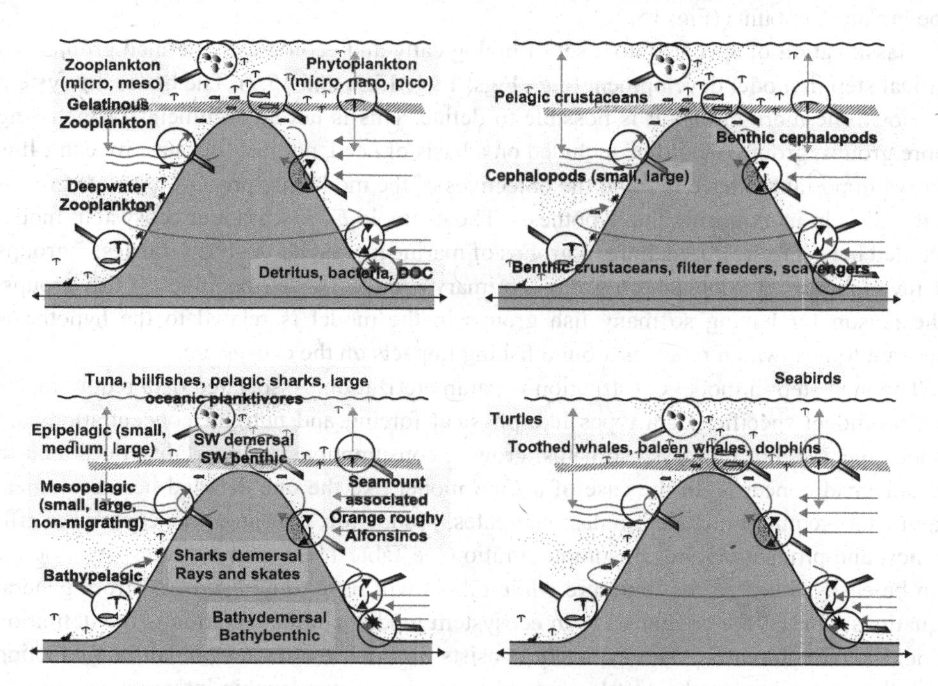

Fig. 15.8 The biological and hydrological components of a generic seamount model. Spyglasses show key biotic components. Orange arrows represent vertical migrations. Blue arrows show water flow (see also Chapter 14).

groups and biomass pools, whereas the higher trophic level vertebrates are treated in much greater detail, often divided into subgroups based around taxonomy, stock structure, ontogeny, and even behaviour.

One of the most difficult things to grapple within very open systems such as seamounts is how to handle ecological processes that occur outside the system but have an impact within the system. For instance, many species spend a variable part of the year or life-cycle away from the seamount. How is consumption or mortality during that period to be handled? How is production brought in *via* the currents to be handled? In the example model given here these complications have been simplified. In a more location-specific model they may have to be considered in careful detail (as was the case for Koslow, 1997; Bulman *et al.*, 2002).

Interaction through predation is not the only process animals experience. Most importantly, fisheries are also removing part of the system. Globally, seamount fisheries around the world have been conducted mainly with technologically developed deepwater trawls, targeting dense resident fish aggregations (Sasaki, 1986; Clark, 1999; Humphreys and Moffitt, 1999; Clark *et al.*, 2000; Koslow *et al.*, 2000; Boyer *et al.*, 2001; Branch, 2001; Hareide *et al.*, 2001; Vinnichenko, 2002; Bergstad *et al.*, 2003; Lack *et al.*, 2003; Chapter 17). Orange roughy (*Hoplostethus atlanticus*), pelagic armourhead (*Pseudopentaceros wheeleri*), blue ling (*Molva dipterygia*), alfonsino (*Beryx splendens*), roundnose grenadier (*Coryphaenoides rupestris*), and redfish (*Sebastes marinus*) have been particularly targeted. Other fisheries take advantage of the aggregation of pelagic species on top of seamounts, such as tuna, billfish, and pelagic sharks. On seamounts close to islands, semi-artisanal fisheries may also develop, and often target resident demersal fish species and invertebrates. Hence, using available information we can identify six types of fisheries that operate on seamounts (Fig. 15.9).

Classification of species into a set of biologically and ecologically defined groups is a critical step in model development (e.g., Figs. 15.1, 15.5, and 15.7). The better the system is known the more groups it is possible to define. This is not an argument for including more groups, groups should be included on a basis of need, not just 'because we can'. It is always important to have in mind the objectives of the modelling process to create groups that will help in exploring the hypothesis. The example *EwE* seamount ecosystem model detailed here (Fig. 15.8), includes 3 groups of marine mammals, seabirds, turtles; 7 groups of invertebrates; 3 zooplankton groups, primary producers, detritus; and 20 fish groups. The reason for having so many fish groups in the model is related to the hypothesis we want to test, which revolves around fishing impacts on the ecosystem.

The next step in model construction is parameterization. Associated data requirements can be model specific. Data types like physical forcing and nutrient concentrations are model specific, but biomass estimates, growth, consumption, and mortality rates tend to be universal concerns. In the case of a *EwE* model like the one detailed here ecological parameters required include biomass estimates, production, consumption, ecotrophic efficiency, and production and consumption ratio (see Table 15.2). Some of these parameters can be easily found in the literature while others will have to be estimated from general equations. One of the crucial steps in ecosystem model parameterization is the definition of an interaction matrix. This typically consists of a diet matrix for predators interacting with their prey, but can be supplemented by a series of non-trophic interactions, such as

Fig. 15.9 Potential fisheries operating around seamounts located close to islands. Those seamounts that are far from any shore may be target of pelagic and deepwater longliners, and deepwater trawlers: (a) deepwater trawl (targeting seamount-associated species, including orange roughy and alfonsinos) (photo courtesy Alan Blacklock, NIWA, New Zealand); (b) swordfish fishery – deepwater longline (targeting bathypelagic and bathybenthic fish species; (c, e) demersal longline (targeting shallow water demersal and benthic fish species); (d) tuna fishery; and (f) small pelagics fishery. Photos b-f courtesy of *Imag*DOP.

the provision of cover against predators. The final basic trophic structure of our example *EwE* seamount model is shown in Fig. 15.10. Within the system, phytoplanktonic primary production and detritus are the sources of energy that sustain zooplanktonic secondary production and benthos. Predator groups benefit from linking benthic and pelagic food chains at the top of the food web.

The final step in model construction is to incorporate the human components. These need not only focus on fisheries–resource interactions, though they are the primary focus of this publication. Fisheries and other anthropogenic data are often incorporated using an interaction matrix as shown in Table 15.3, that includes parameters for the species caught by each fleet, bycatch organisms, prices, costs, and profits. This is supplemented by parameters specific to the processes represented in the particular fleet dynamics or in another exploitation model.

Table 15.2 Parameters of a 37-component trophic model for a potential seamount ecosystem in the North Atlantic. Italicized values, trophic levels and omnivory index were estimated by the model. Parameter values are based on those from a model of seamounts in the Azores.

Group name	Trophic level	Omnivory index	Biomass (t/km²)	Production/ biomass (per year)	Consumption/ biomass (per year)	Ecotrophic efficiency
Detritus	1.00	0.33	100.000	–	–	*0.046*
Phytoplankton	1.00	0.00	7.160	290.00	–	*0.358*
Benthos filter feeders	2.00	0.00	0.595	0.80	9.00	*0.619*
Benthic crustaceans	2.00	0.00	3.858	1.60	10.00	*0.951*
Shallow zooplankton	2.11	0.11	16.684	*11.21*	37.38	*0.774*
Deep zooplankton	2.23	0.21	6.849	*8.70*	29.00	*0.683*
Benthos scavengers	2.28	0.24	3.089	1.83	13.57	*0.868*
Pelagic crustaceans	2.72	0.32	*5.161*	1.45	9.67	0.950
Gelatinous zooplankton	3.08	0.20	*8.893*	0.85	2.00	0.800
Small epipelagic	3.10	0.22	0.859	2.05	19.87	*0.567*
Non-migrating mesopelagic	3.12	0.01	*1.419*	0.50	1.57	0.950
Small migrating mesopelagic	3.37	0.11	2.000	1.98	8.00	*0.982*
Bathybenthic fishes	3.52	0.35	*1.142*	0.20	0.50	0.950
Baleen whales	3.56	0.49	0.123	0.06	5.56	*0.032*
Large oceanic planktivores	3.56	0.34	0.003	0.11	2.07	*0.138*
Medium epipelagic	3.59	0.33	0.113	1.08	10.75	*0.904*
Shallow benthic fishes	3.71	0.33	0.820	0.59	4.70	*0.923*
Cephalopods resident	3.76	0.32	0.189	2.89	10.00	*0.871*
Cephalopods drifting small	3.83	0.23	*0.175*	4.45	16.86	0.950
Rays and skates	3.90	0.44	0.020	0.17	1.50	*0.701*
Beryx spp.	3.91	0.21	1.620	0.06	2.00	*0.501*
Shallow demersal fishes	4.03	0.39	*0.193*	0.66	5.20	0.950
Sea turtles	4.08	0.00	0.001	0.15	3.50	*0.976*
Seamount-associated fishes	4.13	0.16	3.230	0.06	2.20	*0.791*
Bathypelagic	4.15	0.52	*0.030*	0.50	1.48	0.950
Large epipelagic	4.18	0.23	0.014	0.69	5.10	*0.609*
H. atlanticus	4.18	0.23	41.930	0.05	2.00	*0.402*
Bathydemersal fishes	4.18	0.33	*1.278*	0.20	0.60	0.950
Cephalopods drifting large	4.33	0.51	0.001	2.50	10.00	*0.566*
Large migrating mesopelagic	4.34	0.25	*0.002*	0.60	3.55	0.950
Seabirds	4.36	0.26	0.001	0.04	84.39	*0.026*
Sharks benthopelagic	4.39	0.44	0.030	0.51	6.90	*0.137*
Billfishes	4.54	0.12	0.025	0.50	4.20	*0.081*
Dolphins	4.58	0.16	0.040	0.07	11.41	*0.054*
Tunas	4.58	0.12	0.040	0.74	16.29	*0.597*
Sharks pelagic	4.70	0.57	0.015	0.30	3.10	*0.702*
Toothed whales	5.16	0.13	<0.001	0.02	10.27	*0.513*

After parameterization comes calibration in which the model is fitted as best as possible to available data such as biomass or catch time series. For the example *EwE* model being discussed here, a summary of the resulting 'balanced' seamount model and its parameter set is given in Table 15.2. The final total biomass of the modelled ecosystem, excluding detritus, was estimated as 108 t/km². This generic seamount is assumed to support large aggregations of fish, which account for 44% of the total biomass of the system. Primary producers contribute 7.2 t/km² forming 6.7% of the total biomass, whereas at the other extreme toothed whales, pelagic sharks, tunas, and dolphins occupy the highest trophic levels, contributing 0.1% of the total biomass.

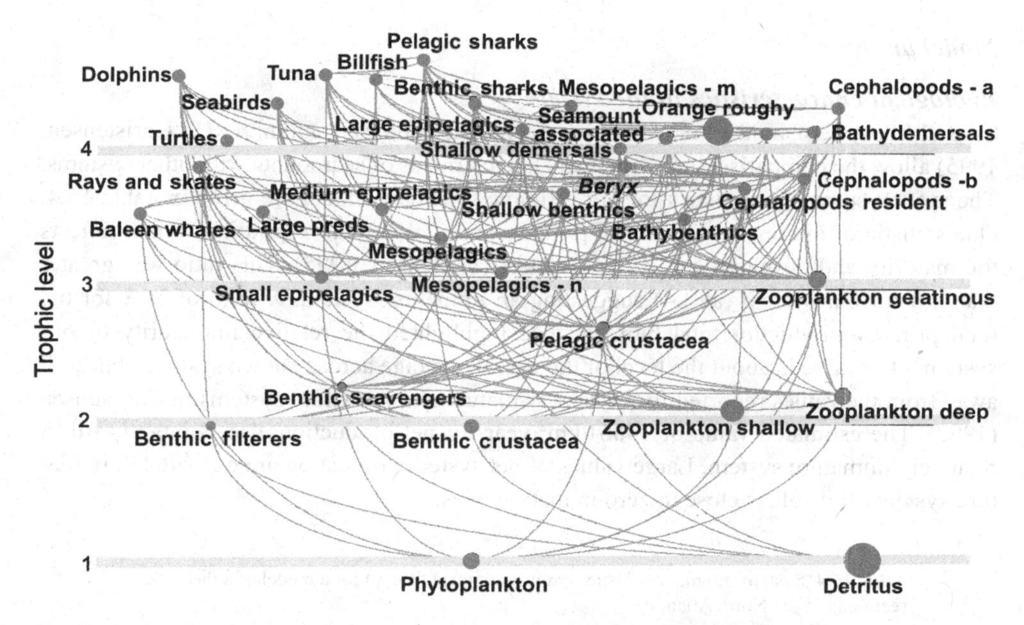

Fig. 15.10 Flow diagram of a seamount ecosystem used in ecosystem simulation modelling. The size of the nodes is roughly proportional to the biomass of each group.

Table 15.3 Fisheries in the example seamount ecosystem model (Morato and Pikcher, 2005). Average landings (t/km^2/year) estimated for the different fisheries considered in the theoretical seamount.

Group name	Landings by fleet (t/km^2/year)						t/year
	BL	DWL	SP	T	SW	DWT	Total
Tunas				0.011			30.0
Billfishes					0.001		2.7
Sharks pelagic	0.001				0.001		5.5
Sharks benthopelagic	0.001					0.001	5.5
Rays and skates	0.002						5.5
Small epipelagic			0.05				136.4
Medium epipelagic			0.01				27.3
Shallow benthic fishes	0.080						218.2
Shallow demersal fishes	0.020						54.6
Seamount-associated fishes		0.001				0.010	30.0
H. atlanticus						0.010	27.3
Beryx spp.	0.005					0.005	27.3
Bathypelagic		0.005				0.001	16.4
Bathybenthic fishes		0.002				0.001	8.2
Bathydemersal fishes	0.001	0				0.001	5.5
Total (t/km^2/year)	0.11	0.008	0.06	0.011	0.002	0.029	600.4
Number of jobs per catch	20	20	9	3	3	1	

Data on number of jobs per catch value is also presented.
BL: bottom longline; DWL: deepwater longline; SP: small pelagic fishery; T: tuna; SW: swordfish; DWT: deepwater trawl.

Model analysis

Ecological characteristics of the system

Several ecosystem indices drawn from theoretical ecology (Odum, 1971; Christensen, 1995) allow the comparison of ecological characteristics of seamounts with other systems. The system summary statistics for our example seamount model are given in Table 15.4. One statistic of interest is the ratio of primary production/total respiration, which reflects the maturity and development of the system (Christensen, 1995). This ratio was greater than 1.0 for our system (at 3.4) suggesting an immature system. A ratio of 19.3 for the total primary production/total biomass also highlighted the relative immaturity of our system – the value is about the level of the overall average across known systems, but well away from the values reported for extremely mature and immature systems in Christensen (1992). The estimated value of 1466 t/km^2/year for net production also points towards a relatively immature system. Large values of net system production are expected in immature systems and values close to zero in mature ones.

Table 15.4 System summary statistics (from ecological theory) for a model of a theoretical seamount in the North Atlantic.

Parameter	Value
Sum of all consumption (t/km^2/year)	1119.9
Sum of all exports (t/km^2/year)	1465.8
Sum of all respiratory flows (t/km^2/year)	610.5
Sum of all flows into detritus (t/km^2/year)	1623.7
Total system throughput (t/km^2/year)	4820
Sum of all production (t/km^2/year)	2361
Mean trophic level of the catch	4.08
Gross efficiency (catch/net primary production)	0.0005
Calculated total net primary production (t/km^2/year)	2076
Total primary production/total respiration	3.4
Net system production (t/km^2/year)	1466
Total primary production/total biomass	19.3
Total biomass/total throughput	0.022
Total biomass (excluding detritus) (t/km^2)	107.6
Total catches (t/km^2/year)	1.09
Connectance index	0.20
System omnivory index	0.23

The net primary production of 2076 t/km^2/year required to sustain the system was similar to the estimated primary production of 2030 t/km^2/year in a North Atlantic *EwE* model (Guénette and Morato, 2001). It was, however, much lower than well recognized highly productive systems, like the Benguela which has a calculated total net primary production of over 7000 t/km^2/year (Heymans *et al.*, 2004). It is interesting to note that while the system is relatively immature and not overly productive it does have a reasonably well-connected web. The system omnivory index value of 0.23 indicates that the model system is fairly stable in its current form, with reasonably high levels of interaction diversity.

The indicators discussed here are just a few examples of ecosystem indices that can be used to consider the final form of systems. They allow for comparisons between systems and for a consideration of system state and type, and can give insights into the system dynamics and existing pressures. For instance, the values indicate that seamounts similar in form to the generic one represented by the model may be relatively ecologically immature. The potential for fisheries and management to cause changes in the state of an ecosystem has been a topic of increasing interest for many years (e.g., Christensen, 1992).

Fisheries in the system

The role of commercial fisheries in the example seamount model is equivalent to a predator occupying a mean trophic level of 4.08; which is at the same level as carnivorous sea turtles, but below that of the largest cephalopods, mesopelagic and benthic species and typical top predator groups, which include seabirds, billfishes, sharks, dolphins, tunas, and toothed whales. Figure 15.11 shows the corresponding mean trophic levels of each fishing fleet and the annual catch. Fisheries for small pelagic fish occupy the lowest

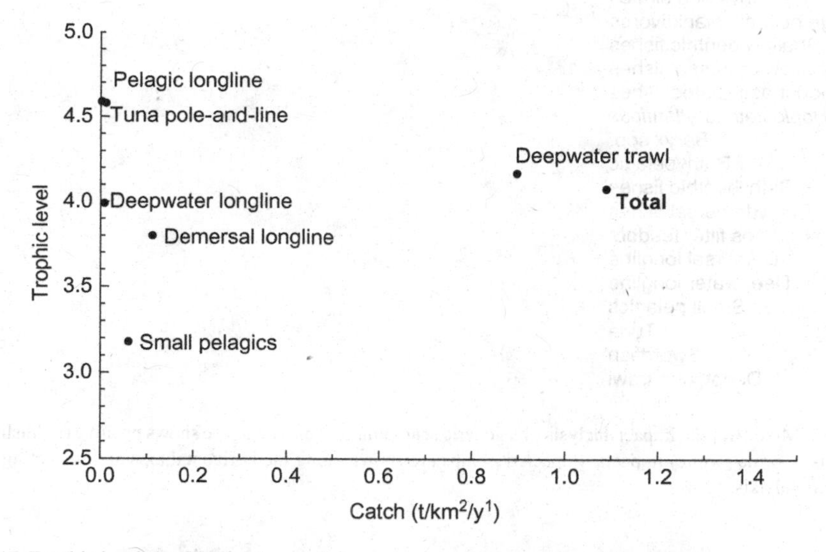

Fig. 15.11 Trophic levels by fleet for the catch from an example seamount model.

trophic levels, matching the trophic level of the small epipelagic and mesopelagic groups. The demersal longline fishery has the next lowest trophic level of 3.80. It is only slightly below the overall fisheries average trophic level at 4.08, which puts it on a footing with small and resident cephalopods, as well as skates and rays. The deepwater longline and trawl fisheries have intermediate trophic levels, which are close to the overall average. The tuna and swordfish fisheries had the highest trophic levels, which is as high as the largest of the top predators represented by the pelagic sharks and toothed whales. This shows that the fisheries have the potential to compete with the biggest predators, which are often the main species of concern (Fig. 15.12). The magnitude of our simulated catch is such that it does not put too much pressure on the system as shown by simulations, which demonstrate stability over a period of 25–50 years.

Fig. 15.12 Mixed trophic impact analysis of a generic seamount model. The figure shows positive (reddish), negative (bluish), or no (white) impacts. Impacted groups are shown along the horizontal axis and impacting groups on the vertical axis.

Reconciling fisheries with conservation on seamounts

Seamount fisheries: 'mining' fish aggregations

Seamount fisheries have recently attracted attention due to their increased importance and recognized impact on ecosystems. However, information about seamount fisheries is very sparse and it is usually quite difficult to make a distinction between deepwater fishing activities in general and those occurring on seamounts. Moreover, fish species living on seamounts are known to occur in other habitats, such as the continental slope. Since catch statistics are rarely spatially allocated, it is difficult to obtain an estimate of the total fisheries occurring on seamounts worldwide, even though seamount fisheries are usually assumed to be very important economically (Chapters 16–19). Seamount fisheries have been mainly conducted using technologically advanced deepwater trawl gear to target dense fish aggregations. Because of the life history of targeted species (Morato *et al.*, 2006b), deepwater fisheries in general, and seamounts fisheries in particular, are usually characterized by a boom and bust sequence (Koslow *et al.*, 2000; Hareide *et al.*, 2001; Chapter 19). In some oceanic islands in the South Pacific, or in the Azores and Madeira archipelagos in the North Atlantic, there has been a long history of artisanal handline fisheries for deepwater species. Currently, these seamount fisheries that operate with semi-industrial longline, handline, and pole-and-line are believed to be sustainable (Chapter 16). The typical boom and bust sequence has not been observed, although symptoms of stock decline have been identified (Santos *et al.*, 1995; Menezes, 2003; Chapter 16).

We may then ask whether seamount fisheries, especially those prosecuted through deepwater trawling, may be sustainable in the long term (Clark, 2001). Several scientific studies (e.g., Hopper, 1995; Merrett and Haedrich, 1997; Moore, 1999; Moore and Mace, 1999; Probert, 1999; Roberts, 2002), environmental agencies (Lack *et al.*, 2003), and governments have strongly advocated an urgent need for implementation of fishing regulations for deepwater fisheries. At least in part, these could be *via* the establishment of MPAs, or a ban on deepwater trawling in sensitive habitats, like seamounts.

Tradeoffs between conservation and fisheries

There is a rising concern about threats to seamount ecosystems in the exclusive economic zones of coastal states and the high seas. Several countries, such as Canada, Australia, New Zealand, and Portugal have begun to undertake action for the protection of such 'fragile' ecosystems. That said, seamount ecosystems remain one of the worst cases of data-limited situations, comprising a true challenge for fishery scientists and managers, who are urged to respond to societal demands for development of new fisheries, under the precautionary approach of the FAO Code of Conduct. Modelling allows preliminary exploration of management regimes such as MPAs of different sizes, trawl bans, or combinations of measures with the current open access situation. While a low risk of depletion and/or adverse ecosystem effects in a model scenario will not necessarily reflect what would happen on the ground, scenarios that show high risk send a clear alarm signal. The models therefore help us to implement a precautionary approach as advocated by the UN Code of Conduct for Responsible Fisheries (FAO, 1995).

Morato and Pitcher (2005) explored the types of fisheries that might be sustainable on seamount ecosystems using used the *EwE* policy search options of Christensen and Walters (2004b). They used a general *EwE* model developed for North Atlantic seamounts, which is identical to the example *EwE* model presented here. The model covers an area of about 3000 km^2, includes 37 functional groups, and is assumed to have a low initial level of exploitation with landings varying from 0.08 t/km^2/year for shallow benthic fishes to 0.001 t/km^2/year for billfishes. In this context, Morato and Pitcher (2005) used *EwE*'s policy optimization features to evaluate the time patterns of relative fishing effort by fishing fleet. The objectives considered were

1. 'Economy', or maximization of net economic value evaluated as total value of the catch.
2. 'Jobs', or employment maximization. This is a social indicator, assumed proportional to gross landed value of catch for each fleet with a different jobs/landed value ratio for each fleet.

3. 'Ecology', or maximized ecological 'stability' measured by assigning a weighting factor to each group based on their longevity, and optimizing for the weighted sum.

The resulting integrated objective function can be thought of as a 'multi-criterion objective', represented as a weighted sum for the three above-mentioned objectives; with the final objective function given by:

$$OBJ = W_V \sum NPV_{ijt} + W_J \sum J_{ijt} + W_E \sum \left(B/P \right)_{ijt} \tag{15.7}$$

where W, weighting factors; V, value; J, jobs; E, ecology; i, fleets; j, species caught; t, time in years; NPV, net present value; and B/P, biomass production ratio, assumed to be proportional to species longevity and thus ecological stability.

Optimal scenario solutions for a range of ecological and economic objective weightings were accessed to search for tradeoffs among objective functions. Surface plots (Fig. 15.13) clearly show that it is not possible to maximize the performance of all three objectives, net economic value, number of jobs, and ecological 'stability', simultaneously. Net economic value (Fig. 15.13a) is maximized with high economic and low ecological weighting factors, while number of jobs (Fig. 15.13b) is maximized with high weightings of both factors. This results in a decrease in the ecological 'stability' of the system. Conversely, to maximize 'ecosystem stability' (Fig. 15.13c), a high weighting must be given to 'ecology' and a low weighting to 'economy'. Not surprisingly, assigning a low weighting factor to 'economy' and a high weighting factor to 'ecology' results in a decrease of the net economic value and the number of jobs, with a corresponding increase in the system's ecological stability. Intermediate weightings produce, in general, intermediate performances for the three objective functions. Interestingly, this form of result is not restricted to seamounts and has been found for other systems, such as a shallow enclosed bay (Fulton and Smith, 2004).

The fishing rates required to achieve different performances of the objective functions (i.e., fishing policies) are presented in surface plots in Fig. 15.14. Maximizing net economic value of the system (lower-right corner of Fig. 15.14) required an increase in fishing rates for most fisheries. In this scenario, optimal fleet configuration would favour deepwater longline (Fig. 15.14a), deepwater trawl (Fig. 15.14b), and pelagic longline (Fig. 15.14c). In contrast, achieving ecological 'stability' in the system (upper-left corner of Fig. 15.14) required most fisheries to lower their fishing rates. In this case optimal fleet configuration will favour small a pelagic fishery (Fig. 15.14d), tuna pole-and-line fishery (Fig. 15.14e), and bottom longline fisheries (Fig. 15.14f).

The ecosystem indicators, mean trophic level of the catch and biodiversity, and total catch estimates derived from the optimal fishing strategies for the overall range of weighting factors for 'ecology' and 'economy' are presented in Fig. 15.15. Total catches were maximized when weighting was high for 'economy' and low for 'ecology' objective functions (Fig. 15.15a), resulting in low mean trophic level of the catch (Fig. 15.15b) and in a loss of biodiversity (Fig. 15.15c).

The analyses showed that simulations with policy objectives that maximize economic performance would favour a fleet configuration based on deepwater trawling, but having a

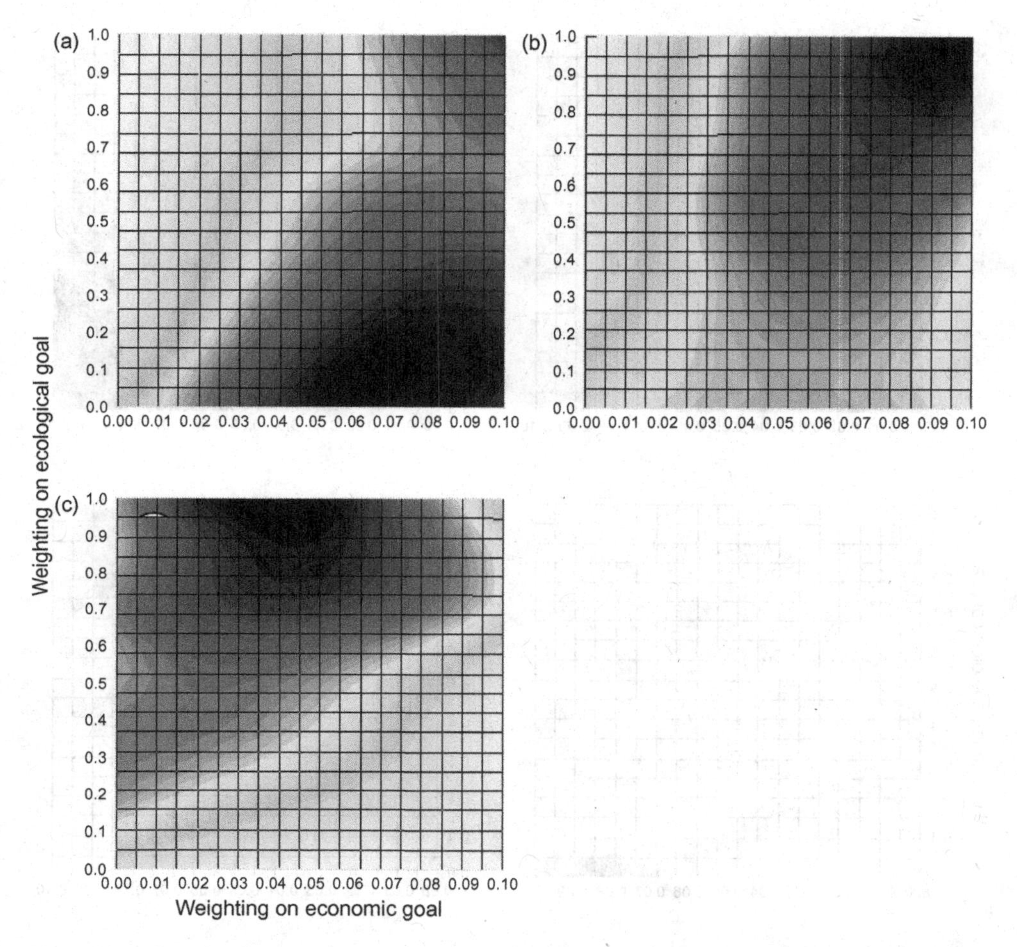

Fig. 15.13 Surface plots from a *EwE* model showing the relative performance of scenarios for a range of weightings of ecological and economic objective functions. Points on the surface represent solutions that optimize objective functions for (a) maximizing net economic value, (b) maximizing number of jobs, and (c) maximizing ecological 'stability'. Scale goes from light blue (low performance) to dark red (high performance). Smooth surface is interpolated.

cost for the ecosystem. Maximizing ecological performance will favour a fleet configuration based on small pelagic and bottom longline fisheries, but would sacrifice total catches and jobs, while maximizing the biomass of long-lived species and biodiversity. The overall study suggested that sustainable seamount fisheries with tolerable ecosystem impacts appeared to be closer to those found by maximizing the 'ecological' objective function.

This discussion has not dealt with spatial management options. It would be relatively easy to consider them, however, using an *Ecospace* extension of the example model presented here. Alternatively, if the same scenario were run using one of the other modelling methods discussed earlier, a range of other management approaches, including spatial management, and monitoring schemes could be considered. Each type of model is suited to answering only certain questions, so it is always good to use as many different models as possible, to cover as many angles of a topic as possible.

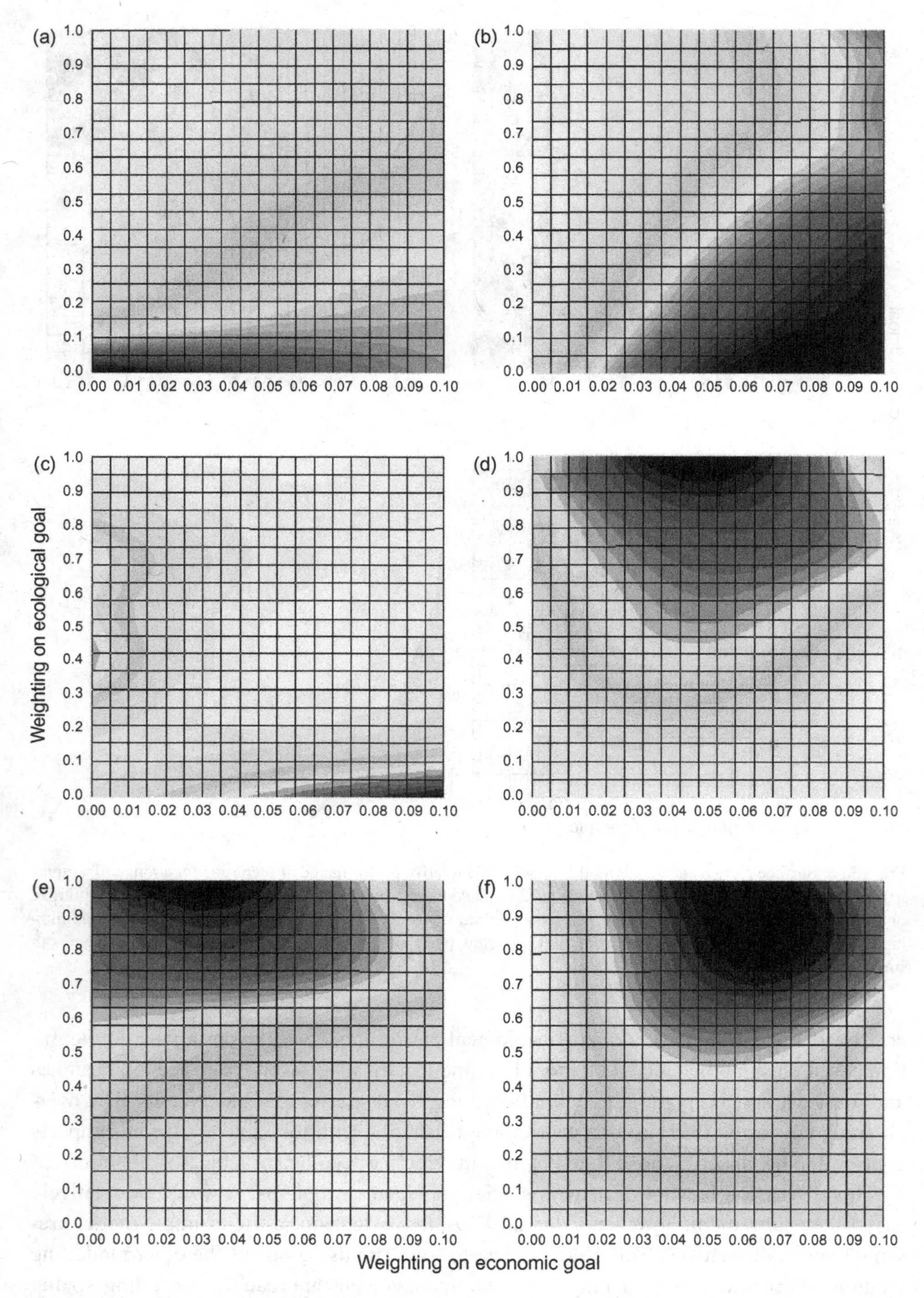

Fig. 15.14 Surface plots from a *EwE* model showing the resulting fishing rates, as proportion of base model rates, of the optimal scenario solutions for a range of weightings of ecological and economic objective functions: (a) deepwater longline, (b) deepwater trawl, (c) pelagic longline, (d) small pelagic fishery, (e) tuna fishery, (f) demersal (bottom) longline. Scale goes from light blue (low proportion of base model rates) to dark red (high proportion of base model rates). Smooth surface is interpolated.

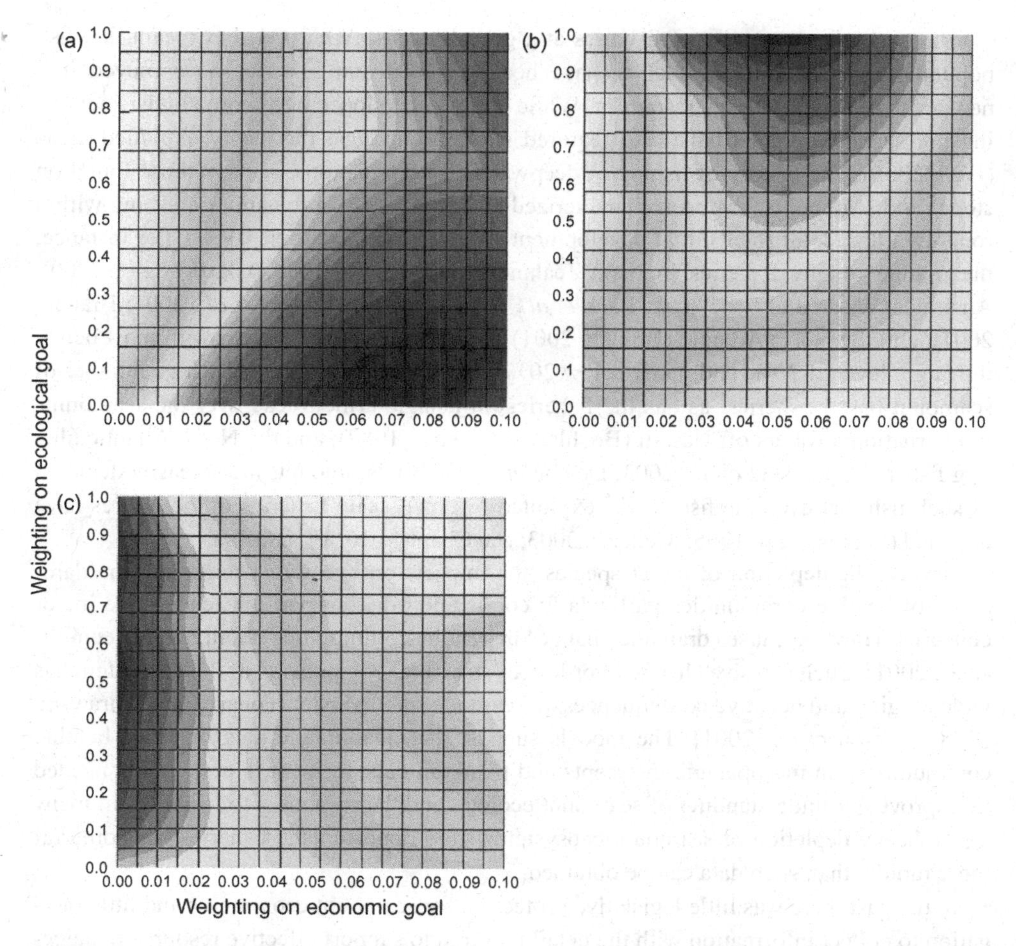

Fig. 15.15 Surface plots from a *EwE* model showing the resulting ecosystem indicators and total fisheries catch for the optimal scenario solutions for a range of weightings of ecological and economic objective functions: (a) total catch, (b) mean trophic level of the catch, and (c) biodiversity index (Q75). Scale goes from light blue (low) to dark red (high). Smooth surface is interpolated.

Models like these can be built in a few months, but refined versions that are robust enough to be used in management can take several years to build and validate.

Conclusions

The complex topography of seamounts can make them localized hotspots of productivity and biodiversity at all trophic levels, both in the pelagic and benthic zones. Seamounts also attract and concentrate marine life from wide surrounding areas making them natural targets for fisheries. Unfortunately, their constrained spatial nature and the life histories (long lived, slow growing, late maturing) of some of the most prominent species at these locations mean they are very vulnerable to over exploitation.

The generally poor state of fisheries at a global level (FAO, 2004) has meant that fish populations on continental shelves have been depleted and fisheries have moved into new areas in search of new resources. While continental slopes have come under increasing pressure, seamounts have also featured strongly amongst the newly exploited areas (FAO, 2005). The history of fishing of deepwater fish populations, such as those found on slopes and seamounts, has been characterized by boom and bust dynamics crashing within roughly a decade of their initial development (Merrett and Haedrich, 1997). For instance, the orange roughy fisheries in New Zealand (Clark, 1995, 1999; Clark *et al.*, 2000), Australia (Wayte and Bax, 2001; Lack *et al.*, 2003), Namibia (Boyer *et al.*, 2001; Branch, 2001), and the North Atlantic (Branch, 2001) have all shown this pattern with most below the target level of 30%, and many below 20%. Serial depletion occurred with a number of seamount-based fisheries; as has the fisheries on pelagic armourhead over the seamounts in international waters off Hawaii (Boehlert and Genin, 1987); and the North Atlantic blue ling fisheries (Bergstad *et al.*, 2003; Devine *et al.*, 2006). Seamounts are so easily depleted of such fish that even the fish stocks exploited by small-scale fisheries in the Azores have declined (Santos *et al.*, 1995; Menezes 2003; see Chapters 16, 17, and 19).

Beyond the depletion of target species, the impacts on bycatch species, and the damaging of benthic communities particularly corals and other suspension feeders are major concerns. Trawling causes dramatic changes in benthic communities (Goni, 1998; Jennings *et al.*, 2001). Such massive change in or loss of structural components of the ecosystem has wide ranging and negative consequences on seamount biodiversity (Fogarty and Murawski, 1998; Jackson *et al.*, 2001). The models suggest that quantitative data on these benthic communities, on mesopelagic intercepts and plankton enhancement is desperately needed to improve our understanding of seamount ecology and the responses to fisheries. In many cases, heavy depletion of seamount ecosystems from opportunistic fisheries has come far more rapidly than such data can be obtained.

In the past there was little legislative protection for seamount ecosystems and little obligation to collect information with the detail required to support effective resource management. Bodies such as the FAO have identified the need to rectify this situation. Progress has only been patchy to date, so supplementary tools are needed to consider the best means of managing seamount fisheries. Ecosystem models are one such tool. Their advantage over single species analysis is that they explicitly allow for evaluation of bycatch, habitat impacts, anthropogenic and environmental interactions, and the implications of alternative policy scenarios and management strategies can be considered. For example, based on the preliminary seamount model discussed here, protection at the level required under ecosystem-based management legislation can only ensue if ecological objectives outweigh economic ones. A switch from trawling to longlining might be called for to minimize disturbance to seabed habitats and associated fauna, but the model suggests that a switch to longlining may actually be detrimental if trawling pressure is not sufficiently reduced, as the longliners can access spatial refuges where stocks are safe from trawling.

The biology, ecology, and life history dynamics of many seamount ecosystem components remain poorly understood. Until these data gaps are filled and more is known about these fragile ecosystems and the long-term impacts of fishing, a precautionary ecosystem-based approach should be adopted to ensure appropriate management of all ecosystem components (FAO, 2005). In some cases, this may require substantial changes in fisheries practices, for example, halting fisheries that are unsustainable or switching to less damaging gears. The judicious use of ecosystem models for forecasts can help guide these decisions and help avoid unexpected consequences of new management actions. Comparisons among the different types of models may encourage greater confidence in their use, although this work is only just beginning.

Acknowledgements

T. Morato acknowledges support from the 'Fundação para a Ciência e Tecnologia' (Portugal, BD/4773/2001) and the European Social Fund through the Third Framework Programme.

References

Ainsworth, C.H. and Pitcher, T.J. (2005) Evaluating marine ecosystem restoration goals for northern British Columbia. In: *Fisheries Assessment and Management in Data-Limited Situations* (eds. Kruse, G.H., Gallucci, V.F., Hay, D.E., Perry, R.I., Peterman, R.M., Shirley, T.C., Spencer, P.D., Wilson, B. and Woodby, D.), pp. 419–38. Alaska Sea Grant, Fairbanks, Alaska.

Baretta-Bekker, J.G. and Baretta, J.W. (eds.) (1997) Special issue: European regional seas ecosystem model II. *Journal of Sea Research*, 38(3/4).

Beattie, A., Sumaila, U.R., Christensen, V. and Pauly, D. (2002) A model for the bioeconomic evaluation of marine protected area size and placement in the North Sea. *Natural Resource Modeling*, 15(4), 413–37.

Bergstad, O.A., Gordon, J.D.M. and Large, P. (2003) Is time running out for deepsea fish? ICES Webpage, http://www.ices.dk/marineworld/deepseafish.asp

Boehlert, G.W. and Genin, A. (1987) A review of the effects of seamounts on biological processes. In: *Seamounts, Island and Atolls* (ed. Keating, B.H.), pp. 319–34. American Geophysical Union, Washington, DC.

Boehlert, G.W. and Sasaki, T. (1988) Pelagic biogeography of the armorhead, *Pseudopentaceros wheeleri*, and recruitment to isolated seamounts in the North Pacific Ocean. *Fishery Bulletin*, 86, 453–66.

Boyer, D.C., Kirchner, C.H., McAllister, M.K., Staby, A. and Staalesen, B.I. (2001) The orange roughy fishery of Namibia: lessons to be learned about managing a developing fishery. In: *A Decade of Namibian Fisheries Science* (eds. Paine, A.I.L., Pillar, S.C. and Crawford, R.J.M.). *South African Journal of Marine Science*, 23, 205–11.

Branch, T.A. (2001) A review of orange roughy *Hoplostethus atlanticus* fisheries, estimation methods, biology and stock structure. In: *A Decade of Namibian Fisheries Science* (eds. Paine, A.I.L., Pillar, S.C. and Crawford, R.J.M.). *South African Journal of Marine Science*, 23, 205–11.

Bulman, C.M. and He, X. (2001) Modelling trophic interactions of the mid-slope demersal fish community off southern Tasmania, Australia. In: Trophic ecology and food web modelling of mid-slope demersal fishes off Southern Tasmania, Australia (ed. Bulman, C.M). PhD Thesis, University of Tasmania, Australia.

Bulman, C.M., Butler, A.J., Condie, S., Ridgway, K., Koslow, J.A., He, X., Williams, A., Bravington, M., Stevens, J.D. and Young, J.W. (2002) A trophodynamic model for the Tasmanian seamounts marine reserve: links between pelagic and deepwater ecosystems. *Final Report to Environment Australia*, CSIRO, Hobart, Tasmania, Australia.

Butterworth, D.S. and Punt, A.E. (1999) Experiences in the evaluation and implementation of management procedures. *ICES Journal of Marine Science*, 56, 985–98.

Cheung, W.L. and Pitcher, T.J. (2005) Designing fisheries management policies that conserve marine species diversity in the Northern South China Sea. In: *Fisheries Assessment and Management in Data-Limited Situations* (eds. Kruse, G.H., Gallucci, V.F., Hay, D.E., Perry, R.I., Peterman, R.M., Shirley, T.C., Spencer, P.D., Wilson, B. and Woodby, D.) pp. 419–38. Alaska Sea Grant, Fairbanks, Alaska.

Christensen, V. (1992) Network analysis of trophic interactions in aquatic ecosystems. PhD Thesis, Royal Danish School of Pharmacy, Copenhagen, Denmark, pp. 55+13 Appendices.

Christensen, V. (1995) Ecosystem maturity – towards quantification. *Ecological Modelling*, 77, 3–32.

Christensen, V. and Pauly, D. (1992) ECOPATH II – a software for balancing steady-state ecosystem models and calculating network characteristics. *Ecological Modelling*, 61, 169–85.

Christensen, V. and Pauly, D. (2004) Placing fisheries in their ecosystem context, an introduction. *Ecological Modelling*, 172, 103–7.

Christensen, V. and Walters, C. (2004a) *Ecopath with Ecosim*: methods, capabilities and limitations. *Ecological Modelling*, 172, 109–39.

Christensen, V. and Walters, C. (2004b) Trade-offs in ecosystem-scale optimisation of fisheries management policies. *Bulletin of Marine Science*, 74(3), 549–62.

Clark, M.R. (1995) Experience with the management of orange roughy (*Hoplostethus atlanticus*) in New Zealand, and the effects of commercial fishing on stocks over the period 1980–1993. In: *Deep-Water Fisheries of the North Atlantic Oceanic Slope* (ed. Hopper, A.G.), pp. 251–66. Kluwer Academic Publisher, Dordrecht, The Netherlands.

Clark, M.R. (1999) Fisheries for orange roughy (*Hoplostethus atlanticus*) on seamounts in New Zealand. *Oceanologica Acta*, 22, 1–10.

Clark, M.R. (2001) Are deepwater fisheries sustainable? The example of orange roughy (*Hoplostethus atlanticus*) in New Zealand. *Fisheries Research*, 51, 123–35.

Clark, M.R., Anderson, O.F., Francis, R.I.C.C. and Tracey, D.M. (2000) The effects of commercial exploitation on orange roughy (*Hoplostethus atlanticus*) from the continental slope of the Chatham Rise, New Zealand, from 1979 to 1997. *Fisheries Research*, 45, 217–38.

Cochrane, K. (2002) Using of ecosystem models to investigate ecosystem-based management strategies for capture fisheries: introduction. In: *The Use of Ecosystem Models to Investigate Multispecies Management Strategies for Capture Fisheries* (eds. Pitcher, T.J. and Cochrane, K.). *Fisheries Centre Research Report* 10(2), pp. 5–10. University of British Columbia, Vancouver, BC, Canada.

Cochrane, K.L., Butterworth, D.S., De Oliveria, J.A.A. and Roel, B.A. (1998) Management procedures in a fishery based on highly variable stocks and with conflicting objectives: experiences in the South African pelagic fishery. *Reviews in Fish Biology and Fisheries*, 8, 177–214.

de la Mare, W.K. (1996) Some recent developments in the management of marine living resources. In: *Frontiers of Population Ecology* (eds. Floyd, R.B., Shepherd, A.W. and De Barro, P.J.), pp. 599–616. CSIRO Publishing, Melbourne, Australia.

DeAngelis, D.L., Shuker, B.J., Ridgeway, M.S., Scheffer, M. (1993) Modeling growth and survival in an age-0 fish cohork. *Transactions of the American Fisheries Society*, 122, 927–41.

Devine, J.A., Baker, K.D. and Haedrich, R.L. (2006) Deep-sea fishes qualify as endangered. *Nature*, 439, 29.

Dower, J., Freeland, H. and Juniper, K. (1992) A strong biological response to oceanic flow past Cobb Seamount. *Deep-Sea Research*, 39, 1139–45.

Dower, J.F. and Mackas, D.L. (1996) 'Seamount effects' in the zooplankton community near Cobb Seamount. *Deep-Sea Research I*, 43, 837–58.

Dower, J.F. and Perry, R.I. (2001) High abundance of larval rockfish over Cobb Seamount, an isolated seamount in the Northeast Pacific. *Fisheries Oceanography*, 10(3), 268–74.

Duplisea, D.E. (2000) The ecological hierarchy, model complexity, predictive ability and robust management. *Report on the Young Scientists Conference on Marine Ecosystem Perspectives, ICES Cooperative Research Report No. 240*. International Council for the Exploration of the Sea, Copenhagen, Denmark, pp. 59–60.

FAO (Food and Agriculture Organization) (1995) *Code of Conduct for Responsible Fisheries*, 41 pp. FAO, Rome, Italy.

FAO (Food and Agriculture Organization) (2004) *The State of World Fisheries and Aquaculture*. FAO, Rome, Italy.

FAO (Food and Agriculture Organization) (2005) *Report on Deep-Sea 2003, An International Conference on Governance and Management of Deep-Sea Fisheries*, Queenstown, New Zealand, 1–5 December 2003. *FAO Fisheries Report No. 772*, Rome, Italy.

Fasham, M.J.R. (1993) Modelling the marine biota. In: *The Global Carbon Cycle* (ed. Heimann, M.), pp. 457–504. Springer-Verlag, Berlin.

Fedosova, R.A. (1974) Distribution of some copepods species in the vicinity of an underwater Hawaiian Ridge. *Oceanology*, 14, 724–7.

Fogarty, M.J. and Murawski, S.A. (1998) Large-scale disturbance and the structure of marine systems: fishery impacts on Georges Bank. *Ecological Applications*, 8(Suppl. 1), S6–S22.

Fonteneau, A. (1991) Monts sous-marins et thons dans l'Atlantique tropical Est. *Aquatic Living Resources*, 4, 13–25.

Francis, R.I.C.C. and Shotton, R. (1997) 'Risk' in fisheries management: a review. *Canadian Journal of Fisheries and Aquatic Sciences*, 54(8), 1699–715.

Fransz, H.G., Mommaerts, J.P. and Radach, G. (1991) Ecological modelling of the North Sea. *Netherlands Journal of Sea Research*, 28(1/2), 67–140.

Fulton, E.A. and Smith, A.D.M. (2004) Lessons learnt from the comparison of three ecosystem models for Port Phillip Bay, Australia. *African Journal Marine Science*, 26, 219–43.

Fulton, E.A., Smith, A.D.M. and Johnson, C.R. (2003) Effect of complexity on marine ecosystem models. *Marine Ecology Progress Series*, 253, 1–16.

Fulton, E.A., Fuller, M., Smith, A.D.M. and Punt, A.E. (2004a) Ecological indicators of the ecosystem effects of fishing: final report. *Australian Fisheries Management Authority Report, R99/1546*.

Fulton, E.A., Parslow, J.S., Smith, A.D.M. and Johnson, C.R. (2004b) Biogeochemical marine ecosystem models II: the effect of physiological detail on model performance. *Ecological Modelling*, 173, 371–406.

Fulton, E.A., Smith, A.D.M. and Johnson, C.R. (2004c) Effects of spatial resolution on the performance and interpretation of marine ecosystem models. *Ecological Modelling*, 176, 27–42.

Fulton, E.A., Smith, A.D.M. and Punt, A.E. (2005) Which ecological indicators can robustly detect effects of fishing? *ICES Journal of Marine Science*, 62, 540–51.

Genin, A. and Boehlert, G.W. (1985) Dynamics of temperature and chlorophyll structures above a seamount: an oceanic experiment. *Journal of Marine Research*, 43, 907–24.

Genin, A., Dayton, P.K., Lonsdale, P.F. and Spiess, F.N. (1986) Corals on seamount peaks provide evidence of current acceleration over deep-sea topography. *Nature*, 322, 59–61.

Genin, A., Haury, L. and Greenblatt, P. (1988) Interactions of migrating zooplankton with shallow topography: predation by rockfishes and intensification of patchiness. *Deep-Sea Research*, 35, 151–75.

Genin, A., Greene, C., Haury, L., Wiebe, P., Gal, G., Kaartvedt, S., Meir, E., Fey, C. and Dawson, J. (1994) Zooplankton patch dynamics: daily gap formation over abrupt topography. *Deep-Sea Research I*, 41(5/6), 941–51.

Goni, R. (1998) Ecosystem effects of marine fisheries: an overview. *Ocean and Coastal Management*, 40, 37–64.

Gray, R., Fulton, E.A., Little, L.R. and Scott, R. (2006) Operating model specification within an agent based framework. *North West Shelf Joint Environmental Management Study Technical Report*, Vol. 16. CSIRO, Hobart.

Grigg, R.W., Malahoff, A., Chave, E.H. and Landahl, J. (1987) Seamount benthic ecology and potential environmental impact from manganese crust mining in Hawaii. In: *Seamounts, Island and Atolls* (ed. Keating, B.H.), pp. 379–90. American Geophysical Union, Washington, DC.

Guénette, S. and Morato, T. (2001) The Azores archipelago, 1997: and Ecopath approach. In: *Fisheries Impacts on North Atlantic Ecosystems: Models and Analyses* (eds. Guénette, S., Christensen, V. and Pauly, D.). *Fisheries Centre Research Reports*, 9(4), pp. 241–70. University of British Columbia, Vancouver, BC, Canada.

Hall, S.J. and Mainprize, B. (2005) Towards ecosystem-based fisheries management. *Fish and Fisheries*, 5, 1–20.

Haney, J.C., Haury, L.R., Mullineaux, L.S. and Fey, C.L. (1995) Sea-bird aggregation at a deep North Pacific seamount. *Marine biology*, 123, 1–9.

Hareide, N.-R., Garnes, G. and Langedal, G. (2001) The boom and bust of the Norwegian longline fishery for redfish (*Sebastes marinus* 'Giant') on the Reykjanes Ridge. Scientific Council Research Documents 01/126. Northwest Atlantic Fisheries Organization, Dartmouth, Canada.

Haury, L., Fey, C., Newland, C. and Genin, A. (2000) Zooplankton distribution around four eastern North Pacific seamounts. *Progress in Oceanography*, 45, 69–105.

Hazin, F.H.V., Zagaglia, J.R., Broadhurst, M.K., Travassos, P.E.P. and Bezerra, T.R.Q. (1998) Review of a small-scale pelagic longline fishery off northeastern Brazil. *Marine Fisheries Review*, 60(3), 1–8.

Heymans, J.J., Shannon, L.J. and Jarre, A. (2004) Changes in the northern Benguela ecosystem over three decades: 1970s, 1980s, and 1990s. *Ecological Modelling*, 172, 175–95.

Holland, K.N., Kleiber, P. and Kajiura, S.M. (1999) Different residence times of yellow fin tuna, *Thunnus albacares*, and bigeye tuna, *T. obesus*, found in mixed aggregations over a seamount. *Fishery Bulletin*, 97, 392–5.

Holling, C.S. (ed.) (1978) *Adaptive Environmental Management and Assessment*. John Wiley and Sons, Inc., Chichester.

Hollowed, A.B., Bax, N., Beamish, R., Collie, J., Fogarty, M., Livingston, P., Pope, J. and Rice, J.C. (2000) Are multispecies models an improvement on single-species models for measuring fishing impacts on marine ecosystems? *ICES Journal of Marine Science*, 57, 707–19.

Hopper, A.G. (ed.) (1995) *Deep-Water Fisheries of the North Atlantic Ocean Slope*. Kluwer Academic Publisher, Dordrecht, The Netherlands.

Humphreys, R. and Moffitt, R. (1999) Western Pacific bottomfish and armorhead fisheries. Unit 17. In: *Our Living Oceans* (ed. NMFS). *Report on the Status of US Living Marine Resources*. US Department of Commerce, NOAA Technical Memorandum NMFS-F/SPO-41, on-line version, http://spo.nwr.noaa.gov/unit17.pdf

Huskin, I., Anadon, R., Medina, G., Head, R.N. and Harris, R.P. (2001) Mesozooplankton distribution and copepod grazing in the subtropical Atlantic near the Azores: influence of mesoscale structures. *Journal of Plankton Research*, 23, 671–91.

Itano, D.G. and Holland, K.N. (2000) Movement and vulnerability of bigeye (*Thunnus obesus*) and yellowfin tuna (*Thunnus albacares*) in relation to FADs and natural aggregation points. *Aquatic Living Resources*, 13(4), 213–23.

IWC (International Whaling Commission) (1992) Report of the Scientific Committee, Annex D. Report of the Sub-Committee on Management Procedures. *Reports of the International Whaling Commission*, 42, 87–136.

Jackson, J.B.C., Kirby, M.X., Berger, W.H., Bjorndal, K.A., Botsford, L.W., Bourgque, B.J., Bradbury, R.H., Cooke, R., Erlandson, J., Estes, J.A., Hughes, T.P., Kidwell, S., Lange, C.B., Lenihan, H.S., Pandolfi, J.M., Peterson, C.H., Steneck, R.S., Tegner, M.J. and Warner, R.R. (2001) Historical overfishing and the recent collapse of coastal ecosystems. *Science*, 293, 629–38.

Jennings, S., Kaiser, M.J. and Reynolds, J.D. (2001) *Marine Fisheries Ecology*. Blackwell Science, London.

Jørgensen, S.E. (1994) *Fundamentals of Ecological Modelling*, 2nd edition. *Developments in Environmental Modelling*, Vol. 19. Elsevier, Amsterdam.

Klimley, A.P., Butler, S.B., Nelson, D.R. and Stull, A.T. (1988) Diel movements of scalloped hammerhead sharks, *Sphyrna lewini* Griffith and Smith, to and from a seamount in the Gulf of California (Mexico). *Journal of Fish Biology*, 33(5), 751–62.

Koslow, J.A. (1996) Energetic and life-history patterns of deep-sea benthic, benthopelagic and seamount-associated fish. *Journal of Fish Biology*, 49, 54–74.

Koslow, J.A. (1997) Seamounts and the ecology of deep-sea fisheries. *American Scientist*, 85, 168–76.

Koslow, J.A., Boehlert, G.W., Gordon, J.D.M., Haedrich, R.L., Lorance, P. and Parin, N. (2000) Continental slope and deep-sea fisheries: implications for a fragile ecosystem. *ICES Journal of Marine Science*, 57, 548–57.

Koslow, J.A., Gowlett-Holmes, K., Lowry, J.K., O'Hara, T., Poore, G.C.B. and Williams, A. (2001) Seamount benthic macrofauna off southern Tasmania: community structure and impacts of trawling. *Marine Ecology Progress Series*, 213, 111–25.

Lack, M., Short, K. and Willock, A. (2003) *Managing Risk and Uncertainty in Deep-Sea Fisheries: Lessons from Orange Roughy*. Traffic Oceania and WWF Endangered Seas Programme, Australia.

Lueck, R.G. and Mudge, T.D. (1997) Topographically induced mixing around a shallow seamount. *Science*, 276, 1831–3.

Mace, P.M. (2001) A new role for MSY in single-species and ecosystem approaches to fisheries stock assessment and management. *Fish and Fisheries*, 2, 2–32.

McLain, R.J. and Lee, R.G. (1996) Adaptive management: promises and pitfalls. *Environmental Management*, 20, 437–48.

Menezes, G.M. (2003) Demersal fish assemblages in the Atlantic Archipelagos of the Azores, Madeira, and Cape Verde. PhD Thesis, Universidade dos Açores, Horta, Portugal.

Merrett, N.R. and Haedrich, R.L. (1997) *Deep-Sea Demersal Fish and Fisheries*. Chapman and Hall, London.

Monteiro, L.R., Ramos, J.R., Furness, R.W. and del Nevo, A.J. (1996) Movements, morphology, breeding, molt diet and feeding of seabirds in the Azores. *Colonial Waterbirds*, 19, 82–97.

Moore, J.A. (1999) Deep-sea finfish fisheries: lessons from history. *Fisheries*, 24, 16–21.

Moore, J.A. and Mace, P.M. (1999) Challenges and prospects for deep-sea finfish. *Fisheries*, 24, 22–3.

Morato, T. and Pitcher, T.J. (2002) Challenges and problems in modelling seamount ecosystems and their fisheries. International Council for the Exploration of the Sea (ICES) CM 2002/M:8, 28 pp.

Morato, T. and Pitcher, T.J. (2005) Ecosystem simulations in support of management of data-limited seamount fisheries. In: *Fisheries Assessment and Management in Data-Limited Situations* (eds. Kruse, G.H., Gallucci, V.F., Hay, D.E., Perry, R.I., Peterman, R.M., Shirley, T.C., Spencer, P.D., Wilson, B. and Woodby, D.), pp. 467–86. Alaska Sea Grant, Fairbanks, Alaska.

Morato, T., Watson, R., Pitcher, T.J. and Pauly, D. (2006a) Fishing down the deep. *Fish and Fisheries*, 7(1), 23–33.

Morato, T., Cheung, W.W.L. and Pitcher, T.J. (2006b) Vulnerability of seamount fish to fishing: fuzzy analysis of life history attributes. *Journal of Fish Biology*, 68(1), 209–21.

Mouriño, B., Fernández, E., Serret, P., Harbour, D., Sinha, B. and Pingree, R. (2001) Variability and seasonality of physical and biological fields at the Great Meteor Tablemount (subtropical NE Atlantic). *Oceanologica Acta*, 24, 167–85.

Nesis, K.N. (1986) Cephalopods of seamounts in the western Indian Ocean. *Oceanology*, 26, 91–6.

Nihoul, J.C.J. and Djenidi, S. (1998) Coupled physical, chemical and biological models. In: *The Sea* (eds. Brink, K.H. and Robinson, A.R.), John Wiley and Sons, Inc., New York.

Odate, T. and Furuya, K. (1998) Well-developed subsurface chlorophyll maximum near Komahashi No. 2 seamount in the summer of 1991. *Deep-Sea Research I*, 45, 1595–607.

Odum, E.P. (1971) *Fundamentals of Ecology*. W.B. Saunders Co., Philadelphia, PA.

Ohkushi, K. and Natori, H. (2001) Living benthic foraminifera of the Hess Rise and Suiko Seamount, central North Pacific. *Deep-Sea Research I*, 48, 1309–24.

Patterson, K., Cook, R., Darby, C., Gavaris, S., Kell, L., Lewy, P., Mesnil, B., Punt, A., Restrepo, V., Skagen, D.W. and Stefansson, G. (2001) Estimating uncertainty in fish stock assessment and forecasting. *Fish and Fisheries*, 2(2), 125–57.

Pauly, D., Christensen, V. and Walters, C. (2000) Ecopath, Ecosim, and Ecospace as tools for evaluating ecosystem impact of fisheries. *ICES Journal of Marine Sciences*, 57, 697–706.

Pauly, D., Christensen, V., Guénette, S., Pitcher, T.J., Sumaila, U.R., Walters, C.J., Watson, R. and Zeller, D. (2002) Towards sustainability in world fisheries. *Nature*, 418, 689–95.

Pinnegar, J.K., Blanchard, J.L., Mackinson, S., Scott, R.D. and Duplisea, D.E. (2005) Aggregation and removal of weak-links in food-web models: system stability and recovery from disturbance. *Ecological Modelling*, 184, 229–48.

Pitcher, T.J. (2000) Fisheries management that aims to rebuild resources can help resolve disputes, reinvigorate fisheries science and encourage public support. *Fish and Fisheries*, 1(1), 99–103.

Pitcher, T.J. (2001) Fisheries managed to rebuild ecosystems: reconstructing the past to salvage the future. *Ecological Applications*, 11(2), 601–17.

Pitcher, T.J. (2005) 'Back to the future': a fresh policy initiative for fisheries and a restoration ecology for ocean ecosystems. *Philosophical Transactions of the Royal Society*, 360, 107–21.

Pitcher, T.J., Buchary, E.A. and Hutton, T. (2002) Forecasting the benefits of no-take human-made reefs using spatial ecosystem simulation. *ICES Journal of Marine Science*, 59, S17–S26.

Pitcher, T.J., Ainsworth, C.H., Buchary, E.A., Cheung, W.L., Forrest, R., Haggan, N., Lozano, H., Morato, T. and Morissette, L. (2005a) Strategic management of marine ecosystems using whole-ecosystem simulation modelling: the 'Back-to-the-Future' policy approach. In: *The Strategic Management of Marine Ecosystems* (eds. Levner, E., Linkov, I. and Proth, J.-M.) Springer, NATO Science Series: IV: Earth and Environmental Sciences 50, pp. 199–258.

Pitcher, T.J., Ainsworth, C.H., Lozano, H., Cheung, W.L. and Skaret, G. (2005b) Evaluating the role of climate, fisheries and parameter uncertainty using ecosystem-based viability analysis. ICES CM 2005/M:24, 1–6.

Plagányi, É.E. and Butterworth, D.S. (2004) A critical look at the potential of *Ecopath with Ecosim* to assist in practical fisheries management. In: *Ecosystem Approaches to Fisheries in the Southern Benguela* (eds. Shannon, L.J., Cochrane, K.L. and Pillar, S.C.). *African Journal of Marina Science*, 26, 261–87.

Probert, P.K. (1999) Seamounts, sanctuaries and sustainability: moving towards deep-sea conservation. *Aquatic Conservation: Marine and Freshwater Ecosystems*, 9, 601–5.

Probert, P.K., McKnight, D.G. and Grove, S.L. (1997) Benthic invertebrate bycatch from a deep-water trawl fishery, Chatham Rise, New Zealand. *Aquatic Conservation: Marine and Freshwater Ecosystems*, 7, 27–40.

Reeves, R.R. and Mitchell, E. (1993) Status of Baird's beaked whale, *Berardius bairdii*. *Canadian Field-Naturalist*, 107(4), 509–23.

Reichert, P. and Omlin, M. (1997) On the usefulness of overparameterized ecological models. *Ecological Modelling*, 95, 289–99.

Rice, J.C. (2000) Evaluating fishery impacts using metrics of community structure. *ICES Journal of Marine Science*, 57, 682–8.

Richer de Forges, B., Koslow, J.A. and Poore, G.C.B. (2000) Diversity and endemism of the benthic seamount fauna in the southwest Pacific. *Nature*, 405, 944–7.

Roberts, C.M. (2002) Deep impact: the rising toll of fishing in the deep sea. *Trends in Ecology and Evolution*, 17, 242–5.

Robinson, L.A. and Frid, C.L.J. (2003) Dynamic ecosystem models and the evaluation of ecosystem effects of fishing: can we make meaningful predictions? *Aquatic Conservation: Marine and Freshwater Ecosystems*, 13, 5–20.

Rogers, A.D. (1994) The biology of seamounts. *Advances in Marine Biology*, 30, 305–50.

Sainsbury, K. and Sumaila, U.R. (2003) Incorporating ecosystem objectives into management of sustainable marine fisheries, including 'best practice' reference points and use of marine protected areas. In: *Responsible Fisheries in the Marine Ecosystem* (eds. Sinclair, M. and Valdimarson, G.), pp. 343–61. Food and Agricultural Organization of the United States/CABI Publishing, Rome.

Sainsbury, K.J. (1988) The ecological basis of multispecies fisheries, and management of a demersal fishery in tropical Australia. In: *Fish Population Dynamics* (ed. Gulland, J.A.), 2nd edition, pp. 349–82. Wiley, New York.

Sainsbury, K.J., Punt, A.E. and Smith, A.D.M. (2000) Design of operational management strategies for achieving fishery ecosystem objectives. *ICES Journal of Marine Science*, 57, 731–41.

Santos, R.S., Hawkins, S., Monteiro, L.R., Alves, M. and Isidro, E.J. (1995) Case studies and reviews. Marine research, resources and conservation in the Azores. *Aquatic Conservation: Marine and Freshwater Ecosystems*, 5, 311–54.

Sasaki, T. (1986) Development and present status of Japanese trawl fisheries in the vicinity of seamounts. In: *The Environment and Researches of Seamounts in the North Pacific* (eds. Uchida, R.N., Hayasi, S. and Boehlert, G.W.). *Proceedings of the Workshop on the Environment and Resources of Seamounts in the North Pacific, NOAA Technical Report NMFS*, 43, 21–30.

Sayer, M.D.J., Magill, S.H., Pitcher, T.J., Morissette, L. and Ainsworth, C.H. (2005) Simulation-based investigations of fishery changes as influenced by the scale and design of artificial habitats. *Journal of Fish Biology*, 67(Suppl. B), 1–26.

Sedberry, G.R. and Loefer, J.K. (2001) Satellite telemetry tracking of swordfish, *Xiphias gladius*, off the eastern United States. *Marine Biology*, 139, 355–60.

Sibert, J., Holland, K. and Itano, D. (2000) Exchange rates of yellowfin and bigeye tunas and fishery interaction between Cross seamount and near-shore fads in Hawaii. *Aquatic Living Resources*, 13(4), 225–32.

Silvert, W.L. (1981) Principles of ecosystem modelling. In: *Analysis of Marine Ecosystems* (ed. Longhurst, A.R.), pp. 651–76. Academic Press, New York.

Tseytlin, V.B. (1985) Energetics of fish populations inhabiting seamounts. *Oceanology*, 25, 237–9.

van Nes, E.H., Lammens, E.H.R.R. and Scheffer, M. (2002) PISCATOR, an individual-based model to analyze the dynamics of lake fish communities. *Ecological Modelling*, 152, 261–78.

Vinnichenko, V.I. (2002) Prospects of fisheries on seamounts. ICES CM 2002/M:32 (poster).

Voronina, N.M. and Timonin, A.G. (1986) Zooplankton of the region of seamounts in the western Indian Ocean. *Oceanology*, 26(6), 745–8.

Walters, C., Christensen, V. and Pauly, D. (1997) Structuring dynamic models of exploited ecosystems from trophic mass-balance assessments. *Reviews in Fish Biology and Fisheries*, 7, 139–72.

Walters, C., Pauly, D. and Christensen, V. (1999) Ecospace: prediction of mesoscale spatial patterns in trophic relationships of exploited ecosystems, with emphasis on the impacts of marine protected areas. *Ecosystems*, 2, 539–54.

Walters, C., Pauly, D., Chistensen, V. and Kitchell, J.F. (2000) Representing density dependent consequences of life history strategies in aquatic ecosystems: EcoSim II. *Ecosystems*, 3, 70–83.

Walters, C., Christensen, V. and Pauly, D. (2002) Searching for optimum fishing strategies for fishery development, recovery and sustainability. In: *The Use of Ecosystem Models to Investigate Multispecies Management Strategies for Capture Fisheries* (eds. Pitcher, T. and Cochrane, K.). *Fisheries Centre Research Reports*, 10(2), 11–15. The University of British Columbia, Vancouver, BC, Canada.

Walters, C.J. (2000) Impacts of dispersal, ecological interactions and fishing effort dynamics on efficacy of marine protected areas: How large should protected areas be? *Bulletin of Marine Science*, 66, 745–57.

Walters, C.J. and Martell, J.D.S. (2004) *Fisheries Ecology and Management*, pp. 399. Princeton University Press, Princeton, NJ.

Wayte, S. and Bax, N. (2001) Orange roughy (*Hoplostethus atlanticus*). Stock Assessment Report 2001 compiled for the South East Fishery Stock Assessment Group, Australian Fisheries Management Authority, Canberra, Australia.

Whipple, S.J., Link, J.S., Garrison, L.P. and Fogarty, M.J. (2000) Models of predation and fishing mortality in aquatic ecosystems. *Fish and Fisheries*, 1, 22–40.

Wilson, R.R. and Kaufmann, R.S. (1987) Seamount biota and biogeography. In: *Seamounts, Island and Atolls* (ed. Keating, B.H.), pp. 355–77. American Geophysical Union, Washington, DC.

Part V

Exploitation, Management and Conservation

Chapter 16
Small-scale fishing on seamounts

Helder Marques da Silva and Mário Rui Pinho

Abstract

This chapter attempts to estimate the significance of worldwide small-scale seamount fisheries. As it is impossible to determine the location of each catch, we have broadened the definition of a seamount and have excluded areas with few seamounts. Small-scale fisheries are defined as those using artisanal fishing vessels and gear. We also had to define the groups of species to include in the analyses, as different groups of seamount fishes behave in different ways. Tuna caught by artisanal fisheries are included as the species are attracted to seamounts and so become a temporary part of the fish community. We estimate that about 150 000–250 000 t of fish are caught worldwide from small-scale fisheries on seamounts. This estimate should be used with caution as there are likely to be unreported catches in some areas, and many islands, such as in the Western Central Pacific, had to be excluded, as catches from such locations include a mix of seamount and coral reef benthic species. We also present some case studies of small-scale seamount fisheries: the Azorean fishery in the early sixteenth century; the Hawaiian fishery; and the Seychelles fishery in the Indian Ocean. Other special cases are the black scabbard fishery of Madeira, in the Atlantic, and the oil fishery and the fishery over the Bowie Seamount, both in the Pacific. In this last the political dimension is of particular interest.

Introduction

A seamount is an undersea mountain, usually of volcanic origin (see Chapter 1). An oceanic island is a special case of a seamount that rises above the surface. Seamounts can usually be found, in varying numbers, around oceanic islands making them accessible to local small-scale fishers. In many cases, traditional fishing methods such as hand lines have been replaced by electric or hydraulic reels. As opposed to deepwater fisheries along the continental margins of some tropical and sub-tropical areas, seamount fisheries developed almost exclusively in the proximity of oceanic islands, at varying depths, although deep water species may often represent a considerable volume of the catches. Although of minor global significance, these fisheries often have high cultural and social as well as economic importance to the island states and regions where they evolved and can still be found.

Our first as sources for data were island states or regions including some continental regions close to seamounts. Ninety-four islands or archipelagos and 20 continental states were identified in this way. While there is no internationally agreed definition of a seamount (see Gubbay, 2002, 2003; Kitchingman and Lai, 2004; ICES, 2005) the definition in the Preface to this book is appropriate as it is based primarily on ecological criteria relevant to fisheries. We define a 'seamount fishery' as one that develops in the vicinity of a seamount as we cannot separate artisanal catch made over the top of a seamount from catch on the slope. While this separation may be necessary to evaluate some physical, chemical or ecological phenomena, our main concern is to identify artisanal catch from such topographic areas. We define a small-scale fishery as one that uses vessels under 30 m using artisanal fishing gears such as handlines, pole-and-line, traps and pots, small nets and short longlines.

We also need to define what we mean by a seamount species. There is a case for broadening the category of seamount species so that interactions among fish communities can be accounted for. As a result, we have not tried to differentiate between permanent residents, temporary residents and seamount visitors, as it would make analysis difficult. We therefore deal with seamount fisheries as a community rather than dealing with each species in turn.

The tunas are the most obvious group of species that are not resident members of seamount communities but which contribute significantly to catches (Fonteneau, 1991).

This chapter presents the first attempt to identify the global occurrence of small-scale seamount fisheries and estimate their significance. It includes some case studies of small-scale seamount fisheries, covering at least one area per ocean with contrasting conditions. Atlantic examples include the Azorean fishery in the early sixteenth century, and the fishery for black scabbardfish off Madeira. In the Pacific, we discuss the Hawaiian fishery, as well as both the oil fishery and over the Bowie Seamount, paying particular attention to the political dimension and describe the Seychelles fisheries in the Indian Ocean.

Global small-scale fisheries

A topographic map of the ocean floor shows where the major seamount assemblages are located, which provides a clue as to where seamount fisheries might be found. A fishery is much more likely to exist when the seamounts are close to inhabited islands (Fig. 16.1). We excluded continental areas from the analysis, since these often have continental shelf platforms and only seldom are near seamount areas. Exceptions are found off the coast of the US and Italy, where medium seamount densities are found.

In a second step, we used information from the 'Sea Around Us Project' (SAUP, 2005) to classify seamount island areas by high, medium and low density (Fig. 16.1). To reduce the chance of incorporating non-seamount catches in our global estimate, low density areas were excluded from the analysis, although we acknowledge that this will result in an underestimation. Seven areas were so identified, one area of high seamount densities in the Western Central Pacific, and six other areas of medium density: Japan, Indian Ocean, New Zealand, Easter, Gulf of Mexico and Mid-Atlantic Ridge.

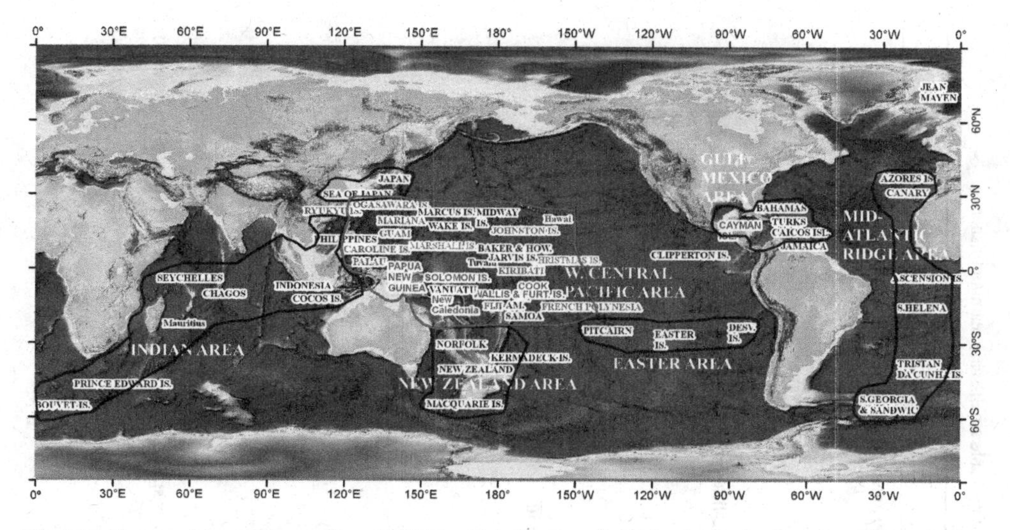

Fig. 16.1 Topographic world map (from NOAA) with indication of all island states/regions classified as having high (red) or medium (black) seamount densities.

Another problem was identified. Although all seven areas can be defined as seamount regions, other types of ecosystem, notably coral reefs also occur and support important fisheries. As the 'Sea Around Us' database includes the percentage of coral reefs in each area, as well as the respective shelf areas, we were able to classify each seamount region by the density of coral reefs. A threshold was defined by the average density of coral reefs in all seven regions previously identified, allowing the separation of 'clean' seamount areas. These were: the Mid-Atlantic Ridge; Easter area; Japan, including the islands of Ryukyu and Ogasawara and the contiguous North Marianas; and Indonesia, Mauritius and Bouvet Island in the Indian Ocean. The Western Central Pacific is not only the area of the highest seamount density, but also the 'noisiest' with respect to possible interactions among seamount and coral reef fisheries. This means that greater care should be given to the analyses of the seamount catch compositions and the allocation of those catches to locations.

A preliminary analysis of the catches from individual seamount areas led us to exclude Japan, including the Sea of Japan and Ryukyu Island, as well as the Philippines, Indonesia and the island of New Zealand. The volume of catches and species composition suggest that industrial and/or large-scale fisheries largely dominate the catches from these areas (Table 16.1).

Some islands were also excluded as they had no recorded catch, although in some cases catches had been by distant water fleets, which are not part of the present chapter. Islands in this category are Bouvet, S. Georgia and Sandwich, Tristan da Cunha, Ascension, S. Helena, Midway, Pitcairn and Clipperton.

Another conclusion of the catch analysis was that while the bulk of Western Central Pacific islands represent the highest seamount densities in the world, they do not constitute a seamount fishing area. This is particularly true for demersal fish, which are mostly caught over the coral reefs that dominate the surrounding shelf areas. Nonetheless, the catch of pelagic species which is largely dominated by tuna, albacore (*Thunnus alalunga*),

Table 16.1 Results from analyses of catch data from 'Sea Around Us Project' (SAUP, 2005). Bold rows represent areas of probable small-scale seamount fisheries.

Country/region	Classification	Local catches (1000 t)	Species	Notes
American Samoa		**4–10**	**Cuttlefish, tuna, seaurchins, scorpaenidae**	**Scorpaenidae catches 9000 t (since 1950s)**
Azores Islands (Portugal)	**Seamount fisheries**	**10–20**	**Tuna, demersals and others**	
British Indian Ocean Territory (UK)		**Up to 4–10**	**Mostly tuna**	
Canary Islands (Spain)		**20–70**		
Christmas Islands (Australia)		**4**	**Several including tuna and tuna-like species**	
Cook Islands (NZ)		0.5	Several, no tuna	Apparently not seamount species
Desventuradas Islands (Chile)		4	Similar to Easter Island	
Easter Islands (Chile)		3–8	Litoral species and shrimp	
Fiji		15	Several	Apparently not seamount species
French Polynesia (France)		**2–5**	**Tuna and sea catfish**	**Sea catfish**
Guam (USA)		**4**	**Tuna and tuna-like species**	
Hawai (USA)		**40**	**Small pelagic, tuna, cuttlefish and scorpaenidae**	
Indonesia	Not small scale	3000		Only industrial?
Jan Mayen Island (Norway)		Peaks >0.1		
Japan (Pacific Ocean Coast)	Not small scale			Only industrial?
Japan (Sea of Japan)	Not small scale			Only industrial?
Jarvis Islands (USA)		**⊂1–5**	**Mainly tuna**	
Johnston Island (USA)		12–15	Mostly cuttlefish. Also some tuna	
Kermadeck Island (NZ)		20	Orange roughy/Oreos	Likely industrial
Macquarie Island (Australia)		0.1–0.8	Small pelagic	
Marcus Islands (Japan)		30	Several including Tuna	Likely industrial
Marshall Islands		**⊂20**	**Tunas**	
Mauritius		**15–20**	**Mainly tuna**	
Micronesia, Federated States of		**<10**	**Mainly tuna**	
New Caledonia (France)		**5**	**Several including *Tuna* and *Beryx* sp.**	**Beryx catches1151 t (over several years)**

(*Continued*)

Table 16.1 *Continued*

Country/region	Classification	Local catches (1000 t)	Species	Notes
New Zealand	Not small scale			Only industrial?
Norfolk Island (Australia)		1–8	Orange roughy and mullets	Industrial and soft bottom?
North Marianas (USA)		**7–9**	**Tuna**	
Palau		**<50**	**Mainly tunas**	
Palmyra Island/Kingman Reef (USA)		,20	**Several including Tuna and Scorpaenidae**	**Identical to Hawai**
Papua New Guinea		**Up to 100**	**Mainly tunas**	Identical to Hawai and Palmyra
Philippines		Up to 1000	Small pelagic, shrimp and tuna	Only industrial?
Prince Edward Island (South Africa)		0.2–0.8	Patagonian tooth fish	Only industrial?
Ryukyu Islands (Japan)	Not small scale	800–1600		Only industrial?
Samoa		0.15–4	Littoral species and small pelagics	Apparently not seamount species
Seychelles		**3–5**	**Several included tuna**	**Industrial tuna**
Solomon Islands		**60**	**Mainly tuna**	
Tuvalu		**12**	**NEI, tuna and tuna-like species**	
Vanuatu		**0.5–2.5**	**Mostly tuna**	
Wake Island (USA)		**4**	**Tuna and tuna-like species**	
Wallis/Futuna Island (France)		**4**	**NEI and tuna**	

bigeye (*Thunnus obsesus*), yellowfin (*Thunnus albacares*) and skipjack (*Katsuwonus pelamis*), caught by hook and line fishing gears, occurs outside of the reef areas, thus over the seamounts that dominate the rough topography of the sea bottom (Table 16.1).

It would be interesting and valuable to discover the volume of seamount catches made worldwide. Although difficult to answer with a high level of precision, a rough estimate may be obtained by the sum of the landings from the island states or regions selected for analysis, excluding those catches of species, or groups of species, typically caught on reef ecosystems (Table 16.1; see also Chapter 18). Doing this, leads to an estimate of about 150 000–250 000 t, half of which is represented by tuna species. These figures are likely conservative as there may well be some significant unidentified small-scale catch around islands where the bulk of catches are from industrial fisheries. Similarly, some of the islands excluded because of high densities of coral reefs, may well have unidentified catches from seamounts which were not included. Excluding areas of low seamount density also probably led to an underestimate of global catch. On the other hand, one might question the reasonability of including tuna and the exclusion of tuna catches would greatly reduce the volume. Table 16.1 provides a rough estimate under the assumption that tuna are seamount species, but also allows an estimate based on a tuna-free total.

The Azores

Early fisheries

A sixteenth century account by Gaspar Frutuoso describes the Azorean fisheries of S. Miguel Island after its discovery in 1427: ‘Five years after the discovery of this island no one yet had a hook. They used to fish with a big chunk of meat tied to a line hanging from the tip of a thin branch. There was so many fish at that time that they used to catch more fish without hooks than is done today with sophisticated gears.’ ‘They used to catch much fish of all kinds with twisted nails; but sometimes with no nails or hooks at all, just with their hands. They could catch fish swimming in the shallows’ (quoted in Frutuoso, 1968).

A further description cited in Frutuoso (1983), refers to Terceira Island and the huge amount of fish that could be caught off the Frades islets: ‘southeast of the these islets there is an undersea elevation. About 3 km from the islet the elevation has 400 fathoms of depth (about 730 m) but further away sailors found a steep rock that could get as shallow as 60 fathoms (about 110 m). This seems to be a tall seamount but no depth soundings have been made beyond this shallow point’ … ‘Sailors agree that there were many mackerel schools around this area. … Over this seamount, there were also many large fish like the wreckfish (*Polyprion americanus*), conger eels (*Conger conger*), six-gill sharks (*Hexanchus griseus*) … porgies (*Pagrus pagrus*), dogfish (*Galeorhinus galeus*). … There were also shortfin mako shark (*Isurus oxyrinchus*), an edible shark, kitefin shark (*Dalatias licha*), dusky grouper (*Epinephelus guaza*), blackspot seabream (*Pagellus bogaraveo*), redfish (*Helicolenus dactylopterus*) …’.

Further describing the coast of Terceira Island, namely Pico do Altar, Gaspar Frutuoso (Frutuoso, 1983) wrote ‘Pico do Altar was used as an orientation point for fishermen of Angra that fished off the north coast of Terceira Island. About 15 km northeast of Pico do Altar fishermen have found a shallow seamount with 50–60 fathoms depth (about 90–110 m). ... Over it there were so many fish of all kinds that fishermen could fill their boats in less than 2 h ... and the fish caught there was easily recognized for being fat because of the impressive amount of food available.'

The descriptions show that the existence of seamounts was already recognized several centuries ago, as were the existence and quality of many fish species associated with those rocky complexes.

In the mid-twentieth century, recognizing the association of high fish concentrations with seamounts a list was published of the seamounts off the coast of Terceira Island of most relevance to pelagic and demersal fish species (Lopes, 1948). Nine seamounts were considered to have abundant stocks of pelagic species, mostly mackerel (*Scomber japonicus*) and horse mackerel (*Trachurus picturactus*), but also jacks (*Seriola* sp.), sardine (*Sardina pilchardus*) and barracuda (*Sphyraena sphyraena*), while 29 seamounts were identified to have demersal fish species, namely wreckfish, redfish, conger eel, blackspot seabream and the forkbeard (*Phycis phycis*). In addition, a list of 251 coastal fishing locations was also presented (Lopes, 1948).

The ecosystem

The Azores archipelago is located on the mid-Atlantic ridge, at the intersection of the European, American and African plates. The archipelago consists of nine volcanic islands forming three groups, running WNW-ESE, at 37–40°N latitude and 25–32°W longitude (Fig. 16.2). The Azores Exclusive Economic Zone (EEZ) comprises only 0.28% of the total surface of the world oceans, but contains 0.4% of the seamounts (SAUP, 2005; Table 16.1), making the Azores a major seamount area.

The land portion of the archipelago is 2344 km^2, but the marine EEZ is almost 1 million km^2 (948 439 km^2). The marine topography is highly irregular, quickly achieving depths of over 1000 m on moving offshore. The continental shelf is non-existent and the bottom is mainly rocky. The Azores Islands have a very steep seafloor, imposing significant constraints on the distribution of shallow water organisms. Around the islands, strata from 0 to 1000 m represent about 1% of the total EEZ area while the seamount areas including knolls, hills or guyots, within the same strata represent about 2% (Martins, 1986; Isidro, 1996; Menezes, 2003). Recent seamount mapping around the Azores by Machete *et al.* (2005), indicates that seamounts represent about 6% of total EEZ. Thus, areas below 1000 m, considered as less productive for fisheries, represent about 94–97% of the total EEZ. While interactions between coastal areas and the different seamounts are not yet very well understood (Pinho, 2003), the structure of the fish communities in both areas is quite similar (Menezes, 2003). The above characterization of the Azores Sea explains the domination of deep water and pelagic species in this ecosystem and supports the assumption that all fish species belong to the same seamount fish community.

Main fisheries of the Azores

The Azores has four main fisheries: (1) the small pelagic (horse mackerel and mackerel) fishery which uses open deck vessels under 12 m with small nets (Fernandes, 1994; Pinho

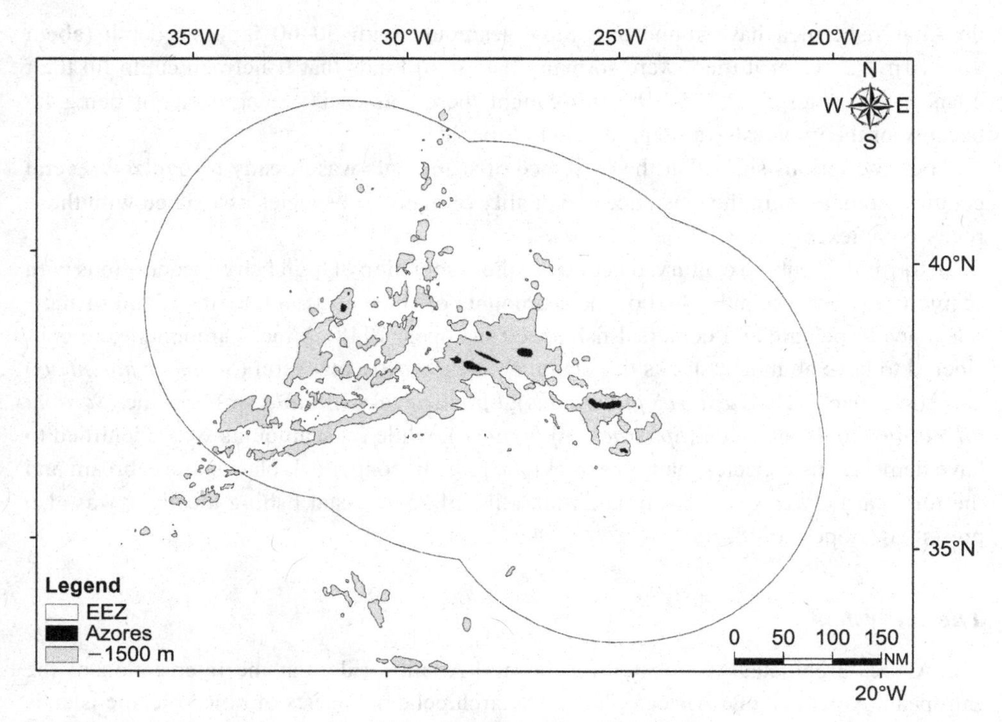

Fig. 16.2 Azores map showing the 12 and 200 mile EEZs and the 1500 m isolines.

et al., 1995; Isidro, 1996); (2) the pole-and-line tuna fishery from March to October which uses boats over 18 m (Pereira, 1995); (3) the demersal fishery, which uses hook-based fishing gears such as handlines and deep longlines and operates from both small open deck vessels under 12 m and closed deck vessels over 12 m (Menezes, 1996) and (4) the swordfish fishery, which was developed during the mid-1980s, using surface longlines, and operates from boats of both demersal fishery categories presented above (Simões, 1995; Silva and Pereira, 1999).

The Azores fleet has three main components: artisanal open deck vessels of mean length under 12 m and a Gross Registered Tonnage (GRT) <10 t, artisanal closed deck vessels, with a mean length smaller than 18 m and a GRT<50 t, and a group of larger vessels, with mean length greater than 18 m but smaller than 30 m and a GRT>50 t (Fig. 16.3).

Landings

Demersal species are the most valuable among Azorean fisheries, although until 1998 they ranked second in weight after the tuna fishery (Pinho and Menezes, 2006). From then on, demersal species became the most important fishery in weight as a result of the significant decrease of the catches of tuna in the Azores (Fig. 16.4). Total landings increased until 1993, with a decadal rate of increase of about 1000 t, and started decreasing afterwards. The demersal fishery generated an annual income between 15 and 18 million Euros, from first sale only, over the period 1998–2003.

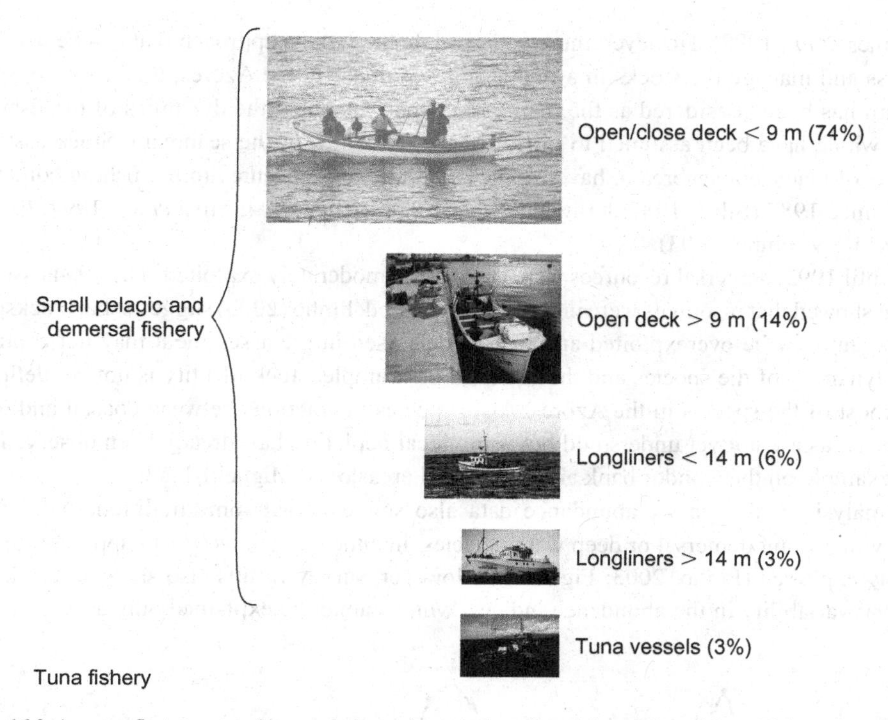

Fig. 16.3 Azorean fleet structure by vessel size and proportion of total fleet (adapted from Pinho and Menezes 2006).

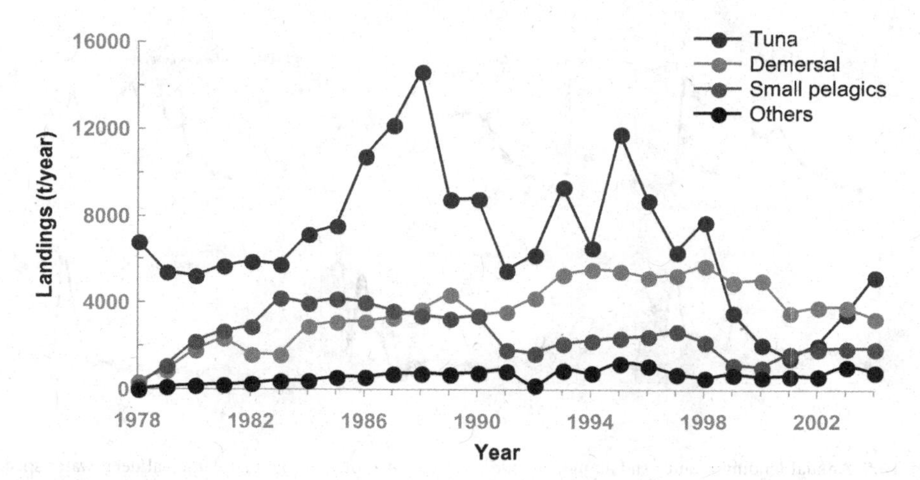

Fig. 16.4 Landings, in weight, by the Azorean fisheries (adapted from Pinho and Menezes, 2006). —●— Demersal; —●— Tuna; —●— Small Pelagics; —●— Others.

Management and state of the fisheries

The Azorean demersal or deep water fishery is a multispecies fishery. Results from diet studies show that predation among demersal species is not observed, suggesting that predator–prey relationships are less important for assessment than technological interactions

(Gomes *et al.*, 1998). However, there is no simple theoretical approach that can be used to assess and manage the stocks in a multispecies context. In the Azores, the blackspot seabream has been considered as the main target species and so the dynamics of the fishery as a whole have been assumed to mirror of the dynamics of the seabream. Stock assessments of blackspot seabream, have been carried out and presented to the fishing community since 1987 (Silva, 1987; Krug and Silva, 1988; Krug, 1994; Silva *et al.*, 1994; Pinho *et al.*, 1999; Pinho, 2003).

Until 1992, demersal resources were considered moderately exploited. The 1994 assessment showed that resources were intensively exploited. Pinho (2003) suggests that blackspot seabream may be overexploited and that models used in the assessment may not capture the dynamics of the species and the fishery. For example, stock identity is not yet defined for most of the species in the Azores. Also, species interactions between coastal and offshore areas are not yet understood but some local depletion has already been observed as for example on the Condor bank and the coastal areas of S. Miguel Island.

Analysis of the survey abundance data also suggests that some traditional commercially important demersal or deep water species, like the alfonsinos (*Beryx* spp.) are intensively exploited (Pinho, 2003; Fig. 16.5). However, survey results also show a very high annual variability in the abundance indices, which cannot be explained only as a result of

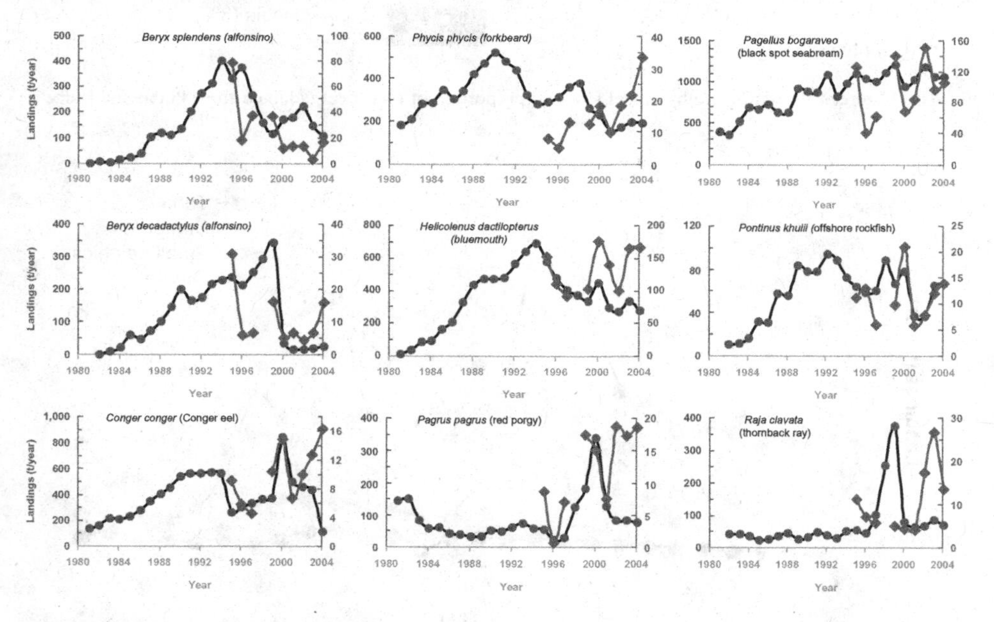

Fig. 16.5 Annual landings (blue) and abundance indices (red) of some commercial demersal/deep water species (RPN: Relative Population Number from the Azorean Spring bottom longline survey) (adapted from Pinho and Menezes, 2006).

fishing effects. This suggests that the Azorean ecosystem is complex. Particular concern must be noted also on the consequences of the pattern of resource exploitation on the sustainability of less abundant species because harvest decisions have usually been based on what is known about target species. This concern has led to local discussions of the need to improve management of the Azorean multispecies fishery (see Pinho, 2003 for details).

Until 1992, demersal resources were considered moderately exploited and no management action was taken. Between 1998 and 2000 some technical measures were implemented, including restrictions on licences based on a minimum threshold landing in value, hook size limit and access restrictions to fishing areas by vessel size and gear type. Coastal Marine Protected Areas (MPAs) have been proposed and implemented under the EU. A 'Natura 2000' network and offshore MPAs have also been proposed under the OSPAR convention (annex V; Afonso and Santos, 2004).

Exploitation of new resources of deep water species, like Mora moro and black scabbardfish (*Aphanopus carbo*), have been encouraged, as a consequence of the intensive fishing of the traditional species. Growth of these fisheries has however been slow as there is no local market for these species (ICES, 2005).

Management measures taken under the Common Fishery Policy of the European Community included introduction of a Total Allowable Catch (TAC) for some deepwater species like black scabbardfish, orange roughy (*Hoplostethus atlanticus*), blackspot seabream and the deepwater sharks *Deania* spp., *Centrophorus* spp., *Etmopterus* spp., *Centroscysmnus*, spp. and *Dalatias licha* (EC Reg. 2340/2002; EC Reg. 2270/2004). Effort restrictions based on a threshold tonnage have been implemented through licensing of the deep water fleet (EC Reg. 2347/2002). Additional and exceptional measures were introduced in the Azores with the creation of a temporary buffer zone of 100 nautical miles around the islands where only Azorean vessels can be operated, so preventing higher fishing intensity from distant water fleets (EC Reg. 1954/2003).

The Hawaiian Islands

The Hawaiian archipelago stretches northwestward over 2778 m, from about 19°N and 155°W to 28°N and 178°W (Fig. 16.6). The seven Main Hawaiian Islands (MHI): Hawaii, Maui, Lanai, Molokai, Oahu, Kauai and Niihau, comprise over 99% of the total land area and have virtually all of the State's population of over 1 million residents. The relatively tiny eroded and submerged islands and reefs extending to the northwest are remnants of high islands that existed millions of years ago (Macdonald *et al.*, 1983). The Hawaiian Islands were reached by colonizing Polynesians, probably from the Marquesas Islands, in about the fourth century SCAD, somewhat later than most other Polynesian islands. The human population reached 200 000 before contact with Europeans in 1778.

Marine ecosystems within the oceanic Pacific can be categorized as coastal, shelf, slope or seamount, and pelagic. Although the pelagic ecosystems are relatively homogeneous, the coastal ecosystems of the Pacific islands include numerous habitat types, and the slope ecosystem is intermediate in complexity (Holthus and Maragos, 1995).

Hawaiian fisheries

The Hawaiian commercial fishing sector includes a wide array of fisheries (Boehlert, 1993). Several fisheries can be identified including tuna fishing using longline, pole-and-line, troll and handlines (Boggs and Ito, 1993; Pooley, 1993; WPRFMC, 2004); bottomfish using mainly hook and line techniques for the grouper–snapper-jack complex (*Epinephelus*

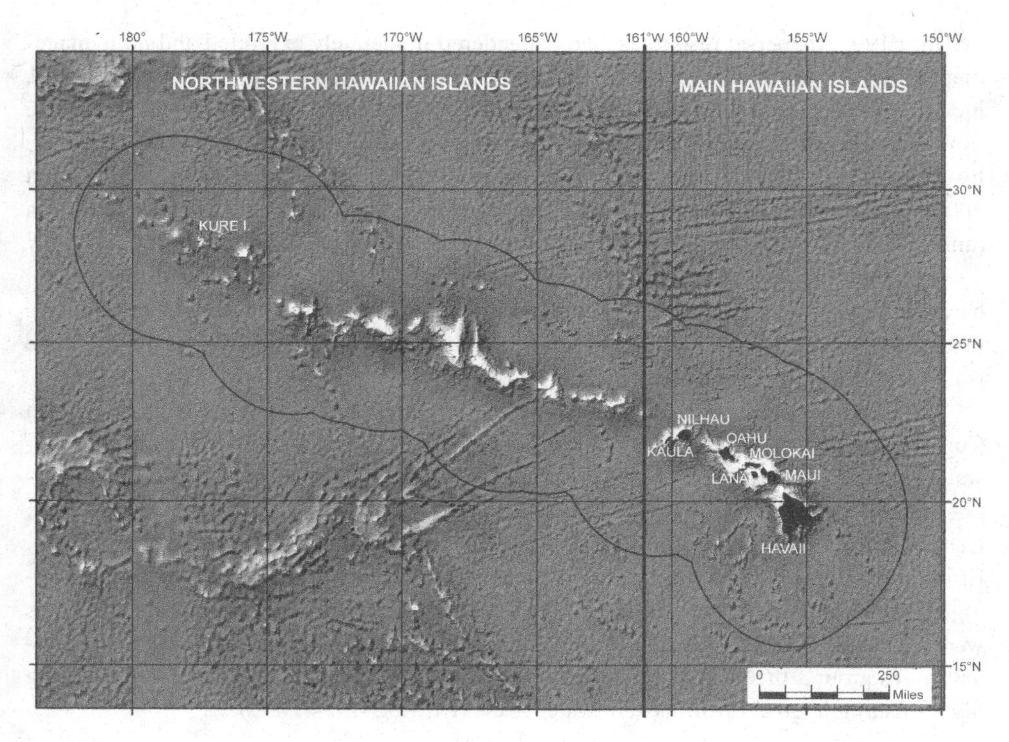

Fig. 16.6 Main and NW Hawaiian Islands and EEZ (data source from ESRI Data and Maps [CD-ROM], 2005; http://www.brocku.ca/maplibrary/digital/ESRIdata2005.htm).

quernus, Etelis spp., *Pristipomoides* spp., *Aprion virescens, Aphareus rutilans, Lutjanus kasmira, Caranx* spp.) (Haight *et al.*, 1993; WPRFMC, 2003); lobster (*Panulinus marginatus*) trapping (Polovina, 1993); net fishing for species such as bigeye scad (*Selar crumenopthlmus*) and mackerel scad (*Decapterus macarellus*). There is also trolling for pelagic species such as marlin (*Makaira mazara, M. indica, Tetrapturus audax*), wahoo (*Acanthocybium solandri*) and mahimahi (*Coryphaena hippurus*). Hawaiian fishing fleets are also usually divided into three overlapping segments: large commercial fisheries; small commercial fisheries and recreational fisheries including part time commercial and subsistence fishing (Pooley, 1993).

Large commercial fisheries include the pole-and-line skipjack fishery and longliners for tuna and swordfish, distant water trollers for albacore and multipurpose vessels for bottomfish and lobsters operating mainly in the Northwest Hawaiian Islands (NWHI). Small-scale fisheries include a wide variety of trollers and handliners, targeting pelagic species such as tunas, billfish and wahoo, and demersal fish. There is some trapping targeting reef fish, bottomfish, and crustaceans and net fishing which targets mid-water scads. These vessels operate almost exclusively in the MHI. The recreational fishery includes vessels similar to the small commercial, charter and dive fishing vessels and also a component of very small vessels. The target species are very varied, including reef species, pelagic species such as tunas, billfishes, mahimahi and wahoo, demersal fish and crustaceans. The fishing methods are also very varied.

Landings

Pelagic landings average between 80% and 90% of total landings of Hawaii. Tunas are the most important resource, dominating the landings in almost all periods of the time series starting in 1948 (Fig. 16.7). The tuna fishery between 1950s and 1970s was dominated by

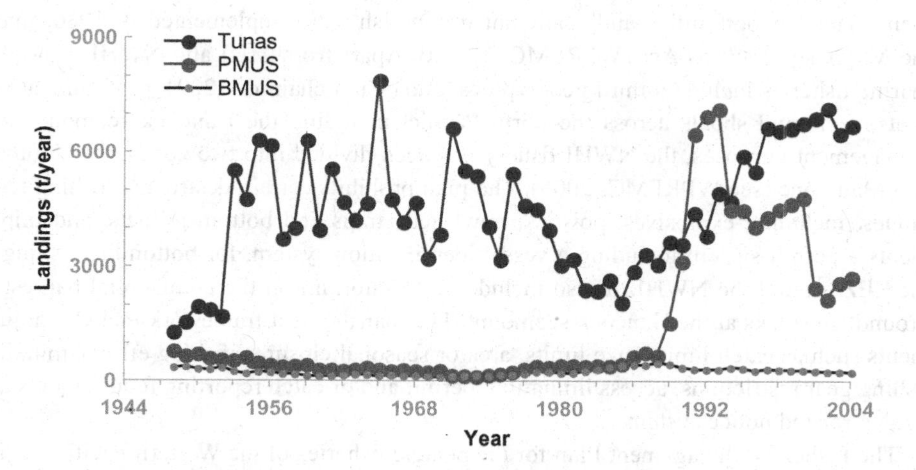

Fig. 16.7 Hawaii annual commercial fishery landings by fishery: —●— PMUS – Pelagic Management Unit Species; —•— BMUS – Bottom Management Unit Species; —●— Tunas (data source from HDAR, Hawaii Division of Aquatic Resources; http://www.pifsc.noaa.gov/wpacfin/hi/dar/Pages/hi_data_1.htm).

the live bait pole-and-line fleets targeting skipjack tuna which was landed mainly for canning. Skipjack landings represented about 50–60% of total during that period, with fluctuations over time. Landings started to decrease during the mid-1970s due to the reduction in effort on this component of the tuna fleet, which followed the closing of the canning industry in 1984. Tuna landings increased again from 1985 as a consequence of important structural changes in local and world markets, introduction of new technologically more advanced and efficient fleets of trollers and longliners, and expansion of the geographic fishing range. This period of expansion, industrialization and diversification of Hawaiian pelagic fisheries coincide with the expansion of the longliners and a decline of pole-and-line, handlines and troll fleets (Boggs and Ito, 1993; Pooley, 1993). The longline fishery is by far the largest pelagic fishery in Hawaii, representing 77% of the total pelagic catch in 2003 and targets mainly tunas, sharks and swordfish (WPRFMC, 2004). Swordfish landings began to increase in 1989 peaked at 5561 t in 1993, and started declining substantially thereafter.

Landings of demersal fish in Hawaii present a cyclic trend with two peaks of about 500 t in the early 1950s and at the end of the 1980s (Fig. 16.7). The fishery declined steadily through 1950s and 1960s as a consequence of decreased effort in the NWHI due to poor fish prices (Haight *et al.*, 1993). The landings started to increase again during 1970s, concentrating mainly in the NWHI. The fishery expanded during 1980s with landings peaking at 500 t in 1988. Landings decreased thereafter due to the introduction of management measures to reduce fishing effort, particularly on the NWHI area.

Management of the fisheries

The Western Pacific Regional Fishery Management Council (WPRFMC) is the authority for managing EEZ fisheries in Hawaii. Bottomfish resources from the MHI area are intensively exploited, with some stocks considered as overexploited (five major BMUS species in 2003 are considered in recruitment overfishing) (WPRFMC, 2003). Fishery management plans for bottomfish and seamount groundfish were implemented in 1986, under the Magnuson-Stevens Act (WPRFMC, 1986a). Apart from MHI and NWHI, Hawaiian marine fisheries include a third geographical subarea, a chain of 2222 km of uninhabited islets, reefs and shoals across the north Pacific, including the Hancock seamount. For management purposes, the NWHI fishery has been divided into two zones, the Hoòmalu and Mau zone (see WPRFMC, 2005). The plan prohibits certain destructive fishing techniques, including explosives, poisons, trawl nets, traps and bottom gillnets, and implements a permit system including a vessel identification system, for bottomfish fishing in the EEZ around the NWHI. It also includes a moratorium on the commercial harvest of groundfish stocks at the Hancock seamount. The management framework includes adjustments such as catch limits, size limits, area or seasonal closures, fishing effort limitation, fishing gear restrictions, access limitation, permit and/or catch reporting requirements and a rules-related notice system.

The Fisheries Management Plan for the pelagic fisheries of the Western Pacific Region was published in 1986 (WPRFMC, 1986b). This FMP regulates the US domestic fisheries for tuna, swordfish, marlin, sharks and other pelagic species in this region. The FMP prohibits drift gillnet fishing within the region's EEZ and foreign longline fishing within certain areas of the EEZ. The plan also establishes a management framework of domestic fisheries that includes closed areas to fishing, a vessel monitoring system and permit limits.

The US has proclaimed the creation of a marine national monument in the NWHI that accords nearly 140 000 square miles of the NW Hawaiian Islands, the Nation's highest form of marine environmental protection. (http://www.hawaiireef.noaa.gov/pdfs/nwhi_factsheet.pdf).

The Seychelles

The Republic of Seychelles comprises 115 widely scattered islands of which 41 are composed of granites and 74 of coralline, extending between 3–10° and 45–58°E in the Western Indian Ocean (Fig. 16.8). Total land area is approximately 443 km^2 with an EEZ of 1 374 000 km^2. Only 3.5% (48 019 km^2) of the EEZ is shallower than 200 m (see Metz, 1995), the rest is over of 1000–1500 m.

Description of the fisheries

The fisheries sector is divided into two distinct categories: traditional fishing by a domestic fleet of some 400 vessels and industrial tuna fishing by foreign vessels, which began to develop in the mid-1970s and has emerged as a major revenue source. Ninety per cent of

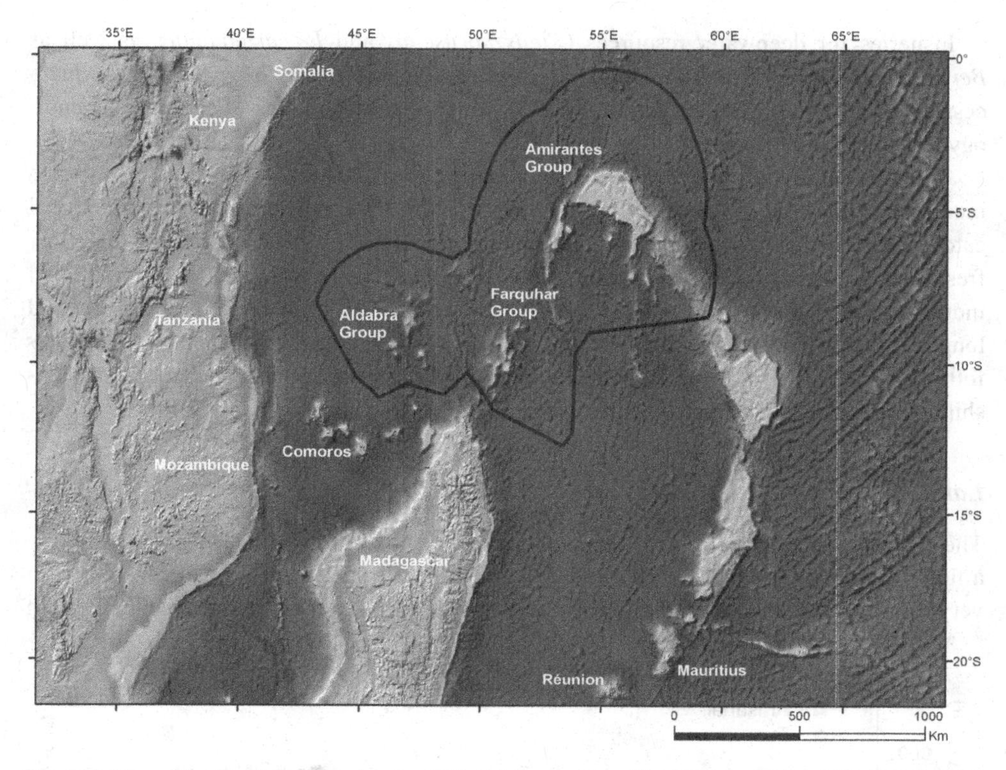

Fig. 16.8 Map of the Seychelles showing the EEZ (data source from ESRI Data and Maps [CD-ROM], 2005; http://www.brocku.ca/maplibrary/digital/ESRIdata2005.htm).

the artisanal fleet consists of open deck boats or small cabin fibreglass boats under 12 m equipped with inboard or outboard engines (15–40 hp), operating within a radius of 50 km of the main islands. A small component of the artisanal fleet, partially decked whalers and schooner vessels, catches deep water species up to 100 km from shore, with fishing trips lasting an average of 8 days. Some domestic offshore operations on banks surrounding the Mahé group and the Amirantes Islands are conducted with handlines from larger boats. Most of the catch is frozen (Mees, 1989; Metz, 1995).

The artisanal fishery is highly seasonal, being intense during the inter-tropical monsoon periods of March–May and October–November, when sea conditions are favourable to fishing. Until the introduction of the schooner vessels in 1974, the fleet was restricted to the near shore fishing grounds on Mahé Plateau. Most of the areas around the coralline islands and atolls are still underexploited (Robinson and Shroff, 2004).

The artisanal fishery targets demersal or semi-demersal species of which the most important are trevally (*Carangoides* spp., 30%); red snapper (*Lutjanus* spp., 13%); jobfish (*Aprion virescens*, 18%); emperors (*Lethrinus* spp., 8%); groupers (*Epinephelus* spp., 4%); rabbitfish (*Siganus* spp., 4%) and mackerel (*Rastrelliger* spp., 2%). The most important fishing technique, which accounts for 60% of the landings, is hook and line for pelagics and demersal, followed by trap fishing (30%) which is mainly for reef-associated species, and nets (10%). The dropline technique, which was introduced in the early 1990s, targets deep water snapper (*Pristipomoides* spp.) at the edge of the Mahé Plateau (FAO, 2001).

Fisheries for deepwater resources (*Etelis coruscans* and *E. carbunculus*, as well as *Beryx splendens* and *B. dacadactylus*) are underdeveloped, largely because they are inaccessible to artisanal fisheries, due to distance from the coast and lack of adequate technology to fish at deeper waters.

A small-scale longline fleet (12–18 m), targeting tunas and swordfish for the fresh fish market, has operated in the oceanic waters since 1995. An export trade market from local catches developed after the opening of the international airport made it possible to deliver fresh fish to Europe and other markets. The offshore fishery consists of an international industrial tuna fishery from purse seiners which are mainly from Spain and France, and longliners from Taiwan and South Korea. This is a highly migratory fishery, with vessels following the fish all over the western Indian Ocean. The majority of the catches are transhipped to reefer vessels destined for Europe and Puerto Rico (Anon, 2002).

Landings

The main resources are tropical tuna species, skipjack, yellowfin and bigeye, fished by a fleet of industrial vessels since 1984 (Fig. 16.9). Before 1984, tuna has been caught in very small quantities in the Seychelles. Two years later this had changed dramatically, with

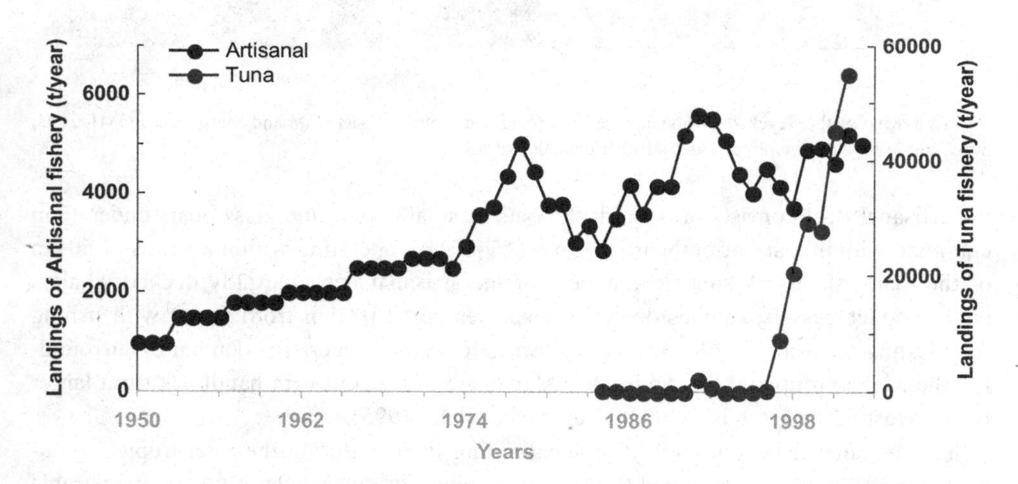

Fig. 16.9 Seychelles annual commercial fishery landings —●— landings of artisanal fishery; —●— landings of tuna fishery (data source from FAO (FAOSTAT 2005 – Fishstatplus – IOTC database; http://www.iotc.org/English/data/databases.php).

approximately 50 foreign purse seiners licensed in the Seychelles. Port Victoria became the most important tuna transhipping port in the Indian Ocean, accounting for more than 80% of the total transshipment for the SW Indian Ocean. Fishing has become the second biggest economic earner, after tourism. A canning factory built in the early 1990s, has grown to process 400 t of tuna/day and employs about 1800 workers.

By contrast, although the local artisanal fleet has increased in size and has been modernized, with the introduction of new fishing techniques, the total catch stabilized around 4500 t during the last two decades (Robinson and Shroff, 2004; Fig. 16.9). The artisanal catch composition has not changed significantly since 1985 with catches dominated by the pelagic carangids, trevally and bludger (*Carangoides gymnostethus*), followed by demersal and bottom species groups like snappers, emperors and groupers.

Landings from Seychelles fisheries stabilized because of the fragile nature of the resources, with certain demersal species currently threatened with overexploitation. In the last few years, management plans have been drawn up, with closed seasons, restricted zones, protected areas, banning of certain gear such as gill nets for sharks and quotas for mother ships working the area. These have helped to stabilize the fishery and avoid overexploitation whilst meeting the local requirement for fish protein, and permitting around 700 of fish to be exported. There is limited scope for large-scale expansion of this fishery, mainly due to variable weather conditions and the fragile nature of the stocks of long-lived species such as groupers (Serranidae) and snappers (Lutjanidae).

Management and state of the fisheries

Coastal resources in Seychelles are managed by the Seychelles Fishing Authority (SFA). Tunas and tuna-like species are managed under the Indian Ocean Tuna Commission (IOTC), a statutory body of the United Nations Food and Agriculture Organization (FAO). The coastal demersal fisheries are considered to be overexploited whilst unexploited potential still exists on the outer Mahé Plateau and Banks off the southern island groups. As a consequence, the expansion of the artisanal fishery is limited to the offshore areas, but fishing pressure on the inshore resources continues (Robinson and Shroff, 2004).

Technical measures implemented or proposed to avoid overexploitation include:

- A ban on demersal trawling in Seychelles waters.
- Introduction of protected areas inside reefs, where fishing is regulated and net fishing prohibited.
- Prohibiting the use of nets for shark fishing.
- Introduction of a closed season for lobster with only a 3-month annual fishing period allowed.
- Restriction on the mesh sizes of traps and the length of gill nets for mackerel fishing.
- Requirements that all fishing vessels and nets must be licensed and marked with a seal and a licence number.
- Special fishing reserves have been introduced on some of the islands in addition to several marine parks around the granitic island.
- Restricted zones for the artisanal fishery on all plateau areas (the Mahé and Amirantes plateaus) and within 22 km of any island.
- Introduction of Vessel Monitoring Systems to allow for better control of the foreign industrial fleets.

Some special cases

This section deals with fisheries from areas that, while mostly of low seamount density, are specially interesting from a social perspective, as in the fishery for black scabbardfish in Madeira; a cultural standpoint, as in Canada's Bowie Seamount; or simply as an historical piece of information, as in the Polynesian oil fishery. These are not intended as case studies for small-scale fisheries on seamounts. Rather, they are simply intended to represent unique cases, mostly dealing with artisanal fisheries which occur (red) in seamount areas irrespective of the density of seamounts.

The black scabbardfish fishery in Madeira

The black scabbardfish is a benthopelagic fish, belonging to the Trichiuridae family, with an extremely elongate and compressed body, living over continental slopes or underwater rises at about 200–1600 m deep (Parin, 1986). The fishery of this species in Madeira dates back to the mid-seventeenth century when Tomás (1635) refers to scabbardfish in his poem 'A Insulana'. It was not until two centuries later that the naturalist Lowe (1843), who lived in Madeira for 50 years, first described this species. Due to its fine flavour, the fishery rapidly expanded, particularly in Câmara de Lobos village, the major centre for this fishery since the 1920s.

Black scabbardfish landings data show three distinct periods of the history of the fishery starting in the late 1930s (Fig. 16.10). The first period with landings between 500 and 2000 t ended in the mid-1980s with a value rising steeply over the last decade. The second period

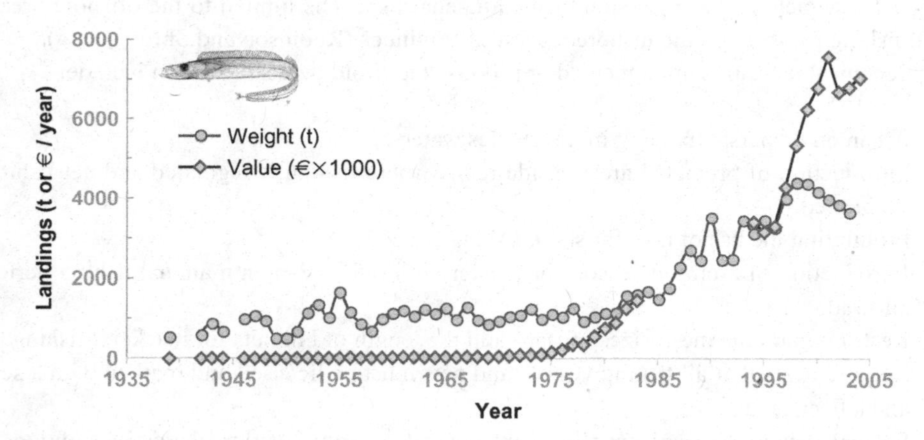

Fig. 16.10 Landings of black scabbardfish (*Aphanopus carbo*), in —○— weight (t) and —◇— value (Euras) since 1930–2004 (data sources from INE (http://www.ine.pt/prodserv/quadros/periodo.asp), FAO-CECAF (http://www.fao.org/fi/statist/fisoft/FISHPLUS.asp) and Ataíde, Personal communication).

was characterized by a steady increase in the volume of fish landed, achieving almost 4500 t in the late 1990s. This period was followed by an increase in the value of the fishery. The last period, although still short (1998–2004), shows a steady decline in the volume of black scabbardfish landed in Madeira, at least partially compensated for by the high value of the fishery (Fig. 16.10).

The 1980s in Madeira saw a change from vertical lines to horizontal drifting longline fishing (Fernandes, 1984; Reis *et al.*, 2001). This, together with the general improvements in fishing boats and equipment, are considered to be the major reasons for the steep increase in landings. In the beginning of the 1980s, there were more than 100 boats specializing in this type of fishing, most of them under 6 m. Nowadays, there is a fleet of some 40 boats fishing for the black scabbardfish, with most of them between 10 and 13 m and a crew of 9 people (Reis *et al.*, 2001).

Pacific oil fishery

Another important deep water seamount fishery, specially in the South Pacific, is the Oilfish (*Ruvettus pretiosus*) fishery that was carried out traditionally throughout Polynesia. The Oilfish belongs to the Gempylidae family which can move between the demersal, bathypelagic, mesopelagic and even the pelagic zones (Merrett and Haedrich, 1997). This rather opportunistic distribution concurs with its behaviour, as a top predator a role characterized by large teeth, firm musculature and fast swimming speed. This fishery, commonly referred to as the Pacific oil fishery, is among the oldest deep sea fisheries worldwide and the species concentrate near oceanic islands of volcanic origin, precisely where seamounts seem to concentrate (see Fig. 16.1).

Traditional fishing for *Ruvettus* in Polynesia is described by Gudger (1927), and discussed by Merrett and Haedrich (1997). Fishing takes place at depths between 150 and 750 m, using wooden hooks often arranged into gangs of a small number of hooks, around 20–36 cm long, frequently baited with flying fish (Exocoetidae family) and weighted by a 1.3–2.3 kg sinker, which is usually a rock.

Canada's Bowie Seamount

The Bowie Seamount Area lies in the vicinity of three Pacific Fisheries Management subareas. The Bowie, Hodgkins and Davidson seamounts lie to the south, in one of these subareas (Fig. 16.11). The Bowie Seamount Area lies within the traditional territory claimed by the Haida Aboriginal people (Canessa *et al.*, 2003). In March 2002, the Haida Nation commenced a court action in which they claim aboriginal title to Haida Gwaii (Queen Charlotte Islands) and the surrounding waters and seabed out to the 200 mile EEZ. Their claim thus includes the Bowie Seamount Area.

Further consultation between governments is required to determine the significance of the Bowie Seamount Area to the Haida and to discuss implications of managing the seamount as an MPA. Other than its significance to the Haida, there are no other known historical or cultural values associated with the Bowie Seamount (Canessa *et al.*, 2003).

Despite the relatively high biomass and biodiversity, seamounts such as the Bowie Seamount can represent delicate and isolated ecosystems making them susceptible to human pressures. In an effort to understand the extent to which even limited fishing activities affect fish stocks and other ecosystem components and to implement appropriate fishery management measures, seamount fisheries have generally been managed as experimental (Canessa *et al.*, 2003).

The Bowie Seamount Area lies in the northern offshore seamount management area and has been fished by the Canadian fleet since 1985 (Murie *et al.*, 1996). Fisheries in the Bowie Seamount Area have primarily existed on a limited basis for groundfish such as

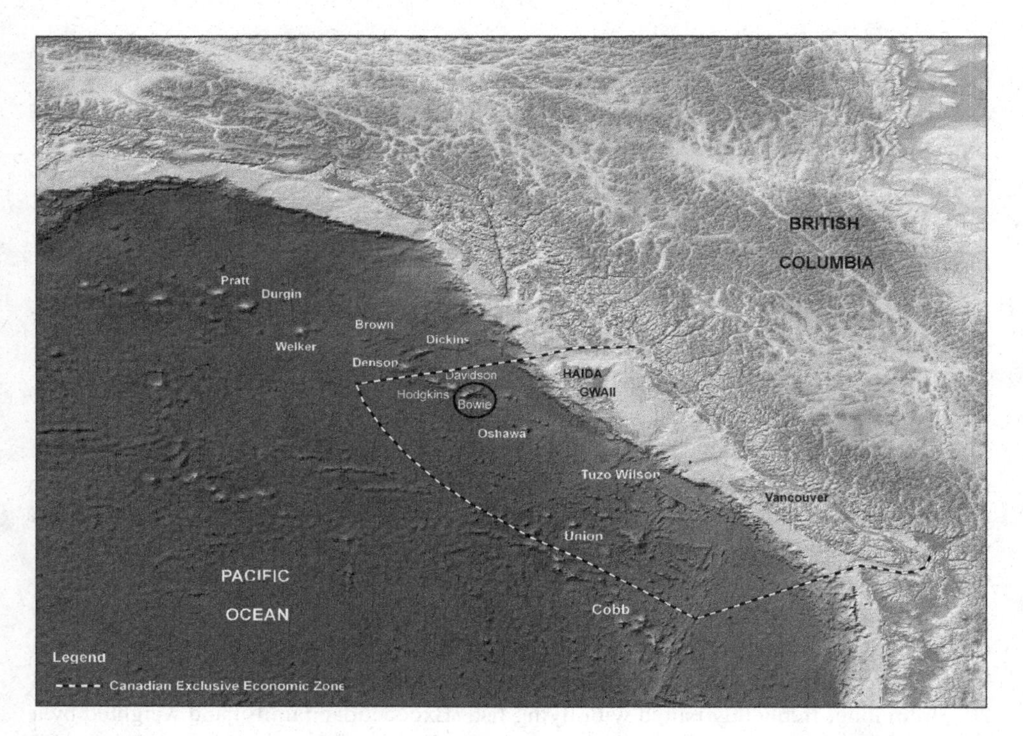

Fig. 16.11 Pacific Fisheries Management Subareas, in the vicinity of the Bowie Seamount Area (data source from: ESRI Data and Maps [CD-ROM], 2005; http://www.brocku.ca/maplibrary/digital/ESRIdata2005.htm).

rockfish (*Sebastes* spp.), sablefish (*Anoplopoma fimbria*), Pacific halibut (*Hippoglossus stenolepis*) and some migratory species such as albacore tuna (*Thunnus alalunga*; DFO, 1997). Test fishing for flying squid (*Ommastrephes bartrami*) has also taken place in the Area (DFO, 1998). Gear types used at the seamounts comprise bottom trawl, jig, longline and traps.

The cumulative catch of sablefish at Bowie Seamount over the period 1987–2000 was approximately 1450t, with an average annual catch over the period of approximately 100t (see Canessa *et al.*, 2003). Since 1990, nine vessels have fished at Bowie Seamount for sablefish executing approximately 45 commercial fishing trips. Records of the number of traps set at Bowie Seamount between 1987 and 2001 show that the depths ranged from 242 to 1326m (mean depth between 1995 and 2000 was approximately 860m (Canessa *et al.*, 2003). The CPUE (catch per unit effort) was significantly greater in 1988 than in 1989, and decreased considerably in 1993 suggesting that the population of sablefish on Bowie Seamount was not sustainable at current fishing levels (Murie *et al.*, 1996).

Rockfishes (*Sebastes* spp.), a group where rougheye rockfish (*S. aleutianus*) formed 90% of the catch, includes at least 10 other species, all of which were taken incidentally in the sablefish fishery until 1992, when a directed fishery was permitted on the seamounts. Coincident with the decline of the sablefish fishery, the rockfish fishery expanded focusing exclusively on Bowie Seamount. Cumulative rockfish catch between 1993 and 2000 was 1430t with an average of 130t year). The target depth was 200–550m (Canessa *et al.*, 2003). There is no specific management plan for the seamount rockfish fishery and the scientific permit for the rockfish fishery for seamounts was closed after the 1999 season, except as bycatch.

Halibut fishing in the Bowie Seamount Area has been conducted since approximately the 1950s. Between 1984 and 1992 there were only five boat landings of fish caught from Bowie Seamount, for a total weight of just under 63 t (Canessa *et al.*, 2003). Halibut fishing techniques include bottom trawls and longline sets.

Discussion

One major shortcoming of the present analysis concerns the definition of seamount catches. What characterizes a small-scale fishery is clear, but a seamount fishery is less well defined. A narrow interpretation of the word 'seamount' would have made the allocation of corresponding catches impossible in the form presented in this chapter. The only way to proceed was to identify areas of high seamount density. It was then assumed that fisheries in that area involved taking some fish from over seamounts. For the fisheries to qualify as small scale, the seamounts in an area had to be near the coast of the regions identified.

Separating seamount ecosystems from coral reef ecosystems was difficult in the Pacific. Some Pacific islands have bigger catches from coral reefs than from seamounts, even though seamounts exist at significant densities. Islands in this group had to be excluded from the analyses, contributing to the conservative nature of the global estimates of small-scale seamount catches.

Froese and Sampang (2004) argue that the closeness of the relationship a species has with a seamount is the defining criterion (i.e., that only demersal species truly belong to seamounts). Should it be the case then that only such permanent residents should be included in seamount catches? We contend that temporary visitors, when mostly caught in aggregations over seamounts; form part of seamount catches. Tuna, even though they are mainly migratory, certainly aggregate over seamounts (Fonteneau, 1991). Tuna taken over a seamount are part of the seamount catch, irrespective of how a seamount species is defined.

The same argument applies to deep sea species. In this case, however, the nature of the species depends on its bathymetric distribution. A seamount that comes close to the surface, or emerges from it to form an island, will necessarily be occupied by shelf species, which once caught become seamount shelf catches. Because most of the fisheries described in this chapter are small scale, the non-pelagic species included in the analysis are all shelf seamount species.

A rough estimate of 150 000–250 000 t of fish caught worldwide from small-scale fisheries on seamounts was estimated this way, half of it represented by tuna. This figure should be considered conservative as there may be unreported catches in some of the studied areas, and many islands, such as in the Western Central Pacific, contained very high densities of seamounts, but where it was impossible to separate demersal fish caught over seamounts from those caught over coral reefs, which were extensive.

Seamount ecosystems seem to exhibit fragile equilibria within groups of species. Diversity tends to be high, consequently species abundance tends to be low, at least locally. Many seamount species, especially deep sea fish, are k-strategists, therefore long lived and with low fecundity. All this adds to the fragility of these ecosystems to exploitation. Nonetheless, small-scale fisheries have existed for centuries in many oceanic islands situated over seamount regions, making a strong case for their long-term sustainability compared to industrial fisheries. In many places, small-scale and industrial fisheries coexist, making it more difficult to relate decreasing catch rates in certain areas to the joint effects of both types of fisheries.

Small-scale seamount fisheries are very important to local island communities. One way to assess this importance is simply to divide the volume of estimated catch by the overall population of those islands and to compare it with the world catch per capita. This exercise leads to roughly 37 kg per capita of fish caught from seamounts only compared to 13 kg of marine fish caught per capita worldwide from all sources. Notice, however, that other small-scale fisheries, not included in the 37 kg, add to the overall socio-economic impact of fisheries in the island nations exploiting seamounts. The relative isolation of many of these islands makes the product of these catches a major source of protein to local population.

In conclusion, there is a strong case for changing TAC and quota management systems. Favourable consideration should be given to allocations to communities that are strongly dependent on small-scale fisheries and have proved through time to be capable of exploiting fish resources in a sustainable way. Application of TAC and quota-based management systems to large-scale commercial or distant water fleets rewards those responsible for overexploitation and does not take into account the low impact and great dependence of many island communities on those fisheries. One possibility is to weigh positively the quota of those closer to the fish resources and, consequently, largely dependent on those fish stocks. This would accord with the 'adjacency' principle of giving greater access and management responsibility to those closer to the fish resources per sections 6.18, 7.2.2c and 12.12 of the Code of Conduct for Responsible Fishing (FAO, 1995). The only way to recover from successive management mistakes and succeed in fisheries management worldwide is to reward responsible fishing and make those irresponsible pay.

References

Anon (2002) *Annual Report for 2002*. Seychelles Fishing Authority (SFA). Victoria, Seychelles: SFA, 2003, 58 pp. http://hdl.handle.net/1834/176 (last accessed August 2005).

Afonso, P. and Santos, R.S. (2004) Implementation of Natura 2000 in the Azores: balancing marine conservation and fisheries management. In: *Marine Protected Areas and Fisheries, Proceedings of the International Expert Workshop* (eds. Ritterhoff, J., Gubbay, S. and Zucco, C.), pp. 95–7. International Academy for Nature Conservation, Isle of Vilm, Germany, 28 June–2 July, 2004, 117 pp.

Boehlert, G.W. (1993) Fisheries and marine resources of Hawaii and the US-associated Pacific islands: an introduction. *Marine Fisheries Review*, 55(2), 3–5.

Boggs, C.H. and Ito, R.Y. (1993) The Hawaii's pelagic fisheries. *Marine Fisheries Review*, 55(2), 69–82.

Canessa, R., Conley, K. and Smiley, B. (2003) Bowie seamount pilot marine protected area: an ecosystem overview. *Canadian Technical Report Fisheries Aquatic Science*, 2461, xi, 85 pp.

DFO (Fisheries and Oceans Canada) (1997) Report of the fifteenth North Pacificalbacore workshop. In: *North Pacific Albacore Workshop* (eds. Shaw, W., Bartoo, N. and Nanaimo, B.C.), 3–5 December 1997, 35 pp.

DFO (Fisheries and Oceans Canada) (1998) *News Release* (NR-PR-98-104E), 8, December 1998.

ESRI Data and Maps [CD-ROM] (2005) Redlands, CA: Environmental Systems Research Institute. http://www.brocku.ca/maplibrary/digital/ESRIdata2005.htm (last accessed July 2006).

FAO (1995) *Code of Conduct for Responsible Fisheries*, 41 pp. FAO, Rome.

FAO (2001) FAO Country Profiles and Mapping Information System. Seychelles: Fishery Country Profile. FAO, January 2001. http://www.fao.org/countryprofiles/index.asp?lang=en&iso3=SYC&subj=6 (last accessed July 21, 2005).

FAOSTAT (2005) Food and Agriculture Organization (FAO), Fisheries Department, FISHSTAT plus: Universal software for fishery statistical time series (Version 2.3), Rome. http://www.fao.org/figis/servlet/static?xml=FIDI_STAT_org.xml&dom=org&xp_nav=3,1,2&xp_banner=fi (last accessed February 2005).

Fernandes, A.M. (1984) Estudo preliminar da dinâmica da população de Aphanopus carbo Lowe, 1839 (peixe-espada-preto), no Arquipélago da Madeira. Faculdade de Ciências de Lisboa. Relatório de estágio, 200 pp.

Fernandes, L.M.R. (1994) Artes de pesca dos Açores. Secretaria Regional De Agricultura E Pescas Dos Açores, 17 pp.

Fonteneau, A. (1991) Monts sous-marins et thons dans l'Atlantique tropical est. *Aquatic Living Resources*, 4, 13–25.

Froese, R. and Sampang, A. (2004) Tasconomy and biology of seamount fishes. In: *Seamouns: Biodiversity and Fisheries* (eds. Morato, T. and Pauly, D.), pp. 25–31. *Fisheries Centre Research Report*, 12(5), 78 pp.

Frutuoso, G. [Palavras prévias de João Bernardo de Oliveira Rodrigues] (1968) Livro segundo das Saudades da Terra/Doutor Gaspar Frutuoso. Ponta Delgada, Instituto Cultural, XXVI, 473 pp.

Frutuoso, G. [Palavras prévias de João Bernardo de Oliveira Rodrigues] (1983) Livro Terceiro das Saudades da Terra/Doutor Gaspar Frutuoso. Ponta Delgada, Instituto Cultural, CLXXVII, 300 pp.

Gomes, T.M., Sola, E., Grós, M.P., Menezes, G. and Pinho, M.R. (1998) Trophic relationships and feeding habitats of demersal fishes from Azores: importance to multispecies assessment. *ICES*, CM:1998/O.

Gubbay, S. (2002) Offshore directory. Review of a selection of habitats, communities and species of the North-East Atlantic. *Report for WWF-UK, North-East Atlantic Programme*, 36 pp.

Gubbay, S. (2003) Seamounts of the North-East Atlantic. *Report for WWF Germany and OASIS*, 38 pp.

Gudger, E. (1927) Wooden hooks used for catching sharks and Ruvettus in South Seas: a study of their variation and distribution. *Anthropological Papers of the American Museum of Natural History*, 28, 199–355.

Haight, W.R., Kobayashi, D.R. and Kawamoto, K.E. (1993) Biology and management of deepwater snappers of the Hawaiian archipelago. *Marine Fisheries Review*, 55(2), 20–7.

Holthus, P.F. and Maragos, J.E. (1995) Marine ecosystem classification for the tropical island Pacific. In: *Marine and Coastal Biodiversity in the Tropical Island Pacific Region* (eds. Maragos, J.E., Peterson, M.N.A., Eldredge, L.G., Bardach, J.E. and Takeuchi, H.F.), Vol. I. Species systematics and information management priorities. Program on Environment, East-West Center, Honolulu, Hawaii, pp. 239–78.

ICES (2005) Report of the working group on deep water ecology (WGDEC). *International Council for the Exploitation of the Sea (ICES)*, CM:2005/ACE: 02, Ref.: E, G, 76 pp.

IOTC (Indian Ocean Tuna Commission) Database. http://www.iotc.org/English/data/databases.php (last accessed June 2005).

Isidro, E.J. (1996) Biology and population dynamics of selected demersal fish species of the Azores archipelago. PhD Thesis, Department of Environmental and Evolutionary Biology, Port Erin Marine Laboratory, University of Liverpool, England, 249 pp.

Kitchingman, A. and Lai, S. (2004) Inferences and potential seamount locations from mid-resolution bathymetric data. In: *Seamounts: Biodiversity and Fisheries* (eds. Morato, T. and Pauly, D.), Vol. 12(5), pp. 7–12. Fisheries Centre Research report, University of British Columbia. Available online: http://www.seaaroundus.org/report/seamounts/05_AKitchingman_Slai/AK_SL_TEXT.pdf (last accessed August 2005).

Krug, H. (1994) Biologia e avaliação do stock açoriano de goraz, Pagellus bogaraveo. PhD Thesis, Department of Oceanography and Fisheries, University of the Azores, Horta. *Arquivos do DOP, Série estudos, No.7/94*, 192 pp.

Krug, H.M. and Silva, H.M. (1988) Virtual population analysis of Pagellus bogaraveo (Brunnich, 1768) from the Azores. *International Council for the Exploitation of the Sea (ICES)*, Demersal Fish Committee, CM:1988/G: 19.

Lopes, F. (1948) A Pesca na Ilha Terceira. In: *Boletim do Instituto Histórico da Ilha Terceira*, N 6, pp. 61–106.

Lowe, R.T. (1843) Notice of fishes newly observed or discovered in Madeira during the years 1840, 1841 and 1842. *Proceedings of the Zoological Society of London*, 11, 81–95. Systematic/nomenclature; Checklist; Geographical distribution.

Macdonald, G.A., Abbott, A.T. and Petersen, F.L. (1983) *Volcanoes in the Sea: The Geology of Hawaii*, 519 pp. University of Hawaii Press, Honolulu.

Machete, M., Morato, T. and Menezes, G. (2005) Modeling the distribution of two fish species in seamounts of the Azores. In: *Deep Sea 2003: Conference on the Governance and Management of Deep Sea Fisheries. Part I: Conference Reports* (eds. Shotton, R.), pp. 182–95. Queenstown, New Zealand, 1–5 December 2003. *FAO Fisheries Proceedings*. No. 3/1. Rome, FAO, 2005, 718 pp.

Martins, J.A. (1986) Potencialidades da ZEE Açoriana. In: *Relatório da VI Semana das Pescas dos Açores* (eds. Secretaria Regional de Agricultura e Pescas, Direcção Regional das Pescas dos Açores), 6(1985), pp. 125–32. Horta, Açores.

Mees, C.C. (1989) Artisanal fishing boats in the Seychelles. *Seychelles Fishing Authority Technical Report*, 49 pp. Seychelles Fishing Authority, Victoria, Seychelles.

Menezes, G.M. (2003) Demersal fish assemblages in the Atlantic archipelagos of the Azores, Madeira and Cape Verde. PhD Thesis, Department Oceanography and Fisheries, University of the Azores, Portugal.

Menezes, G.M.M. (1996) Interacções tecnológicas na pesca demersal dos Açores. Master Thesis, University of the Azores. *Arquivos do DOP, Série Estudos, No. 1/96*, 187 pp.

Merrett, N.R. and Haedrich, R.L. (1997) *Deep Sea Demersal Fish and Fisheries*, 282 pp. Chapman and Hall, London.

Metz, H.C. (eds.) (1995) *Indian Ocean: Five Island Countries*, 3rd edition. Federal Research Division, Library of Congress, Washington, DC. Available online: http://lcweb2.loc.gov/frd/cs/sctoc.html#sc0016 (last accessed August 2005).

Murie, D.J., Saunders, M.W. and McFarlane, G.A. (1996) Canadian trap fishery for sablefish on seamounts in the northeastern Pacific Ocean, 1983–1993. Canadian Manuscript. *Report, Fisheries and Aquatic Sciences No. 2348.*

Parin, N.V. (1986) Trichiuridae. In: *Fishes of the North-East Atlantic and the Mediterranean* (eds. Whitehead, P.J., Bauchot, M.L., Hureau, J.C., Nielsen, J. and Tortonese, E.), Vol. II, 977 pp. UNESCO, Paris.

Pereira, J.G. (1995) A pesca do atum nos Açores e o atum patudo (Thunnus obesus, Lowe 1839) do Atlântico. PhD Thesis, Department Oceanography and Fisheries, University of the Azores, Portugal, 330 pp.

Pinho, M.R. (2003) Abundance estimation and management of Azorean demersal species. PhD Thesis, Department Oceanography and Fisheries, University of the Azores, Portugal, 173 pp.

Pinho, M.R. and Menezes, G. (2006) Azorean deepwater fishery: ecosystem, species, fisheries and management approach aspects. In: *Deep Sea 2003: Conference on the Governance and Management of Deep-sea Fisheries. Part II: Conference Poster Papers and Workshop Papers* (eds. Shotton, R.), pp. 330–43. Queenstown, New Zealand, 1–5 December 2003. Dunedin, New Zealand, 27–29 November. *FAO Fisheries Proceedings*. No. 3/2. Rome, The Food and Agriculture Organization of the United Nations, 2005. 487 pp.

Pinho, M.R., Pereira, J. and Rosa, I. (1995) Caracterização da pesca do isco da frota atuneira Açoreana. *Arquivos do DOP, Série: Estudos, No. 2/95*, 29 pp.

Pinho, M.R., Menezes, G. and Krug, H. (1999) Estado de exploração dos recursos demersais dos Açores – Proposta de gestão. Relatório apresentado ao Workshop: 'Gestão de Pescas', Secretaria Regional de Agricultura Pescas e Ambiente, Direcção Regional das Pescas. Horta 20 de Março de 1999.

Polovina, J.J. (1993) The lobster and shrimp fisheries in Hawaii. *Marine Fisheries Review*, 55(2), 28–33.

Pooley, S.G. (1993) Hawaii marine fisheries: some history, long term trends and recent developments. *Marine Fisheries Review*, 55(2), 7–19.

Reis, S., Sena-Carvalho, D., Delgado, J.H. and Afonso-Dias, M. (2001) Historical overview of the black scabbardfish (Aphanopus carbo Lowe, 1839) fishery in Madeira island. *Deep Sea Fisheries Symposium-Poster, NAFO SCR Doc. 01/103*, Serial Nº N4491.

Robinson, J. and Shroff, J. (2004) The fishing sector in Seychelles: an overview, with emphasis on artisanal fisheries. *Seychelles Medical and Dental Journal*, Special issue, 7(1), November 2004.

SAUP (Sea Around Us Project) (2005) A global database on marine fisheries and ecosystems. www.seaaroundus.org. Fisheries Centre, University British Columbia, Vancouver (British Columbia, Canada). (last accessed July 2005).

Silva, H.M. (1987) Estado dos stocks de goraz e abrótea. In: *Relatório da VII Semana das Pescas dos Açores* (eds. Secretaria Regional de Agricultura e Pescas, Direcção Regional das Pescas dos Açores), 197–9. Horta, Açores.

Silva, A.A. and Pereira, J.G. (1999) Catch rates for pelagic sharks taken by the Portuguese swordfish fishery in the waters around the Azores 1993–1997. *ICCAT, Collective Volume Scientific Papers SCRS/98/168*, XLIX(4),300–305.

Silva, H.M., Krug, H. and Menezes, G. (1994) Bases para a regulamentação da pesca de demersais nos Açores. *Arquivos do DOP, Série estudos, No. 94.*

Simões, P.R. (1995) The swordfish (Xiphias gladius L. 1758) fishery in the Azores, from 1987–1993. *ICCAT, Collective Volume Scientific Papers SCRS/94/109*, XVIV(3), 126–31.

Tomás, M. (1635) Insulana. *Em amberes: Em casa de Ioam meursio impressor*, 1635, 493 pp.

WPRFMC (1986a) Combined fishery management plan, environmental assessment and regulatory impact review for the bottomfish and seamount groundfish fisheries of the Western Pacific Region. *Final Report, Western Pacific Regional Fishery Management Council*, March 1986. Available online: http://www.wpcouncil.org/bottomfish.htm (last accessed July 2005).

WPRFMC (1986b) Fishery management plan for the pelagic fisheries of the Western Pacific Region. *Final Report, Western Pacific Regional Fishery Management Council*, July 1986. Available online: http://www.wpcouncil.org/pelagic.htm (last accessed July 2005).

WPRFMC (2003) Bottomfish and seamount groundfish fisheries of the Western Pacific Region, 2003. *Annual Report, Western Pacific Regional Fishery Management Council*. Available online: http://www.wpcouncil.org/bottomfish.htm (last accessed July 2005).

WPRFMC (2004) Pelagic fisheries of the Western Pacific Region, 2004. *Annual Report, Western Pacific Regional Fishery Management Council*. Available online: http://www.wpcouncil.org/pelagic.htm (last accessed July 2005).

WPRFMC (2005) Final environmental impact statement, bottomfish and seamount groundfish fisheries of the Western Pacific Region. *Final Report, Western Pacific Regional Fishery Management Council*, May 2005. Available online: http://www.wpcouncil.org/bottomfish.htm (last accessed July 2005).

Chapter 17
Large-scale distant-water trawl fisheries on seamounts

Malcolm R. Clark, Vladimir I. Vinnichenko, John D.M. Gordon, Georgy Z. Beck-Bulat, Nikolai N. Kukharev and Alexander F. Kakora

Abstract

Seamounts support a large number and wide diversity of fish species. A number of these species can form aggregations for spawning or feeding, and are the target of large-scale trawl fisheries. Since the 1970s, seamounts throughout the world's oceans have been explored for commercial resources, starting with the Soviet Union and Japan which deployed distant-water fleets around the world. Since then, a large number of other countries have pursued fisheries on seamounts, especially in deepwater. Historical data on seamount trawl fisheries throughout the world's oceans are summarized here together with some information on line and pot fisheries. In particular, detailed information is compiled from the activities of the former Soviet Union. The results indicate that the total cumulative catch from seamount trawl fisheries likely exceeds 2.25 million tonnes. Catch information is presented by region, and individual seamount or seamount group where possible. These histories show that the long-term prospect for large-scale commercial exploitation of fish and invertebrate resources on seamounts is not encouraging.

Introduction

Seamounts support a large number and wide diversity of fish species. A total of almost 800 species have been recorded from seamounts (see Chapter 9; Morato and Pauly, 2004; Morato *et al.*, 2006). Most of these also occur widely on continental shelf and slope habitat, but seamounts can be an important habitat for commercially valuable species which may form dense aggregations for spawning or feeding (Clark, 2001), and on which a number of large-scale fisheries have developed.

Intensive searching for fisheries resources on seamounts around the world's oceans was initiated by the former Soviet Union in the late 1960s and 1970s. Over that period, searches for new fisheries were systematically conducted by offshore trawler fleets (Fig. 17.1) in the Atlantic, Pacific, and Indian oceans. Seamounts with concentrations of fish and invertebrates were found, and commercial fisheries developed in a number of areas (Fig. 17.2). The first major fisheries to be exploited were those of pelagic armourhead (*Pseudopentaceros wheeleri*) in the Pacific ocean, where catches by the USSR and Japanese distant-water fleets rapidly grew. This was followed by the development of fisheries in other parts of the Atlantic and Indian oceans, and established offshore seamounts as important targets for global fisheries. In subsequent decades, other countries such as South Korea, and later China, Cuba, the European Union, Norway, Iceland, Faroe Islands, New Zealand, Australia, and Southern African countries, also developed fisheries on seamounts.

Fig. 17.1 A Soviet 'BMRT' class trawler, about 80 m long and 2000 GRT, typical of those that explored seamount resources throughout the world's oceans in the 1970–1980s (photo Neil Bagley, NIWA).

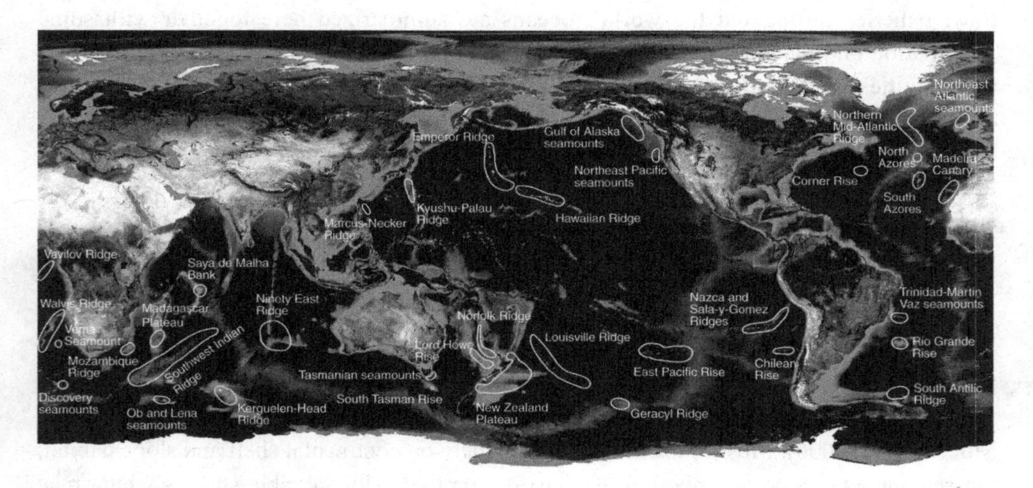

Fig. 17.2 World map showing the distribution of the main seamount and ridge features and fishing grounds referred to in this chapter.

Deepwater trawl fisheries occur widely for a number of species. These include alfonsino (*Beryx splendens*), black cardinalfish (*Epigonus telescopus*), orange roughy (*Hoplostethus atlanticus*), southern boarfish (*P. richardsoni*), macrourid rattails (primarily roundnose grenadier *Coryphaenoides rupestris*), and oreos (several species of the family Oreosomatidae, including smooth oreo *Pseudocyttus maculatus* and black oreo *Allocyttus niger*). Other fisheries occur over seamounts, such as those for pelagic species (mainly tunas), near-bottom fishing for mackerels, and for smaller-scale line fisheries (e.g., black scabbardfish *Aphanopus carbo*).

Many of these fisheries have not been sustainably fished or managed. A number have shown a 'boom-and-bust' pattern, with catches rapidly developing and declining within a decade (e.g., Uchida and Tagami, 1984; Koslow *et al.*, 2000; Clark, 2001; 2002a). This has raised concerns over whether such fisheries can continue (e.g., Roberts, 2002; Gianni, 2004; Stone *et al.*, 2004).

This chapter summarizes available historical data on seamount trawl fisheries throughout the world's oceans, together with some information on line and pot fisheries, and draws conclusions from the patterns in these fisheries about long-term prospects for large-scale commercial exploitation of fish and invertebrate resources on seamounts. This complements the account in Chapter 16 of small-scale artisinal seamount fisheries, and combines with it some information on tuna fisheries on seamounts.

Data sources

Catch summaries have been compiled from a range of publications, fisheries statistics, and data of Soviet, Russian, and Ukrainian scientific research and exploratory cruises. Published reports have provided good information for some seamount areas on fisheries conducted by Japan, New Zealand, Australia, Spain, other EU countries, and Namibia, while personal contacts and data extracted by the authors have been used in some cases to provide 'guestimates' of likely species composition and catch for some seamount regions. Food and Agriculture Organization (FAO) catch statistics were also examined for some areas as a general check on whether catches ascribed to seamounts appeared reasonable, or where no other sources were available. The report by Gianni (2004) on high seas (areas outside of national jurisdiction) fishing in general (not just seamounts) was examined for some areas where much of the high seas catch was thought to be from seamounts. We have focused on bottom trawl fisheries, but have also included comments on midwater trawl, line, and pot fisheries where good data on seamount fisheries are available.

The catch figures are known to be incomplete. There has often been little or no reporting of high seas catches, or misreporting of location if commercially sensitive. We have been unable to source recent Japanese, Chinese, and Korean catches on seamounts. Some catch figures are known to include fish taken from areas of adjacent slope, but a split has not been possible. There is also, at times, a variety of fishing years (not calendar year) used by different countries, and it has not been practical to try and resolve these issues. Even so, the compilation is the most comprehensive attempted to date for seamount fisheries, and at the very least gives a reasonable indication of general catch levels in seamount trawl fisheries.

Global overview of seamount fisheries

Seamount fisheries have taken place on a large number of seamounts throughout the world's oceans. These are prominent in the Pacific Ocean, but also in the southern Indian

Ocean, Mid-Atlantic Ridge (MAR) in the North Atlantic, and off the African coast in the South Atlantic (Fig. 17.3). The main species are listed in Table 17.1.

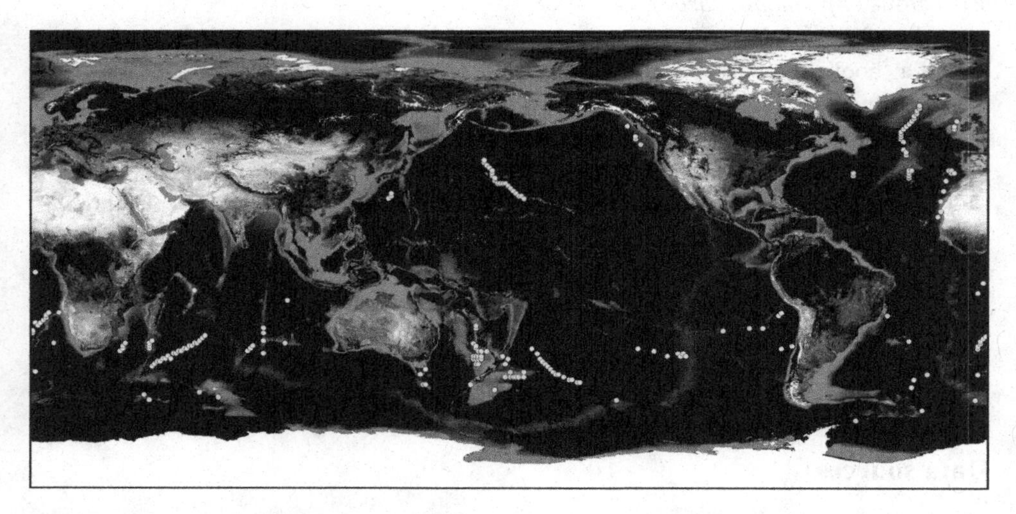

Fig. 17.3 World map showing location of known (incomplete) seamount fishing operations (yellow circles represent a 1 degree square).

Table 17.1 Common names and scientific names of the main commercial fish species on seamounts, and the depth range commonly fished.

Species and code	Scientific name	Main depth range (m)
Alfonsino	*Beryx splendens*	300–600
Black cardinalfish	*Epigonus telescopus*	500–800
Rubyfish	*Plagiogenion rubiginosum*	250–450
Black scabbardfish	*Aphanopus carbo*	600–800
Redbait	*Emmelichthys nitidus*	200–400
Sablefish	*Anoplopoma fimbria*	500–1000
Pink maomao	*Caprodon* spp.	300–450
Southern boarfish	*Pseudopentaceros richardsoni*	600–900
Pelagic armourhead	*Pseudopentaceros wheeleri*	250–600
Orange roughy	*Hoplostethus atlanticus*	600–1200
Oreos	*Pseudocyttus maculatus, Allocyttus niger*	600–1200
Bluenose	*Hyperoglyphe antarctica*	300–700
Redfish	*Sebastes* spp. (*S. marinus, S. mentella, S. proriger*)	400–800
Roundnose grenadier	*Coryphaenoides rupestris*	800–1000
Toothfish	*Dissostichus* spp.	500–1500
Notothenid cods	*Notothenia* spp.	200–600

The largest seamount trawl fisheries have occurred in the Pacific Ocean. In the 1960s–1980s large-scale fisheries for pelagic armourhead and alfonsino occurred on the Hawaiian and Emperor Seamount chains in the North Pacific (Fig. 17.4). In total about 800 000 t of pelagic armourhead was taken and about 80 000 t of alfonsino. In the southwest Pacific, fisheries for orange roughy, oreos, and alfonsino have been large, and continue to be locally important. Orange roughy has also been the target of fisheries on seamounts on the MAR in the North Atlantic, off the west coast of southern Africa, and in the SW Indian Ocean. Roundnose grenadier was an important fishery for the Soviet Union in the North Atlantic, where catches have been over 200 000 t. Smaller fisheries for alfonsino, mackerel, and cardinalfish have occurred on various seamounts in the Mid-Atlantic and off the coast of North Africa. In the Southern Ocean, fisheries for toothfish, notothenids, and icefish can occur on seamounts as well as slope and bank areas. Most of these seamounts are fished with bottom trawls, but several are also subject to midwater trawl and longline fisheries.

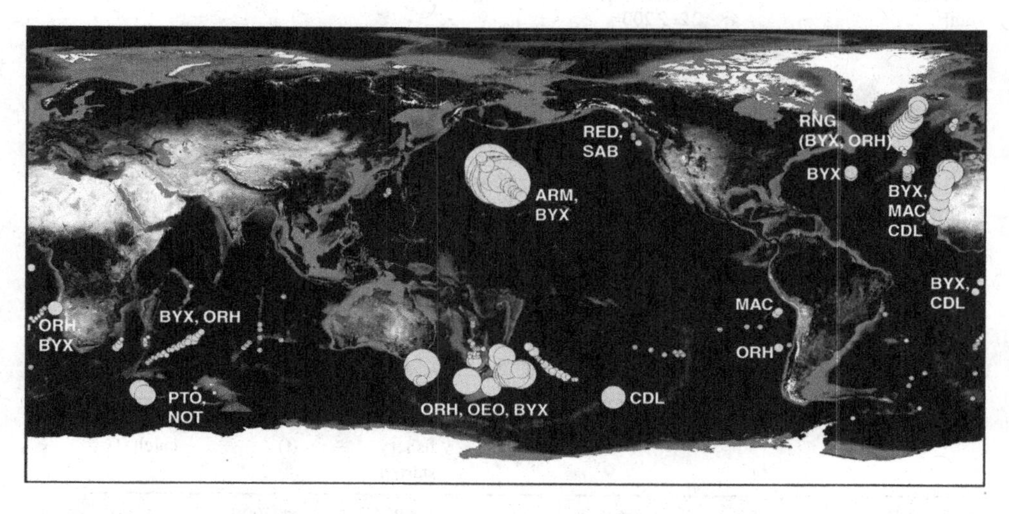

Fig. 17.4 Relative size of historical seamount fisheries. Data are gridded by 1 degree squares. Circle size is proportional to total catch for that grid square, maximum is 85 000 t (see Table 17.2 for codes to the fish species).

In total, the international catch of demersal fishes on seamounts by distant-water fishing fleets is estimated to be over 2.25 million tonnes of fish since the 1960s (Table 17.2, and compare estimates in Chapter 18).

Table 17.2 Summary of total estimated catch (t) of main commercial species from seamounts, major fishing periods, and main gear types used in the seamount fisheries. Species codes are used in Fig. 17.4.

Species	Total historical catch (t)	Main fishery years	Gear type
Alfonsino (BYX)	166 950	1977 to present	Bottom and midwater trawl, some longline
Orange roughy (ORH)	500 700	1978 to present	Bottom trawl
Oreos (OEO)	146 500	1970 to present	Bottom trawl
Cardinalfish (CDL)	52 100	1978 to present	Bottom (and midwater trawl)
Redfish (RED)	54 450	1996 to present	Bottom and midwater trawl
Southern boarfish	9600	1982 to present	Bottom trawl
Pelagic armourhead	803 400	1968–1982	Bottom and midwater trawl (ARM)
Mackerel species (MAC)	148 200	1970–1995	(Bottom) and midwater trawl
Roundnose grenadier (RNG)	232 400	1974 to present	Bottom and midwater trawl
Blue ling	17 000	1979–1980	Bottom trawl

(Continued)

Table 17.2 *Continued*

Species	Total historical catch (t)	Main fishery years	Gear type
Scabbard fish	75000	1973–2002	Bottom and midwater trawl
Sablefish (SAB)	1400	1995 to present	(Bottom trawl), line
Bluenose	2500	1990 to present	Bottom and midwater trawl
Rubyfish	1500	1995 to present	Bottom and midwater trawl
Pink maomao	2000	1972–1976	Bottom and midwater trawl
Notothenid cods (NOT)	36250	1974–1991	Bottom trawl
Toothfish (PTO)	12250	1990 to present	Bottom trawl, longline
Total	2262200		

Seamount fisheries of the Pacific Ocean

The Pacific Ocean contains a large number of seamounts, perhaps as many as 30000 (see Chapters 1 and 2; Wessel, 2001), with many on the Pacific tectonic plate. It is not therefore surprising that large-scale offshore seamount fisheries first developed in the Pacific, as Soviet and Japanese trawling fleets began to explore wide areas in the 1960s (Table 17.3).

Table 17.3 Summary data on seamount fisheries in the Pacific Ocean.

Area	Species	Year fishery started	Total catch (t)	Maximum annual catch (t)
Hawaiian and Emperor Ridges	*Pseudopentaceros wheeleri*	1968	800000	178300
	Beryx splendens	1970	80000	31000
Northeast Pacific seamounts	*Sebastes* spp. *Anoplopoma fimbria* *Trachurus symmetricus* *Lithodes* spp. *Chionoecetes tanneri*	1978	2000	
Kyusmu-Palau Ridge	*Beryx splendens*, *Emmelichthys sruhsakeri*	1970	600	
Geracl Ridge	*Epigonus denticulatus*	1972	11100	8800
Norfolk Ridge	*Caprodon longimanus*	1974	2000	1000
	Beryx splendens	1980	1300	500
	Hoplostethus atlanticus	2000	1800	1000
Lord Howe Ridge	*Hoplostethus atlanticus* *Beryx splendens*	1988	13600	2400
Louisville Ridge	*Epigonus* spp.	1977	12000	12000
	Beryx splendens	1990		
	Hoplostethus atlanticus	1994	37200	11300
	Oreos (*Allocyttus niger*, *Pseudocyttus maculatus*)	1996	2300	1000
New Zealand Plateau	*Hoplostethus atlanticus*	1979	192100	18000
	Oreos (*Allocyttus niger*, *Pseudocyttus maculatus*)	1978	130000	11000
	Beryx splendens	1984	7800	800
	Epigonus telescopus	1989	18000	2300

(*Continued*)

Table 17.3 *Continued*

Area	Species	Year fishery started	Total catch (t)	Maximum annual catch (t)
Tasmanian seamounts	*Hoplostethus atlanticus*	1985	196 400	58 600
	Oreos (*Pseudocyttus maculatus, Neocyttus rhomboidalis*)	1989	7500	2000
South Tasman Rise	*Hoplostethus atlanticus*	1998	10 900	4100
	Oreos (*P. maculatus, N. rhomboidalis*)	1994	3800	1600
Nazca and Sala-y-Gomez Ridges	*Emmelichthys* spp., *Trachurus murphyi*	1978	7200	5100
	Beryx splendens			
	Projasus parkeri	1980	1000	1000
Chilean Rise	*Hoplostethus atlanticus*	1999	8200	1900

North Pacific fisheries

Emperor and Hawaiian Ridges

The Emperor Seamount chain and seamounts of the northern Hawaiian Ridge, northwest of the Hawaiian Islands, comprise numerous features between latitudes 52–20°N, and longitudes 165°E–155°W. There are 18 known seamounts outside the Hawaiian EEZ (Exclusive Economic Zone) with summit depths less than 1000 m.

Pelagic armourhead was not well known as a commercial species until 1967, when Soviet trawlers discovered large aggregations on seamounts in the southern Emperor Seamount chain (Borets, 1975, 1979). The armourhead aggregates over these seamounts to spawn during the last years of their lives, and heavy fishing by the Soviet fleet caught up to 130 000 t of armourhead a year in the early stages of the fishery (see Table 17.4). Most catches

Table 17.4 Catch (t) of the main species caught by USSR and Japanese trawlers on the Emperor Seamounts and northern Hawaiian Ridge, 1968–1982 (Japanese figures from Sasaki, 1986).

Year	Pelagic armourhead		Alfonsino		Others		Total
	USSR	Japan	USSR*	Japan	USSR	Japan	
1968	46 330						46 330
1969	144 916	7410		45		1	152 372
1970	136 206	26 262		600		75	163 143
1971	3178	5546		68		164	8956
1972	79 287	34 826	5985	81		234	120 413
1973	149 919	28 355	31 045	12		191	209 522
1974	19 882	26 284				999	47 165
1975	28 780	21 746	1428			1096	53 050
1976	6910	24 829	220	1726		4535	38 220
1977	603	3448	327	1941		424	6743
1978		875		1645		895	3415
1979		499		5383		975	6857
1980		1837	130	8632	2259	3198	16 056
1981	29	1211	135	7916	4798	1991	16 080
1982	10	524	819	8582	176	763	10 874
1983							
1984			6		456		462
Total	616 020	183 652	40 095	36 631	7689	15 541	899 658

*Includes mirror perch, bigeye, and some other species.

were taken on Kimmei, Milwaukee, Colahan, and Hancock seamounts. These have summit depths between 200 and 400m, with most fishing occurring between 300 and 600m on the peaks and upper flanks. USSR surveys estimated early stock size at between 240000 and 350000t (Borets, 1979; Kulikov and Darnitsky, 1992). Large Japanese trawlers joined the fishery in 1969, and combined catches of the two fleets exceeded 200000t in 1973, before starting to drop rapidly. By 1982, the fishery had ceased because of a lack of aggregations. In subsequent years, the seamounts were often surveyed by USSR exploratory and fishing vessels, but no concentrations of armourhead were found, and catch rates were generally only 100–150kg/haul. Some limited Japanese fishing is thought to occur periodically.

When armourhead catches started to drop dramatically from 1976, the Japanese fleet then shifted its focus to alfonsino (Sasaki, 1986). Alfonsino catches were sometimes large, and taken by both longline and midwater trawl. The Japanese longline and handline fishery for alfonsino started in 1973 at Milwaukee Seamount. Vessels from Korea and Taiwan also worked the area later. Catches are not well known, but in 1975 the Japanese hook and line vessels caught 4000t of groundfish (Seki and Tagami, 1986). The fishery stopped after the US EEZ was declared in 1977.

In addition to armourhead and alfonsino, several other deep-sea fishes were of commercial importance in the area, including mirror perch/dory (*Zenopsis nebulosus*) and broad alfonsino (*B. decadactylus*) which were at times subject to target fisheries. Catches by the USSR fleet were high in some years, reaching several thousand tonnes (Table 17.3). Pearlside (*Maurolicus muelleri*) and mackerel (*Scomber japonicus*) were targets of a midwater trawl fishery between 1979 and 1982. Several thousand tonnes were caught, but neither fishery continued beyond this date (Belyaev, 2003).

The Emperor Seamounts were also subject to a Japanese pole-and-line fishery for albacore tuna (*Thunnus alalunga*) from 1970 into the 1980s. Annual catches between 1970 and 1983 were often 10000–20000t (Yasui, 1986). Extensive tuna fisheries occur around the Hawaiian Islands (see Chapter 10A), including Cross Seamount to the south. Bigeye (*T. obesus*) and yellowfin (*T. albacares*) are caught, with catches around 500t/year (Adam *et al.*, 2003).

Fisheries for red coral (*Corallium secundum*) have occurred on the Emperor Seamount chain. A large bed of this coral at about 400m was discovered by Japanese fishers in 1965, and together with Taiwanese vessels the fishery expanded to a peak of 150t in 1969 (Grigg, 1993). Most vessels in this fishery used various types of tangle-net dredge. Catch levels dropped away until 1978 when a new species of *Corallium* was found in abundance at depths of 900–1500m on other Emperor Seamounts. Over 100 coral boats were involved in this fishery, with catches of almost 300t in 1981. Fisheries have occurred on other seamounts in the area, such as Hancock seamount, but catches have generally been small.

Japanese exploratory surveys were also carried out on mid-central Pacific seamounts southwest of the Emperor Seamount chain. These seamounts are relatively deep, with summit depths of 1200–1500m, and no commercially viable species or concentrations were located (Sasaki, 1986).

Northeast Pacific seamounts

There are a number of scattered offshore seamounts along the western coasts of Canada and the USA in the NE Pacific and Gulf of Alaska (Maloney, 2004). Exploration in this region was pioneered in the 1970s by Soviet and Japanese exploratory research vessels. A total of 26 seamounts with depths less than 1000 m were found and explored (Darnitsky *et al.*, 2001), but catch rates were generally low, the main species being sablefish (*Anoplopoma fimbria*) (Beck-Bulat *et al.*, 1985). Similar results were obtained during Japanese research (Sasaki, 1986) where catches at Patton and Pratt Seamounts were dominated by sablefish, with rougheye rockfish (*Sebastes aleutianus*) and shortspine thornyhead (*Sebastolabus alascanus*). King crabs (*Paralithodes camtschaticus*, *P. platypus*, *Lithodes aequispinus*, and *L. couesi*), snow crab (*Chionoecetes opilio*), and tanner crabs (*C. bairdi* and *C. tanneri*) occurred on eight of nine seamounts surveyed by the USA in the Gulf of Alaska (Hughes, 1981). Crab fisheries have developed, although these operate mainly on the continental shelf and slope, or on the slope around offshore islands. Data on catches from seamounts are not available.

Fish concentrations were found by the Soviet investigations only on Cobb, Union, and Warwick (46–48°N, 130–133°W), and Fiberling seamounts (32–33°N, 127–128°W). On Cobb seamount the most abundant species were rockfish (*Sebastes* spp.), and on Fiberling seamount, California rattail (*Nezumia stelgidolepis*). Catches down to 150 m were dominated by *S. entomelas*, and down to 300 m by *S. borealis* and big-eyed rockfish (*S. zacentrus*). Catch rates from pelagic trawling were 15–60 t/day, which were much higher than from bottom trawling (4–6 t/day). Catch rates by bottom trawl of redfishes on Union seamount ranged between 3 and 15 t/tow. In midwater above the seamounts, concentrations of jack mackerel (*Trachurus symmetricus*) were detected, with catches up to 40 t/haul (Beck-Bulat *et al.*, 1985). Catches of mackerel (*S. japonicus*) have also been taken. Russian activity ceased after the introduction of EEZs in 1978.

Japanese trawling on Cobb and Warwick seamounts for several species of rockfish (harlequin, *S. variegates*; red-stripe, *S. proriger*; rosethorn, *S. helvomaculatus*; and black, *S. melanops*) during 1978 and 1979 caused declines in catch and in the size of individual fish. Catch rates decreased from 2.6 to 0.7 t/h between the 2 years (Sasaki, 1986).

The catch from seamounts in the NE Pacific is poorly known. It is estimated that in the 1970–1980s on Cobb seamount the Soviet exploratory fleet took about 2000 t. In this period, Japanese longliners also fished there with observations made of a daily catch of 5–15 t of redfishes. Longline vessels from other nations also fished occasionally. Trawl and acoustic surveys by the Soviet research vessels estimated the total biomass of redfishes in this area to be about 20 000 t (Stepanenko *et al.*, 2002; Vologdin, 2002; Kodolov and Darnitsky, 2004).

Small fisheries have occurred on more inshore seamounts such as Bowie, off the west coast of Canada. Rockfish (*Sebastes* spp.), sablefish, Pacific halibut (*Hippoglossus stenolepis*), and tuna have been fished. These line fisheries have totalled about 1450 t for sablefish, 1430 t for rockfish, and 63 t for halibut (and unknown for tuna) (WWF, 2003). Exploratory trawling by Canadian vessels was undertaken in 1993 on Bowie, Union, and Cobb seamounts, and catches of *Sebastes* spp. were as high as 16 t in a 10-min tow (Stitt, 1993).

Marcus-Necker Ridge and Magellan seamounts

The Marcus-Necker Ridge is southeast of Taiwan. About 20 seamounts with summit depths less than 1000 m are known. Magellan seamounts make up a large system in the central part of the East-Marianas Deep with 15 peaks shallower than 1500 m (Darnitsky and Kodolov, 2004).

Soviet vessels explored the Marcus-Necker Ridge in 1979, 1983, and 1984, and the Magellan seamounts in 1984. No fishable aggregations were found in either area. Catches were generally less than 100 kg/tow (Orlov, 1991; Darnitsky and Kodolov, 1997).

Kyushu-Palau and Bonin Ridges

Kyushu-Palau Ridge is a chain of seamounts extending from Kyushu Island to the NW extremity of New Guinea. Bonin Ridge is located in the western part of the Philippines Basin. A total of 25 seamounts were explored by Soviet vessels in the early 1970s between 24–30°N and 133–136°W (Borets and Kulikov, 1986). Most effort was directed at the Kyushu-Palau Ridge, where fishable aggregations were found on four seamounts. On the Kita-Kocho and Minami-Kocho seamounts, catches consisted primarily of redbait (*Emmelichthys nitidus*) and 'ruby' fish (*Erythrocles schlegeli*). Alfonsino was the main commercial species on the other seamounts (Darnitsky *et al.*, 2004; Kodolov and Darnitsky, 2004). Fish aggregations were highly variable. Periods with high catches (up to 10–20 t/tow) alternated with times of catches less than 0.1–0.2 t. Borets and Kulikov (1986) estimated yield for the ridge to be 5000–7000 t/year.

Detailed information about the activities of the Soviet/Russian fishing fleet on the Kyushu-Palau and Bonin Ridges are not available. Japanese and Taiwanese commercial vessels worked the area, but again no details are known (Kulikov and Kodolov, 1991).

South Pacific Fisheries

Norfolk Ridge

Exploratory fishing by Soviet and Japanese vessels occurred in the mid-1970s on seamounts of the Norfolk Ridge which runs between New Zealand and New Caledonia. This ridge feature has a large number of peaks which rise to less than 1000 m below the surface, while towards New Caledonia there are more isolated seamounts and guyots.

Soviet exploratory vessels in the 1970s focused on the Wanganella Bank, where aggregations consisted mainly of pink maomao (*Caprodon longimanus*) at depths of 100–250 m (Boldyrev *et al.*, 1981; Boldyrev, 1986). The total catch taken by the exploratory vessels between 1972 and 1975 was about 1000 t. The stock of *C. longimanus* was estimated at 25 000–50 000 t, with a possible yield of 5000–10 000 t/year (Borets and Kulikov, 1986). However, the Soviet fishing fleet did not operate there.

Japanese fishing on the southern section of the Norfolk Ridge in 1976 caught a variety of species. Pink maomao was targeted by a commercial Japanese trawler in 1976, which caught 1000 t over 2 months. Catch per unit effort (CPUE) of the exploratory research vessel decreased from 1.7 to 0.2 t/h over the year (Sasaki, 1986).

In 1980 and 1986, Japanese and French vessels carried out exploratory bottom trawling between 220 and 690 m on the Norfolk Ridge and the Lord Howe Rise. Catches consisted mainly of alfonsino, boarfish (*P. richardsoni* and *P. japonicus*), and ruby snapper (*Etelis carbunculus* and *E. coruscans*). The total catch by a Japanese vessel over 14 days was estimated at 140 t (Grandperrin and Richer de Forges, 1988).

Five seamounts off New Caledonia, ranging in depth from 500 to 750 m at their summit, were fished by bottom longline between 1988 and 1991. Three vessels were allowed to fish inside the New Caledonian EEZ, although only one vessel operated at any given time. The main target was alfonsino, and this comprised 92% of the total catch over the period of 1169 t (Lehodey *et al.*, 1994; Lehodey and Grandperrin, 1996). Although an active research programme has been carried out in the EEZ, with a focus on deepwater species, no further commercial stocks have been found. The alfonsino fishery has not resumed.

Fishing grounds for deepwater species developed in 2000 on the West Norfolk Ridge, which runs NW from the North Island of New Zealand both within and beyond the New Zealand EEZ. Most fishing has occurred on six to seven seamounts or peaks of the ridge. The fishery developed in 2000 within the New Zealand EEZ, and was followed a year later by vessels working the ridge in international waters. Annual catches of orange roughy from 2000 to 2004 ranged from 100 t to about 1000 t (Clark, 2004). Total catch over the period was 1800 t.

Lord Howe Rise

The Lord Howe Rise extends from the NW margin of the Challenger Plateau, off the west coast of New Zealand, out to Lord Howe Island in the western Tasman Sea. The ridge is mostly in international waters, although it does extend into both the Australian and New Zealand EEZs.

Soviet vessels exploring the Lord Howe Rise in the 1970s found high species diversity and low abundance on the shallow Kelso and Keipl seamounts. The main commercial species there, and on the Argo, Nova, and Gifford Guyots was redbait, with catch rates up to several t/h at depths of 300–500 m. The total biomass of redbait on the ridge was estimated at 50 000–75 000 t (Borets and Kulikov, 1986; Polishchuk *et al.*, 1989; Boldyrev and Darnitsky, 1991a). The catch taken by exploratory vessels amounted to about 100 t. Soviet commercial fisheries did not develop.

A major fishery for orange roughy developed on the main ridge top near the south-eastern end of the Lord Howe Rise in 1988, when a Japanese commercial vessel discovered spawning aggregations (Clark and Tilzey, 1996). The fishery expanded rapidly as New Zealand, Australian, Korean, Russian, and Norwegian vessels joined. However, catches declined, and the fishery then progressively shifted to small seamounts on the edges of the Northwest Challenger Plateau, and since the early 1990s has been carried out by only New Zealand and Australian fishers. There are two main fishing areas. The fishing ground on the Lord Howe Rise itself is a broad platform, and not a seamount, although there are a number of very small pinnacles near the major grounds. Fishing areas on the NW margin of the Challenger Plateau have numerous small hills and seamount features (generally only several km^2 in area, and 100–200 m elevation, see Fig. 17.5). Orange roughy catches on these seamounts have varied from 50 to 2400 t/year (based on data from Clark, 2004). The total estimated catch from 1988 to 2004 was 13 600 t. Catch rates have declined on seamounts from peak rates of 4 t/tow in 1996 to between 1 and 2 t/tow in the early 2000s (Clark, 2004). The bycatch in the fishery is small, although some black cardinalfish (*E. telescopus*) have been taken, and bottom trawling directed at alfonsino has occurred on sections of the northern Lord Howe Rise.

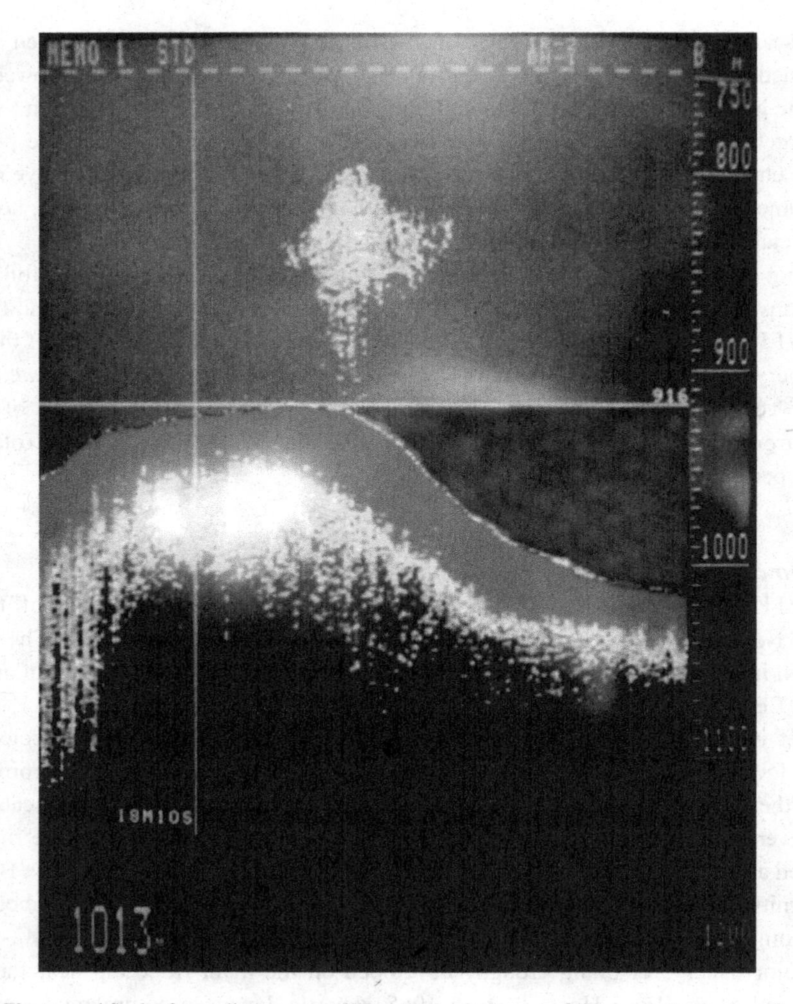

Fig. 17.5 The top and flank of a small seamount off New Zealand, seen on a fishing vessel echo-sounder, showing a mark of orange roughy above the summit (photo Malcolm Clark, NIWA).

New Zealand Plateau

There are about 500 seamounts within the New Zealand EEZ (Rowden *et al.*, 2005). These occur throughout the region, but are prominent along the tectonic plate boundary which runs NE–SW, and features the Kermadec and Colville Ridges to the north, and Macquarie Ridge to the south. There are also a large number of small seamounts on the edge of the slope of the Chatham Rise which runs due east from New Zealand. Fisheries occur on many of these seamounts for a number of commercial species (Table 17.5), including alfonsino, black cardinalfish, smooth oreo, black oreo, bluenose, rubyfish, and, especially, orange roughy (Fig. 17.6).

Fig. 17.6 A catch of orange roughy on deck. A recent and valuable commercial species, the search for orange roughy has involved developing fisheries on many seamounts around the world (photo Malcolm Clark, NIWA).

Table 17.5 Estimated catch (t) of deepwater species by New Zealand vessels from seamounts around New Zealand within the EEZ (1994=1993–1994 NZ fishing year (October–September))

Year	Orange roughy	Oreos	Black cardinalfish	Bluenose	Rubyfish	Alfonsino	Total
1978	0	0					0
1979	52	0					52
1980	643	404					1048
1981	2325	3139					5463
1982	976	1409					2385
1983	1778	979					2757
1984	1173	1358					2531
1985	2557	791					3347
1986	2568	1388					3956
1987	1832	941					2773
1988	5808	1651					7460
1989	6593	2316					8909
1990	7137	4546	164	26	0	106	11 980
1991	11 451	7336	813	70	0	576	20 245
1992	16 775	8552	433	85	0	402	26 246
1993	18 028	10 104	563	60	3	207	28 963
1994	15 613	11 024	1047	95	30	461	28 270
1995	15 815	7175	1749	132	81	573	25 524
1996	13 600	9186	2174	162	84	790	25 995
1997	9265	9861	2276	148	21	516	22 088
1998	7925	7123	937	112	9	555	16 662
1999	8737	7106	885	98	0	425	17 251
2000	8318	6378	1560	163	51	459	16 928
2001	7441	6231	1157	161	0	508	15 498
2002	6942	5199	1097	176	9	467	13 891
2003	7799	4818	1590	263	11	638	15 120
2004	5148	5303	1035	186	0	442	12 113
2005	5817	5071	639	335	89	632	12 582
Total	192 116	129 389	18 117	2272	387	7756	350 038

Data updated from Clark and O'Driscoll (2003), and unpublished MFish records and data.

The proportion of total catch of these species that are taken from seamounts varies and has also changed over time (Clark and O'Driscoll, 2003). Orange roughy catch was mainly from slope grounds in the early 1980s, and increased on seamounts to peak at 50–60% in the mid-1990s, before stabilizing since at around 40%. Oreo catch from seamounts also increased over time, peaking at 50% of the total catch in 1994, and since decreasing to around 30–40%. Cardinalfish catches are largely from seamounts, at 60–70% of the total catch. Bluenose and alfonsino catch from seamounts is 20–40% of the total. Rubyfish is not a major seamount species, at 10–15% of the total catch.

Tasmanian seamounts (Australia)

Around Tasmania are large numbers of small seamounts, some of which host deepwater fish populations. St Helens seamount off the northeastern coast of Tasmania is a spawning site for orange roughy, and south of Tasmania is a field of approximately 70 small seamounts (Pedra Branca/Maatsuyker region), at depths of 1000–2000 m (Hill *et al.*, 1997) which are fishing grounds for orange roughy and black, spiky (*Neocyttus rhomboidalis*) and smooth oreo.

Small catches were taken off several of the southern seamounts in 1987 and 1988, but in 1989 St Helens seamount was discovered, and the Australian orange roughy fishery really began. Landings of orange roughy increased rapidly to over 40 000 t in 1990 (Table 17.6), of which 17 000 t was taken from St Helens. Almost 70 vessels were

Table 17.6 Reported catch (t) of deepwater species by Australian vessels on the Tasmanian seamounts (Eastern and Southern Zones of the SouthEast Trawl management area for orange roughy, 'east' area for oreos, which differs from the orange roughy zone).

Fishing year	Orange roughy (total)	Orange roughy St Helens hill	Spiky oreo	Smooth oreo	Total
1985	64	0			64
1986	664	2			664
1987	663	3			663
1988	2418	137			2418
1989	37 121	1401	36	164	37 321
1990	58 630	16 947	1	441	59 072
1991	26 585	5274	35	837	27 457
1992	31 173	8357	215	1794	33 182
1993	10 637	2879	179	342	11 158
1994	6697	1568	72	530	7299
1995	4116	1571	22	338	4476
1996	2800	970	57	237	3094
1997	2517	1151	72	332	2921
1998	2219	1421	158	186	2563
1999	2128	443	81	97	2306
2000	2307	607	24	122	2453
2001	2180	578	65	401	2646
2002	1751	0	83	338	2172
2003	982	0	82	153	1217
2004	748	0	N/A	N/A	748
Total	196 400	56 309	1182	6312	203 894

Data from Southeast Fishery Stock Assessment Group (plus N. Bax, S. Wayte, Personal communication, St Helens hill data from Wayte and Bax, 2002). N/A: not available.

operating during that year. Catches dropped consistently after that, with stronger management associated with declining stocks. Between 1999 and 2001, the annual catch from St Helens was between 400 and 600t, and the seamount was closed to trawling completely in 2002, with much reduced total allowable catches (TACs) for both Eastern and Southern zones.

South Tasman Rise

The South Tasman Rise is a prominent ridge extending south from Tasmania into the Southern Ocean. It has a series of small peaks near its main summit at about 900m just outside the Australian EEZ. Occasional fishing activity by New Zealand and Australian vessels occurred from the mid-1980s, with from 1989 increased exploratory trawling by Norwegian vessels licenced to fish the southern Australian region. Small catches of oreo were reported (Tilzey, 2000).

The main fishery for orange roughy started in September 1997 by Australian vessels, which were quickly joined by New Zealand fishers, and expanded rapidly. An estimated 3900t was caught in the 1997–1998 fishing year (March–February fishing year). Reported catches were 1700t the following year, and increased to over 4000t in 1999–2000 (Clark and O'Driscoll, 2002; Clark, 2004). One Belizean and three South African registered vessels fished for a period during the 1999 winter, but no other non-Australasian vessels are known to have fished the seamounts. Oreos were previously taken as bycatch in the fishery, with over 1000t in both 1997–1998 and 1998–1999. Catches have dropped markedly since then, to less than 200t during the last 2 fishing years.

Louisville Ridge

The Louisville Ridge is a chain of seamount and guyot features extending southeast from north of New Zealand for over 4000km. It is a 'hotspot' spreading chain, which comprises more than 60 seamounts, most of which rise to peaks of 200–500m from the surrounding seafloor at depths around 4000m. The ridge is entirely in international waters.

Soviet exploratory vessels found 14 seamounts with summit depths less than 1000m during operations in 1977. Quantities of cardinalfish (*E. pectinifer*, *E. denticulatus*, and *E. geracleus*) were found on 2 of them, with daily catches between 10 and 50t. A third seamount produced small catches of alfonsino (Polishchuk *et al.*, 1989, Boldyrev and Darnitsky, 1991a). Exploratory vessels caught a total amount of 10000–12000t of fish. The biomass of cardinalfish was estimated from acoustic surveys at 123000–300000t (Polishchuk *et al.*, 1989), but the Soviet commercial fleet did not work the ridge. Japanese commercial vessels are known to have fished on the Louisville Ridge in the 1980s, with catches of alfonsino and bluenose, but no published information is available on their activities.

The main commercial fishery by New Zealand on the seamounts of the ridge started in the early 1990s for deepwater species such as alfonsino, which were taken by longline. However, a more significant bottom trawl fishery for orange roughy developed in 1993, and catches peaked at over 13000t during 1995 (Clark, 2004). This year saw over 30 trawlers working the seamounts, mainly New Zealand operated vessels (and charters or joint ventures with countries including Korea, Japan, Russia, and Ukraine), but almost 2000t was caught by Australian boats. The number of New Zealand vessels operating in the fishery has generally been between 10 and 15, and catches since 2000 have averaged about 1300 t/year. The New Zealand catch to 2004 is over 34 500 t, with total known catches of 37 200 t. Other nations (e.g., China, Russia, Ukraine, Japan, and Korea) are thought to have fished the ridge at various times, but their catch is unknown.

Catches have been taken from many seamounts along the ridge (at least 15), but the most productive grounds have been those seamounts between 40°S and 45°S. Catches and catch rates have varied between seamounts, and over time, with several seamounts showing a strong decrease in catch rates. The fishery for orange roughy initially occurred throughout much of the year, but after the first 3 years started to concentrate in the months of June–August. Catch rates overall have dropped from peaks of over 3.5 t/tow to levels between 1 and 2 t/tow (Clark, 2004). Black and smooth oreo are a significant bycatch in the fishery, especially in the more southeastern parts of the Louisville Ridge. Their combined catch is about 2300 t (Clark, 2004).

The Geracl Ridge

Fish concentrations on the Geracl Ridge (53–55°S, 139–142°W, southeast of the Louisville Ridge near the Eltanin Fracture Zone) were discovered by a Soviet exploratory vessel in 1972. Two seamounts yielded bigeye cardinalfish (*E. denticulatus* and *E. parini*), with southern blue whiting (*Micromesisteus australis*), and long-nosed grenadier (*Caelorinchus australis*) (Polishchuk *et al.*, 1989; Boldyrev and Darnitsky, 1991b). Trawl and acoustic surveys estimated the total biomass of cardinalfish in 1972–1981 to be 114 000–533 000 t (Boldyrev and Darnitsky, 1982; Boldyrev, 1986). The catch by exploratory vessels totalled about 15 000 t.

A Soviet commercial fleet operated on the ridge in 1972, 1973, and 1980 and catches constituted 8800, 1300, and 1000 t, respectively (Polishchuk *et al.*, 1989). Bottom trawls were mostly used.

East Pacific Rise

There is a chain of 18 seamounts on the western flank of the East Pacific Rise. Fish concentrations were discovered on three seamounts by Soviet exploratory vessels in the 1980s, south of Easter and Pitcairn Islands. Horse mackerel (*T. murphyi*) catch rates in midwater trawls of 10–15 t/h were recorded on one of the seamounts. However, no Soviet/Russian commercial fisheries developed.

New Zealand trawlers also explored the area at the western end of the Challenger Fracture Zone in the late 1990s. Some small catches of orange roughy (about 20 t in total) were reported (Clark and O'Driscoll, 2002).

The Nazca and Sala-y-Gomez Ridges

The Nazca Ridge is located in the SE Pacific between 15–25°S and 76–82°W. Most of the seamounts of the ridge are in international waters, and only its northeastern part is in the 200-mile zone of Peru. In the southwest the Sala-y-Gomez Ridge (24–27°S, 84–102°W) joins with it. There are at least 40 seamounts with a summit depth of less than 1000 m, but the ridges have not been thoroughly explored.

A Soviet fishery for horse mackerel (*T. murphyi*) and redbaits (*E. elongates* and *E. nitidus*) was conducted in 1978 and 1979 with catches of 2100 and 5100 t, respectively.

Pelagic trawls were mainly used, with bottom trawls on some seamounts. Total biomass on the seamounts was estimated at 50 000 t. Concentrations of alfonsino were also found on two seamounts (Anon., 1980, 1985; Dalimaev *et al.*, 1980; Riabikov and Fomin, 1981; Parin *et al.*, 1997).

In 1979–1980, aggregations of rock lobster (*Projasus parkeri*) were surveyed, with biomass on five seamounts estimated at 9000 t (Polishchuk *et al.*, 1989). It is believed a total catch of about 1000 t of rock lobster was taken during research and experimental work.

In the 1980s, sporadic USSR research and fishing was carried out, directed towards the study and development of pelagic fisheries (for horse mackerel, mackerel, and pilchard) outside the EEZs of Peru and Chile. Total fish catch was about 4000 t, mainly horse mackerel and redbait. Slender seasnipe (*Macrorhamphosus* spp.) was also taken in quantity at times (up to 8 t/haul). In 2002–2003, one Russian vessel fished for rock lobsters, with daily catches of about 500 kg (Zhukov, 2005).

There are few data on fishery activities by other countries. A Japanese fleet (10–12 vessels) conducted a fishery for tunas and sharks by pelagic longline both over the seamounts and in the surrounding region, at least during 1978–1979 (Riabikov and Fomin, 1981). In 2003–2005, several Spanish vessels fished for rock lobster on the Nazca Ridge.

The Chilean Rise

The Chilean Rise area (34–46°S, 78–102°W) was briefly surveyed by the former Soviet Union in 1979. A seamount was found with aggregations of cardinal fish (*Epigonus* spp.), but catch rates were less than 5 t/tow.

Joint research voyages between Japan and Chile in 1977–1979 found orange roughy on the southern continental slope. However, it was not until 1998 that spawning aggregations were found on seamounts of the Archipelago Juan Fernandez. Catches were also made at Bajo O'Higgins seamounts (about 160 km off the Chilean coast), near Isla Mocha, and in Punta Sierra just off the coast. The Chilean fishery for orange roughy started in 1999, limited by a 1500 t TAC, which was subsequently increased to 2500 t. The fishery is based on eight main seamount features, five in the Archipelago Juan Fernandez, two at Bajo O'Higgins, and one at Punta Sierra. Catches peaked at 1870 t in 2001, and has reduced to annual catches of around 1200 t during 2003 and 2004. The total catch in this fishery up to 2004 was 8000 t (updated from Young, 2003).

Atlantic Ocean

The Atlantic Ocean topography is dominated by the MAR, with many peaks as well as side ridges, and individual seamounts. Many seamount areas in the Atlantic were explored by Soviet research and exploratory vessels in the 1970s and 1980s, and fisheries developed on most of them (Table 17.7).

North Atlantic Fisheries

Northern MAR

Fisheries on peaks of the northern MAR started in 1973, when dense concentrations of roundnose grenadier (*C. rupestris*) were discovered (Fig. 17.7). The greatest annual catch

Table 17.7 Summary data on seamount fisheries in the Atlantic Ocean.

Area	Species	Year fishery started	Total catch (t)	Maximum annual catch (t)
Northern MAR	*Coryphaenoides rupestris*	1973	232400	22900
	Beryx splendens	1978	5800	1100
	Hoplostethus atlanticus	1993	5100	1300
	Brosme brosme	1996		300
	Sebastes marinus	1996		1000
	Sebastes mentella	1990		
	Molva dypterygia	1979	17000	8000
Northeast Atlantic seamounts	*Hoplostethus atlanticus* *Epigonus telescopus*	2001	9000	5300
Corner Rise	*Beryx splendens*	1976	21500	10200
South Azores seamounts	*Trachurus picturatus, Scomber japonicus, Lepidopus caudatus*	1973	23000	12000
	Beryx splendens	1976	3800	3200
Madeiran–Canarian seamounts	*Trachurus picturatus, Scomber japonicus, Lepidopus caudatus*	1970	190000	46500
Vavilov Ridge	*Beryx splendens, Epigonus denticulatus*	1978	10300	4200
Walvis Ridge	*Beryx splendens*	1976	10000	3900
	Hoplostethus atlanticus	1994	5400	2200
	Pseudopentaceros richardsoni, Epigonus telescopus, oreos			
Vema Seamount	*Jasus tristani*	1964		200
	Decapterus macarellus, Emmelichthys nitidus	1977	500	300
Discovery seamounts	*Maurolicus muelleri*	1983	350	
Rio-Grande Rise	*Beryx splendens*	1982	1500	800
Trinidad-Martin Vaz seamounts	*Caranx* spp., *Decapterus macarellus, Seriola zonata*	1982		
South Antilic Ridge	*Lepidonotothen macroftalmus, Notothenia kempi, Dissostichus eleginoides*	1981	1900	1200

(almost 30000t) in that area was taken by the Soviet Union in 1975, and in subsequent years the catch varied substantially from several hundred tonnes to over 20000t. The fishery declined after the dissolution of the Soviet Union in 1992, and since then there has been a sporadic fishery by vessels from Russia (annual catch estimated at 200–3200t), Poland (500–6700t), Latvia (700–4300t), and Lithuania (data on catch are not available) (ICES, 2004). Grenadier has also been taken as bycatch (<500t/year) in the Faroese orange roughy and Spanish blue ling fisheries. During the entire period of fishing to 2005, the catch of roundnose grenadier from the northern MAR amounted to almost 232 000 t.

Fig. 17.7 A large catch (about 20 t) of roundnose grenadier. Taken on the MAR by the Soviet exploratory trawler 'Rzhev', 1977 (photo Vladimir Vinnichenko, PINRO).

USSR data indicated that roundnose grenadier aggregations may have occurred on 70 seamount peaks of the ridge between 46–62°N, but only 30 of them were commercially important and subsequently exploited. A stock size of up to 900 000 t was estimated in the 1970–1980s (Pavlov *et al.*, 1991; Shibanov, 1998). The fishery is mainly conducted using pelagic trawls although on some seamounts it is possible to use bottom gear. Deepwater redfish (*S. mentella*), orange roughy, black scabbardfish (*A. carbo*), and deepwater sharks are caught as bycatch in the fishery (Kemenov *et al.*, 1979; Anon., 1988a, b, 1990).

The deepwater fisheries off Iceland tend to be on the continental slopes, although a short-lived fishery on spawning blue ling (*Molva dypterygia*) was reported on a 'small steep hill' at the base of the slope near the Westman Islands. The fishery began in 1979, peaked at 8000 t in 1980 and subsequently declined rapidly (Magnússon and Magnússon, 1995). French trawlers fished for blue ling again in 1993 on small seamounts in the southern Reykjanes Ridge, with a catch of 390 t. Iceland caught over 3000 t in 1993, but catches dropped sharply to 300 and 117 t in the next 2 years before fishing stopped until 2000 when Spanish trawlers resumed fishing there with catches up to 1000 t (ICES, 2006).

Orange roughy occurs in Icelandic waters, where in restricted areas of the Reykjanes Ridge it can be abundant on the tops and the slopes of narrow underwater peaks. These are generally difficult to fish, although in 1991 a single trawler made some noteworthy catches of orange roughy off the south coast of Iceland (Bates, 1993). In 1992, the Faroe Islands began a series of exploratory cruises for orange roughy beginning in their own waters and later extending into international waters (Thomsen, 1998). Exploitable concentrations were found in late 1994 (annual catch 260 t) and early 1995 (1040 t), mostly on the MAR. The fishery took place on five features on the MAR and Hatton Bank. Catches peaked in 1996 at 1320 t, and since then have generally been less than 500 t (ICES, 2006).

In the 1980s, a bottom longline fishery developed for tusk (*Brosme brosme*) and northern wolffish (*Anarhichas denticulatus*) on some of the northern MAR seamounts (Prozorov *et al.*, 1985; Zaferman and Shestopal, 1991, 1996). Catches of tusk were taken on 20 seamounts between 51°N and 57°N, with maximum catch rates of 0.8 t/1000 hooks on a seamount named 'Hekate'.

In 1996, a small fleet of Norwegian longliners began a fishery for 'giant' redfish (ocean perch *S. marinus*) and tusk on the Reykjanes Ridge (Hareide *et al.*, 2001). The fishery was mainly conducted close to the summits of seamounts or coral banks and a new type of vertical longline was developed for the fishery. The fishery continued in 1997, but ceased after CPUE decreased markedly. Another species of oceanic redfish (*S. mentella*) can also form aggregations close to the bottom over seamounts of the Reykjanes Ridge (60–62°N). In ICES Sub-area XII, where part of the Reykjanes Ridge is located, large catches (up to 94 000 t/year) have been taken by pelagic trawls, but it has not been possible to estimate the proportion of the catch obtained from over seamounts.

Spanish vessels explored several seamounts on the MAR between 1997 and 2000, and a longline survey was conducted in 2004, but except for sporadic fisheries in the northern area (ICES Division XIVb) there has been a decline in interest (Duran Muñoz *et al.*, 2000; ICES, 2006).

The MAR to the north of the Azores has over 20 seamounts with a depth of less than 1000 m. In 1976, small catches of alfonsino and scabbardfish (*Lepidopus caudatus*) were taken on some seamounts (Vinnichenko, 2002b). A commercial pelagic trawl fishery for alfonsino developed on 'Spectr' seamount in 1977 and this and other seamounts were exploited in 1978 (700 t catch) and 1979 (1100 t catch). No commercial fishing took place during the 1980s but nine exploratory and research cruises by the USSR yielded about 1000 t of alfonsino (Fig. 17.8), with black cardinalfish, orange roughy, black scabbardfish, and silver roughy (*H. mediterrraneus*) (Vinnichenko, 2002b). Aggregations of deepwater crab (*Chaceon affinis*) were also observed (Moskalenko, 1978; Zaferman and Sennikov,

Fig. 17.8 Another day of fishing for alfonsino on seamounts of the MAR by the Soviet exploratory trawler 'Mikhail Verbitsky', 1981 (photo Vladimir Vinnichenko, PINRO).

1991). Russian commercial and exploratory Norwegian, Faroese, and British fishing on more seamounts in the 1990s yielded more than 3000 t of alfonsino, making the total catch from the region about 6000 t. In recent years, there have been no indications of fishable concentrations of alfonsino (Shnar *et al.*, 2005).

Along with deepwater demersal fishes, some epipelagic and mesopelagic species are of commercial interest on the northern MAR seamounts. During the 1970s and 1980s north of the Azores (43–52°N) tuna were regularly taken by Soviet research and exploratory vessels. Albacore (*T. alalunga*) occurred most frequently, with catch rates up to 20 t/haul. Bluefin tuna (*T. thynnus*) and swordfish (*Xiphias gladius*) were also found (Vinnichenko, 1989, 1996). Atlantic saury (*Scomberesox saurus*), shortfinned squid (*Illex illecebrosus*), and Bartrami squid (*Ommastrephes bartrami*) in this area were also of commercial interest (Zilanov, 1975; Volkov *et al.*, 1977; Moskalenko, 1978, Pavlov *et al.*, 1985).

Fishing by a chartered trawler was carried out for deepwater resources on seamounts around the Azores in 2001–2002, and involved a scientific programme as well as commercial exploration. Orange roughy were found on several seamounts during the survey, mainly at depths between 1000 and 1200 m. The total catch during the surveys was 373 t, predominantly orange roughy (92%) with smaller catches of black scabbardfish, smoothheads (*Alepocephalus* spp.), black cardinalfish, and roundnose grenadier (Melo and Menezes, 2002). However, a fishery did not develop.

Northeast Atlantic seamounts

Orange roughy fisheries have also been carried out by French fleets along the continental margin of the British Isles since the early 1990s, and most of the early catches were centred on the Hebridean Seamount (ICES, 2006). This fishery has collapsed, but continues further south mainly around the base and on the slopes of seamounts. An Irish fishery for orange roughy began in 2000 to the west and north of the Porcupine Bank and increased to over 5000 t in 2002 (Shephard and Rogan, 2006). Catch rates started to decrease, EU quotas were introduced, and annual landings since 2003 have been only several hundred tonnes. Total cumulative catch in the fishery amounts to about 9000 t. The fishery is divided into an all-year round catch on the 'flats' and seasonal aggregations on about 20 small hills. The bycatch of the hill fishery was predominantly black cardinal fish.

Some gillnet and longline activity for tusk, ling, and blue ling is known to have occurred on Rosemary Bank, a seamount in the northern Rockall Trough. Catches are unknown.

Corner Rise seamounts

Exploration on the Corner Rise seamounts (a cluster between 34–37°N and 47–53°W) began in 1976–1977, with a focus on three seamounts (Anon., 1988a, 1993). Catches in 1976 totalled more than 10 000 t, which comprised predominantly alfonsino with black scabbardfish, wreckfish, black cardinal fish, barrelfish (*Hyperoglyphe perciforma*), and silver roughy taken in pelagic trawls. A stock size of 80 000 t was estimated (including part of the MAR, Vinnichenko, 1998). In the following year, no stable aggregations were found in the area and the catch decreased to 800 t (Vinnichenko, 1997). Commercial activity resumed in 1987, with a catch of about 2000 t of alfonsino. The fishery on the Rise then ceased until the mid-1990s, when it resumed with variable effort until the end of the century. Since 2000, there has been no significant fishery in this area. Russian research vessels found no stable aggregations in 2003–2004 (Shnar *et al.*, 2005). Efforts by French and Japanese trawlers have also been largely unsuccessful. The total catch taken by the USSR and Russia on the Corner Rise seamounts has amounted to more than 21 000 t.

Some exploratory work was carried out by a Canadian company in 1995. Bottom and midwater trawling resulted in about 80 t of fish being caught, mainly wreckfish (*Polyprion* spp.), alfonsino, and snowy grouper (*Epinephelus niveatus*) (D. Kulka, Department of Fisheries and Oceans, Personal communication). Exploratory crab fishing has also been carried out.

In 1995–1996, the US funded a number of exploratory deepwater fishery vessels to search seamounts of the New England seamount chain for potential deepwater resources. One commercial bottom trawler fished on Bear Seamount, but catches were small. An American longliner is also thought to have attempted some deep-set longlines on the western New England seamounts, and there may currently be some longline activity near Bear Seamount, but no information on target species or catch is available.

South Azores area

A Soviet pelagic and bottom trawl fishery for several species started in 1973 on seamounts south of the Azores. Scabbard fish was initially the main species (10 000 t catch in 1973–1974), with horse mackerel (7000 t), mackerel (2500 t), and later alfonsino (4000 t in 1976–1977). Several seamounts were not fished after the Azores EEZ was declared in 1977, and the fishery continued periodically thereafter with smaller catches (Samokhvalov *et al.*, 1981). Total USSR catch for the period from 1973 to 1987 was 28 000 t (Vinnichenko, 2002b). Commercial trawlers from Russia and some Baltic countries have operated occasionally in the area in recent years, with catches of horse mackerel, mackerel, alfonsino, sea snipe, and deepbodied boarfish (*Antigonia capros*). More detailed information is not available, but fishable concentrations were found by Russian research vessels in 2003–2004 (Shnar *et al.*, 2005).

Fisheries potential for deepwater crabs (*Chaceon* spp.) and longline for various species were identified by USSR research vessels in the 1980s (Zaferman and Sennikov, 1991; Zaferman and Shestopal, 1991) without further development.

In 2001, an Irish commercial longliner carried out exploratory fishing on Atlantis, Plato, Cruiser, Irving, Hyéras, and Great Meteor Seamounts (Nolan, 2004). The target was wreckfish and comprised 45% of the commercial catch from all seamounts.

Madeira–Canary Island seamounts

In the region around Madeira and the Canary Islands, there are at least 10 fishable seamounts, including Ampere, Josephine, Dacia, Canary, Bezymyannaya, Seine, Conception, Gettysburg, Endeavor, and Zvezda. A Soviet fishery for horse mackerel, mackerel, and scabbardfish began on Conception Seamount in 1970, and on the others in 1973–1974. The largest catches in this area occurred in the first few years, with annual catches of 17 800–46 500 t (Fomin *et al.*, 1975). When EEZs were established in 1977, the fishery continued intermittently on Josephine and Ampere seamounts, which are in international waters (Fomin *et al.*, 1980; Vinnichenko and Khlopenyuk, 1983). Catches were over 2000 t in 1992, but otherwise less than 1000 t/year. The main fishing gear was pelagic trawl, although on some seamounts bottom trawl and purse seine were also used.

Portuguese longliners occasionally fished for mackerel, scabbardfish, and tunas on Josephine and Ampere seamounts. The Madeiran fishery for black scabbardfish also exploits the Seine, Lion, and Susan seamounts (Martins and Ferreira, 1995).

South Atlantic fisheries

Vavilov Ridge

The Vavilov Ridge is in the tropical Atlantic between 2–13°S and 6°W–2°E. Soviet research vessels explored the region in 1978, and a commercial fishery developed on six seamounts with summit depths less than 1000 m (Anon., 1988a; Strogalev *et al.*, 1989).

The most active fishery in the area was carried out in 1978, when 4200 t of fish (mainly alfonsino and cardinalfish *E. denticulatus*) were caught on 'Udachnaya' seamount. In the subsequent decade, the area was repeatedly searched by exploratory vessels but concentrations were not generally found, and there was no commercial fishery (Strogalev *et al.*, 1989). A fishery resumed in the second half of the 1990s, with trawlers from Russia, Ukraine, Poland, Norway, and New Zealand, with mixed catches of 500–1300 t/year.

Fishing for tuna in the area is carried out by longlines and purse seine. Catches off seamounts in the 1980s (both on the ridge and closer inshore) amounted to several thousand tonnes per year (Fonteneau, 1991). On one of the seamounts, concentrations of crabs (species unknown) were found and some catches were taken (Anon., 1988a; Strogalev *et al.*, 1989).

Walvis Ridge and Vema Seamount

Vema Seamount, located off the eastern margin of the Walvis Ridge, was first fished in 1964, when commercial quantities of rock lobster (*Jasus tristani*) were found (Simpson and Heydorn, 1965). The fishery was quickly depleted after only 2–3 years (catch levels were unknown, Heydorn, 1969), although it did support one further year of exploitation in 1980 when 80 t of tails were processed (Lutjeharms and Heydorn, 1981).

The Soviet fishery in the Walvis Ridge area commenced in 1968 on Valdivia Seamount, but catches in the early years are unknown (Fomin, 2002). In the mid–late-1970s, the fishing area extended to more seamounts and peaks along the Walvis Ridge, and catches of alfonsino reached almost 4000 t in 1977, with over 2000 t of southern boarfish the same year (Anon., 1983). The fishery then decreased with annual catches generally less than 1000 t. The Valdivia Seamount/Bank near the Walvis Ridge was also fished by an exploratory Japanese trawler in 1979, when 24 tows caught 54 t of fish, mainly southern boarfish and bluemouth (*Helicolenus dactylopterus*) (Sasaki, 1986).

On some seamounts of the ridge in the 1980s, Japanese, South African, and Portuguese vessels occurred from time to time fishing with traps and longlines, as well as trawl (Fomin, 2002). The fishery in this area increased during the latter part of the 1990s, when trawlers from Russia, Ukraine, Poland, Norway, Spain, Namibia, South Africa, and New Zealand operated there fishing for alfonsino and orange roughy.

At the eastern end of the Walvis Ridge, where it abuts the coastline of Namibia and Angola, is a small seamount inside the Namibian EEZ. 'Hotspot' seamount was discovered during exploratory deepwater fishing by a Namibian vessel in 1994. The main target species has been orange roughy. Catches were initially high (1994 catch of 2170 t) but decreased rapidly to only several hundred tonnes per year (Anon., 2005). Black cardinalfish, alfonsino, and several species of oreo (including smooth oreo, warty oreo *A. verrucosus*, and spiky oreo) are taken as bycatch or alternative targets in the fishery on hotspot. Detailed catch data are not available, but annual catches of alfonsino are likely to have been around 1000 t for several years in the mid-1990s.

Trinidad-Martin Vaz seamounts

Soviet vessels explored the Trinidad-Martin Vaz seamounts (20–21°S, 36–39°W) in 1982. Bottom trawling took place on three seamounts in the western part of the chain (Anon., 1988a). Catches on Dogaressa Seamount consisted mainly of jacks (*Caranx* spp.), mackerel scad (*Decapterus macarellus*), amberjack (*Seriola zonata*), *Sphyraena* spp., red snapper (*Lutjanus aya*), sea bass (*Paranthias furcifer*), wreckfish and coney (*Cephalopholis fulva*). Catch rates up to 2–3 t/tow were recorded. Detailed data on catches in this area are not available.

Rio-Grande Rise area

In the Rio-Grande Rise area (28–35°S, 20–38°W), aggregations of alfonsino were found on three seamounts in the 1980s. In 1982, seamounts at the southeastern end of the rise (which the USSR called 'Otdalennaya' and 'Fevralskaya') were fished, and 300 t of alfonsino were taken by both bottom and midwater trawling (Anon., 1988a; Zasel'skii and Pleteshkov, 1988). Small quantities of southern boarfish, barrelfish, and black cardinalfish were also taken (Table 17.8). The fishery stopped after 1984, but resumed in 2000 when a new seamount was discovered to the northwest, named 'Sorokin' (Kakora, 2003, 2005). The total known catch from the Rio-Grande Rise is about 1450 t (Table 17.8). There are no available data after 2002.

Table 17.8 Soviet/Russian catches in the Rio-Grande Rise area from 1982 to 2000 (from logbooks and cruise reports).

Year	Alfonsino	Boarfish	Bluenose	Cardinalfish	Total
1982	306	20	4	3	333
1984	10				10
2000–2001	799	26	38	22	885
2002	327				327
Total	1442	46	42	25	1555

Discovery seamounts

These seamounts, located between 41–44°S and 4°E–3°W, represent a cluster which forms the Discovery Tablemount group. In 1983 aggregations of pearlside, *M. muelleri*, were found on two seamounts. Catches by Soviet research trawlers using pelagic trawls totalled about 350 t, with catch rates of 10–40 t/haul. Between 1984 and 1987, fish density and catch rates decreased, and no commercial fishery was established.

The Japanese carried out exploratory trawling in 1978 and 1979 around Discovery seamount and the nearby Crowford seamounts, McNish seamount, RSA seamount, and Shannon seamount (Sasaki, 1986). Catches were generally small, comprising a mixture of species depending on the seamount and its depth. Catches on Discovery seamount comprised several species of grenadier (*Coryphaenoides* spp.), while 14 t of butterfish (*Hyperoglyphe* sp.) was taken on Crowford seamount. No fisheries have developed.

South Antilic Ridge

The South Antilic Ridge lies SW of the southern part of the MAR (53–62°S, 52–40°W). In late 1980 and early 1981, two seamounts were found on the northwestern part of the ridge, and commercial catches of notothenids (*Lepidonotothen macroftalmus*) with bycatch of *Patagonotothen ramsayi* and Patagonian toothfish (*Dissostichus eleginoides*) were taken by bottom trawl (Anon., 1988a). In 1981, the total catch was about 1200 t, which decreased to 200 t in 1983, and to 20 t in 1984. In 1987, a new seamount was discovered near the South Orkney Islands. Bottom trawl catches totalled 255 t, and consisted of striped-eyed rockcod (*L. kempi*) with bycatch of marbled rockcod (*Notothenia rossi marmorata*) and Patagonian toothfish. In 1989, the catch was about 200 t of notothenids.

A New Zealand vessel explored several seamounts on the Islas Orgadas Ridge (about 43°S, 28°W) in 2001. Small catches of orange roughy were reported, totalling about 50 t. No commercial fishery is known in the area.

Indian Ocean

The Indian Ocean has a number of large ridge features that trend southwards along the east coast of Africa, and south of India, and the main Indian Ocean Ridge which runs southwest from the middle reaches of the Indian Ocean to well south of Africa. These ridges have a large number of peaks which are often of considerable elevation. Most fisheries activity has been on the ridges and seamounts in the southern part of the Indian Ocean (Table 17.9).

Table 17.9 Summary data on seamount fisheries in the Indian Ocean.

Area	Species	Year fishery started	Total catch (t)	Maximum annual catch (t)
Ob and Lena seamounts	*Lepidonotothen squamifrons*, *Dissostichus eleginoides*	1967	34 200	9600
Southwest Indian Ridge	*Beryx splendens*	1980	~20 000	3000
	Emmelichthys nitidus	1980	7000	3200
	Hoplostethus atlanticus	1998	19 000	12 200
	Oreos	1998	3500	2700
Madagascar Ridge	*Pseudopentaceros richardsoni*	1982	800	400
	Epigonus telescopus	1981	300	100
	Beryx splendens	1981		
Mozambique Ridge	*Hoplostethus atlanticus*	2000	?2000	
NinetyEast, Broken, and Mid-Indian Ridges	*Beryx splendens, Emmelichthys nitidus*	1981	600	400

Southwest Indian Ridge region (includes Mozambique Ridge and Madagascar Ridge)

Soviet exploratory fishing on seamounts of the SW Indian Ridge between 20°S and 40°S began in the early 1970s (Romanov, 2003). Concentrations of alfonsino and centrolophid butterfish (including rudderfish (*Centrolophus niger*), bluenose, violet warehou (*Schedophilus velaini*), black ruffe (*S. maculatus*), and New Zealand ruffe (*S. huttoni*)) were found with southern boarfish and cardinal fish (*Epigonus* spp.). Commercial fishing began in 1980 with seven trawlers, using mainly bottom trawl gear. Redbait and rubyfish were also targeted (Romanov, 2003). The total catch was over 6000 t in 1980 (Table 17.10), but decreased to low levels in the mid-1980s before increasing again in the 1990s as alfonsino fisheries developed on new seamounts.

Table 17.10 Soviet/Russian/Ukranian catches (t) on seamounts of the Southwest Indian Ridge and Madagascar Plateau (FAO Statistical Areas 51.01 and 51.02) from 1980 to 2001 (from Romanov, 2003).

Year	Alfonsino	Redbait/rubyfish	Centrolophidae	Mackerel	Cardinal	Southern boarfish	Total catch
1980	20	3252	1831	133			6029
1981	2524	375	1047	255	63		5472
1982	921	268	80	50	80	414	2112
1983	852	356	255	45	27	4	1720
1984	57	40	7	22			152
1985	3	91	4	46			153
1986		9	1				34
1987	2	807	608			9	1472
1988		66	128	312			583
1989	10			26			36
1990		10	4			1	15
1991							10
1992	314	468	828		20	45	1676
1993	462	551	301	127	18		1500
1994	1534	227	732			40	2633
1995	2249	144	485	3	33	54	2970
1996	3079	28	254		53	17	3480
1997	1031	7	440		17	33	1570
1998	859	275	395	15		78	1623
1999	1964	181	753	11		108	3303
2000	1578		360				2015
2001	371	121	299	7		12	810
Total	17 830	7276	8812	1052	311	815	39 368

In 1997–1998, a new fishery developed on the Southwest Indian Ridge for deepwater species. Orange roughy, black cardinalfish, southern boarfish, oreo, and alfonsino were targeted by vessels from Australia and New Zealand, and soon after from South Africa, then Russia, Norway, and the EU in 1999. Fishing was initially successful, with a total catch of demersal fish species over 14 000 t (FAO, 2002; Japp and James, 2005) (Table 17.11), although it is believed that the orange roughy catch alone was over 10 000 t (Anon., 2001). Orange roughy was the main target species, fished at depths of 800–1200 m. The following year saw a dramatic increase in effort, with 45–50 ships from

Table 17.11 Estimated deepwater catches (t) from the Southwest Indian Ocean area (Southwest Indian Ridge and Madagascar Plateau).

Year	Number of nations	Number of vessels	Orange roughy	Oreos	Alfonsino	Southern boarfish	Cardinalfish	Bluenose	Others	Total
1998	1	1			859	78			685	1622
1999	6	8	5210	406	2462	2583	360	30	3475	14526
2000	7	13	12218	2689	6526	2066	1771	15	39412	39413
2001	8	8	1568	358	3471	45	406	28	2090	7966

Data are incomplete (from FAO, 2002).

New Zealand, Australia, Russia, Ukraine,- Mauritius, China, Norway, Namibia, Spain, Portugal, and others under the flag of Panama and Belize. However, the total orange roughy catch was about the same as that estimated the previous year. Catch rates dropped substantially, with the average catch per vessel for the season estimated to have decreased from 1600 t to less than 300 t. The following years saw reduced numbers of vessels, and a shift from orange roughy to alfonsino and rubyfish targets on the Madagascar Plateau, Mozambique Ridge, and Mid-Indian Ridge. Reported catches of orange roughy in FAO Area 51 were much lower than those estimated in Table 17.11: 1265, 711, and 38 t for 2000, 2001, and 2002, respectively. No catches were reported in the official returns to FAO prior to 2000 (FAO, 2003), possibly because of commercial sensitivity of new fishing grounds. Catches of several thousand tonnes are thought to have occurred in recent years. Reported catches of alfonsino from FAO Area 51 were generally 1000–3000 t/year during the 1990s, but have decreased since (FAO, 2003).

Soviet trials in the early 1990s, with droplines and bottom longline on seamounts at depths of 100–800 m, targeted wreckfishes (*Polyprion oxygeneios* and *P. americanus*) (Romanov, 2003). Rock lobsters (*J. lalandii*) were also caught on the tops of seamounts and banks. However, none of these fisheries have developed.

Japanese exploratory trawling took place in 1977 and 1978 further north on the Saya de Malha Bank, west of the Mid-Indian Ridge (Sasaki, 1986). Catches totalled 700 and 350 t in the respective years, mainly comprising scad (*Decapterus* spp.), lizardfish (*Saurida undosquamis*), and butterfly bream (*Nemipterus personii*).

Ob and Lena seamounts

Soviet exploratory fishing was carried out at Ob and Lena seamounts in the early 1970s. In 1974, a commercial fishery for grey notothenid (*L. squamifrons*) developed at depths of 230–270 m. Between one and six trawlers worked the seamounts irregularly. Patagonian toothfish and marbled notothenid were taken as bycatch, but did not form aggregations on the seamounts. Maximum catches totalled over 9000 t in 1986 and over 34 000 t over the period of the Soviet fishery between 1974 and 1991 when fishing stopped.

Japanese vessels also explored Ob and Lena seamounts, as well as banks around the Crozet Islands, in the mid-late 1970s (Sasaki, 1986). Patagonian toothfish, the notothenids *L. squamifrons* and *N. rossii*, and icefish (*Champsocephalus gunnari* and *Chaenichthys rhinoceratus*) dominated catches, but were not regarded as commercial then. There is thought to have been some illegal and unreported fishing activity for Patagonian toothfish in the area in recent years.

Kerguelen-Heard Ridge

Peaks on this ridge around the Kerguelen Islands were explored by Japanese vessels in 1977 and 1978. Catch was 1600t in 1977, and comprised Patagonian toothfish, notothenid cods, and icefishes (Sasaki, 1986). Soviet vessels have also fished for Patagonian toothfish within the EEZ around the Kerguelen Islands (Romanov, 2003).

NinetyEast Ridge (includes Mid-Indian Ridge and Broken Ridge)

Exploratory fishing in deepwater in central and eastern parts of the Indian Ocean has occurred on occasion. Soviet vessels fished during the early 1980s. In 1981, over 400t of rubyfish was caught on the Mid-Indian Ridge, but most effort in subsequent years to 1984 was on Broken Ridge where alfonsino and centrolophids were caught, catches totalling about 120t each (Romanov, 2003) (Table 17.12).

Table 17.12 Soviet/Russian/Ukranian catches (t) on seamounts of NinetyEast Ridge and Broken Ridge (FAO Statistical Area 51.04) from 1980 to 1984 (from Romanov, 2003).

Year	Alfonsino	Redbait/rubyfish	Centrolophidae	Mackerel	Cardinal	Southern boarfish	Total catch
1981	120	442	109	3		14	995
1982	2					2	28
1983							10
1984	6	3					40
Total	128	445	109	3	0	16	

Australian vessels, and some from New Zealand, have also been active in the region. The former explored the ridge areas extensively during the late 1990s and early 2000s. Most effort was on NinetyEast Ridge, where seamounts at depths of 800–1000m were targeted with bottom trawl gear. Exploratory fishing also occurred to the north on parts of the Investigator Ridge, and along Broken Ridge. Details of catches are unknown. The main target species has been orange roughy, but no large catches have been taken to our knowledge.

Discussion

Common characteristics of seamount fisheries

Fisheries on seamounts are subject to additional difficulties and increased commercial risk compared to fishing on the continental shelf and slope. Most seamounts are offshore and located a great distance from the coast. Large vessels are generally required to fish these grounds, and running costs can be high. Catches and catch rates in these areas can show sharp fluctuations; fishing operations are difficult because of hard ground, complex water circulation, and unstable and dynamic fish concentrations (e.g., Kemenov *et al.*, 1979; Sasaki, 1986; Anon., 1988a, 1989, 1993; Vinnichenko, 1998, 2006; Clark, 1999).

Modern deepwater trawls have large bobbin or rockhopper ground gear, which together with advances in navigational and electronic fishing aids since the 1980s, have made trawling on rough seamounts much more feasible than 20 years ago. Small seamounts and trawlable paths can routinely be located and fished. Nevertheless, the hard and rough nature of seamount geology means there are still some limitations on the fisher's ability to bottom trawl on seamounts.

Many of the seamount fisheries described in this chapter have shown similar trends. The highest catches and catch rates are typically observed during the first years of the fishery. Subsequently, these substantially decrease and can remain low over a long period. Often, even relatively small catches (in the range of 500–1000 t) cause lower density and stability of aggregations and consequently reduced catches.

Seamount fisheries are highly variable in their size and duration. In some fisheries, such as those on the Tasmanian seamounts, Emperor, Hawaiian, Southwest Indian Ridges, the northern MAR, and the Madeiran–Canarian area the maximum annual catches were relatively high, tens of thousands of tonnes, sometimes over 100 000 t. However, such catches were generally not sustained for long, and these large-scale fisheries showed a 'boom-and-bust' situation, especially in international waters where there was no management at the time. Trends in some of the major seamount fisheries are plotted in Fig. 17.9. These graphs show the early development of seamount fisheries for pelagic armourhead in the North Pacific, and for a range of species in the northern Atlantic Ocean in the mid-1970s, and the development of deepwater trawl fisheries for orange roughy in the southern hemisphere in the 1980s. Catch histories are affected by a large number of factors, and in some of these fisheries there are effects of EEZ declaration in the late 1970s (e.g., Madeira–Canary), and management measures imposed within EEZs. Nevertheless, they demonstrate the points above about the variability in size of seamount fisheries (note the *y*-axis scales differ between plots) and in their duration. Many fisheries, such as those for armourhead/alfonsino in the North Pacific, alfonsino on Corner Rise, scabbard fish and mackerel around Madeira–Canary Islands and the Azores, had several years of large catches, but then marked reductions in catch, and in most cases the fishery has not resumed, despite regular searches for aggregations. Orange roughy fisheries also typically show strong decreases in catch and catch rates after high levels in the early years of the fishery. For some orange roughy fisheries there appears to have been a continual decline (e.g., Tasmanian seamounts, Hot Spot off Namibia, and South Tasman Rise (not graphed) while for others fisheries have continued at lower catch levels once the initial 'high-catch phase' has passed (e.g., New Zealand seamounts and Louisville Ridge). However, while there is variation in seamount fisheries, it is clear that high-volume trawl fisheries are generally not sustainable. Fisheries that have restricted levels of effort and catch, those that are well regulated inside EEZs, small-scale fisheries (see Chapter 16), or those using other methods (e.g., line and pot) may occur over a longer period.

Fishery potential

Shortage of commercial data and lack of current research prevent an adequate global assessment of the overall status of fish stocks on seamounts. Biomass estimates have been made of some fish species on oceanic seamounts, especially by earlier USSR and Japanese

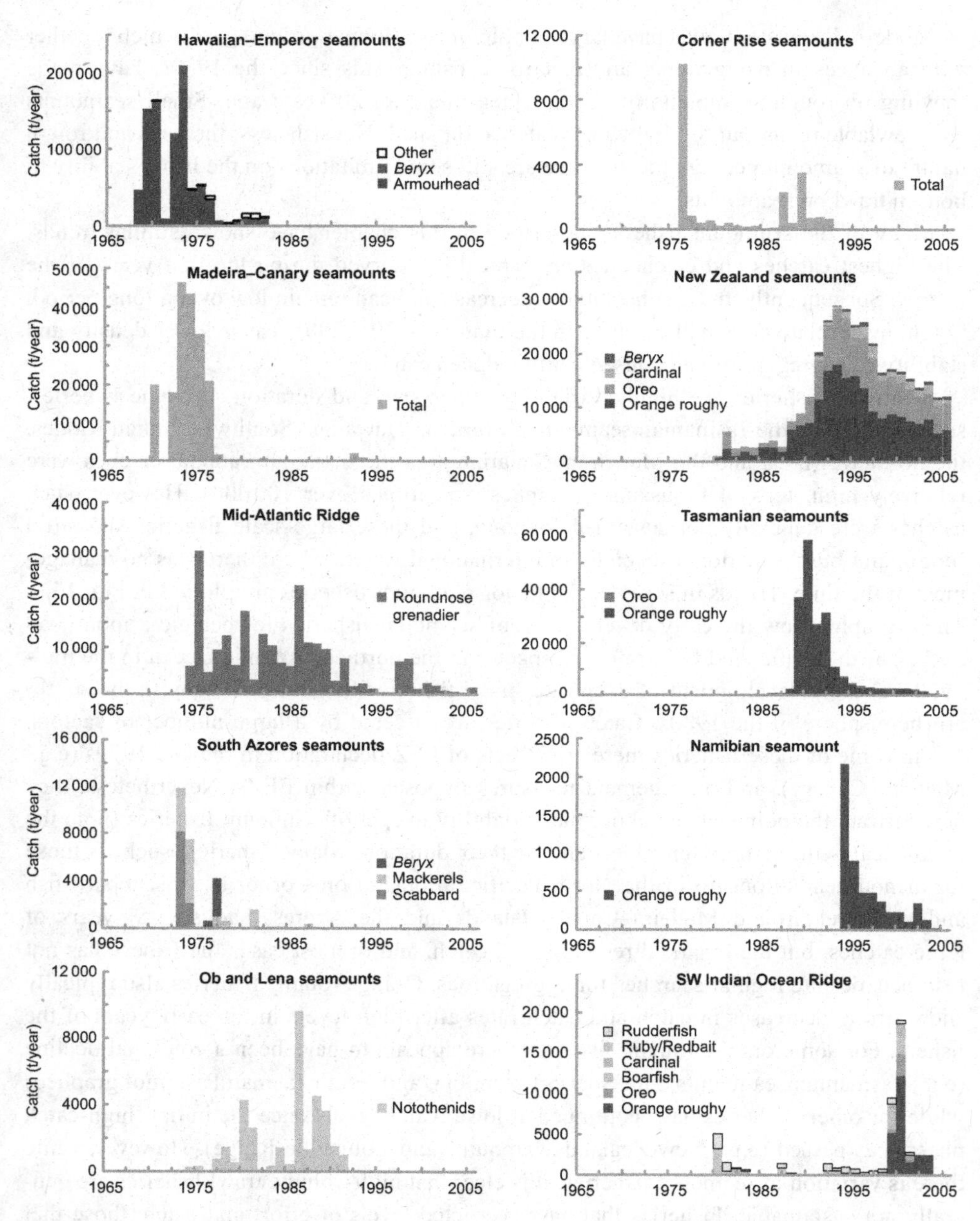

Fig. 17.9 Catches over time in a selection of seamount fisheries around the world, for a variety of species: *y*-axis is estimated catch (t) and *x*-axis is year.

exploratory investigations. Scientific techniques to measure biomass of deepwater species are however often inaccurate, and subject to large degrees of uncertainty (e.g., Clark, 1996). Where estimates have been made, the biomass of many fish stocks on seamounts is relatively low and does not exceed several hundreds of thousands of tonnes for even the most abundant species (e.g., Sasaki, 1986; Vinnichenko, 1998, 2002b). An analysis of New Zealand seamount fisheries for orange roughy suggested that few seamounts can support long-term catches of more than a few hundred tonnes (Clark *et al.*, 2001).

Seamounts host a wide variety of fish species, with varying life history characteristics and productivity levels (see Chapter 9). Stock size and catch vary widely in different areas, depending on the fish production on the seamounts (affected by size, relief and depth of the seamounts, water circulation, and general biological productivity of the area), as well as on the number of seamounts in each area (Boldyrev, 1995). This can be seen in differences between, for example, the Emperor Seamount chain and the northern MAR, compared with Corner Rise and the North Azores area. In each of the first two areas more than 30 'commercial' seamounts are known. Many of them are located in the biologically productive areas of the sub-polar front (Borets and Kulikov, 1986; Anon., 1990). The biomass of the largest aggregation of armourhead on the Emperor–Hawaiian Ridge was estimated at 243–350 000 t (Borets, 1979; Kulikov and Darnitsky, 1992). The biomass of roundnose grenadier at the northern MAR was estimated at 500 000–900 000 t (Pavlov *et al.*, 1991; Shibanov, 1998). By contrast, fisheries on the Corner Rise and in the North Azores area occur on only seven seamounts, and oceanographic conditions there are less productive (Anon., 1993). In the 1970s and 1980s, the total biomass of alfonsino in these two areas was estimated at 50 000–80 000 t (Vinnichenko, 1995, 2002b). The highest annual catch in the North Azores area did not exceed 1200 t and on the Corner Rise 10 200 t.

Most demersal fish species that inhabit seamounts appear to be highly vulnerable to overfishing. Aspects covered in Chapter 9 indicate a low productivity for many deepwater seamount species, and that is borne out by the poor track record of fisheries for species like oreos and orange roughy (e.g., Francis and Clark, 2005). Some pelagic fish (e.g., mackerels) and shallower species (e.g., alfonsino) may be more resilient to heavy fishing, and have been classified as less vulnerable (e.g., Gordon 2005), but are still unlikely to withstand high-catch levels for more than a few years.

There has been intense fishery pressure on the seamounts of the Corner Rise, North Azores area of MAR, Vavilov, Walvis, Southwest Indian, Emperor, and Hawaiian Ridges. Hence, many fish stocks are in a depressed state, and no increase of catch in these areas will be likely within the next few years. Relatively new fishing grounds such as seamounts on the Norfolk, Lord Howe, Louisville, and South Tasman Rise have been targeted for deepwater species such as orange roughy, and are also fully exploited. Some areas of the northern MAR, and some offshore seamounts located in Antarctic waters and the central oceanic regions may have some potential for further exploitation. However, this is likely to be small and short lived if based on deepwater species, and if it occurs without careful management.

Management of seamount fisheries

There has been little regulation of the fisheries in seamount areas until recently, and there are several major problems facing management of these fisheries. Accurate information on catch and effort are important, but reports on the fishing activities in such areas are often incorrect. There is an element of secrecy over new fishing grounds which in the past may have lead to extensive misreporting of catches and locations. Scientific investigations are carried out only occasionally offshore, and a shortage or a complete lack of scientific and fisheries data can significantly increase the probability of overexploitation of seamount resources (especially when the fishing operation is competitive). Even in areas inside EEZs, or offshore regions covered by a Regional Fisheries Management Organisation, seamount fisheries can be short lived. The fine spatial scale needed to research and manage seamount stocks is problematic, serial depletion of fish stocks can happen very fast (Clark, 1999), and combined with the aggregating nature of many species over seamounts, catch rates can be maintained even though biomass is being reduced very quickly (e.g., Clark, 2001). Experience with orange roughy fisheries around New Zealand has highlighted the need to limit the amount of effort that is applied to seamount stocks. Heavy fishing pressure can disperse aggregations (e.g., Clark and Tracey, 1994), and also if uncontrolled lead to vessels 'queueing' for a shot at the hill, and consequently trawl for longer and catch more than is required because of the time it can take to wait for another shot (a situation that occurred in some Australian and New Zealand fisheries).

Management initiatives both within EEZs and on the high seas for seamount fisheries have increased in recent years (see Chapter 20), with closed seamounts, method restrictions, depth limits, individual seamount catch quotas, bycatch quotas, and habitat exclusion areas (e.g., hydrothermal vents). The large geographic areas to be covered present major challenges for international cooperation, and require a collaborative approach between fishing companies, scientists, and managers. However, as discussed in Chapter 20, large-scale sustainable fisheries on seamounts are not possible without them.

Acknowledgements

We gratefully acknowledge data and information provided by N.A. Ivanin, V.V. Paramonov, L.K. Pshenichnov (all YugNIRO), I.P. Shestopal, V.N. Shibanov, M.L. Zaferman (all PINRO), and S.A. Onufrienko (Murmansk Trawl Fleet Company). V.V. Paramonov, S.V. Kornilova, and N.A. Kovalenko also contributed with excellent translation from Russian. The CoML programme on seamounts ('CenSeam') supported a planning workshop for this book, and New Zealand seamount fishery data were made available through research funded by the Foundation for Research, Science and Technology (contract CO1X0508) and Ministry of Fisheries (ORH200503).

References

Adam, M.S., Sibert, J., Itano, D. and Holland, K. (2003) Dynamics of bigeye (*Thunnus obesus*) and yellowfin (*T. albacares*) tuna in Hawaii's pelagic fisheries: analysis of tagging data with a bulk transfer model incorporating size-specific attrition. *Fishery Bulletin*, 101, 215–28.

Anon. (1980) *Fisheries Atlas of the South-Eastern Pacific Ocean for 1979*, 50 pp. Zaprybpromrazvedka, Kaliningrad (in Russian).

Anon. (1983) *Fishery Description of the Walvis Ridge*, 63 pp. AtlantNIRO Central Directorate of Navigation and Oceanography, St. Petersburg (in Russian).

Anon. (1985) *Fisheries Description of the South-Eastern Pacific Ocean*, 154 pp. AtlantNIRO Central Directorate of Navigation and Oceanography, St. Petersburg (in Russian).

Anon. (1988a) Description of seamounts and rises in the fishing grounds of the World Ocean (the open part). In: *The Atlantic and Indian Oceans* (ed. Valchuk, S.V.), Vol. 1, 485 pp Central Directorate of Navigation and Oceanography, St. Petersburg (in Russian).

Anon. (1988b) Fisheries description of the northern part of the Mid-Atlantic Ridge, 179 pp. PINRO Central Directorate of Navigation and Oceanography, St. Petersburg (in Russian).

Anon. (1989) Description of seamounts and rises in the fishing grounds of the World Ocean (the open part). In: *The Pacific Ocean* (ed. Valchuk, S.V.), Vol. 2, 386 pp. Central Directorate of Navigation and Oceanography, St. Petersburg (in Russian).

Anon. (1990) Fisheries description of the western part of the Reykjanes Ridge, 157 pp. PINRO Central Directorate of Navigation and Oceanography, St. Petersburg (in Russian).

Anon. (1993) *Fisheries Description of the North Azores Complex of Seamounts and Corner Rising*, 170 pp. PINRO, Central Directorate of Navigation and Oceanography, St. Petersburg (in Russian).

Anon. (2001) Fleet flops on seamounts: mountains of debt from Indian Ocean. *Fishing News International*, 40(1), 1.

Anon. (2005) Orange roughy (*Hoplostethus atlanticus*). State of the orange roughy stock. Unpublished report prepared by the Namibian Government.

Bates, Q. (1993) Alternatives to the hardship of tightened fishing quotas. *Fishing Boat World*, 4(October), 14.

Beck-Bulat, G.Z., Borets, L.A., Darnitsky, V.B., Panochkin, V.P. and Pashchenko, V.M. (1985) *Results from Fisheries Research on Seamounts in the Northeast Pacific*, 99 pp. TURNIF, Vladivostok.

Belyaev, V.A. (2003) *Ecosystem of the Kurshio Current and Its Dynamics*, 383 pp. Khabarovsk Press, Khabarovsk (in Russian).

Boldyrev, V.Z. (1986) Distribution of ichthyofauna on seamounts in the southwestern part of the Pacific. *Biological Resources of the Pacific*, pp. 520–27. Nauka Press, Moscow (in Russian).

Boldyrev, V.Z. (1995) Study of seamounts. TINRO Report 70, 139–146. TINRO, Vladivostok (in Russian).

Boldyrev, V.Z. and Darnitsky, V.B. (1982) The first cruise of the exploratory vessel 'Darwin' to the southern part of the Pacific. *Fisheries Research in the Southeastern Pacific*, pp. 140–54. VNIRO, Moscow (in Russian).

Boldyrev, V.Z. and Darnitsky, V.B. (1991a) Conditions of living and distribution of fish in the Lord Howe Rise area. *Biological Resources of the World Ocean*, pp. 231–45. VNIRO, Moscow (in Russian).

Boldyrev, V.Z. and Darnitsky, V.B. (1991b) Distribution of fish on seamounts in the Eltanin fracture zone. *Biological Resources of the World Ocean*, pp. 258–75. VNIRO, Moscow (in Russian).

Boldyrev, V.Z., Darnitsky, V.B. and Markina, N.P. (1981) Conditions of living and peculiarities of fish ecology on the Wanganella Bank (Tasman Sea). *Marine Biology*, 2, 15–21 (in Russian).

Borets, L.A. (1975). Some results of studies on the biology of the pelagic armourhead (*Pentaceros richardsoni* Smith). *Investigations of the Biology of Fishes and Fishery Oceanography*, pp. 82–90. TINRO, Vladivostok (in Russian).

Borets, L.A. (1979) Biology and abundance dynamics of pelagic armourhead (*Pentaceros richardsoni* Smith) on the Hawaiian Ridge. Author's Abstract for PhD (Biology). INBYUM, Sevastoplol, 24 pp (in Russian).

Borets, L.A. and Kulikov, M.Yu. (1986) Thalassobathyal. *Biological Resources of the Pacific Ocean*, pp. 505–20. Nauka, Moscow (in Russian).

Clark, M.R. (1996) Biomass estimation of orange roughy: a summary and evaluation of techniques for measuring stock size of a deepwater fish species in New Zealand. *Journal of Fish Biology*, 49(Suppl. A), 114–31.

Clark, M.R. (1999) Fisheries for orange roughy (*Hoplostethus atlanticus*) on seamounts in New Zealand. *Oceanologica Acta*, 22(6), 593–602.

Clark, M.R., Bull, B. and Tracey, D. (2001) The estimation of catch levels for new orange roughy fisheries on seamounts: a meta-analysis of seamount data. *New Zealand Fisheries Assessment Report 2001/75*, 40 pp.

Clark, M.R. (2001) Are deepwater fisheries sustainable? The example of orange roughy. *Fisheries Research*, 51, 123–35.

Clark, M.R. and O'Driscoll, R.L. (2002) Descriptive analysis of orange roughy fisheries in the Tasman Sea outside the New Zealand EEZ: Lord Howe Rise, Northwest Challenger Plateau, and South Tasman Rise from 1986–87 to the end of the 2000–01 fishing year. *New Zealand Fisheries Assessment Report* 2002/59, 26 pp.

Clark, M.R. and O'Driscoll, R.L. (2003) Deepwater fisheries and aspects of their impact on seamount habitat in New Zealand. *Journal of Northwest Atlantic Fishery Science*, 31, 441–58.

Clark, M.R. (2004) Descriptive analysis of orange roughy fisheries in the New Zealand region outside the EEZ: Lord Howe Rise, Northwest Challenger Plateau, West Norfolk Ridge, South Tasman Rise, and Louisville Ridge to the end of the 2002/03 fishing year. *New Zealand Fisheries Assessment Report 2004/51*, 36 pp.

Clark, M.R. and Tracey, D.T. (1994) Changes in a population of orange roughy (*Hoplostethus atlanticus*) with commercial exploitation on the Challenger Plateau, New Zealand. *Fishery Bulletin (US)*, 92, 236–53.

Clark, M.R. and Tilzey, R. (1996) *A Summary of Stock Assessment Information for Orange Roughy Fisheries on the Lord Howe Rise*, 24 pp. Bureau of Resource Sciences, Canberra.

Dalimaev, A.P., Kudrin, B.D., Zheleznyakov, A.A. and Barkan, V.B. (1980) *Fisheries Description of the South-Eastern Pacific Ocean*, 98 pp. Sevrybpromrazvedka, Murmansk (in Russian).

Darnitsky, V.B. and Kodolov, L.S. (1997) Investigations of TINRO-center in the Magellan seamounts area. *Abstracts from the Tenth Conference on Fisheries Oceanography*, pp. 40–41. VNIRO, Moscow (in Russian).

Darnitsky, V.B. and Kodolov, L.S. (2004) Seamount researches in the West Central Tropical Pacific. Part II: Mid-Pacific and Necker Ridges. *Abstracts from the Thirteenth Annual Meeting of the North Pacific Marine Science Organization (PICES)*. Hawaii, USA, p. 202.

Darnitsky, V.B., Stepanenko, M.A. and Vologdin, V.N. (2001) The TINRO field investigations of seamounts in the North-Eastern Pacific. *Abstracts from the Tenth Annual Meeting of the North Pacific Marine Science Organization (PICES)*. Victoria, Canada, pp. 65–6.

Darnitsky, V.B., Belyaev, V.A., Pahorukov, N.P. and Bomko, S.P. (2004) Composition and changeability of the fish community of the central seamounts of the Kyushu-Palau Ridge and some oceanographic feature (Part II). *Abstracts from the Thirteenth Annual Meeting of the North Pacific Marine Science Organization (PICES)*. Hawaii, USA, pp. 173–4.

Duran Muñoz, P., Román, E. and González, F. (2000) Results of a deep-water experimental fishing in the North Atlantic: and example of cooperative research with the fishing industry. ICES CM 2000/W:04, 15 pp.

FAO (2002) Report of the second ad-hoc meeting on management of deepwater fisheries resources of the Southern Indian Ocean. *FAO Fisheries Report No. 677*, 106 pp.

FAO (2003) Fishery statistics – capture production. *FAO Yearbook*, 96(1). FAO, Rome, Italy.

Fomin, G.V. (2002) *Raw Materials of the Thalassobathyal of the Walvis Ridge (Recommendations on Fishery)*, 103 pp. AtlantNIRO, Kaliningrad. (in Russian).

Fomin, G.V., Torin, Yu.A., Bulanenkov, S.K., Nozdrin, Yu.P. and Sinitsyn, V.A. (1975) *Manual on Searching and Fisheries for Fish in the East Atlantic*, 458 pp. Zaprybpromrazvedka, Kaliningrad (in Russian).

Fomin, G.V., Sundakov, A.Z., Akhramovich, A.P., Galimullin, M.G., Vasiljev, G.P., Zalessinsky, L.A. *et al.* (1980) *Manual on Fisheries on Seamounts of the Open Part of the Atlantic Ocean*, 186 pp. Zaprybpromrazvedka, Kaliningrad (in Russian).

Fonteneau, A. (1991) Seamounts and tuna in the tropical Eastern Atlantic. *Aquatic Living Resources*, 4, 13–25.

Francis, R.I.C.C. and Clark, M.R. (2005) Sustainability issues for orange roughy fisheries. *Bulletin of Marine Science*, 76(2), 337–51.

Gianni, M. (2004) *High Seas Bottom Trawl Fisheries and Their Impacts on the Biodiversity of Vulnerable Deep-Sea Ecosystems: Options for International Action.* IUCN, Gland, Switzerland.

Gordon, J.D.M. (2005) Environmental and biological aspects of deepwater demersal fishes. In: *Deep Sea 2003: Conference on the Governance and Management of Deep-Sea Fisheries. Part 1: Conference Papers* (ed. Shotton, R.). *FAO Fisheries Proceedings*, No. 3/1, 70–88.

Grandperrin, R. and Richer de Forges, B. (1988) Chalutages exploratoires sur quelgues monts sous-marins en Nouvelle-Caledonie. *Peche Maritime*, 1325, 752–5.

Grigg, R.W. (1993) Precious coral fisheries of Hawaii and the US Pacific Islands. *Marine Fisheries Review*, 55(2), 50–60.

Hareide, N.-R., Garnes, G. and Langedal, G. (2001) The boom and bust of the Norwegian longline fishery for redfish (*Sebastes marinus* 'Giant') on the Reykjanes Ridge. *NAFO SCR Document 01/126*, 13 pp.

Heydorn, A.E.F. (1969) The South Atlantic rock-lobster *Jasus tristani* at Vema Seamount, Gough Island and Tristan da Cunha. *RSA Division of Sea Fisheries Investigational Report No. 73*, 20 pp.

Hill, P.J., Exon, N.F. and Koslow, J.A. (1997) Multibeam sonar mapping of the seabed off Tasmania: results for geology and fisheries. *Third Australasian Hydrographic Symposium, Special Publication*, 38, 9–19.

Hughes, S.E. (1981) Initial US exploration of nine Gulf of Alaska seamounts and their associated fish and shellfish resources. *Marine Fisheries Review*, 43, 26–33.

ICES (2004) Report of the working group on the biology and assessment of deep-sea fisheries resources (WGDEEP). ICES CM 2004/ACFM:15, 308 pp.

ICES (2006) Report of the working group on the biology and assessment of deep-sea fisheries resources (WGDEEP). ICES CM 2006/ACFM:28, 486 pp.

Japp, D.W. and James, A. (2005) Potential exploitable deepwater resources and exploratory fishing off the South African coast and the development of the deepwater fishery on the south Madagascar Ridge. In: *Deep Sea 2003: Conference on the Governance and Management of Deep-Sea Fisheries, Part 1: Conference Papers* (ed. Shotton, R.). *FAO Fisheries Proceedings*, No. 3/1, 162–8.

Kakora, A.F. (2003) Certain biological features of the slender beryx *Beryx splendens* (Berycidae) on an isolated seamount in the southwest Atlantic. *Journal of Ichthyology*, 43(8), 642–5.

Kakora, A.F. (2005) Alfoncino *Beryx splendens* (Berycidae) of the southwestern Atlantic. *Journal of Ichthyology*, 45(1), 55–61.

Kemenov, V.E., Klimenkov, A.I., Vinnichenko, V.I., Roshupko, A.F. and Garkusha, V.K. (1979) *Manual on the Search and Fishery for Deepwater Fishes on Seamounts of the Mid-Atlantic Ridge*, 82 pp. Sevrybpromrazvedka (Nothern Fish Scouting Organization), MVIMU, Murmansk.

Kodolov, L.S. and Darnitsky, V.B. (2004) Ichthyofauna of seamounts in the North Pacific. *Abstracts from the Thirteenth Annual Meeting of the North Pacific Marine Science Organization (PICES)*, 201 pp. Hawaii, USA.

Koslow, J.A., Boehlert, G.W., Gordon, J.D.M., Haedrich, R.L., Lorance, P. and Parin, N. (2000) Continental slope and deep-sea fisheries: implications for a fragile ecosystem. *ICES Journal of Marine Science*, 57, 548–57.

Kulikov, M.Yu. and Darnitsky, V.B. (1992) Abundance dynamics of benthopelagic fishes in the Northwest Pacific and possible reasons behind that. *Oceanographic Basis of Biological Productivity in the Northwestern Part of the Pacific*, pp. 4–19. TINRO, Vladivostok (in Russian).

Kulikov, M.Yu. and Kodolov, L.S. (1991) *Biological Resources of the Thalassobathyal Zone in the World Ocean*, pp. 168–77. Selected Papers. VNIRO, Moscow (in Russian).

Lehodey, P. and Grandperrin, R. (1996) Age and growth of the alfonsino *Beryx splendens* over the seamounts off New Caledonia. *Marine Biology*, 125, 249–58.

Lehodey, P., Marchal, P. and Grandperrin, R. (1994) Modelling the distribution of alfonsino, *Beryx splendens*, over the seamounts of New Caledonia. *Fishery Bulletin (US)*, 92, 748–59.

Lutjeharms, J.R.E. and Heydorn, A.E.F. (1981) Recruitment of rock lobster on Vema Seamount from the islands of Tristan da Cunha. *Deep-Sea Research*, 28A(10), 1237.

Magnússon, J.V. and Magnússon, J. (1995) The distribution, relative abundance, and the biology of the deep-sea fishes of the Icelandic slope and Reykjanes Ridge. In: *Deep-Water Fisheries of the North Atlantic Oceanic Slope* (ed. Hopper, A.G.), pp. 161–99. Kluwer Academic Publishers, Dordrecht, The Netherlands.

Maloney, N.E. (2004) Sablefish, *Anoplopoma fimbria*, populations on Gulf of Alaska seamounts. *Marine Fisheries Review*, 66(3), 1–12.

Martins, M.R. and Ferreira, C. (1995) Line fishing for black scabbardfish (*Aphanopus carbo* Lowe, 1839) and other deep water species in the Eastern Mid Atlantic to the North of Madeira. In: *Deep-Water Fisheries of the North Atlantic Oceanic Slope* (ed. Hopper, A.G.). Kluwer Academic Publishers, Dordrecht, The Netherlands.

Melo, O. and Menezes, G. (2002) Exploratory fishing of the orange roughy (*Hoplostethus atlanticus*) in some seamounts of the Azores archipelago. ICES CM:2002/M:26, 11 pp.

Morato, T. and Pauly, D. (eds.) (2004) Seamounts: biodiversity and fisheries. *Fisheries Centre Research Reports 12*, xx pp.

Morato, T., Cheung, W.L. and Pitcher, T.J. (2006) Vulnerability of seamount fish to fishing: fuzzy analysis of life history attributes. *Journal of Fish Biology*, 68(1), 209–21.

Moskalenko, V.P. (1978) *Brief Description of Results of the Cruise in the Open Areas of Atlantic from 02 August to 26 December 1977*, 23 pp. Research SRTM-0839 'Ladoga'. VRPO 'Sevryba'; Sevrybpromrazvedka, Murmansk (in Russian).

Nolan, C.P. (ed.) (2004) *A Technical and Scientific Record of Experimental Fishing for Deepwater species in the Northeast Atlantic, by Irish Fishing Vessels in 2001*, Vol. 1(1), 172 pp. Fisheries Resources Series, Bord Iascaigh Mhara (Irish Sea Fisheries Board), Dun Laoghaire, Ireland.

Orlov, A.M. (1991) Preliminary results of fishery studies in the Marcus-Necker Ridge area in the winter of 1983. *Biological Resources of the Thalassobathyal Zone in the World Ocean*, pp. 177–83. Selected Papers. VNIRO, Moscow (in Russian).

Parin, N.V., Mironov, A.N. and Nesis, K.N. (1997) Biology of the Nazca and Sala y Gomez submarine ridges, an outpost of the Indo-West Pacific fauna in the eastern Pacific Ocean: composition and distribution of the fauna, its communities and history. *Advances in Marine Biology*, 32, 145–242.

Pavlov, A.I., Poletaev, V.A. and Sennikov, A.V. (1985) Status, prospects of study and fisheries development of biological resources in the high sea areas of the North Atlantic. *Materials from the All-Union Meeting 'Further Fisheries Development in the High Sea Areas of the World Ocean'*, TSNIITEIRH, pp. 46–52 (in Russian).

Pavlov, A.I., Shibanov, V.N. and Oganin, I.A. (1991) State and prospects of harvesting biological resources in the open areas of the North Atlantic. *Biological Resources in Thalassobathyal Zone of the World Ocean*, pp. 6–15. Selected Papers. VNIRO, Moscow (in Russian).

Polishchuk, V.M., Seniukov, V.L., Serdiuk, O.P. and Sokolov, A.M. (1989) *Fisheries Resources of the Southern Pacific*, p. 75. Sevrybpoisk, Murmansk (in Russian).

Prozorov, A.S., Dvinin, Yu.F., Lisovsky, S.F. and Shestopal, I.P. (1985) *Manual on the Search and Fishery for Tusk with Bottom Line on Seamounts of Open Part of North Atlantic*, 32 pp. PINRO, Murmansk (in Russian).

Riabikov, O.G. and Fomin, G.V. (1981) *Fisheries Potential of the South-Eastern Pacific Ocean*, p. 566. Zaprybpromrazvedka, Kaliningrad (in Russian).

Roberts, C.M. (2002) Deep impact: the rising toll of fishing in the deep sea. *Trends in Ecology and Evolution*, 17, 242–5.

Romanov, E.V. (ed.) (2003) Summary and review of Soviet and Ukrainian scientific and commercial fishing operations on the deepwater ridges of the southern Indian Ocean. *FAO Fisheries Circular No. 991*, 84 pp.

Rowden, A.A., Clark, M.R. and Wright, I.C. (2005) Physical characterisation and a biologically focused classification of 'seamounts' in the New Zealand region. *New Zealand Journal of Marine and Freshwater Research*, 39(5), 1039–59.

Samokhvalov, V.V., Lukjanov, V.A. and Kolchin, V.A. (1981) Preliminary results of studying of lanternfishes of the Atlantic and South Oceans. *Biology and Behaviour of Commercial Fishes on the Seamounts of the Azores Complex*, 121 pp. Zaprybpromrazvedka (Western Fish Scouting Organization), Kaliningrad (in Russian).

Sasaki, T. (1986) Development and present status of Japanese trawl fisheries in the vicinity of seamounts. In: *Environment and Resources of Seamounts in the North Pacific* (eds. Uchida, R.N., Hayasi, S. and Boehlert G.W.). *NOAA Technical Report NMFS*, 43, 21–30.

Seki, M.P. and Tagami, D.T. (1986) Review and present status of handline and bottom longline fisheries for Alfonsin. In: *Environment and Resources of Seamounts in the North Pacific* (eds. Uchida, R.N., Hayasi, S. and Boehlert, G.W.). *NOAA Technical Report NMFS*, 43, 31–5.

Shephard, S. and Rogan, E. (2006) Seasonal distribution of orange roughy (*Hoplostethus atlanticus*) on the Porcupine Bank west of Ireland. *Fisheries Research*, 77, 17–23.

Shibanov, V.N. (1998) Biological foundation of fishing roundnose grenadier. Abstract of Biology Science Candidate Dissertation. VNIRO, Moscow, 24 pp. (in Russian).

Shnar, V.N., Burykin, S.N. and Sirota, A.M. (2005) Fishery-oceanological research carried out by AtlantNIRO in the open North Atlantic Ocean during 2000–2005. *Proceedings of the Twelth International Conference on Fisheries Oceanology*, AtlantNIRO, Kaliningrad, pp. 331–5.

Simpson, E.S.W. and Heydorn, A.E.F. (1965) Vema Seamount. *Nature*, 4994, 249–51.

Stepanenko, M.A., Darnitsky, V.B. and Vologdin, V.N. (2002) On the history of fisheries research by TINRO in the Northeast Pacific. In: *Historical Experience of Scientific and Fisheries Research in Russia*, pp. 179–84. VNIRO, Moscow (in Russian).

Stitt, S. (1993) Deepwater fisheries: Canada's first look down in the deeps – tests on three seamounts. *Fishing News International*, 32(7), 16–17.

Stone, G.S., Madin, L.P., Stocks, K., Hovermale, G., Hoagland, P., Schumacher, M., Etnoyer, P., Sotka, C. and Tausig, H. (2004). Seamount biodiversity, exploitation and conservation. In: *Defying Ocean's End* (eds. Glover, L.K. and Earle, S.A.), pp. 45–70. Island Press, Washington, DC.

Strogalev, V.D., Afnagel, E.L., Shulga, A.D., Pantyukhin, V.L., Zvyagintsev, E.F., Gorlenko, V.A., Khromov, V.E., Sammelberg, H.M. and Kalachev, S.I. (1989) *Possibilities of Fishery on the Oceanic Banks of the Vavilov Rise*, 68 pp. Zaprybpromrazvedka, Kaliningrad (in Russian).

Thomsen, B. (1998) Faroese quest of orange roughy in the North Atlantic. ICES CM 1998/O:31, 8 pp.

Tilzey, R. (2000) South Tasman Rise trawl fishery. In: *Fishery Status Reports 1999: Resource Assessments of Australian Commonwealth Fisheries* (eds. Caton, A. and McLoughlin, K.). BRS, Canberra.

Uchida, R.N. and Tagami, D.T. (1984) Groundfish fisheries and research in the vicinity of seamounts in the North Pacific Ocean. *Marine Fisheries Review*, 46, 1–17.

Vinnichenko, V.I. (1989) *Background for Searching Works on Studying Possibilities of Trawl Fisheries for Tuna in the Open Part of the North Atlantic*, 114 pp. VNTO of the Food Industry, Murmansk (in Russian).

Vinnichenko, V.I. (1995) On commercial stocks of alfonsino in the open North Atlantic. *Abstracts of Reports of the Fourth Russian Fishery Forecasting Conference*. PINRO, Murmansk, p. 32 (in Russian).

Vinnichenko, V.I. (1996) New data on distribution of some species of tuna (Scombridae) in the North Atlantic. *Journal of Ichthyology*, 36(8), 679–81.

Vinnichenko, V.I. (1997) Russian investigations and deep-water fishery on the Corner Rising Seamount in Subarea 6. *NAFO Scientific Council Studies*, No. 30, 41–9.

Vinnichenko, V.I. (1998) Alfonsino (*Beryx splendens*) biology and fishery on the seamounts in the open North Atlantic. ICES CM 1998/O:13, 1–15.

Vinnichenko, V.I. (2002a) Prospects of fisheries on the seamounts. ICES CM:2002/M: 32.

Vinnichenko, V.I. (2002b) Russian investigations and fishery on seamounts in the Azores area. Proceedings of the XVIII e XIX Semana das Pescas dos Acores, Secretaria Regional da Agricultura e Pescas, Azores, pp. 115–29.

Vinnichenko, V.I. (2006) Fisheries resources on the seamounts. *Voprosy rybolovstva* 4(28), 655–669 (in Russian).

Vinnichenko, V.I. and Khlopenyuk, V.F. (1983) *Manual on the Fishery for Mackerel and Hourse Mackerel on the Josephine and Ampere Seamounts*, 28 pp. Sevrybpromrazvedka (Northern Fish Scouting Organization), Murmansk (in Russian).

Volkov, V.V., Gerber, E.M., Semenov, A.P. *et al.* (1977) Status of resources and possibilities of fisheries for squids, Atlantic saury, anchovy, longspine snipefish and flyingfish in the open part of the Atlantic Ocean. *The Atlantic Ocean. Fishsearching Investigations*, Vol. 6, pp. 80–94. Zaprybpromrazvedka, Kaliningrad (in Russian).

Vologdin, V.N. (2002) On the history of hydroacoustic research of TINRO in the north-eastern part of the Pacific. In: *Historical Experience of Scientific and Fisheries Research in Russia*, pp. 39–42. VNIRO, Moscow (in Russian).

Wayte, S. and Bax, N. (Comps.) (2002) Draft stock assessment report 2002-Orange roughy (*Hoplostethus atlanticus*). Report prepared for Australian Fisheries Management Authority, Canberra.

Wessel, P. (2001) Global distribution of seamounts inferred from gridded Geosat/ERS-1 altimetry. *Journal of Geophysical Research*, 106(B9), 19431–41.

WWF (2003) *Management Direction for the Bowie Seamount MPA: Links Between Conservation, Research, and Fishing*, 69 pp. World Wildlife Fund, Canada.

Yasui, M. (1986) Albacore, *Thunnus alalunga*, pole-and-line fishery around the Emperor Seamounts. In: *Environment and Resources of Seamounts in the North Pacific* (eds. Uchida, R.N., Hayasi, S. and Boehlert G.W.). *NOAA Technical Report NMFS*, 43, 37–40.

Young, Z.U. (2003) Fishery of orange roughy in Chile. Unpublished report from Instituto de Fomento Pesquero, 13 pp.

Zaferman, M.L. and Sennikov, A.M. (1991) *Biology and Peculiarities of Behaviour of Deepwater Crab* Geryon quinquedens, pp. 43–5. PINRO, Murmansk (in Russian).

Zaferman, M.L. and Shestopal, I.P. (1991) *Underwater Search for Fish and Deepwater Longlining in the Areas of Underwater Mountains of the North Atlantic*, pp. 50–77. PINRO, Murmansk (in Russian).

Zaferman, M.L. and Shestopal, I.P. (1996) Estimation of tusk abundance on the sea mounts of the North Atlantic from the underwater observation data and long-lining. *Instrumental Methods of Fishery Investigations*, pp. 65–79. PINRO, Murmansk (in Russian).

Zasel'skii, V.S. and Pleteshkov, A.V. (1988) Features of distribution and behaviour of Alfonsino in the area of the Rio Grande Mount in connection with variability of abiotic conditions. *Biological Resources of Thalassobathyal Zone of the World Ocean. Papers at the All-Union Conference on Study of Fishes of the Thalassobathyal of the World Ocean, Rybnoe* (VNIRO, Moscow), p. 23 (in Russian).

Zhukov, V.P. (2005) Feasibility study of organization of fishery in the EEZ of Peru. *Exploration of Aquatic Bioresources in the World Ocean*, pp. 50–52. *Papers of the All-Russian Working Meeting*, Federal Agency on Fisheries of the Russian Federation, Moscow (in Russian).

Zilanov, V.K. (1975) *Biology and Fishery for Atlantic Saury in the North Atlantic*, 117 pp. 'Sevryba'; Sevrybpromrazvedka, Moscow (in Russian).

Chapter 18

Catches from world seamount fisheries

Reg Watson, Adrian Kitchingman and William W. Cheung

Abstract

Even the most remote seamounts are targeted by fisheries and overfishing is a serious concern. Estimating the catches taken from seamounts is difficult as catch statistics are typically reported from large ocean areas and cannot be easily related specifically to seamount habitat, much less individual seamounts. In addition, many species fished from seamounts like orange roughy are also caught on the continental slopes. We employ catch statistics allocated to 30 min spatial cells to estimate catches of 13 primary seamount species, those dependent on seamount habitats, and also of 29 species merely associated with seamounts. Catches of primary seamount species seem to have peaked in the early 1990s and many appear vulnerable. Much uncertainty remains to be resolved in future assessments, including the strength of the affinities of many species to seamounts, and how fishing patterns have shifted from continental slopes to isolated seamounts, especially in the far southern hemisphere.

Introduction

Seamounts represent important features of ocean bottom topography that can have profound effects on local abundance of many commercial species (see Chapters 1, 4–12). Their importance has increased in recent years with the failure of many coastal shelf fisheries. After the introduction and enforcement of 200 nautical mile exclusive economic zones (EEZs) around most nations' productive inshore waters, these richer areas of the high seas have seen an intense interest by both fishing fleets and conservation groups.

The recent history of fisheries on seamounts is one of serial depletion (see Chapter 17) and has made many cautious about the ability of seamount areas to support unlimited exploitation (Hopper, 1995; Merreth and Haedrich, 1997; Froese and Sampang, 2004; Morato *et al.*, 2004). Watson and Morato (2004) showed that seamount fisheries collapsed faster and recovered more slowly than non-seamount fisheries. Koslow (2001) summed it up: 'these fisheries are commonly characterized by a boom-and-bust cycle, whereby a stock (and sometimes the entire species) is fished to commercial extinction within 10 years. Seamount trawl fisheries also have severe impacts on the benthic environment, essentially removing the deepwater reefs and associated fauna from heavily fished areas'.

When there is no clear evidence of serial depletion of species fished, there appears to be evidence that sub-populations are being depleted serially and have reduced genetic diversity (see Chapter 19; Koslow *et al.*, 1997; Clark, 1999).

Assessing the global seamount catch presents two major challenges. The first is deciding what seamounts are (see Chapter 1) and where they are located (see Chapter 2). Secondly, there are a number of lists of seamount species, but it is more controversial to determine which commercial species are truly associated with seamounts. Rogers' (1994) list of more than 70 taxa is useful, but it combines 'primary' species, those that depend on seamounts, with 'secondary' species commonly found on seamounts but not exclusive to them. Many pelagic species, including several tunas such as albacore (*Thunnus alalunga*), are often taken in some parts of the world from seamount areas (see Chapter 10A); however, they are also caught in mid-ocean far from seamounts. These latter species are not considered by some to be seamount species and, arguably, their catches should not be included here. If these 'secondary' seamount species are to be included, it becomes important to have a means of prorating their reported landings so that only that part caught in the proximity of seamounts is included. This is problematic but necessary (see Chapters 9 and 16).

Methods

Data source

Sourcing of fisheries data, as with any reporting of global catches, is problematic but critical. We are fortunate to have global catch data compiled by the Sea Around Us Project (SAUP) of the University of British Columbia (Watson *et al.*, 2004). This was sourced from FAO, Eurostat and a variety of other sources. Modifications to original data include clarification of taxonomic identity of catches and corrections for misreporting. This dataset provides catch for all reported marine taxa from each 30 min by 30 min spatial cell of the world's oceans from 1950 until 2002. We used the third version of this dataset (subsets available online at http://seaaroundus.org/; April 2005), which incorporates published seamount affinities.

Source data for seamount locations include Chapter 2, Kitchingman and Lai (2004), Seamounts Online (http://seamounts.sdsc.edu) (Stocks, 2004a) and NOAA (http://www.ngdc.noaa.gov/mgg/global/global.html).

Seamount affinity

The distribution of taxa plays an important role in the SAUP catch database, notably in the spatial catch allocations. Habitat preference is a major factor affecting a taxon's distribution; procedures based on its association with different habitats including seamounts were developed to enhance predictions. (We use the terms 'taxon' and 'taxa' because global catch data do not always identify the species or genus, but often include a range of taxonomic categories.) The process assumes that relative abundance of a taxon in a spatial grid is partly determined by the area of habitat(s) with which it is associated, but recognizes that the effects of habitats will likely range further. The latter is assumed to be a function of the taxon's body size (maximum length) and its habitat 'versatility'. Thus a large-sized taxon that inhabits a wide range of habitats is more likely to extend its range further from associated habitats (Kramer and Chapman, 1999). Affinities to seamount habitats were investigated using FishBase (Froese and Pauly, 2005) and other sources. For a fuller description of the methodology used to associate commercial taxa to habitats such as seamounts (see http://seaaroundus.org/doc/saup_manual.htm#5).

Primary catches

Primary commercial seamount species were defined as those caught primarily or exclusively on seamounts. These were determined by comparing a number of reference sources such as Rogers (1994) and after discussion with a number of experts in the field. This proved challenging as there is not complete agreement on these species. More worrying still is that several of the taxa reported from some sources (see Chapter 17) are not identified in our global catch database, at least not by the same names. Even less is known about some of the invertebrates that might be harvested (Stocks, 2004b). This could mean that some primary species, including several crustaceans, have been left out, but it would appear that these do not constitute high volume catches, even for seamount species.

Even though some species have been nominated as primary seamount species, it is clear that not all reported landings actually originate from seamount areas. Many taxa, including orange roughy (*Hoplostethus atlanticus*), are often caught on the continental slope, and indeed some individual fisheries are reported only from the slope. It was thus necessary to prorate the reported catches of these species. Two methods were used. The first, and most obvious, is to use the estimates provided by experts when it is available (Fig. 18.1). When more information was available, it was possible to include trends through time. The second method was to intersect the modelled distribution, based partially on seamount affinities, with a map of known seamounts, with a 20 km buffer. The proportion of the global landings occurring in this overlapping area was deemed to be caught in the proximity of seamounts. For two primary seamount species, orange roughy (Fig. 18.2) and the Patagonian toothfish (*Dissostichus eleginoides*) (Fig. 18.3), there is generally a strong overlap. Places with the highest abundances, based on our modelled distributions, generally corresponded well with the buffered areas of known seamount areas. This is hardly surprising as our modelled distributions include a habitat affinity for seamounts which both these species demonstrate. For species like these, that are usually found on seamounts, we estimate that a significant proportion of the global catch should be assigned to seamounts (Table 18.1).

Secondary catches

Several species of tunas and smaller pelagic taxa were included as secondary catches, not found either exclusively or even primarily from seamount habitats. Many seamount species also occur on the continental slope. Many pelagic species are widely distributed and have, at best, enhanced catches in the proximity of seamounts. We extend seamount fisheries beyond deepwater species such as those described by Gordon (2001) and others, to include other species which have some association with seamount locations

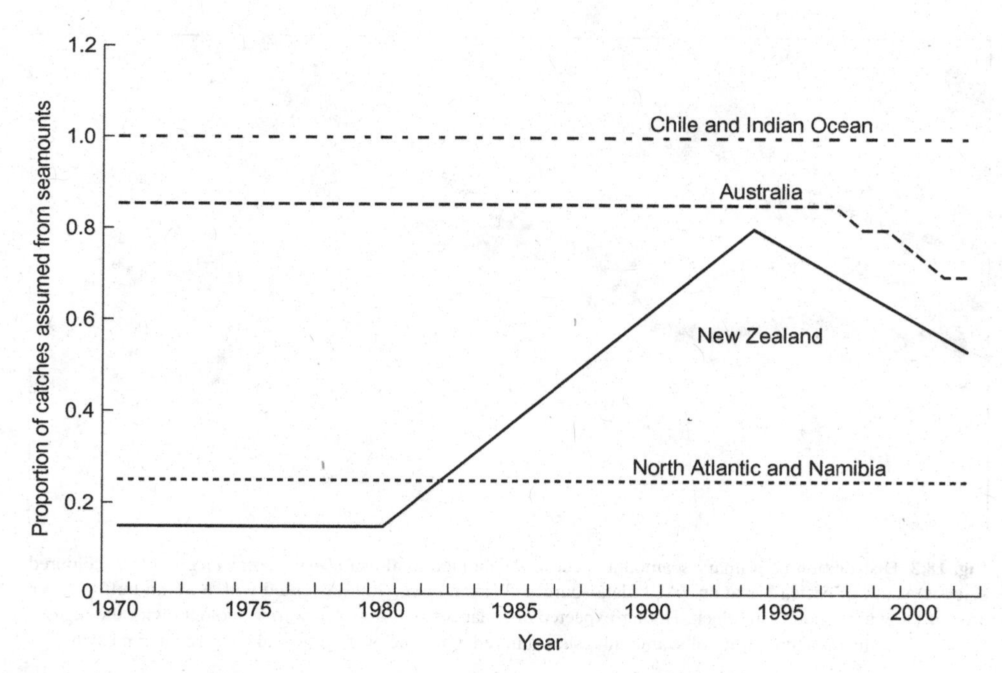

Fig. 18.1 Proportion of catches of primary seamount species: orange roughy (*Hoplostethus atlanticus*) assumed to come from seamounts. New Zealand estimates simplified from Clark and O'Driscoll (2003), Australia Antony Koslow, CSIRO Australia, Personal communication. All other estimates from Malcolm Clark, Deepwater fisheries NIWA, New Zealand, Personal communication.

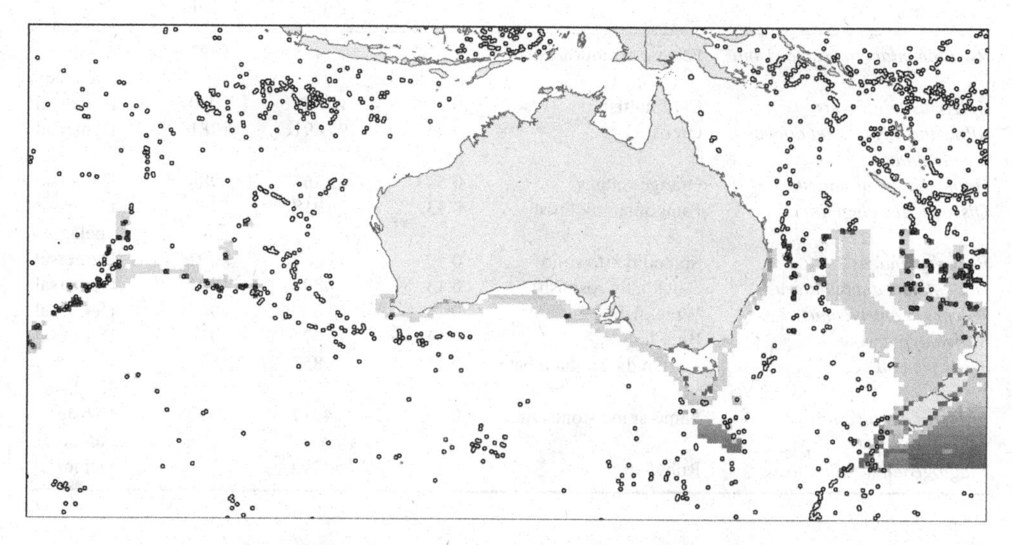

Fig. 18.2 Distribution of primary seamount species: orange roughy (*Hoplostethus atlanticus*) (coloured squares) in the vicinity of Australia based on the SAUP data sources (Watson *et al.*, 2004) overlaid with a 20 km buffer drawn around known seamounts in black. Higher expected abundances are shown in warmer colours with the highest (usually in immediate proximity of seamounts) shown in red. Only the overlap was taken as seamount catch.

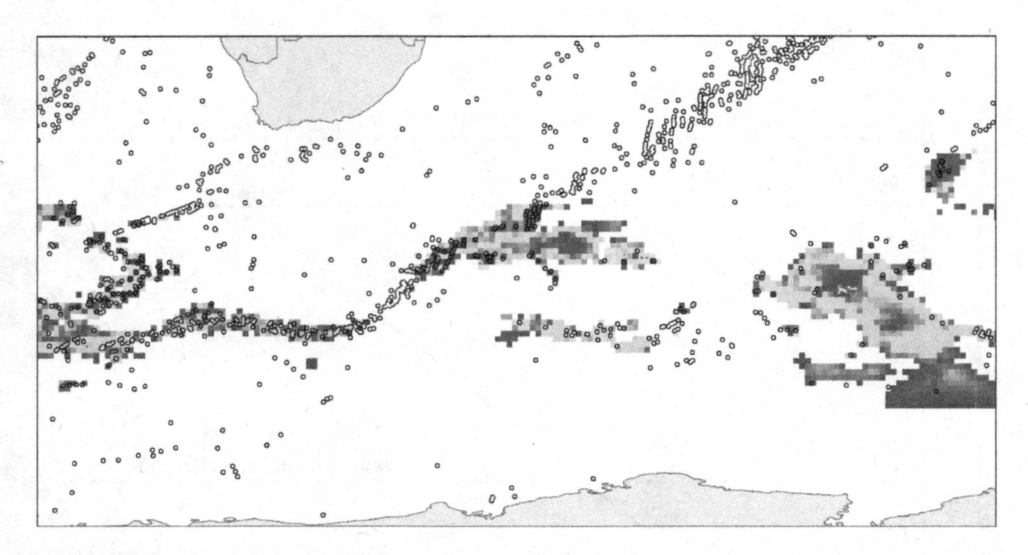

Fig. 18.3 Distribution of primary seamount species: Patagonian toothfish (*Dissostichus eleginoides*) (coloured squares) south of Africa based on SAUP data sources (Watson *et al.*, 2004) overlaid with a 20 km buffer drawn around known seamounts in black. Higher expected abundances are shown in warmer colours with the highest (usually in immediate proximity of seamounts) shown in red. Only the overlap was taken as seamount catch.

Table 18.1 Primary seamount species with proportion of annual landings attributed to seamount, showing magnitude and year of maximum global catch (tonnes).

Species	Common name	Proportion	Maximum annual catch	Year of maximum catch	Group
Pseudopentaceros richardsoni	Pelagic armourhead	1	435	1993	Small pelagic
Epigonus telescopus	Cardinalfish	0.76	10088	2000	Demersal
Allocyttus niger, Pseudocyttus maculates	Oreos	0.76	97491	1981	Demersal
Hoplostethus atlanticus	Orange roughy	0.54	55361	1988	Demersal
Dissostichus eleginoides	Patagonian toothfish	0.23	26918	1992	Large pelagic
Beryx splendens	Splendid alfonsino	0.19	82	2002	Demersal
Helicolenus dactylopterus	Blackbelly rosefish	0.15	4898	1994	Demersal
Polyprion americanus	Wreckfish	0.15	800	1962	Demersal
Molva dypterygia	Blue ling	0.15	10475	1976	Demersal
Jasus tristani	Tristan da Cunha lobster	0.15	857	1972	Lobster and crab
Lithodes aequispina	Same-spine stone crab	0.15	4917	1997	Lobster and crab
Plagiogeneion rubiginosus	Rubyfish	0.15	Nil		Demersal

even if they do not dwell in close proximity to the elevated seafloor that seamounts present. Indeed, we had the same problem distinguishing seamount from non-seamount landings as he reported in separating deepwater and shallower water species. Some of this is based on taxonomic uncertainty in catch reporting, and some is due to the wide distributions of some taxa, and their sometimes weak or uncertain connection with seamounts.

We decided to accept many listed seamount taxa as at least 'secondary' species. Nevertheless, for clarity, and because these taxa have very high annual catches that would overshadow other 'primary' species, we elected to report these separately here. Even more so than with primary seamount species, it was essential to prorate the global catches reported for secondary species. It was assumed that only that catch that was taken in proximity of seamounts (based on a 20 km buffer) was associated with seamounts. For species like albacore tuna (Fig. 18.4) with a very weak association with seamounts, only a small portion of the reported global catch was assigned to seamounts (Table 18.2).

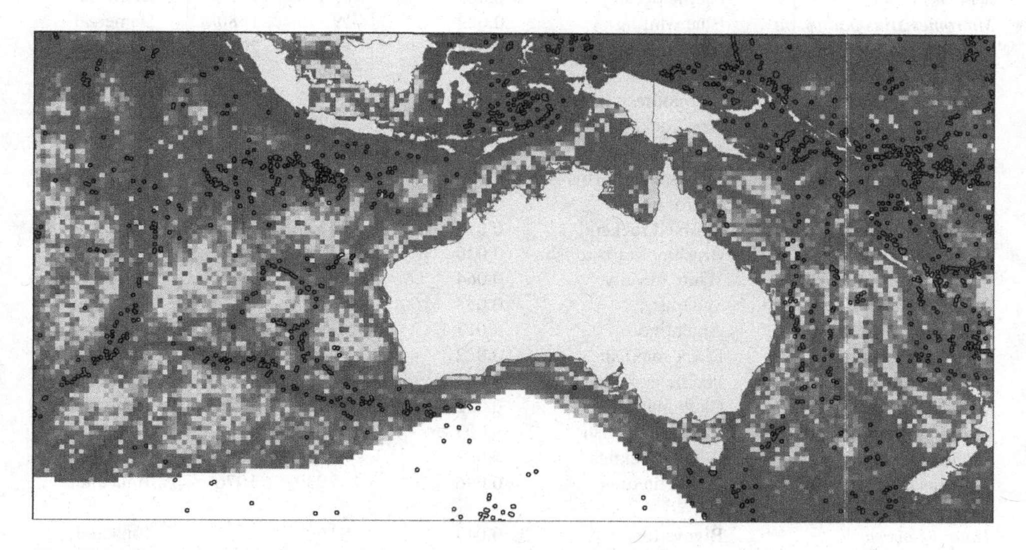

Fig. 18.4 Distribution of secondary seamount species: albacore tuna (*Thunnus alalunga*) (coloured squares) in the vicinity of Australia based on SAUP data sources (Watson *et al.*, 2004) overlaid with a 20 km buffer drawn around known seamounts in black. Higher expected abundances are shown in warmer colours with the highest shown in red. Only the overlap was taken as seamount catch.

One sardine species, *Sardinops sagax*, was particularly problematic. Rogers (1994) lists this sardine as a seamount species under a synonym *Sardinops melanosticta*. This species is, however, extremely abundant and is caught in huge tonnages in many non-seamount areas. In fact, a maximum global catch of 4.5 million tonnes was reported for this species in 1988. It was therefore decided that this species would not be included in our global seamount catches as it would overwhelm all other taxa even if only afforded a 'secondary' seamount species role. Their genuine contribution to seamount fisheries is unknown.

Results

Seamount species

We accepted 13 commercial species as primary seamount species (Table 18.1). Of these, only one, the pelagic armourhead (*Pseudopentaceros richardsoni*), was assumed to come exclusively from seamounts. Seamount catch of other primary species ranged from 76%

Table 18.2 Secondary seamount species with proportion of annual landings attributed to seamount, and magnitude and year of maximum global catch (tonnes).

Species	Common name	Proportion	Maximum annual catch	Year of maximum catch	Group
Trachurus picturatus	Blue jack mackerel	0.036	26 710	1973	Demersal
Epinephelus aeneus	White grouper	0.016	2100	1968	Demersal
Seriola lalandi	Yellowtail amberjack	0.105	4285	1952	Demersal
Sebastes alutus	Pacific ocean perch	0.003	383 900	1965	Demersal
Micromesistius poutassou	Blue whiting	0.022	719 573	1980	Demersal
Caranx ignobilis	Giant trevally	0.042	5527	2002	Demersal
Nemadactylus macropterus	Tarakihi	0.043	14 400	1962	Demersal
Chionodraco hamatus	Chionodraco hamatus	0.007	2	1989	Demersal
Lepidonotothen squamifrons	Grey rockcod	0.021	38 728	1971	Demersal
Promethichthys prometheus	Roudi escolar	0.133	7	2002	Demersal
Sebastolobus alascanus	Shortspine thornyhead	0.100	1616	2000	Demersal
Zenopsis nebulosus	Mirror dory	0.149	695	1992	Demersal
Notothenia rossii	Marbled rockcod	0.012	250 631	1970	Demersal
Saurida undosquamis	Brushtooth lizardfish	0.030	1215	1972	Demersal
Pseudocaranx dentex	White trevally	0.064	6526	1978	Demersal
Anoplopoma fimbria	Sablefish	0.053	50 700	1972	Demersal
Argentina sphyraena	Argentine	0.017	40	1973	Demersal
Sebastes melanops	Black rockfish	0.022	148	2001	Demersal
Trachurus symmetricus	Pacific jack mackerel	0.041	66 462	1952	Small pelagic
Scomber japonicus	Chub mackerel	0.032	1 625 753	1978	Small pelagic
Champsocephalus gunnari	Mackerel icefish	0.003	112 013	1982	Small pelagic
Sebastes entomelas	Widow rockfish	0.001	6886	1997	Small pelagic
Thunnus albacares	Yellowfin tuna	0.096	165 930	1976	Tuna and swordfish
Thunnus obesus	Bigeye tuna	0.044	103 600	1963	Tuna and swordfish
Thunnus alalunga	Albacore	0.042	90 444	1976	Tuna and swordfish
Katsuwonus pelamis	Skipjack tuna	0.039	254 701	1988	Tuna and swordfish
Euthynnus affinis	Kawakawa	0.120	57 899	1989	Tuna and swordfish
Acanthocybium solandri	Wahoo	0.038	5087	1971	Tuna and swordfish
Thunnus thynnus	Northern bluefin tuna	0.020	12 000	1964	Tuna and swordfish

for cardinalfish (*Epigonus telescopus*) and oreos (*Allocyttus niger* and *Pseudosyttus maculates*) down to 15%. Overall 54% of orange roughy were assumed to have come from seamounts globally based on estimates in Fig. 18.1, however, there is considerable variation by year and by area/country. The other primary seamount species ranged from 15% to 23% of the catch associated with seamounts. None of these species reported more than 100 000 tonnes caught in any 1 year. Most were demersal fishes, but there were also two crustaceans. Two species were pelagic. Most species had their maximum annual catch a decade or more ago (see Chapter 17). There was no reported catch for one taxon, the rubyfish (*Plagiogeneion rubiginosus*), in our database, which highlights a reporting problem for minor species.

Twenty-nine taxa were considered to be secondary seamount species (Table 18.2). Most were demersal fishes but there were a few small pelagic fishes, while the remainder were tunas. The proportion of catches associated with seamounts varied from only marginal to about 10%. Maximum annual catch reported also varied widely from an insignificant quantity to one in excess of one million tonnes annually. Several species provided in various lists of seamount species have never been reported in our global catch records, therefore these were not included here. By contrast, some others have such large catches that it is clear that the bulk of their production occurs over much larger areas than seamounts. Caution must be exercised when assuming the role of seamounts in their global catches.

As with primary seamount species, it is clearly unlikely that a fixed annual proportion of any one taxa comes from seamounts. Since catches were first reported for most taxa there undoubtedly has been a trend for fleets to fish more from seamount habitats, especially as the more easily accessed slope and shelf areas became over-exploited.

Annual catches

From the proportion of the global catch of primary and secondary species apportioned to seamounts (Tables 18.1 and 18.2, and also Fig. 18.1), it was possible to plot annual landings (Figs. 18.5 and 18.6). Substantial tonnages of most primary species were not reported until the 1970s. Primary catches were dominated by oreos (Oreosomatidae) and orange roughy. Like many deepwater fish, the orange roughy is a long lived, possibly more than 100 years (Clark *et al.*, 2000), and vulnerable species. Catches of primary species appear to have peaked overall by the early 1990s, by which time it is likely that almost all productive seamounts were accessible to fisheries (see Chapter 17).

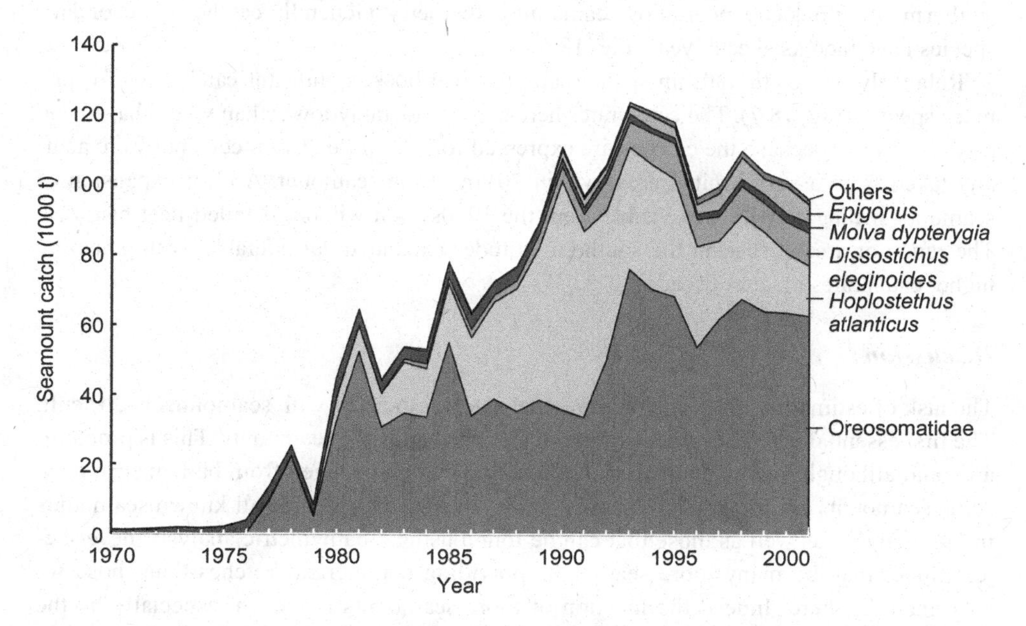

Fig. 18.5 Global catch of primary seamount species (based on data described in Watson *et al.*, 2004).

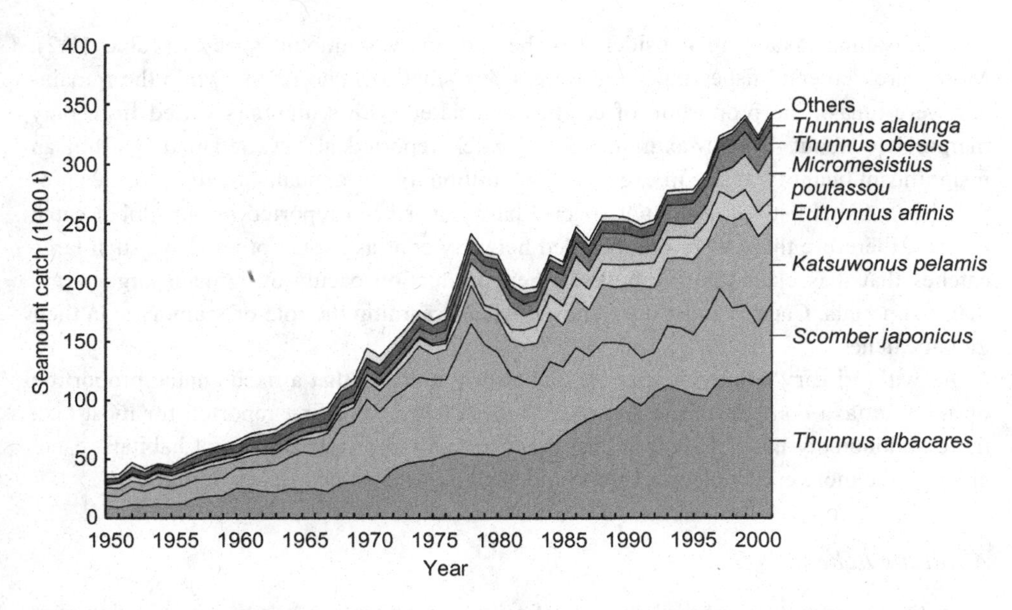

Fig. 18.6 Global catch of secondary seamount species (based on data described in Watson *et al.*, 2004).

Landing of some secondary species such as chub mackerel (*Scomber japonicus*) was reported prior to the mid-1970s (Fig. 18.6). For some species, a portion of the catches reported here are perhaps wrongly attributed to seamounts, however, for some species such as the blue whiting (*Micromesistius poutassou*), significant catches only began in the last few decades. Koslow (2001) reported several examples where some tuna fleets use seamounts. Increased productivity in these areas may improve catches as might local isotherm variations engendered by seamount bathymetry. Generally catches of secondary species have increased each year (Fig. 18.6).

Relatively few of the 30 min spatial cells mapped boast significant catch rates for primary species (Fig. 18.7). The catch rates here are considerably lower than would have been observed at sea because the catches are expressed for the entire 30 min cells but were actually taken from a very limited area, within 20 km of the seamount. A clear expansion of seamount fishing activity was visible from the 1970s even without detailed fleet histories. The range of areas fished in the southern latitudes expanded considerably, with generally higher catch rates in latter decades.

Discussion

The task of estimating the global commercial catch associated with seamounts is difficult. The first essential is to know the location of all possible fished seamounts. This is problematic, and although much about seamount location can be deduced from bathymetric data, some seamounts are too small to show up. We have tried to include all known seamounts in our analysis, as well as those that can be found using a bathymetric analysis, nevertheless, there may be many more seamounts providing commercial catches than those we have included here. Indeed the location of more seamounts is known, especially by the world's submariners, as well as conservation and fishing groups, but for many reasons

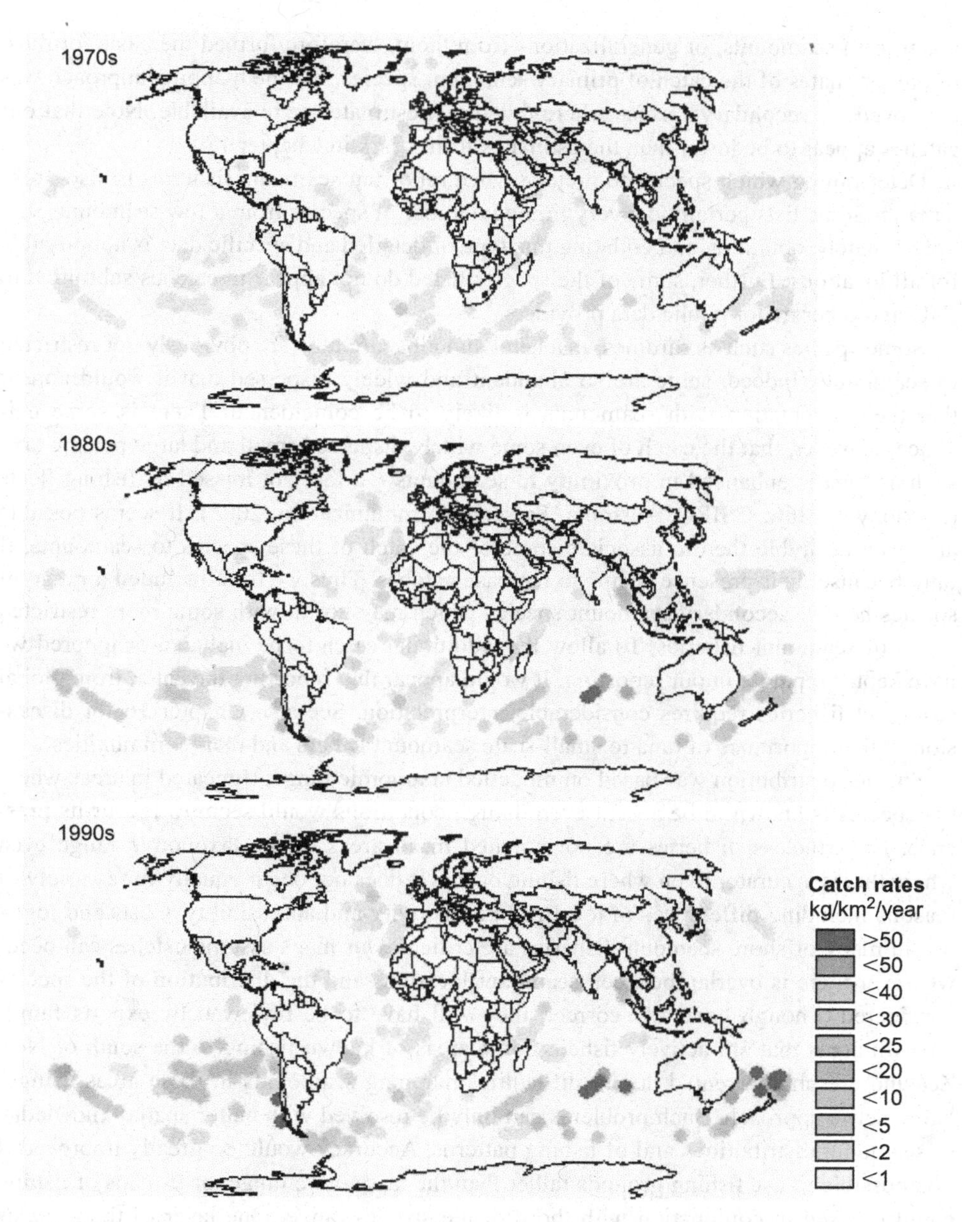

Fig. 18.7 Global catch rates of primary seamount species by decade for 30 min spatial cells which include identified seamounts.

these locations are not made public (see Chapter 17). We are aware that seamounts in higher latitudes appear to be under-represented including some fished commercially south of New Zealand. The method we employed therefore has a possibility of being an underestimate, particularly as incomplete knowledge of seamount locations limits an analysis of overlap between seamounts and the species' abundance distribution. For this and other reasons we choose to use direct estimates such as that from Clark and O'Driscoll (2003) when they were available. Direct estimates of the proportion of regional or national catch taken from seamounts, or generalizations from them, therefore formed the basis for most of our estimates of the catch of primary seamount species, but the mapping approach was employed for secondary species where few direct estimates were available. Note that our catches appear to be lower than the estimates from Clark in Chapter 17.

Determining which species to include when reporting seamount fisheries is also problematic. Some lists pertain to a very limited number of species from a few seamounts, but unfortunately, obtaining and collating this type of detailed and specific data is not possible for all locations. Further, some of the species listed do not appear in catches submitted to FAO and other major public data providers.

Some species such as sardines, mackerel and albacore tuna are obviously not restricted to seamounts, indeed, some are so abundant and widely dispersed that it would appear that their association with seamounts is likely to be coincidental. There is some evidence, however, that the catch of even some widely dispersed small and large pelagic taxa such as tuna is enhanced in proximity to seamounts – a fact not missed by fishing fleets (Anthony Koslow, CSIRO, Australia, Personal Communication, 2005). It seems possible and even desirable then to associate some of the catch of these species to seamounts, if only because their presence seems to increase catches. Thus we have included a range of species here as secondary seamount species which may not fit with some more restricted views of seamount fisheries. To allow this additional catch to be included or ignored we have kept it separate in our reporting. It would appear that reporting the catch from global seamount fisheries requires considerable interpretation. See also Chapter 16 for discussion of the importance of tuna to small-scale seamount fishers and their communities.

Species distribution was based on modelled taxonomic range, truncated in areas where the species is known to be absent even though conditions would seem to favour its presence. Nevertheless, fisheries are not pursued in all areas of the taxonomic range even where this is accurate. Even where fishing occurs it does not occur equally for a variety of reasons including differences in localized productivity and accessibility. Costs and logistics in many offshore seamount fisheries are critical. Our maps assume catches can occur wherever there is overlap between seamount locations and the distribution of the species considered. Though generally correct, they will have to be reviewed by experts familiar with areas that are actively fished. Some areas of known fishing to the south of New Zealand are unrepresented due to difficulties mapping seamounts in these areas using a bathymetric approach. Such problems can only be resolved with better spatial knowledge of seamount distributions and of fishing patterns. Accuracy would be greatly improved if it is possible to use fishing grounds rather than the taxonomic range, or if maps of fishing could be used in conjunction with those of the species range. One layer of the analysis could present where the species occurs while another, acting on that layer, would show where the distribution is actually fished that year. Modelling fishing fleets would appear to be a logical way of integrating all available information.

As our work on associating the distribution of marine animals with habitats such as seamounts improves (see Chapter 12a), we are likely to improve our ability to define the degree of seamount affinity, and indeed dependence of some commercial species for these habitats. Seamounts, of course, are highly variable in profile and subject to a number of definitions. Some likely do not approach a depth where their effects are important, especially to pelagic species. The tops of some are so deep as to be barely within the productive photic zone. Some are very isolated, while others are found in groups. Many aspects of their shape, arrangement and location can affect their suitability for supporting fisheries. Therefore not all seamounts are equally productive for commercial fisheries. Future work will characterize their productivity better, following on from work by Clark and O'Driscoll (2003) and others, and improve global estimates for the fisheries they support (see Chapters 4, 5 and 14).

We have taken a simplistic approach to calculating the proportion of the global catch of seamount-associated species. One obvious omission is chronological changes in fishing patterns. Though we have incorporated a time series for the proportion of catch of orange roughy and oreos, the two major primary taxa taken from seamounts annually and often taken as bycatch in former fisheries, we were unable to access this information for all locations or for all species. Some fisheries pursued taxa we have accepted as 'seamount' species only on the continental slope long before fleets moved, most likely from necessity, into deeper waters to take them from seamounts. Declaration of exclusive EEZs in the 1970s closed coastal waters to many fleets at a time when the technology was finally available to find and use deeper and distant fishing locations such as seamounts. It was our intention to try and reconstruct this chronology on a fishery-by-fisheries basis for our analysis, so that our seamount proportions would reflect these historical changes. This was not possible for many, but is still feasible for several major species such as orange roughy. Ignoring this trend is likely to have the effect of overestimating seamount catches in early years (pre-mid-1970s) and underestimating them in more recent years.

In an era of increasing concern about the need for fisheries regulation on the high seas, and about the sustainability of deepwater/cold water species, it is important to explore trends in resource use. Seamounts, and the species that use them, represent a resource that has been discovered by fishing fleets, and in some cases quickly exhausted. They are often offshore and difficult to patrol even if they fall within a management jurisdiction. Because of the difficulties involved, it will take many new approaches in concert to reconstruct what has been happening to these valuable fisheries. Of course, catches are best put in context if they pertain to stocks rather than global values for a species, especially for potentially widely separated and potentially endangered populations. For all its weaknesses, we hope that our attempt can help stimulate further discussion and work including integrated plans to protect seamount fishes from over-exploitation, especially on the high seas.

Acknowledgements

The authors are grateful for suggestions, support and patience provided by Telmo Morato, Tony Koslow and Malcolm Clark. The authors thank G. Nowara and an anonymous editor for their assistance. The authors were supported by the Sea Around Us Project, which is funded by the Pew Charitable Trusts of Philadelphia, USA.

References

Clark, M. (1999) Fisheries for orange roughy (*Hoplostethus atlanticus*) on seamounts in New Zealand. *Oceanologica Acta*, 22, 593–602.

Clark, M. and O'Driscoll, R. (2003) Deepwater fisheries and aspects of their impact on seamount habitat in New Zealand. *Journal of Northwest Atlantic Fisheries Science*, 31, 441–58.

Clark, M.R., Anderson, O.F., Francis, R.I.C.C. and Tracey, D.M. (2000) The effects of commercial exploitation on orange roughy (*Hoplostethus atlanticus*) from the continental slope of the Chatham Rise, New Zealand, from 1979 to 1997. *Fisheries Research*, 40, 217–38.

Froese, R. and Pauly, D. (eds.) (2005) FishBase.World Wide Web electronic publication. www.fishbase.org, version (15 May/2006).

Froese, R. and Sampang, A. (2004) Taxonomy and biology of seamount fishes. In: *Seamounts: Biodiversity and Fisheries* (eds. Morato, T. and Pauly, D.), pp. 25–32. *Fisheries Centre Research Reports*, 12(5), 25–32.

Gordon, J.D.M. (2001) Deep-water fish and fisheries. In: Managing risks to biodiversity and the environment on the high sea, including tools such as marine protected areas – scientific requirements and legal aspects. *Proceedings of the Expert Workshop held at the International Academy for Nature Conservation Isle of Vilm*, Germany, 27 February–4 March, 2001 (eds. Thiel, H. and Koslow, J.A.), pp 31–8.

Hopper, A.G. (ed.) (1995) *Deep-Water Fisheries of the North Atlantic Oceanic Slope*. Kluwer Academic Publishers, Dordrecht, The Netherlands. 420 pp.

Kitchingman, A. and Lai, S. (2004) Inferences on potential seamount locations from mid-resolution bathymetric data. In: *Seamounts: Biodiversity and Fisheries* (eds. Morato, T. and Pauly, D.), pp. 25–32. *Fisheries Centre Research Reports*, 12(5), 7–12.

Koslow, A.J. (2001) Fish stocks and benthos of seamounts. In: Managing risks to biodiversity and the environment on the high sea, including tools such as marine protected areas – scientific requirements and legal aspects. *Proceedings of the Expert Workshop Held at the International Academy for Nature Conservation Isle of Vilm*, Germany, 27 February–4 March, 2001 (eds. Thiel, H. and Koslow, J.A.), pp. 43–54.

Koslow, J.A., Bax, N.H., Bulman, C.M., Kloser, R.J., Smith, A.D.M. and Williams, A. (1997) Managing the fishdown of the Australian orange roughy resource. In: *Developing and Sustaining World Fisheries Resources: The State of Science and Management: 2nd World Fisheries Congress Proceedings* (eds. Hancock, D.A., Smith, D.C., Grant, A. and Beumer, J.P.), pp. 558–62. CSIRO Publishing, Collingwood, Victoria.

Kramer, D.L. and Chapman, M.R. (1999) Implications of fish home range size and relocation for marine reserve function. *Environmental Biology of Fishes*, 55, 65–79.

Merrett, N.R. and Haedrich, R.L. (1997) *Deep-Sea Demersal Fish and Fisheries*. Chapman and Hall, London, UK.

Morato, T., Cheung, W.W.L. and Pitcher, T.J. (2004) Vulnerability of seamount fish to fishing: fuzzy analysis of life-history attributes. In: *Seamounts: Biodiversity and Fisheries* (eds. Morato, T. and Pauly, D.), pp. 25–32. *Fisheries Centre Research Reports*, 12(5), 51–60.

Rogers, A.D. (1994) The biology of seamounts. *Advances in Marine Biology*, 30, 305–50.

Stocks, K. (2004a) Seamounts online: an online resource for data on the biodiversity of seamounts. In: *Seamounts: Biodiversity and Fisheries* (eds. Morato, T. and Pauly, D.), pp. 25–32. *Fisheries Centre Research Reports*, 12(5), 13–16.

Stocks, K. (2004b) Seamount invertebrates: composition and vulnerability to fishing. In: *Seamounts: Biodiversity and Fisheries* (eds. Morato, T. and Pauly, D.), pp. 25–32. *Fisheries Centre Research Reports*, 12(5), 17–24.

Watson, R. and Morato, T. (2004) Exploitation patterns in seamount fisheries: a preliminary analysis. In: *Seamounts: Biodiversity and Fisheries* (eds. Morato, T. and Pauly, D.), pp. 25–32. *Fisheries Centre Research Reports*, 12(5), 61–5.

Watson, R., Kitchingman, A., Gelchu, A. and Pauly, D. (2004) Mapping global fisheries: sharpening our focus. *Fish and Fisheries*, 5, 168–77.

Chapter 19
Impacts of fisheries on seamounts

Malcolm R. Clark and J. Anthony Koslow

Abstract

Seamounts are the focus of many commercial fisheries, and most of them have some impact either on the target species, associated bycatch species, or the benthic communities and habitat. Longlines, gillnets, traps, and pots can all have some effect on seafloor habitats, but bottom trawling is the best known for causing considerable impacts. In this chapter, we review the effects of fishing operations and known physical impacts on the seafloor, infauna, finfish, and epibenthic invertebrate fauna. Indirect effects of fishing, such as sediment re-suspension and mixing, and discharge of processing waste are also considered. There are few published studies on seamounts specifically, but the general effects of fishing in other benthic habitats often apply equally to seamounts. Fishing has clear consequences for the structural complexity of benthic habitats and can alter species composition, abundance, size structure, age composition, growth rates, reproductive output, and other biological parameters in the seamount ecosystem. Human-induced changes are likely to be more intense, and occur over a shorter time period, than natural events, especially in deepwater. Changes in the type of fishing gear can reduce impacts, but in many cases the most effective option for preventing excessive damage to seamount communities is protection from bottom trawling.

Introduction

Seamounts are often sites of high productivity, and the focus of important commercial fisheries based on fish and crustacean species that form aggregations in association with them (see Chapters 16 and 17, where the major target species and locations are described). Seamounts are generally regarded as fragile habitat, which can be readily impacted by heavy trawl gear (Keating *et al.*, 1987; Rogers, 1994; Probert, 1999). The effects of trawl gear have long been known (see reviews by Jones, 1992; Dayton *et al.*, 1995; Dorsey and Pederson, 1998; Jennings and Kaiser, 1998; Hall, 1999) but, along with a dramatic increase in the intensity of fishing in the twentieth century, there has been an expanding scientific literature (e.g., de Groot, 1984; Hutchings, 1990; Collie *et al.*, 1997; Auster and Langton, 1999; Hall, 1999, Koslow *et al.*, 2001) providing evidence of changes in faunal communities and habitats exposed to trawling. These studies have been carried out mainly in shallow water (e.g., Wattling and Norse, 1998; Hall, 1999; Kaiser and de Groot, 2000) and few studies have looked specifically at seamounts. In this chapter we evaluate these studies and draw widely on other research on fishing impacts, especially in deeper water, which could apply also to seamounts. The fishing activities considered are primarily those aimed at demersal/semi-demersal species; pelagic fisheries are not considered here directly (see Chapter 10).

First, we describe the types of fishing gear deployed on seamounts, mainly trawls, with some longlines, and potting/trapping for crustaceans. Summaries and reviews of the effects of fishing gear on the seafloor are given in Johnson (2002), NREFHSC (2002), and National Research Council (2002). Descriptions of the various gear types may be found in Gabriel *et al.* (2005).

Fishing gear on seamounts

Bottom otter trawl gear and midwater trawl gear used on the seafloor

Otter trawls comprise a net with a wide horizontal opening that tapers to a closed cod end; contact is made with the seafloor by the trawl doors ('Otter boards'), sweeps, bottom bridle, wing-end assemblies, and the groundrope and chain extensions. A groundrope is attached to the bottom edge of the trawl net: on rough seamount features it often consists of large rubber or steel discs, rollers, or bobbins, and can weigh several tonnes. The cod end can also drag along the seafloor. Effects on the benthos may, therefore, extend in a band the width of the trawl doors. Midwater trawls can also be fished very close to the seafloor, with chains and weights actually touching the bottom. The relative severity of effects on the seafloor will vary with different trawl designs and the way that the net is rigged for bottom contact.

Bottom longlines

Bottom longlines have a long set-line anchored to the seafloor, often furnished with thousands of hooks, weights to keep it near the bottom, and floats to hold it off the seafloor. Bottom contact is made by the anchors, weights, and the line itself, together with snoods (also called gangions; short lines carrying a hook that branch off the main longline) and hooks, which can scrape against the seafloor.

Sink/anchor gillnets

Gillnets are occasionally deployed on seamounts: they comprise a floated headline anchored at each end, the main webline netting, and a bottom weighted leadline, which can scrape the bottom. The netting is often monofilament, making it an almost invisible wall in which swimming fish become meshed. Several nets are normally strung together and can be several kilometres in length.

Pots and traps

A wide variety of designs of pots and traps are deployed for crustaceans and some fish species. They are generally round or square and can be large and heavy (up to 2 m or more across in the case of king crab pots in Alaska). Pots are weighted to keep them stable on the seafloor.

Tangle nets

Fishing for corals on seamounts has often been carried out using tangle nets/tangle net dredges. This method has several configurations, but essentially has bunches of netting tied onto stones or weights. This gear is towed slowly across the bottom, the coral branches become entangled in the mesh, and the coral base is broken off by the stone weight.

Fishing operations on seamounts

In the 1980s and 1990s there were numerous changes in the technology of fishing that have had a huge effect on the success of fisheries operating on seamounts, but have brought serious consequences in terms of impacts on seamount habitats.

Fishing gear designed to fish over rough seafloor includes 'Bobbin rigs', which became standard on larger vessels working offshore on seamounts (Fig. 19.1a); steel or rubber bobbins up to 0.6 m in diameter raise the footline clear of the bottom and reduce the chance of damage, and also enable the rig to 'roll' over rocks, boulders, and other obstructions on the seafloor. Later, 'Rockhopper' gear (Fig. 19.1b) was introduced for a similar purpose, comprising large solid rubber discs, designed to 'bounce' (not roll) over protrusions. For use on seamounts, smaller trawl nets were developed with cut-away lower wings that reduce the risk of snagging on rough and hard bottom.

Possibly the greatest influence on seamount fisheries has advances in navigation and electronic technology. The advent of global positioning systems (GPS) in the 1990s has meant that vessels can easily locate and position themselves over a seamount in the middle of the ocean with an accuracy that was hardly dreamt of a few decades ago. Established safe trawl lines can be worked repeatedly, and foul areas avoided. Computer displays update vessel position and seafloor bathymetry from depth data that are captured and plotted automatically. Echosounder technology has improved considerably, and split-beam and multibeam transducers, with scanning and net-borne sonars, have made it much easier to locate both seamount features, and aggregations of fish associated with them (Simmonds and Maclennan, 2005). However, the slopes of seamounts are often hard and rugged, and gear can readily be damaged; this problem has been addressed by the development of net monitoring systems and automatic winch controls. Electronic sensors allow the optimization of multiple gear parameters, including doorspread, wingtip distance, depth, headline height, groundrope contact, and degree of net fullness, all giving greater control and greatly enhancing the effectiveness of seamount fishing operations.

Extent of fishing on seamounts

There are extensive fisheries on seamounts and a large number of seamounts are, or have been, fished (see Chapters 16, 17, and 18). Although it is difficult to know the total number of seamounts that are fished globally, and in small areas, the effort applied to seamounts may be considerable.

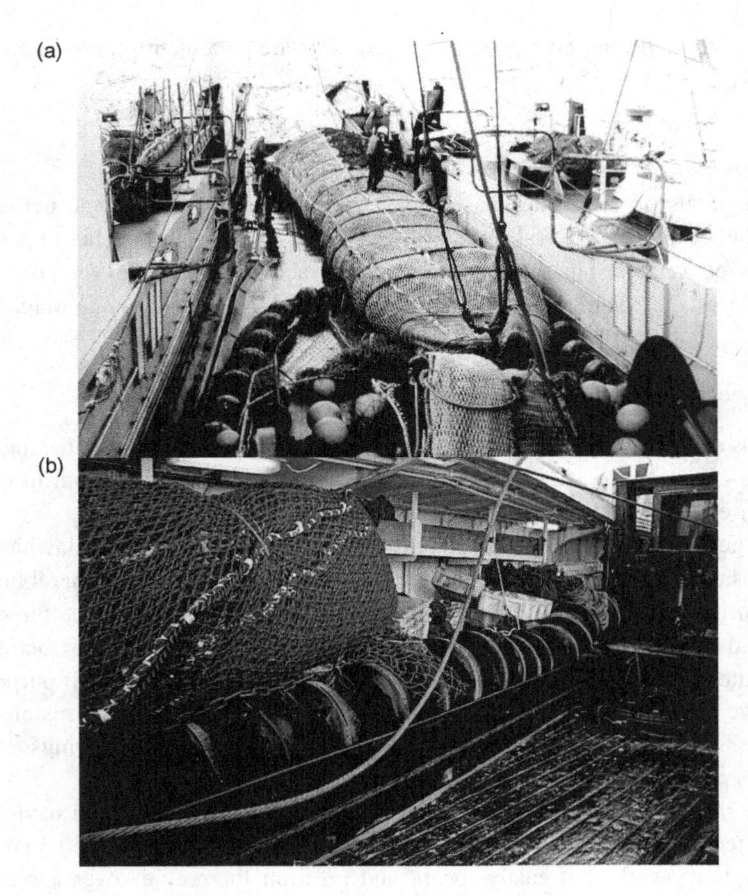

Fig. 19.1 Trawl gear groundrope commonly used in commercial fisheries on seamounts: (a) bobbin rig and (b) rockhopper (photos Malcolm Clark and Neil Bagley, NIWA).

Around New Zealand, major deepwater fisheries occur on seamounts for orange roughy, oreos, black cardinalfish (*Epigonus telescopus*), and alfonsino (Clark, 1999; Clark and O'Driscoll, 2003). Although the fisheries for orange roughy and oreos also occur extensively on the gentle contours of the continental slope, over time they have become more focused on seamounts. Orange roughy catch from seamounts increased from 20% to over 70% of New Zealand landings between the 1980s and mid-1990s (Clark and O'Driscoll, 2003). Commercial vessels actively sought seamount features (Chapter 17; Fig. 19.2), until over 80% of the known seamount features at orange roughy depths were being exploited (Clark and O'Driscoll, 2003).

Trawling intensity on individual features can be very high. Soviet fishing effort for pelagic armourhead on a relatively few seamounts in the southern Emperor-northern Hawaiian Ridge system was some 18 000 trawler days over the period 1969–1975 (Borets, 1975). Koslow *et al.* (2001) and Clark and O'Driscoll (2003) reported effort levels of hundreds to several thousand trawls carried out on small seamount features in orange roughy fisheries around Australia and New Zealand, and O'Driscoll and Clark (2005) recorded a median total tow length of trawls on seamounts of 130 km per square km of seamount

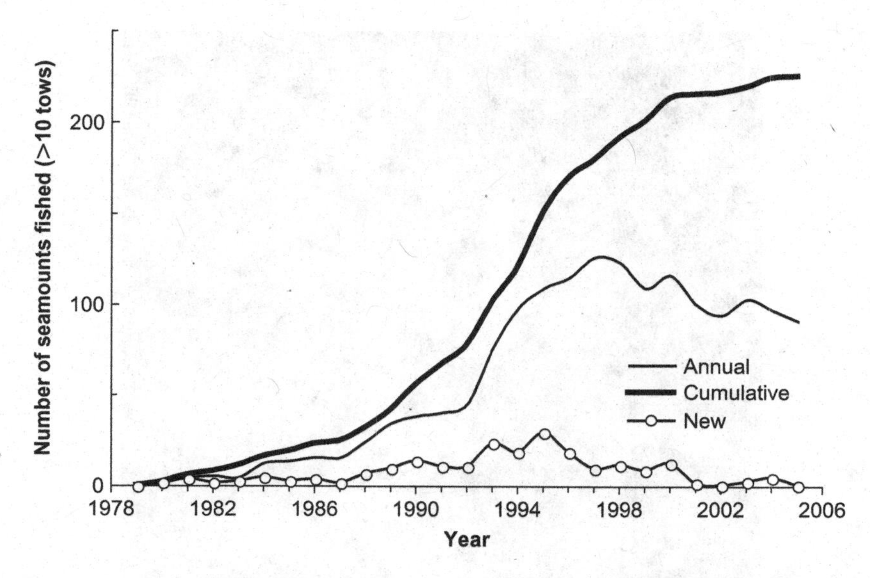

Fig. 19.2 Changes over time in the number of New Zealand seamounts being commercially fished (cumulative is the total number fished, annual is the number fished in any given year, and new is the number of previously unfished that are exploited in a year) (updated from Clark and O'Driscoll, 2003).

area. The distribution of trawl effort varies between seamounts, from locations where a few sites are routinely fished, to where more scattered trawling in all directions is evident (Clark and O'Driscoll, 2003; O'Driscoll and Clark, 2005).

Effects of fishing gear

Fishing gears that contact the seabed have many different effects on the seafloor, the benthic fauna, and associated fish stocks.

Physical impact

The physical effects on seamount habitat of fishing gear are generally the same as recorded for other habitats, and have been reviewed in a number of papers (e.g., Jones, 1992; Kaiser and de Groot, 2000; Johnson, 2002; Kaiser, 2002; National Research Council, 2002).

With bottom trawl gear, the trawl doors dig into the substrate, and create gouges and grooves (Fig. 19.3). The doors can be heavy (over 2000 kg) and are often tilted back to run on their trailing corner, which increases their impact. Towing speeds can be slow as the trawl is fished down-slope on the flanks of the feature, and occasionally doors can 'fall-over', crushing biota. Doors also scrape across the surface, but most of this action is by the sweeps, bridles, and ground gear. This can affect a distance of 100–200 m between the doors, depending on the length of sweeps. The trawl doors and sweeps/bridles together create a sediment cloud, which increases the sediment load above the bottom. This in turn reduces light levels, and can smother small sessile animals as the sediment resettles. The ploughing effect of the doors and ground gear can also 'churn' up the bottom, re-stratifying

Fig. 19.3 Trawl door grooves on soft sediment on top of a seamount (photo courtesy of NIWA).

the sediment layers near the surface and mixing surface organic material with subsurface anaerobic conditions.

Longline and gillnets can cause damage while deploying/hauling, especially if there are strong currents: lines may swing about in the current, or when fish pull on the hooks, which can drag the line across the bottom. Gillnets can also move around, especially if meshed fish try to escape (NREFHSC, 2002). Corals and sponges can be damaged when gillnet or longline gear is hauled (e.g., ICES, 2000), for example in *Lophelia* beds off Norway (Fossa *et al.*, 2000, 2002).

Pots and traps are considered less damaging than mobile gear because they are stationary (more or less) and contact a smaller area of seafloor (e.g., Eno *et al.*, 2001). In rough weather, especially if the buoy lines are relatively short, they tend to 'bounce' along the bottom, increasing the area of impact (Johnson, 2002). If they are linked by trotlines, these can further snag and damage fauna when deploying or retrieving gear.

'Ghost-fishing' can occur with most gear types (Brown *et al.*, 2005), for example when animals are trapped by lost gillnets or when pots become snagged on the bottom. Some lobster fisheries require traps to have an escape opening that becomes available when a retaining device decomposes over time, creating a large opening and allowing animals to escape in future. Some countries routinely collect lost nets (e.g., South Korea).

The physical impact of bottom gear will depend on the type, weight, and set-up of the fishing gear, but also to a large extent on the nature of the substrate and the frequency of disturbance. The extensive literature on the impact on different bottom types is described and summarized in a number of reports and papers (e.g., Dorsey and Pederson, 1998; Kaiser and de Groot, 2000; Johnson, 2002; National Research Council, 2002; NREFHSC, 2002;

Thrush and Dayton, 2002). Harder substrates (e.g., gravel, cobble, boulders) were often viewed as more susceptible to effects of fishing than softer ones (e.g., sand, mud). The degree of gouging and sediment re-suspension may be less on hard bottom, but the scraping effect can be more severe for sessile erect fauna, which may only occur on the exposed hard bottom. Stability of substrate is also a factor. Fauna on stable substrates (such as mud, gravel, coral) are more adversely affected by bottom contact than those on less consolidated coarse sediment (e.g., sand) (Collie *et al.*, 2000). Kaiser and Spencer (1996a) and Kaiser *et al.* (1998) also found the effects of trawling less in mobile sediments. Sediment stability can be affected by the removal of larger epibenthic fauna where such animals as corals, sponges, polychaetes can act to stabilize the sediment at the seafloor surface.

Substrate type can be highly variable on seamounts. Multibeam and photographic surveys on several New Zealand seamounts revealed a patchy mosaic of sediment composition (e.g. Rowden *et al.*, 2005). Koslow *et al.* (2001) found considerable variability both within and between seamounts off Tasmania. In general, the sides have thin layers of soft sediment because of the slope, and current movement, but the areas around the summit and bases can have extensive areas of soft bottom. This means that the impact of trawl gear can vary in different parts of the seamount.

Effects on infauna

In areas of seafloor with soft sediment, much of the biodiversity is to be found in the top 5–8 cm of the sediment (Giere, 1993). Gear that digs into the sediment can clearly have an impact through direct crushing of buried fauna, compacting the substrate through increased weight, or, conversely, by stirring up the sediment, dislodging animals, and later leaving a suspended sediment cloud that slowly settles.

Bottom trawl gear exerts considerable downwards pressure on the substrate due to the weight of the ground gear and net. We are not aware of any work measuring the depth of penetration of the ground gear into the superficial sediment with the type of fishing gear used on trawlers working offshore seamounts. The gear itself can vary considerably, depending on the size of the vessel, as can the physical nature of the seamount (whether hard and rough, or relatively smooth and 'trawl-friendly'). Studies on beam trawls in Europe (Fonteyne *et al.*, 1998) indicate a direct effect of 3–4 cm, although this is affected by the size of the trawl and weight of tickler chains rigged. Other studies have suggested penetration to at least 6 cm (Bergman and Hup, 1992). Otter trawls (of the type used in the *Nephrops* fishery) were observed to have less physical effect, although damage to burrow entrances may have an impact. Infaunal bivalves (e.g., *Arctica islandica*) and heart urchins (*Echinocardium* spp.) have been shown in several studies to suffer high mortality from repeated trawling (e.g., summaries in Lindeboom and de Groot, 1998; Hall, 1999). The effects may vary depending upon the depth and hydrographic conditions on the seamount.

Effects on epifauna

The effects of bottom fishing gear on epifaunal fish and invertebrates living on or near the seafloor of seamounts are much better documented than for infauna. These include major depletions of target species and populations and changes in community composition.

Target fish stocks

The history of commercial fishing operations on seamounts is generally one of 'boom and bust' (see Chapter 17). The biological characteristics of many seamount-associated fish species makes them vulnerable to heavy exploitation (see Chapter 9), and they are poor candidates for large-scale commercial fisheries (see Chapter 17).

Finfish fisheries

Pelagic armourhead (*Pseudopentaceros wheeleri*) were not well known until 1967, when Soviet trawlers discovered large aggregations on seamounts in the southern Emperor Seamount Chain, northwest of the Hawaiian Islands, where adults aggregate to spawn. They live on the seamounts permanently during the last years of their lives, making them extremely vulnerable to serial depletion, whereby the fish that live on each seamount can be heavily fished and depleted in turn. Soviet trawlers, later joined by Japanese vessels, landed between 50 000 and 200 000 t of armourhead from the seamounts each year (Borets, 1975; Sasaki, 1986). After 10 years, and over 800 000 t of cumulative catch, the fishery was all but gone (Chapter 17; Koslow, 2007): the fleet then shifted its focus to alfonsino (*Beryx splendens*) (Sasaki, 1986).

Orange roughy off New Zealand and Australia are also well-known seamount-based fisheries that have been overexploited. Fishing for orange roughy on the New Zealand continental slope dates from the mid–late-1970s, with rapid expansion to a number of fishing grounds in the 1980s (Clark, 1995). As some stocks became fully or overexploited, active exploration for new areas often involved searching for seamount features, which were found throughout the exclusive economic zone (EEZ) and beyond (Clark, 1999; see Chapter 17). The catch from seamounts peaked at over 70% of the total New Zealand orange roughy catch in 1994 (Clark and O'Driscoll, 2003), and has since been relatively stable at about 60% (O'Driscoll and Clark, 2005). Many fisheries targeting seamount stocks showed strong signs of serial depletion, moving from one seamount to another (Clark, 1999), and most stocks were driven to low levels within 10 years or less of the fishery development (Clark, 2001; Francis and Clark, 2005). Off Australia, there has been a similar picture of rapid development, several years of large catches, and then strong decreases as stocks have been depleted. The main Australian stock around Tasmania was fished from the mid-1980s, but catches increased rapidly with the discovery of spawning aggregations in 1989. The fishery was almost unregulated for a short time, and catches exceeded 40 000 t in 1990 from the one stock. Management appeared to be working for a while with catch limits matching catch (Koslow *et al.*, 1997), but by the late 1990s the stock was estimated to be as low as 10% of its original size (although estimates varied depending upon assumptions; Bax, 2000; Koslow, 2007). This story has been repeated with a seamount ('Hot Spot') stock off Namibia (Branch, 2001).

Two species of oreosomatids are also fished around New Zealand on the continental slope and seamounts. Black oreo (*Allocyttus niger*) and smooth oreo (*Pseudocyttus maculatus*) are both targeted and taken as bycatch in fishing for orange roughy, and average annual landings have been about 20 000 t. Around 50% of the New Zealand catch of both species combined is taken from seamount features (Clark and O'Driscoll, 2003). Commercial catch per unit effort in the main fishery area containing seamounts on the Chatham

Rise decreased for both species during the 1980s, and through the 1990s for black oreo, but not for smooth oreo which has stabilized (Sullivan *et al.*, 2005). Although the overall stock on the Chatham Rise does not appear to have been overexploited, several smaller stocks on seamounts elsewhere (e.g., Puysegur Bank) have seen strong declines in catch rate to very low levels (McMillan *et al.*, 2002).

Seamount fisheries may last even shorter periods of time than experienced with pelagic armourhead and orange roughy stocks. Catch rates of rockfish (*Sebastes* spp.) off Cobb Seamount were reported to decline rapidly to 25% after 1 year, and fish size also decreased (Sasaki, 1986). Pink maomao (sea bass, *Caprodon longimanus*) also showed rapid declines in stock size on seamounts in the northern Tasman Sea. Japanese analyses of catch and effort data suggested that fishing by one trawler over a 6-week period, with catches of 1000 t, may have depleted the stock to one-sixth its original size (Sasaki, 1986). Soviet fishing for alfonsino on the Corner Rise seamounts in 1976 is also thought to have depleted the stock substantially, and catches the following year were poor (Vinnichenko, 1997; see Chapter 17).

Small-scale artisanal fisheries on seamounts, however, can be sustainable (see Chapter 16). Fisheries in the South Pacific for oilfish (*Ruvettus pretiosus*) and in the North Atlantic around Madeira for the black scabbardfish (*Aphanopus carbo*) were carried out by handlonglining for centuries, although are now changing with the introduction of modern fishing methods (see accounts in Merrett and Haedrich, 1997; Chapter 16).

Invertebrate fisheries

Similar problems occur with invertebrate fisheries. Rock lobster (*Jasus tristani*) was discovered on Vema Seamount off the west coast of southern Africa in the early 1960s, and a fishery rapidly developed (Simpson and Heydorn, 1965). The stock was depleted in the mid-1960s, and although research showed no recovery by 1979 (Lutjeharms and Heydorn, 1981a), a commercial fishery did re-establish briefly in 1980, but the stock was once again quickly overexploited (Lutjeharms and Heydorn, 1981b).

Coral fisheries have shown a similar boom-and-bust pattern. The best known example from seamounts is the extraction of red coral (*Corallium secundum*) from the Emperor Seamount Chain in the western Pacific. Japanese fishers discovered a large bed of this coral at about 400 m in 1965, and together with Taiwanese vessels, the fishery expanded to a peak of 150 t in 1969 (Grigg, 1993). Most vessels used various types of tangle net dredge (see above), although in shallow water SCUBA is used, and off Hawaii, submersibles selectively harvest coral. Catches fell until 1978 when a new species of *Corallium* was found in abundance at depths of 900–1500 m on the Emperor seamounts. Over 100 coral boats were involved in this fishery, with catches of almost 300 t in 1981. Fisheries have occurred on other seamounts in the area, such as Hancock seamount, but catches have generally been small. Exploration, discovery, exploitation, and depletion characterize these fisheries (Grigg, 1989). Most of the species involved are slow growing, and have low rates of recruitment which makes them vulnerable to overfishing. Exploration still continues, as shown by surveys of Cross Seamount (Grigg, 2002), but coral exploitation is now more carefully managed.

King crab (*Paralithodes camtschaticus*, *P. platypus*, *Lithodes aequispinus*, *L. couesi*), snow crab (*Chionoecetes opilio*), and tanner crabs (*C. bairdi*, *C. tanneri*) occur on seamounts in the Gulf of Alaska (Hughes, 1981), although the fisheries operate mainly on the continental shelf and slope, or on the slope around offshore islands. The resources, and their fisheries, fluctuate widely. There were two major collapses of the king crab fishery, off Kodiak in 1969, and in Bristol Bay in 1982, apparently due to a combination of heavy fishing (Dew and McConnaughey, 2005) and changing environmental conditions (Witherell and Harrington, 1996). As with other seamount fishes, crab stocks appear vulnerable to localized depletion.

It is evident from these examples that seamount fisheries pose some major management problems. The formation of aggregations, either permanent, or seasonally associated with spawning or feeding, makes several deepwater species commercially attractive, yet most of them have low productivity and hence low sustainable yields (see Chapter 9). Overexploitation is the norm, even where there have been active management regimes and research activities.

Non-target fish bycatch

A large number and variety of fish and invertebrate species may be caught in bottom trawling operations on seamounts (see Chapter 9; Parin *et al.*, 1997; Koslow and Gowlett-Holmes, 1998; Grandperrin *et al.*, 1999; Fock *et al.*, 2002; Clark and Roberts 2003; Moore *et al.*, 2003; Tracey *et al.*, 2004). Fluctuations and shifts in bycatch composition over time with heavy fishing have been shown in a number of major fishing areas, such as Georges Bank and the North Sea, although findings have varied for different fish communities (e.g., Greenstreet and Rogers, 2000). Trawling on seamounts, as elsewhere, results in a decline of all associated species, although over time some species may show an increase. For example, off New Zealand, rattails and dogfish have increased in some orange roughy fisheries, although the reasons are unknown (Clark *et al.*, 2000b; Livingston *et al.*, 2003).

Although seamount communities can be composed of many species, seamount trawl fisheries can be relatively 'clean' in terms of bycatch when they target aggregations. Using observer data, Clark *et al.* (2000a) and Anderson *et al.* (2001) found the bycatch of non-target fish species to be 5–10% of catch in deepwater fleets fishing for orange roughy and oreos. Similarly, about 5% was recorded in an orange roughy fishery on seamounts south of Tasmania (Anderson and Clark, 2003), where oreos (Oreosomatidae), rattails (Macrouridae), and dogfishes (Squalidae) were the main bycatch. The bycatch of sharks in seamount trawl fisheries can be much less than in longline fisheries (e.g., Gordon, 1999).

Benthic invertebrates

Bottom trawling and dredging commonly reduce habitat complexity and invertebrate biodiversity (e.g., Sainsbury, 1987; Auster *et al.*, 1996; Collie *et al.*, 1997; Cryer *et al.*, 2002), where benthic invertebrates comprise unwanted bycatch (e.g., Jennings and Kaiser, 1998; Hall, 1999; Kaiser and de Groot, 2000; Johnson, 2002). Stocks (2004) noted a high proportion of 'emergent' sessile upright epifauna extending above the seafloor on seamounts, such as corals, sponges, anemones, and crinoids. These animals are more vulnerable than organisms found on the softer sediments that characterize the general expanses of shelf, slope, and abyssal plain.

Corals, sponges, and other sessile suspension feeders often dominate benthic communities on continental margins, canyons, mounds, and other environments where currents are accelerated (Freiwald *et al.*, 2004). The impacts of fishing activities, mostly trawling but also longlining and gillnetting, have now been documented on a wide range of such communities. Between 30% and 50% of the *Lophelia* reefs off the coast of Norway, the most extensive in the world, have been partially or totally damaged by trawling (Fossa *et al.*, 2002). Impacts on *Lophelia* reefs have also been observed from trawl grounds to the west of Scotland (Roberts *et al.*, 2000), off western Ireland (Hall-Spencer *et al.*, 2002), and in the northern Rockall Trough (ICES, 2002, 2003) (see Chapter 9). In the Northwest Atlantic, a video survey of the Northeast Channel between Georges and Browns Banks recorded trawl damage to gorgonian corals (*Paragorgia arborea*, *Primnoa resedaeformis*, and *Acanthogorgia armata*) on 29% of transects (Mortensen *et al.*, 2005). *Oculina* reefs off Florida have been extensively damaged by trawling, even after designation as protected areas (Reed *et al.*, 2005). In the North Pacific, substantial gorgonian coral and sponge bycatch is recorded from trawling within the coral and sponge 'garden' habitat along the Aleutian Islands (Shester and Ayers, 2005) and on unique hexactinellid sponge reefs in British Columbia (Ardron, 2005). Many corals and sponges are long lived and slow growing, meaning their recovery from trawling is slow (see Chapter 8; Wilson, 1979; Hall-Spencer *et al.*, 2002; Tracey *et al.*, 2003).

The impacts of trawling on the benthic fauna of seamounts have been studied most intensively off Australia and New Zealand, where large deepwater fisheries for species such as orange roughy have raised concerns about the effects of bottom trawling. Koslow and Gowlett-Holmes (1998) and Koslow *et al.* (2001) used a towed camera sled and epibenthic sled to survey the benthic macrofauna of a group of small seamounts south of Tasmania, Australia that were being, or had been, extensively trawled in the orange roughy fishery. The seamounts are relatively small, with elevations of 300–600 m, at depths from 660 to 1700 m. There were strong differences in faunal composition and distribution on fished and unfished seamounts. Corals, especially the reef-building *Solenosmilia variabilis*, dominated the unfished or lightly fished seamounts at moderate depths, urchins the deeper seamounts, while shallower heavily fished sites were relatively depauperate. Fished seamounts typically had half the species and hosted a 7-fold lower biomass of benthic invertebrate species (Table 19.1). Although the depth distributions of fished and unfished seamounts were not identical, the contrast in faunal community composition was so strong that trawling was deemed responsible, in particular by stripping coral cover from the seamounts.

A similar 'compare and contrast' study was carried out on the Chatham Rise off New Zealand in 2001 (Clark and O'Driscoll, 2003; Rowden *et al.*, 2004), an area heavily trawled for orange roughy since the early 1990s. Half of the eight study seamounts (peaks 750–1000 m) were considered 'unfished' (less than 10 trawls) and the other 'fished' (40–1500 trawls). Benthic macroinvertebrates at each seamount were sampled using towed camera and epibenthic sleds, and the presence of habitat-forming fauna (e.g., live corals), substrate type, and indications of trawling (e.g., trawl door marks) were noted. Fisheries catch-effort data were examined to determine the amount and distribution of bottom trawling effort. Results showed the benthic communities of 'fished' and 'unfished' seamounts were significantly different; 'unfished' seamounts possessed a relatively large

Table 19.1 Comparison of mean sample weight and number of species from lightly and heavily fished seamounts off Tasmania (from Koslow et al., 2001).

	Lightly fished (n = 11)	Heavily fished (n = 11)
Biomass (kg)	7.0 (±5.8)	1.1 (±3.4)
Number of species	20.1 (±3.6)	8.7 (±6.3)

amount of live coral habitat comprising *S. variabilis* and *Madrepora oculata* (predominantly on the peaks), whilst 'fished' seamounts possessed relatively little coral habitat (Figs. 19.4 and 19.5) and bore six times more physical indications of trawling, in direct relation to the amount of effort on individual seamounts (Clark and O'Driscoll, 2003). Multivariate analyses revealed that the variability observed in macroinvertebrate assemblage composition between the study seamounts could in part be explained by the relative occurrence of live coral and coral rubble, both structural habitats, although the differences in species diversity between fished and unfished seamounts were not as marked as found by Koslow *et al.* (2001).

Probert *et al.* (1997) examined the invertebrate bycatch on small seamounts of the northeastern Chatham Rise during a research survey using commercial trawl gear. They

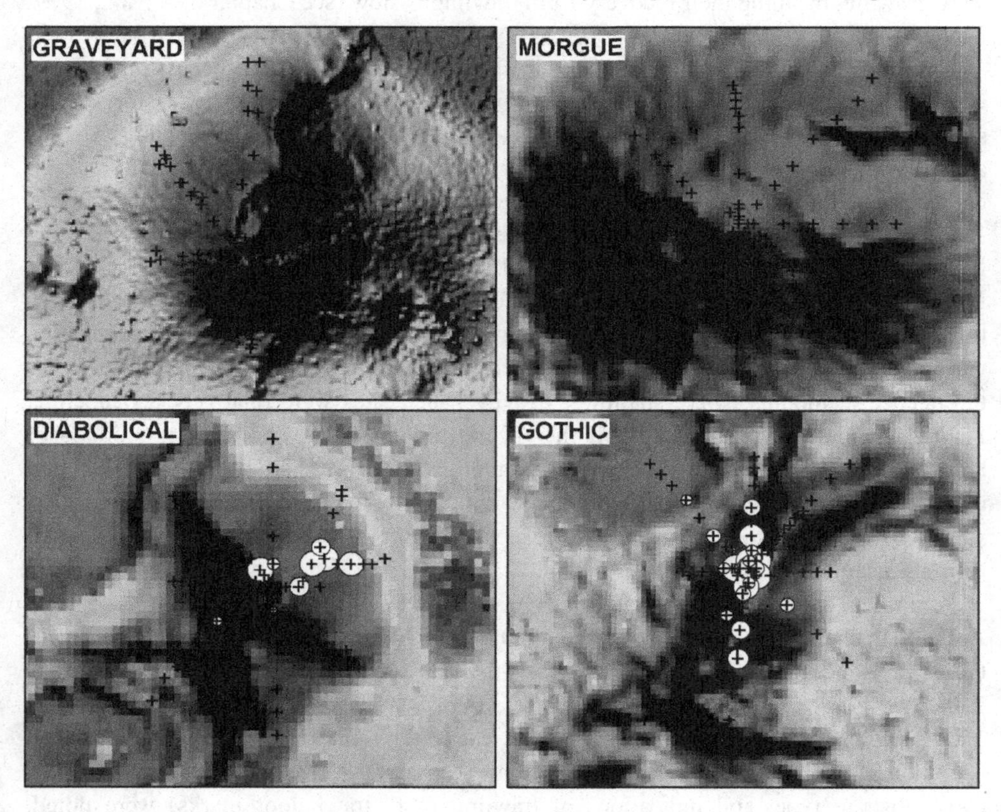

Fig. 19.4 Distribution of live coral cover on four seamounts off the east coast of New Zealand. The upper panels are fished seamounts, the lower panels are unfished. The crosses indicate the position of camera stations, and the expanding symbols show the relative amount of seafloor in the image that is covered by coral (based on Clark and O'Driscoll, 2003).

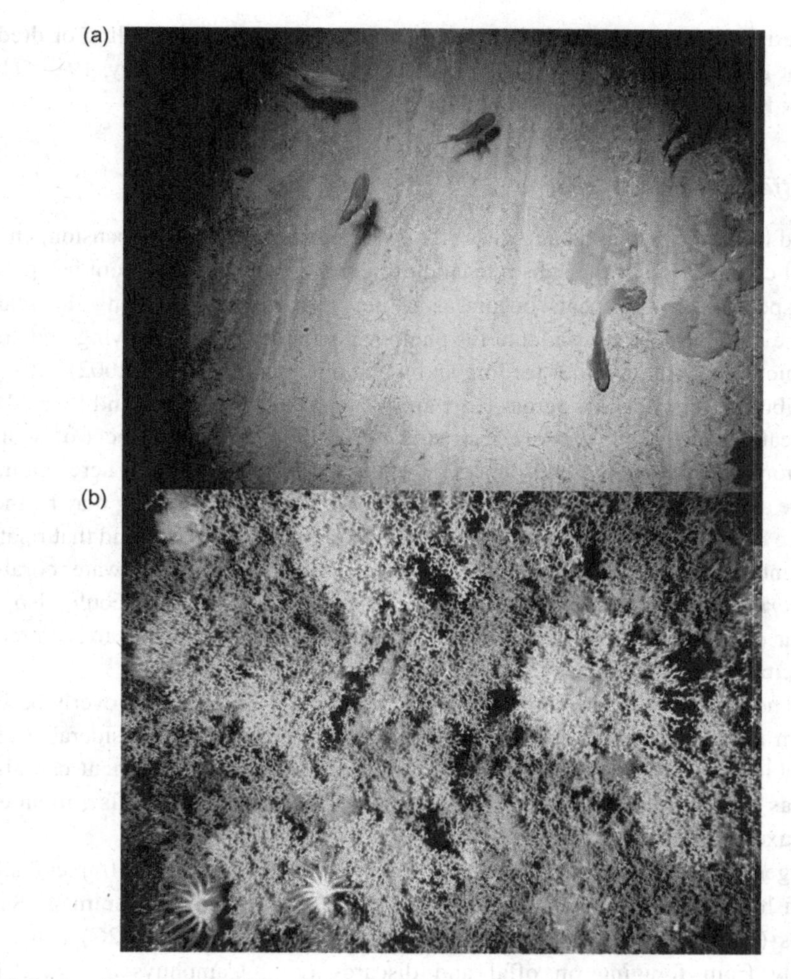

Fig. 19.5 Seafloor images from a (a) heavily fished seamount and (b) lightly fished seamount off the east coast of New Zealand (photos courtesy of NIWA).

found differences in species composition and abundance between the seamounts and adjacent slope (soft sediment), and between two seamount groups. However, no specific results could be attributed to fishing.

Anderson and Clark (2003) analysed observer data from vessels in the early years of a new orange roughy fishery on seamounts of the South Tasman Rise. In the first year of fishing an estimated 1750 t of coral was taken in an orange roughy catch of 3900 t. The amounts of coral bycatch subsequently declined, thought to be a combination of a decrease in abundance as it was removed from the seafloor and a response by vessel skippers to avoid areas of coral abundance due to net damage.

The Australian and New Zealand studies described above have inferred that differences in benthic invertebrate communities were caused by fishing. Comparing fished with unfished areas of similar habitat characteristics is a common approach in impact studies (e.g., Riemann and Hoffman, 1991; Collie *et al.*, 1997), but results are less definitive than comparing benthic communities before-and-after experimental trawling or dredging in areas not commercially fished (e.g., van Dolah *et al.*, 1987; Sainsbury, 1988; Thrush *et al.*, 1995; Kaiser and Spencer, 1996b; Sainsbury *et al.*, 1997).

Indirect effects of fishing gear

Indirect and less well-known fishing gear effects include sediment re-suspension, changes in chemical composition of the substrate, and offal and bycatch discard from fish processing. Re-suspension of sediments occurs as fishing gear is dragged along the seafloor, reducing the amount of light available for photosynthetic organisms, burying and smothering benthic biota, and affecting feeding and metabolic rates (Johnson, 2002). Alteration and redistribution of sediments across large areas can occur (e.g., Black and Parry, 1999), with implications for fauna that are restricted to a certain habitat. Effects of sediment re-suspension are site specific and depend strongly on substrate type. Where seamounts are offshore and deep, storm stresses may be relatively weak, and trawling may be the primary source of suspended sediment (Churchill, 1989). Rogers (1999) found that relatively small amounts of sediment settling on the bottom can smother small cold water corals like *Lophelia pertusa* and prevent expansion or recovery of the colony. This could also occur with similar coral genera such as *Solenosmilia* and *Goniocorella*, which are common on southern hemisphere seamounts (Cairns, 1995; Freiwald *et al.*, 2004).

There is no information for seamounts on the mixing of sediments and overlying water, but this can alter their chemical makeup of the sediment and have considerable effects in deep stable waters (Rumohr, 1998). Chemical release from the sediment can also be enhanced, as shown for phosphate in the North Sea, although this was offset to an extent by lower fluxes following trawling (ICES, 1992).

Discharged processing waste from factory trawlers (e.g., heads, guts, frames), as well as whole fish that are lost or discarded at the surface, can affect other animals. Seabird populations (which can be associated with seamounts; see Chapter 12C) can benefit substantially from foraging on offal and discards (e.g., Camphuysen *et al.*, 1993; Furness, 1996): seabirds consumed about 50% of discards in the North Sea (Camphuysen *et al.*, 1993). The remainder sinks to the seabed where it becomes available to midwater and benthic predators and scavengers. The attraction of both seabirds and marine mammals (in particular seals) has become an important issue worldwide. Although they may benefit from increased food, they are also at risk from lines, hooks, and wires, and can drown when trapped in trawls or taking bait or catch off longline hooks. For example, sperm whales take blackcod (*Anoploma fimbria*) off longlines on Canada's west coast (Anon, 2006), and killer whales and colossal squid strip toothfish off lines in the Antarctic (Oliver, 2003; Kock *et al.*, 2005). This predator attraction is an on-going problem for responsible fisheries, leading to devices such as bird-scarers, seal-exclusion grids, and regulations on offal discharge. Such discharge can reach the seafloor at considerable depths. Offal from a New Zealand fishery for hoki (*Macruronus novaezelandiae*) could reduce oxygen levels at 800 m depth (Livingston and Rutherford, 1988) and alter community composition (Grange, 1993). Hence, if factory trawlers remain over a seamount while processing, there could be such an effect depending upon depth and current flows affecting the descent and scattering of the offal.

Effects on seamount community structure

Dramatic shifts in fish community composition with heavy fishing are commonly reported (e.g., Hempel, 1978; Daan, 1980; Grosslein *et al.*, 1980; Gulland and Garcia, 1984; Pauly, 1988; Fogarty and Murawski, 1998), and shallow-water studies have shown that intensive trawling may alter benthos population and community structure for long-time periods (e.g., Rumohr and Krost, 1991; Hall *et al.*, 1994). Seamount fisheries may reasonably be expected to cause changes in structural complexity, biodiversity, endemism, size structure, trophic relationships, and genetic composition.

Structural complexity

Many long-lived epifaunal animals perform a structural role within benthic communities, providing a microhabitat for other species (e.g., Nalesso *et al.*, 1995). The loss of such structures can affect survivorship and re-colonization of associated fauna and has prompted analogies with clear-cutting of forests on land (e.g., Wattling and Norse, 1998; Wattling, 2005). Probert *et al.* (1997) observed that corals were relatively abundant on hill features of the Chatham Rise compared with surrounding 'flat' slopes, and proposed they had an important role in increasing the complexity of benthic habitat: trawling damage would consequently depress local biodiversity. Both Australian (Koslow *et al.*, 2001) and New Zealand (Clark and O'Driscoll, 2003; Rowden *et al.*, 2004) studies have attributed differences in species number and composition between fished and unfished features to reduction of large structure-forming corals.

Colonies of cold water corals like *L. pertusa* and *Goniocorella dumosa* provide niches for many animals including fishes, other scleractinians, stylasterids, bryozoans, stoloniferans, sponges, polychaetes, ophiuroids, asteroids, bivalves, gastropods, crabs, and anemones (e.g., Cairns, 1995; Fossa *et al.*, 2000; Krieger and Wing, 2002; Freiwald *et al.*, 2004): Rogers (1999) recorded over 850 species found in association with beds of *L. pertusa*.

Disturbance can cause dominant species to alter from large sessile types (e.g., corals, hydroids, sponges) to small, fast growing, opportunistic colonizing species, scavengers, and juveniles: on Chatham Rise fished and unfished seamounts were distinguished in this way (Rowden *et al.*, 2004) and by differences in the relative frequency and abundance of scavenging gastropods, which may benefit from regular bottom disturbance. Clark and Hill (2001) reported higher abundance of pagurid crabs on trawled sections of a small seamount off the east coast of New Zealand.

Fishing activities provide two main sources of food for benthic scavengers: first from discards, bycatch, and offal that sink to the seafloor; and secondly from towed gear that digs up, damages, or displaces benthic fauna in the path of the gear. Scavengers of bathyal communities are likely to include lysianassid amphipods, isopods, decapod crustaceans, polychaetes, and various fish (Britton and Morton, 1994). They provide pulses of organic matter to localized areas of the seabed, and scavengers can move considerable distances to take advantage of this important source of food in the deep sea (Dayton and Hessler, 1972). A number of studies have demonstrated an increase in fish density following trawling (e.g., Hall *et al.*, 1993; Kaiser and Spencer, 1994, 1996b; Demestre *et al.*, 2000), although the duration of a response can be short lived (Demestre *et al.*, 2000), affected by the extent of continued trawling (Ramsay *et al.*, 2000), and it can be quite different for similar species (Ramsay *et al.*, 1996). Switches in diet can accompany increased feeding on trawled grounds (e.g., Hughes and Croy, 1993; Kaiser and Spencer, 1994).

A reduction in the age composition and size structure of species may occur with fishing. Trawl gear can be size selective, and larger, older fish are often taken more than smaller younger fish. The result is a reduction in the size and age spectra of exploited populations (Royce, 1972; Haedrich, 1995; King, 1995; Rice and Gislason, 1996; Large *et al.*, 1998). Community structure can change to dominance by species or size classes that turn over faster and consume smaller prey (Merrett and Haedrich, 1997; Christensen, 1998; Pauly *et al.*, 1998). Increased growth rates have been observed with some species (e.g., sole; Millner and Whiting, 1996). Reproductive output can be reduced, with lower fecundity and viability of eggs (Berkeley *et al.*, 2004). Trophic relationships also change with a shift in community structure, as predator–prey balances change (Fogarty and Murawski, 1998; Jennings *et al.*, 2001). An increase in piscivorous predators in the Georges Bank ecosystem is thought to impede recovery of depleted groundfish (Fogarty and Murawski, 1998).

There do not appear to be any empirical studies on the impact of deepwater fisheries on predator or prey populations of the target species. However, changes in the fish biomass aggregated over seamounts likely affect local prey abundance (see Chapter 15; Koslow *et al.*, 2000). Species such as pelagic armourhead and orange roughy require production from an area 10 times larger than their home range (Tseitlin, 1985; Koslow, 1997), so that reduction of their biomass may increase prey abundance (Clark *et al.*, 2000b).

The balance of effort between various gear types can affect changes in community composition. Philippart (1998) found that otter trawlers in the southeastern North Sea caught relatively more fish than invertebrates, while beam trawlers caught proportionately more invertebrates.

The selectivity of fishing can also lead to long-term changes in community composition and characteristics. Body size and shape, age composition, maturation, reproductive rates, growth rates, and behavioural traits all come under selective pressure through exploitation (Law, 2000; Law and Stokes, 2005). Genetic changes have been thought likely, but difficult to prove (Nelson and Soule, 1987) although evidence is now increasing that loss of genetic diversity can be caused by fishing (Hauser *et al.*, 2002; Hutchinson *et al.*, 2003; Law and Stocks, 2005). Smith *et al.* (1991) reported a loss of genetic diversity in several New Zealand orange roughy populations and attributed this to fishing pressure and a heavy reduction in the biomass of the fish.

Substrate type and gear design are two important elements determining the nature and extent of bottom trawling impacts on benthic fauna. On hard bottom, results from single tows have been mixed. Tilmant (1979) reported very high damage to stony corals caused by shrimp trawling (80% crushed or torn loose), sponges (50%), and soft corals (38%). In contrast, Van Dolah *et al.* (1987) noted damage to individuals of most species immediately after trawling, but only one species (a barrel sponge) showed a marked decrease in density. Research off Alaska by Freese *et al.* (1999) showed a 40–90% decrease in density for sponges and anthozoans after trawling. In work off eastern Australia 5–20% of the fauna was removed with each trawl, and after 13 tows 70–90% of the initial benthos was gone (Burridge *et al.*, 2003). These studies together indicate that, depending on the ground gear used, a single pass with a trawl may not have a devastating effect, but repeated trawling can strip areas severely.

Endemism

Following the initial work of Wilson and Kaufman (1987), high levels of endemism have been recorded on some seamounts (endemism is here defined as species found on only one seamount, or a seamount chain; see Chapters 7 and 13) with values ranging from less than 5% to over 50%. It is difficult to generalize, given that in many areas sampling has been inadequate to capture many macrofaunal groups, the number of seamounts sampled may have been small, and there was little or no sampling of the area of surrounding slope (see Chapter 7). Nevertheless, if endemism is a significant feature of seamounts, localized intensive fishing pressure can have a severe impact on species adapted to an isolated structure. These species may have limited ability to recruit eggs/larvae over large distances, and their growth and reproductive rates might also be low.

'Relict' faunas – lineages previously thought long extinct – have also been discovered on seamounts (see Chapters 7, 8, and 13; Kelly and Rowden, 2001; Schlacher *et al.*, 2003). In the same way as endemic species can be vulnerable to fishing pressure because they are so localized, these relict faunas can also be significantly affected by fishing activities.

Impacts of fishing gear on the benthic substrate and sessile fauna are readily appreciated, but longer term effects on ecosystem productivity are difficult to evaluate. Jennings *et al.* (2001) estimated a 75% reduction in overall infaunal productivity on fished grounds compared with unfished, even though production per unit biomass was greater in the fished areas because of the shift to smaller organisms. Seamounts are known to be sites of fish spawning, but many do not appear to be nursery grounds for small fish. Adult orange roughy, for example, still return to the seamounts for spawning during winter months, and so habitat changes have not obviously affected adult behaviour. However, nothing is known of benthic–pelagic interactions and linkages, and there may be longer term changes. Tuck *et al.* (1998) undertook a 3-year study documenting changes between fished and unfished benthic infauna. Initially they observed only changes in the relative abundance of species, and it took over a year for total species richness to start to decline in the trawled areas.

Natural vs fishing disturbance

Natural physical disturbances can influence benthic fauna; these include storms, hurricanes, landslides, ice scour, low-salinity excursions, and large temperature changes (Witman, 1998). Storms increase wave-induced water motion and can dislodge organisms when the hydrodynamic forces exceed the attachment strength of animals such as sponges, corals, echinoderms, mussels, and ascidians. Severe storms can literally scour 10s of square metres, but are rare events. Generally these effects are uncommon at depths greater than 30m (Riedl, 1971). Monitoring of sites in the Gulf of Maine over a decade showed few effects of storms at these depths (Witman, 1998). Most offshore seamounts that are fished commercially have summit depths greater than 50m, and so natural effects of weather are unlikely to be significant. In areas of high latitude, seamounts may be affected by iceberg scour, which can be a major natural occurrence to depths of 500m (Barnes and Lien, 1988).

Natural changes in community composition or density can also occur through biological disturbance. Predation by sea stars, sea urchins, nudibranchs, and other larger mobile animals can strip patches of mussels, algae, corals, and sponges (Witman, 1998), but little is known about such impacts on seamount fauna. Fish aggregations can have a localized effect (e.g., cod on polychaetes and amphipods; Witman and Cooper, 1983), but most seamount fishes are not benthic predators (see Chapters 9 and 14).

The community structure of many habitats reflects the occurrence of natural 'intermediate events' such as large storms that occur infrequently. Stable habitats, which are not much affected, tend to have climax communities with successful competitors dominating the fauna (Field, 2005), a situation that appears appropriate for deep offshore seamounts in the absence of fishing. The effects of external physical disturbance may be greater in these environments where communities are not adapted to frequent natural perturbations (Rogers *et al.*, 1998).

Active volcanism can have a substantial impact, with the destruction of hydrothermal vent communities documented on the East Pacific Rise at 9° N (Haymon *et al.*, 1993). Unlike sites impacted by terrestrial volcanism, sites of submarine volcanism and venting are re-colonized within a few years by vent organisms, such as tubeworms, as observed at the 9° N site and at Axial Seamount on the Juan de Fuca Ridge (van Dover, 2000). The impact of volcanoes and the recovery of more typical coral-dominated seamount faunas are not known. In the New Zealand region, Rumble III and Monowai seamounts have recently undergone eruptions, with strong physical effects on the shape and structure of the seamount. The absence of before-and-after surveys means there is no record of changes in faunal composition or distribution, but a large sector of Monowai Seamount completely slumped, which must have buried or displaced most of the fauna living on the flanks of the volcano. The tsunami on Boxing Day 2004 in the Indian Ocean also likely had considerable effects at depth, although recovery from an earlier tsunami in the region was reported to be quite rapid (BCN, 1999).

Climate change, with the global effects of sea-level rise making the seafloor relatively deeper, an increase in water temperatures, and the slow acidification of the oceans, will undoubtedly have a slow impact on seamount communities over future decades. However, for the same reasons that fish aggregate in a localized area, so too do fishers concentrate their effort in finding and fishing seamounts leading to higher levels of disturbance from fishing than from any natural cause.

Methods of reducing impacts

The challenge of reducing the impact of fishing gear is not specific to seamounts as it is a major concern around the world in most habitats. Seamounts are especially problematic because of sessile fauna like corals and the low productivity of many deepwater seamount species. Impacts can be reduced through modification to fishing gear and by the application of management measures (see Chapters 14 and 20). Bottom trawls are the principal cause of damage. Various technical modifications can reduce bycatch, including changes in mesh size, the use of separator panels, sorting grids, various types of mesh, and escape openings (Van Marlen, 2000; Kennelly and Broadhurst, 2002). Such changes have been proven effective in reducing bycatch/non-target catch in many fisheries.

Reducing the impact on benthic animals, sessile invertebrates in particular, is more difficult. Corals and sponges are fragile structures; video footage from New Zealand shows coral being easily damaged by a relatively light towed camera frame. Replace that with a bottom trawl rig weighing several tonnes and there is little question of the damage that can be done. Changing gear type may be an option in some fisheries. If the shape and nature of the seamount allows, then use of midwater gear close to the bottom may be effective in catching some species such as alfonsino or armourhead. However, experience has shown that this approach does not work with orange roughy. In addition, midwater trawling on steep-sided seamounts, or where the bottom is rugged and uneven, causes expensive net damage. Replacement of trawling with a line fishery does not work for orange roughy, although it could be a management option for some seamount fisheries.

If gear type cannot be changed, then the groundrope could be lightened, trawl doors balanced (weight distribution less concentrated on the lower part of the door) so they do not dig in as much, and sweeps shortened so the trawl covers less area. The use of catch sensors when towing into aggregations can also help reduce tow times, as hauling can begin as soon as the catch is large enough, rather than towing for longer. Design of the ground gear should also be investigated. In some deepwater fisheries (e.g., orange roughy) there has tended to be a shift from bobbin rigs to rockhopper gear, which is generally lighter being designed to drag and bounce, rather than roll. Its impacts may differ, but may not necessarily be reduced. The likely application of such methods will vary with target species and seamount, and operationally may entail a trade-off between reduced bottom impact and catch rate of the fish species.

Closure of seamounts and other cold water coral habitat to fishing has proved the most effective means to reduce impacts. Marine protected areas and other forms of closure have now been instituted off Norway and for the Darwin Mounds in the Northeast Atlantic, with similar protection being sought for seamounts off the Azores and Madeira, as well as Bowie Seamount off Canada's west coast (see Chapters 16 and 17). Some NE Atlantic gorgonian coral and *Lophelia* grounds have been closed to fishing in the Canadian Atlantic, and *Oculina* reefs off Florida are similarly protected, although the efficacy of this closure has proven limited. Some seamounts off southern Tasmania and around New Zealand are now protected, and the trawling 'footprint' has been frozen in areas of coral and sponge habitat around the Aleutians in the North Pacific (Freiwald *et al.*, 2004). Recent area closures in the North Atlantic (NAFO) and North Pacific (USA-Alaska) include seamounts. Trawling is banned within 100 nautical miles of the Azores, which serves to protect seamounts, important for the local artisanal fisheries, from these impacts (see Chapter 16).

Discussion

The effects of fishing are similar across many habitat types but, despite a paucity of seamount-specific studies, there are five important aspects of seamounts and their communities:

1. Seamounts can host aggregations of fish, which can be vulnerable to overexploitation, rapid depletion, and be slow to recover from fishing.
2. Target trawling on seamounts is often very localized. Seamounts are often small in area, and therefore the density of tows per seamount area can be high.
3. Heavy bottom trawl gear is used to tow on the rough and hard bottom which is often characteristic of seamounts.

4. Many seamount fisheries occur offshore, and so are carried out by large powerful vessels with the ability to work large gear, catch, and process large amounts of fish, and stay at sea for long periods, leading to high levels of fishing effort.
5. The invertebrate fauna is often dominated by large, slow-growing, sessile organisms, which are especially vulnerable to damage by fishing gear.

There is little doubt about the physical impact that fishing has had on seamount communities and habitat. However, although medium and long-term changes have been described in several ecosystems, which have been studied over decades, it is still difficult to attribute changes directly to fishing (Rogers *et al.*, 1998), partly because of the poor spatial resolution of much fisheries data (Jennings and Reynolds, 2000), which makes the 'burden of proof' an issue (Dayton, 1998). Knowledge of biodiversity, endemism, structure, function, and productivity of seamount ecosystems is poor. In most cases science is still in a descriptive phase, rather than a process–function stage, and so the effects of disruption are difficult to measure, despite being critical to assessing risk and conservation priorities. In particular, almost nothing is yet known of recovery processes on seamounts impacted by fishing.

The productivity of seamounts is generally regarded as high (but see Chapters 5, 9, and 14), but the fisheries have typically overexploited them. Conservative management practices are needed for many species, especially as most deep sea fisheries have only developed in the last few decades, and have occurred against a backdrop of limited, if any, knowledge of long-term natural fluctuations in species abundance. Given how little we know about the composition and functioning of benthic communities, and about linkages between fisheries and benthic fauna, there is a clear and strong message to adopt a conservative approach, and manage seamounts as fragile ecosystems with some form of precautionary and spatial management. Gear modifications cannot significantly ameliorate trawling impacts, and area closures to trawling at present provide the most effective means to conserve seamount ecosystems.

Acknowledgements

Some unpublished data on New Zealand seamount fisheries and impacts were made available from research funded by the Foundation for Research, Science and Technology (contract CO1X0508) and Ministry of Fisheries (project number ZBD2000/04).

References

Anderson, O.F. and Clark, M.R. (2003) Analysis of bycatch in the fishery for orange roughy, *Hoplostethus atlanticus*, on the South Tasman Rise. *Marine and Freshwater Research*, 54, 1–10.

Anderson, O.F., Gilbert, D.J. and Clark, M.R. (2001) Fish discards and non-target catch in the trawl fisheries for orange roughy and hoki in New Zealand waters for the fishing years 1990–1991 to 1998–1999. *New Zealand Fisheries Assessment Report*, 2001/16, 57 pp.

Anon (2006) Draft MSC Assessment: the United States North Pacific Sablefish fishery. Draft report No. 2 prepared by Scientific Certification Systems Inc. Available www.msc.org/assets/docs

Ardron, J.A. (2005) *Protecting British Columbia's Corals and Sponges from Bottom Trawling*, 22 pp. Living Oceans Society, British Columbia, Canada.

Auster, P.J. and Langton, R.W. (1999) The effects of fishing on fish habitat. *American Fisheries Society Symposium*, 22, 150–87.

Auster, P.J., Malatesta, R.J., Langton, R.W., Watling, L., Valentine, P.C., Donaldson, C.L.S., Langton, R.W., Shepard, A.N. and Babb, I.G. (1996) The impacts of mobile fishing gear on seafloor habitats in the Gulf of Maine (Northwest Atlantic): implications for conservation of fish populations. *Reviews in Fisheries Science*, 4, 185–202.

Barnes, P.W. and Lien, R. (1988) Icebergs rework shelf sediments to 500 m off Antarctica. *Geology*, 16, 1130–33.

Bax, N.J. (comp) (2000) Stock assessment report 2000: orange roughy (*Hoplostethus atlanticus*). Unpublished report prepared for the South East Fishery Stock Assessment Group, 81 pp.

BCN (Biodiversity Conservation Network) (1999) *Evaluating Linkages Between Business, the Environment, and Local Communities. Final Stories from the Field*. Biodiversity Conservation Network, Biodiversity Support Program, Washington, DC, USA.

Bergman, M.J.N. and Hup, M. (1992) Direct effects of beam trawling on macrofauna in a sandy sediment in the southern North Sea. *ICES Journal of Marine Science*, 49, 5–11.

Berkeley, S.A., Chapman, C. and Sogard, S.M. (2004) Maternal age as a determinant of larval growth and survival in a marine fish, *Sebastes melanops*. *Ecology*, 85(5), 1258–64.

Black, K.P. and Parry, G.D. (1999) Entrainment, dispersal, and settlement of scallop dredge sediment plumes: field measurements and numerical modeling. *Canadian Journal of Fisheries and Aquatic Sciences*, 56, 2271–81.

Borets, L.A. (1975) *Some Results of Studies on the Biology of the Boarfish* (*Pentaceros richardsoni* Smith). *Investigations of the Biology of Fishes and Fishery Oceanography*, pp. 82–90. TINRO, Vladivostok.

Branch, T.A. (2001) A review of orange roughy *Hoplostethus atlanticus* fisheries, estimation methods, biology and stock structure. In: *A decade of Namibian Fisheries Science* (eds. Payne, A.I.L., Pillar, S.C. and Crawford, R.J.M.), *South African Journal of Marine Science*, 23, 181–203.

Britton, J.C. and Morton, B. (1994) Marine carrion and scavengers. *Annual Reviews in Oceanography and Marine Biology*, 32, 369–434.

Brown, J., Macfadyen, G., Huntington, T., Magnus, J. and Tumilty, J. (2005) Ghost fishing by lost fishing gear. *Report to DG Fisheries and Maritime Affairs of the European Commission*. Fish/2004/20. Institute for European Environmental Policy/Poseidon Aquatic Resource Management Ltd. 151 pp.

Burridge, C.Y., Pitcher, C.R., Wassenberg, T.J., Poiner, I.R. and Hill, B.J. (2003) Measurement of the rate of depletion of benthic fauna by prawn (shrimp) otter trawls: an experiment in the Great Barrier Reef, Australia. *Fisheries Research*, 60, 237–53.

Cairns, S.D. (1995) The marine fauna of New Zealand: Scleractinia (Cnidaria: Anthozoa). *New Zealand Oceanographic Institute Memoir*, 103, 210 pp.

Camphuysen, C.J., Ensor, K., Furness, R.W., Garthe, S., Huppop, O., Leaper, G., Offringa, H. and Tasker, M.L. (1993) Seabirds feeding on discards in winter in the North Sea. Netherlands Institute for Sea Research. *NIOZ-Rapport*, 1993–1998, 142 pp.

Christensen, V. (1998) Fishery-induced changes in a marine ecosystem: insight from models of the Gulf of Thailand. *Journal of Fish Biology*, 53, 128–42.

Churchill, J.H. (1989) The effect of commercial trawling on sediment resuspension and transport over the Middle Atlantic Bight continental shelf. *Continental Shelf Research*, 9(9), 841–64.

Clark, M.R. (1995) Experience with the management of orange roughy (*Hoplostethus atlanticus*) in New Zealand, and the effects of commercial fishing on stocks over the period 1980–1993. In: *Deep-Water Fisheries of the North Atlantic Oceanic Slope* (ed. Hopper, A.G.), pp. 251–66. Kluwer, The Netherlands.

Clark, M.R. (1999) Fisheries for orange roughy (*Hoplostethus atlanticus*) on seamounts in New Zealand. *Oceanologica Acta*, 22(6), 593–602.

Clark, M.R. (2001) Are deepwater fisheries sustainable? – the example of orange roughy (*Hoplostethus atlanticus*) in New Zealand. *Fisheries Research*, 51, 123–35.

Clark, M. and Hill, A. (2001) Effects of fishing on deepwater seamounts: Ritchie Bank study shows localised effects of trawling on benthos. *Seafood New Zealand*, 9(4), 58–9.

Clark, M.R and O'Driscoll, R.L. (2003) Deepwater fisheries and aspects of their impact on seamount habitat in New Zealand. *Journal of Northwest Atlantic Fisheries Science*, 31, 441–58.

Clark, M.R. and Roberts, C.D. (2003) NORFANZ marine biodiversity survey uncovers mysteries of the deep. *Aquatic Biodiversity and Biosecurity*, 5, 1.

Clark, M.R., Anderson, O.F. and Gilbert, D.J. (2000a) Discards in trawl fisheries for southern blue whiting, orange roughy, hoki, and oreos in waters around New Zealand. *NIWA Technical Report*, 71, 73 pp.

Clark, M.R., Anderson, O.F., Francis, R.I.C.C. and Tracey, D.M. (2000b) The effects of commercial exploitation on orange roughy (*Hoplostethus atlanticus*) from the continental slope of the Chatham Rise, New Zealand, from 1979 to 1997. *Fisheries Research*, 45(3), 217–38.

Collie, J.S., Escanero, G.A. and Valentine, P.C. (1997) Effects of bottom fishing on the benthic megafauna of Georges Bank. *Marine Ecology Progress Series*, 155, 159–72.

Collie, J.S., Hall, S.J., Kaiser, M.J. and Poiner, I.R. (2000) A quantitative analysis of fishing impacts on shelf-sea benthos. *Journal of Animal Ecology*, 69, 785–98.

Cryer, M., Hartill, B. and O'Shea, S. (2002) Modification of marine benthos by trawling: toward a generalization for the deep ocean? *Ecological Applications*, 12(6), 1824–39.

Daan, N. (1980) A review of replacement of depleted stocks by other species and the mechanisms underlying such replacement. *Conseil International pour l'Exploration de la Mer, Rapports et Proces-Verbaux des Reunions*, 177, 405–21.

Dayton, P. (1998) Reversal of the burden of proof in fisheries management. *Science*, 279, 821–2.

Dayton, P.K. and Hessler, R.R. (1972) Role of biological disturbance in maintaining diversity in the deep sea. *Deep-Sea Research*, 19, 199–208.

Dayton, P.K., Thrush, S.F., Agardy, M.T. and Hofman, R.J. (1995) Viewpoint: environmental effects of marine fishing. *Aquatic Conservation*, 5, 205–232.

de Groot, S.J. (1984) The impact of bottom trawling on benthic fauna of the North Sea. *Ocean Management*, 9, 177–90.

Demestre, M., Sanchez, P. and Kaiser, M.J. (2000) The behavioural response of benthic scavengers to otter-trawling disturbance in the Mediterranean. In: *Effects of Fishing on Non-Target Species and Habitats: Biological, Conservation and Socio-Economic Issues* (eds. Kaiser, M.J. and de Groot, S.J.), pp. 121–9. Blackwell Scientific Publications, Oxford, UK.

Dew, C.B. and McConnaughey, R.A. (2005) Did bottom trawling in Bristol Bay's red king-crab broodstock refuge contribute to the collapse of Alaska's most valuable fishery? [Symposium abstract] In: *Benthic Habitats and the Effects of Fishing* (eds. Barnes, P.W. and Thomas, J.P.), *Proceedings of the Symposium on Effects of Fishing Activities on Benthic Habitats: Linking Geology, Biology, Socioeconomics and Management. American Fisheries Society Symposium*, 41.

Dorsey, E.M. and Pederson, J. (eds.) (1998) *Effects of Fishing Gear on the Sea Floor of New England*. Conservation Law Foundation, Boston, MA, USA.

Eno, N.C., MacDonald, D.S., Kinnear, J.A.M., Amos, S.C., Chapman, C.J., Clark, R.A., Bunker, F.P.D. and Munro, C. (2001) Effects of crustacean traps on benthic fauna. *ICES Journal of Marine Science*, 58, 11–20.

Field, M.E. (2005) Living with change: response of the sea floor to natural events. In: *Benthic Habitats and the Effects of Fishing* (eds. Barnes, P.W. and Thomas, J.P.), *Proceedings of the*

Symposium on Effects of Fishing Activities on Benthic Habitats: Linking Geology, Biology, Socioeconomics and Management. American Fisheries Society Symposium, 41.

Fock, H., Uiblein, F., Koster, F. and von Westernhagen, H. (2002) Biodiversity and species–environment relationships of the demersal fish assemblage at the Great Meteor Seamount (subtropical NE Atlantic), sampled by different trawls. *Marine Biology*, 141, 185–99.

Fogarty, M.J. and Murawski, S.A. (1998) Large-scale disturbance and the structure of marine systems: fishery impacts on Georges Bank. *Ecological Applications* (Suppl.), 8(1), S6–S22.

Fonteyne, R., Ball, B., Kujawski, T., Munday, B.M., Rumohr, H. and Tuck, I. (1998) Physical impact. In: *Impact II: The Effects of Different Types of Fisheries on the North Sea and Irish Sea Benthic Ecosystems* (eds. Lindeboom, H.J. and de Groot, S.J.). *NIOZ-Rapport*, 1998 – 1, 404 pp.

Fossa, J.H., Mortensen, P.B. and Furevik, D.M. (2000) Lophelia-korallrev langs Norskekysten forekomst og tilstand. *Fisken og Havet*, No. 2, 94 pp.

Fossa, J.H., Mortensen, P.B. and Furevik, D.M. (2002) The deepwater coral *Lophelia pertusa* in Norwegian waters: distribution and fishery impacts. *Hydrobiologia*, 471, 1–12.

Francis, R.I.C.C. and Clark, M.R. (2005) Sustainability issues for orange roughy fisheries. *Bulletin of Marine Science*, 76(2), 337–51.

Freese, L., Auster, P.J., Heifetz, J. and Wing, B.L. (1999) Effects of trawling on seafloor habitat and associated invertebrate taxa in the Gulf of Alaska. *Marine Ecology Progress Series*, 182, 119–126.

Freiwald, A., Fossa, J.H., Grehan, A.J., Koslow, J.A. and Roberts, J.M. (2004) *Cold-Water Coral Reefs: Out of Sight-No Longer Out of Mind*. UNEP, UK. http://www.unep-wcmc.org

Furness, R.W. (1996) A review of seabird responses to natural or fisheries-induced changes in food supply. In: *Aquatic Predators and their Prey* (eds. Greenstreet, S.P.R. and Tasker, M.L.), pp. 168–73. Blackwell Scientific Publications, Oxford, UK.

Gabriel, O., Lange, K., Dahm, E. and Wendt, T. (2005) *Von Brandt's Fish Catching Methods of the World*, 4th edition, 53 pp. Blackwell Scientific Publications, Oxford, UK.

Giere, O. (1993) *Meiobenthology, the Microscopic Fauna in Aquatic Sediments*. Springer-Verlag, Berlin.

Gordon, J.D.M. (1999) Management considerations of deep-water shark fisheries. In: *Studies of the Management of Elasmobranch Fisheries* (ed. Shotton, R.C.). *FAO Fisheries Technical Paper*, No. 378, 774–818.

Grandperrin, R., Auzende, J.M., Henin, C., LaFoy, Y., Richer de Forges, B., Seret, B., Van de Beuque, S. and Virly, S. (1999) Swath-mapping and related deep-sea trawling in the south-eastern part of the economic zone of New Caledonia. *Proceedings of the 5th Indo-Pacific Fish Conference* (eds. Noumea, S.B. and Sire, J.Y.), Society Francais Ichtyologie, Paris, pp. 459–68.

Grange, K. (1993) Hoki offal dumping on the continental shelf: a preliminary benthic assessment. *New Zealand Marine Sciences Society Review*, 35, 15.

Greenstreet, S.P.R. and Rogers, S.I. (2000) Effects of fishing on non-target fish species. In: *Effects of Fishing on Non-Target Species and Habitats* (eds. Kaiser, M.J. and de Groot, S.J.), 399 pp. Blackwell Scientific Publications, Oxford, UK.

Grigg, R.W. (1989) Precious coral fisheries of the Pacific and Mediterranean. In: *Marine Invertebrate Fisheries: Their Assessment and Management* (ed. Caddy, J.F.), pp. 637–45. John Wiley and Sons, New York, USA.

Grigg, R.W. (1993) Precious coral fisheries of Hawaii and the U.S. Pacific Islands. *Marine Fisheries Review*, 55(2), 50–60.

Grigg, R.W. (2002) Precious corals in Hawaii: discovery of a new bed and revised management measures for existing beds. *Marine Fisheries Review*, 64(1), 13–20.

Grosslein, M.D., Langton, R.W. and Sissenwine, M.P. (1980) Recent fluctuations in pelagic fish stocks of the northwest Atlantic Georges bank region, in relation to species interactions. *Conseil*

International pour l'Exploration de la Mer, Rapports et Proces-Verbaux des Reunions, 177, 374–404.

Gulland, J.A. and Garcia, S. (1984) Observed patterns in multispecies fisheries. In: *Exploitation of Marine Communities* (ed. May, R.), pp. 155–90. Springer Verlag, Berlin, Germany.

Haedrich, R.L. (1995) Structure over time of an exploited deepwater fish assemblage. In: *Deepwater Fisheries of the North Atlantic Oceanic Slope* (ed. Hopper, A.G.), pp. 70–96. Kluwer Academic Publishers, Dordrecht.

Hall, S.J. (1999) *The Effects of Fishing on Marine Ecosystems and Communities*, 274 pp. Blackwell Scientific Publications, Oxford, UK.

Hall, S.J., Robertson, M.R., Basford, D.J. and Heaney, S.D. (1993) The possible effects of fishing disturbance in the northern North Sea: an analysis of spatial patterns in community structure around a wreck. *Netherlands Journal of Sea Research*, 31, 201–8.

Hall, S.J., Raffaelli, D. and Thrush, S.F. (1994) Patchiness and disturbance in shallow water benthic assemblages. In: *Aquatic Ecology, Scale, Pattern and Processes* (eds. Hildrew, A.G., Giller, P.S. and Raffaelli, D.), pp. 333–75. Blackwell Scientific Publications, Oxford, UK.

Hall-Spencer, J., Allain, V. and Fossa, J.H. (2002) Trawling damage to Northeast Atlantic ancient coral reefs. *Proceedings of the Royal Society of London B*, 269, 507–11.

Hauser, L., Adcock, G.J., Smith, P.J., Bernal Ramirez, J.H. and Carvalho, G.R. (2002) Loss of microsatellite diversity and low effective population size in an overexploited population of New Zealand snapper (*Pagrus auratus*). *Proceedings of the National Academy of Science, USA*, 99(18), 11742–7.

Haymon, R.M., Fornari, D.J., Von Damm, K.L., Lilley, M.D., Perfit, M.R., Edmond, J.M., Shanks III, W.C., Lutz, R.A., Grebmeier, J.M., Carbotte, S., Wright, D., McLaughlin, E., Smith, M., Beedle, N. and Olson, E. (1993) Volcanic eruption of the mid-ocean ridge along the East Pacific Rise at 9° 45-52′N: direct submersible observations of seafloor phenomena associated with an eruption event in April, 1991. *Earth and Planetary Science Letters*, 119, 85–101.

Hempel, G. (1978) North Sea fisheries and fish stocks – a review of recent changes. *Conseil International pour l'Exploration de la Mer, Rapports et Proces-Verbaux des Reunions*, 173, 145–67.

Hughes, R.N. and Croy, M.I. (1993) An experimental analysis of frequency-dependent predation (switching) in the 15-spined stickleback, *Spinachia spinachia*. *Journal of Animal Ecology*, 62, 341–52.

Hughes, S.E. (1981) Initial US exploration of nine Gulf of Alaska seamounts and their associated fish and shellfish resources. *Marine Fisheries Review*, 43, 26–33.

Hutchings, P. (1990) Review of the effects of trawling on macrobenthic epifaunal communities. *Australian Journal of Marine and Freshwater Research*, 41, 111–20.

Hutchinson W.F., van Oosterhout, C., Rogers, S.I. and Carvalho, G.R. (2003) Temporal analysis of archived samples indicates marked genetic changes in declining North Sea cod (*Gadus morhua*). *Proceedings of the Royal Society of London – Biological Sciences*, 270, 2125–32.

ICES (1992) Report of the Study Group on Ecosystem Effects of Fishing Activities. *ICES, CM*:1992/G11, 144 pp.

ICES (2000) Report of the Working Group on Ecosystem Effects of Fishing Activities. *ICES, CM*: 2000/ACE 2.

ICES (2002) Report of the Study Group on Mapping the Occurrence of Cold-Water Corals, *ICES, CM*: 2002/ACE 5.

ICES (2003) Report of the Study Group on Cold-Water Corals (SCCOR) for the Advisory Committee on Ecosystems (ACE), *ICES, CM*: 2003/ACE 2.

Jennings, S. and Kaiser, M.J. (1998) The effects of fishing on marine ecosystems. *Advances in Marine Biology*, 34, 201–351.

Jennings, S. and Reynolds, J.D. (2000) Impacts of fishing on diversity: from pattern to process. In: *Effects of Fishing on Non-Target Species and Habitats* (eds. Kaiser, M.J. and de Groot, S.J.), 399 pp. Blackwell Scientific Publications, Oxford, UK.

Jennings, S., Pinnegar, J.K., Polunin, N.V.C. and Warr, K.J. (2001) Impacts of trawling disturbance on the trophic structure of benthic invertebrate communities. *Marine Ecology Progress Series*, 213, 127–42.

Johnson, K.A. (2002) A review of national and international literature on the effects of fishing on benthic habitats. *NOAA Technical memorandum NMFS-F/SPO-57*, 72 pp.

Jones, J.B. (1992) Environmental impact of trawling on the seabed: a review. *New Zealand Journal of Marine and Freshwater Research*, 26, 59–67.

Kaiser, M. (2002) Modification of marine habitats by trawling activities: prognosis and solutions. *Fish and Fisheries*, 3, 114–36.

Kaiser, M.J. and de Groot, S.J. (eds.) (2000) *Effects of Fishing on Non-Target Species and Habitats*, 399 pp. Blackwell Scientific Publications, Oxford, UK.

Kaiser, M.J. and Spencer, B.E. (1994) Fish scavenging behaviour in recently trawled areas. *Marine Ecology Progress Series*, 112, 41–9.

Kaiser, M.J. and Spencer, B.E. (1996a) The effects of beam trawl disturbance on infaunal communities in different habitats. *Journal of Animal Ecology*, 65, 348–58.

Kaiser, M.J. and Spencer, B.E. (1996b) The behavioural response of scavengers to beam trawl disturbance. In: *Aquatic Predators and their Prey* (eds. Greenstreet, S.P.R. and Tasker, M.L.), pp. 116–23. Blackwell Scientific Publications, Oxford, UK.

Kaiser, M.J., Edwards, D.B., Armstrong, P.J., Radford, K., Lough, N.E. and Jones, H.D. (1998) Changes in megafaunal benthic communities in different habitats after trawling disturbance. *ICES Journal of Marine Science*, 55(3), 353–61.

Keating, B.H., Fryer, P., Batiza, R. and Boehlert, G.W. (eds.) (1987) Seamounts, islands, and atolls. *Geophysical Monograph*, 43, 405 pp.

Kelly, M. and Rowden, A. (2001) Rock sponges on seamounts: relicts from the past. *NIWA Biodiversity Update*, 4, 3.

Kennelly, S. and Broadhurst, M. (2002) Bycatch begone: changes in the philosophy of fishing technology. *Fish and Fisheries*, 3, 340–55.

King, M. (1995) *Fisheries Biology, Assessment and Management*. Fishing News Books, Oxford, UK.

Kock, K-H., Purves, M.G. and Duhamel, G. (2005) Interactions between cetacean and fisheries in the Southern Ocean. International Whaling Commission Scientific Committee Document, SC/57/O13. Available www.iwcoffice.org/_documents/sci_com

Koslow, J.A. (1997) Seamounts and the ecology of deep-sea fisheries. *American Scientist*, 85, 168–76.

Koslow, J.A. and Gowlett-Holmes, K. (1998) The seamount fauna off southern Tasmania: benthic communities, their conservation and impacts of fishing. (Final report to Environment Australia and the Fisheries Research Development Corporation.)

Koslow, J.A., Bax, N.J., Bulman, C.M., Kloser, R.J., Smith, A.D.M. and Williams, A. (1997) Managing the fishdown of the Australian orange roughy resource. In: *Developing and Sustaining World Fisheries Resources: The State of Science and Management* (eds. Hancock, D.A., Smith, D.C., Grant, A. and Beumer, J.P.), pp. 558–62. *Proceedings of the Second World Fisheries Congress*, CSIRO Publishing, Collingwood, Australia.

Koslow, J.A., Boehlert, G.W., Gordon, J.D.M., Haedrich, R.L., Lorance, P. and Parin, N. (2000) Continental slope and deep-sea fisheries: implications for a fragile ecosystem. *ICES Journal of Marine Science*, 57, 548–57.

Koslow, J.A., Gowlett-Holmes, K., Lowry, J.K., O'Hara, T., Poore, G.C.B. and Williams, A. (2001) Seamount benthic macrofauna off southern Tasmania: community structure and impacts of trawling. *Marine Ecology Progress Series*, 213, 111–25.

Koslow, T. (2007) *The Silent Deep: The Discovery, Ecology and Conservation of the Deep Sea.* University of New South Wales Press, Sydney, Australia.

Krieger, K.J. and Wing, B.L. (2002) Megafauna associations with deepwater corals (*Primnoa* spp.) in the Gulf of Alaska. *Hydrobiologia*, 471, 83–90.

Large, P.A., Lorance, P. and Pope, J.G. (1998) The survey estimates of the overall size composition of the deepwater fish species on the European continental slope, before and after exploitation. *ICES, CM*: 1980/O: 24.

Law, R. (2000) Fishing, selection, and phenotypic evolution. *ICES Journal of Marine Science*, 57, 659–68.

Law, R. and Stokes, K. (2005). Evolutionary impacts of fishing on target populations. In: *Marine Conservation Biology: The Science of Maintaining the Sea's Biodiversity* (eds. Norse, E.A. and Crowder, L.B.), pp. 232–46. Island Press, Washington.

Lindeboom, H.J. and de Groot, S.J. (eds.) (1998) Impact II: the effects of different types of fisheries on the North Sea and Irish Sea benthic ecosystems. *NIOZ-Rapport*, 1998–1, 404 pp.

Livingston, M. and Rutherford, K. (1988) Hoki wastes on west coast fishing grounds. *Catch*, 15(2), 16–17.

Livingston, M.E., Clark, M.R. and Baird, S.J. (2003) Trends in incidental catch of major fisheries on the Chatham Rise for fishing years 1989–1990 to 1998–1999. *New Zealand Fisheries Assessment Report*, 2003/52.

Lutjeharms, J.R.E. and Heydorn, A.E.F. (1981a) The rock lobster *Jasus tristani* on Vema Seamount: drifting buoys suggest a possible recruiting mechanism. *Deep-Sea Research*, 28A(6), 631–6.

Lutjeharms, J.R.E. and Heydorn, A.E.F. (1981b) Recruitment of rock lobster on Vema Seamount from the islands of Tristan da Cunha. *Deep-Sea Research*, 28A(10), 1237.

McMillan, P.J., Coburn, R.P., Hart, A.C. and Doonan, I.J. (2002) Descriptions of black oreo and smooth oreo fisheries in OEO 1, OEO 3A, OEO 4, and OEO 6 from 1977–1978 to the 2000–2001 fishing year. *New Zealand Fisheries Assessment Report*, 2002/40.

Merrett, N.R. and Haedrich, R.L. (1997) *Deep-Sea Demersal Fish and Fisheries*. Chapman and Hall, London, UK.

Millner, R.S. and Whiting, C.L. (1996) Long-term changes in growth and population abundance of sole in the North Sea from 1940 to the present. *ICES Journal of Marine Science*, 53, 1185–95.

Moore, J.A., Vecchione, M., Collette, B.B., Gibbons, R., Hartel, K.E., Galbraith, J.K., Turnipseed, M., Southworth, M. and Watkins, E. (2003) Biodiversity of Bear Seamount, New England seamount chain: results of exploratory trawling. *Journal of Northwest Atlantic Fisheries Science*, 31, 363–72.

Mortensen, P.B., Buhl-Mortensen, L., Gordon Jr., D.C. Fader, G.B.J., McKeown, D.L. and Fenton, D.G. (2005) Effects of fisheries on deep-water gorgonian corals in the Northeast Channel, Nova Scotia (Canada). In: *Benthic Habitats and the Effects of Fishing* (eds. Barnes, P.W. and Thomas, J.P.), *Proceedings of the Symposium on Effects of Fishing Activities on Benthic Habitats: Linking Geology, Biology, Socioeconomics and Management. American Fisheries Society Symposium*, 41.

Nalesso, R.C., Duarte, L.F.L., Rierozzi, I. and Enumo, E.F. (1995). Tube epifauna of the polychaete *Phyllochaetopterus socialis* Claparede. *Estuarine and Coastal Shelf Science*, 41, 91–100.

National Research Council (2002) *Effects of Trawling and Dredging on Seafloor Habitat.* National Academy Press, Washington, DC, USA.

Nelson, K. and Soule, M. (1987) Genetical conservation of exploited fishes. In: *Population Genetics and Fisheries Management* (eds. Ryman, N. and Utter, F.), pp. 345–68. University of Washington Press, Seattle, USA.

NREFHSC (Northeast Region Essential Fish Habitat Steering Committee) (2002) Workshop on the effects of fishing gear on marine habitats off the northeastern United States, October 23–25, 2001, Boston, MA. *Northeast Fisheries Science Center Reference Document*, 02–01, 71 pp.

O'Driscoll, R.L. and Clark, M.R. (2005) Quantifying the relative intensity of fishing on New Zealand seamounts. *New Zealand Journal of Marine and Freshwater Research*, 39, 839–50.

Oliver, P. (2003) Colossal squid a formidable customer. *New Zealand Herald*, 3 April 2003.

Parin, N.V., Mironov, A.N. and Nesis, K.N. (1997) Biology of the Nazca and Sala y Gomez submarine ridges, and outpost of the Indo-West Pacific fauna in the Eastern Pacific Ocean: composition and distribution of the fauna, its communities and history. *Advances in Marine Biology*, 32, 145–242.

Pauly, D. (1988) Fisheries research and the demersal fisheries of southeast Asia. In: *Fish Population Dynamics* (ed. Gulland, J.A.), pp. 329–48. Academic Press, New York, USA.

Pauly, D., Christensen, V., Dalsgaard, J., Froese, R. and Torres Jr., F. (1998) Fishing down marine food webs. *Science*, 279, 860–63.

Philippart, C.J.M. (1998) Long-term impact of bottom fisheries on several bycatch species of demersal fish and benthic invertebrates in the southeastern North Sea. *ICES Journal of Marine Science*, 55, 342–52.

Probert, P.K. (1999) Seamounts, sanctuaries and sustainability: moving towards deep-sea conservation. *Aquatic Conservation*, 9, 601–5.

Probert, P.K., McKnight, D.G. and Grove, S.L. (1997) Benthic invertebrate bycatch from a deep-water fishery, Chatham Rise, New Zealand. *Aquatic Conservation*, 7, 27–40.

Ramsay, K., Kaiser, M.J. and Hughes, R.N. (1996) Changes in hermit crab feeding patterns in response to trawling disturbance. *Marine Ecology Progress Series*, 144, 63–72.

Ramsay, K., Kaiser, M.J., Rijnsdorp, A.D., Craeymeersch, J.A. and Ellis, J. (2000) Impact of trawling on populations of the invertebrate scavenger *Asterias rubens*. In: *Effects of Fishing on Non-Target Species and Habitats: Biological, Conservation and Socio-Economic Issues* (eds. Kaiser, M.J. and de Groot, S.J.), pp. 151–62. Blackwell Scientific Publications, Oxford, UK.

Reed, J.K., Shapard, A.N., Koenig, C.C., Scanlon, K.M. and Gilmour Jr., R.G. (2005) Mapping, habitat characterization, and fish surveys of the deep-water *Oculina* coral reef Marine Protected Area: a review of historical and current research. In: *Cold-Water Corals and Ecosystems* (eds. Freiwald, A. and Roberts, J.M.), pp. 443–65. Springer-Verlag, Berlin.

Rice, J. and Gislason, H. (1996) Patterns of change in the size spectra of numbers and diversity of the North Sea fish assemblage, as reflected in surveys and models. *ICES Journal of Marine Science*, 53, 1214–25.

Riedl, R. (1971) Water movement: animals. In: *Marine Ecology* (ed. Kinne, O.), Vol. 1, pp. 1123–40. Wiley-Interscience, London.

Riemann, B. and Hoffman, E. (1991) Ecological consequences of dredging and bottom trawling in the Limfjord, Denmark. *Marine Ecology Progress Series*, 69, 171–8.

Roberts, J.M., Harvey, S.M., Lamont, P.A., Gage, J.D. and Humphery, J.D. (2000) Seabed photography, environmental assessment and evidence for deep-water trawling on the continental margin west of the Hebrides. *Hydrobiologia*, 441, 173–83.

Rogers, A.D. (1994) The biology of seamounts. *Advances in Marine Biology*, 30, 305–50.

Rogers, A.D. (1999) The biology of *Lophelia pertusa* (Linnaeus 1758) and other deep-water reef-forming corals and impacts from human activities. *International Review of Hydrobiology*, 84, 315–406.

Rogers, S.I., Kaiser, M.J. and Jennings, S. (1998) Ecosystem effects of demersal fishing: a European perspective. In: *Effects of Fishing Gear on the Sea Floor of New England* (eds. Dorsey, E.M. and Pederson, J.), pp. 68–78. Conservation Law Foundation, Boston, MA, USA.

Rowden, A.A., Clark, M.R. and O'Shea, S. (2004) The influence of deepwater coral habitat and fishing on benthic faunal assemblages of seamounts on the Chatham Rise, New Zealand. *ICES, CM*: 2004/AA: 09.

Rowden, A.A., Clark, M.R. and Wright, I. (2005) Physical characterisation and a biologically focussed classification of 'seamounts' in the New Zealand region. *New Zealand Journal of Marine and Freshwater Research*, 39, 1039–59.

Royce, W.F. (1972) *Introduction to the Fishery Sciences*. Academic Press, New York.

Rumohr, H. (1998) Information on impact of trawling on benthos in Kiel Bay. Annex to Eighth Report of the Benthos Ecology Working Group. *ICES, CM*: 1989/L: 19, 80.

Rumohr, H. and Krost, P. (1991) Experimental evidence of damage to benthos by bottom trawling with special reference to *Arctica islandica*. *Meeresforschung*, 33, 340–45.

Sainsbury, K.J. (1987) Assessment and management of the demersal fishery on the continental shelf of northwestern Australia. In: *Tropical Snappers and Groupers: Biology and Fisheries Management* (eds. Polovina, J.J. and Ralston, S.), pp. 465–503. Westview Press, Boulder, CO, USA.

Sainsbury, K.J. (1988) The ecological basis of multispecies fisheries and management of a demersal fishery in tropical Australia. In: *Fish Population Dynamics* (ed. Gulland, J.A.), 2nd edition, pp. 349–82. John Wiley & Sons, London.

Sainsbury, K.J., Campbell, R.A., Lindholm, R. and Whitelaw, A.W. (1997) Experimental management of an Australian multispecies fishery: examining the possibility of trawl-induced habitat modification. In: *Global Trends: Fisheries Management* (eds. Pikitch, E.L., Huppert, D.D. and Sissenwine, M.P.), pp. 107–12. American Fisheries Society Symposium 20. Bethesda, MD, USA.

Sasaki, T. (1986) Development and present status of Japanese trawl fisheries in the vicinity of seamounts. In: *Environment and Resources of Seamounts in the North Pacific* (eds. Uchida, R.N., Hayasi, S. and Boehlert, G.W.), *NOAA Technical Report NMFS*, 43, 21–30.

Schlacher, T.A., Schlacher-Hoenlinger, M.A., Richer De Forges, B. and Hooper, J.A. (2003) Elements of richness and endemism in sponge assemblages on seamounts. Unpublished presentation, 10th Deep Sea Biology Symposium. Coos Bay, Oregon, August 25–9.

Shester, G. and Ayers, J. (2005) A cost effective approach to protecting deep-sea coral and sponge ecosystems with an applicaton to Alaska's Aleutian Islands region. In: *Cold-Water Corals and Ecosystems* (eds. Freiwald, A. and Roberts, J.M.), pp. 1151–69. Springer-Verlag, Berlin.

Simmonds, J. and Maclennan, D. (2005) *Fisheries Acoustics*. 438 pp. Fisheries and Aquatic Resources Series 10, Blackwells, Oxford, UK.

Simpson, E.S.W. and Heydorn, A.E.F. (1965) Vema seamount. *Nature*, 4994, 249–51.

Smith, P.J., Francis, R.I.C.C. and McVeagh, M. (1991) Loss of genetic diversity due to fishing pressure. *Fisheries Research*, 10, 309–16.

Stocks, K. (2004) Seamount invertebrates: composition and vulnerability to fishing. In: *Seamounts: Biodiversity and Fisheries* (eds. Morato, T. and Pauly, D.), pp. 17–24. University of British Columbia. *Fisheries Center Research Report*, 12(5).

Sullivan, K.J., Mace, P.M., Smith, N.W.McL., Griffiths, M.H., Todd, P.R., Livingston, M.E., Harley, S.J., Key, J.M. and Connell, A.M. (2005) Report from the Fishery Assessment Plenary, May 2005: stock assessments and yield estimates. Unpublished report available from NIWA, Private Bag 14-901, Wellington, New Zealand.

Thrush, S.F. and Dayton, P.K. (2002) Disturbance to marine benthic habitats by trawling and dredging: implications for marine biodiversity. *Annual Review of Ecology and Systematics*, 33, 449–73.

Thrush, S.F., Hewitt, J.E., Cummings, V.J. and Dayton, P.K. (1995) The impact of habitat disturbance by scallop dredging on marine benthic communities: What can be predicted from the results of experiments? *Marine Ecology Progress Series*, 129, 141–50.

Tilmant, J.T. (1979) Observations on the impacts of shrimp roller frame trawls operated over hard-bottom communities, Biscayne Bay, Florida. *US National Park Services Report, Series. No.* P-553, 23 pp.

Tracey, D.M., Neil, H., Gordon, D. and O'Shea, S. (2003) Chronicles of the deep: ageing deepwater corals in New Zealand. *Water and Atmosphere*, 11(2), 23–4.

Tracey, D.M., Clark, M.R., Bull, B. and MacKay, K. (2004). Fish species composition on seamounts and adjacent slope in New Zealand waters. *New Zealand Journal of Marine and Freshwater Research*, 38, 163–82.

Tseitlin, V.B. (1985) The energetics of fish populations inhabiting seamounts. *Oceanology*, 25, 237–9.

Tuck, I.D., Hall, S.J., Robertson, M.R., Armstrong, E. and Basford, D.J. (1998) Effects of physical trawling disturbance in a previously unfished Scottish sea loch. *Marine Ecology Progress Series*, 162, 227–42.

Van Dolah, R.F., Wendt, P.H. and Nicholson, N. (1987) Effects of a research trawl on a hard-bottom assemblage of sponges and corals. *Fisheries Research*, 5, 39–54.

Van Dover, C.L. (2000) *The Ecology of Deep-Sea Hydrothermal Vents.* Princeton University Press, Princeton, NJ.

Van Marlen, B. (2000) Technical modifications to reduce the bycatches and impacts of bottom fishing gears. In: *Effects of Fishing on Non-Target Species and Habitats* (eds. Kaiser, M.J. and De Groot, S.J.), pp. 198–216. Blackwell Scientific Publications, Oxford, UK.

Vinnichenko, V.I. (1997) Russian investigations and deep water fishery on the Corner Rising Seamount in Subarea 6. *NAFO Scientific Council Studies*, 30, 41–9.

Wattling, L. (2005) The global destruction of bottom habitats by mobile fishing gear. In: *Marine Conservation Biology: The Science of Maintaining the Sea's Biodiversity* (eds. Norse, E.A. and Crowder, L.B.), pp. 198–210. Island Press, Washington.

Wattling, L. and Norse, E.A. (1998) Disturbance of the seabed by mobile fishing gear: a comparison to forest clearcutting. *Conservation Biology*, 12, 1180–97.

Wilson, J.B. (1979) 'Patch' development of the deep-water coral *Lophelia pertusa* (L.) on Rockall Bank. *Journal of the Marine Biological Association of the United Kingdom*, 59, 165–77.

Wilson, R.R. and Kaufman, R.S. (1987) Seamount biota and biogeography. In: *Seamounts, Islands and Atolls* (eds. Keating, B.H., Fryer, P., Batiza, R., Boehlert, G.W.), pp. 355–77.*Geophysical Monograph*, 43.

Witherell, D. and Harrington, G. (1996) Evaluation of alternative management measures to reduce the impacts of trawling and dredging on Bering Sea crab stocks. In: *High Latitude Crabs: Biology, Management, and Economics. Alaska Sea Grant Report*, 96–02, 15–22.

Witman, J.D. (1998) Natural disturbance and colonization on subtidal hard substrates in the Gulf of Maine. In: *Effects of Fishing Gear on the Sea Floor of New England* (eds. Dorsey, E.M. and Pederson, J.), pp. 30–37. Conservation Law Foundation, USA, 160 pp.

Witman, J.D. and Cooper, R.A. (1983) Disturbance and contrasting patterns of population structure in the brachiopod *Terebratulina septentrionalis* from two subtidal habitats. *Journal of Experimental Marine Biology and Ecology*, 73, 57–79.

Chapter 20

Management and conservation of seamounts

P. Keith Probert, Sabine Christiansen, Kristina M. Gjerde, Susan Gubbay and Ricardo S. Santos

Abstract

This chapter highlights features of seamount ecosystems of particular significance to their conservation and management, examines the impact of human activities, especially fishing, on seamounts, and analyses management issues and options. We list globally agreed targets for seamount conservation and relevant international and regional tools, institutions and activities. And, as conservation depends on national implementation, national programmes relevant to seamount conservation are discussed and recommendations presented for action to improve the conservation and management of seamounts at global, regional, and national levels.

Introduction: features of seamounts

Seamount ecosystems exhibit a number of distinct features that make them of considerable interest to marine scientists, resource managers, and those with an interest in marine conservation (Fig. 20.1) because they are potentially vulnerable to fishing and other anthropogenic disturbances. Given the targeting of seamount resources by world fisheries (see Chapter 17), the conservation of seamount ecosystems has become a matter of major concern (see Chapter 3; Probert, 1999; Alder and Wood, 2004; Johnston and Santillo, 2004; Stone *et al.*, 2004).

Scarcity, age, and isolation

Seamounts (as defined in the Preface) are widespread, abundant topographic features of the world's oceans, occurring within national, regional, and international jurisdictions (see Chapter 2). But deriving accurate information on the abundance and distribution of seamounts is difficult, given the problems of bathymetric resolution, methodology, logistics, and cost (see Chapters 1 and 2). There may be over 100 000 seamounts of elevation of 1 km or more, with about half in the Pacific and most of the remainder in the

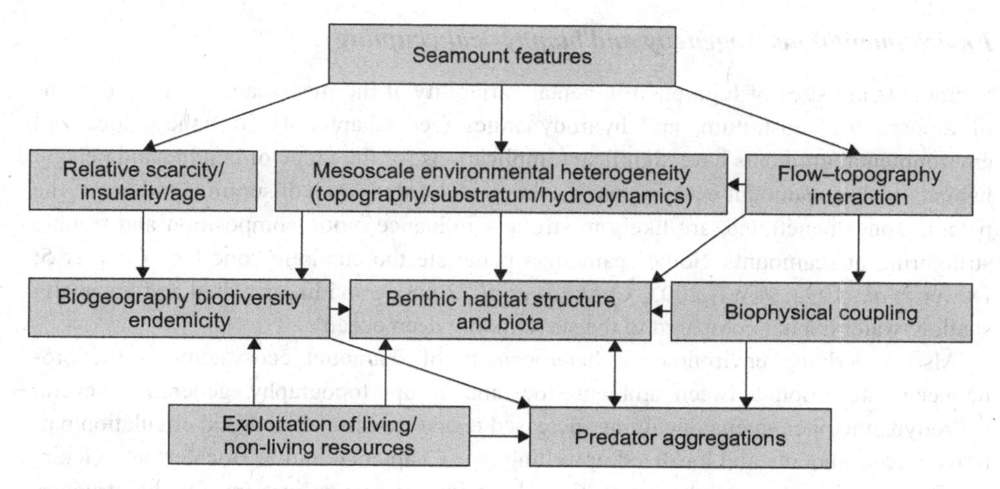

Fig. 20.1 Diagram showing features of seamounts relevant to conservation and management.

Atlantic and Indian oceans (Chapter 1). Despite their abundance, seamounts occupy a small proportion of ocean floor. Rowden *et al.* (2005) estimate that the ~800 seamounts of ⩾100 m elevation within the New Zealand region (24–57°S, 157°E–167°W) account for only about 2.5% of the region's seafloor area. Seamounts are also very diverse (e.g., in terms of size, water depth, elevation, location, overlying water mass characteristics), so certain seamount habitats may be poorly represented in a region.

Although seamounts are relatively young geological features, reflecting the tectonic history of ocean basins, they can vary considerably in age (see Chapter 1). Seamounts in a linear chain produced serially from a mantle hotspot can range from less than 1 million to tens of millions of years in age. Relative age and degree of isolation may strongly influence biotic composition and biogeographic affinity of seamounts (see Chapter 13). Given the evidence for marked regionality in the species composition of seamounts, even individual or small groups of seamounts may assume considerable importance for their distinctive biota. A number of studies have indicated elevated levels of endemism on seamounts; for instance, that up to one-third of benthic macro and megafaunal species from seamounts in the Tasman and SE Coral Seas may be endemic (Richer de Forges *et al.*, 2000; Koslow *et al.*, 2001). High levels of endemism may in part be related to seamount species often having direct development or a short planktonic larval stage, and seamount circulation patterns (see below) tending to retain planktonic larvae (see Chapter 5). But apparent high endemism may be influenced by sampling intensity and relative differences in species abundances (see Chapter 7).

Whether seamounts in general support significantly higher or lower levels of diversity compared to non-seamount habitat is unclear (see Chapter 13). Both may be the case, but at present not enough seamounts have been studied in detail. Seamounts of a given region can be characterized in terms of environmental variables to assess the range, distribution, and relative abundance of seamount types represented. This provides a strong indication of the likely seamount biodiversity of an area and a basis for allocating research effort and informing management (Rowden *et al.*, 2005).

Environmental heterogeneity and biophysical coupling

Seamounts are sites of high environmental variability at the mesoscale, notably in terms of topography, substratum, and hydrodynamics (see Chapter 4), and these localized environmental gradients have significant implications for the range of benthic and pelagic habitats within seamount ecosystems. Elevation and water depth of summit, and hence the pelagic zones penetrated, are likely to strongly influence biotic composition and trophic structuring at seamounts. Some seamounts penetrate the euphotic zone (see Chapter 5; Dower *et al.*, 1992; WWF, 2003; Cardigos *et al.*, 2005), providing isolated and unusually shallow-water habitat compared to the surrounding deep ocean.

Also underlying environmental heterogeneity of seamount ecosystems is the pronounced interaction between ambient flow and abrupt topography, generating several hydrodynamic phenomena, including increased residence time of enclosed circulation patterns over seamounts and localized upwelling (see Chapters 4 and 5; Boehlert and Genin, 1987). Such flow–topography interactions have important implications for the structure and functioning of seamount biota. Pelagic and benthic communities of seamounts are often of high biomass compared to those of surrounding areas, though to what extent this reflects a local increase in productivity is unclear. The residence time of upwelled nutrient enriched water generally appears insufficient for enhancement of primary productivity to occur over seamounts themselves. Instead, seamount ecosystems are likely to be fuelled more by advected particulate organic matter and plankton and vertical entrapment of mesopelagic migrators, leading to localized aggregations of fish and higher predators (see Chapters 4 and 6). Physical processes affecting biota may extend considerable distances downstream of seamounts. Given the spatial and temporal complexities of physical processes it is hard to generalize about a 'seamount effect' (see Chapter 4).

Aggregation of fish at seamounts is a conspicuous result of this biophysical coupling. Eight hundred species of fish have been reported from seamounts (see Chapter 9; Froese and Sampang, 2004; Morato *et al.*, 2004) of which a number, although not restricted to seamounts, characteristically form aggregations (*sensu* Koslow, 1996) over seamounts. Best known are those species targeted by fisheries, such as orange roughy (*Hoplostethus atlanticus*), alfonsinos (*Beryx* spp.), pelagic armourhead (*Pseudopentaceros wheeleri*), Patagonian toothfish (*Dissostichus eleginoides*), oreos (Oreosomatidae), and rockfishes (*Sebastes* spp.) (see Chapter 9; Rogers, 1994). Seamount aggregating fishes tend to be strongly K-selected (long lived, slow growing, late maturing, low fecundity), rendering them highly vulnerable to exploitation (see Chapter 9; Koslow, 1996; Morato *et al.*, 2006). Higher predators recorded in increased numbers at seamounts include tunas, billfishes, sharks, marine mammals, and seabirds (see Chapter 10A and B; Chapter 12A, B, and C; Yen *et al.*, 2004). Analyses of the distribution of oceanic predators indicate biodiversity hotspots associated with prominent topographic features, including seamounts; oceanographically, such features are characterized by mesoscale habitat complexity and as key feeding areas (Worm *et al.*, 2003; Cheung *et al.*, 2005). The predator component of seamount food webs comprises resident species as well as shorter term aggregators and visitors, illustrating the trophic complexity of seamount ecosystems (see Chapter 14). The distinction also has important management implications; species that tend to be resident are likely to be more vulnerable to fishing impacts than visitors, at least for fishing focused on seamounts.

Benthic habitats and biotas

Seamount benthic environments encompass a wide range of bathymetry, topography, and substrata (Rogers, 1994; Koslow *et al.*, 2001). Hard substrata (atypical of the deep sea) and enhanced particulate flux from flow–topography interactions provide for benthic assemblages where large sessile suspension feeders are conspicuous. Prominent megafauna on seamounts often includes stylasterid, gorgonarian, antipatharian, and sclaractinian corals (see Chapter 8; Boehlert and Genin, 1987; Probert *et al.*, 1997; Koslow *et al.*, 2001; Piepenburg and Muller, 2004), with some corals forming extensive reefs (Rogers, 1999). This sessile epifauna provides habitat for numerous associated species (Koslow *et al.*, 2001). For example, nearly 900 species have been recorded living on or in *Lophelia pertusa* reefs in the NE Atlantic (Rogers, 1999). Costello *et al.* (2005) recorded 25 fish species associated with *Lophelia* reefs of which 17 were of commercial importance. Fish species richness and abundance were far higher on the reef than on the surrounding seabed (92% of species and 80% of individuals were associated with the reef). Fish may, however, respond to deep-water corals as a type of complex habitat rather than associating specifically with coral (Auster, 2005). Deep-sea corals and other sessile epifauna of seamounts can be very long lived, of the order of hundreds of years (see Chapters 7 and 8; Tracey *et al.*, 2003; Thresher *et al.*, 2004; Andrews *et al.*, 2005), signifying very prolonged post-disturbance recovery where such organisms contribute importantly to benthic habitat structure.

Sediment benthos of seamounts is poorly known (see Chapter 7; Levin *et al.*, 1991b; Rogers, 1994), but factors such as intensified near bottom flow, sediment texture and mobility, organic input, and penetration into the oxygen minimum or euphotic zones can strongly influence community structure (e.g., Levin *et al.*, 1991a, 1994; Dower *et al.*, 1992; Wishner *et al.*, 1995; Thistle *et al.*, 1999; Heinz *et al.*, 2004). Xenophyophores – large sediment-agglutinating protozoans – are often conspicuous on seamounts (Levin, 1994) and may provide macrofaunal habitat (Levin *et al.*, 1991b). Further detailed studies are needed to clarify the role of the benthos on seamounts.

Major concerns

Fisheries impacts in particular highlight a pressing need to address the conservation and management of seamounts. Large-scale fisheries for seamount-aggregating species date from the mid-1960s and typically follow boom-and-bust cycles as successive stocks have been targeted and collapsed (see Chapters 17 and 18; Watson and Morato, 2004). Life history characteristics of deep-water fishes in general render them highly susceptible to overexploitation (see Chapter 9) and call into question the sustainability of such fisheries (Merrett and Haedrich, 1997; Clark, 2001; Roberts, 2002; Francis and Clark, 2005). Several major orange roughy stocks off New Zealand were serially depleted to a few per cent of their estimated virgin biomass in 10 years, in some cases in less than 5 years, but sustainable yields are estimated at only 1–2% of virgin biomass level per year (see Chapters 17 and 19; Clark *et al.*, 2000; Clark, 2004).

It is unclear what effect such large-scale removal of biomass may have on seamount ecosystems, and particularly whether depletion might significantly alter fish community composition or threaten seamount obligates. It may be too early for identifiable changes, given the recent development of the fisheries and the prevailing life history traits of deep-water fishes (Koslow *et al.*, 2000). Seamount fisheries began in earnest some 20 years ago. But as orange roughy become sexually mature at about 35, a full reproductive cycle has yet to occur. Size-selective fishing can have various effects on size and age distribution and reproductive output of exploited populations (e.g., Jennings *et al.*, 2001; Birkeland and Dayton, 2005), but there is as yet limited information on the response of seamount species (e.g., Clark *et al.*, 2000). Conditions for successful recruitment also remain unknown; nor have impacts of seamount fisheries on predator and prey populations of target species been examined. Modelling studies will help in understanding the wider effects of fisheries on seamount ecosystems (see Chapters 14 and 15).

Attention has focused on impacts of bottom trawling on seafloor habitat and benthos of seamounts. Effects of bottom trawling on benthos at bathyal depths are still poorly understood, but are generally expected to be more severe than for shelf areas, given the life history characteristics of deeper-water species and the slow, decadal recovery times (see Chapter 19; Jones, 1992; Dayton *et al.*, 1995). Of particular concern is the impact of trawling on large sessile epifauna that can contribute importantly to structuring seabed habitat, especially if such species are seamount obligates. Benthic habitats differ considerably in structural complexity and vulnerability to disturbance from fishing gear (Auster, 1998), but emergent epifauna provide complex habitat that is especially susceptible to damage from fishing (Collie *et al.*, 2000). Fishing impact will also depend on trawling intensity relative to seamount area. For New Zealand seamounts, O'Driscoll and Clark (2005) estimated a median fishing intensity of about 130 km of trawled seabed per km^2 of seamount area, but as much as 17 400 km of tows per km^2 for some seamounts.

Deep-water, coral-dominated benthos is highly vulnerable to physical disturbance (see Chapters 7, 8, and 19; Probert *et al.*, 1997; Rogers, 1999; Koslow *et al.*, 2001; Hall-Spencer *et al.*, 2002). Trawling can result in a large bycatch of coral on previously unfished or lightly fished seamounts, and coral-dominated habitat may be severely degraded by fishing pressure. For the orange roughy fishery in the South Tasman Rise, total coral bycatch (dominated by the reef-forming *Solenosmilia variabilis*) fell from about 1750 to 100 t/year between 1997–1998 and 2000–2001 (Anderson and Clark, 2003). For seamounts south of Tasmania, trawling effectively removed corals and other suspension feeders from heavily fished seamounts at depths of up to 1400 m. Benthic biomass of samples from unfished seamounts was 106% greater than from heavily fished seamounts, and the number of species per sample was 46% greater (Koslow *et al.*, 2001). Loss of coral and other long-lived, slow-growing sessile epifauna that structure benthic habitat could have major long-term implications for seamount biodiversity, especially for isolated, distinctive faunas.

Fishing may disrupt important interactions between pelagic and benthic components of seamount ecosystems. Depletion of seamount aggregating fish may reduce an important source of organic detritus to seamount benthos (Koslow *et al.*, 2001). The integrity of benthic habitat structure may also be important for deep-water fishes. North Atlantic *Lophelia* reefs are important fish habitat and may function as feeding, breeding, and nursery grounds (Costello *et al.*, 2005). Protection of reefs is important for maintaining fish biodiversity as well as fisheries.

Precious corals – gorgonacean and antipatharian species highly valued for jewellery – have been targeted and often overexploited, mainly on North Pacific seamounts (see Chapter 17; Grigg, 1993; Rogers, 1994). Grigg (1984) estimated the maximum sustainable yield for a Hawaiian seamount black coral at only 3.5% of biomass per year. Seamounts may also in time be exploited for cobalt-rich ferromanganese crusts and metallic sulphides (Grigg *et al.*, 1987; Rogers, 1994; Glasby, 2002; Rona, 2003), and possibly for pharmacologically active metabolites (Laurent and Pietra, 2004).

Seamounts support distinctive ecosystems, unusual for the range of bathymetry, substrata, and hydrodynamic conditions they typically encompass, and in the degree of biophysical coupling between the flow field and pelagic and benthic components of the biota. They are also relatively scarce environments in areal extent and, depending on their degree of isolation, may support biota that are biogeographically distinct and with significant endemicity. Flow–topography interactions mean that seamounts tend to be sites of high biomass, most conspicuously in supporting dense aggregations of fish. Enhanced particulate flux and abundant hard substrata favour benthic communities dominated by large sessile epifauna. Seamount communities are of high conservation interest, heightened by their vulnerability to intensive fishing. It is important, therefore, to examine management and conservation tools applicable to seamounts, conservation initiatives at national, regional, and international level, and future strategies.

Management issues for seamounts

Management of human activities affecting biodiversity on and around seamounts raises six major issues:

1. Deep-sea bottom fishing (presently to 2000 m) has been identified as the major threat to seamount communities and habitats.
2. Addressing fisheries impacts is critical to conservation of seamount biodiversity. This requires dealing with destructive gear types, access rights, enforcement, and surveillance.
3. Management objectives need to
   (a) incorporate natural variability on both short and very long timescales;
   (b) work within, and potentially across, the major jurisdictional divisions of ocean space;
   (c) be highly precautionary (given the state of knowledge and nature and extent of human impacts on seamounts);
   (d) target recovery and restoration as well as prevention of degradation;
   (e) be within a wider management framework; this may be regional as well as global.
4. No single management model is applicable to all seamounts. Measures are likely to range from activity-based restrictions to marine protected areas (MPAs), and regulation of activities beyond the immediate vicinity of seamounts.
5. Sustainable, economically viable fisheries may not be possible for some species on seamounts, so fisheries management will have to adjust. Likely knock-on effects on fishing activity elsewhere will also need to be considered.

6. In the short term, there is little likelihood of cataloguing every seamount and detailing their key characteristics. Action can most usefully be directed at regions and clusters of seamounts within fishing depth, and seamounts that have been studied in detail. Grouping seamounts according to their general characteristics can help to prioritize conservation action.

These issues are familiar to those concerned with the management of activities affecting inshore and shallow-water features of biodiversity interest (e.g., Kenchington, 1990; Norse, 1993). Management of activities affecting seamounts should, therefore, be seen as an extension of existing programmes inside Exclusive Economic Zones (EEZs), but with a greater focus on the high seas. The main differences are in the legal and administrative frameworks supporting such management (described below), rather than in the management tools.

Approaches and tools for management of seamounts

Two complementary categories of management measures are likely to be most appropriate for seamounts: (1) site-specific measures, particularly the establishment of MPAs and regulation of human activities therein; and (2) activity-specific measures to regulate human activities on a wider geographic scale.

Site-specific management tools

MPAs incorporating a variety of management options have become an important component of marine conservation programmes worldwide (Gubbay, 1995; Agardy, 1997; Kelleher, 1999; Salm *et al.*, 2000; Sobel and Dahlgren, 2004). Increasingly, this has included consideration of offshore MPAs (Thiel and Koslow, 2001; Gjerde and Breide, 2003; Unger, 2004). MPAs are now recognized as a valuable tool for management of activities around seamounts (e.g., WSSD, 2002; CBD, 2004; Schmidt and Christiansen, 2004), and, to date, some 84 MPAs worldwide include seamounts (Alder and Wood, 2004).

Defining the boundary of management areas

Many early MPAs cover relatively small coastal areas, but large areas are needed for management of biodiversity at an ecosystem scale. This is the case for seamounts, which span a wide depth range and with influences extending tens of kilometers across the surrounding ocean, and where management should be at seamount scale, an approach demonstrated by the few seamount MPAs that extend over large areas (e.g., Lord Howe Island, Australia, and Bowie Seamount, Canada, pilot MPA).

Management zones

Both vertical and horizontal zoning schemes have been introduced for seamount MPAs to support management options ranging from strict protection (e.g., Tasmanian seamounts, and Formigas and Dollabarat Bank, Azores) to multiple use (e.g., Saba National Marine Park, Netherlands Antilles). Horizontal zoning, a standard practice in many MPAs, can provide for a range of activities. For example, the Tasmanian and Lord Howe Island reserves both have a highly protected zone (equivalent to IUCN Category 1a) and Managed Resource Zones (IUCN Category IV) (IUCN, 1994). Vertical zoning, a newer idea, is seen in the Tasmanian Seamounts Marine Reserve (Commonwealth of Australia, 2002). It may be appropriate given the bathymetric range of many seamounts and if activities at the surface have only limited effect on seabed features. Vertical zoning could, therefore, be used to separate management measures.

Management planning

For both offshore and coastal MPAs, there is a need to focus on managing users rather than resources, and for regulators and users to cooperate. MPA management plans need clear objectives to facilitate actions, and to enable progress and effectiveness to be evaluated (Pomeroy *et al.*, 2004). Documenting management intentions and inviting consultation make the process more open, essential in gaining the support of users, even for remote sites. Linking existing initiatives and sharing resources and information will also be critical to management, given that no single agency will have the full suite of powers to establish, manage, monitor, and enforce seamount MPAs. This may be a relatively new area of operation for parties involved, so the roles of different organizations in relation to seamount location (i.e., territorial waters, EEZ, or high seas) need to be clarified.

Activity-related management tools

Fisheries are currently the main activity at seamounts (see Chapters 16–18). These range from longlining to trawling and encompass commercial, artisanal, and recreational fishing. Options for fisheries management on seamounts range from prohibition of all fishing or specific gear types, to regulation of activity through measures such as effort control, gear limitations, quotas, and licences. Measures which also contribute to enforcement and accurate data on fishing activity include observer schemes and mandatory logbooks supporting fisheries audits. Environmental assessment and regulations guiding exploratory/expanding fisheries could also be used before the start of a new fishery and to review existing fisheries.

Mineral extraction and bioprospecting may in time become significant activities on seamounts (Glasby, 2002; Arico and Salpin, 2005). Licensing arrangements will be critical to managing such activity on a precautionary basis and could be used in combination with exclusion zones that act as reference areas.

Management of shipping activity may also be appropriate to support seamount conservation, particularly where seamounts extend into shallow water, with measures relating to ship routing, discharges, and designation of sensitive areas. Wider measures, for example, to control waste disposal from ships, ship design criteria to enhance safety, and preventing the transfer of alien species in ballast water are also relevant, although not specific to seamounts.

Seamounts that rise close to the surface and are accessible may become important for tourism, such that boating, anchoring, and diving may need to be considered (e.g., D. João de Castro Bank, Azores, and Saba Marine Park, Antilles).

Common management issues

Regulation

Regulation of activities at seamounts is complex. National legislation with provisions for establishing and managing MPAs or particular activities may exist. But many seamounts are remote, there is often incomplete understanding of all that is required to safeguard such areas, and many of the management issues need to be tackled at an international as well as a local level.

Within national waters, general MPA regulations could provide the legal framework for designation. Such legislation will usually require a lead agency to coordinate action. This relationship is one of the most important elements of legislation for MPAs, because all aspects of management of such sites will rely on cooperation between regulatory and interested parties. This is important when dealing with international activities such as fishing and shipping where proposed measures must be agreed through regional or international bodies, and where countries cannot work effectively in isolation.

Non-MPA legislation will also be needed to achieve management objectives at particular sites. General biodiversity and fisheries management measures, as well as area-specific measures such as setting water-quality standards and specifying permitted fishing effort and permitted fishing gear, will give important additional support to the management of activities which affect seamounts. Using a broad base of regulation will also reinforce the point that both general and site-specific measures are needed for effective management.

Voluntary measures, such as Codes of Practice, have been used to complement statutory management measures for some activities. They have been introduced to support the management of activities in coastal waters as well as the high seas, for example, the FAO Code of Conduct for Responsible Fisheries (FAO, 1995). Another approach is to set up formal agreements, such as memoranda of understanding and concordats.

Stakeholder involvement

The importance of involving the general public and local communities in the establishment and operation of MPAs and other management tools is widely accepted (Kelleher, 1999; Roberts and Hawkins, 2000; DEFRA, 2004). This is more difficult for offshore sites where there is no resident 'local community' and where user groups may come from far afield. The perception of offshore resources as common or public property held in trust by a coastal state can limit public interest. Nevertheless, it is possible to identify a 'constituency of interest' and effort must be made to involve such groups in management.

Designation of seamount MPAs beyond the 12 nm territorial sea or 200 nm EEZ, but on the legal continental shelf (which can include the slope and margin beyond the physical continental shelf), may require consultation with other regional and international bodies to regulate activities such as shipping or fishing. International shipping is regulated by the International Maritime Organization (IMO). For European Union members, the Common Fisheries Policy (CFP) provides for access of member states to all Community waters between 12 nm (or less) and 200 nm offshore, thus protective measures for fisheries must be adopted at the regional level. However, in other nations, responsibility for fisheries out to the limits of the EEZ generally rests with national authorities. In some 60 nations, the legal continental shelf may extend beyond 200 nm, and hence the coastal state retains the right to control exploitation of sedentary living resources. The right to control exploitation also implies the right to regulate activities such as bottom fishing that may impact on those sedentary resources (Molenaar, 2005). However, to secure the widest possible cooperation, it may be advisable for the coastal nation to consult with the relevant regional fishery management organization before adopting rules with regards to bottom fishing on the extended continental shelf. This will require coordination between the government departments responsible for regulating, assessing, and monitoring these activities at a national level before discussions with these European and international bodies take place.

Involvement of casual users in the establishment and operation of potential MPAs or other management measures on seamounts will be essential. Consultation with stakeholder groups will help to achieve this, with information posted in appropriate publications. Another option would be to establish an offshore liaison committee with the role of identifying the main groups to be contacted at national and international level and the most appropriate methods for doing this for all the offshore seamount MPAs being considered in a region or EEZ, rather than liaising on a site-by-site basis.

Education and interpretation

The need for education about seamounts and their conservation is, if anything, greater than for inshore waters. There is still much ignorance about the importance of the oceans and of the need to use marine resources wisely, especially when sites proposed for protection are far removed from daily life and less accessible. Much can be done to further education and interpretation of offshore sites. The materials produced, such as leaflets, books, videos, and displays, may be similar to those for other types of marine management, but greater effort is needed to take such information to people rather than providing it only when users and visitors come to the site or make enquiries. Information about offshore sites needs to be available from a range of sources such as aquariums, museums, research institutes, and schools, as well as in places frequented by the main user groups. There are also considerable opportunities through the use of electronic media, remote sensing imagery, and the internet (e.g., Centro de Interpretação Marinha Virtual, www.cimvazores.info, last accessed 30 November 2006).

Compliance and enforcement

The virtual impossibility of directly observing all infringements means that most MPA programmes emphasize compliance rather than coercion. This will certainly be the case for offshore MPAs at seamounts, albeit backed by a regulatory framework. To achieve this it will be essential to provide users with information about regulations, including the reasons for them. Another important element should be to work with agencies that already have a regulatory role in the area. Enforcement of regulations is essential to the success and credibility of any management measure. The difficulties of achieving this in the marine environment are well known and likely to be even more problematic in the remote locations of many seamounts where infringements are harder to observe and where a rapid response to problems is unlikely. Both water-based and aerial surveillance are used in these circumstances and can be supported by satellite monitoring. Vessel monitoring systems (VMSs) that automatically transmit the position of a fishing vessel to enforcement agencies are now mandatory for a number of fisheries (e.g., required by vessels permitted to take orange roughy adjacent to the Tasmanian Seamounts Reserve, Commonwealth of Australia, 2002).

Research and monitoring

Measures to support the conservation of seamounts can also facilitate research and monitoring programmes that improve understanding of marine systems and of management requirements for conservation. Such work would benefit from being coordinated at a national level, particularly for seamounts, where it is generally more difficult to establish long-term programmes and where permission may need to be sought from both national governments and international organizations. Mechanisms for sharing and accessing data, for streamlining procedures to work in such areas, and for coordinating research effort should be set up, as well as links to research programmes in EEZs where appropriate. Monitoring, evaluation, and review are important to the successful management of activities impacting on seamounts. Monitoring must be linked to review procedures, so that any necessary changes are made to management measures.

Global seamount conservation goals, tools, and strategies

Global agreements and goals for action

A substantial body of resolutions and agreements has evolved in recent years with the potential to improve the conservation and management of seamount species and ecosystems throughout the world's oceans. However, significant efforts, including new legal agreements, are required for seamount ecosystems beyond national jurisdiction.

The 1982 UN Convention on the Law of the Sea (UNCLOS) is fundamental as it provides for jurisdictional zones in the world's oceans, including EEZs, where states have sovereign rights extending out to 200 nm (see International tools and institutions below and Supplementary materials online). A number of coastal states have a legal continental shelf extending beyond 200 nm, wherein states retain exclusive rights to exploit and explore mineral and sedentary living resources such as corals, sponges, and crabs. It is estimated that nearly half of seamounts of over 1000 m elevation are within EEZs (Alder and Wood, 2004). No study has yet been undertaken to identify what percentage of seamounts is located on legal continental shelves. Such information would help to clarify who is primarily responsible for management and conservation of the sedentary living resources on seamounts.

International seabed beyond 200 nautical miles, or the legal continental shelf, whichever is furthest, is known as the 'area'. Activities related to seabed mining are overseen by the International Seabed Authority, which is currently developing regulations for prospecting and exploration of polymetallic sulphides and cobalt crusts that may occur on seamounts. Deep seabed mining, when it occurs, will be subject to international regulations to protect the marine environment. By contrast, deep-sea fishing that impacts the international seabed has no such constraint as UNCLOS negotiators did not foresee that activities besides mining might affect the deep seabed. Thus, there are no international regulations or bodies responsible for protecting the deep seabed environment from human activities other than mining (Arico and Salpin, 2005).

Since 2003, a growing number of states, scientists, environmental, conservation, and fish worker organizations have been calling on the United Nations General Assembly

(UNGA) to adopt a resolution for a moratorium on high seas bottom trawling until effective conservation and management regimes can be put in place (e.g., a 'Scientists' Statement' signed by more than 1100 scientists from 69 countries, MCBI, 2004). In response, UNGA in 2004 adopted a resolution calling for urgent action to protect deep-sea biodiversity on the high seas, though not for a global moratorium. The resolution also called for a review of state and regional action to protect vulnerable marine ecosystems and consideration in 2006 of what further measures may be necessary. Support for this resolution was endorsed in 2005 by the FAO Committee of Fisheries, the major global intergovernmental forum where fisheries issues are discussed and recommendations adopted. Also in 2004, UNGA established a working group to examine conservation of biodiversity in areas beyond national jurisdiction. The European Union has proposed that this should develop an implementing agreement to UNCLOS to address the conservation and management of biodiversity in areas beyond national jurisdiction.

Seamounts were a key topic for discussions at the 2004 and 2006 Conference of Parties of the Convention on Biological Diversity (CBD). Parties designated seamount and cold-water corals as conservation priorities and called for their protection through short-, medium-, and long-term measures. In 2004, the CBD Parties also established a working group to explore options for the establishment of MPAs in areas beyond national jurisdiction (COP VII/28), and ecological criteria for high seas MPAs were considered at a meeting in 2006.

Fundamental goals and targets for marine conservation agreed at the 2002 World Summit on Sustainable Development (WSSD) in Johannesburg commit governments to

1. Encourage the application by 2010 of the ecosystem approach (WSSD Joint Plan of Implementation (JPOI), para. 30(d)).
2. Maintain the productivity and biodiversity of important and vulnerable marine and coastal areas, including in areas within and beyond national jurisdiction (para. 32(a)).
3. Develop and facilitate the use of diverse approaches and tools, including the ecosystem approach, the elimination of destructive fishing practices, the establishment of MPAs consistent with international law and based on scientific information, including representative networks by 2012 and time/area closures for the protection of nursery grounds and periods (para. 32(c)).
4. Develop national, regional, and international programmes for halting the loss of marine biodiversity, including in coral reefs and wetlands (para. 32(d)).
5. A commitment to achieve by 2010 a significant reduction in the current rate of loss of biological diversity (para. 44).

Achievement of these goals and targets would shelter seamounts in a framework of ecosystem-based management throughout the oceans. A full toolbox of approaches would be in use everywhere, including the elimination of destructive fishing practices such as bottom trawling on seamounts, the establishment of MPA networks to protect specific areas from a broader spectrum of human uses (consistent with international law and based on science), and the use of closed areas, both permanent and temporary, on seamounts to protect spawning grounds and nursery areas for juvenile stages. All human activities would be managed consistent with programmes to prevent loss of biodiversity.

The following sections tackle the difficult question of how we achieve these goals and targets. What legal tools and institutions may be used to enhance the conservation and management of seamount ecosystems? Where these fall short, what are the most promising avenues to broaden the scope of these instruments? What are the best strategies for achieving this for seamounts within and beyond national jurisdiction?

International tools and institutions

The following international tools and institutions are described more fully in online materials, www.seamountsbook.info. *The UN General Assembly* sets global oceans policy. *The UNCLOS* prescribes the basic rights and duties of states, including an obligation to protect and preserve all areas of the marine environment. *The International Seabed Authority* is charged with administering and distributing the benefits from minerals related activities in the seabed beyond national jurisdiction, deemed the 'common heritage of mankind' under UNCLOS. The 1995 *Agreement on Highly Migratory and Straddling Fish Stocks* is the major international instrument designed to elaborate and implement the vague provisions of UNCLOS. The *UN Food and Agriculture Organization* (FAO) has been requested to develop technical guidelines on deep-sea fisheries and on the design, implementation, and testing of MPAs and to assist member states to establish representative MPA networks by 2012 (FAO, 2005). The *FAO Code of Conduct* (FAO, 1995) calls for ecosystem-based management, a precautionary approach, environmentally safe and selective fishing gear and practices, minimal waste, bycatch and impacts on other species, and protection of critical habitats such as reefs, nursery, and spawning grounds.

The 1992 *Convention on Biological Diversity* sets out a framework for conservation, sustainable use and equitable benefit sharing of biodiversity, with an emphasis on *in situ* protection through protected areas and ecosystem-based management. *The Convention on Migratory Species* requires 'Range States' to protect listed migratory species including sperm whales, sea turtles, sea birds and small cetaceans, and their habitat.

Regional tools and actions

Many regional agreements and institutions have been developed to elaborate and supplement the UNCLOS regime in their respective regions. Each regional organization is generally empowered to establish conservation and/or environmental protection measures for a defined geographic area. Only four regional seas agreements cover areas beyond national jurisdiction, whereas most regional fisheries management organizations (RFMOs) specifically cover such regions (Fig. 20.2), but may also cover areas within national jurisdiction. Boundaries for RFMOs generally reflect the geographic range of the stocks or species concerned (CBD, 2005).

The following section describes the three regions – North East Atlantic, Mediterranean, and the Southern Ocean – where regional seas organizations and RFMOs have been most active in marine biodiversity conservation and deep-sea fisheries management beyond national jurisdiction. Several other regional initiatives are also discussed.

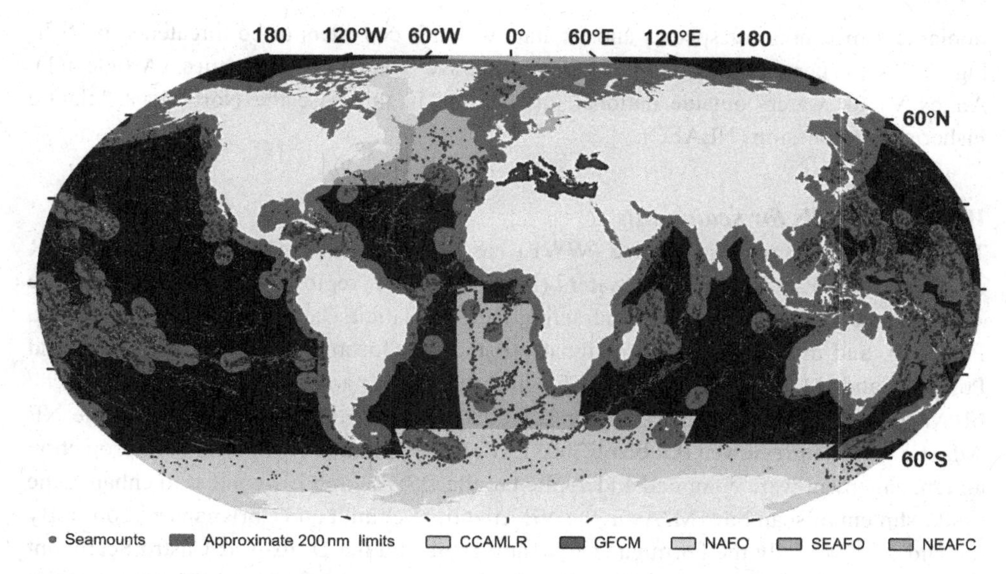

Fig. 20.2 RFMOs that have competence to regulate deep-sea fisheries. Dark blue, areas that lack coverage by an RFMO with the competence to manage deep-water fisheries on the high seas. Coloured areas (in light blue, purple, green yellow, and orange), where RFMOs have a mandate to regulate deep-water fisheries. CCAMLR, Convention for the Conservation of Antarctic Marine Living Resources; GFCM, General Fisheries Commission for the Mediterranean; NAFO, North West Atlantic Fisheries Organization; SEAFO, South East Atlantic Fisheries Organization; NEAFC, North East Atlantic Fisheries Commission (*Source*: P. Hoagland, WHOI). The boundaries shown are purely indicative, have no jurisdictional basis and do not imply any expression of any opinion of the publishers or authors concerning the legal status of any country, territory or area, or concerning the delimitation of its frontiers or boundaries.

North East Atlantic

The OSPAR Convention

The 1992 OSPAR Convention (Convention for the Protection of the Marine Environment of the North-East Atlantic) extends from the latitude of the Strait of Gibraltar to the North Pole and west to Greenland, more than half of which is beyond national jurisdiction. OSPAR has 15 'Contracting Parties' including the EU. Annex V, adopted in 1998, requires parties to undertake individual and collaborative measures to protect ecosystems and conserve biological diversity. Fisheries and shipping are specifically excluded from OSPAR's remit.

In 2003, Parties to OSPAR adopted a declaration to protect cold-water corals throughout the OSPAR maritime area, as well as a list of threatened and/or declining species and habitats that includes seamounts, *L. pertusa* reefs, deep-sea sponge grounds, and species such as orange roughy, loggerhead (*Caretta caretta*), and leatherback (*Dermochelys coriacea*) turtles (OSPAR, 2004). They also recommended establishing an 'ecologically coherent network of well-managed' MPAs by 2010 (OSPAR, 2003). Selection criteria for these MPAs include prioritizing for species and habitats on the OSPAR list. Parties have agreed to include areas beyond national jurisdiction as components of the MPA network. Therefore, there are various instruments to select and designate seamount MPAs in the OSPAR maritime area. A serious obstacle for OSPAR is its lack of competence to implement measures for species and habitats which it considers to be threatened by fishing. OSPAR is permitted only to notify the respective 'competent authorities' (Article 4(1), Annex V) for waters outside national jurisdiction, in this case the North East Atlantic Fisheries Commission (NEAFC).

WWF proposals for seamounts

The World Wide Fund for Nature (WWF) has proposed several seamounts to OSPAR as candidate MPAs in territorial waters (Formigas Bank, regional marine reserve in the Azores since 1988), in EEZs of Contracting Parties (Galicia Bank, Spain; Gorringe Bank, Portugal), and in waters beyond national jurisdiction (Josefine Bank, between mainland Portugal and Madeira Portuguese EEZs) (http://www.ngo.grida.no/wwfneap/Projects/MPAmap.htm, last accessed 21 September 2006). Information on seamounts in the NE Atlantic has been reviewed (Gubbay, 2003). Global experiences with seamount protection and management were compiled and applied to the OSPAR and EU context to enhance the establishment of seamount MPAs in the NE Atlantic (Schmidt and Christiansen, 2004). By the end of 2006, only the Formigas/Dollabarat Bank and the D. João de Castro Seamount have been nominated (by Azores/Portugal) for inclusion in the OSPAR MPA network, due to be finalized by 2010 (see section below on national action). No other seamounts or offshore banks are expected to be listed in the near future, as most Contracting Parties are struggling with selecting and designating sites for the Natura 2000 network in territorial waters, which is an EU commitment (see below). WWF has also prepared a justification for the Lucky Strike as an offshore MPA and the Rainbow Hydrothermal Vent Field for a high seas MPA. In 2006, these two deep-sea sites, together with Menez Gwen (see also Santos *et al.*, 2003), were proposed by Portugal/Azores as offshore MPAs. Both OSPAR and WWF are proceeding with the evaluation seamounts and/or other deep-sea habitats as potential MPAs beyond national jurisdiction in the NE Atlantic.

North East Atlantic Fisheries Commission

The NEAFC has the competence to regulate fisheries through various mechanisms, including gear restrictions and closed areas, in the waters beyond national jurisdiction in the NE Atlantic. This competence is limited to activities related to the conservation of 'fishery resources' (Article 1, para. 2). Consequently, NEAFC does not have a clear mandate to establish closed areas to protect seamounts or cold-water corals solely for their biodiversity. Widening the scope of NEAFC to include all living marine resources and the effect of fisheries on other parts of the ecosystem are being examined (NEAFC, 2005a).

NEAFC has responded to growing international pressure to protect cold-water corals and seamounts. Understanding that it should be more proactive pending the expansion of its legal mandate, NEAFC agreed in 2004 to close four seamounts to fishing with demersal trawl and static gear for 3 years, two on the Mid-Atlantic Ridge, and two in the ocean basins to the east and west, as well as a section of the Reykjanes Ridge (NEAFC, 2005b) (Fig. 20.3). On the other hand, proposals from OSPAR and Norway to close the western slopes of Rockall Bank and part of the Hatton Bank to fisheries were not accepted, both areas being under significant fishing pressure. ICES (the body advising NEAFC, OSPAR, and European Commission) subsequently advised that closures were necessary to protect

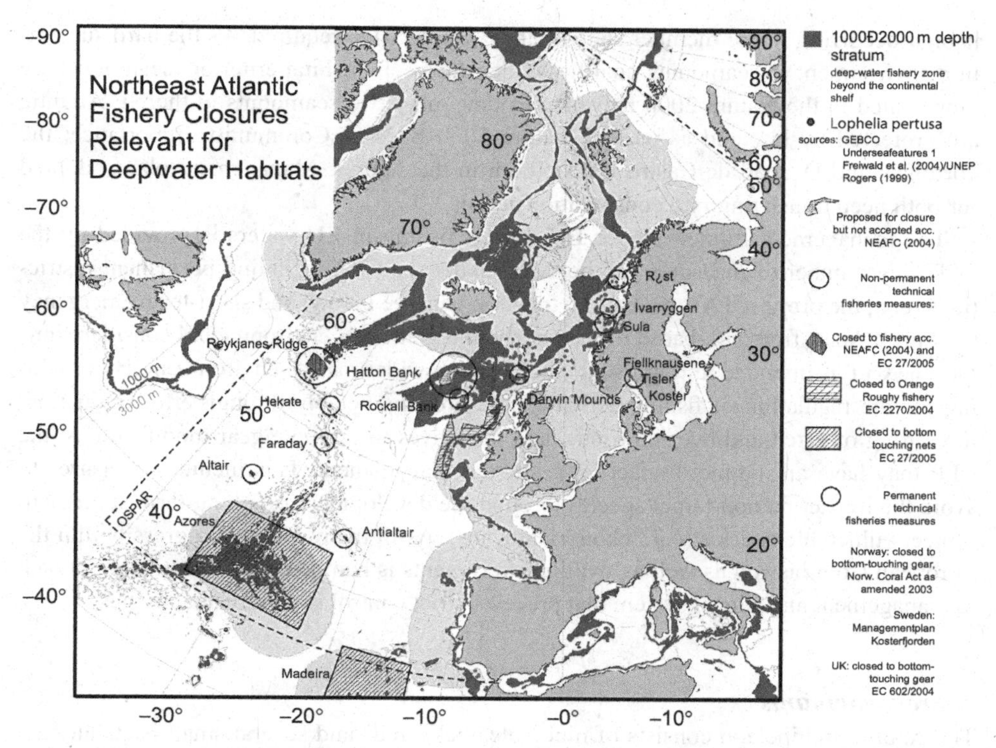

Fig. 20.3 Map of fishery closures in the northeast Atlantic of benefit to deep-water habitats. Key to shading at right (*Source*: WWF International redrawn by Les Gallagher *Imag*DOP).

corals in these two areas from fishing impacts (ICES, 2005). This is consistent with the ICES advice that the deep-water fishery is unsustainable and that, 'the only proven method of preventing damage to deep-water biogenic reefs from fishing activities is through spatial closures to towed gear that potentially impacts the bottom' (ICES, 2002).

European Union

A key environmental objective agreed by EU Heads of State and Government Council at Göteburg, Sweden, June 2001, is 'to halt the decline of biodiversity by 2010'. The EU biodiversity policy is developing in the context of international commitments and targets (e.g., those agreed at WSSD, CBD Conference of the Parties 7, and by the regional conventions for the NE Atlantic (1992 OSPAR Convention), Baltic (1992 Helsinki Convention), and Mediterranean (1995 Barcelona Protocol)). Since 2002, the European Commission has sought to develop an ecosystem-based approach to promote the sustainable use of the seas and conservation of marine ecosystems, paying special attention to sites of high biodiversity.

The legal framework for establishing a European network of protected areas, Natura 2000, consists of two major instruments: the EU Birds and Habitats Directives (79/409/EEC, 92/43/EEC). The Habitats Directive lists marine habitats and species for which protected areas have to be designated. As a rule, a minimum of 20% of the respective habitat occurring in the member state (Hab 97/2 Rev. 4) is required. As the hard substratum environment of seamounts qualifies under the 'reefs' habitat criterion, seamounts are represented in the Natura 2000 network. To date, only two seamounts in the NE Atlantic are protected by law and as part of Natura 2000 Sites of Community Importance: the Formigas and D. João de Castro Banks, both in the Azores. Management plans adopted for both need translating into conservation action.

The management framework for EU vessels fishing in EU waters is provided by the CFP. Any number of measures can be taken under the CFP, including permanent restrictions (e.g., the proposed Azores bottom trawling ban, see below) and short-term emergency measures. Apart from site-based measures, the integration of environmental considerations into the CFP is intended to improve its focus on the wider marine environment. By providing for the regulation of fishing activity more broadly, notably through effort reduction, the setting of catch and bycatch limits, and incentives to promote gear modifications, the CFP may have an indirect impact on seamount management. This includes measures to avoid the impact on non-target species through the development of a long-term strategy to protect vulnerable species (e.g., chondrichthyans). As in many places, progress within the European Commission as well as within governments is hampered by a sectoral approach to management and a legal system that prevents strict control over human activities.

Azores seamounts

The Azores archipelago consists of nine volcanic islands and several small islets along a tectonic zone in the mid-NE Atlantic (37–40°N, 25–32°W) (Fig. 20.4). The archipelago has the largest EEZ in the NE Atlantic, more than 1 million km^2, with only 8% shallower than 1500 m, predominantly on the slopes of the islands and on emerging summits of seamounts and offshore banks. Some 140 seamounts are known in the Azores, and fishing focuses on these features (see Chapter 16; Machete *et al.*, 2005).

The marine environment of the Azores and its EEZ is of considerable conservation and biological interest, largely because of its isolation and relatively pristine condition due to pursuit of traditional rather than industrial fishing (see Chapter 16; Santos *et al.*, 1995). Nevertheless, Azorean waters have been affected by increasing human activity. As a result of ecological deterioration over recent decades, various legislative measures have been taken to preserve marine species and habitats. In 1988, the Azores government designated an MPA of 377 km^2 to enclose the Formigas Islets and Dollabarat Bank and to 1800 m water depth. This constituted one of the first seamount MPAs. The only fishing activity then allowed was demersal and pelagic hook fishing with open deck boats smaller than 14 m and pole and line for tuna, but in 2003 the MPA was increased to 595 km^2 and the restrictions extended to all fisheries. The maximum depths remained more or less the same.

With the application of the EC Birds and Habitats Directives in the Archipelago, 17 marine Sites of Conservation Importance (SCIs) and 13 marine Special Protection Areas (SPAs) were designated (Council Regulation (EC) No. 11/2002). These include the D. João de Castro seamount and the Formigas/Dollabarat complex, but the maximum depth for the SCIs is 200 m. In mid-2006, all sites were integrated in a Sectoral Plan for Natura 2000 in the Azores (Decreto Legislativo Regional No. 20/2006/A) which establishes the scope and framework of conservation measures. However, whilst Natura 2000 has been the main driver for management and conservation of marine habitats in the region, it is limited in scope (Santos, 2004). Anticipating a revision of Natura 2000 to include habitats beyond 200 m deep, the Azores government has recently proposed Lucky Strike and Menez Gwen as two new sites for the network, under the classification of reefs.

An important aspect of the conservation of Azorean deep-sea resources is that the Azores government and the fishing sector decided to support the artisanal fishery to ensure long-term sustainable use of the resource rather than industrializing the fishery. Prohibition of bottom trawling and deep gillnetting, as well as pole-and-line fishing and

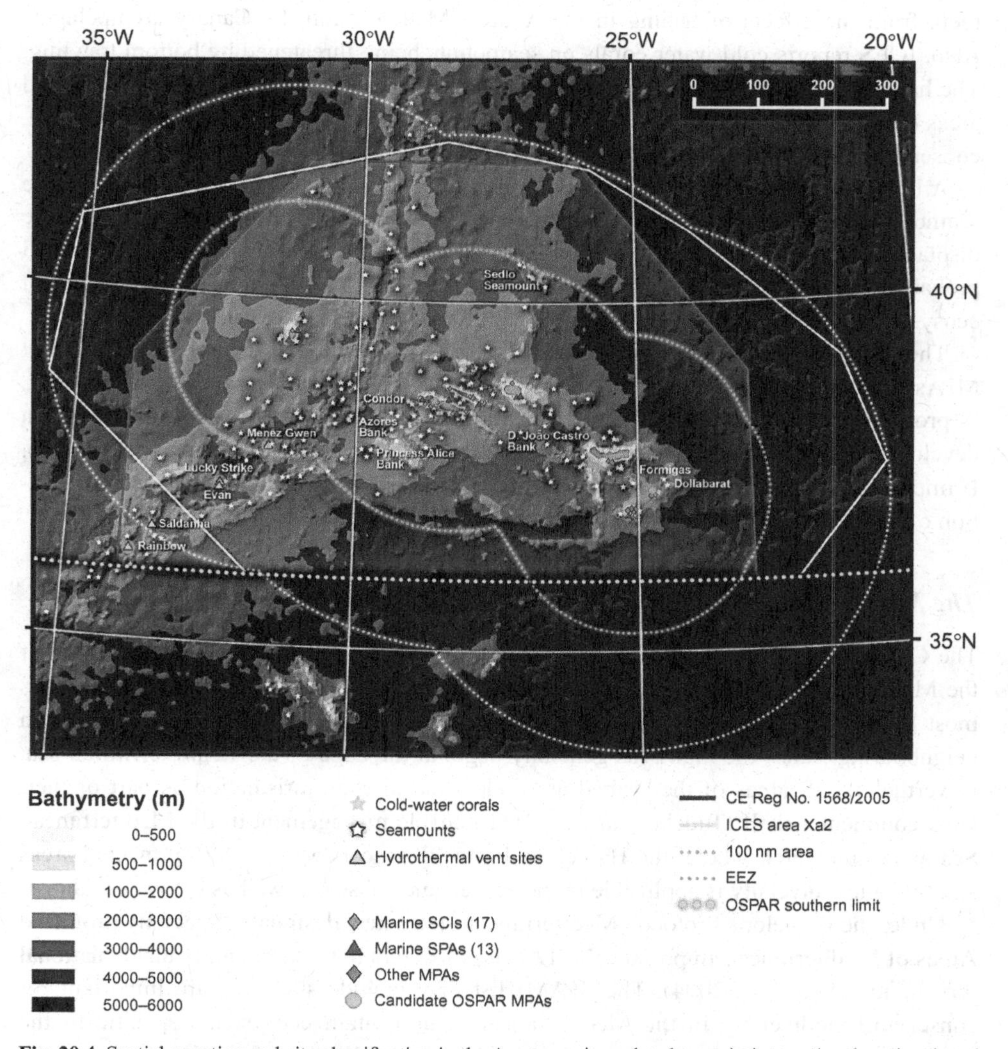

Fig. 20.4 Spatial zonation and site classification in the Azores region related to main international, national, and regional conventions and laws (e.g., ICES, OSPAR, the CFP 100 and 200 nm regulations, the EC Reg. 1568/2005 for the protection of deep-sea habitats, Natura 2000 habitat, and birds directive SIC and SPA sites, regional OSPAR MPAs) and indication of known location of OSPAR priority deep-sea habitats: cold-water corals (A. Henriques *et al.* based on several sources, DOP database), seamounts (Morato *et al.*), and hydrothermal vent sites.

reduced pelagic longlining, has resulted in a relatively low level of exploitation compared to that of surrounding waters.

Until the end of 2003, management of fishing resources in the 200 nm zone around the Azores was the responsibility of the Azores government. Special access agreements under the EU CFP reserved the fishing rights exclusively to Portuguese vessels in the Azorean EEZ. Revised legislation (Council Regulation (EC) No. 1954/2003) now allows the European deep-water fleet to fish in the outer 100 nm from the islands. Bottom trawling has been temporarily banned since 2004 (Council Regulation (EC) No. 1811/2004) on the grounds that the previous management regime forbade it. An extended regulation was adopted a year later (Council Regulation (EC) No. 1568/2005) to protect deep-water coral reefs from the effects of fishing in the Azores, Madeira, and the Canary archipelagos. Also, ICES reports cold-water corals on seamounts being threatened by bottom trawling. The hard bottom habitat hosting the corals requires designation of Natura 2000 protected areas (see above) and is on the OSPAR list of threatened habitats (see above). Therefore, conservation action has to be taken.

A ban on bottom trawling will not prevent an increase in fishing effort on and near the seamounts, since large statistical rectangles used for setting of fishing quotas allow for the displacement of effort of foreign vessels into the former Azorean fishing zone. This will probably result in the usual pattern of consecutive overfishing of seamounts, with concurrent ecosystem effects, and loss of fishing opportunities for the local community.

The Azores government is going forward with the designation of several more seamount MPAs using the Natura 2000 and OSPAR framework within a regional MPA network. A proposal to manage Sedlo seamount in the north of the Azores as a MPA has been developed under the EC research project OASIS (Gubbay, 2005; Neumann *et al.*, 2005). If implemented and well enforced this network will be of major importance to conservation of species and habitats of the northern Mid-Atlantic Ridge.

The Mediterranean

The Convention for the Protection of the Marine Environment and the Coastal Region of the Mediterranean 1976, as amended in 1995, applies throughout the Mediterranean. As most Mediterranean coastal states have not declared EEZs, due to boundary issues with neighbouring states, the high seas generally begin at the edge of the 12 nm territorial sea. Nevertheless, all areas of the seabed are under coastal state jurisdiction as part of their legal continental shelf (Tudela *et al.*, 2004). To enable management of the Mediterranean Sea as an integrated whole, the Barcelona Protocol concerning specially protected areas and biological diversity is applicable to the seabed and subsoil as well as the waters above.

Under the Barcelona Protocol, Mediterranean states may designate 'Specially Protected Areas of Mediterranean Importance' (SPAMIs) in waters both within and outside national jurisdiction (Scovazzi, 2004). The SPAMI list may include sites that are important for conserving biodiversity in the Mediterranean, that contain ecosystems specific to the Mediterranean or the habitats of endangered species, or that are of special interest at the scientific, aesthetic, cultural, or educational levels (Article 8.2). SPAMIs beyond national jurisdiction must be agreed by consensus. Once agreed, all Parties to the Barcelona Protocol are bound to comply with management measures and may neither approve nor undertake any actions that might harm them. Again, the Protocol excludes fisheries regulation. One high seas SPAMI already exists, the Pelagos Sanctuary for Mediterranean Marine Mammals (accepted 2001). About half the Sanctuary area is located in international waters beyond national jurisdiction, the rest is within internal maritime and territorial waters of France, Italy, and Monaco (in total 84 000 km^2).

WWF and IUCN strategy for seamount protection

There are substantial numbers of seamounts in the Gulf of Lions, Alboran Sea, eastern Tyrenian Basin, south of the Ionian Abyssal Plain, and in the Levantine seas. Biological data on most of these are scarce, and the best-known seamount is Eratosthenes south of Cyprus. WWF and IUCN proposed a two pronged approach for the conservation of biodiversity in the Mediterranean: firstly, a trawling ban in waters below 1000 m depth, which was adopted in 2005 by the General Fisheries Commission for the Mediterranean (the relevant RFMO); and secondly, inclusion of the best-known Mediterranean seamounts in a representative network of MPAs plus, on a precautionary basis, those of presumed biodiversity importance and most threatened by the potential development of human activities (Tudela *et al.*, 2004).

Southern Ocean

Antarctic Protocol on Environmental Protection

The 1959 Antarctic Treaty area covers the continent and surrounding seas south of 60°S. Territorial disputes have been set aside, so for the most part the marine regions are regulated as if the high seas begin at the shoreline. The 1991 Protocol on Environmental Protection and its five annexes provide a broad remit to the Antarctic Treaty Parties to conserve and protect the environment and its biota. Annex V on Area Management provides for two categories of protected area: Antarctic Specially Protected Areas (ASPAs) and Antarctic Specially Managed Areas (ASMAs). To date, six ASPAs have been established that are fully marine. Two of these are quite large, the Western Bransfield Strait (900 km^2) and Eastern Dallman Bay (580 km^2), and have benthic fauna of particular interest (CBD, 2005). The Environmental Protocol covers most human activities. The principal exception is fishing, which is regulated by the Commission for the Conservation of Antarctic Marine Living Resources.

Convention on the Conservation of Antarctic Marine Living Resources

Of the 30 plus RFMO agreements, the Convention on the Conservation of Antarctic Marine Living Resources (CCAMLR), adopted in 1981, is one of the few to reflect ecosystem-based management. This approach applies to the conservation and rational use of marine living resources out to the limits of the Treaty area. It extends beyond fisheries to cover all associated and dependent ecosystems and species. Recognizing that the Southern Ocean ecosystem is poorly understood, difficult in terms of management and of extremely high conservation value, the Convention calls for cautious and conservative use. The measures taken include conventional fishery closures, precautionary catch limits, bycatch minimization measures, and an ecosystem monitoring programme.

The CAMLR Commission has recently initiated action to establish a comprehensive system of MPAs. In 2005 CCAMLR agreed on the need to develop a strategic approach to MPA design and implementation throughout the CCAMLR area, in harmony with measures taken under the Antarctic Treaty and the Madrid Environmental Protection Protocol (Grant, 2005). CCAMLR agreed on some basic parameters for MPAs with the general goal of maintaining biodiversity and ecosystem processes that could include the protection of the following:

1. Representative areas.
2. Scientific areas to assist with distinguishing between the effects of harvesting and natural ecosystem changes, and to provide opportunities for understanding of the Antarctic marine ecosystem in areas not subject to human interference.
3. Areas potentially vulnerable to impacts by human activities, to mitigate those impacts, and/or ensure sustainability of the rational use of marine living resources.

In November 2006 CCAMLR adopted a Conservation Measure restricting the use of bottom trawling gears in high seas areas of the Southern Ocean for a 2-year period. The measure also calls for a review of the use of bottom trawling gear in 2007 based on scientific criteria to determine what constitutes, 'significant harm to benthos and benthic species' (CCAMLR Conservation Measure 22-05).

North West Atlantic

The Discovery Corridor Concept of the Department of Fisheries and Oceans Canada (DFO), represents an innovative way to focus research and education on a cross section of the Atlantic continental margin. The long-term objective is to provide advanced knowledge on conservation concepts and establish a network of 'biodiversity reference areas' encompassing both unique and representative areas. These areas could provide controls against which to measure changes in non-reference areas (Centre for Marine Biodiversity, 2004). The Gulf of Maine Pilot Corridor, a trans-boundary area along the Canadian–US border, extends from coastal areas to abyssal plains and includes the New England Seamount chain. Four seamounts within the US EEZ off the southern edge of Georges Bank (Bear, Physalia, Retriever, and Mytilus seamounts) and the minor topographic rises surrounding them, were proposed for designation as 'Habitat of Particular Concern' to the New England Fisheries Management Council in March 2005 (Auster and Watling, 2005). As these seamounts are presently not fished for benthic or demersal species, this is meant to be a precautionary measure to ensure long-term conservation of sensitive benthic habitats, including a highly diverse and abundant cold-water coral fauna.

North East Pacific

Bowie Seamount Pilot MPA

Bowie Seamount is located 180 km west of Haida Gwaii (the Queen Charlotte Islands) in Canada's NE Pacific. It rises to within 25 m of the surface from a depth of over 3000 m and is a productive ecosystem, rich in rockfish (*Sebastes* spp.) and benthic organisms

(McDaniel *et al.*, 2003; WWF, 2003). Bowie's shallow habitat is uncommon in the open ocean and its distance from the coast makes it relatively isolated ecologically; the only targeted commercial fishery is for sablefish (*Anoplopoma fimbria*).

In 1998, DFO identified Bowie Seamount as a pilot MPA, and the area of interest has since expanded to include the neighbouring Hodgkins and Davidson Seamounts. But since establishing the pilot MPA, the process has stalled for a number of reasons. The Haida First Nation supports the site, but discussion about shared management of the area with the Government of Canada is tied to larger rights and title issues, making it complex to develop an agreement specifically around Bowie. Parties continue to work towards resolving the issue.

Davidson Seamount

Davidson Seamount is about 120 km southwest of Monterey, California, outside the southern part of the Monterey Sanctuary. It rises 2300 m from the seafloor with a summit at 1300 m water depth. Two government entities are considering measures that could protect the seamount from bottom trawling and other potential threats. The National Marine Sanctuary Program has proposed expanding the Monterey Bay Sanctuary to include the seamount and to prohibit fishing on it. The Sanctuary Program has given the Pacific Fishery Management Council, a regional fishery advisory body, and the option of drafting any fishing regulation. If successful, the Sanctuary Program will have the authority to provide comprehensive protection to the Davidson Seamount ecosystem. Another option is for federal fishery management agencies to adopt measures to minimize adverse effects of fishing on essential fish habitat of groundfish, as required under the Magnuson-Stevens Fishery Conservation and Management Act (16 USC 1853 Section 303(a)(7)). The Pacific Fishery Management Council and National Marine Fisheries Service (NMFS), the agency responsible for managing fisheries in the US 200 nm zone, are considering proposed amendments to the Pacific Groundfish Fishery Management Plan that would prohibit bottom trawling or bottom fishing on Davidson Seamount. NMFS is scheduled soon to make a final decision on those proposals, which have strong public support.

Alaska

In 2005, almost 1 million km^2 of seafloor, including seamounts, off the Aleutian Islands, USA, became exempt from bottom trawling, including 380 km^2 banned to all bottom gear in the deep-sea coral and sponge gardens in the Aleutian Islands. In the Gulf of Alaska, 7200 km^2 of seafloor are now closed to bottom trawling (www.fakr.noaa.gov/npfmc/current_issues/HAPC/HAPCmotion205.pdf and http://www.fakr.noaa.gov/npfmc/current_issues/HAPC/HAPCmaps205.pdf, last accessed 22 September 2006). The regulation was a compromise: the industry can keep its traditional trawling grounds, but cannot expand to those areas now closed because of their conservation status. Trawlers will also be allowed to continue netting fish in Aleutian areas historically yielding the best catches.

Indonesian Seas

In the Sulu-Sulawesi, one of the deepest regional seas (over 4300 m), seamounts are a feature of deep, yet nearshore, waters (Kahn, 2003), and so can often be managed under national or even district maritime jurisdictions relating to the establishment of protected areas or management of fisheries. Some nearshore seamounts are of tourism interest. In the rest of the eastern Indonesian Seas (Flores-Banda and into Bismarck-Solomon Seascapes) seamounts are plentiful in the deep basins, often rising >2000–3000 m. Regionally, seamounts are specifically identified as oceanic habitats important for cetaceans, billfish, and tunas (Fortes *et al.*, 2003). Several assessments are underway in Indonesia to include seamounts in MPA sites and in ecoregional/seascape planning. Indonesia's new National Committee for Ocean Conservation is developing options for improved seamount management.

Australia

In 1999, Australia declared the Tasmanian Seamounts Marine Reserve (370 km^2), 100 km south of Tasmania, to protect 15 seamounts representative of the region. The management plan details a vertical zoning into a 'strict nature reserve' (IUCN Category Ia) from 500 m water depth to 100 m below the seabed (to account for any extractive operations) and a management zone in the upper 500 m of the water column (IUCN Category VI). In the lower zone, fishing and extractive operations are not permitted, and an access permit is required to enter the reserve at or below 500 m water depth. In the upper 500 m, access to commercial pelagic fleets using non-trawl methods is by permit (Commonwealth of Australia, 2002). A research and monitoring programme, integral to the implementation of the management plan, has yet to be established.

New Zealand

Rapid development of New Zealand seamount fisheries during the 1990s, and increasing awareness of the nature of seamounts and impact of fishing prompted the Ministry of Fisheries to close 19 seamounts in 2001, and develop a seamount management strategy (Brodie and Clark, 2004). The closures were declared after 2 years of negotiations with the fishing industry. Although the closed seamounts are relatively deep and unfished, the industry questioned the legality of the closures. Further research and monitoring of the closed seamounts is planned.

High seas seamount conservation

Strategies for seamount conservation in areas beyond national jurisdiction include making better use of existing institutions and tools as well as further action where these fall short (de Fontaubert, 2001).

High seas MPAs

States can adopt MPAs and other conservation measures for seamounts in national waters, but in areas beyond national jurisdiction; this can be done only in cooperation with other states, based on scientific information and consistent with international law. A Ten-Year Strategy for the development of a system of high seas MPA networks developed by IUCN, government, and NGO experts (IUCN, 2004) highlights short-, medium-, and long-term steps that states, NGOs, scientists, and others can pursue to establish high seas MPAs (see CBD, 2005, for more detailed information on high seas MPAs). A number of regional instruments already authorize the establishment of protected or strictly managed areas in waters beyond national jurisdiction. Thus in the NE Atlantic, Mediterranean, and Southern Ocean, states could include high seas seamounts within regional networks of protected areas. Protective measures would be binding on regional participants, though not on non-Parties from outside the region. To protect a seamount area or cluster from adverse impacts of shipping, seabed mining, or fishing on the high seas, it is generally necessary to obtain the consent of the relevant international organization or regional fisheries body. The IMO may designate 'Special Areas' where vessel discharges are more stringently regulated, and Particularly Sensitive Sea Areas (PSSAs), which alert the maritime community to the need for a high level of caution. PSSAs also serve as an umbrella for the adoption of other IMO routing, reporting, and discharge-related measures. The ISA is authorized to set aside areas as 'preservation reference zones' in which no mining shall occur. The FAO is now developing guidelines for the use of MPAs for fisheries-related purposes, and most RFMOs have the ability to adopt closed areas and other specially managed areas for conservation of target fish stock and sometimes for broader biodiversity purposes (CBD, 2005).

In addition to the sector-specific approach, there are five mechanisms under existing international law through which countries can agree to regulate their activities to protect a high seas area:

1. Through the Convention on Migratory Species and its provision for sub-agreements, states can identify and protect seamounts that serve as important breeding or feeding habitats or are located along migration corridors for species such as sea turtles, marine mammals, and seabirds.
2. States can agree to pursue a new agreement to protect a specific seamount or cluster, e.g., the agreement between Canada, France, UK, and US to preserve the wreck of the *Titanic*.
3. Regional seas conventions and bodies that cover areas beyond national jurisdiction could establish cooperative programmes with other organizations that focus on specific areas. For example, regional seas organizations and RFMOs can agree to adopt complementary programmes and regulations aimed at protecting biodiversity in specific seamount areas.
4. Nations and regional and international organizations could develop cooperative programmes to protect seamount areas that straddle national and international waters.
5. MPAs can also be based on non-binding agreements and codes of conduct. Non-legal designations such as UNESCO Biosphere Reserves or PSSAs could be used to highlight the significance of the area and to encourage the adoption of protective measures for shipping and other impacts. Voluntary codes of conduct can also be used to guide marine scientific research, cable laying, bioprospecting, and other potentially harmful activities. These can be adopted by a relevant international organization (e.g., the Intergovernmental Oceanographic Commission) or through professional associations.

In summary, there are opportunities for states to use existing legal agreements and to develop agreements both voluntary and binding among themselves to protect specific seamount areas. This requires strong political support from key states, clear identification of potential candidates, and high-quality scientific information. The cooperation of all relevant user groups is essential and would be difficult without a legally binding agreement on enforcement.

Pursuing seamount conservation through RFMOs

RFMOs could provide the most direct route to establishing protective measures for seamount ecosystems beyond national jurisdiction. Under the Fish Stocks Agreement, RFMOs should be able to enforce gear restrictions, as well as closed or strictly regulated areas, to protect seamounts from harmful fishing activities. Of over 30 RFMOs, few have a mandate for managing deep-water fisheries. Even fewer have measures to prevent destruction of vulnerable ecosystems under their remit (see above; High Seas Task Force, 2005). RFMO competence often extends only to pelagic fishing activities and only permits them to regulate activities to conserve target fish stocks, not living marine resources or biodiversity in general (Molenaar, 2005).

In recent years, the global call for implementing conservation measures has become ever louder. Thus more protective measures, such as those adopted by CCAMLR, NEAFC, and the General Fisheries Commission for the Mediterranean, can be expected. However, the small scale of areas protected in most regions compared to the areas where bottom trawling is still permitted suggests that RFMOs other than CCAMLR remain reluctant to act proactively. Most still prefer to act only when a strong link can be demonstrated between seamounts and essential habitat for commercial fisheries (and the impacts on existing fisheries are minimal). This requires a high level of scientific data and certainty, elements lacking for the vast majority of seamounts and associated fisheries.

Efforts are required to improve the capacity of RFMOs to ensure that: (1) ecosystem-based management is applied, including the elimination of destructive fishing practices and the use of MPAs; (2) all areas and all fisheries are regulated; and (3) fishery management organizations are effective, transparent, participatory, and accountable. This would include establishing new management bodies or expanding the legal authority of RFMOs or other bodies to enable them to adopt conservation measures as envisaged by the Fish Stocks Agreement to protect biodiversity (including on seamounts) wherever high seas fishing activities occur. Many are also proposing some form of institutionalized oversight of the performance of RFMOs to increase their accountability (Lodge and Nandan, 2005). States have so far agreed only to a self-assessment by RFMOs, not the independent audit that is required for full transparency and accountability.

In the broader context of high seas fisheries governance, efforts are also required to eliminate the overcapacity of fishing vessels and the excessive killing power of current fishing technology, end illegal, unreported, and unregulated fisheries and establish mechanisms for effective high seas enforcement (Gjerde, 2005).

Integrated seamount conservation and management in international waters

Efforts to ensure integrated, comprehensive, and precautionary management for all the high seas are essential. Seamounts are just one example of high seas and deep ocean biodiversity impacted by human activities. Efforts to establish MPA networks for seamounts in areas beyond national jurisdiction will be stymied if outside influences undermine them, or if other threats to high seas ecosystems and biodiversity are not controlled.

An implementing agreement to UNCLOS proposed by the EU could improve global efforts at conservation and sustainable use of marine biodiversity in areas beyond national jurisdiction. Such an agreement could ensure regulation of all human activities, including destructive fishing practices, provide mechanisms for the establishment and enforcement of an integrated network of MPAs, and improve coordination between existing regulatory frameworks and bodies with competence over specific areas or activities. This remains a key topic for discussion within the United Nations.

Conclusions

Few seamount ecosystems have been studied in detail, and there is an urgent need to greatly increase our understanding of biology and ecology over the full spectrum of seamount morphology and oceanographic conditions. There is wide scientific consensus that seamount ecosystems are under threat, and that effective management strategies are urgently needed. Seamount conservation worldwide lags behind recognition of the problem and behind measures for coastal areas. Questions regarding deep-water fisheries on seamounts were raised in the early 1990s (by ICES, Northwest Atlantic Fisheries Organization), but concern has gathered momentum. Nevertheless, there are still calls for more proof of damage to seamount ecosystems. Information on sensitivity and vulnerability of seamount ecosystems points to the necessity of taking a highly precautionary approach to managing activities close to seamounts.

Pelagic fisheries in waters beyond national jurisdiction are covered by some agreements, though not necessarily successfully. Demersal fisheries on the other hand are more or less unregulated. RFMOs, where in place, and competent to regulate demersal fisheries, have only recently had the mandate to act on the basis of environmental concerns. Ecosystem-based management of human activities, in particular fisheries, is required which aims at long-term sustainable fishing based on non-destructive practices.

Governance solutions are needed because of the complexity or absence of a regulatory framework, as in waters beyond national jurisdiction. There is a need to establish a new UN treaty and authority to implement the requirements of UNCLOS for the conservation and sustainable use of marine biodiversity in areas beyond national jurisdiction. This could ensure precautionary regulation of all human activities and provide for the establishment and enforcement of an integrated network of MPAs. Such a move could also help in coordinating the actions of existing regulatory frameworks and bodies with competence over specific areas or activities (e.g., FAO and RFMOs for fisheries and ecosystem conservation and management; IMO for attention to high seas areas; ISA for deep seabed mining and environmental protection), and make codes of conduct stronger and more explicit.

Many management instruments are known (e.g., MPAs, closed areas, site-based effort control, licensing, gear restrictions), but they have not been widely applied to seamounts, nor have their effects been properly tested at seamounts. Although existing fisheries management tools are available, management units may need to be reduced to reflect the seamount scale of fisheries. Also, fundamental fisheries management questions, such as subsidies, may mask the real productivity of deep-sea fisheries and prolong the economic viability of fishing depleted stocks.

Policy and management must be supported by a robust body of scientific information from seamount researchers (hence the role of SeamountsOnline and Census of Marine Life on Seamounts (CenSeam) projects, see Stocks *et al.*, 2004). Lack of knowledge due to lack of research and insufficient knowledge due to the vastness and complexity of the area to be investigated should not hamper action towards conservation and sustainable management. The precautionary approach has to be the guiding principle for management, so that environmental vulnerability is assessed before any new activity starts.

Recommendations

An array of short-, medium-, and long-term options will be needed to achieve the 2002 WSSD's goals of (1) ecosystem-based oceans and fisheries management by 2010, (2) prohibition of destructive fishing practices, (3) establishment of MPA networks by 2012, and (4) fisheries closed areas. The short-term options are essential to ensure there is still something left to protect when longer-term reforms are enacted.

Short term

States, scientists, and conservation organizations to

1. Work for the immediate development and adoption of precautionary regulations and measures to protect, conserve, and sustainably manage human activities at seamounts to prevent deliberate or accidental damage.
2. Work for adoption of a UNGA resolution for an interim prohibition on high seas bottom trawling until scientists have adequately identified and studied seamounts, and a more comprehensive protective regime has been put in place.
3. Develop an exchange of information and data with the sectors and industry associations that potentially operate at seamounts.

Coastal states to

1. Identify and protect seamounts in their waters from harmful practices (as is their obligation under UNCLOS Articles 192, 193, 194.5).
2. Invest more research, including the use of modern technologies for habitat mapping and visual inspection of seamounts, in particular where they are far offshore.

Scientists, scientific, and conservation organizations to fund and cooperate in developing and improving the SeamountsOnline database and maps to make it the central global seamounts database.

Coastal states and the scientific community to strengthen their cooperation, exchange of expertise, and logistical support for more efficient seamount research.

Medium term

States to use existing legal agreements and develop voluntary agreements to protect specific seamount areas. Concerned states can take the first steps and encourage other states to join in, or work in collaboration with many. States, scientists, and conservation organizations to

1. Implement the FAO Code of Conduct in waters under their national jurisdiction, and states and RFMOs with respect to high seas fisheries. This could significantly improve deep-sea fisheries management and protection of seamount ecosystems.
2. Support proposals for protective and precautionary measures for seamounts through RFMOs, and to work within regional seas agreements to elevate the priority of seamount conservation, including through MPAs.
3. Promote full implementation of the Fish Stocks Agreement, as well as its extension or a similar agreement to discrete high seas fish stocks, complemented by a global mechanism to oversee RFMO performance.
4. Promote reforms within RFMOs as well as more broadly in high seas fisheries governance to enable resource management bodies to adopt ecosystem-based management. This would address the need to eliminate destructive fishing practices, reduce bycatch, and adopt MPAs (and closed areas) for both fisheries conservation and biodiversity protection.
5. Promote and conduct impact assessments prior to the start of human activities directly or indirectly affecting seamount communities and habitats and take the necessary measures to prevent, reduce, or eliminate adverse effects.
6. Become more heavily involved in the development of regulations at the International Seabed Authority for exploration and prospecting of polymetallic sulphides and cobalt crusts on seamounts. As required by UNCLOS, such regulations should prevent pollution, protect and conserve the natural resources, and prevent damage to the marine biota.

Long term

States, scientists, and conservation organizations to promote the development of an implementing agreement to UNCLOS for the conservation and sustainable use of marine biodiversity in areas beyond national jurisdiction. This could ensure regulation of all human activities and elimination of destructive fishing practices, provide mechanisms for the establishment and enforcement of an integrated network of MPAs, and better coordinate the actions of existing regulatory frameworks and bodies with competence over specific areas or activities.

Coastal states need to ensure that an adequate representation of the range of seamount types is included in national and regional networks of MPAs.

Acknowledgements

We thank Benjamin Kahn (APEX Environmental), Karen Garrison (NRDC), and Michele Patterson (WWF Canada) for kindly providing information; and reviewers for useful comments on earlier drafts.

References

Agardy, T.S. (1997) *Marine Protected Areas and Ocean Conservation.* R.G. Landes Co., Georgetown, Texas and Academic Press, San Diego, CA.

Alder, J. and Wood, L. (2004) Managing and protecting seamount ecosystems. In: *Seamounts: Biodiversity and Fisheries* (eds. Morato, T. and Pauly, D.). Fisheries Centre, University of British Columbia, Canada. *Fisheries Centre Research Report*, 12(5), 67–75.

Anderson, O.F. and Clark, M.R. (2003) Analysis of bycatch in the fishery for orange roughy, *Hoplostethus atlanticus*, on the South Tasman Rise. *Marine and Freshwater Research*, 54, 643–52.

Andrews, A.H., Cailliet, G.M., Kerr, L.A., Coale, K.H., Lundstrom, C. and DeVogelaere, A.P. (2005) Investigations of age and growth for three deep-sea corals from the Davidson Seamount off central California. In: *Cold-Water Corals and Ecosystems* (eds. Freiwald, A. and Roberts, J.M.), pp. 1021–38. Springer-Verlag, Berlin/Heidelberg.

Arico, S. and Salpin, C. (2005) *Bioprospecting of Genetic Resources in the Deep Seabed: Scientific, Legal and Policy Aspects*. United Nations University, Institute of Advanced Studies, Yokohama, Japan.

Auster, P.J. (1998) A conceptual model of the impacts of fishing gear on the integrity of fish habitats. *Conservation Biology*, 12, 1198–203.

Auster, P.J. (2005) Are deep-water corals important habitats for fishes? In: *Cold-Water Corals and Ecosystems* (eds. Freiwald, A. and Roberts, J.M.), pp. 747–60. Springer-Verlag, Berlin/Heidelberg.

Auster, P. and Watling, L. (2005) The New England Seamounts HAPC. In: *Habitat Area of Particular Concern Candidate Proposal Submission Form* (ed. Christiansen, S.). New England Fishery Management Council, Newburyport, MA.

Birkeland, C. and Dayton, P.K. (2005) The importance in fishery management of leaving the big ones. *Trends in Ecology and Evolution*, 20, 356–8.

Boehlert, G.W. and Genin, A. (1987) A review of the effects of seamounts on biological processes. In: *Seamounts, Islands and Atolls* (eds. Keating, B., Fryer, P., Batiza, R. and Boehlert, G.). *Geophysical Monograph*, 43, 319–34.

Brodie, S. and Clark, M.R. (2004) The New Zealand seamount management strategy – steps towards conserving offshore marine habitat. In: *Aquatic Protected Areas: What Works Best and How Do We Know*? (eds. Beumer, J.P., Grant, A. and Smith, D.C.). *Proceedings of the World Congress on Aquatic Protected Areas*, Cairns, 2002. Australian Society for Fish Biology, Australia, pp. 664–73.

Cardigos, F., Colaço, A., Dando, P.R. *et al.* (2005) Shallow water hydrothermal vent field fluids and communities of the D. João de Castro Seamount (Azores). *Chemical Geology*, 224, 153–68.

CBD (Convention on Biological Diversity) (2004) Decision VII/5. *Marine and Coastal Biological Diversity*. UN Document UNEP/CBD/COP/7/21, pp. 134–77.

CBD (Convention on Biological Diversity) (2005) *The International Legal Regime of the High Seas and the Seabed Beyond the Limits of National Jurisdiction and Options for Cooperation for the Establishment of Marine Protected Areas (MPAs) in Marine Areas Beyond the Limits of National Jurisdiction.* Ad Hoc Open-Ended Working Group on Protected Areas, First Meeting, Montecatini, Italy, June 13–17, 2005. UNEP/CBD/WG-PA/I/INF/2 (April 28, 2005).

Centre for Marine Biodiversity (2004) *Report of a Workshop on the Discovery Corridor Concept and Its Applicability.* St Andrews Biological Station – January 13–14, 2004. Available at http://www.marinebiodiversity.ca/en/corridor_work.html (last accessed September 21, 2006).

Cheung, W., Alder, J., Karpouzi, V. *et al.* (2005) *Patterns of Species Richness in the High Seas.* Secretariat of the Convention on Biological Diversity, Montreal. Technical Series, No. 20, 31 pp.

Clark, M. (2001) Are deepwater fisheries sustainable? The example of orange roughy (*Hoplostethus atlanticus*) in New Zealand. *Fisheries Research*, 51, 123–35.

Clark, M. (2004) Case Study 1: Seamounts and Deep Sea Fisheries in New Zealand. *Presentation at Fisheries, Oceanography and Society: Deep-Sea Fisheries: Ecology, Economics and Conservation*. Woods Hole Oceanographic Institution and New England Aquarium, September 12–14, 2004. Available at http://www.whoi.edu/institutes/oli/activities/symposia-deepsea-program.htm (last accessed September 21, 2006).

Clark, M.R., Anderson, O.F., Francis, R.I.C.C. and Tracey, D.M. (2000) The effects of commercial exploitation on orange roughy (*Hoplostethus atlanticus*) from the continental slope of the Chatham Rise, New Zealand, from 1979 to 1997. *Fisheries Research*, 45, 217–38.

Collie, J.S., Escanero, G.A. and Valentine, P.C. (2000) Photographic evaluation of the impacts of bottom fishing on benthic epifauna. *ICES Journal of Marine Science*, 57, 987–1001.

Commonwealth of Australia (2002) *Tasmanian Seamounts Marine Reserve Management Plan*. Environment Australia, Canberra, Australia.

Costello, M.J., McCrea, M., Freiwald, A. *et al.* (2005) Role of cold-water *Lophelia pertusa* coral reefs as fish habitat in the NE Atlantic. In: *Cold-Water Corals and Ecosystems* (eds. Freiwald, A. and Roberts, J.M.), pp. 771–805. Springer-Verlag, Berlin/Heidelberg.

Dayton, P.K., Thrush, S.F., Agardy, M.T. and Hofman, R.J. (1995) Environmental effects of marine fishing. *Aquatic Conservation: Marine and Freshwater Ecosystems*, 5, 205–32.

de Fontaubert, A.C. (2001) Legal and political considerations. In: *The Status of Natural Resources on the High-Seas* (eds. WWF/IUCN). WWF/IUCN, Gland, Switzerland.

DEFRA (Department for Environment, Food and Rural Affairs) (2004) *Review of Marine Nature Conservation. Summary of Working Group Report to Government*. DEFRA, London.

Dower, J., Freeland, H. and Juniper, K. (1992) A strong biological response to oceanic flow past Cobb Seamount. *Deep-Sea Research A*, 39, 1139–45.

FAO (Food and Agriculture Organization of the United Nations) (1995) *Code of Conduct for Responsible Fisheries*, 41 pp. FAO, Rome, Italy.

FAO (Food and Agriculture Organization of the United Nations) (2005) *Deep Sea Fisheries*. Committee on Fisheries. *Twenty-Sixth Session*. Rome, Italy, March 7–11, 2005. COFI/2005/6.

Fortes, M., Djohani, R. and Kahn, B. (eds.) (2003) Introducing the Regional Action Plan (RAP) to strengthen a resilient network of effective marine protected areas in Southeast Asia 2002–2012: technical background and RAP description. Final Report to the IUCN World Commission for Protected Areas (WCPA), Gland, Switzerland, 128 pp.

Francis, R.I.C.C. and Clark, M.R. (2005) Sustainability issues for orange roughy fisheries. *Bulletin of Marine Science*, 76, 337–51.

Froese, R. and Sampang, A. (2004) Taxonomy and biology of seamount fishes. In: *Seamounts: Biodiversity and Fisheries* (eds. Morato, T. and Pauly, D.). Fisheries Centre, University of British Columbia, Canada. *Fisheries Centre Research Reports*, 12(5), 25–31.

Gjerde, K.M. (2005) Editor's introduction: moving from words to action. *International Journal of Marine and Coastal Law*, 20, 323–44.

Gjerde, K.M. and Breide, C. (2003) *Towards a Strategy for High Seas Marine Protected Areas: Proceedings of the IUCN, WCPA and WWF Experts Workshop on High Seas Marine Protected Areas*, Malaga, Spain, January 15–17, 2003. IUCN, Gland, Switzerland.

Glasby, G.P. (2002) Deep seabed mining: past failures and future prospects. *Marine Georesources and Geotechnology*, 20, 161–76.

Grant, S.M. (2005) Challenges of marine protected area development in Antarctica. In: *High Seas Marine Protected Areas* (eds. Kelleher, G. and Gjerde, K.M.). IUCN, Gland, Switzerland. *Parks*, 15(3), 40–47.

Grigg, R.W. (1984) Resource management of precious corals: a review and application to shallow water reef building corals. *Pubblicazioni della Stazione Zoologica di Napoli I: Marine Ecology*, 5, 57–74.

Grigg, R.W. (1993) Precious coral fisheries of Hawaii and the US Pacific Islands. *Marine Fisheries Review*, 55, 50–60.

Grigg, R.W., Malahoff, A., Chave, E.H. and Landahl, J. (1987) Seamount benthic ecology and potential environmental impact from manganese crust mining in Hawaii. In: *Seamounts, Islands and Atolls* (eds. Keating, B., Fryer, P., Batiza, R. and Boehlert, G.). *Geophysical Monograph*, 43, 379–90.

Gubbay, S. (ed) (1995) *Marine Protected Areas. Principles and Techniques for Management*. Chapman and Hall, London.

Gubbay, S. (2003) *Seamounts of the North-East Atlantic*. OASIS, Hamburg: WWF-Germany, Frankfurt am Main.

Gubbay, S. (2005) *Toward the Conservation and Management of the Sedlo Seamount*. OASIS, Hamburg, 34 pp.+Appendices.

Hall-Spencer, J., Allain, V. and Fossa, J.H. (2002) Trawling damage to Northeast Atlantic ancient coral reefs. *Proceedings of the Royal Society of London B*, 269, 507–11.

Heinz, P., Ruepp, D. and Hemleben, C. (2004) Benthic foraminifera assemblages at Great Meteor Seamount. *Marine Biology*, 144, 985–98.

High Seas Task Force (2005) High Seas Governance. Meeting of the High Seas Task Force. Paris, March 9, 2005. HSTF/09. HSTF, Paris. Available at http://www.high-seas.org/ (last accessed September 21, 2006).

ICES (International Council for the Exploration of the Sea) (2002) Report of the ICES Advisory Committee on Ecosystems, 2002. *ICES Cooperative Research Report 254*, 129 pp.

ICES (International Council for the Exploration of the Sea) (2005) Report of the Working Group on Deep-Water Ecology (WGDEC), ICES Headquarters, Copenhagen, March 8–11, 2005. ICES CM 2005/ACE:02, 76 pp.

IUCN (World Conservation Union) (1994) *Guidelines for Protected Area Management Categories*. IUCN, Gland, Switzerland.

IUCN (World Conservation Union) (2004) Ten-year high seas marine protected areas strategy: a ten-year strategy to promote the development of a global representative system of high seas marine protected area networks. Summary Version as agreed by Marine Theme Participants at the *fifth World Parks Congress*, Durban, South Africa, September 8–17, 2003.

Jennings, S., Kaiser, M.J. and Reynolds, J.D. (2001) *Marine Fisheries Ecology*. Blackwell Science Ltd, Oxford.

Johnston, P.A. and Santillo, D. (2004) Conservation of seamount ecosystems: application of a marine protected areas concept. *Archive of Fishery and Marine Research*, 51, 305–19.

Jones, J.B. (1992) Environmental impact of trawling on the seabed: a review. *New Zealand Journal of Marine and Freshwater Research*, 26, 59–67.

Kahn, B. (2003) The Indo-Pacific Marine Corridors of Eastern Indonesia. Ecological Significance for Oceanic Cetaceans and other Large Migratory Marine Life and Implications for MPA networks in Southeast Asia. Paper presented at the *fifth IUCN World Parks Congress (WPC) Marine Stream* workshop *Scaling Up to Build MPA Networks: Benefits for Fisheries and Endangered Species*. Durban, South Africa, September 8–17, 2003.

Kelleher, G. (1999) *Guidelines for Marine Protected Areas*. IUCN, Gland, Switzerland and Cambridge, UK.

Kenchington, R.A. (1990) *Managing Marine Environments*. Taylor & Francis, New York.

Koslow, J.A. (1996) Energetic and life-history patterns of deep-sea benthic, benthopelagic and seamount-associated fish. *Journal of Fish Biology*, 49(Suppl. A), 54–74.

Koslow, J.A., Boehlert, G.W., Gordon, J.D.M., Haedrich, R.L., Lorance, P. and Parin, N. (2000) Continental slope and deep-sea fisheries: implications for a fragile ecosystem. *ICES Journal of Marine Science*, 57, 548–57.

Koslow, J.A., Gowlett-Holmes, K., Lowry, J.K., O'Hara, T., Poore, G.C.B. and Williams, A. (2001) Seamount benthic macrofauna off southern Tasmania: community structure and impacts of trawling. *Marine Ecology Progress Series*, 213, 111–25.

Laurent, D. and Pietra, F. (2004) Natural-product diversity of the New Caledonian marine ecosystem compared to other ecosystems: a pharmacologically oriented view. *Chemistry and Biodiversity*, 1, 539–94.

Levin, L.A. (1994) Paleoecology and ecology of xenophyophores. *Palaios*, 9, 32–41.

Levin, L.A., Huggett, C.L. and Wishner, K.F. (1991a) Control of deep-sea benthic community structure by oxygen and organic matter gradients in the eastern Pacific Ocean. *Journal of Marine Research*, 49, 763–800.

Levin, L.A., McCann, L.D. and Thomas, C.L. (1991b) The ecology of polychaetes on deep seamounts in the eastern Pacific Ocean. *Ophelia*, Suppl. 5, 467–76.

Levin, L.A., Leithold, E.L., Gross, T.F., Huggett, C.L. and Dibacco, C. (1994) Contrasting effects of substrate mobility on infaunal assemblages inhabiting 2 high-energy settings on Fieberling Guyot. *Journal of Marine Research*, 52, 489–522.

Lodge, M.W. and Nandan, S.N. (2005) Some suggestions towards better implementation of the United Nations Agreement on Straddling Fish Stocks and Highly Migratory Fish Stocks of 1995. *International Journal of Marine and Coastal Law*, 20, 345–79.

Machete, M., Morato, T. and Menezes, G. (2005) Modelling the distribution of two fish species on seamounts of the Azores. *FAO Fisheries Proceedings*, 3/1, 182–95.

MCBI (Marine Conservation Biology Institute) (2004) Scientists' Statement on Protecting the World's Deep-Sea Coral and Sponge Ecosystems. Available at http://www.mcbi.org/what/dscstatement.htm/ (last accessed September 21, 2006).

McDaniel, N., Swanston, D., Haight, R., Reid, D. and Grant, G. (2003) Biological Observations at Bowie Seamount, August 3–5, 2003. Preliminary Report prepared for Fisheries and Oceans Canada, Unpublished, 25 pp.

Merrett, N.R. and Haedrich, R.L. (1997) *Deep-Sea Demersal Fish and Fisheries*. Chapman and Hall, London.

Molenaar, E.J. (2005) Addressing regulatory gaps in high seas fisheries. *International Journal of Marine and Coastal Law*, 20, 533–70.

Morato, T., Cheung, W.W.L. and Pitcher, T.J. (2004) Additions to Froese and Sampang's checklist of seamount fishes. In: *Seamounts: Biodiversity and Fisheries* (eds. Morato, T. and Pauly, D.). *Fisheries Centre Research Reports*, 12(5), Appendix 1: 1–6. Fisheries Centre, University of British Columbia, Canada.

Morato, T., Cheung, W.W.L. and Pitcher, T.J. (2006) Vulnerability of seamount fish to fishing: fuzzy analysis of life-history attributes. *Journal of Fish Biology*, 68, 209–21.

NEAFC (North East Atlantic Fisheries Commission) (2005a) Report of the meeting of the Working Group on the future of the North East Atlantic Fisheries Commission, London, April 26, 2005, 13 pp.

NEAFC (North East Atlantic Fisheries Commission) (2005b) *Recommendation IV from the 23rd Annual Meeting*. NEAFC Recommendation for the protection of vulnerable deep-water habitats by Denmark (in respect of the Faroe Islands and Greenland), Estonia, the European Community, Iceland, Norway and Poland.

Neumann, C., Christiansen, S. and Christiansen, B. (eds.) (2005) *Seamounts and Fisheries – Conservation and Sustainable Use*. First OASIS Stakeholder Workshop, Horta, Azores, April 1–2, 2004. OASIS/WWF.

Norse, E.A. (ed.) (1993) *Global Marine Biological Diversity: A strategy for Building Conservation into Decision Making*. Island Press, Washington, DC.

O'Driscoll, R.L. and Clark, M.R. (2005) Quantifying the relative intensity of fishing on New Zealand seamounts. *New Zealand Journal of Marine and Freshwater Research*, 39, 839–50.

OSPAR (OSPAR Commission for the Protection of the Marine Environment of the North-East Atlantic) (2003) OSPAR Recommendation 2003/3 on a Network of Marine Protected Areas. Meeting of the OSPAR Commission, Bremen, June 23–27, 2003. Annex 9 (Ref. § A-4. 44a). OSPAR, London.

OSPAR (OSPAR Commission for the Protection of the Marine Environment of the North-East Atlantic) (2004) Initial OSPAR List of Threatened and/or Declining Species and Habitats (Reference Number: 2004-06). OSPAR, London.

Piepenburg, D. and Muller, B. (2004) Distribution of epibenthic communities on the Great Meteor Seamount (North-East Atlantic) mirrors pelagic processes. *Archive of Fishery and Marine Research*, 51, 55–70.

Pomeroy, R.S., Parks, J.E. and Watson, L.M. (2004) *How Is Your MPA Doing? A Guidebook of Natural and Social Indicators for Evaluating Marine Protected Area Management Effectiveness.* IUCN, Gland, Switzerland and Cambridge, UK.

Probert, P.K. (1999) Seamounts, sanctuaries and sustainability: moving towards deep-sea conservation. *Aquatic Conservation: Marine and Freshwater Ecosystems*, 9, 601–5.

Probert, P.K., McKnight, D.G. and Grove, S.L. (1997) Benthic invertebrate bycatch from a deep-water trawl fishery, Chatham Rise, New Zealand. *Aquatic Conservation: Marine and Freshwater Ecosystems*, 7, 27–40.

Richer de Forges, B.R., Koslow, J.A. and Poore, G.C.B. (2000) Diversity and endemism of the benthic seamount fauna in the southwest Pacific. *Nature*, 405, 944–7.

Roberts, C.M. (2002) Deep impact: the rising toll of fishing in the deep sea. *Trends in Ecology and Evolution*, 17, 242–5.

Roberts, C.M. and Hawkins, J.P. (2000) *Fully-Protected Marine Reserves: A Guide.* WWF Endangered Seas Campaign, Washington, DC, USA and Environment Department, University of York, UK.

Rogers, A.D. (1994) The biology of seamounts. *Advances in Marine Biology*, 30, 305–50.

Rogers, A.D. (1999) The biology of *Lophelia pertusa* (Linnaeus 1758) and other deep-water reef-forming corals and impacts from human activities. *International Review of Hydrobiology*, 84, 315–406.

Rona, P.A. (2003) Resources of the sea floor. *Science*, 299, 673–4.

Rowden, A.A., Clark, M.R. and Wright, I.C. (2005) Physical characterisation and a biologically focused classification of 'seamounts' in the New Zealand region. *New Zealand Journal of Marine and Freshwater Research*, 39, 1039–59.

Salm, R.V., Clark, J.R. and Siirila, E. (2000) *Marine and Coastal Protected Areas: A Guide for Planners and Managers*, 3rd edition. IUCN, Gland, Switzerland and Cambridge, UK.

Santos, R.S. (2004) Introduction to Topic 3: Marine Protected Areas: Natura 2000 and beyond (OSPAR) – The Azores, pp. 10–12 and Summary of discussions on Topic 3: Marine Protected Areas: Natura 2000 and beyond (OSPAR) – The Azores, pp. 27–29. In: *Electronic Conference on Biodiversity Research That Matters*! (eds. Heip, C.H.R. *et al.*) Summary of discussions, November 15–26, 2004. Flanders Marine Institute, Oostende, Belgium.

Santos, R.S., Hawkins, S., Monteiro, L.R., Alves, M. and Isidro, E.J. (1995) Marine research, resources and conservation in the Azores. *Aquatic Conservation: Marine and Freshwater Ecosystems*, 5, 311–54.

Santos, R.S., Colaço, A. and Christiansen, S. (eds) (2003) Planning the Management of Deep-Sea Hydrothermal Vent Fields MPAs in the Azores Triple Junction (Workshop proceedings). *Arquipélago – Life and Marine Sciences*, Suppl. 4, 64 pp.

Schmidt, S. and Christiansen, S. (2004) *The Offshore MPA Toolbox. Implementing Marine Protected Areas in the North-East Atlantic Offshore: Seamounts – A Case Study*, 55 pp. OASIS, Hamburg and WWF Germany, Frankfurt am Main.

Scovazzi, T. (2004) Marine protected areas on the high seas: some legal and policy considerations. *International Journal of Marine and Coastal Law*, 19, 1–17.

Sobel, J.A. and Dahlgren, C.P. (2004) *Marine Reserves: A Guide to Science, Design, and Use.* Island Press, Washington, DC.

Stocks, K.I., Boehlert, G.W. and Dower, J.F. (2004) Towards and international field programme on seamounts within the Census of Marine Life. *Archive of Fishery and Marine Research*, 51, 320–27.

Stone, G.S., Madin, L.P., Stocks, K. *et al.* (2004) Seamount biodiversity, exploitation and conservation. In: *Defying Ocean's End: An Agenda for Action* (eds. Glover, L.K. and Earle, S.A.), pp. 43–70. Island Press, Washington, DC.

Thiel, H. and Koslow, J.A. (eds.) (2001) Managing risks to Biodiversity and the Environment on the High Sea, including Tools Such as Marine Protected Areas. Scientific Requirements and Legal Aspects. *Proceedings of the Expert Workshop*, Isle of Vilm, Germany, February 27–March 4, 2001. Federal Agency for Nature Conservation. Bonn, Germany. *BfN-Skripten*, 43.

Thistle, D., Levin, L.A., Gooday, A.J., Pfannkuche, O. and Lambshead, P.J.D. (1999) Physical reworking by near-bottom flow alters the metazoan meiofauna of Fieberling Guyot (northeast Pacific). *Deep-Sea Research I*, 46, 2041–52.

Thresher, R., Rintoul, S.R., Koslow, J.A., Weidman, C., Adkins, J. and Proctor, C. (2004) Oceanic evidence of climate change in southern Australia over the last three centuries. *Geophysical Research Letters*, 31, L07212.

Tracey, D., Neil, H., Gordon, D. and O'Shea, S. (2003) Chronicles of the deep: ageing deep-sea corals in New Zealand waters. *Water and Atmosphere*, 11(2), 22–4.

Tudela S., Simard, F., Skinner, J. and Guglielmi, P. (2004) The Mediterranean deep-sea ecosystems: a proposal for their conservation. *The Mediterranean Deep-Sea Ecosystems: An Overview of Their Diversity, Structure, Functioning and Anthropogenic Impacts, with a Proposal for Conservation*, pp. 39–47. IUCN, Málaga and WWF, Rome.

UNGA (United Nations General Assembly) (2004) United Nations General Assembly Resolution 59/25. Adopted on November 17, 2004 (Doc. A/RES/59/25, of January 17, 2005).

Unger, S. (2004) *Managing Across Boundaries. The Dogger Bank – A Future International Marine Protected Area*. WWF Germany, Frankfurt am Main.

Watson, R. and Morato, T. (2004) Exploitation patterns in seamount fisheries: a preliminary analysis. In: *Seamounts: Biodiversity and Fisheries* (eds. Morato, T. and Pauly, D.). Fisheries Centre, University of British Columbia, Canada. *Fisheries Centre Research Reports*, 12(5), 61–5.

Wishner, K.F., Ashjian, C.J., Gelfman, C. *et al.* (1995) Pelagic and benthic ecology of the lower interface of the eastern tropical Pacific oxygen minimum zone. *Deep-Sea Research I*, 42, 93–115.

Worm, B., Lotze, H.K. and Myers, R.A. (2003) Predator diversity hotspots in the blue ocean. *Proceedings of the National Academy of Sciences of the United States of America*, 100, 9884–8.

WSSD (World Summit on Sustainable Development) (2002) *Plan of Implementation*. WSSD, Johannesburg, NY.

WWF (World Wide Fund for Nature) (2003) Management direction for the Bowie Seamount MPA: links between conservation, research, and fishing. Final Report prepared for WWF Canada by Axys Environmental Consulting Ltd.

Yen, P.P.W., Sydeman, W.J. and Hyrenbach, K.D. (2004) Marine bird and cetacean associations with bathymetric habitats and shallow-water topographies: implications for trophic transfer and conservation. *Journal of Marine Systems*, 50, 79–99.

Chapter 21

The depths of ignorance: an ecosystem evaluation framework for seamount ecology, fisheries and conservation

Tony J. Pitcher, Telmo Morato, Paul J.B. Hart, Malcolm R. Clark, Nigel Haggan and Ricardo S. Santos

Abstract

Seamounts are some of the least known habitats on the planet. Indigenous people and early navigators knew of some of them, but we are only just finding out that there may be 100 000 large seamounts and up to a million smaller features. Seamounts are steep-sided underwater volcanoes with a geological life history. The physical characteristics can generate upwelling of nutrients, the formation of density cones or the retention of water masses. These hydrological phenomena may lead to local enhancement of primary production. However, a more important mechanism appears to be the trapping of small migrating organisms, both zooplankton and mesopelagic organisms, over the summit and flanks, depending on the depth of the peak. As a result, larger mobile sea creatures visit seamounts to feed on the concentrations of small organisms. Species of seabirds, sharks, tuna, billfish, sea turtles and marine mammals can aggregate at seamounts to 'raid the larder', and sometimes to spawn. Seamount biota, especially fishes but also corals, present an attractive target for human exploitation. Small-scale artisanal fisheries from oceanic island chains have, for generations, taken advantage of life on nearby seamounts and have proven sustainable over long periods. Large-scale fisheries, in contrast, have a poor record of sustainability, often causing serial and serious depletion of fish on seamounts. Unregulated distant water fleets overexploited many high seas seamount areas in the 1970s and 1980s, and catch data from this period is only now being fully assembled and analysed. Trawl gear destroys delicate and long-lived benthic organisms such as cold water corals and sponges. Seamounts need some protection from trawling and other fishing, and rational management if they are to provide sustainable fisheries as well as serve as reservoirs of abundance and biodiversity; 'islands in the deep'. We present a two-part ecosystem evaluation framework (EEF) for seamounts by pulling together information from the preceding chapters. Part A scores the extent of our knowledge about individual seamounts: a more detailed version might express the extent of local enhancement of biomass and biodiversity. Part B assesses the severity of a range of threats, mainly from human fisheries, to the abundance and diversity of living organisms found at individual seamounts.

Introduction

The smallest rock in the tropical seas, by giving a foundation for the growth of innumerable kinds of seaweed and compound animals, supports likewise a large number of fish. The sharks and the seamen in the boats maintained a constant struggle which should secure the greater share of the prey caught by the fishing-lines. I have heard that a rock near the Bermudas, lying many miles out at sea, and at a considerable depth, was first discovered by the circumstance of fish having been observed in the neighbourhood.

Charles Darwin (1839), Voyage of the Beagle, Chapter 1,
St Jago – Cape de Verd Islands.

The apparently featureless ocean is packed with many tens of thousands of large seamounts, formed from extinct volcanoes. Seamounts ecosystems are islands of abundance, biomass and biodiversity in the oligotrophic deep ocean, but their discovery is quite a recent story. In the quote above, Darwin is clearly describing a shallow seamount and its associated fish fauna; as was his wont Darwin correctly guessed that flat-topped seamounts ('guyots') provide a base for coral reef formation. Darwin's 'discovery' would have been old news to indigenous people, who, as in the case of Polynesian navigators, had a phenomenal knowledge of ocean features and currents. They may also have been attracted by birds feeding on forage fish[1] forced to the surface by predators, a process termed 'trophic mediation' which we will consider again in the case of corals and sponges.

Seamounts entered the European scientific canon on 2 July, 1869, when the Josephine Bank in the eastern north Atlantic was found and named by the Swedish Navy Corvette *Josephine* (Chapter 3). Research on seamount ecology began in earnest in the 1950s, with the application of underwater sonar and the development of survey devices such as Doppler and scanning sonar (Simmonds and MacLennan, 2005), that can measure currents and the movement of fish and small mesopelagic organisms. Since then, the ubiquity and pivotal role of these 'islands in the deep' has begun to emerge and provides the basis for the 21 chapters of this book.

Physics and geology

Seamounts are undersea mountains characterised by their height above the surrounding abyssal plain, depth of the peak below the surface and to some extent, steepness of slope. Nearly all seamounts are underwater volcanoes: they represent about 20% of global volcanic extrusions so that their distribution relates directly to spatial and temporal variations in intra-plate volcanic activity. Since they tend to occur in island arcs, where their location intercepts global water currents their geomorphology enhances the trapping of water masses in several ways. Seamounts have unique 'magnetic signatures', which may contribute to their location and use as rest stops and 'cafés' for sharks, whales and other migrants.

[1] This raises the question of what is, and is not, a 'seamount species'. Sardines (*Sardinops sagax*) are often found over seamounts, but were excluded from catch in Chapter 18, as their global catch of 4.5 million tonnes would swamp all else. The presence of sardines over seamounts would attract birds and hence early navigators.

Seamounts obstruct currents and thus enhance tidal dissipation. Several mechanisms can enhance upwelling of nutrients. Formation of Taylor caps and wake effects may trap water masses and create upwelling currents. Taylor columns (Chapter 4) are formed by the effect of the Earth's rotation on directional current flow split by a seamount. Intriguingly, the great red spot of Jupiter may have been the first Taylor column to be described (Chapter 3). Stratification of water may turn the spinning column into a Taylor cap (or cone). Taylor caps form, or not, over a particular seamount depending on width, local current and tidal flow, height and the Coriolis force, which varies by latitude. The Rossby and Burger numbers can forecast where a Taylor cap may form (Chapter 4), some seamounts always producing Taylor cones, others only intermittently or not at all.

Taylor cones encourage a doming of water density layers, which in turn creates two physical effects, each with biological implications. First, below the thermocline, isolated topographic features may produce large vertical displacements in the density gradient so that small deep seamounts can cause significant local current structures so that deep seamounts over about 0.5 km in height are especially important in producing density-layered domes of water. Secondly, density domes over shallow seamounts may reach the euphotic zone and have powerful effects on local plankton abundance including increased vertical mixing. In addition, tides can generate similar density cones and wave spin-offs over seamounts (Chapter 4), and eddies like this can be formed by several different mechanisms. Chapters 4 and 5 outline considerable field evidence for all of these processes, although no one seamount may exhibit all of them, and some seamounts may show none.

Seamount biota

There is therefore a range of hydrological features that may lead to higher local production and biomass, although not all of them may be observed at any one seamount, and they may operate intermittently or for most of the time. Although oceanographic processes such as Taylor columns have the potential to hold water over seamounts for periods of several days, it is unlikely that planktonic communities over seamounts are significantly and permanently different from those in the surrounding ocean. There is evidence that seamounts can cause upwelling of nutrients, with consequent enhancement of the rate of primary production. Seamounts that come close to the surface may well cause eddies downstream of the current with resulting differences in the planktonic community from surrounding open ocean areas. A few very shallow seamounts with macrophytes host significant communities. Higher primary production may be expected over some seamounts, but has not been widely observed and may not be as widespread a phenomenon as was once thought. Indeed, modelling (Chapter 12) supports this empirical evidence, which suggests that enhanced local primary production would rarely be sufficient or widespread enough to account for the high biomass of both resident and visiting organisms often found at seamounts.

Enhanced current flows on the tops and flanks of seamounts may encourage higher abundances of sessile filter feeding organisms such as corals and sponges, provided they are within range of sources of settling larvae for colonies to be founded. The fixed nature of benthic animals makes it more likely that seamount communities are different from the surrounding ocean. If the seamount is very tall, the benthic community of the upper slopes will be very different from the abyssal benthic community, mainly because physical conditions are so different. Distance from a continental shelf may also determine how endemic the benthic fauna is. It appears that the invertebrate benthic communities of seamounts are more often not similar to those found on nearby shelves (Chapter13). However, some communities are very different from those elsewhere in the nearby region, as shown in Chapters 7 and 9 where a few seamounts in the SW Pacific appear to have high levels of benthic invertebrate and coral endemism. Many of the organisms on seamounts are long lived and slow growing, making the community relatively unproductive and vulnerable to exploitation and damage.

Benthic-structuring organisms such as corals and sponges can live hundreds of years and can be extremely slow growing. The interdependence of seamount fish and other motile species with benthic-structuring organisms is unknown, but the 'feed rest' hypothesis (see Chapters 5, 6, 8 and 10A; Genin, 2004) suggests they may be important to both the safety and energy budgets of these and other fish. The role of benthic structure in the spawning and rearing of commercial and other species is equally dark but likely significant. Chapter 8 raises serious concerns about the ability of corals destroyed by trawling to recover, or to recolonise under current and anticipated climate regime.

For many benthic animals on seamounts a major problem is to evolve a life cycle that ensures that sufficient new recruits join the local populations. With water passing continuously over the top of the seamount, and being held up for only 2–3 weeks by Taylor processes and by eddies, planktonic larval stages have to be short enough to ensure that there will be a shallow surface to settle on. Evidence presented in Chapter 13 shows that for species of both invertebrates and fish that live over shallow seamounts, larvae are in the plankton for less time than is true for species that live in equivalent shelf and slope waters.

There is strong evidence that vertically migrating zooplankton can be advected over seamounts at night then and so prevented from returning to the depths. These trapped plankton concentrations around and over seamounts will attract fish that feed on plankton and may well enhance food supplies to sessile filter feeders anchored to the seamount. In this way a seamount will have the same influence on the plankton community as would an island or an oceanic front between two water masses. Given that temporary concentrations of plankton occur sufficiently often, a seamount can become a location that attracts planktivorous fish and their predators. The consistency with which plankton concentrations are found will determine how permanent these aggregations of fish are. There is evidence that tuna, sharks, turtles and some marine mammals and seabirds spend time feeding around seamounts, although it is unlikely that they remain over one seamount for more than a few days (Chapter 10A and B; Chapter 12A–C).

Where all of these factors act together, it is possible that some seamounts may generate sufficient settling detritus to encourage a resident detritivore community (such as rock lobsters), and hence attract resident small fish and cephalopods (Chapter 14).

Small-scale seamount fisheries

Many seamounts are close to inhabited islands, such as the Azores or Hawaii. People on these islands have a long history of fishing over nearby seamounts, as detailed in

Chapter 16. Many of the small-scale fisheries exploit deepwater fish using relatively simple gear and labour intensive equipment. This means that fishing pressure is relatively low and such fisheries have proved to be sustainable over long-time periods. Although the global catch of fish from small-scale seamount fisheries is estimated to be about 250000t, this is probably a very poor estimate. In many locations local landings by artisanal fishers are from a variety of locations in particular from shallow tropical coral reefs. There is no good way in which to separate the landings from those that come from local seamounts. Nevertheless, the cultural, social and economic importance of artisanal seamount fisheries are locally very significant. Very often, fishing is one of the only occupations that is available for the inhabitants of an island so that many people will be dependent on the catches from small-scale seamount fisheries, as indicated by the *per capita* annual consumption estimate of 37kg from seamount fisheries compared to a global average of 13kg (Chapter 16).

Large-scale seamount fisheries

Large-scale trawl fisheries on offshore seamounts have a much more recent history than do the island-based artisanal fisheries. They developed from 1960s to 1970s when large trawlers from primarily the USSR and Japan searched the world's oceans for fisheries resources. Some high-volume fisheries were developed for deepwater species like pelagic armourhead (*Pseudopentaceros wheeleri*), alfonsino (*Beryx splendens*), orange roughy (*Hoplostethus atlanticus*), oreos (Oreosomatidae) and grenadiers (*Coryphaenoides rupestris*). The cumulative catch from these fisheries is estimated at over 2 million tonnes (Chapter 17). These fisheries have not proven sustainable, in many cases lasting a decade or less (Chapters 17 and 18). In part this is because deepwater species often have biological characteristics that make them less productive and more vulnerable to overfishing than shallower shelf or slope species (Chapter 9), but also the development of fishing technology in the late twentieth century that allowed seamount features (and their fish) to be consistently located and successfully fished.

Impacts on ecosystems: biota/habitats

The effects of fishing on seamounts are basically the same as on other habitats. In Chapter 19 the types of fishing gear and their effects are described, and there is nothing specific to seamounts for most of them. However, seamounts are often the only features at suitable depths for commercial species of demersal fish in large areas of the abyssal ocean plains – true 'islands in the deep'. Because fish can be localised over and around seamounts, so are their fisheries. Studies off New Zealand and Australia document very high levels and densities of bottom trawling on small seamounts for orange roughy, and hence the impacts can be highly concentrated. Technology has contributed to this, with development of global positioning systems (GPS) in 1980s–1990s making it possible for seamounts to be easily and consistently located, and with design of trawl gear that can fish rough, hard seafloor and the introduction of factory and freezer trawlers that can stay at sea for long periods and process large quantities of fish. Trawling, and its associated impacts on seamount fish and habitat, can operate 24h a day, 7 days a week. Without management, fish stocks and fragile seamount ecosystems are highly vulnerable.

Management issues at seamounts

While a large undersea mountain 5–10 km across may seem rather obvious feature, on a global scale the numbers and locations of many seamounts remain unknown. For example in 2005, an American submarine ran into an uncharted Pacific seamount (Fig. 21.1), and in 2004 the Norwegian RV *G.O. Sars* involved in the CoML project MAR-ECO identified a set of topographic features at the Mid-Atlantic Ridge between the Azores and Iceland that were either not charted or mistakenly represented in available bathymetry. Determining the number and location of seamounts presents several problems. Precise hydrographic sonar surveys have charted many ocean areas, especially the exclusive economic zones (EEZs) of developed nations and areas of strategic military importance, yet it would take thousands of years to survey all of the world oceans. Short cuts are the use of satellite altimeters or the use of detection algorithms on approximations of ocean bathymetry. Analyses presented in this book support an estimate of over 100 000 large seamounts, with many smaller ones waiting to be discovered; about 60% are located in the Pacific basin. Complete mapping of the world's seamounts thus has important ramifications for marine geophysics, physical oceanography, marine ecology and fisheries conservation.

Fig. 21.1 Many seamounts remain uncharted. On 8 January 2005, the US nuclear submarine *San Francisco* en route to Brisbane Australia for a port visit, ran into an underwater mountain at 35 knots about 350 miles south of Guam leaving one sailor dead: a number of sailors were awarded medals for bravery in dealing with over 70 wounded. The seamount is now named the 'San Francisco seamount'.

Management instruments

Seamount ecosystems exhibit a number of distinct features that make them of considerable interest to marine scientists, resource managers and those with an interest in marine conservation because they are particularly susceptible to fishing both as habitats, but also due to the high vulnerability of seamount aggregating fish (Morato *et al.*, 2006; Chapter 9).

Thus, for seamounts within EEZs, compliance with management and issues of precaution in setting quotas (FAO, 1995, 1996) are more critical when compared with those applied to continental slope fisheries. For this reason it is probably safer to set up exclusion zones for fisheries that are damaging to seamount biota, as they are easier to enforce than the quotas and effort limitation regulations commonly employed in the management of continental slope fisheries.

Marine protected areas (MPAs) could be the most practical way of managing seamounts, but could they be effectively policed? The same question can be applied to trawl bans or any other restriction on the way fishing is carried out. As has been shown in several chapters in the book, seamount communities often have fixed structures, such as cold water corals, that are very vulnerable to trawling. Without an effective way of limiting fishing, these communities will be destroyed very quickly and would then take tens to hundreds of years to re-establish, if further trawling could be prevented. For some seamount areas around oceanic islands with a large EEZ, it might be feasible to create a large MPA area around the islands within which only local, artisanal fisheries are allowed.

In some areas under national jurisdiction, unique regulations are coming in force. Of special interest is the case of the Azores seamounts. In this region, an autonomous region of Portugal and part of the European Union, small-scale artisanal fisheries have been practiced for many years. Deep-sea trawling and other deep-sea nets have been forbidden here for many years. In 2005, the revised EU Common Fisheries Policy (CFP) created a huge area, encompassing the greater portion of the EEZ, where deep-sea trawl and other deleterious deep-sea fisheries are excluded. This was extended to the two other Macaronesian archipelagos of the EU, Madeira (Portugal) and the Canaries (Spain). Also, in recognition of the distinctive features of the seamount fauna, the CFP is applied differently from other regions (see Chapter 20 for details). In fact, the 'non-trawl area' created (see Fig. 20.4) can be considered as an MPA. But, even for the Azores it has not been possible to reach agreement to extend the MPA far enough to include all the seamounts that are known in the Azores EEZ. The negative aspect of the CFP applied to the Azores is that it opened the area between the 100 and 200 miles to non-local fleets and to EC central management, leading to a huge increase (the fleet has increased from around 10 ships to around 150) of pelagic long liners whose main target is the swordfish (*Xiphias gladius*), but that are having high impacts in other species like pelagic sharks (Chapter 10B) and sea turtles (Chapter 12B). No proper impact study is under way.

The fact that more than half of the world's seamounts are outside EEZs presents a serious challenge (see Chapters 2 and 20). Pitifully few are covered by international agreements and/or conventions established in view to regulate and manage fisheries in this high seas areas (see Fig. 20.2). They are therefore highly prone to pirate fisheries and policing is a major problem.

Many international bodies and conventions now recognise the particular threat to these habitats. In Europe, OSPAR (Convention for the Protection of the Marine Environment of the North-East Atlantic) includes seamounts, and a set of other associated habitats (e.g., *Lophelia pertusa* reefs and deep-sea sponge aggregations) and species (e.g., orange roughy) in an initial list of threatened and/or declining species and habitats (OSPAR, 2004). This creates obligations for the contracting parties towards the conservation of these habitats and species.

No single management model is applicable to all seamounts. Measures are likely to range from activity-based restrictions to MPAs, and regulation of activities beyond the immediate vicinity of seamounts. Given the failure of many traditional management processes in fisheries worldwide (Pitcher, 2001, 2005; Pauly *et al.*, 2002; Pauly and Maclean, 2003; but see Hilborn, 2006), MPAs and moratoria for some fishing technologies (e.g., deep-sea trawls) are of growing interest both in EEZs as also as in the high seas.

The high seas pose some complicated jurisdictional problems since, under the Law of the Sea, it is not technically possible for a state to create MPAs on the high seas. However the international community is discussing this issue and some NGOs have already made indicative applied proposals (Roberts *et al.*, 2006). At present (2007), the main effort is concentrated on trying to obtain a Moratorium to the UN on 'deep-sea bottom trawl fishing on the high seas until legally binding regimes for the effective conservation and management of fisheries and the protection of biodiversity on the high seas can be developed, implemented and enforced by the global community' (see www.savethehighseas.org).

There are few effective ways to solve the high seas policing problem. Other remote areas of the world show similar problems. Policing the Patagonian toothfish (*Dissostichus eleginoides*) fishery in the Antarctic has been poor despite using observers on board fishing boats. In the case of the toothfish, a remedy is to monitor landings of the species and to require skippers to inform the management agency as to where the fish were caught. This approach is possible for a single species, but it would be difficult to implement a ban on landings of seamount fish as the fish communities on these structures are not necessarily unique (Chapter 9). How would one know that a given species had or had not come from a seamount? The alternative is to have observers on board vessels fishing beyond EEZs, but they are subject to considerable pressure being the only representative of the management authority on board. It would hardly be feasible to have inspection vessels monitoring activity over remote seamounts, so it would be very difficult to prevent single vessels cruising the oceans fishing over seamounts at will, as was done in the 1970s by vessels from the former USSR and Cuba (see Chapter 17). One glimmer of hope lies in the development of remote operated vehicles, in particular 'ocean gliders' that can remain at sea for 6–12 months, travel 1000s of kilometres and carry a variety of sensors. Current cost for a glider is ~$70 000. One possible use is to listen for ship activity and report by satellite. A pilot project to deploy 'Slocum' gliders, named for the first man to sail alone around the world (http://www.physics.mun.ca/~glider/) and manufactured by Webb Research (http://www.webbresearch.com/index.htm) at Canada's Bowie Seamount met with some technical difficulty, but the technology holds a lot of promise (Dale Gueret, Integrated Coastal Management Coordinator, Personal Communication, 2006). The University of Washington in the US has a similar 'sea glider' (www.apl.washington.edu/projects/seaglider/new_note_seaglider.html).

We have no doubt that MPAs in some form will be a major tool in seamount conservation and restoration. In addition to the Azores, Madeira and Canary Islands in the Atlantic, there is already some protection for a few seamounts in the Pacific off New Zealand and Australia, while discussions continue in Canada and the US. As discussed above, policing will be hard, but enforcement and vigilance might be facilitated by space technologies including satellite tracking of fishing vessels.

As shown in the next section, many seamounts we know something about are often heavily exploited and underexplored. More knowledge about how seamount ecosystems work will aid future management, but of course we cannot wait around for that knowledge to be gained. In fact, we do not need more information to appreciate that most seamount ecosystems need protection now to recover, and really the only way to do that is to ban all trawling. The critical task is to find a way of implementing such a ban in an effective way.

Conclusion: an EEF for seamounts

This chapter is a summary of the principal findings presented in the 20 earlier chapters of this book. But we also present a novel synthesis in the form of an EEF that attempts to evaluate the status of seamounts worldwide.

Table 21.1 sets out the schema. The first part derives from the EEF analysis set out in Chapter 14 and describes the important attributes, including geological, hydrological

Table 21.1 The EEF for knowledge of seamount attributes according the scheme presented in Chapter 14. Each seamount in the analysis is scored using the table below. Knowledge of each attribute is scored on a scale of 1 (completely unknown) to 4 (very well known).

	Knowledge score	Notes
Oceanographic factors Depth of peak		0–4 depending on how well known the measure is
Depth of surrounding ocean		
Height of peak		
Slope of seamount		
Proximity to shelf		
Proximity to neighbour seamounts		
Ocean currents link to shelf		
Ocean currents to neighbour seamounts		
Taylor cap forms		
Overall oceanographic knowledge status		
Ecological factors Macrophytes present		
Corals present		
Larval settlement regime		
Nutrient upwelling occurs		
Phytoplankton enhancement		
Zooplankton enhancement		
Deep scattering layer organisms entrapped		
Settled filter feeders		
Zooplankton migrates in feeding range		
Predators/grazers present		
Detritus build-up present		
Detritivores present		
Small resident invertebrate predators		
Small resident fish predators		
Resident cephalopods		
Aggregating deep-sea fish		
Visiting fish predators		
Visiting elasmobranch predators		
Visiting marine turtles		
Visiting mammal predators		
Visiting seabird predators		
Overall ecological knowledge status		

and biotic features, that are found on seamounts and may contribute to their enhancement of local biomass and biodiversity. Not all of these features occur on all seamounts and so this part of the evaluation framework scores their presence and likely magnitude. The actual values of each attribute will determine the extent of any local enhancement effect on the food web, and indeed that is the focus of Chapter 14, but, in this version of the EEF we score how much we know about these seamount features so that scanning the EEF immediately reveals the location and depth of our ecological ignorance.

Figure 21.2 applies the EEF for some Azores and Canadian seamounts as an example. Colour coding indicates the extent of our knowledge. Green means values of an attribute (e.g., peak depth) are accurately measured and known, through pale green and orange, meaning that values are indirectly estimated or inferred, to red, which means completely

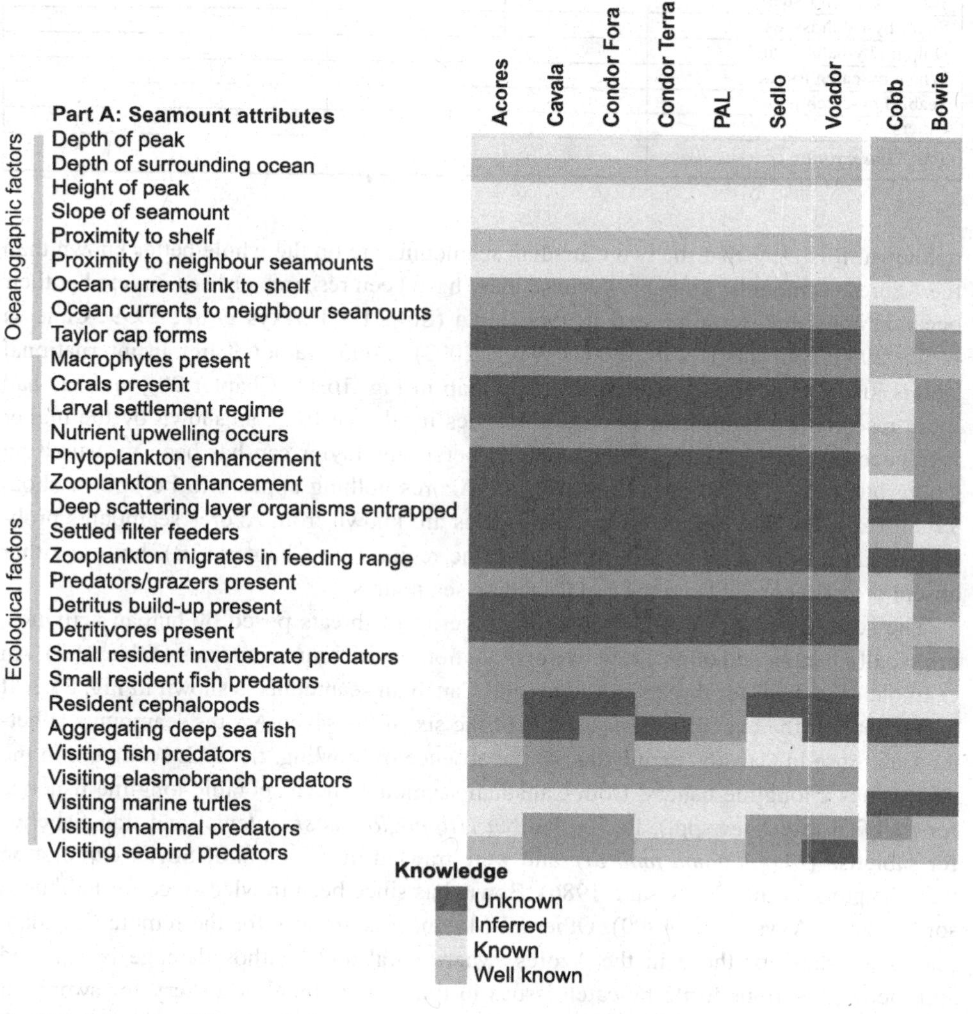

Fig. 21.2 Knowledge EEF example for seven Azores and two Canadian seamounts.

Table 21.2 The EEF for threats to seamounts. Each seamount in the analysis is scored using the table below. Threats posed by each fishery or factor are scored on a scale of 0 (no threat) to 10 (severe threats). Uncertainty in each score may be taken account of in the overall analysis.

	Status score	Limits		Notes
		Low	High	
Fisheries				
Trawl fishery				Presence and status on scale 0 (none) to 10 (severe)
Longline fishery				
Handline fishery				
Purse seine fishery				
Others				
Total fisheries status				
Conservation concerns				
Corals and benthos damage				
Turtle by-catch issues				
Shark by-catch issues				
Dolphin by-catch issues				
Whale by-catch issues				
Seabird by-catch issues				
Others				
Total conservation concern status				

unknown. It is clear that the two Canadian seamounts are on the whole better known than the Azores seamounts; probably because there have been research cruises directed at their oceanography and, to some extent, their biota (Stitt, 1993; Axys *et al.*, 1999; Canessa *et al.*, 2003; McDaniel *et al.*, 2003; WWF, 2003). Cobb seamount lies in international waters just outside the Canadian EEZ (see map in Fig. 16.11; Chapter 20), and has had more research attention than Bowie, which lies inside the EEZ, as shown by the higher incidence of green shading. For example, a persistent Taylor cap has been measured on Cobb, but only inferred over Bowie: in the Azores nothing appears to be known about Taylor caps. On the other hand, visiting turtles are known from Azores seamounts probably because of a turtle tracking project in the region (see Chapter 12B), but are either absent or not recorded from the two Canadian seamounts.

The second part of the EEF scores the severity of threats posed by human activities, principally fishing and other extractive exploitation of seamount resources (Table 21.2). An example of an EEF for threats to Azores and Canadian seamounts is shown in Fig. 21.3. It is evident that the overall fisheries status of the six of the seven Azores seamounts is better than those in Canada, mainly due to the absence of trawling: the PAL seamount in the Azores has a longline fishery. Both Canadian seamounts have sporadic longline fisheries for rockfish (*Sebastes* spp.), Pacific halibut (*Hippoglossus stenolepis*) and trap fisheries for sablefish (*Anoplopoma fimbria*); and were trawled in the years of Soviet exploration (see Chapters 17 and 20; Sasaki, 1986). Bowie has since been trawled since for halibut to some extent (Axys *et al.*, 1999). Other EBM concerns are less for the remote Canadian seamounts than for those in the Azores, where coral and benthos damage is reported and there are serious turtle by-catch issues in the surface longline fishery for swordfish (Morato *et al.*, 2001).

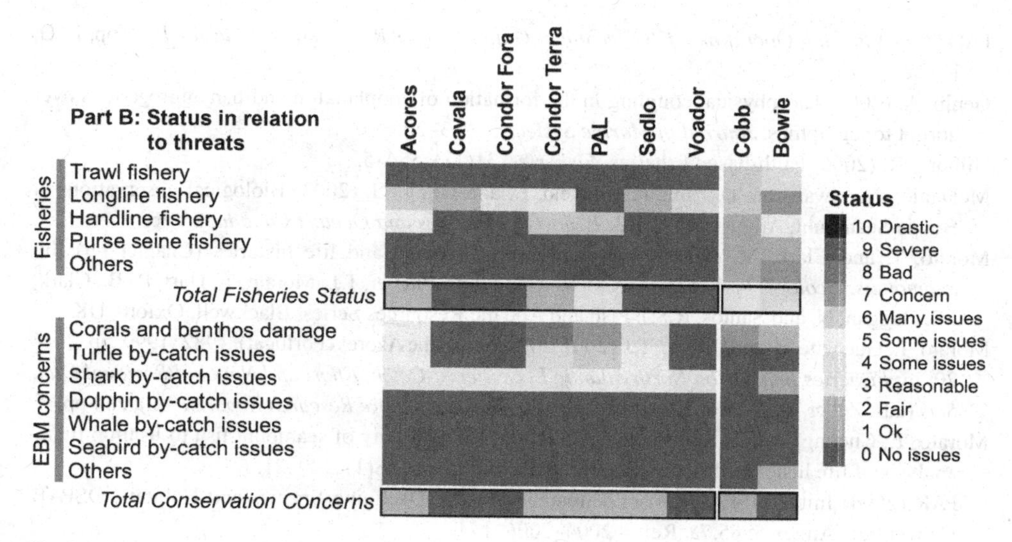

Fig. 21.3 Threats EEF example for seven Azores and two Canadian seamounts.

Conclusions

Seamounts are important islands of biodiversity in the surrounding ocean. Seamount fish and benthic-structuring organisms such as corals and sponges can have very long lives. The extent to which 'seamount fishes' rely on benthic structure for shelter, feeding, spawning and rearing is unknown, but is likely significant. On the darkside, we present a history of serial depletion, destruction of benthos and handwringing about research costs and toothless international instruments.

On the bright side, we are now aware of these fragile but fascinating habitats. We have a better sense of fishing levels that seamount populations might withstand. We can draw on indigenous and artisanal fisheries for elements of sustainability. There is a growing international will to ban high seas trawling (with the exception of Canada and Iceland). MPAs have been set up and more are in the works. There are proven vessel monitoring systems, remote sensing satellite systems and promising low-cost innovations such as ocean gliders. It may be that the depth of our knowledge is starting to exceed that of our ignorance.

References

Axys, C., Fee, F. and Dower, J. (1999) *The Bowie Seamount Area: A Pilot Marine Protected Area in Canada's Pacific Waters*, 72 pp. Fisheries and Oceans Canada, Vancouver.

Canessa, R.R., Conley, K.W. and Smiley, B.D. (2003) Bowie Seamount marine protected area: an ecosystem overview report. *Canadian Technical Report of Fisheries and Aquatic Sciences*, 2461, 85 pp.

Darwin, C. (1839) *Journal of Researches into the Geology and Natural History of the Various Countries Visitied by H.M.S. Beagle.* Henry Colburn, London, UK.

FAO (1995) *Code of Conduct for Responsible Fisheries*, 41 pp. FAO, Rome.

FAO (1996) *Fishing Operations. FAO Technical Guidelines for Responsible Fisheries 1*, 26 pp. FAO, Rome.

Genin, A. (2004) Bio-physical coupling in the formation of zooplankton and fish aggregations over abrupt topographies. *Journal of Marine Systems*, 50, 3–20.

Hilborn, R. (2006) Faith-based fisheries. *Fisheries*, 31(11), 554–5.

McDaniel, N., Swanston, D., Haight, R., Reid, D. and Grant, G. (2003) Biological observations at Bowie Seamount, August 3–5, 2003. *Report to Fisheries and Oceans Canada*, 25 pp.

Morato, T. and Clark, M. (2007) Seamount fishes: ecology and life histories (Chapter #9). In: *Seamounts: Ecology, Fisheries and Conservation* (eds. Pitcher, T.J., Morato, T., Hart, P.J.B., Clark, M., Haggan, N. and Santos, R.S.). Fish and Aquatic Resources Series, Blackwell, Oxford, UK.

Morato, T., Guénette, S. and Pitcher, T.J. (2001) Fisheries of the Azores (Portugal), 1982–1999, pp. 214–20. In: *Fisheries Impacts on North Atlantic Ecosystems: Catch, Effort and National/Regional Data Sets* (eds. Zeller, D., Watson, R. and Pauly, D.). *Fisheries Centre Research Reports*, 9(3), 254 pp.

Morato, T., Cheung, W.W.L. and Pitcher, T.J. (2006) Vulnerability of seamount fish to fishing: fuzzy analysis of life history attributes. *Journal of Fish Biology*, 68(1), 209–21.

OSPAR (2004) Initial OSPAR List of Threatened and/or Declining Species and Habitats. OSPAR Convention, Annex 5, §5.7a, Ref. # 2004–2006, 1–4.

Pauly, D. and Maclean, J. (2003) *In a Perfect Ocean: The State of Fisheries and Ecosystems in the North Atlantic Ocean*, 160 pp. Island Press, USA.

Pauly, D., Christensen, V., Guénette, S., Pitcher, T.J., Sumaila, U.R., Walters, C.J., Watson, R. and Zeller, D. (2002) Towards sustainability in world fisheries. *Nature*, 418, 689–95.

Pitcher, T.J. (2001) Fisheries managed to rebuild ecosystems: reconstructing the past to salvage the future. *Ecological Applications*, 11(2), 601–17.

Pitcher, T.J. (2005) 'Back to the future': a fresh policy initiative for fisheries and a restoration ecology for ocean ecosystems. *Philosophical Transactions of the Royal Society B: Biological Sciences*, 360, 107–21.

Roberts, C.M., Mason, L. and Hawkins, J.P. (2006) *Roadmap to Recovery: A Global Network of Marine Reserves*, 60 pp. Greenpeace International, The Netherlands.

Sasaki, T. (1986) Development and present status of Japanese trawl fisheries in the vicinity of seamounts. In: *Environment and Resources of Seamounts in the North Pacific* (eds. Uchida, R.N., Hayasi, S. and Boehlert, G.W.). *NOAA Technical Report NMFS*, 43, 21–30.

Simmonds, J.E. and MacLennan, D.N. (2005) *Fisheries Acoustics: Theory and Practice*, 437 pp. Blackwell, Oxford, UK.

Stitt, S. (1993) Deepwater fisheries: Canada's first look down in the deeps – tests on three seamounts. *Fishing News International*, 32(7), 16–17.

WWF (2003) *Management Direction for the Bowie Seamount MPA: Links Between Conservation, Research, and Fishing*, 69 pp. World Wildlife Fund, Canada.

Glossary

'a a' Typically a higher viscosity basaltic lava flow, which cools to produce a very rough surface (term originating from Hawaiian English).

Abyssal plain An extensive, flat region of the ocean bottom from 4000 to 7000 m comprising fine sediments or oozes; the upper abyssal plain (2000–4000 m) is also often referred to as the continental rise.

Acoustic Doppler current profiler (ADCP) An acoustic current meter that estimates water currents from the backscatter of an acoustic beam from passive particles in the water, such as plankton.

Agent-based model (ABM) Like an individual-based model but aggregate groups of several different types (e.g., schools or habitat patches). See '*InVitro* model'.

Anti-cyclonic flow Flows driven by changes in pressure gradient in combination with the Coriolis force. Anti-cyclonic flow is a flow that moves around a region of high pressure. In the Northern Hemisphere the flow has the high pressure to the right of the flow direction and to the left of the flow in the southern hemisphere. Cyclonic flow is a flow with low pressure at its centre, and flow directions are opposite to that for anti-cyclonic flow.

Antipatharia Black corals; corals with a hard axial skeleton (typically black or brown) that can form tall whip-like or tree-like colonies. The skeleton is used to make jewellery and decorative objects.

Asexual reproduction Reproduction without the production of gametes. Can be used to produce new individuals or for growth of a colony. Offspring are usually identical.

Asymptotic length A parameter of the von Bertalanffy growth function (VBGF), expressing the mean length the fish of a given stock would reach if they were to grow for an infinitely long period.

Atlantis model Ecosystem box-model containing sub-models representing each step in the management strategy and adaptive management cycles (biophysical, industry, monitoring, assessment, management, socio-economic drivers).

Baroclinic The type of water motion that is driven by internal pressure gradients within the ocean caused by horizontal changes in density. Baroclinic flows have a vertical change in the velocity structure (i.e., a vertical current shear).

Basalt The most common volcanic rock produced by seafloor spreading; also the source of most oceanic volcanoes.

Bathyal Pertaining to the seabed at continental slope depths.

Bathydemersal fish Living and feeding on the bottom below 1000 m.

Bathymetry Measurement of the depth of the ocean floor; the ocean equivalent of topography.

Bathypelagic fish Living or feeding in open waters at depths between 1000 and 4000 m.

Benthic Living on or pertaining to the bottom of an ocean or other water body.

Benthopelagic fish Living and feeding near the bottom up to 1000 m as well as in midwaters.

Biodiversity The variety of living things in an area, measurable from genetic to ecosystem diversity.

Biogeochemical model Model based on nutrient cycling (e.g., nitrogen) and dynamics in addition to trophic and non-trophic interactions.

Biogeography The scientific study of the geographic distribution of organisms.

Bioherm Structure, such as a reef, that is constructed by one or more species and forms the habitat for a community of associated organisms.

Biome A regional-scale ecosystem or life zone characterized by a distinct assemblage of living organisms.

Biotope An area of uniform environmental (physical) conditions providing habitat for a specific assemblage of plants and animals.

Black coral See 'Antipatharia'.

Broadcast spawning Shedding of eggs and sperm into the water column. Fertilization often takes place externally to the body.

Brooding Retaining developing eggs and larvae internally or externally to the body for a period of time prior to releasing them.

Budding Splitting off part of the body of an animal to form a new individual or part of a colony. Occurs as part of the growth process in corals or as a means of asexual reproduction.

Burger number The Burger number is a non-dimensional number that quantifies whether the Earth's rotation or the vertical density stratification is most important controlling the resultant flow characteristics. High B values will indicate strong stratification conditions, and very low B values low stratification conditions more akin to the homogeneous ocean case.

Bycatch The catch of non-target species in fishing.

Calcareous Having a high concentration of calcium carbonate, usually in the form of the minerals, calcite or aragonite.

Cenozoic The current geologic era, which began 664 million years ago and continues to the present.

Cnidaria Phylum of animals that include corals, sea anemones, jellyfish, box jellyfish, sea firs, and siphonophores. Characterized by stinging cells called cnidae or nematocysts. Animals are radially symmetrical with a single entrance to the gut through the mouth. Many form colonies by asexual reproduction.

Cold-water coral reef Reef formed by corals lacking symbiotic algae, generally in deep, cold water. Found on the continental margins, in fjords and on seamounts. See also 'Antipatharia'.

Community An assemblage of organisms that occurs together at a location at the same time that interact with each other.

Conspecifics Individuals belonging to the same species.

Continental margin Edge of the continental plate.

Continental shelf All waters between the zone influenced by the rising and falling of the tides to the shelf break. Depths are generally down to 200 m, although markedly deeper in the Antarctic (600 m).

Continental slope The edge of the continental plate where the seabed descends from the shelf-break to the abyssal plains.

Convergent plate boundary The boundary between two colliding tectonic plates. In the oceans, the oldest (and densest) plate will subduct beneath the other plate, resulting in the formation of an island arc.

Coral skeleton Matrix of calcium carbonate secreted by corals to support colony. May be massive and formed of aragonite (stony corals). Some groups have a tough flexible axial skeleton made of a proteinaceous material called gorgonin (gorgonians). Some corals have small structures embedded in the coral tissue formed of calcite, known as sclerites (octocorals).

Cretaceous The longest of the major geologic periods (from ~146 to 65 Ma).

Cryptic species Animals that are difficult to observe through camouflage, subterranean lifestyle, transparency, or mimicry.

Cyclonic flow See 'Anti-cyclonic flow'.

Deep bodied fish Laterally compressed fish (flattened from side to side) and having deep bodies (dorso-ventrally elongated).

Deep scattering layer (DSL) Zooplankton and small swimming animals such as shrimp and fish that migrate towards the sea surface at night and return to deep water at daybreak. These animals are visible as reflective layers using acoustic equipment (echsounders). DSLs occur throughout the world's oceans.

Demersal fish Sinking to or lying on the bottom up to 1000 m; living on or near the bottom and feeding on benthic organisms.

Detritivores Organisms that feed on dead or decaying organic matter.

Direct larval development Having no free larval stage.

Divergent plate boundary The boundary between two separating tectonic plates. The void created is filled with new crust created by the resulting mid-ocean spreading ridge.

DNA barcoding Use of genetic information, specifically DNA nucleotide sequences, to help with the identification of species.

Ecopath with Ecosim and Ecospace (EwE) Trophodynamic ecosystem model suite: *Ecopath* creates a static mass-balanced snapshots of the biomass pools in an ecosystem; *Ecosim* is a time varying model using the Ecopath as input; and *Ecospace* incorporates both spatial and temporal dimensions into the model.

Ecosystem engineer A species that, by its presence, alters conditions so as to provide habitat for a community of other species.

Ecotone The boundary or transition between two adjacent ecosystems.

Effusive eruption Volcanic eruptions below sea level affected by water pressure, which may counteract the gas pressure that otherwise might dictate an explosive eruption.

Ekman dynamics Dynamics associated with the flow near surface or bottom boundaries whereby frictional forces become important and break the geostrophic balance between pressure and Coriolis forces. As one approaches the seabed, currents will tend to rotate anti-clockwise (in the Northern Hemisphere).

Endemic Native and restricted to a particular area. For seamounts, endemic species are generally defined as those found only on a single seamount or seamount chain and nowhere else in the ocean.

Endemism The proportion of species in a community found only in a locality or geographic area.

Epifauna Organisms living on or just above a surface, such as the seabed. May be attached or mobile.

Euphotic zone Upper layer of the ocean that receives enough sunlight to support photosynthesis.

Eustatic sea level changes Changes in the relative sea level due to changes in the amount of water in the oceans; typically driven by climatic changes.

Explosive eruption A violent eruption in which the pressurized gases and water vapour in the magma can no longer be contained. The pressure of the overburden decreases as the magma rises to the surface, and when erupting these forces greatly exceed the viscous forces of the lava.

Fecundity The potential reproductive capacity of an organism or population; number of eggs an animal produces during each reproductive cycle.

Fertilization Fusion of gametes (eggs and sperm) to form a zygote.

Filter feeder An organism that obtains food by straining seawater using a specialized feeding structure. Food may include small planktonic organisms, eggs, and larval stages of other animals and organic particles.

Flexural deformation Vertical deformation and isostatic adjustment of the seafloor due to the bending of the elastic lithosphere under the weight of a tectonic load such as a seamount.

Food consumption The amount of food consumed by a fish usually expressed as a percentage of body weight. Also the amount of food consumed by a population.

Foraging-arena dynamics Theory of predator–prey interactions, which appears to match small-scale predatory–prey interactions. Prey typically remain in refuges with only a percentage of the prey population venturing out to feed (and thus being available to predators) at any one time, this vulnerable prey component can rise steeply with population density as refuge sites are filled and overflow.

Forearc Refers to the region between the deep trench and the island arc associated with the subduction zone.

Fuzzy expert system An expert system that uses a collection of fuzzy membership functions and rules, instead of Boolean logic, to reason about data.

Geoid Equipotential surface which closely approximates the mean sea surface. The gravitational vector is normal to the geoid. The geoid can be inferred from measurements of gravity or it can be measured by satellite altimetry.

Growth rate Change over time of the body mass of an organism; a linear dimension of size (e.g., total length) may be used instead of weight, to express growth, as long as this linear dimension can be conveniently related to weight.

Guyot Seamount with a truncated, flat top surface.

Habitat The place where an organism lives and which provides it with requirements to live, grow, and reproduce.

Hermaphroditism Ability of some species to express the reproductive organs of both sexes. This may occur simultaneously or sequentially (first one sex, then the other).

Hermatypic species Species that can form bioherms (see above).

High seas Area of the ocean that lies outside State's exclusive economic zones.

Holocene A geological period that extends from the present day back to about 11 500 calendar years. The Holocene is the fourth and last epoch of the Neogene period.

Hotspot hypothesis A hypothesis which states that seamount chains are formed as plate tectonic motions move oceanic plates over stationary or slow-moving mantle plumes.

Hydrocoral Erect or encrusting corals with a calcareous skeleton known as a coenosteum. This is covered in small pores that house the coral polyps.

Hydrozoa The groups that comprise hydrocorals.

Hyperstability of catch rates The case when catches stay high as abundance drops. It can be expected, for example, in almost any fishery targeting spawning aggregations.

Individual-based model (IBM) Model where individuals are followed as discrete entities rather than modelling entire populations as a single unit; heterogeneity and stochasticity are important features of IBMs (in contrast to the assumptions of homogeneity in mean-field models).

Infauna Any of a diverse group of aquatic animals that live within marine and fresh water sediments.

Intratentacular budding Corals that produce new polyps by forming a bud within a polyp. Part of the process of growth of many cold-water stony corals.

InVitro model An agent-based hybrid model that incorporates a sub-model for each major step in the adaptive management cycle (biophysical, industry, monitoring, assessment, management, socio-economic drivers) for a wide range of human activities (including fisheries, tourism, oil and gas or salt production); the sub-models tie classical dynamic (mean field and metapopulation) models with decision-based agents. See 'Agent-based model'.

Island arc Arcuate alignment of volcanic islands and seamounts formed by magma resulting from the subduction of an oceanic plate beneath another oceanic plate. The subducted slab partly melts and generates a basaltic magma that penetrates the overriding plate and forms volcanoes.

Isopycnal A layer of constant density, or density surface in the ocean.

Isostatic adjustment Vertical adjustment (sinking or uplift) of land masses so as to maintain a constant gravitational pressure at the base of the plates; the geologic equivalence of buoyancy in water.

Keystone species A species that exerts control on the abundance of others by altering community or habitat structure, usually through predation or grazing, and usually to much greater extent than might be surmised from its abundance.

Larvae Development stages of many marine organisms that occur after the egg. Larvae are unlike the adult form and must metamorphose before assuming adult characteristics.

Leicithotrophic development Type of development exhibited by marine animals that involves the production of a large yolky egg which hatches into a non-feeding larva.

Life history The entire life cycle of a species including reproduction, larval development, growth, and mortality.

Longevity Life span; oldest fish ever recorded for a species or stock.

Macrofauna Small benthic animals, such as segmented worms, that live in or associated with sediments and are retained by 0.25–1 mm mesh sieves.

Magnetic signature The distinctive perturbation of the Earth's magnetic field attributable to a seamount and likely detectable by sharks, turtles, tunas, and other migrating organisms.

Management strategy evaluation (MSE) The simulation of the elements (or at least the major elements) of a management strategy and an evaluation of the performance of alternative strategies with regard to specific performance measures and objectives defined *a priori*.

Mantle plume Ascending column of hot mantle believed to originate at the core–mantle boundary at a depth of 2900 km. As such plumes interact with the lithosphere they form volcanic provinces called hotspots.

Mean-field model Model where the variables are assumed to be internally homogeneous (e.g., if modelling a biological population then all individuals are assumed to be identical).

Megafauna Large animals visible to the naked eye and discernible on photographs or video imagery.

Meiofauna That part of the microfauna which inhabits algae, rock fissures, and superficial layers of the muddy sea bottom and measure 0.1–1 mm.

Mesopelagic The pelagic zone below the euphotic zone, in the mid-depths of ocean waters, generally 200–1000 m.

Miocene Past time period from 238 million years ago to 53 million years ago.

Model calibration When a model is fit to available appropriate data (e.g., biomass or catch time series or maps). Also termed 'model tuning'.

Monte Carlo simulation When a solution set is derived by repeatedly running a model, drawing parameters or stochastic parameters randomly from a specified distribution.

Natural mortality That component of total mortality caused by natural causes such as predation, diseases, senility, pollution, etc.

Nekton Aquatic animals that can swim against prevailing currents, such as fish, squid, and whales.

Nektonic Living in open water, with active movement.

Nematocyst The stinging cell of a jellyfish or other cnidarian.

Oceanography The study of the ocean, embracing and integrating all knowledge pertaining to the ocean's physical boundaries, the chemistry and physics of sea water, and marine biology.

Octocorals Diverse group of corals. Name is derived from the fact that the polyps have eight tentacles. Include soft corals and gorgonians.

Oligotrophic waters Waters relatively low in nutrients leading to low levels of primary productivity and unable to support much plant life, such as the open oceans and some lakes.

Oocytes Eggs which have not yet been fertilized (female gametes).

Open system When the system has exchanges with the world beyond its boundaries (e.g., the material brought to or flushed from seamounts by currents means that they are open systems).

Oxygen minimum zone Zone of low dissolved oxygen content, usually between 500 and 1000 m water depth.

'*Pahoehoe*' Typically a lower-viscosity basaltic lava flow, which cools to produce a smooth and ropy surface morphology (term originating from Hawaiian English).

Parameterization Entering the values for the model parameters (e.g., growth and mortality rates) so that the resulting model represents the system of interest.

Pelagic Referring to fish and other organisms that live in the water column and not in association with the seafloor or coastlines. Can include the larval stages of benthic species.

Pelagic fish Living and feeding in the pelagic environments.

Phenotype The observable or visible characteristics of an organism; the expression of genes in an individual.

Phytoplankton Small plants that drift in the ocean.

Plankton Small, free-floating aquatic organisms that cannot swim against the prevailing currents.

Planktonic See 'Plankton'.

Planktotrophic larvae Free, pelagic larvae that can feed while in the water column.

Planula Larva produced by corals. Often covered in tiny bristles (cilia) used for locomotion.

Pleistocene Past time period from 18 million to 10000 years ago. Associated with the most recent ice ages.

Polyp Feeding structure of a coral, generally consisting of a mouth, surrounded by tentacles loaded with stinging cells.

Population A group of individuals of the same species living in a particular area or ecosystem.

Primary catches Catch of primary seamount species.

Primary production The conversion of carbon dioxide to carbohydrates within an ecosystem by photosynthesis or chemosynthesis.

Primary seamount species Species that are caught exclusively in proximity to seamounts.

Pycnocline The region of rapid vertical change in water density. It is often associated with the thermocline. A seasonal pycnocline will develop associated with the summer heating regime and development of a seasonal thermocline.

Recruitment The addition of new individuals to an habitat (seamounts, in Chapter 13); or process whereby larval and juvenile animals become part of the reproductively active adult population or to the fished component of the stock.

Reef accretion The secretion of a complex framework of calcium carbonate by corals and other animals.

Reef associated fish Living and feeding on or near coral reefs.

Reef destruction Natural processes that erode the calcium carbonate framework of coral reefs, such as burrowing by worms and dissolution by sponges.

Regeneration Re-growth of parts of an individual or colony that is damaged or destroyed.

Relative vorticity The tendency for a parcel of fluid to rotate (in addition to the Coriolis forces). Relative vorticity is associated with a horizontal change in velocity which gives rise to the rotational forces involved.

Rossby number A non-dimensional number that quantifies the importance of the Earth's rotation in determining the characteristics of the current. A small value of Rossby number ($\ll 1$) indicates that rotational forces are important.

Scleractinia Stony corals. Large, usually colonial corals with a massive calcareous skeleton called a corallum. Polyps also have a skeleton known as the corallite, often recognizable by radially arranged partitions known as septae.

Sclerochronology Study of growth rings to determine past climatic variation.

Seamount affinity The degree to which a species requires the seamount habitat.

Seamount aggregating fishes Fishes that form large aggregations around seamount and that are the main target for seamount fisheries.

Seamount fishery A fishery operating on top or around a seamount.

Seamount fishes Fishes that live on seamounts.

Seamount species Species living on and caught on seamounts (including permanent residents, temporary residents, or seamount visitors).

Secondary catches Catch of secondary seamount species.

Secondary circulation Minor circulation patterns that are set up in response to the dominant flow pattern.

Secondary seamount species Secondary species that are caught in abundance in proximity to seamounts but certainly not exclusively.

Sediment Small mineral particles such as sand, mud, or carbonate ooze.

Self-fertilization Sexual reproduction whereby a single individual produces both eggs and sperm.

Serpentinite Metamorphic rock whose olivine minerals (iron and magnesium silicates) have been oxidized and hydrolysed. This conversion increases the mineral volume by 40%. Thus, mud volcanoes can form as serpentinized rocks that are squeezed out of the crust.

Sessile organism A species that, as an adult, lives attached to the seabed and unable to move away from the point of attachment.

Shelf break Where the seabed shows an increase in gradient as it plunges towards into the deep sea, marking the seaward edge of the continental shelf.

Shield-building stage Main phase of the construction of seamounts and oceanic islands when huge quantities of effusive lava flows from a shield-like edifice.

Slab pull forces When a tectonic plate subducts it is heavier than the surrounding mantle and thus exerts a pull at the end of the plate. This force can be transmitted through the elastic plate and is believed to be one of the major driving forces that move the tectonic plates.

Small-scale fisheries A fishery using small vessels (usually under 30 m) and artisanal fishing gears such as handlines, pole and line, traps and pots, small nets and small longlines.

Solitary coral Stony coral that is formed by a single polyp.

Species distribution The geographical distribution of species usually dictated by latitude, depth, temperature, proximity to required habitats.

Species evenness A component of species diversity that measures how evenly individuals are distributed among species. An area dominated by many individuals of one or a few species and having few individuals of most other species would have low evenness.

Species richness A count of the number of species in an area.

Spring–neap cycle A periodic change in the range of tidal excursions due to the coupling of the Earth–Moon and Earth–Sun system creating the tidal-generating forces. Spring tides occur when the effect of the Earth–Moon and Earth–Sun system add up (close to new and full Moon) and neap tides when they oppose each other. Spring tides have a greater range of tidal elevations and neap tides a minimum tidal range. The cycle occur at approximately a fortnightly period.

Stress (in geological terms) Plate boundary force or gravitational force per unit area of rock.

Suspension feeder See 'filter feeder'.

Symbiotic algae Algae that live in the tissues of animals where they photosynthesize and supply the host with nutrients. Include zooxanthellae.

System omnivory index (SOI) The average omnivory index (the variance of the trophic levels of a predator's prey) of all predators in the system weighted by the logarithm of each consumer's food intake, this index gives a measure of the diversity of the predator–prey relationships in the system as a whole.

Taxon A grouping used by taxonomists based on evolutionary similarities.

Taxonomy The classification of life into categories according to evolutionary relationships.

Taylor column/cone Features associated with the changes in vorticity that occur when a steady flow encounters an isolated obstacle. Under certain conditions, water will divert around the obstacle and an anti-cyclonic flow pattern may develop over the seamount.

Teleplanic Larvae capable of dispersing over long distances (usually long-lived larvae).

Tidal rectification Tidal rectification is a result of non-linear interaction of a tidal current with steep bathymetry, which, by a combined effect of topographic acceleration and bottom friction, produces asymmetry in tidal transport during a tidal cycle.

Trophic focusing A process by which prey from large volumes is accumulated or trapped in a relatively confined area.

Trophodynamic Dynamics of trophically connected groups of species (foodwebs).

Trophodynamic model Model based on food webs and energy budgets, no (or very few) non-trophic interactions or processes are included.

Vesicular lava A particular texture of volcanic rock that contains small cavities formed by the expansion of gas and water vapour during eruption.

Virgin biomass A fish stock in its natural condition before anyone has fished it.

Vulnerability (intrinsic) The relative extinction risk naturally inherent to marine fishes disregarding other external factors such as fishing intensity or destructive coastal activities (e.g., pollution, coastal development).

Zoanthidea Animals that resemble colonial sea anemones. They have no skeleton but can incorporate sand, sponge spicules, and other debris into the body wall.

Zooplankton Floating and drifting small animals that have little power of independent horizontal movement.

Zooxanthellae Unicellular algae that live in the tissues of corals. The zooxanthellae and corals form a symbiosis whereby each derive benefit from living together. The algae use the waste products of the coral and in turn provide the coral with nutrients derived from photosynthesis.

Zooxanthellate species Corals that harbour zooxanthellae in their tissues.

Subject index

Author index

Species index

Printed in the United States
By Bookmasters